**Transition Metal
Catalysed Reactions**

INTERNATIONAL UNION OF PURE AND APPLIED CHEMISTRY

Transition Metal Catalysed Reactions

A 'Chemistry for the 21st Century' Monograph

EDITED BY

SHUN-ICHI MURAHASHI

Department of Chemical Science and Engineering
Graduate School of Engineering Science
Osaka University
Machikaneyama 1–3
Toyonaka
Osaka 560-853
Japan

AND

STEPHEN G. DAVIES

Dyson Perrins Laboratory
University of Oxford
South Parks Road
Oxford
OX1 3QY
UK

b

Blackwell
Science

© 1999 International Union of Pure and
Applied Chemistry and published for them by
Blackwell Science Ltd
Editorial Offices:
Osney Mead, Oxford OX2 0EL
25 John Street, London WC1N 2BL
23 Ainslie Place, Edinburgh EH3 6AJ
350 Main Street, Malden
  MA 02148 5018, USA
54 University Street, Carlton
  Victoria 3053, Australia
10, rue Casimir Delavigne
  75006 Paris, France

Other Editorial Offices:
Blackwell Wissenschafts-Verlag GmbH
Kurfürstendamm 57
10707 Berlin, Germany

Blackwell Science KK
MG Kodenmacho Building
7–10 Kodenmacho Nihombashi
Chuo-ku, Tokyo 104, Japan

First published 1999

Set by
Excel Typesetters Co., Hong Kong

Printed and bound in Great Britain
by MPG Books Ltd, Bodmin

The Blackwell Science logo is a
trade mark of Blackwell Science Ltd,
registered at the United Kingdom
Trade Marks Registry

DISTRIBUTORS

Marston Book Services Ltd
PO Box 269
Abingdon, Oxon OX14 4YN
(*Orders*: Tel: 01235 465500
        Fax: 01235 465555)

USA
Blackwell Science, Inc.
Commerce Place
350 Main Street
Malden, MA 02148 5018
(*Orders*: Tel: 800 759 6102
         781 388 8250
         Fax: 781 388 8255)

Canada
Login Brothers Book Company
324 Saulteaux Crescent
Winnipeg, Manitoba R3J 3T2
(*Orders*: Tel: 204 837-2987)

Australia
Blackwell Science Pty Ltd
54 University Street
Carlton, Victoria 3053
(*Orders*: Tel: 3 9347 0300
        Fax: 3 9347 5001)

A catalogue record for this title
is available from the British Library

ISBN 0-632-05126-4

Library of Congress
Cataloging-in-publication Data

Transition metal catalysed reactions / edited by Shun-ichi
  Murahashi and Stephen G. Davies.
      p.   cm. — (A 'chemistry for the 21st century'
  monograph)
    Includes bibliographical references.
    ISBN 0–632–05126–4
    1. Catalysis.   2. Transition metal catalysts.
  I. Murahashi, Shun-ichi.   II. Davies, Stephen G.
  III. Series.
  QD505.T73   1999
  547′.1395—dc21   98-44683
                                                        CIP

For further information on
Blackwell Science, visit our website:
www.blackwell-science.com

Contents

Colour plates fall between pages 22 and 23

Contributors

A. ALEXAKIS *Université de Genève Departement de Chimie Organique, 30 quai Eznert-Ansermet, 1211 Genève Switzerland* [303]

H. ALPER *Department of Chemistry, University of Ottawa, 10 Marie Curie, Ottawa, Ontario K1N 6N5, Canada* [261]

I.P. BELETSKAYA *Department of Chemistry, Moscow State University, 119899 Moscow, Russia* [29]

S. BRÄSE *Institut für Organische Chemie der Georg-August Universität Göttingen, Tammannstrasse 2, 37077 Göttingen, Germany* [99]

J.M. BROWN *Dyson Perrins Laboratory, South Parks Road, Oxford OX1 3QY, UK* [465]

A. BRUNEAU *Laboratoire de Chimie de Coordination et Catalyse, UMR 6409, CNRS-Université de Rennes, Campus de Beaulieu, F-35042 Rennes, France* [391]

M. CATELLANI *Dipartimento di Chimica Organica e Industrial dell'Università, Viale delle Scienze, I-43100 Parma, Italy* [169]

A.V. CHEPRAKOV *Department of Chemistry, Moscow State University, 119899 Moscow, Russia* [29]

A. DE MEIJERE *Institut für Organische Chemie der Georg-August Universität Göttingen, Tammannstrasse 2, 37077 Göttingen, Germany* [99]

P.H. DIXNEUF *Laboratoire de Chimie de Coordination et Catalyse, UMR 6409, CNRS-Université de Rennes, Campus de Beaulieu, F-35042 Rennes, France* [391]

H. DOUCET *Dyson Perrins Laboratory, South Parks Road, Oxford OX1 3QY, UK* [465]

M.P. DOYLE *Department of Chemistry, University of Arizona, Tucson, Arizona 85723, USA* [289]

B. EL ALI *Chemistry Department, King Fahd University of Petroleum & Minerals, Dhahran 31261, Saudi Arabia* [261]

E. FERNANDEZ *Dyson Perrins Laboratory, South Parks Road, Oxford OX1 3QY, UK* [465]

D.C. FORBES *Department of Chemistry, University of South Alabama, Mobile, Alabama 36688, USA* [289]

J.P. GENET *Ecole Nationale Supérieure de Chimie de Paris, Laboratoire de Synthèse Sélective Organique et Produits Naturels, UMR CNRS 7573, 11 rue Pierre et Marie Curie, 75231 Paris Cedex 05, France* [55]

R. GRIGG *Molecular Innovation, Diversity and Automated Synthesis (MIDAS) Centre, School of Chemistry, Leeds University, Leeds LS2 9JT, UK* [81]

T. HAYASHI *Department of Chemistry, Graduate School of Science, Kyoto University, Sakyo, Kyoto 606-01, Japan* [405]

H.E. HEERES *Dyson Perrins Laboratory, South Parks Road, Oxford OX1 3QY, UK* [465]

G. HELMCHEN *Organisch-Chemisches Institut, Universität Heidelberg, D-69120 Heidelberg, Germany* [241]

W.A. HERRMANN *Anorganisch-chemisches Institut der Technischen Universität Munchen, Lichtenbergstraße 4, D-85747 Garching b. München, Germany* [375]

M.W. HOOPER *Dyson Perrins Laboratory, South Parks Road, Oxford OX1 3QY, UK* [465]

T. HOSOKAWA *Department of Environmental Systems Engineering, Faculty of Engineering, Kochi University of Technology, Tosayamada, Kochi 782-8502, Japan* [329]

D.I. HULMES *Dyson Perrins Laboratory, South Parks Road, Oxford OX1 3QY, UK* [465]

T. IKARIYA *Department of Chemical Engineering, Faculty of Engineering, Tokyo Institute of Technology, 2-12-1 O-okayama, Meguro-ku, Tokyo 152, Japan* [1]

Y. ITO *Department of Synthetic Chemistry and Biological Chemistry, Graduate School of Engineering, Kyoto University, Kyoto 606-8501, Japan* [419]

F. KAKIUCHI *Department of Applied Chemistry, Faculty of Engineering, Osaka University, Suita, Osaka 565-0871, Japan* [195]

F.I. KNIGHT *Dyson Perrins Laboratory, South Parks Road, Oxford OX1 3QY, UK* [465]

S. KUDIS *Organisch-Chemisches Institut, Universität Heidelberg, D-69120 Heidelberg, Germany* [241]

F.E. KÜHN *Anorganisch-chemisches Institut der Technischen Universität München, Lichtenbergstraße 4, D-85747 Garching b. München, Germany* [375]

T.P. LAYZELL *Dyson Perrins Laboratory, South Parks Road, Oxford OX1 3QY, UK* [465]

S. LEMAIRE-AUDOIRE *Ecole Nationale Supérieure de Chimie de Paris, Laboratoire de Synthèse Sélective Organique et Produits Naturels, UMR CNRS 7573, 11 rue Pierre et Marie Curie, 75231 Paris Cedex 05, France* [55]

B.H. LIPSHUTZ *Department of Chemistry, University of California, Santa Barbara, CA 93106, USA* [317]

G.C. LLOYD-JONES *Dyson Perrins Laboratory, South Parks Road, Oxford OX1 3QY, UK* [465]

XIYAN LU *Shanghai Institute of Organic Chemistry, Chinese Academy of Sciences, 354 Fenglin Lu, Shanghai 200032, China* [133]

SHENGMING MA *Shanghai Institute of Organic Chemistry, Chinese Academy of Sciences, 354 Fenglin Lu, Shanghai 200032, China* [133]

I.I. MOISEEV *N.S. Kurnakov Institute of General and Inorganic Chemistry, Russian Academy of Sciences, Leninsky Prospekt 31, 117907, Moscow GSP-1, Russia* [343]

A. MORTREUX *Laboratoire de Catalyse Hétérogène et Homogène, Groupe de Chimie Organique Appliquée, URA CNRS402 ENSC Lille, USTL, Bât. C7, 59652 Villeneuve d'Ascq Cedex, France* [159]

S.I. MURAHASHI *Department of Chemical Science and Engineering, Graduate School of Engineering Science, Osaka University, Machikaneyama 1–3, Toyonaka, Osaka 560-8531, Japan* [329]

S. MURAI *Department of Applied Chemistry, Faculty of Engineering, Osaka University, Suita, Osaka 565-0871, Japan* [195]

R. NOYORI *Department of Chemistry and Research Center for Materials Science, Nagoya University, Chikusa, Nagoya 464-8602, Japan* [1]

M.T. REETZ *Max-Planck-Institut für Kohlenforschung, Kaiser-Wilhelm-Platz 1, 45470 Mülheim an der Ruhr, Germany* [207]

M. SAVIGNAC *Ecole Nationale Supérieure de Chimie de Paris, Laboratoire de Synthèse Sélective Organique et Produits Naturels, UMR CNRS 7573, 11 rue Pierre et Marie Curie, 75231 Paris Cedex 05, France* [55]

V. SRIDHARAN *Molecular Innovation, Diversity and Automated Synthesis (MIDAS) Centre, School of Chemistry, Leeds University, Leeds LS2 9JT, UK* [81]

H. STEINHAGEN *Organisch-Chemisches Institut, Universität Heidelberg, D-69120 Heidelberg, Germany* [241]

M. SUGINOME *Department of Synthetic Chemistry and Biological Chemistry, Graduate School of Engineering, Kyoto University, Kyoto 606-8501, Japan* [419]

A. SUZUKI *Department of Chemical Technology, Kurashiki University of Science and the Arts, Kurashiki 712-8505, Japan* [441]

H. YAMAMOTO *Graduate School of Engineering, Nagoya University, CREST, Japan Science and Technology Corporation (JST), Chikusa, Nagoya 464-8603, Japan* [225]

A. YANAGISAWA *Graduate School of Engineering, Nagoya University, CREST, Japan Science and Technology Corporation (JST), Chikusa, Nagoya 464-8603, Japan* [225]

Preface

From practical, economic and environmental standpoints, transition-metal-catalysed reactions are set to dominate the chemical industry in the 21st century. These reactions will have an impact on the production of fine chemicals, pharmaceuticals, agrochemicals, polymers, etc. It is not surprising therefore that the field of transition metal catalysis has been, and will remain, at the forefront of both the academic and industrial research arenas.

Transition metal catalysis offers the possibility of achieving complex organic synthesis transformations that combine complete efficiency (100% yield) with complete chemical and stereochemical control (one product only) while minimizing or even eliminating reagents, waste products and solvents. This chemical Utopia is achievable but will require an ever more sophisticated understanding of the interactions of transition metal species and their substrates, investigations of which will continue well into, if not throughout, the 21st century. This monograph offers a snapshot of some of the progress that has been made to date in a few selected areas of transition metal catalysis while providing tantalizing glimpses of what is still to be achieved.

Solvents play an enormous part in chemical reactions, profoundly influencing the rate and selectivity of catalysts. Ikariya and Noyori demonstrate that catalysis of organic reactions in supercritical fluids, rather than conventional organic solvents, can offer enormous advantages in terms of high reactivity, high selectivity and operational simplicity. Beletskaya and Cheprakov show the advantages in terms of reactivity and turnover of changing from phosphine-based catalysts in organic media to phosphine-free systems in water. The advantages of the aqueous medium in palladium-catalysed reactions is emphasized further by Genet, Savignac and Lemaire-Audoire.

Transition metal catalysis continues to have a major impact in the area of carbon–carbon bond-forming processes. Grigg and Sridharan illustrate how palladium catalysts facilitate the assembly of complex hetero- and carbacycles from simple building blocks via cascade reactions involving molecular queuing processes. Meijere and Bräse describe how domino-type synthetic sequences involving palladium-catalysed cross-coupling reactions lead to complex structures in single operations involving multiple carbon–carbon bond formations. Lu and Ma show how palladium catalysis transforms simple acyclic allylic 2-alkynoate esters to α-alkylidene-γ-butyrolactones. Mortreux discusses the influence of the ligand environment on the reactivity and selectivity of nickel-catalysed carbon–carbon bond-forming processes with olefins and dienes as substrates.

Aromatic substitution involving the formation of carbon–carbon bonds has been revolutionized by transition metal catalysis. Catellani gives a general account of such substitution reactions, while Kakiuchi and Murai describe highly efficient methods for the addition of unactivated aromatic carbon–hydrogen bonds across olefins, and Reetz discusses new palladium catalysts for Heck reactions of unreactive aryl halides.

Allyl groups are precursors to many other functional groups and this, coupled with the ready availability of both nucleophilic and electrophilic allyl equivalents, makes allylation reactions one of the most studied areas of transition metal catalysis. Cross-couplng allylations and the allylations of carbonyl compounds, including the enantioselective addition of allyl stannanes to aldehydes in the presence of a chiral Lewis acid, are described by Yanagisawa and Yamamoto. The enantioselective catalysis of allylic substitutions with palladium complexes of phosphooxazolidines is discussed by Helmchen, Steinhagen and Kudis.

Ali and Alper describe the synthesis of lactones and lactams via ring expansions and cyclizations where the carbonyl group in the product is derived from carbon monoxide or an equivalent thereof. Doyle and Forbes describe the enantioselective cyclopropanations, carbon–hydrogen insertion and ylides of diazoketones catalysed by rhodium complexes. Recent advances in organocopper chemistry are addressed by Alexakis, who discusses copper(I)-promoted asymmetric transformations, and by Lipshutz, who demonstrates the synthetic power and potential of cyanocuprates.

The use of organotransition metal complexes to promote the formation of carbon–oxygen bonds is an area of enormous potential. Palladium-catalysed oxidations of alkenes are discussed by Hosokawa and Murahashi. New strategies for hydrogen peroxide activation are reviewed by Moiseev, while organic oxidations promoted by organorhenium oxides as catalysts are detailed by Herrmann and Kühn. Dixneuf and Bruneau illustrate the use of ruthenium catalysts to promote the addition of carboxylic acids across alkynes to generate vinyl esters and ethers.

The formation of carbon–silicon bonds is covered by Hayashi, who summarizes the development of chiral monodentate phosphine ligands to promote the asymmetric hydrosilylation of olefins catalysed by palladium. Suginome and Ito describe the transition-metal-catalysed addition of disilanes across olefins, which involve the formation of two carbon–silicon bonds in one reaction.

Transition metal catalysts can be used to promote the formation of organoboron compounds via carbon–boron bond formation, as well as the reactions of organoboranes to produce carbon–carbon bonds. Suzuki reviews both these areas of organoboron chemistry. Finally, the mechanism and synthetic potential of catalytic hydroboration are detailed by Brown *et al.*

Transtion-metal-catalysed reactions are already established as an important part of the synthetic chemist's arsenal. It is hoped that this monograph will serve as a catalyst to encourage the revelation, throughout the 21st century, of their as yet untapped real potential for synthesis.

Stephen G. Davies, *Oxford*
Shun-Ichi Murahashi, *Osaka*

1 Organic Reactions in Supercritical Fluids

TAKAO IKARIYA* and RYOJI NOYORI†

*Department of Chemical Engineering, Faculty of Engineering, Tokyo Institute of Technology 2-12-1 O-okayama, Meguro-ku, Tokyo 152, Japan, † Department of Chemistry and Research Center for Materials Science, Nagoya University, Chikusa, Nagoya 464-8602, Japan

1 Introduction

Organic reactions are performed largely in liquid solvents, in which all reactants dissolve to achieve homogeneity and to facilitate mass transport. Solvent molecules interact with solute molecules via various kinds of non-covalent forces such as van der Waals' forces, which consist of a dispersion force, charge–charge, charge–dipole or dipole–dipole interaction, and hydrogen bonding. These interactions play a central role in the effects of solvents on reaction rate and selectivity so one must consider carefully the properties of solvents, such as the solubility parameter, molar volume, polarity and cage effects, in addition to the molecular structures in order to interpret the reaction kinetics and mechanisms [1].

Figure 1.1 schematically illustrates the reaction phases: liquid, supercritical and gas. Solvent molecules surround a solute molecule to form a solvent shell due to the intermolecular forces between solute and solvent molecules. In the liquid phase, solvent molecules are packed densely to form a solvation sphere, where there is continuous rotation or exchange of the solvent shell molecules. On the other hand, in the gas phase, molecules have high mobility and interact without the formation of rigid aggregation of these molecules. The properties of supercritical fluids (SCFs) can vary continuously between those of liquids and gases, even though the inherent molecular structures of the fluids are maintained. Near the critical point especially, molecules in the supercritical phase are not uniformly distributed in space, but the solvent molecules aggregate around the solute by intermolecular solvent–solute interactions to form clusters, where the aggregated molecules are in dynamic equilibrium with free solvent molecules. The state of the solvation depends strongly on the density of the SCF and differs from that in the liquid solution or gas phase.

The solvent effects, molecular interactions and phase behavior in SCFs or liquid solutions are evaluated quantitatively under high-pressure and high-temperature conditions by means of spectroscopic analysis such as infrared (IR), near-IR, Raman, visible, ultraviolet (UV) or NMR spectroscopy [2, 3]. Solvatochromism, indicating a solvent-induced change in the electronic absorption spectrum of a solute, can be used to establish empirical solvent polarity [4]. The solvatochromic scales are used to correlate solvent effects on the rate constant in homogeneous liquid phases.

2 Properties of SCFs

The critical point of a pure compound is the endpoint of the liquid–gas line in the phase diagram (Fig. 1.2) and the point at which the liquid and gas phases become indistinguishable. The supercritical region of the diagram is the one at temperatures higher than the

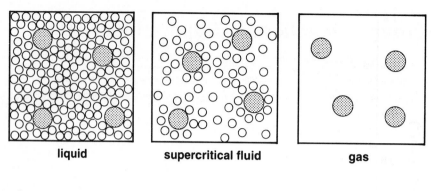

: solute   ○ : solvent

Figure 1.1. Schematic reaction phases.

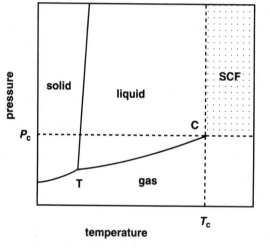

compound	T_c (°C)	P_c (atm)
H_2	−239.9	12.8
C_2H_4	9.9	50.5
Xe	16.6	57.7
CHF_3	25.9	47.8
CO_2	31.0	72.9
C_2H_6	32.3	48.2
CH_3OH	240.0	78.5
H_2O	374.2	218.3

T: triple point, C: critical point

Figure 1.2. Phase diagram of CO_2 and critical data of simple compounds.

critical temperature (T_c). Supercritical fluids exist as a single phase, combining some of the advantageous gas-phase properties, such as high miscibility with other gases, high diffusivity and relatively weak molecular association, with liquid-phase properties such as the ability to dissolve and transport organic compounds. The critical data of carbon dioxide (CO_2), for example, are: $T_c = 31$ °C and critical pressure (P_c) = 72.9 atm. Plate 1.1 shows photographs of the interior of the window-equipped reaction vessel under various conditions: the left photograph shows gently boiling liquid and gaseous CO_2 at 30 °C and close to the critical pressure; the middle photograph illustrates the apparent thickening of the meniscus as the temperature is raised to the critical temperature of 31 °C; and the right photograph shows only one fluid phase, supercritical CO_2 (scCO_2), above the critical temperature, at 32 °C.

Properties such as density (Fig. 1.3) are continuous above the T_c and discontinuous below it. Thus intermediate densities, which are impossible below the T_c, can be obtained in the supercritical region. Table 1.1 compares the density, viscosity and diffusivity of SCFs with those of typical liquids and gases. The properties of any compounds above their criti-

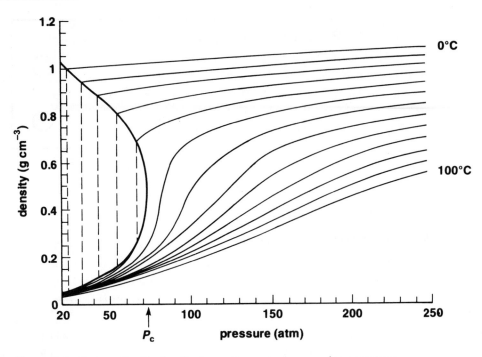

Figure 1.3. Carbon dioxide density depending on pressure and temperature.

Table 1.1. Typical liquid, supercritical fluid (SCF) and gas properties

	Liquid	SCF	Gas
Density (g cm^{-3})	1	0.1–0.5	10^{-3}
Viscosity (Pa·s)	10^{-3}	10^{-4}–10^{-5}	10^{-5}
Diffusivity (cm^2 s^{-1})	10^{-5}	10^{-3}	10^{-3}

cal points differ greatly from those below these points. For example, compared to liquid water, scH$_2$O ($T_c = 374\,°C$ and $P_c = 218\,atm$) is much less polar, can dissolve organics [5] and has a low dielectric constant ($\varepsilon = 6$ at the critical point but $\varepsilon = 90$ at the freezing point [5]) and a large dissociation constant ($pK_w = 8$ at the critical point but $pK_w = 15$ at the freezing point [6]). Supercritical CHF$_3$ ($T_c = 25.9\,°C$, $P_c = 47.8\,atm$) has a tunable dielectric constant that increases with pressure from $\varepsilon \approx 2$ at the critical point to $\varepsilon = 7$–8 at 200 atm [7]. Near the critical point, particularly in the ranges $T/T_c = 1.0$–1.1 and $P/P_c = 1$–2, these properties vary greatly with only minor increases in temperature or pressure because of high compressibility. This region offers intriguing phenomena but sometimes practical difficulties for experiments and industrial applications.

Supercritical CO$_2$, a commonly used SCF, has a large body of published phase and solubility data for binary mixtures with organic solutes [8]. Solvatochromic studies have found that the dipolarity and polarizability of scCO$_2$, as measured by the Kamlet–Taft π* parameter, is slightly lower (−0.1 at c 160 atm and 40 °C) than that of alkanes (0.0) [9]. Similarly, the polarity, measured on the $E_T(30)$ scale, is 31–33, comparable to the value of 31 for alkanes [4, 10, 11]. Supercritical CO$_2$ can dissolve a wide range of organic compounds. Per-

manent gases such as H_2 are sparingly soluble in liquid solvents, but they are highly miscible with $scCO_2$. For example, the concentration of H_2 in a supercritical mixture of 85 atm of
H_2 and 120 atm of CO_2 at 50 °C is 3.2 M, while the concentration of H_2 in tetrahydrofuran
(THF) under the same pressure is merely 0.4 M [12]. Even fluoropolymers that are soluble only in chlorofluorocarbons (CFCs) dissolve in $scCO_2$, indicating that the latter can
replace environmentally hazardous halogenated solvents [13]. The high solubility is probably due to a specific interaction between the fluorinated groups and CO_2. Factors that
increase the solubility of organic solutes in $scCO_2$ or scC_2H_6 include volatility, non-polarity
and the absence of unsaturation or protic functional groups [4, 8–11, 14]. For many compounds, the solubility limit can be raised by adding a small amount of a volatile cosolvent,
usually a polar or protic compound such as small alcohols, acetone or amines (Fig. 1.4)
[15, 16]. Water as a cosolvent of $scCO_2$ dramatically changes the performance of the solvent
[17]. Fourier transform infrared (FTIR) spectroscopy reveals that water in $scCO_2$ forms
microemulsions stabilized by an ammonium carboxylate perfluoropolyether surfactant,
and that water in microemulsion (bulk water, aggregated) is distinguishable from free
water dissolved in $scCO_2$ (Fig. 1.5) [18]. Experimental data concerning the phase behavior
of organic compounds have been accumulating. However, complete phase diagrams are
available for only a few binary mixtures of solutes and $scCO_2$ or scH_2O. At this moment,
therefore, one must determine the phase behavior at operating conditions to elucidate the
SCF effects [19]. Supercritical fluids are used as media in research and industry for extraction and chromatography, where their unique physicochemical properties are utilized for
the separation and purification of low-volatility and thermolabile substances. The decaf-
feination of coffee and tea and the extraction of hops, spices and drugs are now operating
on an industrial scale.

Supercritical fluids also offer many advantages as reaction media. Improved performance or novel behavior of stoichiometric organic reactions, particularly in $scCO_2$ and
scH_2O, has been investigated extensively [20–25]. Replacement of conventional liquid solvents by SCFs can also increase the rate and change the selectivity of homogeneously
catalysed reactions for the following reasons:

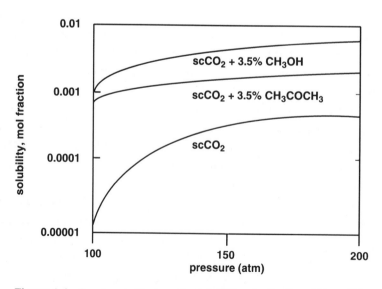

Figure 1.4. Cosolvent effects on the solubility of salicylic acid in $scCO_2$ at 55 °C.

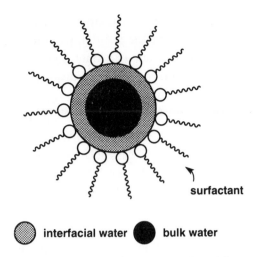

surfactant

interfacial water bulk water

Figure 1.5. Schematic representation of the water environments in microemulsions in scCO$_2$. Conditions: CO$_2$ (148 atm, 32 °C, $\rho = 0.92$ g cm^{-3}) and 1.4 wt.% surfactant with respect to CO$_2$.

1 The high solubility of reactant gases in SCFs [2, 26–31].

2 Rapid diffusion of solutes into, out of and within the supercritical phase.

3 Weakening of the solvation around reacting species in comparison to that in the liquid phase.

4 Local clustering of solutes or solvents resulting in an appreciable increase in the local concentration of solute or solvent molecules.

5 Reduction of the cage effect in radical reactions.

6 Large compressibility near the critical point resulting in very large negative activation volumes [32, 33].

As described in recent review articles, many reports on organic reactions in SCFs have been published, examining unique solvent effects of these media on reaction performance [3, 24, 25]. However, examples of organometallic homogeneous catalysis in SCFs are still rare. This review will concentrate on the recent development of organic reactions, coordination chemistry and homogeneous catalysis using transition-metal-based complexes in SCFs.

3 Organic reactions in SCFs

Supercritical fluids have unique physical properties that are tunable by small changes in pressure and/or temperature and by doping a small amount of a cosolvent. These properties offer the possibility of dramatic changes in reaction rates. In addition, one must consider the potentially strong pressure effects on the rate constant, which can be understood in terms of transition-state theory. Near the critical point, SCFs are highly compressible so that the density varies from gas-like to liquid-like over three orders of magnitude. These density changes can elongate the average intermolecular distance by a factor of 10 in comparison to that in the ordinary liquid phase. The size of clusters and the stoichiometry of the solvate complexes can be determined by spectroscopic methods [34].

In 1946, Toriumi first reported an unusual rate enhancement for the oxidation of ammonia to ammonium nitrate—an inorganic reaction—around the critical point of ammonia [35]. Since then, there have been a number of reports on the chemical reactions in SCFs as unorthodox reaction media to examine the possibility of tuning the reactivity and selectivity or improving the productivity. The rate of a kinetically controlled reaction is influenced significantly by the local environment of the statistical structure of solvents around the reactant molecules. For example, in the Diels–Alder reaction of isoprene and maleic anhydride with a characteristic large negative activation volume, the rate constants obtained in the near-critical region are much larger than those in a higher pressure region or in liquid solvents (Fig. 1.6) [32, 36, 37]. The activation volume $\Delta V^{\ddagger}$, representing primarily a difference in the solvation of reactants and transition state, can be estimated to be $-4000\,cm^3\,mol^{-1}$ at 35 °C and 74 atm, where CO_2 is highly compressible, while at 296 atm it is $-55\,cm^3\,mol^{-1}$, the value of which is comparable to that for the reaction in ethyl acetate, $-37.4\,cm^3\,mol^{-1}$ [1, 33]. Similarly, as illustrated in Fig. 1.7, the rate constant of unimolecular decomposition of α-chlorobenzyl methyl ether increases by an order of magnitude as the pressure decreases from 59 to 44 atm at around the critical point of 1,1-difluoroethane. The $\Delta V^{\ddagger}$ value is as low as $-6000\,cm^3\,mol^{-1}$ near the critical point, but its value is $-72\,cm^3\,mol^{-1}$ at a higher pressure region of $P/P_c = 6.1$, similar to that observed in conventional liquid solvents [32]. This rate enhancement is attributed to the increase in the local density at near-critical conditions, which is well correlated by solvatochromic data obtained for phenol blue in this SCF.

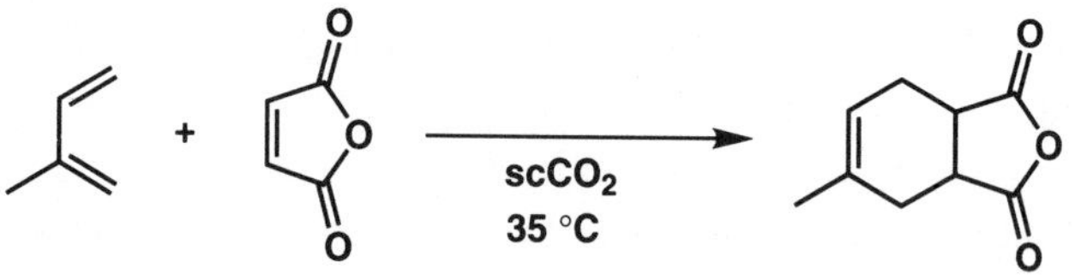

pressure atm	$-\Delta V^{\ddagger}$ $(cm^3\,mol^{-1})$
75	4000
200	70

Figure 1.6. Diels–Alder reaction of isoprene and maleic anhydride.

$\Delta V^{\ddagger} = -6000\,cm^3\,mol^{-1}$

proposed transition state

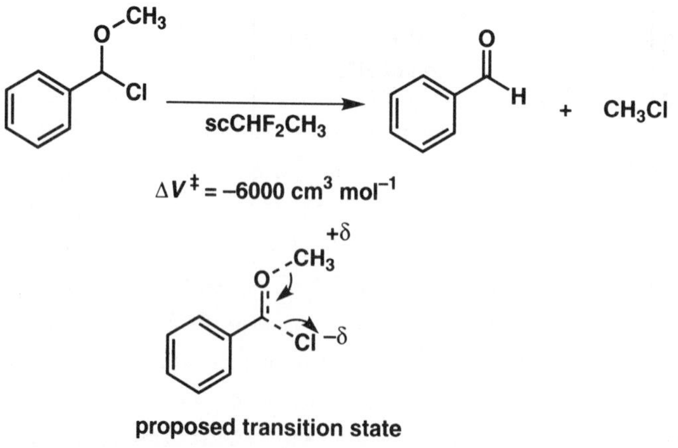

Figure 1.7. Thermal decomposition of α-chlorobenzyl methyl ether.

Solute–solute clustering, which results in an increase in local concentration of reactants, often facilitates bimolecular reactions in a near-critical point region. The hydrogen atom transfer from 2-propanol or 1,4-cyclohexadiene to triplet-state benzophenone, generated by laser flash photolysis in $scCO_2$, scC_2H_6 or $scCHF_3$, proceeds via the same mechanism as that in liquids. The reaction rates obtained in these SCFs are the same order of magnitude as those in liquids, which are approximately three orders of magnitude below diffusion control. As shown in Fig. 1.8, the second-order rate constant decreases sharply to about one-quarter with an increase in pressure above the critical pressure [38, 39]. Because the thermodynamic pressure effect on the rate constant suggests an increase in the rate constant with pressure, the observed unusual pressure effect on the reaction rate at a near-critical point is probably due to an increase in the local concentration of hydrogen donors around the triplet benzophenone.

In the esterification of phthalic anhydride with methanol in $scCO_2$, a 25-fold decrease in the bimolecular rate constant is observed when the pressure is increased from 96 to 164 atm (Fig. 1.9) [40, 41]. The extent of the contribution of the thermodynamic pressure effect suggested by transition-state theory on the rate is not significant: less than a twofold decrease for the increase in the pressure tested. The higher rate constants observed at lower pressures are interpreted as a considerable increase in the local concentration of methanol around anhydride near the critical point. Spectroscopic studies for methanol around the solute in $scCO_2$ show that the local composition increase is of the magnitude that would be required to create the observed discrepancy [41].

The rate of the Michael addition of piperidine to methyl propiolate in $scCHF_3$ or scC_2H_6 at 37 °C depends on fluid density [42] (Fig. 1.10). The quantitative study of the reaction kinetics in scC_2H_6 was unsuccessful due to the low reaction rate. The rate constant in

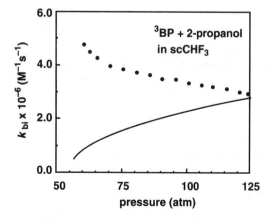

Figure 1.8. Reaction of triplet benzophenone and 2-propanol or 1,4-cyclohexadiene in SCFs.

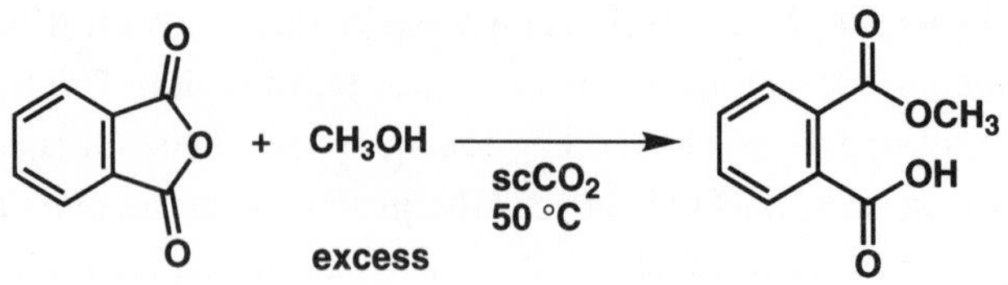

pressure atm	rate constant $dm^3 \, mol^{-1} \, min^{-1}$
96	3.48×10^{-2}
164	1.38×10^{-3}

Figure 1.9. Esterification of phthalic anhydride with CH_3OH in $scCO_2$.

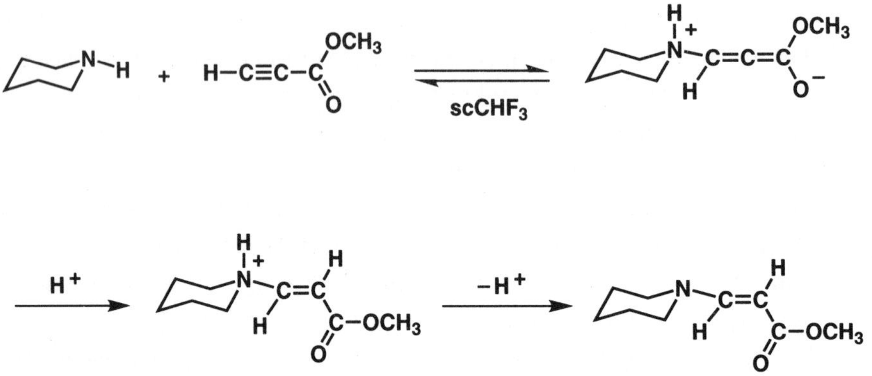

Figure 1.10. Michael addition of piperidine to methyl propiolate in $scCHF_3$.

$scCHF_3$, a polar fluid, is linearly related to pressure except near the critical point. The observed rate constant dependence on pressure is attributed to solvent polarity at higher pressures, as would be consistent with the stabilization of a highly polar transition state. The unusually large rates observed near the critical points are interpreted in terms of solvent–solute clustering in $scCHF_3$.

The enhancement of the local solvent density around a solute in SCFs often retards a rapid reaction such as annihilation of a triplet state of benzophenone [38]. The pyrene excimer formation in $scCO_2$ and scC_2H_4 proceeds at diffusion control near and far from the critical point via the same mechanism as that in liquids. The lifetime of the excimer fluorescence in these SCFs can be lengthened over solution behavior due to a protective environment for the excimer created by solvent clustering. On the other hand, in $scCHF_3$ the reaction is about 10 times below diffusion control in the near-critical region due to the enhancement of the ground-state pyrene–solvent interaction. Supercritical CHF_3 does not affect the lifetime of the excimer, indicating that the cluster sizes of CO_2 and C_2H_4 are larger than that of CHF_3 [43, 44].

In addition to rate control, the selectivity or product distribution in the chemical reactions can also be tuned by using SCFs: through the differing magnitude of solvent effects on parallel reaction pathways; and by the mechanistic change resulting from perturbation in the nature of the SCFs with pressure or doping a cosolvent. For example, as shown in Fig. 1.11, photodimerization of isophorone gives three major products, an *anti*-configurated head-to-head dimer and *anti*- and *syn*-configurated head-to-tail isomers [45]. The product

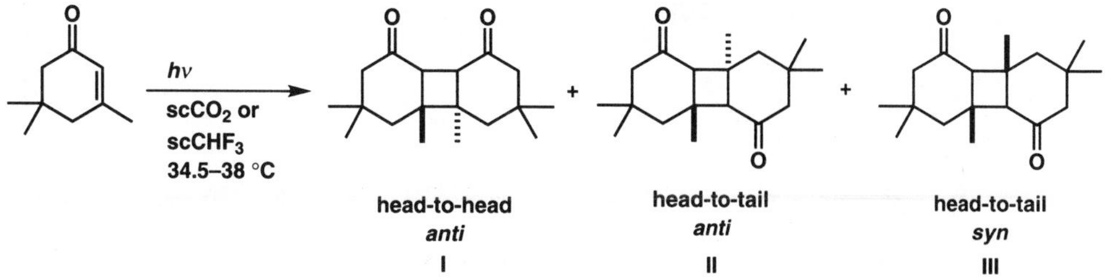

medium	pressure (atm)	ε	head-to-head:head-to-tail (I:II+III)	anti:syn (II:III)
scCO₂	88–280	1.34–1.35	0.09–0.1	2.00–3.95
scCHF₃	56–270	2.5–8.4	0.74–1.0	2.56–3.32

Figure 1.11. Photodimerization of isophorone in SCFs.

distribution varies as a function of the pressure of the supercritical media, the ratio of the head-to-head to total head-to-tail dimers being considerably higher in $scCHF_3$ than in $scCO_2$. This change in the selectivity is attributed to the polarity of the fluid, significantly affecting the polar transition state as well as solvent reorganization. The stereoselectivity is influenced largely by solvent reorganization, whereas the regiochemistry is affected by both solvent polarity and solvent reorganization. The dielectric constant of $scCHF_3$ varies from 2.5 to 8.4 at 34.5 °C with pressure [7], while the dielectric constant of $scCO_2$ does not change significantly over the entire pressure range.

Cosolvents capable of interacting with solute molecules also sensitively tune the properties of SCFs, offering the possibility of reaction control. In the tautomeric equilibrium of the Schiff base 4-(methoxy)-1-(N-phenylforminidoyl)-2-naphthol shown in Fig. 1.12, the enol form is stabilized by intramolecular hydrogen bonding in non-polar solvents, while in polar solvents, such as alcohols, the keto form is more stabilized because of intermolecular hydrogen bonding between the carbonyl oxygen of the keto form and the hydroxy group of solvent alcohols. The equilibrium position of the Schiff-base tautomerism is tunable by modifying pure scC_2H_6 with less than 2 mol.% of protic cosolvents [16]. Alcohols capable of hydrogen bonding interaction lead to a significant shift of this equilibrium from mostly the enol form in the absence of the cosolvent to the keto tautomer (60%) in a mixture of scC_2H_6 and alcohol. The best cosolvent choice is $(CF_3)_2CHOH$ because of its stronger proton donor ability relative to ethanol.

Cage effects of SCFs on radical reactions have been well investigated compared to those in conventional solvents. Because SCFs are less viscous than liquid solvents, a small cage effect might be expected in SCF media, which can directly affect rapid reactions of reactive species in close proximity in a solvent cavity. For example, the initiation efficiency of 2,2′-azobis(isobutyronitrile) (AIBN) for radical polymerization in $scCO_2$ is 1.5 times greater than in benzene, suggesting a weaker cage effect (Fig. 1.13) [4, 46]. As shown in Fig. 1.14, the Norrish type I photolysis of 1-(p-methylphenyl)-3-phenyl-2-propanone in $scCO_2$ and scC_2H_6 gives a statistical mixture of the dibenzyls formed by random coupling of the benzyl and p-methylbenzyl radicals, indicating that no solvent cage effect is operative in these SCFs even near their critical points [47]. Similarly, as shown in Fig. 1.15, photo-initiated free-radical chlorination of cyclohexane in $scCO_2$ and conventional solvents

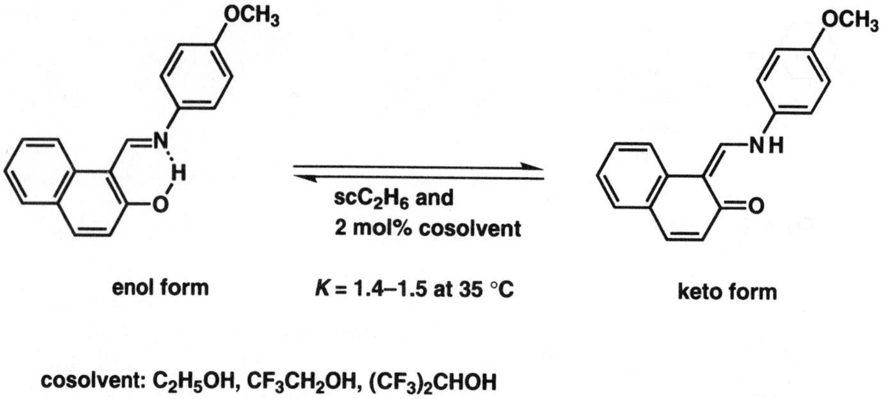

cosolvent: C₂H₅OH, CF₃CH₂OH, (CF₃)₂CHOH

Figure 1.12. Tautomeric equilibrium of a Schiff base in a mixture of scC$_2$H$_6$ and alcohol.

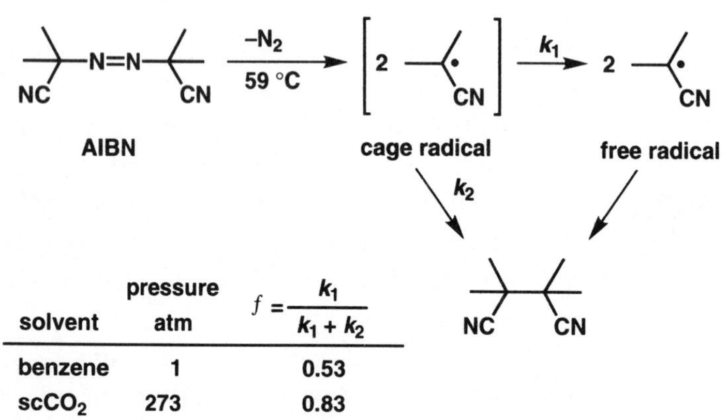

solvent	pressure atm	$f = \dfrac{k_1}{k_1 + k_2}$
benzene	1	0.53
scCO₂	273	0.83

Figure 1.13. Thermal decomposition of 2,2'-azobis(isobutyronitrile) (AIBN) in scCO$_2$.

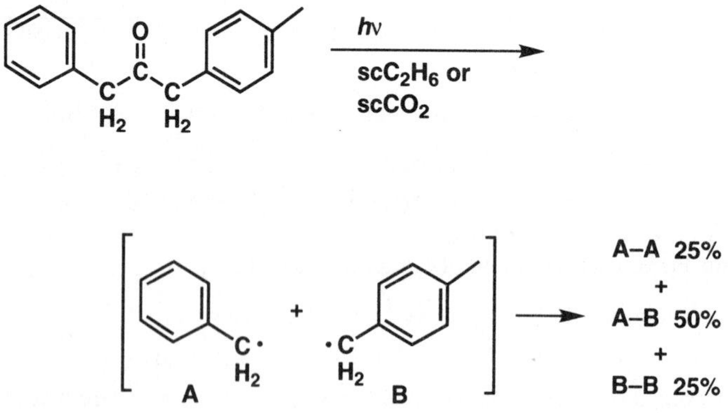

A–A 25%
+
A–B 50%
+
B–B 25%

Figure 1.14. Photolysis of dibenzyl ketones in scC$_2$H$_6$ or scCO$_2$.

provides a mixture of the monochloride (M), generated from the cage-escape reaction, and the polychlorides (P) formed by in-cage abstraction of hydrogen atoms from cyclohexyl chloride. The dependence of the M/P ratio on the inverse of the viscosity is linear over a range of viscosities from conventional solvents to scCO$_2$, indicating no enhanced cage

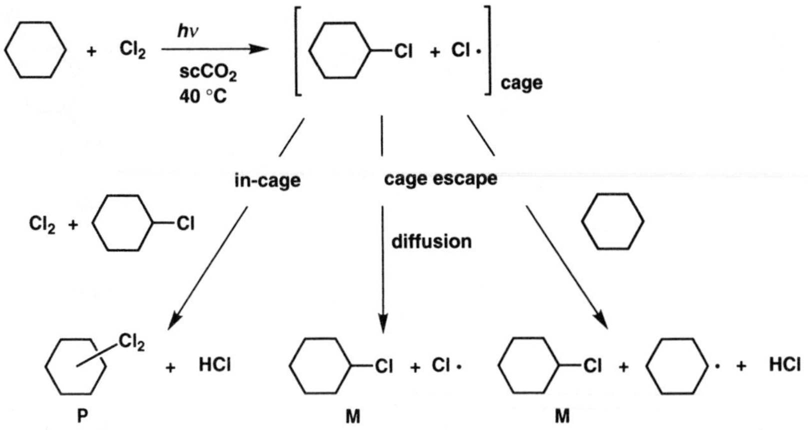

Figure 1.15. Photoinitiated free-radical chlorination of cyclohexane in $scCO_2$.

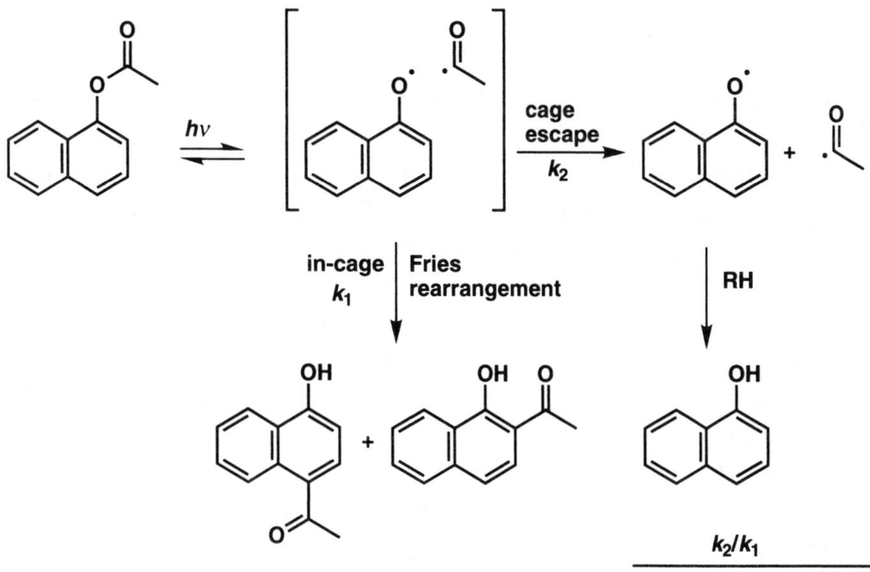

solvent	temp (°C)	76	80	110 (atm)
liqCO$_2$	27	4.1	4.1	4.1
scCO$_2$	47	5.2	13.5	4.0

Figure 1.16. Photo-Fries rearrangement of 1-naphthyl acetate.

effect even near the critical point of $scCO_2$ [48]. The product ratio of M/P is affected only by solvent viscosity.

On the other hand, the clustering between a solute molecule and solvent molecules causes an enhancement of the cage effect due to an increase in local density of a solvent molecule about a radical rather than the bulk density. The photo-Fries rearrangement of naphthyl acetate (Fig. 1.16) is known to proceed via a short-lived singlet radical pair with a lifetime of c 25 ps to form naphthol by abstraction of hydrogen atoms in the absence of a solvent cage. The photolysis in $scCO_2$ containing 2-propanol provides the products arising

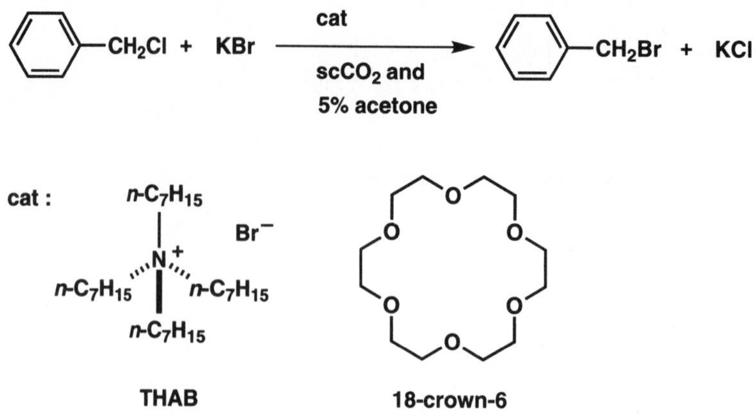

Figure 1.17. Phase-transfer catalysis in scCO$_2$.

from an in-cage reaction of the geminate radical pair and products from a cage-escape reaction [49]. The ratio of the in-cage product to cage-escape product is invariant with pressure except near the critical point, where the in-cage rearrangement product is predominantly formed because of a significant cage effect.

Phase-transfer catalysis (PTC) is an important and effective method for conducting heterogeneous reactions, including both ionic reactants and non-polar molecules, in synthetic organic chemistry. Supercritical fluids are extremely attractive media for PTC because the diffusivity and mass-transfer ability of solutes between phases are gas-like. Common phase-transfer catalysts such as quaternary ammonium salts or cyclic ethers are generally not soluble in scCO$_2$ but dissolve in this medium doped with a suitable cosolvent. For example, the solubility of tetraheptylammonium bromide (THAB) in a mixture of scCO$_2$ and 5% acetone is of the order of 10^{-5} mole fraction and that of 18-crown-6/KBr complex is of the order of 10^{-7}–10^{-8} mole fraction at c 200 atm and 50–75 °C [50]. These catalysts effect the nucleophilic displacement of benzyl chloride with a bromide ion in scCO$_2$ containing acetone, although no reaction takes place in the absence of the catalysts. This is the first example of successful PTC between a solid phase and SCF (Fig. 1.17) [50].

4 Coordination chemistry in SCFs

Although very few quantitative data have been reported on the solubility of transition metal complexes in SCFs, neutral complexes with carbonyl (CO), cyclopentadienyl, porphyrin, acetylacetonate and other chelating ligands have been reported to be soluble to some extent in scCO$_2$ [51, 52]. For example, as shown in Plate 1.2, Fe$_3$(CO)$_{12}$ is soluble in scCO$_2$ at 35 °C and 100 atm, providing a green-colored fluid (left photograph). As the pressure is raised to 200 atm, the green color of the fluid deepens (middle photograph). When the pressure is decreased to atmospheric pressure, black-green crystals are deposited on the walls and the windows of the reaction vessel (right photograph). We confirmed that a trimethylphosphine-containing complex, RuCl$_2$[P(CH$_3$)$_3$]$_4$, is soluble in scCO$_2$, while the triphenylphosphine analog is sparingly soluble [28, 53]. The solubility of metal complexes in scCO$_2$ is also increased by using a lipophilic, fluorinated anion [B{3,5-(CF$_3$)$_2$C$_6$H$_3$}$_4$]$^-$ as the counter ion of cationic Rh complexes [54], using perfluoroalkyl-substituted arylphosphines as ligands [55] or using polyfluorinated alkanols as cosolvents [56].

The complete miscibility of gaseous compounds such as N_2 and H_2 with SCFs allows the preparation of novel organometallic compounds that are otherwise difficult to form [2, 57]. For example, as illustrated in Fig. 1.18, ultraviolet (UV) photolysis of a solution of a Re carbonyl complex and N_2 in scXe ($T_c = 16.7\,°C$, $P_c = 57.6\,atm$) causes substitution of the CO ligands to generate unique N_2 complexes of Re [29, 58]. In a similar manner, a previously unknown H_2–Mn complex with a non-classical structure has been synthesized by selective substitution of the CO ligand. Carbonyl complexes of Cr or Fe also can be used as starting materials in the new methodology [30, 31]. Because Xe possesses no vibrational characteristics and hence is completely transparent to IR, near-IR and UV radiation, spectroscopic analysis in scXe offers an excellent characterization methodology. These products were identified spectroscopically but would not be as stable were it not for the high concentrations of N_2 or H_2 possible in the SCFs [30].

This photochemical method has been used for the formation of a series of organometallic noble gas compounds—$W(CO)_5Ar$, $M(CO)_5Kr$ and $M(CO)_5Xe$ (M = Cr, Mo, W)—in the supercritical state of noble gases at ambient temperature [59]. The time-resolved infrared (TRIR) technique in the SCF offers a general method to study the formation of short-lived complexes and the interaction of weakly coordinating noble gas ligands with metal centers. As shown in Fig. 1.19, flash photolyses of $M(CO)_6$ in scAr ($T_c = -122.1\,°C$, $P_c = 48.3\,atm$), scKr ($T_c = -63.6\,°C$, $P_c = 54.2\,atm$) or scXe in the presence of CO generate noble-gas-coordinated complexes $M(CO)_5L$ (L = Ar, Kr, Xe). Similarly, $M(CO)_5(CO_2)$ is also obtainable in scCO$_2$. The TRIR spectra of a reaction mixture of $M(CO)_5L$ and other gases reveal that the reactivity of these complexes for the substitution is $Kr > Xe \approx CO_2$. Surprisingly, $M(CO)_5Xe$ complexes that remained unisolable at low temperature have proved stable even at room temperature. They react with CO via a dissociative mechanism to give $M(CO)_6$. The second-order rate constants for the reaction of $W(CO)_5Kr$ or

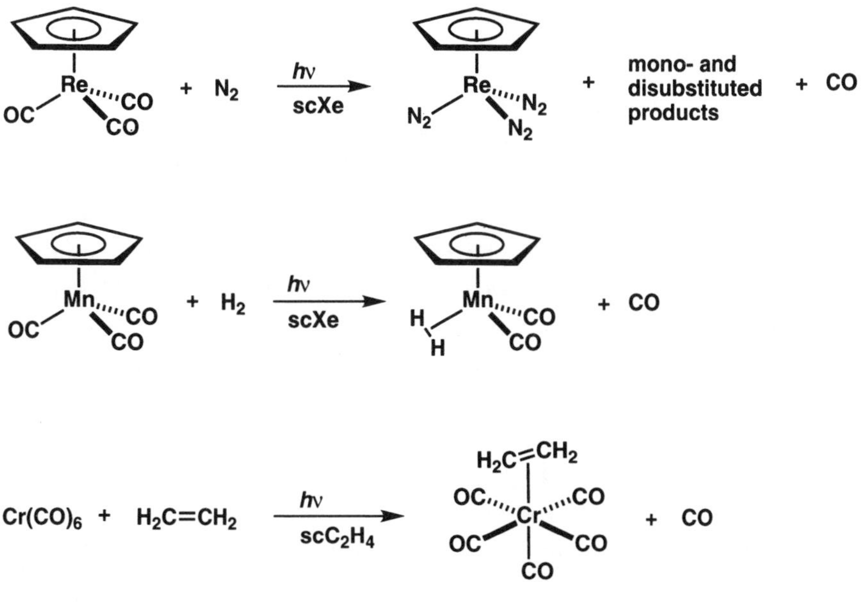

Figure 1.18. Formation of N_2, η^2-H_2 and C_2H_4 complexes in SCFs.

$$M(CO)_6 \xrightarrow[\text{scKr}]{h\nu} M(CO)_5 + CO$$

$$M(CO)_5 + Kr \xrightarrow{\text{scKr}} M(CO)_5Kr$$

$$M(CO)_5Kr + CO_2 \longrightarrow M(CO)_5(CO_2) + Kr$$

$$M(CO)_5Kr + CO \longrightarrow M(CO)_6 + Kr$$

M = Cr, Mo, W

Figure 1.19. Photolysis of $M(CO)_6$ in scKr and reactions of $M(CO)_5Kr$ with CO_2 and CO.

$W(CO)_5Xe$ with CO at 25 °C are 75×10^7 and 2.0×10^6 mol^{-1} dm^3 s^{-1}, respectively. Similarly, $CpRe(CO)_2Kr$ or $CpRe(CO)_2Xe$ can be obtained from the reaction of $CpRe(CO)_3$ and noble gas ligands in scKr or scXe. The complex $CpRe(CO)_2Xe$ is less reactive toward substitution reactions with CO than $W(CO)_5Xe$, and the rate constant for the CO reaction is 4.8×10^3 mol^{-1} dm^3 s^{-1} [59].

Activation of the C–H bonds of C_2H_6 is possible under supercritical conditions (Fig. 1.20). For example, UV irradiation of $Cp^*Ir(CO)_2$ [$Cp^* = C_5(CH_3)_5$] in H_2-doped scC_2H_6 ($T_c = 32.4$ °C, $P_c = 48.2$ atm) generates a mixture of $Cp^*Ir(CO)(H)_2$ and $Cp^*Ir(CO)H(C_2H_5)$, where the scC_2H_6 is both the reactant and the solvent [60]. The high concentration of H_2 in the SCF greatly increases the yield but, even in the presence of D_2, $Cp^*Ir(CO)H(C_2H_5)$ and not $Cp^*Ir(CO)D(C_2H_5)$ is formed. Interestingly, the C–H activation of C_2H_6 is almost completely suppressed in a mixture of scC_2H_6 and polar CHF_3 even in the presence of H_2, although the reaction of $Cp^*Ir(CO)_2$ with H_2 occurs efficiently in scCHF_3. This is probably due to the stronger coordination ability of CHF_3, rather than H_2, on the central Ir atom. These results indicate that the intermediate complex $Cp^*Ir(CO)H_2$ effects more efficiently the C–H bond activation than does $Cp^*Ir(CO)_2$ (Fig. 1.22).

The C–H bond activation of methane is achieved in scCO_2 or scXe but not effectively, whereas in scCH_4 ($T_c = -77.5$ °C, $P_c = 39.5$ atm) at 22 °C effective C–H bond activation occurs with the formation of $Cp^*Ir(CO)_2H(CH)_3$ because of the high concentration of CH_4 under these conditions [60]. Similarly, the C–H bonds of polyethylene (PE) are activated to form $Cp^*Ir(CO)_2H$(polymer) with a characteristic Ir–H vibration band at 2162 cm^{-1}, wherein scCO_2 is used to carry $Cp^*Ir(CO)_2$ into the PE film and irradiation results in C–H bond activation (Fig. 1.21). The unreacted Ir complex can be removed subsequently by washing with scCO_2 [61]. This reaction exploits the ability of SCF to diffuse rapidly into and out of polymers, which property also allows scCO_2 to carry small complexes into PE, a potentially useful matrix for photochemical studies of coordination complexes [62]. Impregnation of PE with $Mn_2(CO)_{10}$ is not normally possible but is achievable by carrying $MnH(CO)_5$ into the PE with scCO_2 and then irradiating the complex to convert it into $Mn_2(CO)_{10}$ with loss of H_2 [63]. Reaction of C_2H_4 and $Cp^*Ir(CO)_2$ in scXe provides predominantly $Cp^*Ir(CO)H(CH=CH_2)$ in addition to two possible ethylene complexes: $Cp^*Ir(CO)(\eta^2\text{-}C_2H_4)$ and $Cp^*Ir(\eta^2\text{-}C_2H_4)_2$ (Fig. 1.21) [60]. A new complex, $Cr(CO)_5(C_2H_4)$, is also isolable by dissolving $Cr(CO)_6$ in scC_2H_4 in a flow reactor system, passing it through a UV photolysis cell and then venting it into a collection vessel [64].

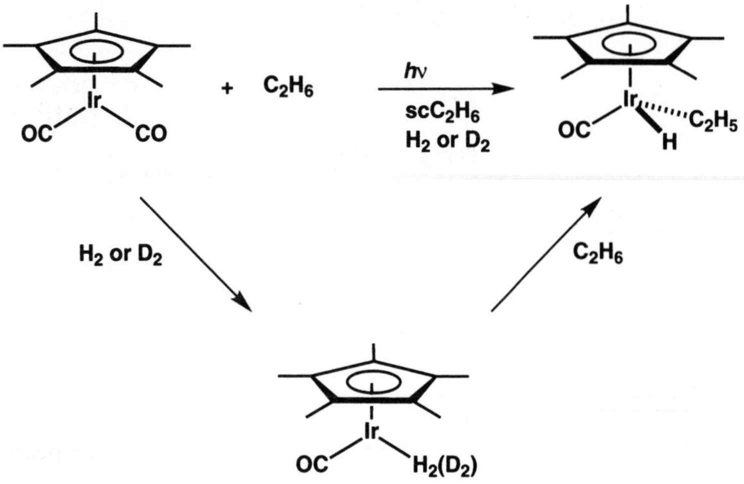

Figure 1.20. Reactions of Cp*Ir(CO)$_2$ and ethane in scC$_2$H$_6$.

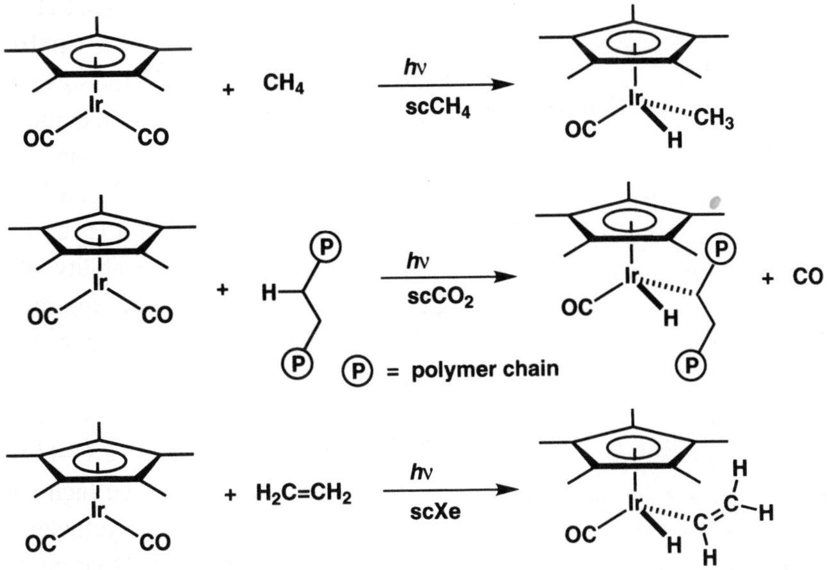

Figure 1.21. Reactions of Cp*Ir(CO)$_2$ with methane, polyethylene and ethylene in SCFs.

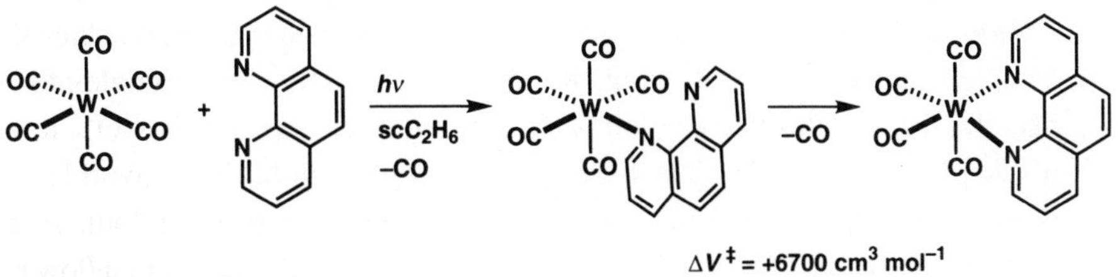

$$\Delta V^{\ddagger} = +6700 \text{ cm}^3 \text{ mol}^{-1}$$

Figure 1.22. Reaction of W(CO)$_6$ with 1,10-phenanthroline in scC$_2$H$_6$.

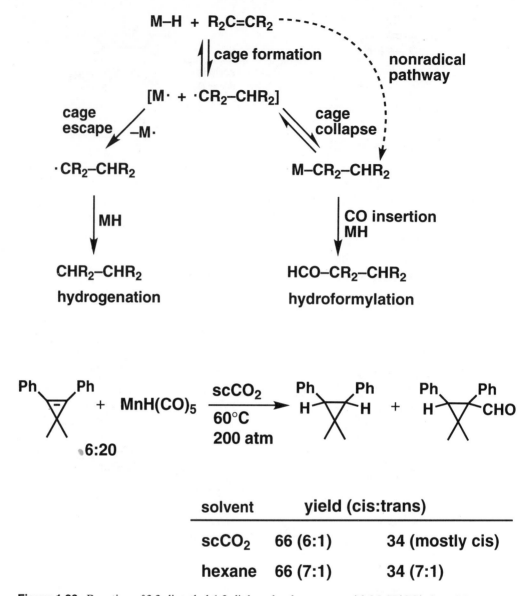

Figure 1.23. Reaction of 3,3-dimethyl-1,2-diphenylcyclopropene with MnH(CO)$_5$ in scCO$_2$.

Ligand substitution in metal carbonyl complexes in SCFs has been investigated to examine the polarity and large compressibility in the near-critical region that affect the rate of organometallic reactions. As illustrated in Fig. 1.22, laser flash photolysis of a mixture of W(CO)$_6$ and 1,10-phenanthroline (phen) in scC$_2$H$_6$ or scCO$_2$ results in the formation of a ring-closed W(CO)$_4$(phen) complex through the ring-opened W(CO)$_5$(phen*) complex that is generated from the extremely reactive species W(CO)$_5$ and the phen ligand. The ring-closure reaction is the rate-limiting step and the volume of activation is 6700 cm^3 mol^{-1} in scC$_2$H$_6$, just above the critical point [65]. This value is about three orders of magnitude larger than those in the higher pressure region that are seen in liquid solvents. The large positive activation volume indicates that in the highly compressible near-critical region a huge repulsive interaction of the solvent and solute is operating toward the transition state of reaction, while the reactant is more solvated by solvent–solute clustering.

We studied the stoichiometric reaction of $MnH(CO)_5$ and activated olefins in order to test the solvent cage effect of $scCO_2$ on the reaction performance [66]. No cage effect is expected for reactions that are much slower than cage escape [21]. However, for very rapid reactions such as those involving radical pairs, the possibility of altered selectivity or rate is considered. Stoichiometric and catalytic hydrogenation and hydroformylation of activated olefins with Mn and Co carbonyls in liquid solvents are generally believed to occur with a radical-pair mechanism as shown in Fig. 1.23. In this mechanism, hydroformylation requires a solvent cage to accommodate the alkyl and metal radicals. A solvent with a weak cage effect should increase selectivity for hydrogenation rather than hydroformylation. However, a test reaction (Fig. 1.23) showed comparable selectivity in hexane and $scCO_2$, indicating either that the cage effects in the two media are comparable or, more likely, that non-radical mechanisms are involved.

5   Homogeneous organometallic catalysis in SCFs

Molecular catalyses are usually conducted in liquid solvents. Use of homogeneous organometallic catalysts of adjustable molecular structures allows high selectivity for the production of a range of small to large molecules, including chiral compounds [67, 68]. Catalysis, however, often suffers from low reaction rates resulting from solvent effects, including a strong solvation to the catalyst and slow mass transfer, as well as low concentration of gaseous reactants in the liquid phase. An extreme case that has no such negative solvent effects is the gas-phase heterogeneous reactions that use solid catalysts. Heterogeneous catalyses are widely used in industry because they normally allow high-rate reactions and easy separation of the solid catalysts from the products, in spite of the difficulty in selectivity or tunability.

The ideal catalyst would have the best of both worlds: high reactivity, high selectivity and operational simplicity. To create an ideal reaction system the molecular catalyst must be dissolved but have a weak solvent shell, slow mass transfer or diffusion must be eliminated and reactants must be in high concentrations. Adopting SCFs as reaction media may allow molecular catalysis to meet these goals [69]. They offer a great opportunity to speed up catalytic reactions involving gaseous reactants such as H_2 and CO, to tune selectivity in the homogeneous catalysis, and to replace conventional organic solvents with environmentally benign solvents such as CO_2.

Simultaneous use of SCFs as both reaction medium and reactant leads to novel, highly selective catalytic chemistry. An early example is seen in the 1966 patent for the polymerization of scC_2H_4 or scC_3H_6 with a soluble $AlCl(C_2H_5)_2/TiCl_4$ catalyst system (Fig. 1.24) [70]. The homogeneous metallocene-catalysed polymerization of scC_2H_4 to form linear low-density polyethylene has been commercialized [71].

Two different groups, in Germany and Japan, have independently and simultaneously studied homogeneously catalysed fixation reactions of CO_2 in the supercritical state. Reetz and co-workers [72] reported a coupling reaction of $scCO_2$ and 3-hexyne catalysed by a Ni complex, giving tetraethyl-2-pyrone (Fig. 1.25) with ~5 TON (turnover number, defined as moles of product per mole of catalyst). The conversion, unfortunately, was lower than in benzene at $120\,°C$ [73].

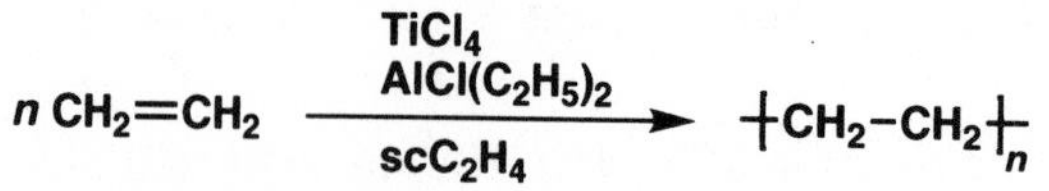

$$n\ CH_2{=}CH_2 \xrightarrow[\text{scC}_2\text{H}_4]{\substack{\text{TiCl}_4 \\ \text{AlCl(C}_2\text{H}_5)_2}} +CH_2{-}CH_2+_n$$

Figure 1.24. Polymerization of scC_2H_4.

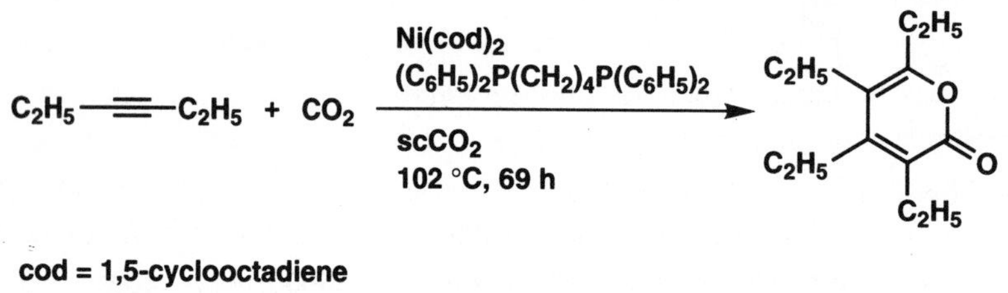

$$C_2H_5{-}{\equiv}{-}C_2H_5\ +\ CO_2 \xrightarrow[\substack{\text{scCO}_2 \\ 102\ °C,\ 69\ h}]{\substack{\text{Ni(cod)}_2 \\ (\text{C}_6\text{H}_5)_2\text{P(CH}_2)_4\text{P(C}_6\text{H}_5)_2}}$$

cod = 1,5-cyclooctadiene

Figure 1.25. Nickel(0)-catalysed coupling reaction of 3-hexyne in $scCO_2$.

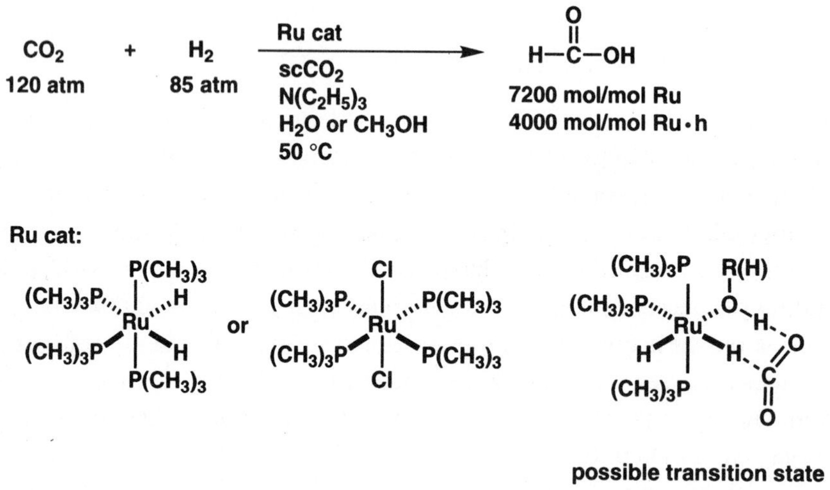

$$CO_2\ +\ H_2 \xrightarrow[\substack{\text{scCO}_2 \\ \text{N(C}_2\text{H}_5)_3 \\ \text{H}_2\text{O or CH}_3\text{OH} \\ 50\ °C}]{\text{Ru cat}} H{-}\overset{\overset{\displaystyle O}{\|}}{C}{-}OH$$

7200 mol/mol Ru
4000 mol/mol Ru·h

Ru cat:

possible transition state

Figure 1.26. Homogeneous hydrogenation of CO_2 to formic acid in $scCO_2$.

We studied the efficient homogeneous hydrogenation of $scCO_2$ to formic acid with $RuH_2[P(CH_3)_3]_4$ or $RuCl_2[P(CH_3)_3]_4$ as catalyst and water as a promoter (Fig. 1.26) [28, 74]. The trimethylphosphine–Ru complexes are the best catalyst choice for this reaction because of their high solubility in this medium. The catalytic efficiency of 7200 TON and initial rate or TOF (turnover frequency, TON per hour) of 1400 are greater than any previously reported for reactions under subcritical conditions. The conventional triphenylphosphine-based catalyst $RuH_2[P(C_6H_5)_3]_4$ is less active, possibly because of its low solubility in $scCO_2$. The base, $N(C_2H_5)_3$, is necessary to combat the unfavorable thermodynamics of the reaction by forming a complex of HCO_2H and $N(C_2H_5)_3$. Addition of a trace amount of an alcohol as a cosolvent to pure $scCO_2$ results in a significant improvement in the rate of the reaction, with a TOF exceeding 4000 [12]. The great enhancement of the reaction rate is attributed to the hydrogen-bond stabilization of the transition state as shown in Fig. 1.26.

As illustrated in the photographs of Plate 1.3, the reaction system is homogeneous at the start of the reaction (left), but, as the reaction proceeds, complexes of HCO_2H and $N(C_2H_5)_3$ precipitate as insoluble liquids (right). When the catalysis is conducted in two phases from the start by using a large amount of the amine, water, THF or CH_3CN as a

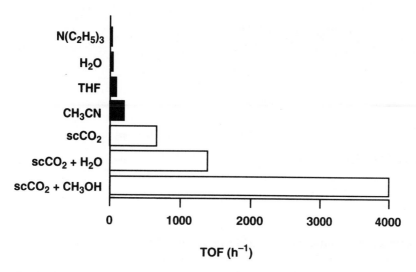

Figure 1.27. Solvent effect on the rate of hydrogenation of $scCO_2$. TOF = turnover frequency. Conditions: 50 °C, 83 atm H_2, 120 atm CO_2, 0.1 mmol H_2O or CH_3OH.

liquid solvent, the reaction proceeds slowly as shown in Fig. 1.27. The decrease in the reaction rate is attributed to dissolution of the Ru catalyst in the liquid phase where the reaction proceeds. The conspicuous rate enhancement demonstrates the benefits of the homogeneous supercritical phase.

Other industrially significant reactions were also tested. First, as shown in Fig. 1.28, addition of methanol in amounts below the solubility limit to $scCO_2$ caused the esterification of the formic acid product to methyl formate [75]. The methanol has two roles, acting as a kinetic promoter of the hydrogenation of CO_2 and as the esterification agent of the resulting HCO_2H. Furthermore, use of secondary or primary amines or ammonia in place of triethylamine resulted in very high yields of formamides. For example, *N,N*-dimethylformamide (DMF) is produced from dimethylamine and CO_2 with up to 420 000 TON and in 99% selectivity [12, 53]. The reaction takes place in two steps: hydrogenation to $[NH_2(CH_3)_2]^+[HCO_2]_-$ followed by thermal dehydration. The formamide synthesis differs from the syntheses of formic acid and methyl formate in that two phases are present from the start. The dimethylamine and CO_2 form an insoluble liquid carbamate salt. Because the Ru catalyst is insoluble in this liquid salt and also in the aqueous product phase that forms in a later stage of the reaction (Fig. 1.29), the hydrogenation occurs in the supercritical phase. The syntheses of methyl formate and DMF in $scCO_2$ have far greater TON and TOF than those reported for the same reactions in liquid solvents. The high rates and productivity are due to the high concentration of H_2 in $scCO_2$, rapid mass transfer between phases and the high reactivity of the catalyst in the supercritical phase. The high TON obtained demonstrates the long catalyst lifetime possible under these conditions.

Even CO_2-insoluble Ru complexes with bidentate phosphine ligands effect efficiently the formation of DMF and methyl formate at 100 °C via hydrogenation of CO_2 under super- and subcritical conditions [76]. The solubility of the Ru complexes might be increased at the elevated temperature even in a subcritical CO_2/H_2 fluid doped with dimethylamine or methanol as a cosolvent, because the latter is capable of strong interaction with the metal complexes. When hydrogen becomes more readily available by electrolysis of water using solar or hydroelectric energy, this efficient and non-toxic CO_2 chemistry could replace some of the current CO-based processes.

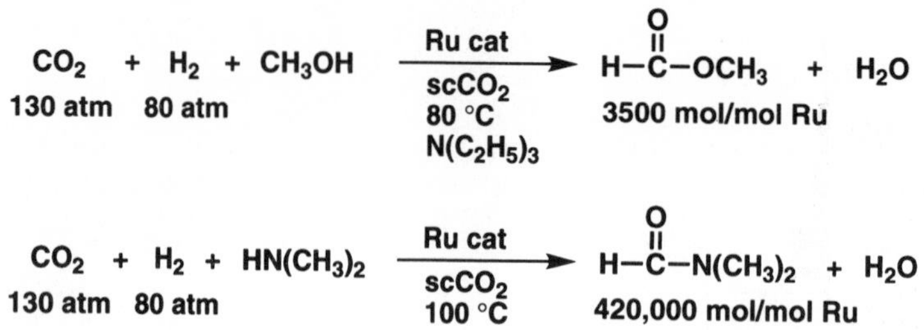

CO2 + H2 + CH3OH
130 atm 80 atm
Ru cat
scCO2
80 °C
N(C2H5)3
H–C(=O)–OCH3 + H2O
3500 mol/mol Ru
CO2 + H2 + HN(CH3)2
130 atm 80 atm
Ru cat
scCO2
100 °C
H–C(=O)–N(CH3)2 + H2O
420,000 mol/mol Ru

Figure 1.28. Formation of methyl formate and *N,N*-dimethylformamide (DMF) via hydrogenation of $scCO_2$.

supercritical phase
liquid phase
CO2, H2
Ru catalyst
R2NH
[NR2H2]+[HCO2]−
[NR2H2]+[NR2CO2]−
CO2, H2
Ru catalyst
DMF
[NR2H2]+[HCO2]−
H2O, DMF
CO2, H2
Ru catalyst
DMF
H2O, DMF
early
intermediate
final

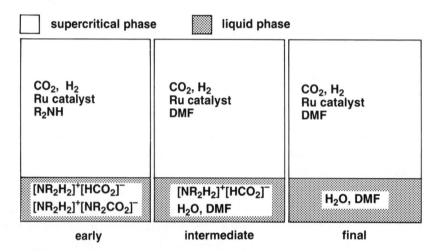

Figure 1.29. Composition of the phases during the formation of *N,N*-dimethylformamide (DMF) from dimethylamine, H_2 and $scCO_2$ ($R = CH_3$).

Supercritical fluids can also be used profitably as inert media for homogeneous catalysis, including hydrogenation, hydroformylation, isomerization and oxidation. An important potential application is the asymmetric hydrogenation of olefinic substrates for the synthesis of chiral fine chemicals [67, 68]. We found that $scCO_2$ is a promising practical medium for the asymmetric hydrogenation of unsaturated carboxylic acids. As shown in Fig. 1.30, hydrogenation of tiglic acid catalysed by Ru complexes of 2,2′-bis(diphenylphosphino)-1,1′-binaphthyl (BINAP) derivatives proceeds efficiently in $scCO_2$ to give an optically active saturated carboxylic acid. With $Ru(O_2CCH_3)_2$[(*S*)-H_8-binap] as catalyst, (*S*)-2-methylbutanoic acid, a useful fruit flavor, is obtainable in 81% enantiomeric excess (ee) at 50 °C [56]. This selectivity is lower than that obtained in protic methanol (82% ee), but superior to that in hexane, an aprotic solvent (73% ee). Addition of $CF_3(CF_2)_6CH_2OH$, a fluorinated alcohol, as a cosolvent increases both the conversion and selectivity of hydrogenation, giving the product in up to 89% ee. Importantly, formic acid is not generated as a byproduct from the reaction of H_2 and CO_2 because of the absence of any base. The reaction takes place in the homogeneous phase, where tiglic acid, the product and H_8-BINAP–Ru complex are soluble. When the reaction is conducted in $scCHF_3$ with pressure-tunable polarity, the ee value of the product is increased from 83 to 90% with an

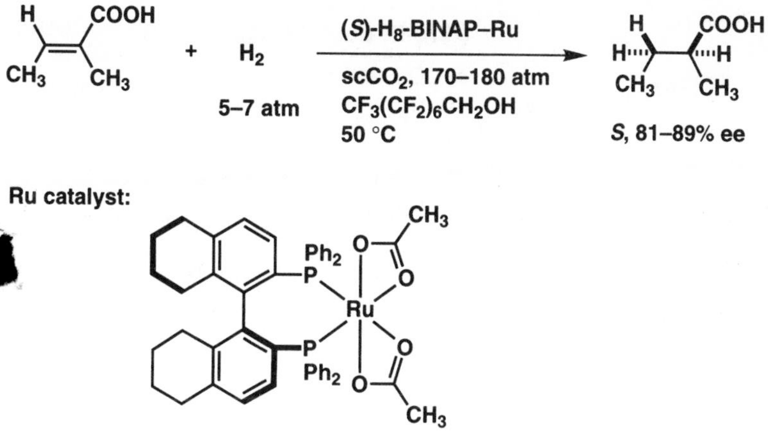

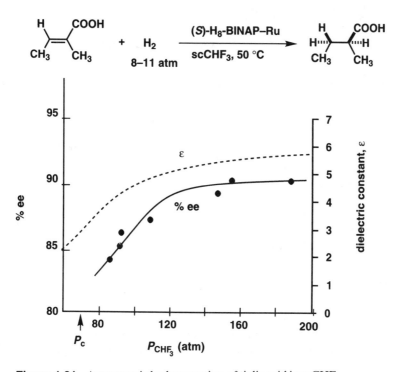

Figure 1.30. Asymmetric hydrogenation of tiglic acid with (*S*)-H$_8$-BINAP–Ru in scCO$_2$.

Figure 1.31. Asymmetric hydrogenation of tiglic acid in scCHF$_3$.

increase in pressure from 82 to 188 atm. As shown in Fig. 1.31, this change in the selectivity correlates well with the change in the dielectric constant of solvent as observed in conventional liquid solvents (J. Xiao, P.G. Jessop, T. Ikariya and R. Noyori, unpubl. data).

Burk and Tumas investigated the asymmetric hydrogenation of α-enamides with Rh complexes of 1,2-bis(*trans*-2,5-diethylphospholano)benzene (DuPHOS) in scCO$_2$. The reaction gives α-amino acid derivatives quantitatively with up to 99.7% ee (Fig. 1.32) [54]. Ultraviolet–visible solubility studies revealed that these chiral Rh complexes are soluble in scCO$_2$ under the reaction conditions. The selectivity achieved in scCO$_2$ is very high and comparable to those in conventional solvents. In particular, the ee values observed in the

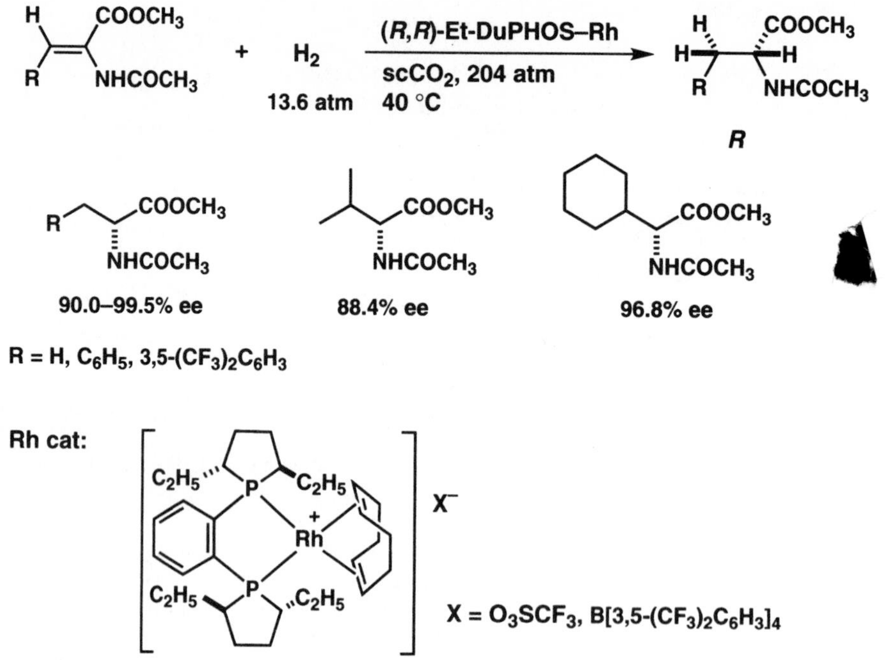

Figure 1.32. Asymmetric hydrogenation of enamides with chiral Rh complexes in scCO$_2$.

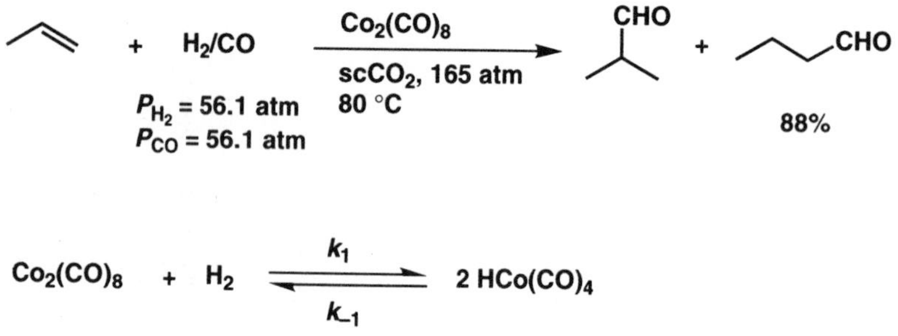

Figure 1.33. Propylene hydroformylation catalysed by Co$_2$(CO)$_8$ in scCO$_2$.

reaction of β,β-disubstituted enamides in scCO$_2$ are higher than those achieved in conventional liquid solvents. The hydrogenation is faster than in hexane or methanol, probably due to the high concentration of H$_2$ in scCO$_2$.

Rathke and co-workers [27] found the first homogeneous hydroformylation in scCO$_2$. The reaction of propylene with a Co catalyst (Fig. 1.33) occurs at a slightly lower rate than in hydrocarbon solvents such as methylcyclohexane and heptane, but gives a higher selectivity for the desired linear aldehyde, butanal. The major advantage of using SCFs is the elimination of gas-to-liquid mass-transfer problems. The narrow linewidths of the [59]Co NMR spectra allow the detection of Co$_2$(CO)$_8$, HCo(CO)$_4$ and even RCOCo(CO)$_4$ in scCO$_2$ [27, 77]. High-pressure NMR studies with the reaction revealed that equilibration between Co$_2$(CO)$_8$/H$_2$ and HCo(CO)$_4$ in scCO$_2$ occurs reproducibly. The equilibrium constants $K_{eq} = 0.025$ and $K_p = 8.8 \times 10^{-4}$ M atm^{-1} obtained in scCO$_2$ at 80 °C are in close agreement with those in heptane, and the rate constants of forward and reverse reactions for hydrogenation of Co$_2$(CO)$_8$ are also comparable to those measured in typical hydroformy-

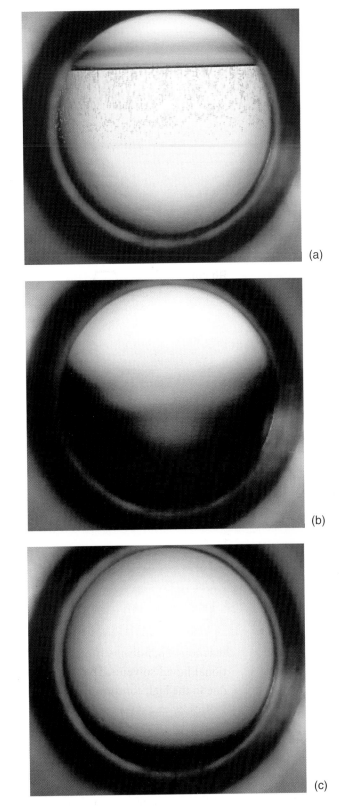

(a)

(b)

(c)

Plate 1.1. Photographs showing phase behaviour of pure CO_2.

[*Facing page 22*]

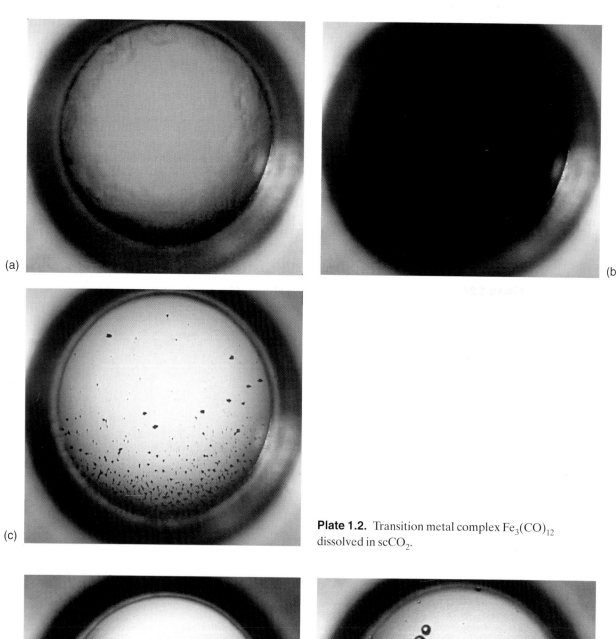

(a)

(b)

(c)

Plate 1.2. Transition metal complex $Fe_3(CO)_{12}$ dissolved in $scCO_2$.

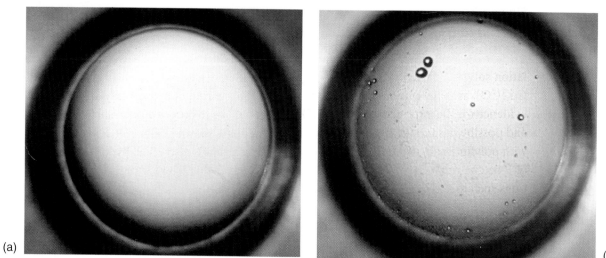

(a)

(b)

Plate 1.3. Photographs showing the interior of the reaction vessel for hydrogenation of $scCO_2$. Initial reaction conditions: H_2 (85 atm), $N(C_2H_5)_3$ (5 mmol), H_2O (0.1 mmol) and Ru complex (3 µmol) in $scCO_2$ (total 216 atm) at 50 °C. The drops are liquid complexes of HCO_2H and $N(C_2H_5)_3$.

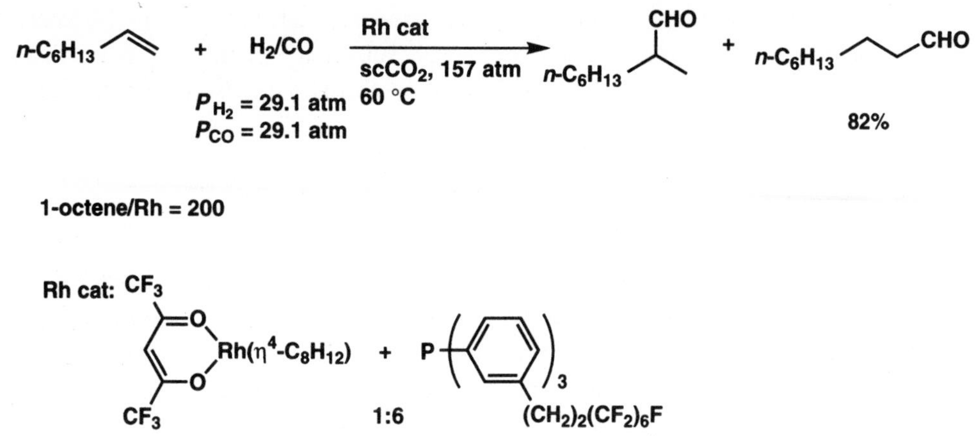

Figure 1.34. Rhodium(I)-catalysed hydroformylation of 1-octene in scCO$_2$.

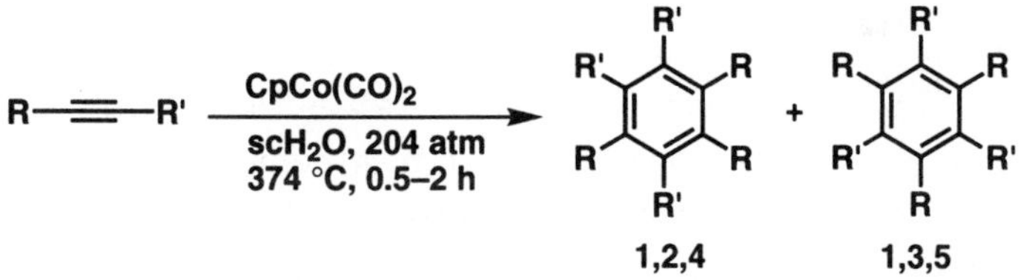

alkyne	% yield	benzene isomer 1,2,4:1,3,5
n-C$_4$H$_9$—≡—H	>95	1:3
C$_6$H$_5$—≡—H	>95	1:6
CH$_3$—≡—CH$_3$	5	–

Figure 1.35. Cobalt(I)-catalysed alkyne cyclotrimerizations in scH$_2$O.

lation solvents. Surprisingly, the rate of hydrogen atom transfer between CoH(CO)$_4$ and Co$_2$(CO)$_8$ is many orders of magnitude higher than the rate of hydroformylation. This is evidence for the intermediacy of Co(CO)$_4$ radicals in the hydrogen atom transfer reaction and possibly also in the hydroformylation itself [78]. As shown in Fig. 1.34, Rh complexes with polyfluoroalkyl-substituted triarylphosphines in scCO$_2$ also effect hydroformylation of 1-octene to give a linear aldehyde in good yield and with 82% selectivity. No side reactions such as hydrogenation or the formation of other isomeric aldehydes were observed [55].

Many studies on these subjects have cited the beneficial effect of the high solubility of H$_2$ in scCO$_2$. The effect was first reported in the AlBr$_3$-catalysed isomerization of n-hexane [23]. This reaction in scCO$_2$ was fivefold more selective for isomerization over cracking than in neat hexane. The selectivity was improved further when H$_2$ was dissolved in the SCF.

Despite the high T_c of H$_2$O, cyclotrimerization of alkynes to substituted benzene derivatives is promoted homogeneously by CpCo(CO)$_2$ in scH$_2$O (Fig. 1.35) [79]. Thermal decomposition of the complex is not a problem even at the temperature of scH_2O. The catalytic efficiency is only 1 TON, however. Selectivity for the 1,2,4-isomer of the product is comparable to that found in organic solvents. In comparison, liquid H_2O at 140 °C was an inferior medium causing significant side reactions.

6   Concluding remarks

Supercritical fluids have proved to be unique and very practical media for chemical reactions, providing a third reaction phase besides conventional gas and liquid phases. The reaction performance can be tuned easily by changing only the temperature and pressure. The utility of SCFs is not limited to chemical reactions described herein. Potential advantages of SCFs as reaction media have been demonstrated in uncatalysed reactions [3, 20–25,80,81], heterogeneously catalysed reactions [82], enzyme-catalysed reactions [83–85] and polymerization [3,23,80]. In addition, the following recent topics illustrate the special benefits. First, use of SCFs allows the elimination of solvent residues and wastes and facile separation of reactants, catalysts and products after reaction by selective precipitation. These characteristics offer a great opportunity for rapid synthesis, workup, purification and analysis of reaction products containing a radionuclide with a short half-life such as ^{11}C ($t_{1/2} = 20.3$ min). In fact, as shown in Fig. 1.36, highly efficient synthesis of ^{11}C-labeled monosubstituted guanidines has been achieved in $scNH_3$ ($T_c = 132.6$ °C, $P_c = 111.3$ atm) by reacting aliphatic or aromatic amines and [^{11}C]cyanogen bromide followed by treatment of ammonia. The syntheses are performed within 10 min and with high reproducibility. The ^{11}C-labeled products are used for both *in vivo* positron emission tomography (PET) studies and *in vitro* assays in modern bioscience [86].

Furthermore, the supercritical conditions of CO_2 can be used for its fixation to useful chemicals catalysed by pure organic compounds, as shown in Fig. 1.37 (S. Iwasa, T. Ikariya and R. Noyori, unpubl. data). Reaction of propargylic alcohols and $scCO_2$ in the presence of a trialkylphosphine as a catalyst gives an excellent yield of cyclic carbonates. The reaction efficiency is superior to that in solution phase [87]. The TON reaches 1200 and the TOF exceeds 400. The sufficient concentration of CO_2 as well as the high reactivity of the ion-pair intermediate in $scCO_2$ are responsible for such high efficiency.

The heterogeneously catalysed Fischer–Tropsch reaction proceeds smoothly under supercritical conditions. The observed enhancements of reaction rate and catalyst lifetime are due partly to higher diffusivities between the SCF and the catalyst surface and reduced coke deposition, which are benefits having no parallel in homogeneous catalysis. The greater diffusivity in an SCF can increase the rate of diffusion-controlled reactions in the liquid phase, one example being enzyme-promoted reactions in $scCO_2$ or $scCHF_3$. Fluoropolymers can be produced efficiently in $scCO_2$, as exemplified by the AIBN-initiated homogeneous free-radical polymerization of 1,1-dihydroperfluorooctyl acrylate to give CO_2-soluble fluoropolymers. In $scCO_2$, the reaction occurs with a 1.5-fold higher initiation efficiency than in benzene, but with a lower reaction rate due to the weaker cage effect of $scCO_2$ [13]. Inert $scCO_2$ can avoid the environmental problems caused by chlorofluorocarbons (CFCs). DeSimone has cleverly used soluble fluoropolymers in small amounts as surfactants in the polymerization of methyl methacrylate [88] or acrylamide [89] in $scCO_2$, resulting in greater rates and uniformity of product particle size, respectively.

Molecular catalysis using homogeneous organometallic complexes is now directed

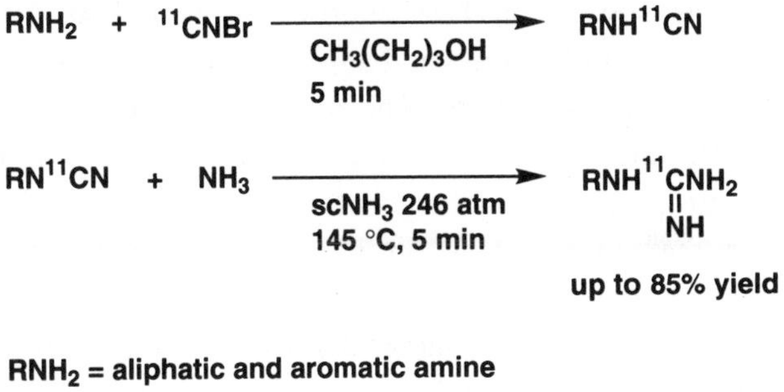

$$RNH_2 + {}^{11}CNBr \xrightarrow[\substack{CH_3(CH_2)_3OH \\ 5\ min}]{} RNH^{11}CN$$

$$RN^{11}CN + NH_3 \xrightarrow[\substack{scNH_3\ 246\ atm \\ 145\ °C,\ 5\ min}]{} RNH\underset{\underset{NH}{\parallel}}{^{11}C}NH_2$$

up to 85% yield

RNH$_2$ = aliphatic and aromatic amine

Figure 1.36. Synthesis of ^{11}C-labeled monosubstituted guanidines in scNH$_3$.

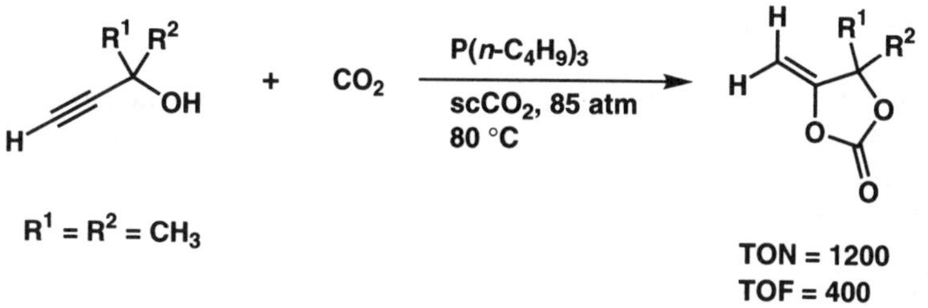

R^1 = R^2 = CH$_3$

TON = 1200
TOF = 400

Figure 1.37. Trialkylphosphine-catalysed reaction of propargylic alcohol and scCO$_2$.

toward synthesis of fine chemicals and materials. Catalysis in SCFs could be of great help because of high reactivity, high selectivity and operational simplicity. It is certain that researchers, either in academia or industry, will adopt SCFs as media for many other reactions and broaden the scope of this exciting new technique. Molecular catalysis in SCFs with high rates, high selectivities and high productivities could lead to perfect chemical reactions.

The industrial outlook for SCFs as reaction media is bright, mainly because of the beneficial environmental effect of dispensing with organic liquid solvents. In particular, scH$_2$O is a very attractive medium for a waste-clean technology such as complete oxidation of organic wastes, including pulp and paper mill sledges, hazardous organics and biological wastes from the pharmaceutical industry. The application of scH$_2$O as a reaction medium has been reviewed extensively [23, 25]. Supercritical fluids are also promising solvents for dyeing processes, solid precipitation, coating processes and extraction and separation of polymeric compounds. High pressure operations would be facilitated in large-scale production processes and, especially, continuous flow systems for which SCFs are well suited. Favorable mass transfer between phases or control of phase behavior facilitates the catalytic hydration of olefins. Reaction of 2-butene and water in the presence of an aqueous heteropolyacid as a catalyst at 200 °C and 200 atm in scC$_4$H$_8$ proceeds effectively to produce 2-butanol [90]. The use of supercritical conditions can overcome the equilibrium problems of this hydration reaction; the product, 2-butanol, can be separated effectively from the catalyst solution by extraction with scC$_4$H$_8$ because of the high solubility of 2-butanol in scC$_4$H$_8$. This process is now working on a large scale of 40000 t yr^{-1} at Idemitsu Petrochemical in Japan.

Chemical transformations in SCFs thus have enormous potential advantages. The possibilities of greater or even outstanding rates and adjustable selectivities will motivate the research of homogeneous catalysis in SCFs. The number of reactions tested so far is still very small, however. Because phase behavior and solubility data for multicomponent systems are scarce, these properties at reaction conditions should be studied carefully by experiments. In addition, the solvent effects on the performance of catalysed reactions should be elucidated. In principle, solute solubility, polarity, solvent cage, solvation and clustering are readily controllable with the change of pressure and temperature. Although a relatively small number of commercial-scale plants are in operation, the chemistry in SCFs is full of promise. Both fundamental research on such a unique phase and its application will exploit new chemical science and technology in SCFs.

7 References

1 Reichardt C. *Solvents and Solvent Effects in Organic Chemistry.* Weinheim: VCH, 1990.
2 Poliakoff M, Howdle SM, Kazarian SG. *Angew Chem Int Ed Engl* 1995; **34**: 1275–95.
3 Kaupp G. *Angew Chem Int Ed Engl* 1994; **33**: 1452–5.
4 Reichardt C. *Chem Rev* 1994; **94**: 2319–58.
5 Shaw RW, Brill TB, Clifford AA, Eckert CA, Franck CA. *Chem Eng News* 1991; **26**: 6–39.
6 Tawa GJ, Pratt LR. *J Am Chem Soc* 1995; **117**: 1625–8.
7 Kamat SJ, Beckman EJ, Russell AJ. *J Am Chem Soc* 1993; **115**: 8845–6.
8 Bartle KD, Clifford AA, Jafar SA, Shilstone GF. *J Phys Chem Ref Data* 1991; **20**: 713–56.
9 Sigman ME, Lindley SM, Leffler JE. *J Am Chem Soc* 1985; **107**: 1471–2.
10 Ikushima Y, Saito N, Arai M. *J Phys Chem* 1992; **96**: 2293–7.
11 Hyatt JA. *J Org Chem* 1984; **49**: 5097–101.
12 Jessop PG, Hsiao Y, Ikariya T, Noyori R. *J Am Chem Soc* 1996; **118**: 344–55.
13 DeSimone JM, Guan Z, Elsbernd CS. *Science* 1992; **257**: 945–7.
14 Dandge DK, Heller JP, Wilson KV. *Ind Eng Chem Prod Res Dev* 1985; **24**: 162–6.
15 Ekart MP, Bennett KL, Ekart SM *et al. AIChE J* 1993; **39**: 235–48.
16 Dillow AK, Hafner KP, Yun SLJ *et al. AIChE J* 1997; **43**: 515–24.
17 Johnston KP, Harrison KL, Clarke MJ *et al. Science* 1996; **271**: 624–6.
18 Clarke MJ, Harrison KL, Johnston KP, Howdle SM. *J Am Chem Soc* 1997; **119**: 6399–406.
19 Renslo AR, Weinstein RD, Tester JW, Danheiser RL. *J Org Chem* 1997; **62**: 4530–33.
20 Subramaniam B, McHugh MA. *Ind Eng Chem Process Des Dev* 1986; **25**: 1–12.
21 Brennecke JF. In: Kiran E, Brennecke JF, eds. *Supercritical Fluid Engineering Science Fundamentals and Applications, ACS Symp Ser 514.* Washington: American Chemical Society, 1993: 201–19.
22 Clifford AA. In: Kiran E, Sengers JMHL, eds. *Supercritical Fluids, Fundamentals for Application of the NATO ASI Series E.* Dordrecht: Kluwer Academic, 1994: 449–79.
23 Savage PE, Gopalan S, Mizan TI, Martino CJ, Brock EE. *AIChE J* 1995; **41**: 1723–78.
24 Eckert CA, Knutson BL, Debenedetti PG. *Nature (London)* 1996; **383**: 313–18.
25 Clifford T, Bartle K. *Chem Ind* 1996: 449–52.
26 Kramer GM, Leder F. *US Patent 3,880,945,* filed July 18, 1973, issued April 29, 1975.
27 Rathke JW, Klingler RJ, Krause TR. *Organometallics* 1991; **10**: 1350–55.
28 Jessop PG, Ikariya T, Noyori R. *Nature (London)* 1994; **368**: 231–2.
29 Howdle SM, Grebenik P, Perutz RN, Poliakoff M. *J Chem Soc Chem Commun* 1989: 1517–19.
30 Howdle SM, Healy MA, Poliakoff M. *J Am Chem Soc* 1990; **112**: 4804–13.
31 Howdle SM, Poliakoff M. *J Chem Soc Chem Commun* 1989: 1099–101.
32 Johnston KP, Haynes C. *AIChE J* 1987; **33**: 2017–26.
33 Kim S, Johnston KP. In: Squires TG, Paulaitis ME, eds. *Supercritical Fluids, Chemical and Engineering, Principles and Applications, ACS Symp Ser 329.* Washington: American Chemical Society, 1987: 42–55.

34 Kajimoto O, Yamasaki K, Honma K. *Faraday Discuss Chem Soc* 1988; **85**: 65–75; Kajimoto O, Fitakami M, Kobayashi T, Yamasaki K. *J Phys Chem* 1988; **92**: 1347–52; Morita A, Kajimoto O. *J Phys Chem* 1990; **94**: 6420–25.

35 Toriumi T, Sakai J, Kawakami T, Osawa D, Azuma M. *J Soc Chem Ind Jpn* 1946; **49**: 1–6.

36 Knutson BL, Dillow AK, Liotta CL, Eckert CA. In: Hutchenson WK, Foster NR, eds. *Innovations in Supercritical Fluids, Science and Technology*. Washington: American Chemical Society, 1995: 166–78.

37 Paulaitis ME, Alexander GC. *Pure Appl Chem* 1987; **59**: 61–8.

38 Roberts CB, Zhang J, Chateauneuf JE, Brennecke JF. *J Am Chem Soc* 1992; **114**: 8455–63.

39 Roberts CB, Brennecke JF, Chateauneuf JE. *AIChE J* 1995; **41**: 1306–18.

40 Ellington JB, Brennecke JF. *J Chem Soc Chem Commun* 1993: 1094–5.

41 Ellington JB, Park KM, Brennecke JF. *Ind Eng Chem Res* 1994; **33**: 965–74.

42 Rhodes TA, O'Shea K, Bennett G, Johnston KP, Fox MA. *J Phys Chem* 1995; **99**: 9903–8.

43 Roberts CB, Zhang J, Chateauneuf JE, Brennecke JF. *J Am Chem Soc* 1993; **115**: 9576–82.

44 Zagrobelny JA, Bright FV. *J Am Chem Soc* 1992; **114**: 7821–6.

45 Hrnjez BJ, Mehta AJ, Fox MA, Johnston KP. *J Am Chem Soc* 1989; **111**: 2662–6.

46 Guan Z, Combes JR, Menceloglu YZ, DeSimone JM. *Macromolecules* 1993; **26**: 2663–9.

47 O'Shea KE, Combes JR, Fox MA, Johnston KP. *Photochem Photobiol* 1991; **54**: 571–6.

48 Tanko JM, Suleman NK, Fletcher B. *J Am Chem Soc* 1996; **118**: 11958–9.

49 Andrew D, Des Islet BT, Margaritis A, Weedon AC. *J Am Chem Soc* 1995; **117**: 6132–3.

50 Dillow AS, Yun SLJ, Suleiman D *et al*. *Ind Eng Chem Res* 1996; **35**: 1801–6.

51 Warzinski RP, Lee C-H, Holder GD. *J Supercrit Fluids* 1992; **5**: 60–71.

52 Lagalante AF, Hansen BN, Bruno TJ. *Inorg Chem* 1995; **34**: 5781–5.

53 Jessop PG, Hsiao Y, Ikariya T, Noyori R. *J Am Chem Soc* 1994; **116**: 8851–2.

54 Burk MJ, Feng S, Gross MF, Tumas W. *J Am Chem Soc* 1995; **117**: 8277–8.

55 Kainz S, Koch D, Baumann W, Leitner W. *Angew Chem Int Ed Engl* 1997; **36**: 1628–30.

56 Xiao J, Nefkens SCA, Jessop PG, Ikariya T, Noyori R. *Tetrahedron Lett* 1996; **37**: 2813–16.

57 Poliakoff M, Howdle S. *Chem Br* 1995; **31**: 118–21.

58 Banister JA, Georege MW, Grubert S *et al*. *J Organomet Chem* 1994; **484**: 129–35.

59 Sun XZ, George MW, Kazarian SG, Nikiforov SM, Poliakoff M. *J Am Chem Soc* 1996; **118**: 10525–32; Sun XZ, Grills DC, Nikiforov SM, Poliakoff M, George MW. *J Am Chem Soc* 1997; **119**: 7521–5.

60 Banister JA, Cooper AI, Howdle SM, Jobling M, Poliakoff M. *Organometallics* 1996; **15**: 1804–12.

61 Jobling M, Howdle SM, Poliakoff M. *J Chem Soc Chem Commun* 1990: 1762–3.

62 Clarks MJ, Howdle SM, Jobling M, Poliakoff M. *J Am Chem Soc* 1994; **116**: 8621–8.

63 Clarke MJ, Howdle SM, Jobling M, Poliakoff M. *Inorg Chem* 1993; **32**: 5643–4.

64 Banister JA, Howdle SM, Poliakoff M. *J Chem Soc Chem Commun* 1993: 1814–15; Banister JA, Lee PD, Poliakoff M. *Organometallics* 1995; **14**: 3876–85.

65 Ji Q, Eyring EM, van Eldik R *et al*. *J Phys Chem* 1995; **99**: 13461–6.

66 Jessop PG, Ikariya T, Noyori R. *Organometallics* 1995; **14**: 1510–13.

67 Noyori R. *Science* 1990; **248**: 1194–9.

68 Noyori R. *Asymmetric Catalysis in Organic Synthesis*. New York: Wiley, 1994.

69 Jessop PG, Ikariya T, Noyori R. *Science* 1995; **269**: 1065–9.

70 Cottle JE. *US Patent 3,294,772*, filed June 17, 1963, issued December 27, 1966.

71 Folie B, Radosz M. In: Brunner G, Perrut M, eds. *Proceedings of the 3rd International Symposium on Supercritical Fluid*, Vol. 3. Nancy, France: Institut National Polytechnique de Lorraine 1994: 281–6.

72 Reetz MT, Konen W, Strack T. *Chimia* 1993; **47**: 493.

73 Inoue Y, Itoh Y, Kazama H, Hashimoto H. *Bull Chem Soc Jpn* 1980; **53**: 3329–33.

74 Jessop PG, Ikariya T, Noyori R. *Chem Rev* 1995; **95**: 259–72.

75 Jessop PG, Hsiao Y, Ikariya T, Noyori R. *J Chem Soc Chem Commun* 1995: 707–8.

76 Kröcher O, Köppel RA, Baiker A. *J Chem Soc Chem Commun* 1996: 1497–8; Kröcher O, Köppel RA, Baiker A. *J Chem Soc Chem Commun* 1997: 453–4.

28 T. IKARIYA AND R. NOYORI

77 Rathke JW, Klingler RJ, Krause TR. *Organometallics* 1992; **11**: 585–8.

78 Klingler RJ, Rathke JW. *J Am Chem Soc* 1994; **116**: 4772–85.

79 Jerome KS, Parsons EJ. *Organometallics* 1993; **12**: 2991–3.

80 Buback M. *Angew Chem Int Ed Engl* 1991; **30**: 641–53; Buback M. In: Kiran E, Sengers JMHL, eds. *Supercritical Fluids, Fundamentals for Application of the NATO ASI Series E*. Dordrecht: Kluwer Academic, 1994: 481–97.

81 Tanko JM, Blackert JF. *Science* 1994; **263**: 203–5.

82 Yokota K, Fujimoto K. *Ind Eng Chem Res* 1991; **30**: 95–100.

83 Aaltonen O, Rantakylä M. *Chemtech* 1991; **21**: 240–48.

84 Russell AJ, Beckman EJ, Chaudhary AK. *Chemtech* 1994; **24**: 33–7.

85 Claudhary AK, Kamat SV, Beckman EJ *et al. J Am Chem Soc* 1996; **118**: 12891–901.

86 Jacobson GB, Westerberg G, Markides KE, Långström B. *J Am Chem Soc* 1996; **118**: 6868–72.

87 Fournier J, Bruneau C, Dixneuf PH. *Tetrahedron Lett* 1989; **30**: 3981–2; Journier JM, Fournier J, Bruneau C, Dixneuf PH. *J Chem Soc Perkin Trans 1* 1991: 3271–4.

88 DeSimone JM, Maury EE, Menceloglu YZ *et al. Science* 1994; **265**: 356–9; Shaffer KA, Jones TA, Canelas DA, DeSimone JM. *Macromolecules* 1996; **29**: 2704–6.

89 Adamsky FA, Beckman EJ. *Macromolecules* 1994; **27**: 312–14.

90 Fukuzato R. In: McHugh MA, ed. *Proceedings of the 2nd International Symposium on Supercritical Fluid*. Baltimore: Johns Hopkins University, 1991: 196.

2 Palladium-catalysed Reactions: from the Art of Today to a Common Tool of Tomorrow

IRINA P. BELETSKAYA and ANDREI V. CHEPRAKOV

Department of Chemistry, Moscow State University, 119899 Moscow, Russia

1 Introduction

Over the last two decades palladium-catalysed reactions have virtually invaded modern organic synthesis, giving an almost unlimited variety of new methods for a straightforward construction of the most complex molecules [1]. However, as always happens with rapidly growing areas, the frontiers are being pushed ahead much faster than the routine and less interesting work of tidying up the already conquered areas, so that the whole community of chemists may enjoy the fruits. Why do many researchers, particularly the chemists working for industry, still prefer traditional reactions described on the faded pages of famous preparative manuals to these fascinating reactions with palladium complexes advertised almost as a universal glue to painlessly tie together any given subunits of the desired structure?

Palladium catalysis is being developed in two major directions: towards complex synthesis where selectivity is the major issue, while the efficiency of catalysts may be completely neglected as the value of compounds being synthesized by far exceeds the cost of catalyst; and towards common organic preparations with the perspective of industrial applications, in which catalyst efficiency often becomes the major issue. For a budget-minded laboratory, as well as for technology, reactions requiring 10 mol.% palladium plus three to four times as much of expensive phosphine ligands are prohibited. Therefore, in spite of a huge synthetic potential, palladium catalysis is still quite rarely used in the large-scale applications. Large loads of palladium catalyst bring forward yet another serious difficulty associated with the purification of products from residual palladium [2], especially those targeted for medicinal or optical applications, because higher initial concentrations in reaction mixtures would require more purification steps to reduce the residual content to appropriate levels.

The aim of this chapter is to discuss the approaches appearing to solve problems associated with mainstream phosphine-mediated catalysis. We tried to include research that helps to reveal the resources of palladium reactivity by focusing attention on those studies that gave an apparent simplification of procedures and marked increase of reactivity and catalytic efficiency, in a hope that an attempt to establish general trends in these, often unconnected, studies may help to elucidate growth points for further development.

2 Overcoming the deficiencies of phosphine-mediated catalysis

Most known catalytic cycles involve the transformation Pd(II) → Pd(0). As far as Pd(0) is believed not to form stable complexes in solution in the absence of strongly coordinating ligands, the majority of which are mono- and bidentate phosphines, it seemed natural to assume that the design of new ligands is the only key to solving the challenges of palladium catalysis. Nevertheless, even in the early papers it had been shown occasionally that sometimes the reactions can be catalysed effectively by palladium salts and weak complexes without the addition of phosphines, although initially such examples were regarded as particular cases relevant only for highly reactive substrates. Later on, this phenomenon was identified as a separate incarnation of palladium reactivity, and such *phosphine-free* (this term is more accurate than an ambiguous and provocative term, *ligandless*, often used in earlier literature) catalysis emerged as a useful alternative approach. Indeed, besides being expensive and unrecoverable reagents, phosphines can have a negative influence on both reaction rate and selectivity. Any step in catalytic transformations requires at least two free coordination sites on the palladium atom, and in most elementary steps involved, the two remaining ancillary ligands must possess a *cis*-configuration. In common phosphine-mediated catalysis the ligands are either added *in situ* or a preformed form of catalyst precursor is used. In any case, the species involved are labile and enter all possible ligation/de-ligation equilibria, as well as *cis–trans* isomerizations (cf. [3], and references therein), while only a few of the species formed in such heavily populated systems can behave as actual catalysts.

2.1   *Rigid complexes as highly efficient catalysts*

Conventional phosphine-mediated catalysis uses phosphines as labile ligands involved in common ligation/de-ligation equilibria. Less-reactive aryl bromides and especially chlorides require high temperatures at which phosphines become inefficient ligands due to several reasons, including the following: ligation equilibria shift to the left, and therefore a large excess of phosphine is needed to avoid the build-up of non-ligated species leading to the nucleation of metal particles; phosphines undergo an oxidative addition to Pd(0), and thus enter the catalytic cycle as reagents leading to scrambling of substituents, and fast depletion of ligands [4]; and use of even higher excess of ligands is both uneconomical and has a negative influence on activity as extra ligands block free coordination sites needed for catalytic processes.

Thus, in a search for higher reactivity, common phosphine-mediated catalysis must be superseded by methods involving either new types of ligands forming highly reactive but stable complexes, or new forms of palladium able to operate in the absence of tightly coordinated ligands. All these new methods can be regarded safely as phosphine-free catalysis, because their main characteristic feature is to stop relying on ligation/de-ligation equilibria with free phosphines, which are always considered as detachable ligands, even if some particular phosphine complex is used as a precursor. New ligands remain strongly bound in the coordination sphere of palladium, so although some of them are derived from phosphines, de-ligation does not occur, and unbound phosphines do not emerge to lead to side processes deteriorating the catalytic activity.

The development of new ligands brought very promising results in recent years, because the catalysts developed in this approach all share very high thermal stability (many of these new catalysts withstand prolonged heating above 200 °C, even in the presence of air, and extremely high catalytic activity in the processes involving oxidative addition to aryl bromides and even chlorides.

First, interesting chemistry was revealed for one of the most famous ligands of palladium catalysis, tris-(*o*-tolyl)phosphine, which readily gave a cyclopalladation product (the

Herrmann catalyst) when treated with palladium acetate. The readiness of formation of this complex allows it to be supposed that it might be involved in reactions catalysed by palladium salts in the presence of tris-(*o*-tolyl)phosphine.

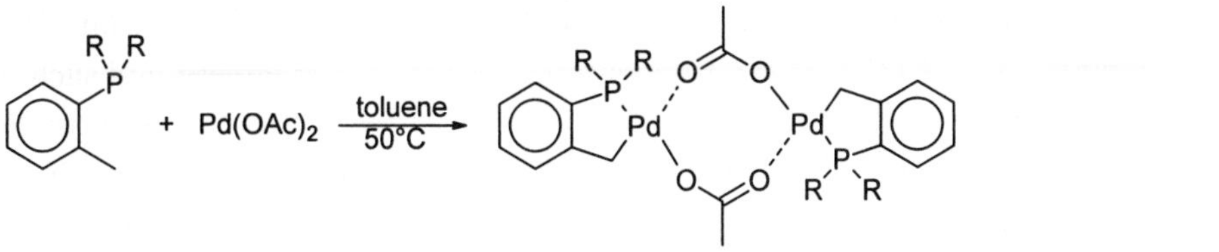

The complex showed very high catalytic activity in Heck and Suzuki reactions with aryl bromides and even with aryl chlorides bearing electron-withdrawing substituents. Turnover numbers as high as 10 000–100 000 were not uncommon in the reactions with more reactive aryl bromides [5, 6]. Moreover, such ligands retain an ample reserve of modification in the aryl rings, with the goal of development of enantioselective catalysts, catalysts suitable for anchoring at solid supports, recyclable hydrophilic or solubilizable amphiphilic catalysts, etc. [7]. The Herrmann catalyst (HC) has proved to be a superior catalyst for some other reactions, including Stille coupling [8] and intramolecular vinyl–aryl [9] and aryl–aryl coupling [10].

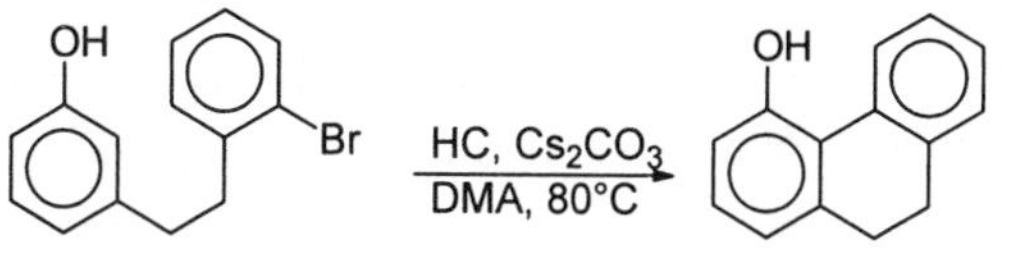

The latter example clearly shows that the palladacycle complex is capable of catalysing reactions with heavily crowded intermediates.

Phosphapalladobicycles obtained by the cyclopalladation of chelating diphosphines provide another example of this approach, and show that the outstanding catalytic performance and stability is not an exclusive feature of the Herrmann catalyst.

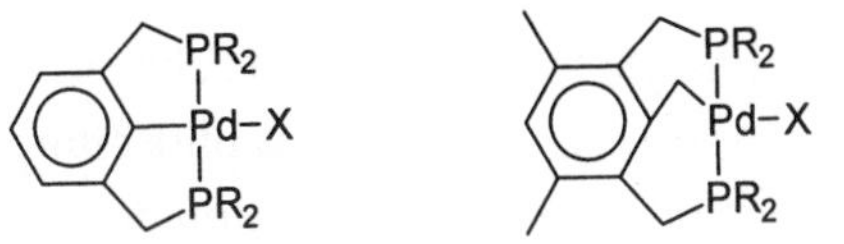

where R = *i*-Pr, *t*-Bu; X = OCOCF$_3$

These complexes were reported as highly efficient catalysts for the Heck reaction (>100 000 catalytic cycles for the reaction of PhBr with methyl acrylate; 500 000 cycles for the similar reaction with PhI), with exceptional stability under very harsh conditions (prolonged heating to 140–180 °C even in air, despite the fact that the ligands used are highly donor phosphines that are themselves extremely susceptible to oxidation). As soon as there was no evidence for de-chelation of the arms during the catalytic process, it was claimed that catalysis by these complexes may not involve the usual Pd(II) → Pd(0) flip-flop, but rather is a new Pd(IV) → Pd(II) process started by the oxidative addition of haloarene to palladobicycle [11].

Phosphapalladocycles are not the only example of stable palladium complexes capable of driving catalytic cycles, including the oxidative addition to C—X bonds. Heterocyclic carbene complexes of palladium like the one below, obtained easily by the reaction of 1,3-dimethylimidazolium iodide with $Pd(OAc)_2$ [12], were shown to be excellent catalysts for the Heck reaction with aryl bromides and aryl chlorides bearing electron-acceptor substituents (NO_2, CHO).

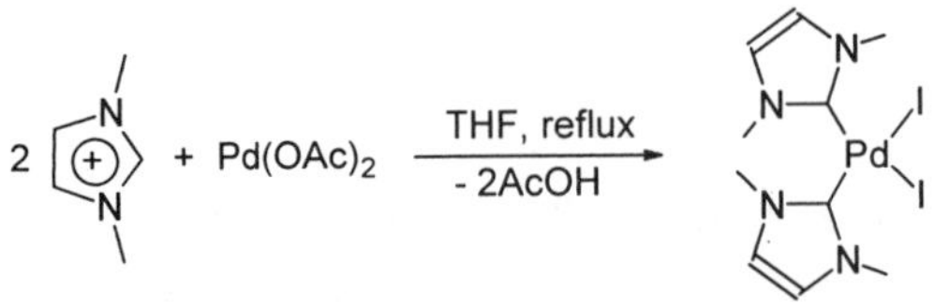

It is interesting that in this case the most efficient and stable complex was the simplest one shown above bearing only monodentate ligands, while more rigid palladocycles formed by a bridged 3,3'-dimethyl-1,1'-methylenediimidazolium salt were less stable towards heat and oxidation.

2.2   *Aqueous catalysis vs phosphine-free catalysis*

A high sensitivity of palladium reactivity to the fine details of the ligand effect was described by Milstein *et al.* [13], who revealed two distinct modes of the carbopalladation pathway. The first is initiated by the de-ligation of a neutral phosphine ligand to give a neutral intermediate that then undergoes addition to olefin, while the second is initiated by the de-ligation of halide ion leading to a cationic palladium complex. Although these findings were made in a study with specific chelating and highly donor trialkylphosphine ligands, the conclusions seem to be highly relevant to the distinctions of phosphine-complex-mediated vs phosphine-free catalysis. For example, in a general case of a Heck arylation catalytic cycle, two pathways may be distinguished: the interplay of normal path A depends on the ligation/de-ligation equilibria of neutral ligands such as phosphines; and the more polar path B, which can occur even in the absence of such ligands, may account for the trends in catalytic activity and the effects of media and additives.

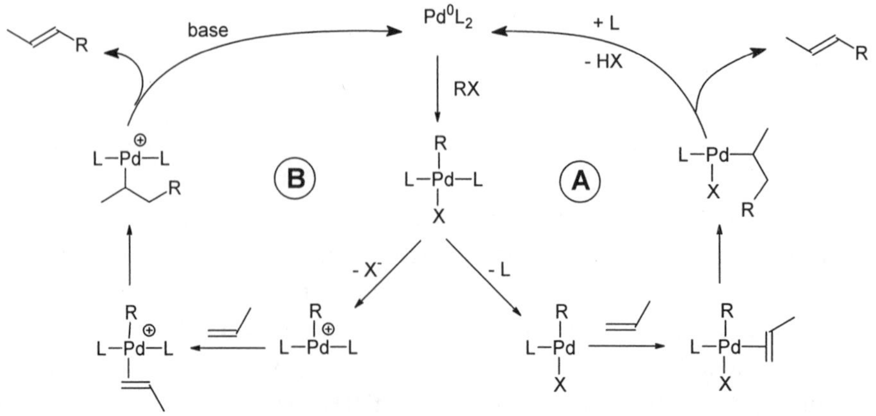

Polar solvents, notably water, thus facilitate the de-ligation of ionic halide ligands; dramatic accelerating effects are exerted by the addition of phase-transfer catalysts with chloride counter ions (Bu_4NCl, Ph_4PCl, etc.; cf. [14]) or even alkali metal chlorides (cf. [15]) but inhibition is caused by added iodide ions, which has been noted in many reactions involving the oxidative addition of organic halides to Pd(0) and further transformations of RPdX intermediates. This effect is probably accounted for by ligand exchange to give an RPdCl intermediate with a weaker chloride ligand more prone to de-ligation (resembling the famous *special salt effect* in nucleophilic aliphatic substitution). The promotive effect of soft Lewis acids (e.g. derivatives of Ag or Tl and possibly of copper salts), which in various palladium-catalysed reactions is often explained by the formation of organocopper intermediates but has never been proven experimentally, thus lends soft electrophilic assistance to the de-ligation of halide.

We shall try to establish a common basis for the whole range of phenomena associated with a more polar pathway in catalytic cycles. This pathway is the only one possible in phosphine-free catalysis. Moreover, although phosphine-free catalysis and aqueous catalysis seem not to be logically associated with each other, it is hardly a coincidence that a good proportion of known phosphine-free cases are either done in aqueous media or exploit the effect of phase-transfer agents, the main function of which is to facilitate ionization in less polar media.

In this context the discovery by Reetz *et al.* of an unprecedented extremely efficient catalytic system, which holds the reactivity record in a Heck reaction with the least reactive substrates, aryl chlorides, looks especially intriguing and brilliantly highlights the huge reserve of reactivity unleashed by a proper choice of ligand environment and promoters. The system is actually an ingenious compromise between phosphine-mediated and phosphine-free catalysis. A simple complex $PdCl_2$ $(MeCN)_2$, which is itself rapidly reduced to unreactive palladium black on heating with olefins, gave an efficient and stable catalyst upon the addition of an excess of Ph_4PCl salt [16].

where X = H, Me, CHO; NMP = *N*-methylpyrrolidone

The role of Ph_4PCl can be dual. First of all, it serves as an effective scavenger of Pd(0) formed by the reduction by olefin, and thus affords a single phosphine ligand keeping reduced palladium from the nucleation of inactive metal particles, but not overloading available coordination sites, which would be inevitable if free phosphine were present.

$$Pd(0) + Ph_4PCl \longrightarrow (Ph_3P)PhPdCl \rightleftharpoons [(Ph_3P)PhPdCl]_2$$

As a penalty, the catalysts always contain a phenyl group, which can be transferred to the product, thus leading to a scrambling of aryl groups in the case of substituted aryl halides. It is interesting that the catalytic system can be activated further with a parallel increase of regioselectivity (suppression of the formation of 1,1-diarylethylenes) by the addition of *N,N*-dimethylglycine, which can serve both for the solvation of chloride ion in an aprotic solvent or as an ancillary ligand for palladium that is similar to acetate, which is well known to accelerate Heck arylation [17], but with an additional bidentate coordination mode.

Going back to the Herrmann catalyst, the functioning of this apparently rigid complex, at least in the reactions involving hydridopalladium intermediates, may resemble the system treated above. There is evidence that this phosphapalladocycle is not so rigid and that de-chelation may occur due to intramolecular reductive elimination [18]. In the absence of free phosphine ligand this process should lead to a highly reactive monophosphine complex.

R = *o*-tolyl

Meanwhile, in the presence of free tris(*o*-tolyl)phosphine the Herrmann catalyst may be a precursor of simple bis- and tris-phosphine complexes PdL_n ($n = 2, 3$), and thus such cases may be described as phosphine-mediated catalysis.

2.3   *Suzuki cross-coupling: a success story of aqueous phosphine-free catalysis*

Among the other cross-coupling protocols, the Suzuki reaction using boronic acids and similar organoboron compounds as a carbanion synthon is, where applicable, the most promising method because it is highly efficient, produces no toxic byproducts (compared to Stille cross-coupling with organotins, for example, which have a wider scope but give such toxic waste that its application to anything other than research purposes is debatable) and is tolerant to aqueous media. It is not surprising, therefore, that this reaction indeed benefited most from the application of phosphine-free aqueous methods and, vice versa, that the potential of phosphine-free palladium catalysis was first unambiguously unveiled for cross-coupling with boronic acids. It should be noted that the original Suzuki procedure from the beginning employed aqueous systems, because reaction with boronic acids requires the presence of bases, such as alkali metal carbonates or hydroxides, introduced to reaction mixtures as aqueous solutions. However, the original method required phosphine complexes of palladium, often taken in very high amounts (in some cases up to 25–30 mol.% palladium). It is quite clear that such a low efficiency effectively disabled this method for possible application in large-scale synthesis. Aqueous phosphine-free catalysis allowed this deficiency to be overcome, allowing catalyst efficiency and selectivity to be increased dramatically.

First, it was shown that the reaction of arylboronic acids with water-soluble organic halides can be performed at room temperature in the presence of simple palladium salts ($PdCl_2$, $Pd(OAc)_2$, etc.) and inorganic bases ($NaOH$, Na_2CO_3, K_2CO_3, K_3PO_4, etc.) to give the respective cross-coupling products in near to quantitative yields [19].

where X = Br, I; Y = H, *p*-F, etc., Z = *m*-, *p*-OH, *o*-, *m*-, p-COOH.

Cross-coupling with water-insoluble aryl halides can be performed in aqueous organic solvents also in the absence of phosphines. Further studies revealed that the high performance of phosphine-free systems is not a casual effect, and that phosphines have a negative influence on both rates and selectivity of Suzuki reactions. Palladium complexes without phosphine ligands were all excellent catalysts for cross-coupling of boronic acids or cyclic boronates with water-insoluble aryl iodides or bromides. As low as 0.02 mol.% of palladium was sufficient for reaction to give high yields of biaryls [20].

where "Pd" = $Pd(OAc)_2$, $[(\eta^3\text{-}C_3H_5)PdCl]_2$, $Pd_2(dba)_3$; solv = THF, Me_2CO, MeCN; Y, Z = H, p-NO_2, o-Me, p-CF_3

On the other hand, the addition of Ph_3P or the use of the preformed complex *trans*-$PhPdI(Ph_3P)_2$ resulted in incomplete conversions and moderate yields. The basicity of the reaction media was another key factor determining the rate of coupling. Best results were achieved for pH > 9, which is sufficient to transform most of the boronic acid into the more reactive tetracoordinate boronate anion.

Besides inhibiting the reaction, the addition of phosphines may result in a decrease of selectivity, first of all due to scrambling of substituents. This phenomenon is particularly important for sterically hindered or highly donor aryl halides, because triphenylphosphine may be involved competitively in oxidative addition to Pd(0) to give its phenyl group for further steps. In the context of the conventional procedure, the scrambling can be partially suppressed by lowering the load of palladium complex, although generally the rate of reaction becomes too low when the amount of palladium is lower than 1 mol.%. Another way is to use phosphines with bulky donor aryls, such as tris-(2,4,6-trimethoxyphenyl)phosphine or trimesitylphosphine [21], although this approach is quite expensive. These problems are effectively circumvented by running the process in the absence of phosphines in aqueous solvent [22]. Under such conditions even the reaction with sterically hindered and highly donor reagents can be run at room temperature to give 80–96% yield of the respective product.

Another useful variation of phosphine-free Suzuki reactions uses a heterogeneous system with neat water as solvent and tetrabutylammonium chloride as promoter and phase-transfer catalyst. Aryl bromides were shown to give higher yields than aryl iodides, because of inhibition by the liberated iodide ion leading to incomplete conversions [23].

Aqueous phosphine-free catalysis is particularly effective for cross-coupling with tetraarylborate salts. The readily available $NaBPh_4$ is a highly reactive reagent showing enormous turnover numbers (beyond 250 000) and capable of transferring all four phenyl groups provided that the pH of the aqueous base used is high enough to keep the boron atom in a tetracoordinate state throughout all four steps [24]. The reaction is applicable to water-soluble chloroarenes. All sorts of phosphine ligands (chelating or monodentate, water-soluble or hydrophobic) strongly inhibit the process [25].

An important role of an ancillary ligand that emerges from the leaving group of the substrate involved in the oxidative addition is well exemplified by the results on Suzuki coupling with iodonium or arenediazonium salts. Cross-coupling of iodonium salts with boronic acids readily proceeds both in water and in organic solvents at room temperature. The reaction can be catalysed by phosphine-free $Pd(OAc)_2$ or by $Pd(PPh_3)_4$. The phosphine-free process is faster but slightly less selective, giving several per cent of the product of oxidative homocoupling of boronic acid. Unlike cross-coupling with halides or triflates, the reaction can be performed in the absence of base, which can be accounted for by enhanced reactivity of the arylpalladium intermediate lacking the halide ligand towards transmetallation with even non-activated tricoordinate organoboron compounds [26]:

$$ArIPh^+BF_4^- + Pd(0) \xrightarrow{-PhI} ArPd^+BF_4^- \xrightarrow{Ar'B(OH)_2} ArPdAr' \xrightarrow{-Pd(0)} ArAr'$$

The reaction of arenediazonium salts with boronic acids gives high yields of cross-coupling products only in the absence of both bases and phosphines, which exhibit an apparent detrimental effect due to extensive formation of byproducts [27].

The interest for efficient and clean methods of cross-coupling is further inspired by a need to develop effective approaches to rigid polymers with conjugated backbones, which possess a number of interesting properties allowing application to light-emitting electronic devices, non-linear optics, molecular wires, etc. Such polymers, the most widely known of which is poly-p-phenylene (PPP), which contain conjugated sequences of aromatic rings with or without intermediate olefinic or acetylenic bonds, are an appealing target for palladium-catalysed reactions involving substrates bearing two or more reactive termini (cf. [28], and references therein). The development of phosphine-free and aqueous procedures may bring special advantages for the preparation of such materials. Indeed, the side-effects of phosphine-mediated catalysis are particularly insidious for the preparation of rigid conjugate polymers because the involvement of the phosphine aryl groups may lead to structural defects or uncontrolled termination of polymeric chains owing to the fact that microelectronic or optical devices require high and reproducible purity and consistency of materials. On the other hand, the introduction of groups rendering solubility in water or amphiphilic properties to such polymers improves the prospects for processing (into films, multilayered structures, mesophases, etc.) because such basic polymers are insoluble and infusible, which makes their exciting inherent properties very hard to obtain.

The preparation of water-soluble poly-*p*-phenylenes was achieved by Suzuki coupling of dibromides bearing hydrophilic groups with diboronic acids. The reactions are carried out in aqueous dimethyl formamide (DMF) in the presence of either the water-soluble complex Pd(TPPTS)$_3$ (where TPPTS = *m*-trisulfonated triphenylphosphine [29] or Pd(OAc)$_2$. The reaction in phosphine-free mode is faster, and thus the use of either of the catalysts allows the rate of polymerization to be controlled, which is reflected in the molecular mass distribution and other properties [30].

$$R = \text{\textbackslash}\,SO_3Na$$

2.4 *The Heck reaction*

Among various palladium-catalysed C–C bond formation reactions, the Heck reaction is the only major method that has a clear industrial perspective because it opens a potentially practicable route to such products as stilbenes or cinnamates, which are already widely applied as dyestuffs, UV screens, etc., which particularly stimulates research targeted at the increase of catalytic reactivity and efficiency, simplification of procedures and decrease of overall cost. Indeed, phosphine-free and aqueous methods have been established as a very promising approach for the Heck reaction, although many aspects of this chemistry still need further clarification. The Heck reaction can be accomplished under phase-transfer catalysis (PTC) conditions [31] with inorganic carbonates as bases under very mild conditions even at room temperature, which allows it to be applied to substrates such as methylvinylketone, which cannot survive the conventional conditions of Heck arylation (action of base at high temperature). Later it was shown that water and aqueous organic solvents can be used successfully for carrying out Heck reactions catalysed by simple palladium salts in the presence of inorganic bases such as K$_2$CO$_3$, Na$_2$CO$_3$, NaHCO$_3$ or KOH [32,33]. As in the case of Suzuki coupling under similar conditions, huge turnover numbers exceeding 100 000 can be achieved for water-soluble aryl iodides. However, the addition of (*o*-MeC$_6$H$_4$)$_3$P ligand is required for aryl bromides.

X = I, Br; Z = H, *p*-Cl, *p*-MeO, *p*-Me, *p*-Ac, *p*-NO$_2$, *p*-CHO, *p*-OH, *m*-COOH, etc.

Aqueous DMF is the solvent of choice for a large scope of both aryl iodides and olefins under phosphine-free conditions with K$_2$CO$_3$, NaOAc or other alkaline salts as bases.

It was shown that the addition of 25–30 vol. % of water to DMF exerts a dramatic accelerating influence on the rate of the Heck reaction and shortens the latent period, which is commonly accounted for by the pre-reduction of a Pd(II) precursor to the reactive Pd(0) complex (A.V. Cheprakov and I.P. Beletskaya, unpubl. data).

The combination of phase-transfer catalysts and water gave another useful method for Heck arylations, proposed by Jeffery. In this procedure the reaction is carried out in a heterogeneous system composed of substrates, water, alkali metal carbonate and tetrabutylammonium salt with hydrophilic anion (chloride, bisulfate, etc.), often in the absence of phosphines. The reaction in such systems can be carried out even at room temperature, which is particularly important for application to the synthesis of compounds that cannot endure harsh conditions of conventional phosphine-mediated Heck reactions in anhydrous solvents [34]. A major inconvenience of the original Jeffery's procedure is that it is actually a case of liquid–liquid phase transfer, and thus it can be applied only if organic reagents can form a liquid phase. With solid reagents, an organic solvent can be added. It must be noted here that both water and ammonium salt have a positive effect on the rate and catalytic efficiency, but there is no evidence that there is a synergism of these effects. Tetrabutylammonium salts with small hydrophilic anions, as was hypothesized above, are not solely phase-transfer catalysts but, rather, may serve as promoters helping to increase the reactivity of the intermediate palladium species during the catalytic cycles. Thus, the action of such salts is retained in homogeneous environments. Such an approach was applied to the arylation of N-acetamidoacrylic acid to afford an easy route to well-known precursors of phenylalanines obtained by asymmetric hydrogenation [35].

Aqueous Heck arylation in the presence of terabutylammonium salts is especially attractive for application towards the synthesis of sensitive natural molecules and their analogs. The modification of pyrimidines and other nitrogen-containing heterocycles occurring in nucleotides was investigated using the model reaction with 2,3-dihydrofurane [36]. Besides clear preparative advantages (milder conditions, higher yields, enhanced selectivity), the aqueous method unexpectedly suppressed double bond migration, which is a well-known cause of poor selectivity in common Heck arylations of olefins having the ability to isomerize through the addition–elimination mechanism.

The arylation of protected glycals opens a route to C-nucleosides.

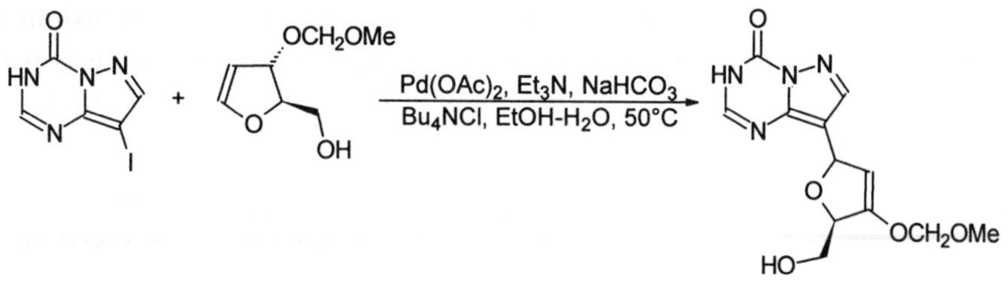

This is not a sole example of the suppression of double bond migration in phosphine-free and aqueous palladium catalysis, thus the arylation of allylphosphonates under conventional anhydrous conditions leads to the predominant formation of rearranged products. The same reaction in aqueous DMF can be run at lower temperatures, and leads exclusively to γ-arylallylphosphonates [37].

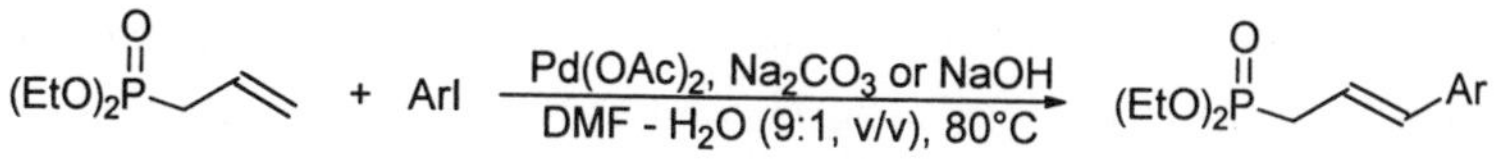

The intramolecular version of the Heck reaction is extensively used for the construction of five- and six-membered rings. Common methods usually lead to *exo*-cyclization, giving a smaller ring of two possible products with an exocyclic double bond. However, there are clear indications that the use of aqueous or phosphine-free methods may favor *endo*-cyclization. Thus, the ring closure mode changed from *exo* to *endo* on changing the system from the conventional anhydrous procedure to aqueous, although in this case the reaction in the latter was mediated by the water-soluble phosphine TPPTS (see below) [38].

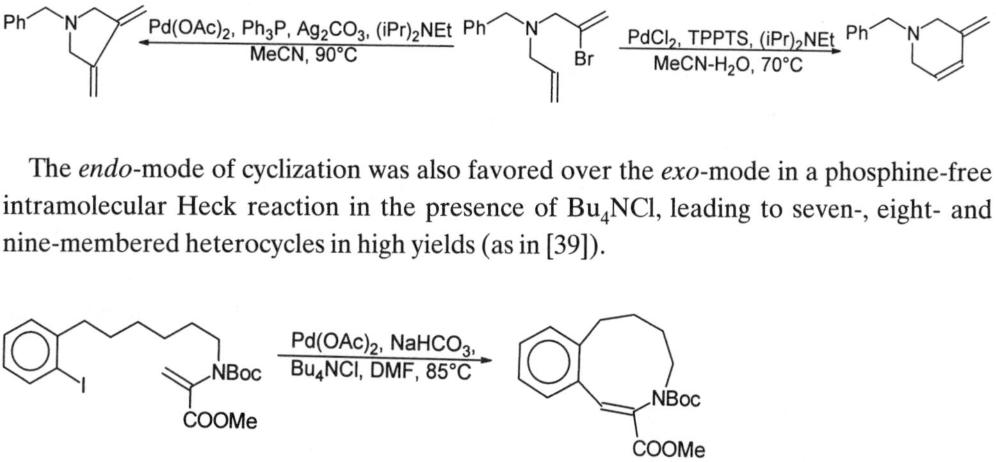

The *endo*-mode of cyclization was also favored over the *exo*-mode in a phosphine-free intramolecular Heck reaction in the presence of Bu₄NCl, leading to seven-, eight- and nine-membered heterocycles in high yields (as in [39]).

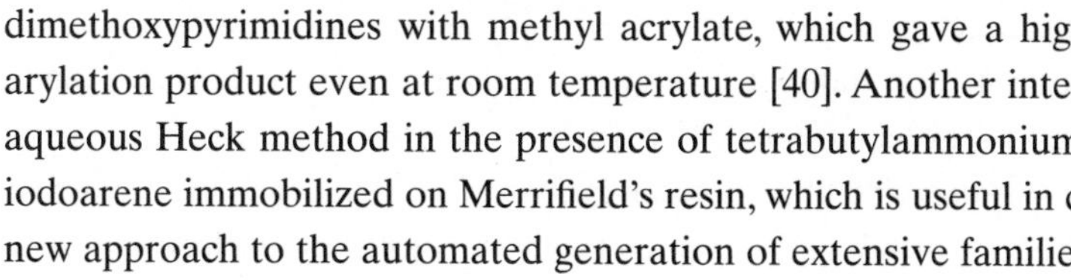

The aqueous Heck method was applied for the reaction of 5-iodo-2,4-dimethoxypyrimidines with methyl acrylate, which gave a high yield of the respective arylation product even at room temperature [40]. Another interesting application of the aqueous Heck method in the presence of tetrabutylammonium salt is the reaction with iodoarene immobilized on Merrifield's resin, which is useful in combinatorial synthesis, a new approach to the automated generation of extensive families of analogous molecules

[41]. Mild conditions account for enhanced tolerance towards various functional groups, which is necessary for a given method to be applicable for combinatorial synthesis.

Z = COOEt, CN, $CONH_2$, $CONR_2$, CHO etc.

3 Reactions in microheterogeneous systems: solubilization and clusters; the concept of a nanoreactor

The development of phosphine-free catalysis and aqueous media has already brought about procedures with markedly enhanced efficiency and selectivity, as well as the simplification of techniques with a positive impact on safety, but there are still a lot of problems to be overcome before going further in this direction. The first is the clarification of how palladium can function in catalytic cycles in the apparent absence of supporting ligands. Without such knowledge, the development of phosphine-free methods shall remain an adventurous hunt in the dark. The deactivation of palladium catalysts occurs mainly through the formation of inactive sediments consisting of particles of metallic palladium, the notorious *black*. It is evident that in order to understand how such deactivation can be avoided, it is necessary first to investigate which factors control the nucleation and growth of metallic particles passing through the stages of small clusters, nanometer-scale particles, crystals, etc. As the physical chemistry of such phenomena is extremely complex and relevant data are still lacking, a thorough study of such phenomena shall wait for the 21st century. However, some data that help to highlight the importance of such phenomena are already available. As far as we know, small palladium clusters characterized as distinct molecular entities were not investigated as catalysts in such processes, although in oxidation or hydrogenation reactions, for example, such catalysts gave promising results [42]. Several recent papers describe the catalytic activity of palladium nanoparticles obtained by different methods in catalytic C–C bond formation reactions.

3.1 *Palladium nanoparticles*

Palladium does not readily form sols in homogeneous solutions, as compared to gold, for example. The reduction of solutions of palladium salts or complexes usually leads to poly-dispersed precipitates known as *palladium black*, possessing poorly reproducible variable properties and thus not suitable for catalysis, although occasionally a certain residual catalytic activity has been reported for such sediments. In order to form stable and reproducible palladium sols, palladium salts are reduced in the presence of either surfactants or polymers, the molecules of which form protective layers on the surface of metallic nanoparticles to prevent their further growth. Because of the nature of such protective shells, the majority of known protected palladium colloids are liosols capable of redispersion in organic solvents. Palladium colloids can be obtained and stored in dry form and redispersed when needed in the appropriate solvent, thus actually being highly reactive and, particularly importantly, *reproducible* forms of palladium black.

A whole family of such colloids is obtained by a general method consisting of a fast reduction of palladium salts by superhydride $MBEt_3H$ or in an electrochemical cell in the presence of various anionic, cationic or zwitter-ionic surfactants [43]. Such colloids are efficient catalysts in Heck and Suzuki reactions with reactive aryl bromides at high temperatures in dimethylacetimide solutions, giving high turnover numbers and yields of respective products [44,45]. Bimetallic clusters Pd/Ni (1:3) can be used in place of pure palladium, showing even higher catalytic activity, and can be used to economize on the precious metal. The main disadvantage of such systems is poor stability of the sol under such conditions. The protective shield of surfactant molecules is easily stripped by hot organic solvent, which leads to deactivation of the catalyst due to the redistribution of palladium between unprotected nanoparticles and the formation of polydispersed palladium black.

A more stable palladium sol, which functions as a true continuous-flow nanoreactor loaded by reagents and unloaded from products through mass transfer between the interior and the bulk environment, was described quite recently [46]. The sol is generated in the core of micelles of amphiphilic block copolymer, such as polystyrene-b-poly-4-vinylpyridine.

P4VP   PS

The idea is to obtain a microstructure partially resembling an inverted micelle, but much more persistent because the block copolymer folds so that the polar part composed of vinylpyridine residues forms a micelle core with polystyrene tails growing outwards, thus imparting micelle solubility in hydrophobic solvents such as toluene. The hydrophilic core solubilizes palladium salts and reduction with fresh superhydride $LiBEt_3H$ gives a fine sol encapsulated in polymeric micelles. In spite of a thick protective layer of polystyrene, such sols turn out to be quite efficient, such as in Heck arylation of styrene with reactive aryl bromides, provided that hydrophobic base NBu_3 is used, which is capable of diffusion to the core where the reagents are located.

Here, the reaction is forced to occur in a nanoreactor built by a copolymer micelle because the compartmentalization of all participating species is unambiguous. The encapsulated catalyst has an outstanding stability and fully retains activity after 50 000 catalytic cycles (this number is, in fact, higher because interior atoms of palladium cluster cannot take part in the reaction and the turnover number is estimated from the net amount of palladium in a sol) at 140 °C in a fairly aggressive environment and taking into account that the reaction leads to the formation of NBu_3HBr. It is noteworthy that pyridine residues in the micelle cores themselves may, and should, act as the base required by the Heck reaction mechanism, although the accumulation of bound protons and bromide counter ions in the nanoreactors might have destroyed the internal structure. Thus, tributylamine in this system acts as a pseudophase-transfer agent to carry away the liberated HBr. However, as the reaction proceeds, the concentration of NBu_3HBr in the solvent grows and progressively more of it must be solubilized in the micelle core, eventually and inevitably ruining this fine structure.

The formation of a byproduct salt in the Heck reaction, in cross-coupling, in allylic substitution and in carbonylation is often overlooked during the design of new catalytic systems, although this factor, which may seem non-essential at first glance, is one of the main problems when catalytic efficiency and re-use of catalyst are the issue. In this respect, processes such as hydrogenation and hydroformylation, which produce no such byproducts, are better suited for the design of recyclable catalysts and therefore may boast industrial implementations of this idea. The encapsulation of palladium clusters in copolymer micelles has another serious drawback because the recovery and recycling of such state-of-the-art encapsulated sol from a reaction mixture is not an easy, if altogether achievable, task.

3.2 *Catalysis in solubilized systems*

Besides the attempts to study the catalytic activity of preformed palladium clusters, the study of microheterogeneous systems has yet another incentive. Successful application of aqueous solvents for palladium-catalysed reactions has highlighted the main limitation of homogeneous media with a high content of water. Such methods can be applicable in most cases only to water-soluble reagents. Because one of the goals pursued by the development of aqueous catalysis was the increase of environmental and technological safety, it must be stressed that the use of aqueous organic media in which the content of organic components is kept high, while solving the solubility restrictions, does nothing for the safety factors. The design of aqueous systems with water as the predominant component and suitable for wider ranger of organic reagents led to the exploitation of the *solubilization* phenomena. It is well known that aqueous solutions of surfactants are capable of enhancing the solubility of hydrophobic organic compounds (as well as the solubility of water and aqueous solutions in hydrophobic liquids often generically referred to as *oils*, both for brevity and to pay tribute to the tradition). The molecules of solubilizate naturally do not form a true solution, but rather are absorbed by the micelles. Common ionic surfactants (like sodium dodecylsulfate or cetyltrimethylammonium bromide, i.e. those consisting of a single ionic head group and a single non-branched hydrocarbon tail) form spherical micelles having very low solubilization capacities, because typically a single micelle built of several tens of surfactant molecules takes a few (1–5) molecules of solubilizate. Ionic species are either being attracted or repelled by the interfacial layer formed by head groups and their environment, depending on the relative charges of surfactant and ions in solution. So, due to a considerable increase (or depletion) of the local concentration of reagents in the micelle, many reactions, especially those involving ionic reagents, are markedly accelerated or retarded. This phenomenon—micellar catalysis [47]—is well known, although its practical application for carrying out reactions on a preparative scale is limited by the very small solubilization capacities of simple micelles. In order to increase the solubilizing ability of the microheterogeneous system, the hydrophile–lipophile balance (HLB) of a surfactant system must be adjusted to counter balance the hydrophilic and hydrophobic interactions. The adjustment of HLB is performed by a number of means, including the use of special surfactants or compound surfactant/cosurfactant systems. Further fine balancing is performed by varying the temperature and the concentration of electrolytes. The resulting microheterogeneous media are characterized by very low interfacial tension between aqueous and oil phases, and are capable of solubilizing considerable amounts of water-insoluble organic compounds. Unlike macroemulsions, such media, the most well known of which are called *micro*emulsions, are thermodynamically stable systems, that are formed spontaneously and do not require continuous mechanical agitation.

Such systems can be applied successfully in palladium-catalysed reactions with water-insoluble reagents. The simplest of media of this type are formed by a combination of single-tail ionic surfactant and short-chain aliphatic alcohol as a cosurfactant dispersed in water in an approximate surfactant ROH water molar ratio = of $1:4:100–200$. After fine balancing using an electrolyte that can act simultaneously, e.g. as a base needed for the majority of catalytic reactions under discussion, these media solubilize organic reagents in amounts roughly equivalent to those commonly used in organic preparations in more conventional media ($0.1–1 \, mol^{-1}$).

Such media can be used for all major types of palladium-catalysed reactions, including carbonylation of aryl halides, hydrogenolysis, acetylenic coupling, N-arylation, Heck and Suzuki reactions, Wacker oxidation of olefins, etc. [48–50]. Most reactions studied so far can be performed in the phosphine-free mode with high efficiency.

Moreover, the study of Heck arylation and Suzuki cross-coupling provided the first experimental evidence on the involvement of palladium clusters in phosphine-free catalysis. These reactions in balanced solubilized systems turned out actually to run through the fast formation of stable palladium sols. Unlike the studies of the catalytic activity of palladium sols treated above, where the nanoparticles were obtained via an independent way and then transferred to the reaction media, in solubilized systems nanoparticles are formed and survive, and thus perform catalytic functions in the native environment. The catalytic process itself proceeded in the presence of palladium colloid, which underwent no further evolutions during the reaction. By carrying out the reactions in systems built around various types of surfactants (anionic carboxylates, phosphonates, sulfates, sulfonates with variable chain lengths, as well as cationic alkyltrimethylammonium salts) the performance of the catalytic system was shown to change in parallel with the ability of the media to assist the formation of colloid and sustain it. The reactions showed very high turnover numbers (up to 50 000), which is not surprising because only a fraction of palladium atoms located at the surface of the cluster actually react, and thus lowering the initial concentration of palladium leads only to the formation of smaller particles having less wasted palladium atoms in cluster interiors.

Thus, we may argue that the microphase of the solubilized system acts as another example of a self-sustained nanoreactor, the role of which is performed by the interfacial layers and surfactant-stabilized Pd clusters. Aqueous and oil phases act as feedstocks of hydrophilic and hydrophobic reagents, respectively, as in Heck arylation, for example. Unlike simple micelles, the solubilized systems possess much wider margins of stability and thus are quite tolerant to a profound change of the composition of the reacting system during the process.

Moreover, the study of Heck and Suzuki reactions in balanced microemulsions afforded experimental evidence in favor of the hypothesis that phosphine-free palladium catalysis is accounted for by the formation of palladium clusters. The factors that help to stabilize reactive clusters and prevent further aggregation leading to unreactive blacks (which are likely to consist of relatively large crystalline particles with rather small surfaces) act in favor of phosphine-free reactions. Thus, tetraalkylammonium salts, which are ubiquitous in phosphine-free methods, may have yet another function by absorbing at the surface of the emerging nanoparticles and forming a supporting shell for them.

Solubilized systems can be formed also by non-ionic surfactants, the derivatives of oligo-ethyleneglycols: $R(OCH_2CH_2)_nOH$. The balancing of such systems does not require a cosurfactant but is performed by varying the number of ethylene oxide residues in the surfactant molecule, with further fine balancing achieved by temperature variation in the vicinity of the so-called phase inversion point, at which the nature of the system goes from more-hydrophilic to more-hydrophobic. Non-ionic surfactants and similar molecules exhibit another intriguing capability to affect the local transport of ions, thus facilitating mass transfer of hydrophilic reagents to the interfacial nanoreactor [48, 51].

4   Recyclable catalytic systems

Water and aqueous systems help to realize technical goals associated with heterogenization of homogeneous catalysis. From a practical viewpoint, homogeneous catalysis has a fatal drawback in that the catalyst is used only for a single run. None of the catalysts used for truly homogeneous processes can be recovered from the reaction mixtures in a reactive form suitable for immediate re-use. Thus, the cost of the catalyst, which is often quite high, adds to the cost of the product. This can be appropriate only for very complex and expensive products (e.g. chiral compounds with high enantiomerism), but is prohibitive for the application of such processes to the vast majority of other chemicals. Besides, homogeneous processes are unsuitable for application in more economical continuous-flow reactors. The heterogeneous processes, however, for which catalyst recovery and continuous-flow operation are inherent, are often hopeless in areas where selectivity is the major issue. It is obvious that combining what is good in each of the types of catalysis while leaving behind what is deficient must be an appealing challenge. It should only be added that increasing catalytic efficiency (the yield of product per load of catalyst) is an approach that is roughly equivalent to recycling. Indeed, any method of recycling cannot have an absolute recovery, and some of the catalyst is still lost at each cycle. For example, increasing the efficiency (turnover number) by 100 times (going from 5 mol.% of catalyst load to 0.05 mol.%) in a non-recyclable process is equivalent to having a recyclable catalyst with total recovery of 99% (summed up for a whole lifetime of a recycled catalyst), which is an enviable value rarely achieved in real processes. As we have seen above, the application of novel rigid complexes and of some phosphine-free systems in aqueous homogeneous or solubilized media affords turnover number exceeding 50000–100000. Such efficiency in a single run already by far exceeds values those achievable from any real recycling procedure.

There are two main approaches for the design of recyclable catalysts: running the process in a biphasic (liquid–liquid or liquid–solid) system in which the catalysts are physically bound to one of the phases; and chemically binding the catalyst to the carrier molecule possessing specific physical properties, which enables the separation of the whole unit by methods applicable to the carrier. Both approaches have been applied already to palladium catalysis.

4.1 Liquid–liquid biphasic catalysis

Liquid–liquid biphasic catalysis relies on catalysts held in one of the phases, while the products accumulate in the other phase. After the reaction the phases are separated, either directly or, in the case of systems prone to emulsification, by applying various technical tricks such as passing the reaction mixture through semi-permeable membranes that allow only one phase to pass through (cf. [52]). The simplest and most obvious example of such systems uses water as the catalyst-containing phase, although there are non-aqueous biphasic systems: the most famous industrial realization of biphasic technology is the Shell Higher Olefin Process, wherein the recyclable polar phase containing the catalyst is immiscible with the products and is claimed to be butyleneglycol.

In order to make the recycling truly efficient, it is important not to allow the palladium to be reduced to metal and to have it quantitatively bound to a complex with such a high solubility in its own phase that any leaching to the other phase can be neglected. In order to achieve this, highly hydrophilic ligands need to be employed, the most important of which has been, and still is, tris(m-sulfonatophenyl)phosphine used as its trisodium salt (TPPTS).

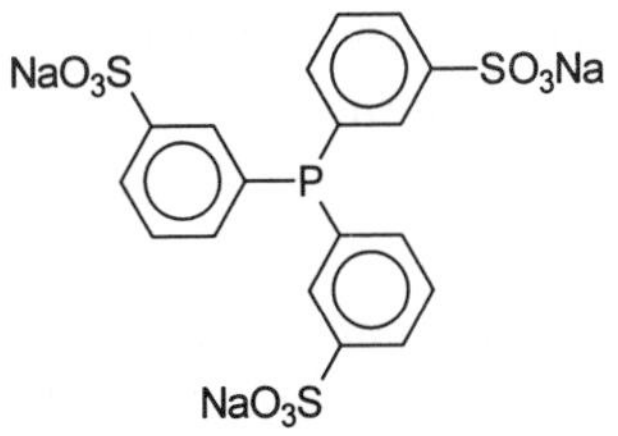

TPPTS 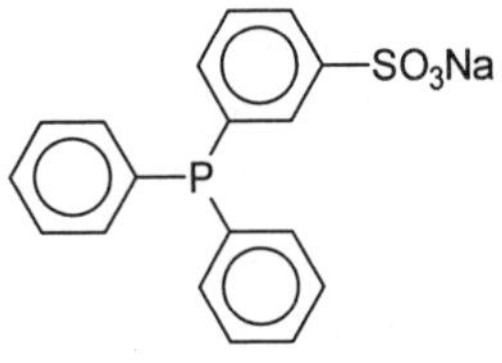   TPPMS

It should be noted that the cheaper and more readily available monosulfonated triphenylphosphine (TPPMS) cannot serve as a reliable substitute for TPPTS because the former ligand, as well as its complexes and intermediates formed in catalytic processes, may bear considerable amphiphilicity, i.e. the ability to aggregate and leach into the organic phase.

Both TPPTS and TPPMS ligands are quite popular as water-soluble analogs of Ph_3P in phosphine-mediated reactions in aqueous media for reasons that have nothing to do with recycling. The incentives for using such catalysts are manifold. Although, as already mentioned, the use of aqueous organic solvents apparently solves no environmental or safety problems, such media are required for applying the catalytic reactions towards highly hydrophilic molecules, e.g. for modification of natural compounds and their analogs, which may be incompatible with organic solvents due to poor solubility or destruction of inherent conformations (denaturing). To mention only a few important contributions, aqueous media and water-soluble palladium catalysts with hydrophilic phosphines were particularly effective for acetylenic coupling of iodoarenes with terminal acetylenes under very mild (*biocompatible*) conditions [53], in contrast with conventional non-aqueous acetylenic coupling, which often requires prolonged heating in refluxing organic solvents. The other example is a cross-coupling with organostannates $[RSn(OH)_{3+n}]^{n-}$, generated *in situ* from $RSnCl_3$ in aqueous alkaline solutions [54], thus affording a modification of the Stille reaction that avoids the formation of toxic organotins as side-products, offering a competitive method for Suzuki cross-coupling.

$$RSnX_3 \xrightarrow{OH^-/H_2O} [RSn(OH)_{3+n}]^- \xrightarrow{ZC_6H_4Hlg,\ PdCl_2\ or\ PdCl_2(TPPMS)_2} Ar\text{-}R$$

True biphasic systems were applied mostly for Tsuji–Trost reactions. Palladium-catalysed allylic substitution is a powerful and extremely versatile method for which, unlike cross-couplings, Heck reactions and other basic methods, no reliable and highly efficient phosphine-free procedures have been realized. The reactions are usually carried out in the presence of considerable amounts of palladium catalyst (4–20 mol.%) supported by phosphine ligands. Therefore, biphasic catalysis with a re-usable catalyst phase is the only practical way to lower the cost of this important procedure. Indeed, allylic acetates and carbonates readily react with various C-, N-, S-nucleophiles in biphasic systems composed of nitriles and aqueous solutions of Pd complex with TPPTS ligands [55]. This method was applied for very mild and selective deprotection of allyl and allyloxycarbonyl derivatives, in which allyl is transferred from the protected molecule onto a small nucleophile such as Et_2NH. The method is easily tunable to distinguish quantitatively the various types of allylic protective groups [56].

The other reaction closely related to allylic substitution that benefited from the application of biphasic techniques is the telomerization of butadiene with water to obtain 2, 7-octadienol-1 as a main product, which is a valuable synthetic intermediate for the production of plasticizers and monomers. Because butadiene possesses considerable solubility in water, the reaction can be run in a conventional biphasic mode with aqueous solution of palladium complex with TPPTS [57]. The selectivity and yield in this reaction is markedly increased in the presence of carbon dioxide.

Conventional biphasic techniques require the use of hydrophilic phosphine ligands to hold the catalyst in the recyclable aqueous phase. Thus, it seems that phosphine-free catalysis is altogether not suitable for the biphasic method. However, this limitation can be broken by a judicious choice of liquid phase that can serve both for the separation of catalyst from product and to provide the support for palladium species. An interesting example of such a system was discovered recently. One of the liquid phases is composed of a melt of 1-*n*-butyl-3-methylimidazolium tetrafluoroborate (BMI·BF$_4$), which is an excellent solvent for various palladium compounds, including Pd(OAc)$_2$, η^3-allyl complexes, etc. This liquid is miscible both with water and hydrophobic compounds at elevated temperatures, so the reaction actually runs in homogeneous solution without mass-transfer limitations. Below −5 °C the mixture separates into distinct phases, with the BMI·BF$_4$ phase holding all the catalyst for easy recycling. The system itself provides methylimidazole ligands for palladium, preventing the formation of inactive Pd metal due to the following process involving the oxidative addition of Pd(II) to a C—N bond, which accounts for the transformation of the salt (BMI)$_2$PdCl$_4$ to bis(1-methylimidazole)palladium dichloride in the presence of water [58].

This reaction leads to the desired telomer with rather high selectivity even in the absence of CO$_2$. It is quite interesting to note that imidazolium salts were reported earlier to react with palladium salts to give carbene complexes, which are highly reactive catalysts for Heck arylation (cf. [12]).

This process is an example of systems that undergo a controlled reversible transition between biphasic and monophasic states. The development of such systems opens new possibilities for phase-separation catalysis. Currently, a few such systems are known, although their application for palladium-catalysed reactions is practically unknown. Besides the liquids, which show behavior known as *the miscibility gap*, any implementation of this approach must also rely on a catalyst that has preferential solubility in one of the phases. The most appraised example of this approach is so-called *fluorous biphasic* catalysis, which uses the perfluorohydrocarbons as the catalyst carrier phase [59]. These liquids exhibit temperature-controlled miscibility with, for example, aromatic hydrocarbons. In order to make the catalysts preferentially soluble in the *fluorous* phase, similar to the design of water-soluble ligands by modification of known ligands by hydrophilic groups, long perfluoroalkyl chains (so-called *fluorous ponytails*) are attached to ligands (cf. [60] for the design of fluorophilic cyclopentadienyl ligands; cf. [61] for the system used for oxidation catalysed by ruthenium or nickel complexes of perfluorinated 1,3-diketone; cf. [62] for the design of chelating ligands for nickel and ruthenium complexes). We shall definitely not wait long for the application of fluorous biphasic catalysis for palladium-catalysed reactions, at least for those reactions not requiring the use of additional reagents such as bases, e.g. hydrogenation or Stille coupling.

It should be noted also that the properties of perfluorocarbon liquids resemble those of supercritical carbon dioxide, the solvent of high interest for chemical technology of the 21st century.

True biphasic systems suffer from the limitations of rate by mass transfer through a relatively small interface, which makes them highly dependent on the efficiency of stirring (this phenomenon resembles the behavior of the conventional liquid–liquid phase-transfer processes). In order to overcome this limitation, ideas derived from partition chromatography were successfully introduced. Liquid phase (most often the aqueous phase) is distributed over the surface of the solid support, making a thin layer with a huge surface. The catalyst, which is usually a complex with hydrophilic phosphines such as TPPTS, is distributed in the supported phase by a simple impregnation of dry carrier with the solution of metal complex. The amount of supported phase must be large enough for it to form liquid film in which the catalyst possesses a certain two-dimensional mobility. This aspect distinguishes the supported liquid phase catalysts (SLPC, or SAPC if *liquid* stands for *aqueous*) from the common supported catalysts, because the latter are immobilized with the carrier and have no own mobility. Thus, in principle, SLPC must be more efficient and, besides, they are much simpler and cheaper. However, such catalysts must be very sensitive to the composition of reaction mixtures, particularly to electrolytes that are partitioned into the supported phase and rapidly demolish this fine structure. Thus, these systems are suitable only for reactions that produce no ionic byproducts, primarily hydroformylation and hydrogenation [63]. As was described above, allylic substitution catalysed by water-soluble palladium complex with TPPTS ligand can be run in the absence of base, thus excluding the influence of electrolyte, the nature and concentration of which continuously vary during the reaction. The reaction with allylic carbonates is especially suitable for implementation using SAPC, because the only byproduct accumulating in the reaction mixture here, the alcohol, is truly unobtrusive. The reaction of cinnamylethyl carbonate with two representative substrates, ethyl acetoacetate and morpholine, was realized with $Pd(TPPTS)_3$ complex supported on both controlled-porosity glass (silica with a narrow distribution of pore sizes) and non-porous silica bearing an aqueous layer [64]. This study is exemplary because it shows all the positive and negative aspects of the use of SAPC in palladium-catalysed reactions.

Both reactions readily run with the SAPC at 80 °C and show higher selectivity towards dialkylation and hydrolysis of the carbonate ester than the respective reactions done under aqueous monophasic and biphasic conditions, all other factors held constant. If run in MeCN solvent the reaction catalysed by SAPC showed lower activity than the same reaction run in homogeneous solution in aqueous MeCN. Owing to the miscibility of this solvent with water, the supported catalyst cannot retain a film of liquid water and the complex may be held only by absorption on the surface of silica. This leads to extensive leaching of palladium and fast degradation of activity, making recycling inefficient. On the other hand, the reaction in PhCN showed higher catalytic activity and excellent reproducibility in tests with recycled catalyst. Some decrease of activity of the re-used catalyst can be attributed to the dehydration of the catalysts due to leaching of water. The catalyst, nevertheless, can be revitalized by rehydration, effected by the simple addition of water. The activity of catalyst is highly dependent on the amount of absorbed water, rapidly growing with higher water content until the wetting volume is reached, after which extra water is readily removed from the support, leading to increased leaching of palladium and degradation of activity. Interestingly enough, no essential dependence on the porosity of silica was noted.

Glycols can be used in place of water both in liquid–liquid biphasic and in SLPC catalysis. In SLPC catalysis, glycols can provide a more stable supported phase than water, less prone to wear due to the dissolution in the bulk phase. Both TPPMS and TPPTS complexes of palladium were dissolved in ethylene glycol supported over controlled porosity glasses to provide a good catalyst both for Heck reactions with reactive aryl iodides and for allylic substitution [65]. It should be noted that SLPC have poor perspectives for reactions requiring harsh conditions, such as Heck arylations with aryl bromides and chlorides, because the supported liquid phase is too weak to withstand high temperatures.

Thus, the SAPC/SLPC concept, though quite narrow in scope, is, where applicable, a very useful approach, allowing easy and very efficient recycling of catalyst and making feasible the implementation in continuous-flow reactors, for which the conventional biphasic technique is unsuitable.

Yet another way to generate fine palladium particles and overcome their further aggregation and deactivation was sought as a compromise between the supported aqueous-phase catalysis and cluster catalysis [66], and applied to carry out Heck arylations and acetylene cross-couplings in the presence of palladium supported on porous glass. In this method porous glass was first impregnated with an aqueous solution of $[Pd(NH_3)_4]Cl_2$ and then the supported palladium complex was reduced by $NaBH_4$ to obtain palladium metal immobilized in pores. Such catalysts are no longer sensitive to water content and stray electrolytes and have been used in the phosphine-free mode to achieve arylation and acetylenic coupling under rather drastic conditions, e.g. conventional or microwave heating in air to as high as 180 °C for Heck reactions and 140 °C for acetylenic couplings. The latter required no copper co-catalyst, e.g.:

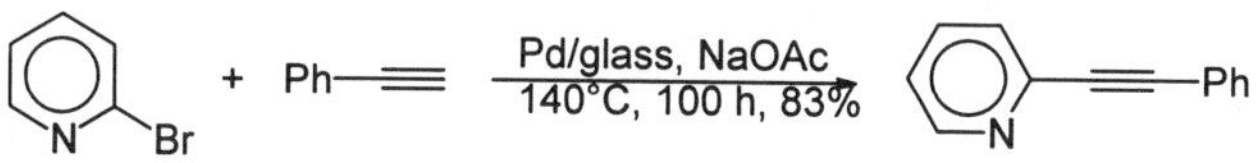

4.2   Immobilized catalysts

Immobilization of palladium catalysts on polymeric resins so far has not produced any spectacular results. The problems here are manifold, including low thermal stability of the polymers that are usually employed for such purposes (mostly various linear cross-linked polyolefins), which leads to the fast degradation of expensive catalyst and prevents recycling, as well as very low rates due to both the limited diffusion of reactants in polymers and the low concentration of catalytic centers that are bound only to end-cap residues. Several attempts to enhance the stability of supports and to increase the load of palladium per molecule of polymer by introducing the ligating molecules inside the polymeric molecule were made recently. For example, a resin containing 1,10-phenanthroline units was developed and used as a support for palladium in Heck arylations of acrylamide to give cinnamamides, which can be converted further to arylacetic aldehydes [67].

Polyimide resins are famous for their outstanding thermal stability. A support made of such a resin containing cyano groups capable of binding palladium was used for preparing recyclable palladium catalyst for Wacker oxidation of olefins [68]. Recycling tests showed considerable leaching of palladium, which is not surprising because nitrile complexes are rather weak.

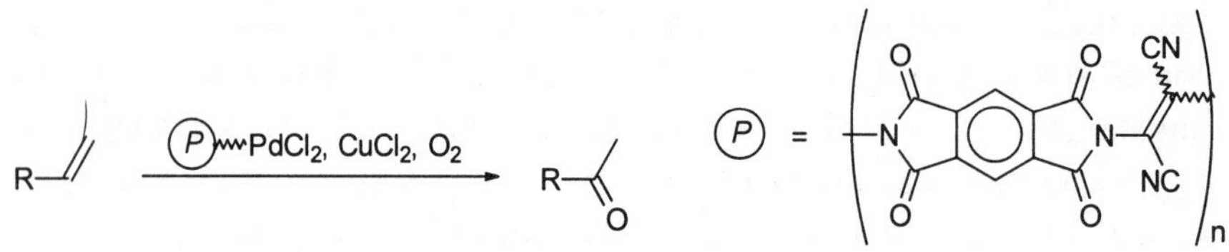

Immobilization of bidentate phosphine ligands on resin was observed to lead to a significant boost of the catalytic activity in comparison with reactions with the same ligand detached from the polymeric support [69]. This phenomenon may be ascribed also to restricted diffusion that prevents Pd(0) centers from meeting together and the formation of clusters.

A further step forward takes advantage of a new class of macromolecules, the dendrimers. Commercially available dendrimer based on 1,4-diaminobutane (DAB) bearing peripheral amino groups was derivatized with diphenylphosphine (PPh$_2$) to give a macromolecule with 16 chelating diphosphine units, DAB-*dendr*-[N(CH$_2$PPh$_2$)$_2$]$_{16}$ [70], which can bind transition metals (e.g. palladium, nickel, rhodium) to give complexes. If the loading of metal is complete, well-characterized complexes are obtained, that also differs dendritic ligands from other macromolecules.

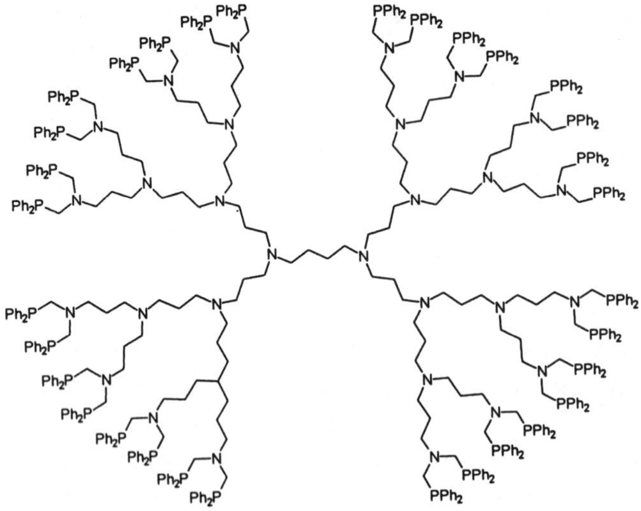

Dendritic molecules like the one above are well known to possess structures close to spherical, thus being actually well-defined monomolecular micelles, which gives yet another analogy with the nanoreactor concept because the whole dendrimer unit charged with palladium operates like a micelle with an interfacial layer emulated by the external diphenylphosphino groups, or like a nanoparticle in which all valuable atoms are located at the edges while the interior is filled with inert material.

Like other macromolecules, dendrimers can be precipitated from solutions by the addition of weakly interacting solvents, making them useful for recycling purposes. A very strong advantage of dendrimeric support over other kinds of organic polymers is a full exposure of catalytic centers to the environment. Because the whole macromolecule is inside the particle, diffusion limitations are not important.

Dendrimeric catalyst DAB-*dendr*-[N(CH$_2$PPh$_2$)$_2$PdMe$_2$]$_{16}$ showed about three times higher catalytic activity than parent complex *n*-PrN(CH$_2$PPh$_2$)$_2$PdMe$_2$ in the Heck arylation of stilbene with bromobenzene carried out in DMF at 130 °C in the presence of NaOAc as base, due to the higher stability of the former. Indeed, the deactivation of palladium catalyst in catalytic reactions is due to the formation of large metallic particles, the process initiated by encounters of Pd(0) species. It is evident that the encounter rate is much higher for small molecules than for macromolecules, due to both diffusional and entropic considerations. The recycling of dendritic catalyst is effected by a simple precipitation by ether, and the recovered catalyst retains the activity.

Advances in the chemistry of materials led to the discovery of a number of new types of regularly structured inorganic polymers with captivating properties, including high thermal stability, reproducible three-dimensional structure and availability of reactive groups for modification with organic residues. Unlike such conventional solid supports as charcoal, silica, alumina, etc., which are charged by reagents or catalysts using simple impregnation-and-drying procedures that rely on physical absorption, which is very hard to control, novel materials are modified via more predictable processes such as ion exchange or chemical bonding.

Thus, a well-known very cheap natural ion exchanger, montmorillonite clay, can be charged by Pd(II) and Cu(II) to provide a recyclable catalyst for Heck arylations with aryl bromides [71]. A synergism of palladium and copper was observed. It is quite interesting that the main function of montmorillonite is soft Lewis acidity, a property that can assist in the de-ligation of ancillary non-phosphinic ligands, analogous to the effect of metal cations (see above).

An interesting family of regular layered porous materials is generated based on zirconium hydroxide forming networks built of ZrO$_6$ octahedra. Such networks are modified by inclusion of other oxoanions using a principle roughly analogous to semiconductor doping, when the additive has a different valency, thus generating a valuable defect. The zirconium hydroxide structure can be doped with tri- and tetracoordinate oxoanions, such as phosphates, phosphites or even organic phosphonates, allowing the introduction of an interesting structural pattern of regular pores but also simultaneously the attachment of organic residues to produce a three-dimensional modified inorganic polymer [72].

This catalyst, reduced with Et_3SiH, showed high activity in Heck reactions of iodoarenes with methylacrylate. Although it is extremely sensitive to oxidation, the catalyst fully retained its initial activity after being used several times provided that all manipulations were carried out in an inert atmosphere, which demonstrates an exceptional stability of the structure of the inorganic polymer. Even more interesting is the finding that reactions with this catalyst reveal a specific type of selectivity towards the size of substrates, very similar to that typical for reactions with the participation of host–guest complexes, e.g. cyclodextrins. The use of the latter was recently offered as a way to promote rhodium-catalysed hydroformylation of long-chain olefins in biphasic systems [73], but this approach is definitely too expensive for practical catalysis. The development of inorganic materials capable of molecular recognition, but much less expensive and easily recoverable, may have a strong impact on the development of recyclable catalysts. Moreover, unlike cyclodextrins, which provide only several predefined cavity sizes depending on the number of glucose units in the ring, zirconium hydroxyphosphites afford materials with variable porosity depending on the amount and nature of spacer anions.

Other novel materials used for the construction of solid recyclables are the *mesoporous molecular sieves*. Such materials, discovered quite recently [74], provide an ingenious link between microstructures formed by surfactants and silicate frameworks. Silicate ions are gathered from the solution by interfacial layers of various microstructures formed by surfactants—simple micelles, swollen micelles of solubilized systems, hexagonal phases (spontaneous assemblies of rod-like micelles formed in concentrated solutions of surfactants), etc.—which serve as templates for the assembly of inorganic frameworks. After calcination and removal of surfactant, the remaining polysilicate material, being actually a postmortem mask of the microstructured template, is a porous material with enormous surface area.

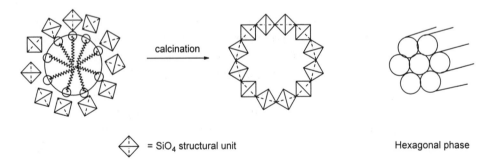

calcination

◇ = SiO_4 structural unit   Hexagonal phase

Owing to the extreme versatility of surfactant systems controlled by a number of easily varied parameters, the composition and morphology of the resulting materials are easily adjustable: e.g. pore sizes can be varied continuously over a broad range of values. Catalytically active sites can be added either during the preparation of the material or later. The first report on the use of such material grafted with palladium for Heck arylation has been reported recently [75]. Palladium metal was deposited onto the hexagonal mesoporous zeolite MCM-41 by vacuum deposition of the volatile complex $(\eta^3-C_3H_5)(C_5H_5)Pd$ with subsequent reduction by hydrogen. The catalyst showed high activity in the test reaction of *p*-bromoacetophenone and some other bromoarenes with butyl acrylate, although the lifetime of the catalyst was limited due to agglomeration of palladium and the formation of large polydispersed particles.

5 References

1 Tsuji J. *Palladium Reagents and Catalysts. Innovations in Organic Synthesis.* Chichester: Wiley, 1995.

2 Rosso VW, Lust DA, Bernot PJ *et al. Org Process Res Dev* 1997; **1**: 311–14.

3 Casado AL, Espinet P. *Organometallics* 1998; **17**: 954–9.

4 Herrmann WA, Brossmer C, Oefele K, Beller M, Fischer H. *J Organomet* 1995; **491**: C1–4.

5 Herrmann WA, Brossmer C, Öfele K *et al. Angew Chem Int Ed* 1995; **34**: 1844–8.

6 Beller M, Fischer H, Herrmann WA, Öfele K, Brossmer C. *Angew Chem Int Ed* 1995; **34**: 1848–9.

7 Beller M, Fischer H, Herrmann WA, Brossmer C. *US Patent 5698755*, filed June 21, 1995, issued December 16, 1997.

8 Louie, J, Hartwig JF. *Angew Chem Int Ed* 1996; **35**: 2359–61.

9 Hennings DD, Iwasa S, Rawal VH. *Tetrahedron Lett* 1997; **38**: 6379–82.

10 Hennings DD, Iwasa S, Rawal VH. *J Org Chem* 1997; **62**: 2–3.

11 Ohff M, Ohff A, van der Boom ME, Milstein D. *J Am Chem Soc* 1997; **119**: 11687–8.

12 Herrmann WA, Elison M, Fischer J, Köcher C, Artus GRJ. *Angew Chem Int Ed* 1995; **34**: 2371–4.

13 Ben-David Y, Portnoy M, Gozin M, Milstein D. *Organometallics* 1992; **11**: 1995–6; Portnoy M, Ben-David Y, Milstein D. *Organometallics* 1993; **12**: 4734–5; Portnoy M, Ben-David Y, Rousso I, Milstein D. *Organometallics* 1994; **13**: 3465–79.

14 Larock RC, Yum EK, Yang H. *Tetrahedron* 1994; **50**: 305–21.

15 Larhed M, Andersson CM, Hallberg A. *Tetrahedron* 1994; **50**: 285–304.

16 Reetz MT, Löhmer G, Schwickardi R. *Angew Chem Int Ed* 1998; **37**: 481–3.

17 Amatore C, Jutand A, Meyer G. *Organometallics* 1995; **14**: 5605–13.

18 Hartwig JP. *SYNLETT* 1997: 329–40.

19 Bumagin NA, Bykov VV, Beletskaya IP. *Izv Akad Nauk SSSR Ser Khim* 1989: 2394 (English translation: *Bull Acad Sci USSR Div Chem Sci* 1989; **38**: 2206).

20 Wallow TI, Novak BM. *J Org Chem* 1994; **59**: 5034–7.

21 O'Keefe DF, Dannock MC, Marcuccio SM. *Tetrahedron Lett* 1992; **33**: 6679–80.

22 Campi EM, Jackson WR, Marcuccio SM, Naeslund CGM. *J Chem Soc Chem Commun* 1994: 2395.

23 Badone D, Baroni M, Cardamone R, Ielmini A, Guzzi U. *J Org Chem* 1997; **62**: 7170–3.

24 Bumagin NA, Bykov VV, Beletskaya IP. *Metalloorg Khim (Russ)* 1989; **2**: 1200.

25 Bykov VV, Bumagin NA, Beletskaya IP. *Bull Russ Acad Sci* 1995; **340**: 775–8.

26 Kang SK, Lee HW, Jang SB, Ho PS. *J Org Chem* 1996; **61**: 4720–4.

27 Sengupta S, Bhattacharyya S. *J Org Chem* 1997; **62**: 3405–6.

28 Goodson FE, Novak BM. *Macromolecules* 1997; **30**: 6047–55.

29 Wallow TI, Novak BM. *J Am Chem Soc* 1991; **113**: 7411–12.

30 Kim S, Jackiw J, Robinson E *et al. Macromolecules* 1998; **31**: 964–74.

31 Jeffery T. *J Chem Soc Chem Commun* 1984: 1287–9.

32 Bumagin NA, More PG, Beletskaya IP. *J Organomet Chem* 1989; **371**: 397–401.

33 Bumagin NA, Andryukhova NP, Beletskaya IP. *DAN SSSR* 1990; **313**: 107–9.

34 Jeffery T. *Tetrahedron Lett* 1994; **35**: 3051–4.

35 Carlstroem AS, Frejd T. *Acta Chem Scand* 1992; **46**: 163–71.

36 Zhang HC, Daves GD. *Organometallics* 1993; **12**: 1499–500.

37 Demik NN, Kabachnik MM, Novikova ZS, Beletskaya IP. *Zh Org Khim (Russ)* 1995; **31**: 64–8.

38 Lemaire-Audoire S, Savignac M, Dupuis C, Genet JP. *Tetrahedron Lett* 1996; **37**: 2003–6.

39 Gibson SE, Guillo N, Middleton RJ, Thuilliez A, Tozer MJ. *J Chem Soc Perkin Trans 2* 1997: 447–55.

40 Basnak I, Takatori S, Walker RT. *Tetrahedron Lett* 1997; **27**: 4869–72.

41 Hiroshige M, Hauske JR, Zhou P. *Tetrahedron Lett* 1995; **36**: 4567–70.

42 Moiseev II. *J Organomet Chem* 1995; **488**: 183; Kaneda K, Fujii M, Morioka K. *J Org Chem* 1996; **61**: 4502–3.

43 Bönnemann H, Braun G, Brijoux W *et al. J Organomet Chem* 1996; **520**: 143–62.

44 Reetz MT, Breinbauer R, Wanninger K. *Tetrahedron Lett* 1996; **37**: 4499–502.

45 Beller M, Fischer H, Kühlein K, Reisinger CP, Herrmann WA. *J Organomet Chem* 1996; **520**: 257–9.

46 Klingelhöfer S, Heitz W, Greiner A *et al. J Am Chem Soc* 1997; **119**: 10116–20.

47 Fendler JH, Fendler EJ. *Catalysis in Micellar and Macromolecular Systems*. New York: Academic Press, 1975.

48 Cheprakov AV, Ponomareva NV, Beletskaya IP. *J Organomet Chem* 1995; **486**: 297–300.

49 Davydov DV, Beletskaya IP. *Russ Chem Bull* 1995; **44**: 1141.

50 Cheprakov AV, Lomakina MA, Tsarkova LA, Beletskaya IP. *Chem Int J* 1998 (In press.)

51 Trost BM, Radinov R. *J Am Chem Soc* 1997; **119**: 5962–3.

52 Livingston JR. *US Patent 5 302 750*, filed May 25, 1993, issued April 12, 1994.

53 Casalnuovo AL, Calabrese JC. 1990; **112**: 4324–30; Amatore C, Blart E, Genêt, JP *et al. J Org Chem* 1995; **60**: 6829–39; Dibowski H, Schmidtchen FP. *Tetrahedron Lett* 1998; **39**: 525–8.

54 Roshchin AI, Bumagin NA, Beletskaya IP. *Tetrahedron Lett* 1995; **36**: 125–8; Rai R, Aubrecht KB, Collum DB. 1995; **36**: 3111–14.

55 Safi M, Sinou D. *Tetrahedron Lett* 1991; **32**: 2025–8; Blart E, Genêt Jp, Safi M, Savignac M, Sinou D. *Tetrahedron* 1994; **50**: 505–14.

56 Genêt JP, Blart E, Savignac M *et al. Tetrahedron* 1994; **50**: 497–503; Lemaire-Audoire S, Savignac M, Pourcelot G, Genêt JP, Bernard JM. *J Mol Catal A* 1997; **116**: 247–58.

57 Monflier E, Bourdauducq P, Courtier J, Kervennal J, Mortreux A. *J Mol Catal* 1995; **97**: 29.

58 Dullius JEL, Suarez PAZ, Einloft S *et al. Organometallics* 1998; **17**: 815–19.

59 Horváth IT, Rábai J. *Science* 1994; **266**: 72–5.

60 Hughes PH, Trujillo HA. *Organometallics* 1996; **15**: 286–94.

61 Klement I, Lütjens H, Knochel P. *Angew Chem Int Ed* 1997; **36**: 1454–6.

62 Kleijn H, Jastrzebski JTBH, Gossage RA *et al. Tetrahedron* 1998; **54**: 1145–52.

63 Arhancet JP, Davis ME, Merola JS, Hanson BE. *J Catal* 1989; **121**: 327–39.

64 Santos S, Tong Y, Quignard F *et al. Organometallics* 1998; **17**: 78–89.

65 Tonks L, Anson MS, Hellgardt K *et al. Tetrahedron Lett* 1997; **38**: 4319–22.

66 Li J, Mau AWH, Strauss CR. *J Chem Soc Chem Commun* 1997: 1275–6.

67 Zhuangyu Z, Yi P, Honwen H, Tsi-yu K. *Synthesis* 1991: 539–42.

68 Ahn JH, Sherrington DC. *Macromolecules* 1996; **29**: 4164–5.

69 Wang PW, Fox MA. *J Org Chem* 1994; **59**: 5358–64.

70 Reetz MT, Lohmer G, Schwickardi R. *Angew Chem Int Ed* 1997; **36**: 1526–9.

71 Ramchandani RK, Uphade BS, Vinod MP *et al. J Chem Soc Chem Commun* 1997: 2071–2.

72 Villemin D, Jaffrè PA, Nechab B, Courivand F. *Tetrahedron Lett* 1997; **37**: 6581–4.

73 Monflier E, Fremy G, Castanet Y, Mortreux A. *Angew Chem Int Ed* 1995; **34**: 2269–71.

74 Sayari A. *Chem Mater* 1996; **8**: 1840–52.

75 Mehnert CP, Ying JY. *J Chem Soc Chem Commun* 1997: 2215–16.

3 Palladium(0)-catalysed Reactions in Aqueous Medium and Synthetic Applications

JEAN PIERRE GENET, MONIQUE SAVIGNAC and SANDRINE LEMAIRE-AUDOIRE

Ecole Nationale Supérieure de Chimie de Paris, Laboratoire de Synthèse Sélective Organique et Produits Naturels,
UMR CNRS 7573, 11 rue Pierre et Marie Curie, 75231 Paris Cedex 05, France

1 Introduction

Over the past three decades, the use of transition metal catalysts has undergone considerable progress in the field of organic synthesis [1]. As the preparation of more and more complex molecules requires highly selective reactions under milder conditions, the need for developing new catalytic transformations has become a major aim for organic chemists. Among the large panel of transition metals frequently used, palladium, discovered in 1803 by Wollatson [2], has a predominant role due to the variety of transformations with which it is associated [3]. Moreover, organopalladium complexes are fairly stable and non-toxic, which explains the recent development of efficient industrial processes [4] including palladium-promoted steps. Many important palladium-catalysed reactions, such as oxidation of alkenes [5], carbonylation [6] and telomerization of conjugated dienes [7], have found various applications in both academic research and industry. Moreover, one of the major contributions of zerovalent palladium catalysts in organic synthesis is the formation of carbon–carbon bonds through many coupling reactions. The vinylation and arylation of alkenes, known as the Heck reaction [8], as well as sp²–sp and sp–sp couplings [9] and Suzuki-type reactions [10], promoted by zerovalent palladium species, were used as key steps in the preparation of elaborated molecules. The catalytic version of π-allylpalladium chemistry developed with carbon nucleophiles and heteronucleophiles is an indispensable technology in fine organic synthesis [1], which provides highly regioselective and stereoselective [11] formation of carbon–carbon and carbon–heteroatom bonds (Scheme 3.1).

To date, synthetic organic chemists have used palladium catalysts in a number of selective transformations. Many reviews and books have been published involving the use of palladium catalysts [1,3,12,13]. Nevertheless, one of the major drawbacks of homogeneous metal catalysis lies in the separation of the reaction products from the catalyst, which requires costly procedures. A solution to this problem entails anchoring the catalyst on an organic or inorganic polymer. Another elegant alternative consists of using water-soluble ligands that, once complexed to the metal, make the catalyst poorly soluble in organic media [14]. The industrial application [15] using sulfonated phosphines in rhodium complexes in a biphasic aqueous system has already indicated the wide scope of this type of catalyst. However until recently, palladium-promoted reactions in aqueous media with water-soluble catalysts remained unexplored. The development of water-soluble catalysts offers several advantages for industrial production: easy separation of the product from the catalyst, high reactivity and recycling of catalyst. In this chapter, an attempt is made to provide recent developments and applications of palladium catalysis in aqueous medium.

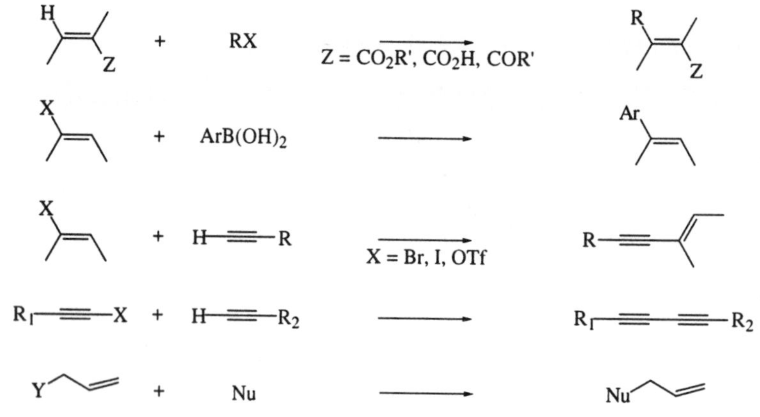

Scheme 3.1.

2   The catalysts

In organic synthesis, two types of palladium compounds, namely Pd(II) and Pd(0) complexes, are used. As shown in Table 3.1, only a few palladium derivatives are used as catalysts in aqueous medium. It has been shown that the commercially and relatively inexpensive palladium acetate without phosphine ligands catalyses cross-coupling reactions in aqueous medium or neat water [16]. Sulfonated triphenylphosphines are used as water-soluble ligands, with which Pd goes into aqueous phase and the catalytic reaction proceeds therefore in water. A preformed water-soluble Pd(0) complex has been prepared using diphenyl(*m*-sulfonated phenyl)phosphine (TPPMS) [17]. This fully characterized catalyst has been used successfully in various cross-coupling reactions. Another useful and versatile water-soluble catalytic system was generated *in situ* from a mixture of Pd(OAc)$_2$ and *m*-trisulfonated triphenylphosphine (TPPTS). This mixture afforded a Pd(0) complex spontaneously. One TPPTS ligand reduces Pd(II) to Pd(0) and this transformation has been established through a series of kinetic and ^{31}P-NMR experiments [18]. This system is an excellent catalyst for various cross-coupling reactions (Heck, Sonogashira, Suzuki) as well as π-allylpalladium chemistry [19, 20].

3   Reactions

3.1   *Vinylation and arylation of alkenes: the Heck reaction*

3.1.1   INTERMOLECULAR REACTIONS

The coupling of vinyl or aryl halides with various alkenes (a process known as the Heck reaction) under anhydrous conditions is a powerful tool in organic chemistry. Intensive investigations have permitted further developments in the synthetic application of the Heck reaction. In particular, the introduction of tetrabutyl ammonium [21], silver(I) or thallium(I) [22] salts and the use of organotriflates [23] have brought some improvements.

The use of water as a reaction medium for transition-metal-catalysed reactions is very attractive for organic synthesis, for both economical and safety reasons. The pioneering

Table 3.1. Catalysts used in aqueous medium

$$Pd(OAc)_2 + nTPPTS \xrightarrow{\text{TPPTS oxide}} Pd(TPPTS)n\text{-}1$$

Palladium catalysts	Applications	Remarks	Ref.
Pd(OAc)$_2$; PdCl$_2$	Heck reaction; carbonylation; reduction of aryl halides	Without phosphine ligands	[16]
Pd(0)(TPPMS)$_3$ from Na$_2$PdCl$_4$ + TPPMS TPPMS: 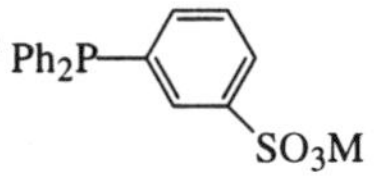 Ph$_2$P—⟨benzene ring⟩—SO$_3$M M = Na$^+$, K$^+$	Heck, Suzuki and Sonogashira cross-coupling	Palladium zerovalent preformed from Na$_2$PdCl$_4$; air-sensitive; well characterized (X-ray, NMR)	[17]
Pd(0)TPPTS from Pd(OAc)$_2$ + TPPTS TPPTS: 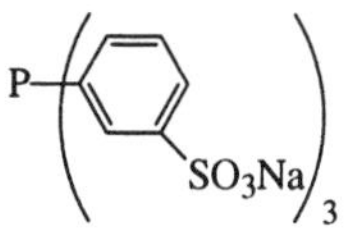 (P—⟨benzene ring⟩—SO$_3$Na)$_3$	Heck, Suzuki and Sonogashira sp–sp coupling π-Allyl-Pd chemistry	Easy to handle *In situ* formation of palladium zerovalent	[18] [19, 20]
Pd(dba)$_2$ or Pd(dba)$_3$ + TPPTS	π-Allyl chemistry	Less convenient than Pd(OAc)$_2$ + TPPTS	[20]

The TPPTS ligand is much more soluble in water than the monosulfonated ligand (TPPMS) (1100 g L^{-1} at 20 °C instead of 80 g L^{-1}). The extremely high solubility of this phosphine ensures the hydrophilic properties of the palladium zerovalent catalyst. dba = dibenzylidenacetone.

work of Beletskaya [16] has shown that the coupling of acrylic acid and acrylonitrile with aryl halides was successful in neat water with good yields (Scheme 3.2). The reaction proceeded selectively in the presence of a mineral base but still at elevated temperature. In the presence of potassium acetate, the reaction was performed at lower temperature and accomplished at a higher rate. Reactions involving water-insoluble substrates and phosphine ligands can be performed efficiently in water in the presence of a combination of an alkali metal carbonate and a quaternary ammonium salt [24].

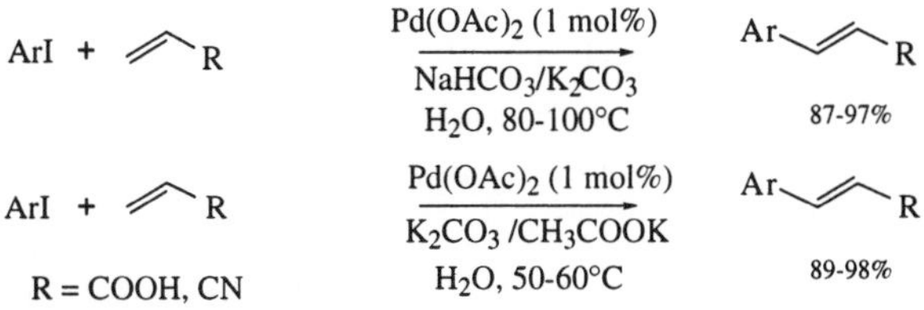

ArI + ⟍R → Pd(OAc)$_2$ (1 mol%) / NaHCO$_3$/K$_2$CO$_3$ / H$_2$O, 80-100°C → Ar⟍⟍R 87-97%

ArI + ⟍R → Pd(OAc)$_2$ (1 mol%) / K$_2$CO$_3$ /CH$_3$COOK / H$_2$O, 50-60°C → Ar⟍⟍R 89-98%

R = COOH, CN

Scheme 3.2.

Vinylations and arylations of alkenes were examined by using a catalyst prepared *in situ* from 2.5% Pd(OAc)$_2$ with 5% TPPTS in an aqueous–organic phase. The first assays were realized with bromide derivatives in different mixed solvents and with potassium acetate as base. Poor yields were obtained in coupling products. Then, the reactivity of iodo substrates was evaluated in Heck coupling in a solvent consisting of water and either acetonitrile or methanol. Different bases were tested, with the best results being observed with triethylamine. The couplings under these aqueous conditions were generally complete within several hours at a temperature of 25–60 °C. α,β-Unsaturated esters or acids were obtained with good to excellent yields (Scheme 3.3).

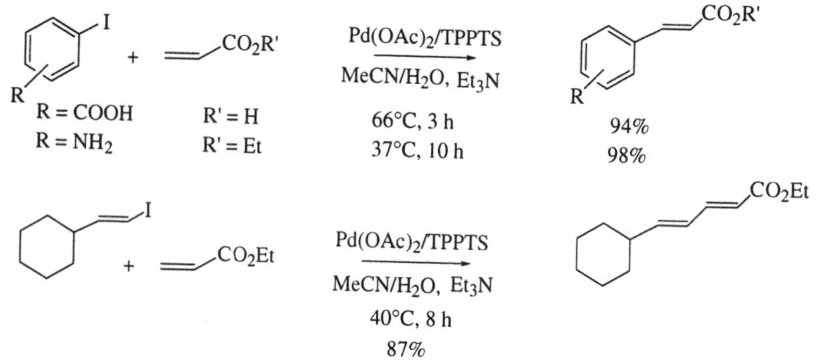

R = COOH   R' = H   66°C, 3 h   94%
R = NH$_2$ R' = Et   37°C, 10 h   98%

40°C, 8 h
87%

Scheme 3.3.

Under the same conditions, the coupling of aryl iodides with cyclic alkenes (3,4-dihydropyran, cyclohexene and cyclopentene) proceeded at room temperature to give a single aryl adduct in 70–91% yield (Scheme 3.4). With 3,4-dihydropyran, it is noteworthy that the conditions afforded solely the thermodynamic product.

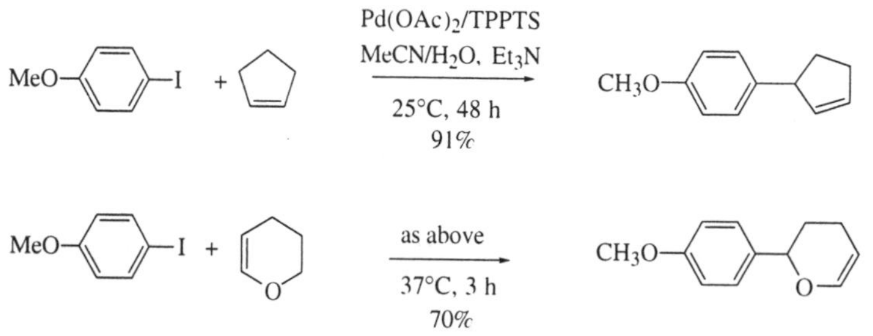

25°C, 48 h
91%

37°C, 3 h
70%

Scheme 3.4.

Thus, Pd(0)/TPPTS was a good catalyst for sp^2–sp^2 coupling under mild conditions; crude products were obtained in pure form by simple filtration.

3.1.2 INTRAMOLECULAR HECK REACTIONS

For a long time, applications of the Heck reaction in the construction of cyclic systems were focused on the preparation of heterocyclic compounds. Only recently, complicated ring systems with different functionalities were assembled. Intramolecular Heck reactions are therefore important key steps in numerous natural product syntheses [25].

Using the Pd(OAc)$_2$/TPPTS catalyst in a homogeneous aqueous phase of water–acetonitrile, the *o*-iodo-*N*-allylaminobenzene cyclized very rapidly at room temperature to produce the 3-methylindole in quantitative yield. In the same way, benzofuran was easily prepared [19] (Scheme 3.5).

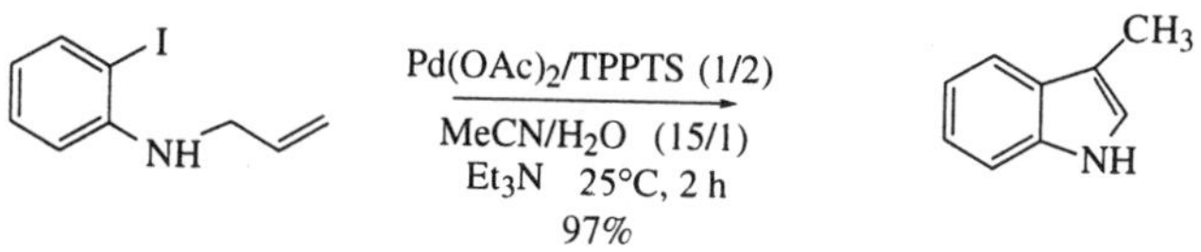

Scheme 3.5.

Both iodide and bromide amino precursors derived from cyclohex-2-ene underwent smooth cyclization to give the corresponding fused bicyclic products in good yields. In some cases, PdCl$_2$ as the Pd(II) salt was more efficient; triethylamine or diisopropylethylamine could be used as base [26] (Scheme 3.6).

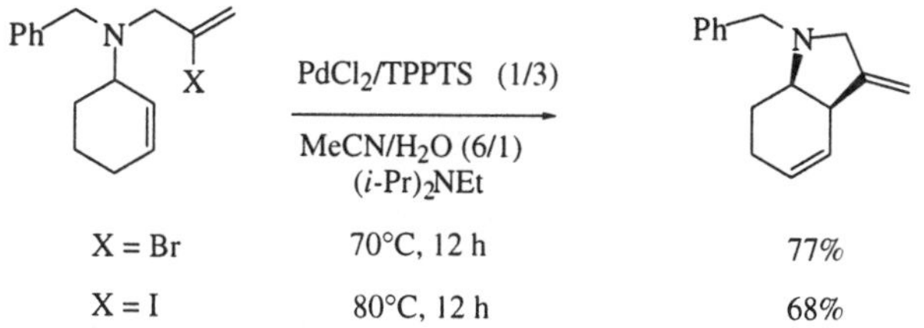

Scheme 3.6.

When the halogen was branched on the internal position of the double bond, only *cis* heterocycles or carbocycles were formed, whereas an equimolar mixture of *cis* and *trans* diastereomers was observed when the iodine atom was on the terminal carbon of the double bond (Scheme 3.7).

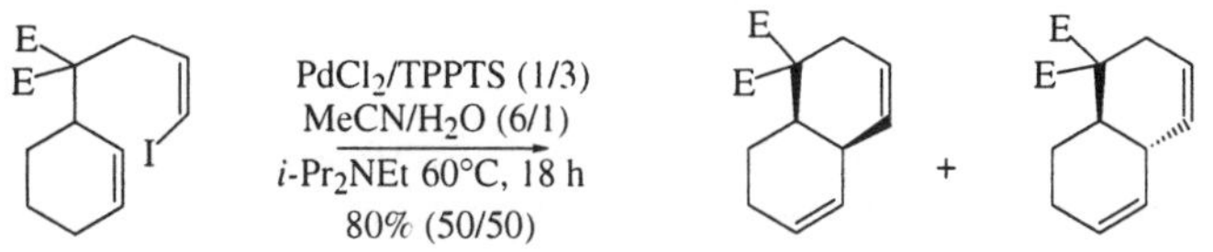

Scheme 3.7.

Moreover, when substrates were likely to cyclize via *exo* or *endo* processes, *exo*-cyclization was commonly observed in many cases. However, the regioselectivity strongly depends on the catalytic species, and a change from 5-*exo* to 6-*endo* preference was reported recently for the intramolecular carbopalladation of 1,6-enynes by modifying the catalyst [27].

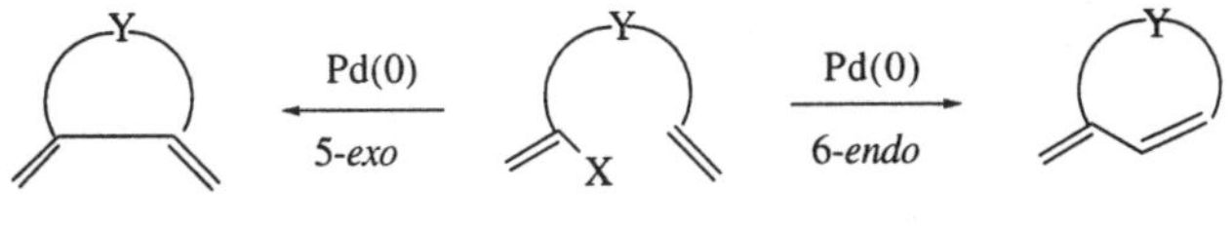

In anhydrous conditions in the presence of Pd(OAc)$_2$, PPh$_3$, AgCO$_3$ and diiso-propylethylamine, 5-*exo*-trig cyclization was observed; meanwhile under phase-transfer conditions, a tendency to reverse the *exo* preference was observed. Finally, the hydrosoluble Pd(0) catalyst has afforded a total inversion of selectivity in favor of 6-*exo* (Table 3.2).

These results were in agreement with the cyclization of highly functionalized enamides, where the unexpected *endo* process was observed using R$_4$N+Cl- conditions [28].

Table 3.2. Regioselectivity of inframolecular Heck reaction

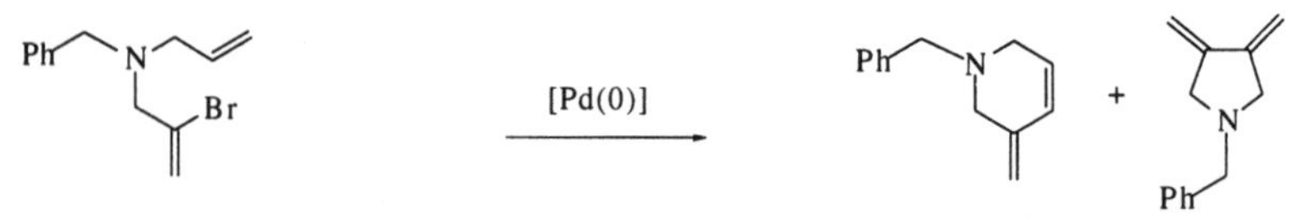

[Pd(0)]	Base	Salt	Solvent	T (°C)	t (h)	A	B	Yield (%)
5% Pd(OAc)$_2$ PPh$_3$	(*i*-Pr)$_2$NEt (1.2 equiv)	AgCO$_3$	MeCN	90	20	14	86	60
5% Pd(OAc)$_2$ PPh$_3$	K$_2$CO$_3$ (2.8 equiv)	*n*-Bu$_4$NCl	DMF, εH$_2$O	40	12	47	53	70
10% PdCl$_2$ TPPTS	(*i*-Pr)$_2$NEt (1.2 equiv)	—	MeCN/H$_2$O (6:1)	70	24	93	7	65

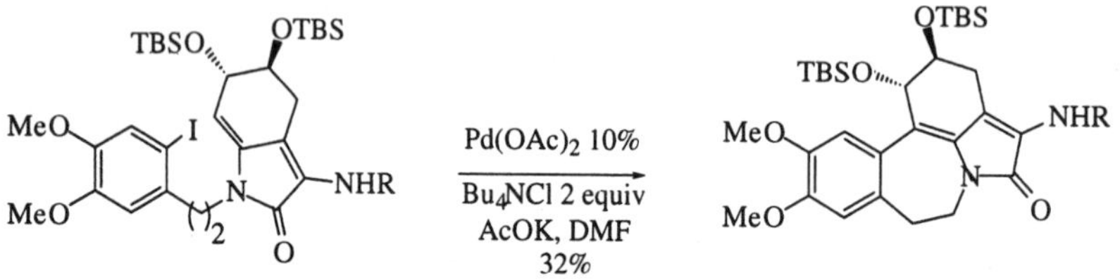

Nitrogen substrates that are likely to undergo both 5-*exo*-trig and 6-*endo*-trig cyclization were submitted to the Pd(0)/TPPTS conditions. With both bromide or iodide derivatives, the *endo* products were generally isolated between 50 and 80 °C (Table 3.3).

This 6-*endo* regioselectivity was also observed with the dimethylmalonate derivative, which cyclized to the six-membered carbocycle (Scheme 3.8).

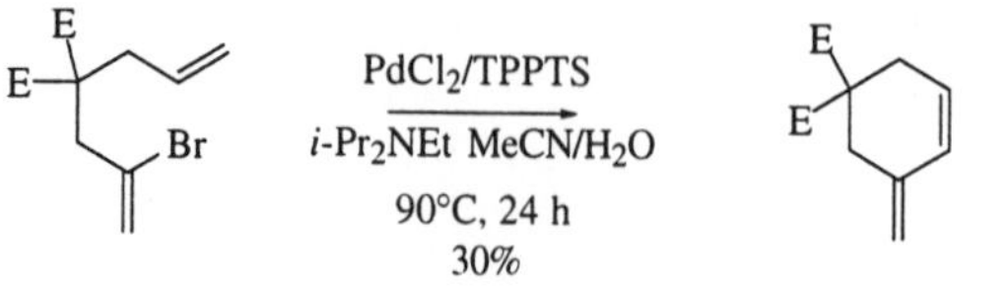

Scheme 3.8.

Table 3.3. 6-endo palladium cyclization in aqueous medium

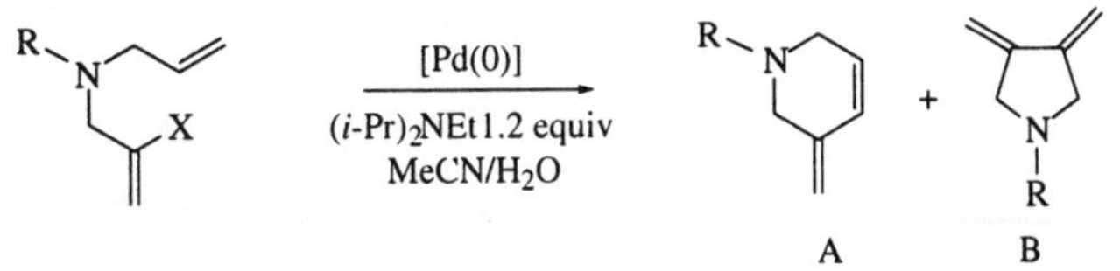

R	X	[Pd(0)] catalyst	T (°C)	t (h)	A	B	Yield (%)
PhCH$_2$	Br	PdCl$_2$/TPPTS	70	24	93	7	65
PhCH$_2$	I	PdCl$_2$/TPPTS	65	14	96	4	61
⬊CH$_2$	I	Pd(OAc)$_2$/TPPTS	80	10	90	10	79
⬊CH$_2$	I	PdCl$_2$/TPPTS	70	10	100	0	80
Ts	I	PdCl$_2$/TPPTS	Reflux	24	100[a]	0	70

[a] With double bond isomer.

A mechanism could be proposed based on an oxidative addition followed by insertion of palladate and then *syn* elimination of HPdX:

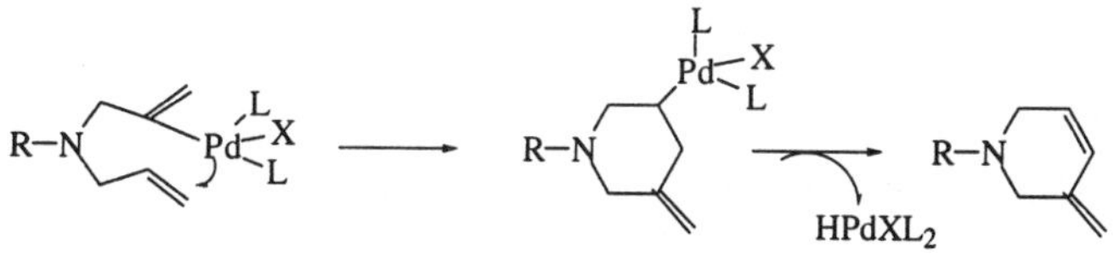

An alternative mechanism began with the more usual *exo*-cyclization followed by cyclopropanation, cyclopropylcarbinyl-to-homoallyl rearrangement and β-dehydropalladation [8].

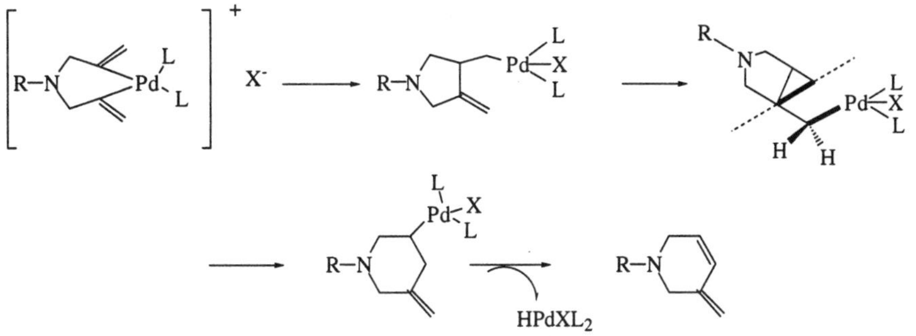

These aqueous conditions have therefore offered the opportunity to reverse the usual *exo* process, affording the *endo* products. By proper choice of reaction medium, it is moreover possible to obtain either *endo* or *exo* ring closure starting from the same precursor.

3.2 Cross-coupling reactions of terminal alkynes with vinyl and aryl halides: synthetic applications

Palladium-catalysed reactions of terminal alkynes with aryl or vinyl halides are usually conducted in *non-aqueous media* with a base as scavenger for hydrogen halide. Copper(I) iodide is a particularly effective co-catalyst, allowing the reaction to occur under mild conditions. This sp²–sp coupling was employed extensively in numerous key steps of antibiotic, antitumor and lipoxine syntheses [29].

Recently, Casalnuovo [17] has developed a hydrosoluble catalyst $Pd(TPPMS)_3$ that is able to couple a wide range of acetylenes with unprotected nucleosides, nucleotides and amino acids in 50% aqueous acetonitrile solution with a CuI promoter. The reactions were generally conducted at room temperature with complete conversion of organic iodide and yields ranging from 50 to 95%. Other types of alkylation reactions, such as the Heck reaction, were catalysed by this complex.

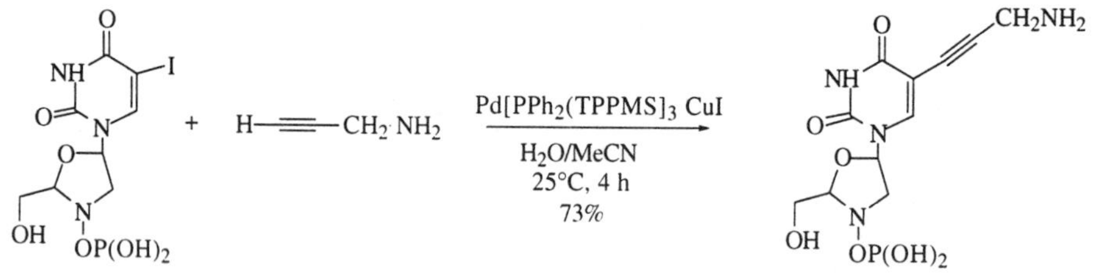

The synthetic versatility of this methodology was illustrated in an alternative synthesis of T-505, part of a family of chain-terminating nucleotide reagents used in automated DNA sequencing and labeling. In the commercial syntheses of these reagents, the acetylene coupling reaction was carried out prior to the introduction of the hydrophilic triphosphate and fluorescein dye groups owing to the insolubility of $Pd(PPh_3)_4$ in water. With the hydrosoluble catalyst the C–C coupling reaction was conducted in the final step, with the 5-iododideoxyuridine 5′-triphosphate and the unprotected fluorescein dye in 50% yield.

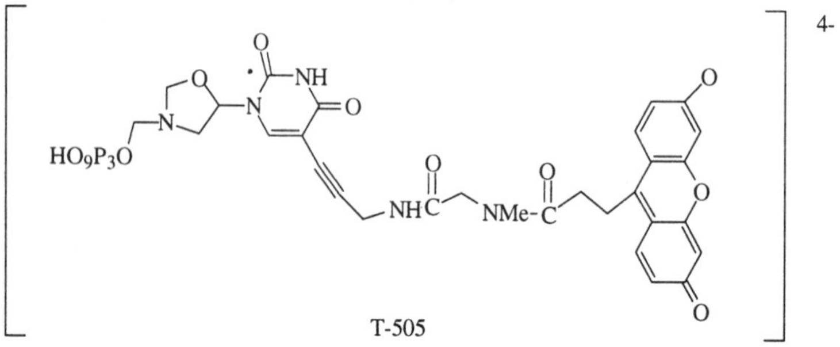

T-505

The $Pd(OAc)_2$/TPPTS catalyst efficiently catalysed couplings of iodoaromatics and vinyl halides with a variety of terminal alkynes in an acetonitrile–water solution (6:1) at room temperature in a few hours *without any Cu(I) promoter*. High yields were observed and, interestingly, in these cross-coupling reactions the catalyst was tolerant of a wide range of functionalities such as sulfur, which is not poisonous for the catalyst (Table 3.4).

Based on these sp²–sp couplings an approach to the tricyclic structure of taxanes has been accomplished in our laboratory [30]. The key elements of our strategy involved a sequential Sonogashira reaction between the protected iodo ketone and trimethyl-silylacetylene and then, after deprotection of the trimethylsilyl group, with 3-iodocyclohexenone to generate the two-carbon C-9 and C-10 of the taxane skeleton. The first cross-coupling failed under aqueous conditions. Interestingly, standard Sonogashira conditions afforded the coupling product in 99% yield. The second cross-coupling was carried out using aqueous conditions (Scheme 3.9).

Table 3.4. Sonogashisa type coupling with water soluble palladium(0) catalyst

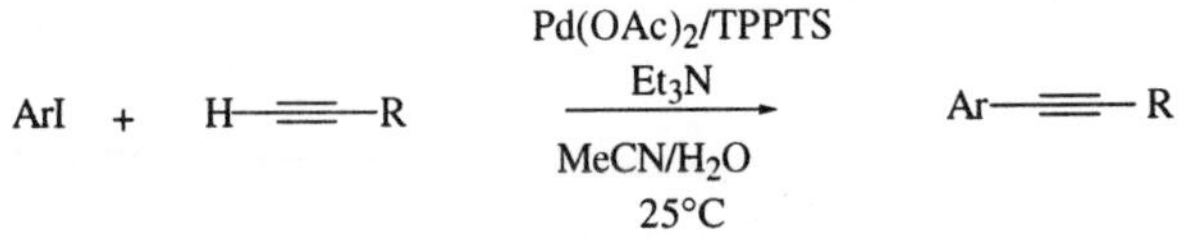

Ar	R	t (h)	Yield (%)
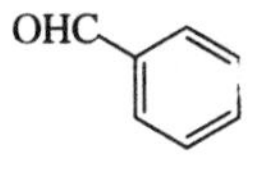 OHC (benzene ring)	SiMe$_3$	3	80
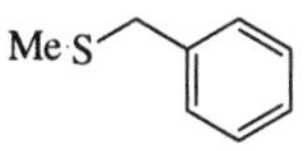 Me-S-CH$_2$ (benzene ring)	SiMe$_3$	18	85
 (thiophene ring, S)	CH$_2$OH	3	99

Reaction: ArI + H—≡—R $\xrightarrow[\substack{\text{MeCN/H}_2\text{O} \\ 25°\text{C}}]{\substack{\text{Pd(OAc)}_2/\text{TPPTS} \\ \text{Et}_3\text{N}}}$ Ar—≡—R

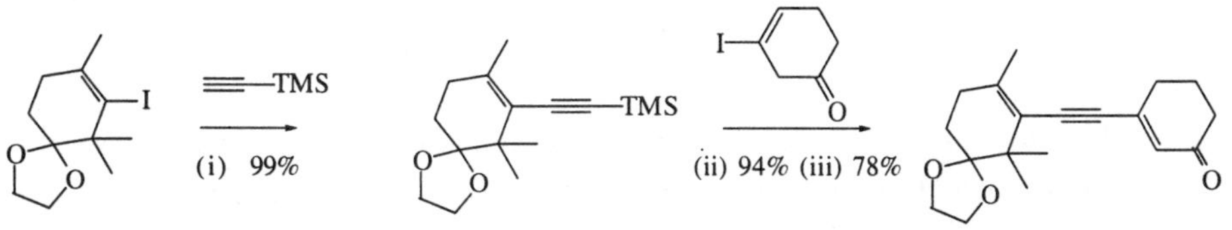

(i): 5% Pd(OAc)$_2$/Ph$_3$P)$_3$P 1/4 (ii): 1.1 equiv TBAF/THF ; (iii): 5% Pd(OAc)$_2$/TPPTS 1/2, MeCN/H$_2$O (6/1)

2.5 equiv Et$_3$N, 1h, 40°C

Scheme 3.9.

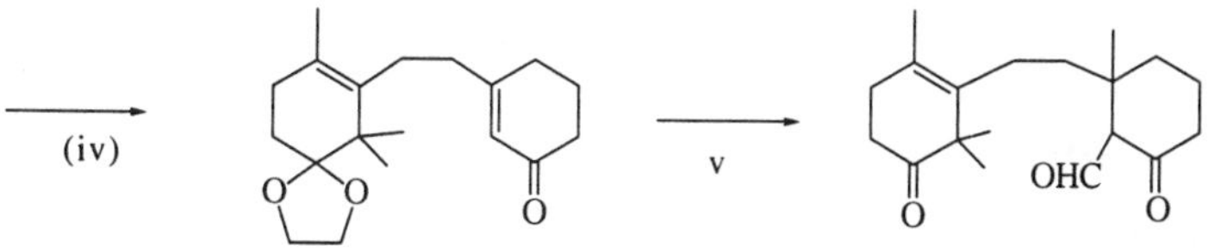

(iv) H$_2$/5% Pd/C/Py, 10% (v) 1.5 equiv Me$_2$CuLi/Et$_2$O, 0°C then 7 equiv BF$_3$Et$_2$O, 7 equiv (MeO)$_3$CH, −78°C

Scheme 3.10.

After complete hydrogenation of the triple bond, conjugate addition and enolate trapping under the reaction conditions shown in Scheme 3.10 afforded the desired compound in 30% yield.

The synthesis of indoles and furans via the two-step sp²–sp coupling/intramolecular cyclization sequence has been studied for a long time. In the beginning, harsh reaction conditions were used in the presence of Cu(I) species. Recent developments in this type of reaction have been conducted under anhydrous conditions, with high temperatures and a phase-transfer agent generally required to give the cyclized products in moderate to good yields. In the presence of the water-soluble catalyst Pd(OAc)$_2$/TPPTS, this sequential two-step reaction has proceeded under very mild conditions (room temperature to 65°C) *without Cu(I)*[18]. Thus, the 2-iodoaniline was coupled with hex-1-yne in the presence of 2.5 mol.% Pd(OAc)$_2$, 5 mol.% TPPTS and triethylamine, leading initially to the expected alkyne, which was then partially cyclized *in situ* at 65°C to give the 2-butylindole in 56%

yield. When the amino function was activated by a trifluoroacetyl group, a complete cyclization was observed. Benzofuran derivatives were prepared with the same strategy in moderate to good yields (Scheme 3.11).

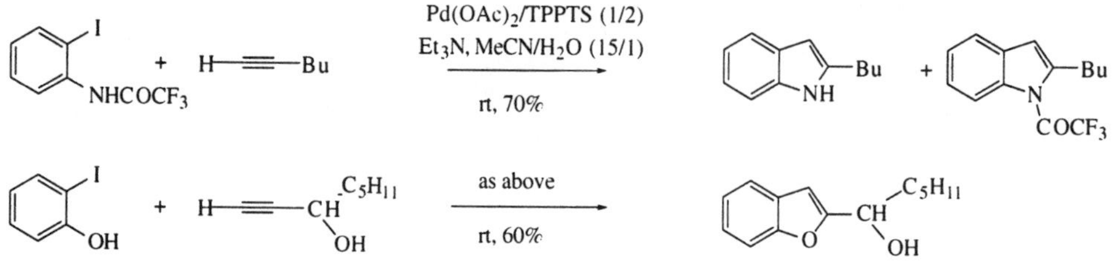

Scheme 3.11.

The σ-organopalladate intermediate underwent an intramolecular nucleophilic attack on the sp carbon to yield the cyclized product.

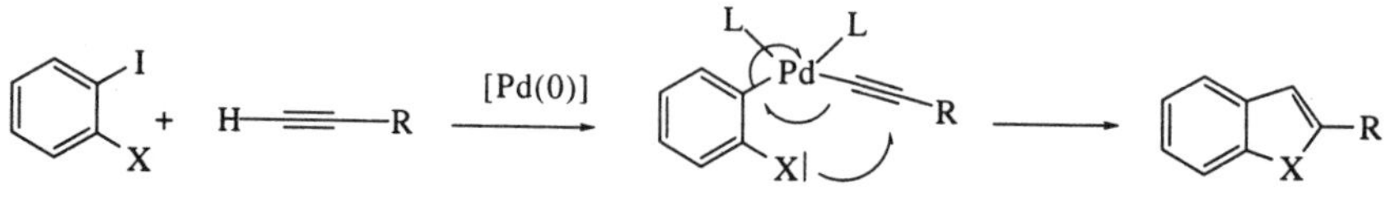

This method is very easy to handle because the catalyst and the base are eliminated by simple water treatment affording, after extraction, very clean crude products.

To demonstrate the synthetic versatility of this technique, an alternative synthesis of eutypine, an antibacterial substance isolated from the culture medium of *Eutypa lata*, was developed [18] (Scheme 3.12). The fungus *E. lata* attacks several woody species and it is responsible, in particular, for a vineyard disease known as eutyopis that has damaged wine-growings in France during the past decade. As shown in Scheme 3.12, the cross-coupling in the presence of Pd(0)/TPPTS proceeded in quantitative yield.

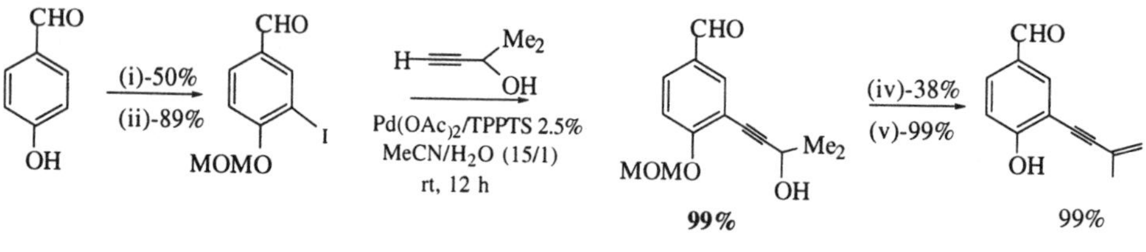

(i)-I_2/KI, Me_2NH, 3 h; (ii)-$CH_2(OMe)_2P_2O_5$, CH_2Cl_2, 5 h; (iv)-$POCl_3$, pyridine, 0 °C→rt, 18 h; (v)-HCl, AcOH, 8 h

Scheme 3.12.

3.3 *Alkyne–alkyne cross-coupling*

Symmetrical diyne molecules have been synthesized by an sp–sp coupling reaction for a long time by use of stoichiometric or catalytic quantities of Cu(I) [31]. More recently, Pd(0) in association with Cu(I) or Cu(II) was used for this type of homocoupling reaction under mild conditions in almost quantitative yield [32]. However, only symmetrical diynes could be prepared. The synthesis of unsymmetrical diynes was described in the 1950s by Cadiot and Chodkiewicz [33], who developed an sp–sp coupling between acetylenic bromides and terminal alkynes or alkynyl cuprates in organoaqueous medium (Scheme 3.13).

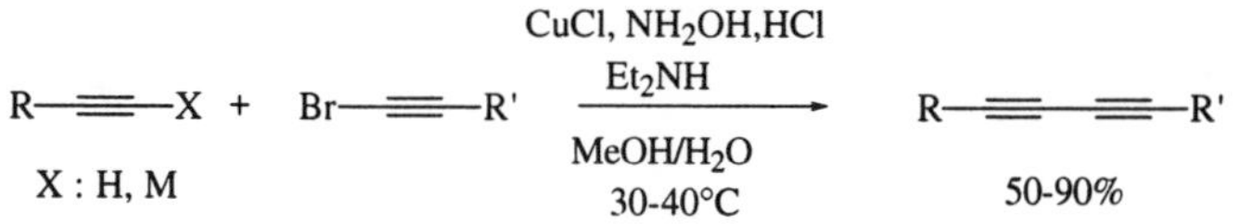

$$R\!-\!\!\equiv\!\!-X + Br\!-\!\!\equiv\!\!-R' \xrightarrow[\substack{MeOH/H_2O \\ 30\text{-}40°C}]{\substack{CuCl, NH_2OH, HCl \\ Et_2NH}} R\!-\!\!\equiv\!\!-\!\!\equiv\!\!-R'$$

X : H, M   50-90%

Scheme 3.13.

Table 3.5. Synthesis of unsymmetrical diynes

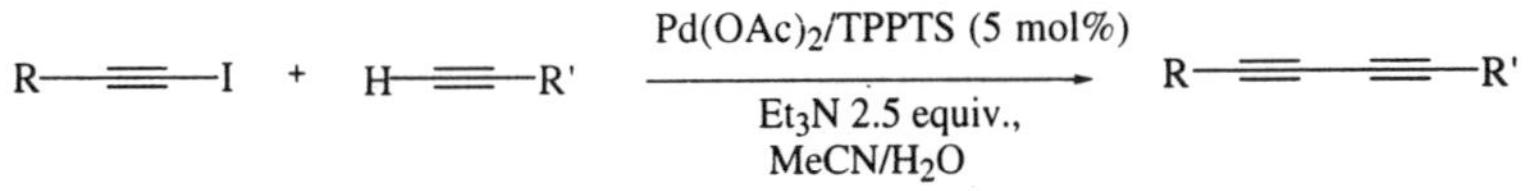

$$R\!-\!\!\equiv\!\!-I + H\!-\!\!\equiv\!\!-R' \xrightarrow[\substack{Et_3N\ 2.5\ equiv., \\ MeCN/H_2O}]{Pd(OAc)_2/TPPTS\ (5\ mol\%)} R\!-\!\!\equiv\!\!-\!\!\equiv\!\!-R'$$

R	R'	T	t (h)	Yield (%)
Bu	Ph	Room temp.		49
Bu	$CH\!-\!Me_2$ OH	Room temp.	1.5	60
Me_3Si	$CH\!-\!Me_2$ OH	Room temp.	0.5	57
$(Et)_2\!-\!CH$ NH_2	$CH\!-\!C_5H_{11}$ OH	35 °C	12	65

Very few sp–sp coupling methods using organopalladium catalysts have been described in the literature [34]. The water-soluble catalyst formed *in situ* from $Pd(OAc)_2$ and TPPTS again has proved its efficiency in these sp–sp coupling reactions *without any Cu(I) promotor.* In aqueous acetonitrile (1:6) and in the presence of triethylamine (2.5 equiv.), various alkynyl iodides, easily prepared from the corresponding alkynes [18], have been coupled with functionalized terminal alkynes under very mild conditions. The competitive homocoupling reaction was only observed with phenylacetylene in 14% yield (Table 3.5). The moderate yields were essentially due to the instability of these functionalized diynes, which easily decompose during purification on silica gel.

Thus, under very mild conditions, the water-soluble catalyst allowed sp–sp cross-coupling without any Cu(I). Furthermore, silylated diynes could be useful nucleophilic precursors for the preparation of new materials.

3.4   *Reactions of organo halides with boronates and boronic acids: Suzuki coupling*

The palladium-catalysed cross-coupling reaction of organometallic reagents with halides has become an important synthetic methodology for regio- and stereoselective bond formation between unsaturated carbon atoms. Among these organometallic reagents, alkenyl and aryl boranes or boronic acids have been used extensively. This coupling, discovered by Suzuki, was catalysed in organic solvents by $Pd(PPh)_4$ in the presence of two equivalents of an inorganic base [10]:

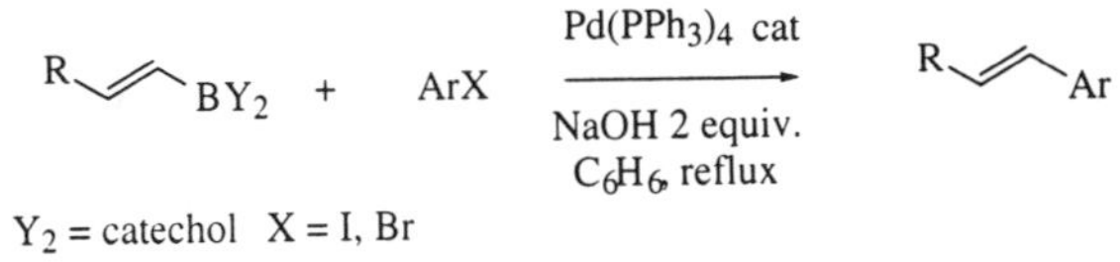

$$R\diagup\!\!\diagdown BY_2 + ArX \xrightarrow[\substack{NaOH\ 2\ equiv. \\ C_6H_6\ reflux}]{Pd(PPh_3)_4\ cat} R\diagup\!\!\diagdown Ar$$

Y_2 = catechol   X = I, Br

Table 3.6. Base effect of suzuki cross-coupling reactions

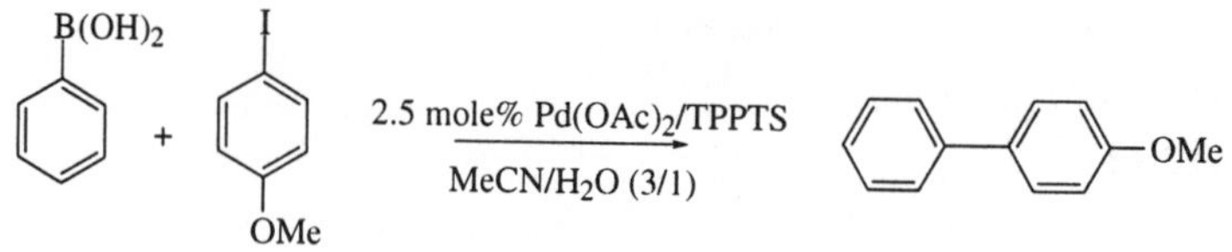

Base	t (h)	Starting material (%)	Yield (%)
K_2CO_3, $Ba(OH)_2$	1.8–24	69–95	—
Et_3N	72	0	64
i-Pr_2NH	72	6	84

The Pd/TPPMS catalyst [17] has been successfully employed in the coupling of various aryl iodides or bromides and substituted phenylboronic acids. The reactions were carried out in a single basic aqueous phase or in a biphasic solution and were generally complete within several hours at 80 °C. It was possible to recycle the catalyst in the case of 4-bromopyridine (Scheme 3.14).

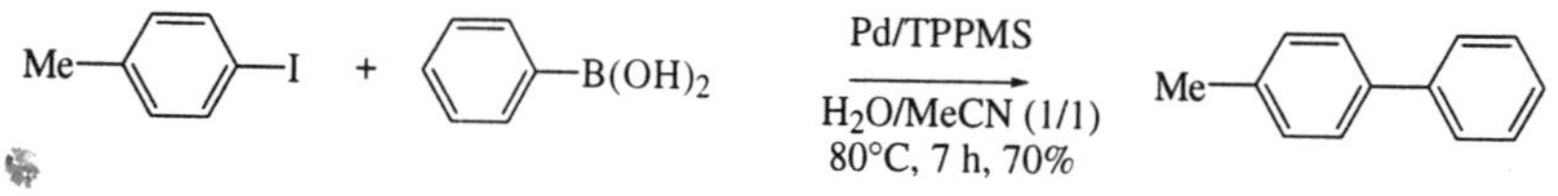

Scheme 3.14.

Investigation of the cross-coupling of phenyl boronic acid with *p*-iodo anisole has shown that in acetonitrile–water (3:1) with catalytic amounts of hydrosoluble catalyst (2.5 mol.%), the inorganic bases such as K_2CO_3, $Ba(OH)_2$ or Cs_2CO_3 were inefficient, while triethylamine and diisopropylamine led to the biphenyl product in good yield [35]. A higher yield was observed with diisopropylamine compared to triethylamine, so subsequent cross-coupling reactions were only performed in the presence of diisopropylamine (Table 3.6).

Under these optimized conditions, the coupling of functionalized alkenyl boronic acids or esters with iodo esters or iodo ketones afforded the functionalized dienes or trienes in good to excellent yields at room temperature (Scheme 3.15).

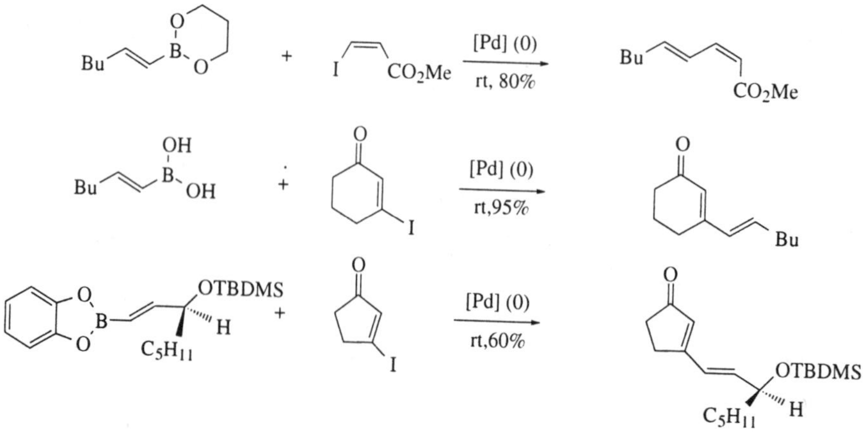

Scheme 3.15.

The nature of the boronic derivatives, boronic acid or ester did not affect the yield. Moreover, a high stereoselectivity was observed. Thus, the reaction of (1E,3Z)-penta-1,3-dienyl-1,3,2-dimethyl oxaborolane with (Z)-methyl-3-iodopropenoate afforded the (2Z,4E,6Z)-methylocta-2,4,6-trienoate in 70% yield. In the same way, the coupling of (E)-hex-1-enyl-boronic acid with 3-iodocyclohexenone gave (E)-3-(hex-1-enyl)cyclo-hexenone. Interestingly, under standard conditions, the stereogenic center of an alkenyl boronate was not affected during the reaction with 3-iodocyclopentenone.

The Suzuki coupling performed with the water-soluble catalyst Pd(OAc)$_2$/TPPTS was therefore very efficient with high stereoselectivities: the mild conditions used have some advantages for the preparation of α,β-unsaturated esters with Z-geometry; in addition, the catalyst was cleanly removed.

3.5 *Reactions of 1,6-enynes*

The cycloisomerization of 1,6-enynes was developed extensively by Trost in the mid-1980s [27]. These reactions were carried out by using Pd(0) or Pd(II) in association with a catalytic amount of phosphine and a carboxylic acid. They afforded efficient access to 1,3-dienes that could be engaged in subsequent Diels–Alder reactions. Incorporation of the dienophile in the chain promoted polycyclizations. The ratio of 5-*exo*- versus 6-*endo*-cyclization depends on the ligand.

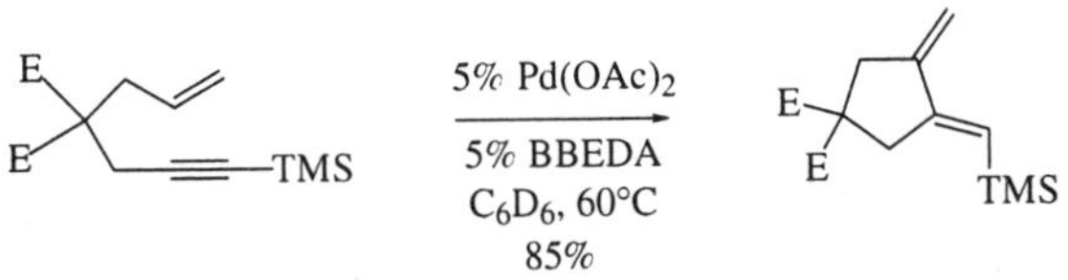

In the development of hydrosoluble palladium catalyst reactions, the cycloisomerization of 1,6-enynes was a challenge. The readily available cinnamyl propargyl ether was used to determine the reaction conditions. Using PdCl$_2$/TPPTS, we observed in 7 h at 60 °C the formation of 3-hydroxytetrahydrofuran derivative as a single diastereomer in 44% yield. Better results were obtained by using a homogeneous mixture of dioxane–water (6:1) and in the absence of acetic acid. To our knowledge, this carbohydroxypalladation reaction has never been reported (Scheme 3.16).

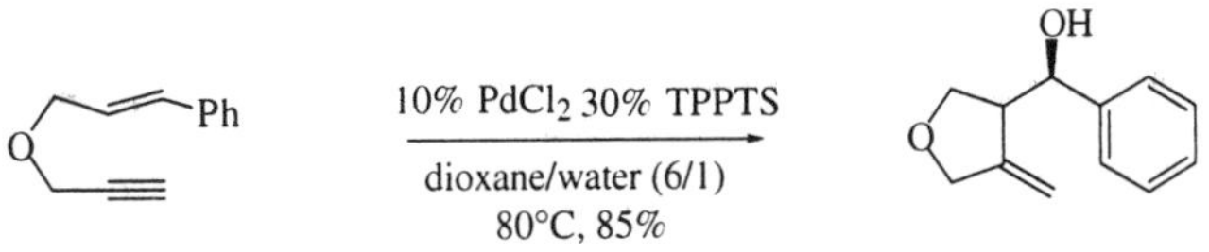

Scheme 3.16.

These conditions were adopted as our standard conditions and applied to various enynyl ethers [36]. When the triple bond was α-substituted, the corresponding hydroxyte-trahydrofuran was formed with good yield and stereoselectivity but required a longer time (Scheme 3.17).

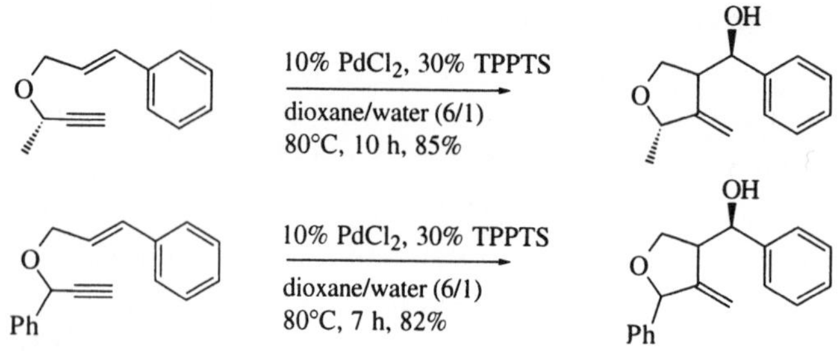

Scheme 3.17.

When the phenyl group was *ortho*-substituted, a single diastereomer was isolated in 47% yield. The phenyl group could be substituted by a thienyl group and carbocycles could be also obtained with moderate yield (Scheme 3.18).

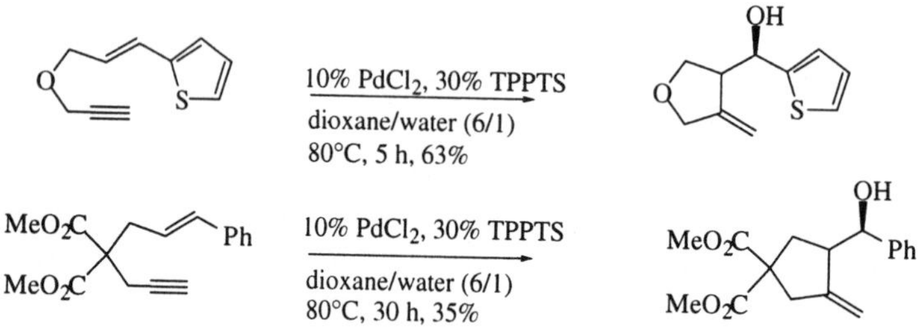

Scheme 3.18.

A mechanism could be proposed beginning with a *syn* addition of an H-Pd-OH species on the triple bond of the substrate, following cyclization and reductive elimination:

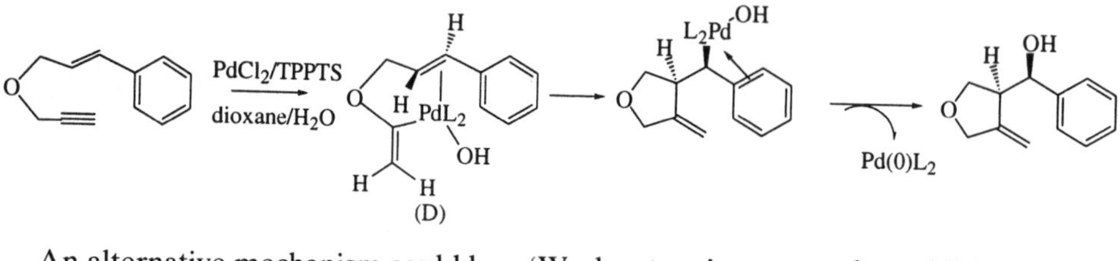

An alternative mechanism could be a 'Wacker-type' process: after addition of Pd(0) on the triple bond, the nucleophilic attack of water was *anti* to the palladate. Thus, the relative stereochemistry would be opposite to that observed in the previous mechanism.

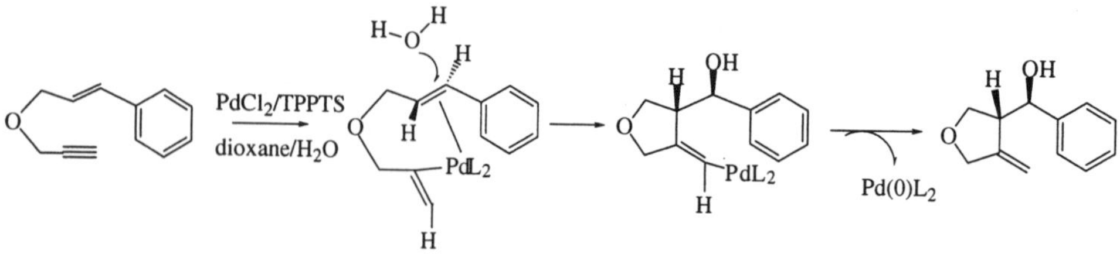

The precise mechanism is not yet established. However, the ease of availability of acyclic substrates has made this atom-economical reaction a very practical approach for the construction of polycyclic skeletons such as podophyllotoxins. Thus the latter seems more appropriate, since we have established by X-ray analysis that the stereochemistry of both hydrogen and hydroxyl groups are *syn*. (J. C. Galland, unpublished data).

3.6 *Reactions of allylic compounds via π-allylpalladium complexes*

3.6.1 ALLYLATION OF STABILIZED CARBON NUCLEOPHILES AND VARIOUS HETERONUCLEOPHILES

Allylic substitution is well known and probably the most widely used palladium reaction in organic syntheses.

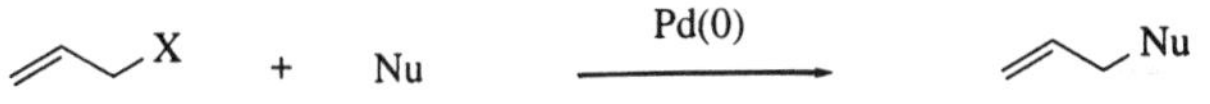

Scope of the reaction. Allylation was carried out using the palladium associated with TPPTS [19, 20]. The useful allylic allylation under neutral conditions was possible using allylic carbonates and vinyloxiranes. The reaction was performed in the absence of base and good yields were obtained with compounds such as acetoacetate and acetylacetone. Cinnamyl carbonate gave a single regio- and stereoisomer with *E* configuration. Alkylation of 3,4-epoxybut-1-ene regioselectively led to the formation of a mixture of *E* and *Z* isomers (85:15). Alkylation of acetate required the presence of a base such as triethylamine or, better, DBU (1,8-diazabicyclo[5,4,0]undec-7-ene) (Scheme 3.19).

Some nucleophiles other than carbon-stabilized nucleophiles were allylated. Secondary amines and primary amines (*n*-butylamine, 2,2-diethyl propargylamine, methylbenzyl-amine, etc.) reacted with cinnamyl acetate using the Pd(0)/TPPTS system in acetonitrile–water (Scheme 3.20).

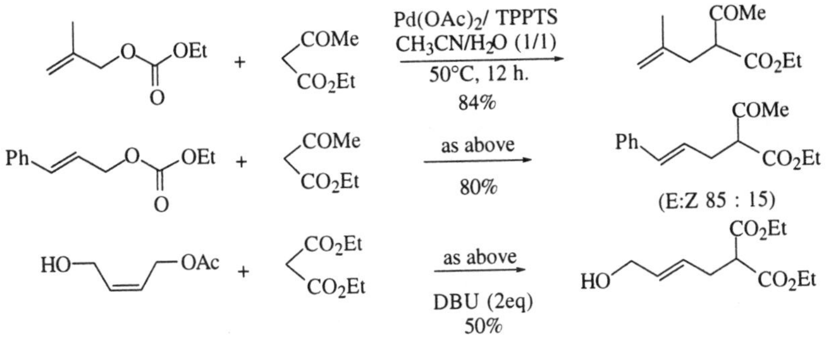

Scheme 3.19.

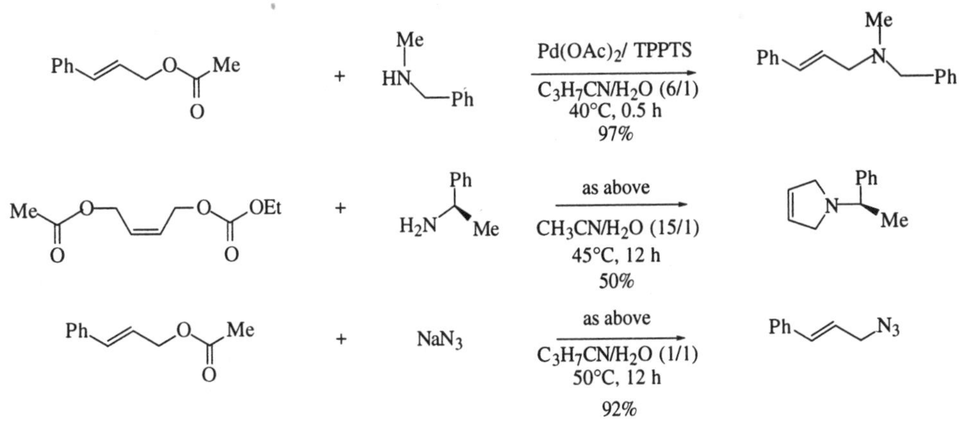

Scheme 3.20.

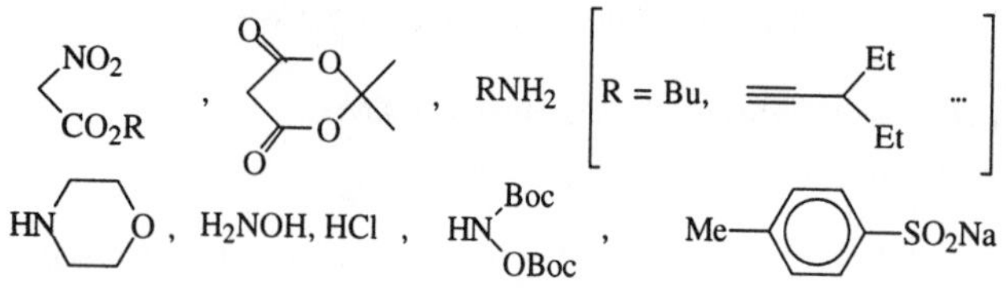

Scheme 3.21.

Table 3.7. Catalyst recycling in biphasic medium

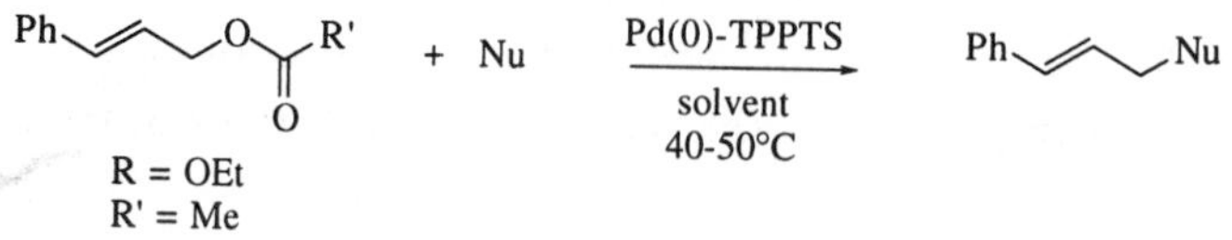

substrat	nucleophile	solvent	time(h)	yield(%)
first run	CO$_2$Et / COMe	PhCN/H$_2$O	50	94
first recycling	"	PhCN/H$_2$O	50	93
second recycling	"	PhCN/H$_2$O	50	90
first run	HN–Ph / Me	*n*-PrCN/H$_2$O	5	95
first recycling	"	*n*-PrCN/H$_2$O	2	95
second recycling	"	*n*-PrCN/H$_2$O	2	95
first run	Me–⬡–SO$_2$Na	*n*-PrCN/H$_2$O	12	95
first recycling	"	*n*-PrCN/H$_2$O	12	95

Other stabilized carbonucleophiles as well as heteronucleophiles reacted equally well using the Pd(0)/TPPTS system as shown in Scheme 3.21.

Recycling the catalyst. As expected, the use of water-soluble Pd(0)/TPPTS ligand allowed easy separation of the catalyst from the product during the work-up. Interestingly, in aqueous solutions using butyronitrile or benzonitrile as cosolvent, after completion of the reaction, the catalyst stays in the aqueous phase and can be recycled without substantial loss of activity. Some examples are summarized in Table 3.7 under these conditions. All types of nucleophiles, including carbon as well as heteronucleophiles, allowed easy recycling [20].

3.6.2   INTRAMOLECULAR REACTIONS OF ALKENES

Intermolecular insertion of alkenes into π-allylpalladium have been reported recently. On the other hand, the intramolecular version has been developed extensively by Oppolzer [37] and has proceeded smoothly to give a wide range of cyclic compounds after β-elimination. Numerous useful synthetic examples have been reported.

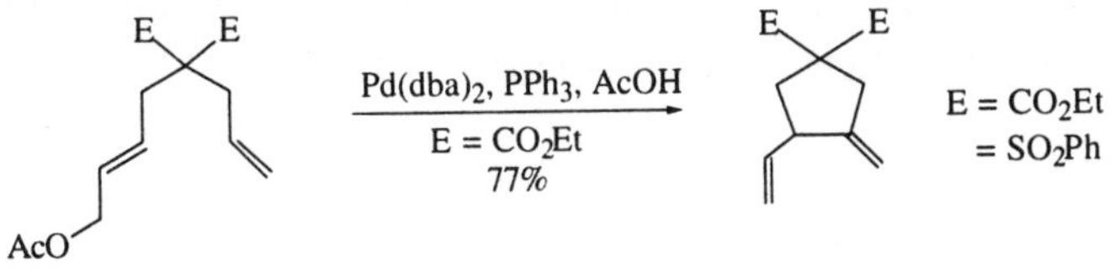

This intramolecular ene-palladium reaction has been realized under milder conditions in aqueous medium using water-soluble Pd(0)/TPPTS catalyst without any additive (acetic acid) and in better yield (J.C. Galland, unpubl. data) (Scheme 3.22).

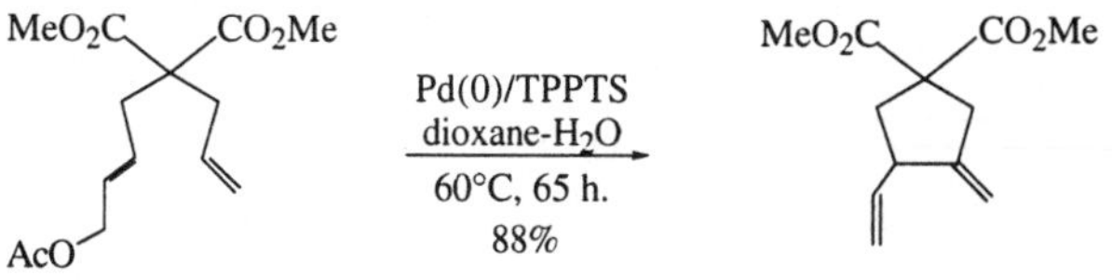

Scheme 3.22.

The following mechanism could be proposed on the basis of intramolecular insertion of alkenes into π-allylpalladium(II) complexes [38].

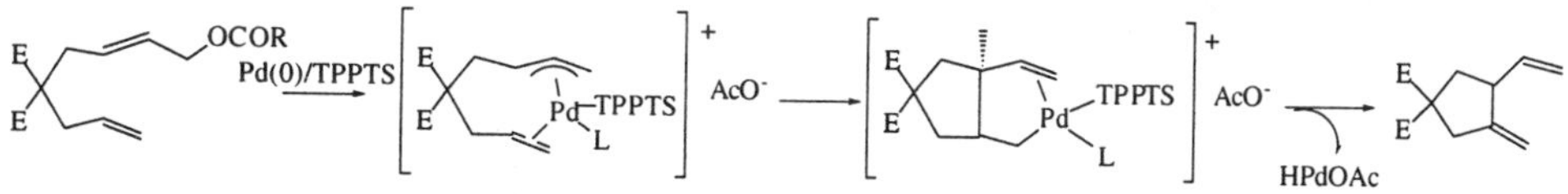

4 Protecting group chemistry

4.1 *Principles*

Among the usual protecting groups for amino, hydroxyl and carboxylic functions, the allyloxycarbonyl (Alloc) and allyl moieties were largely developed over the past 20 years after a methodology using π-allylpalladium complexes was introduced for their cleavage [39].

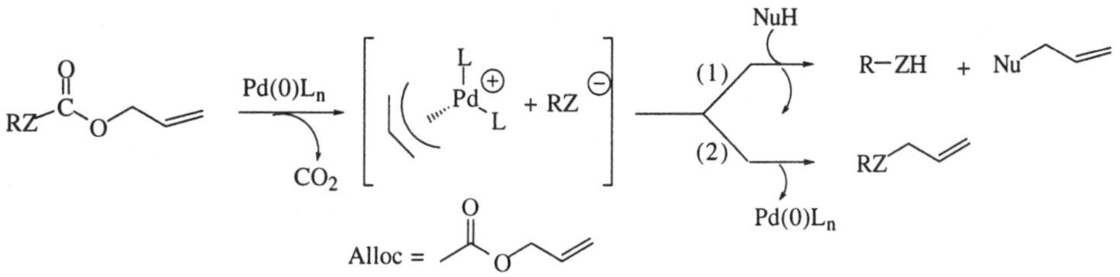

In the literature, various conditions involving different allyl scavengers, such as formic acid [40], morpholine [41], tributyltin hydride [42] or potassium 2-ethylhexenoate [43], have been reported in anhydrous media. Nevertheless, these systems suffered some limitations, particularly for the deprotection of secondary amines, which led to the competitive *N*-allylation (path 2). Although recent progress was achieved using silylated amines [44] or phenyltrihydrosilane [45] as nucleophiles, a simple and inexpensive method for the cleavage of allylcarbamates derived from secondary amines is of great interest. The water-soluble catalyst generated *in situ* from Pd(OAc)$_2$ and the sulfonated phosphine TPPTS

[18] allowed smooth and selective removal of allyl and Alloc groups in the presence of diethylamine as allyl scavenger and in homogeneous (MeCN/H$_2$O) or biphasic (*n*-PrCN/H$_2$O) media.

4.2 *Deprotection of alcohols and carboxylic acids: recycling*

The use of 2% Pd(0) species and excess diethylamine as an allyl-trapping agent led to the fast deprotection of primary and secondary alcohols in homogeneous medium. Under biphasic conditions (*n*-PrCN/H$_2$O), the reaction of protected Alloc menthol proceeded smoothly. Taking advantage of a two-phase system, the water-soluble catalyst could be recycled up to 10 times without loss of efficiency, providing a major asset from an industrial viewpoint [46]. In addition, the use of diethylamine as a cheap allyl scavenger is very attractive because both the excess of nucleophile and the *N*-allyldiethylamine byproduct are simply removed by evaporation, affording very clean crude products after aqueous workup. The same procedure was also successful for the cleavage of allylic esters in homogeneous medium (Scheme 3.23).

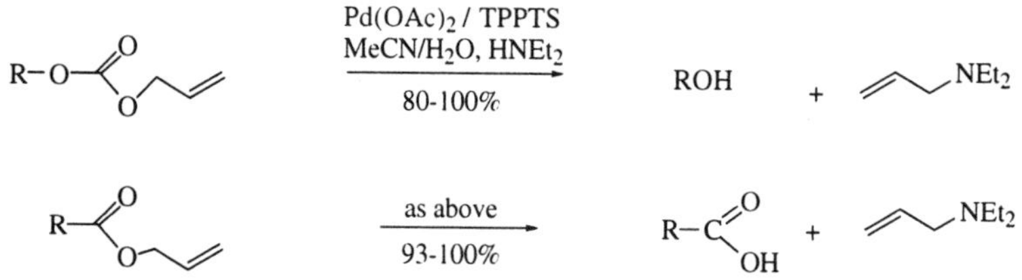

Scheme 3.23.

The mild conditions were compatible with other sensitive molecules such as cephalosporin, which was deprotected using 5% catalyst with 93% yield. Moreover, other protecting groups remained untouched under these conditions, as shown in the deprotection of a highly polyfunctional molecule, the precursor of the west part of ambruticine (Scheme 3.24).

4.3 *Deprotection of primary and secondary amines: recycling*

The cleavages of allylcarbamates derived from primary amines on treatment with 2% Pd(0) catalyst and 2.2 equivalents of nucleophile have been realized smoothly, with short reaction times. However, when *N*-Alloc-*N*-methylbenzylamine was allowed to react under the same conditions, the competitive *N*-allylation process occurred preferentially to give a 30:70 mixture of free amine and side-product. Anticipating that the reaction may be more selective in a biphasic system, the deprotection was carried out in *n*-PrCN/H$_2$O (6:1) medium with 5 mol.% Pd(0) catalyst. Under these conditions, the free secondary amine was recovered quantitatively without any undesired *N*-allylated product (Scheme 3.25).

The use of a two-phase system therefore offered an efficient alternative for the total removal of allylcarbamates derived from secondary amines, avoiding the competitive *N*-allylation. It is reasonable to think that in such a biphasic medium there is almost no contact between the catalyst present in the aqueous phase and the deprotected substrate liberated in the organic layer, resulting in an enhanced selectivity toward the deprotective

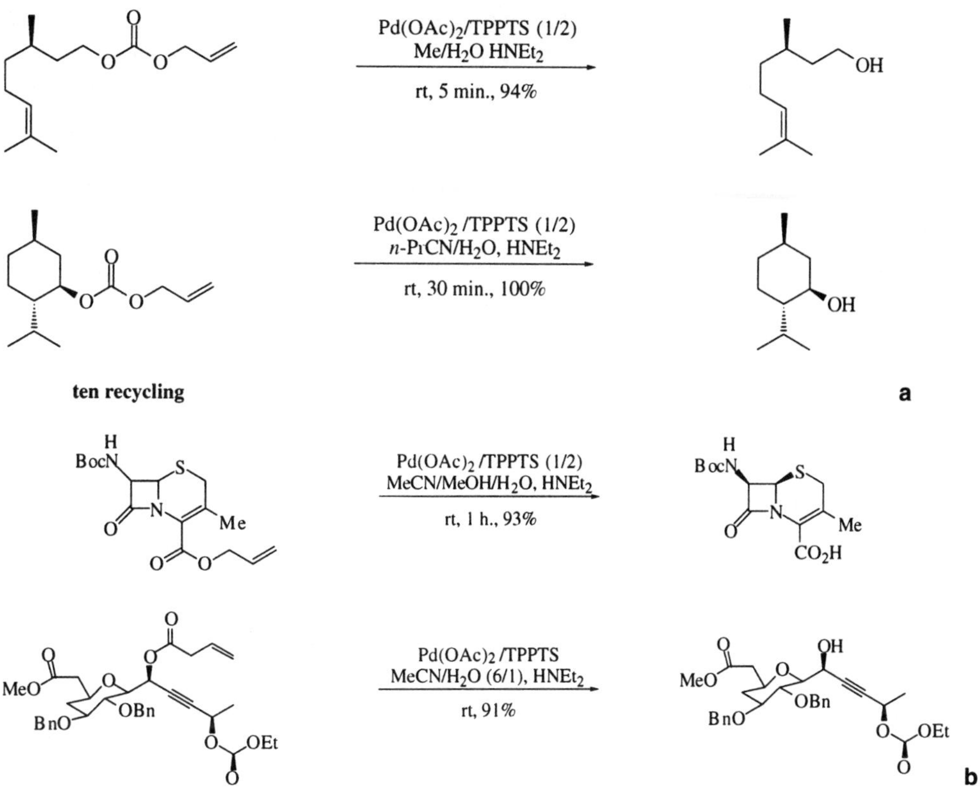

Scheme 3.24.

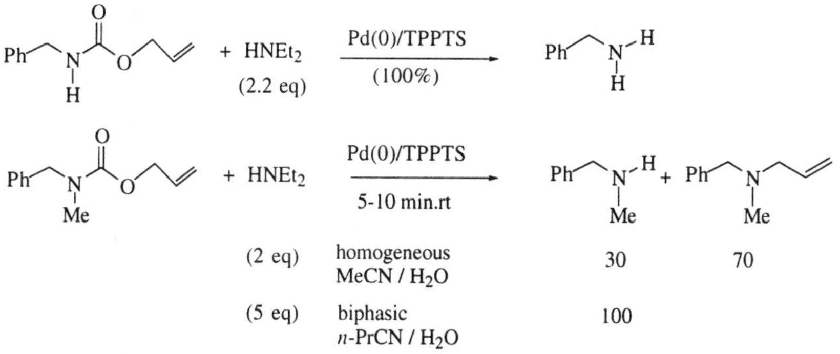

Scheme 3.25.

process. Other protected secondary amines such as (1R,2S)-N-allyloxycarbonylephedrine and N-allyloxycarbonyl-L-proline have reacted equally well upon treatment with a fivefold excess of HNEt$_2$.

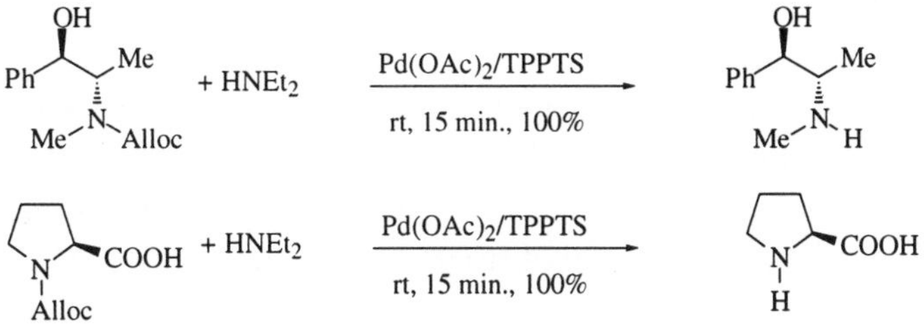

This efficient and inexpensive methodology thus has allowed the removal of allyl and allyloxycarbonyl groups from various substrates and the particularly mild conditions were compatible with polyfunctionalized molecules. Moreover, both the Pd(0) catalyst and the *N*-allyl-diethylamine byproduct were easily separated from the free alcohols, amines and carboxylic acids, which were recovered in almost pure form.

4.4   *Chemoselective removal of allylic protecting groups*

The synthesis of elaborated molecules requires the use of various protecting groups that can be cleaved independently at different stages of the sequence.

The usually accepted mechanism for palladium-promoted removal of allyl carbonates and carbamates involves oxidative addition of the palladium zerovalent species on the allyl group in the first step, leading to the formation of a π-allylpalladium(0) intermediate. We anticipated that substitution on the terminal position of the allyl group could disfavor the oxidative addition step by steric interactions between the substituents and the palladium.

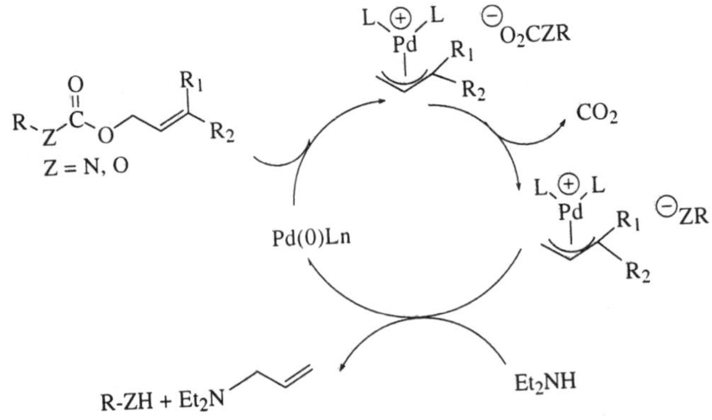

In order to determine the different factors that could allow chemoselective cleavage of allylic moiety using the water-soluble Pd/TPPTS catalyst, we first compared the rate of deprotection of several phenylacetic allyl esters under homogeneous and biphasic conditions. Using 2% catalyst in homogeneous (CH_3CN/H_2O) medium, the reaction time increased from 5 to 75 min while the protecting moiety went from allyl to cinnamyl and finally to dimethylallyl(3-methylbut-2-enyl). Lowering the catalyst rate to 0.5%, dimethyl-allylphenylacetate remained almost untouched after 8 h. Moreover, in a biphasic system only the allyl group was removed, whereas mono- and disubstituted allylic esters were not cleaved (Table 3.8).

Based on these results we envisage obtaining selectivity in a homogeneous medium between allyl and dimethylallyl groups by using a small quantity of Pd(0) species and in a biphasic system between allyl and substituted allylic groups. It thus seems possible to adjust these conditions in order to deprotect selectively the doubly protected bifunctional substrates.

Indeed, the allyloxycarbamate of isonipecotic acid was cleaved selectively and quantitatively under homogeneous conditions in the presence of 1% Pd(0) without affecting the dimethylallyl carboxylate. The resulting monodeprotected product was then deprotected using a higher amount of catalyst (3–5 mol.%) (Scheme 3.26).

Table 3.8. Solvent effect in selective allyl groups cleavage

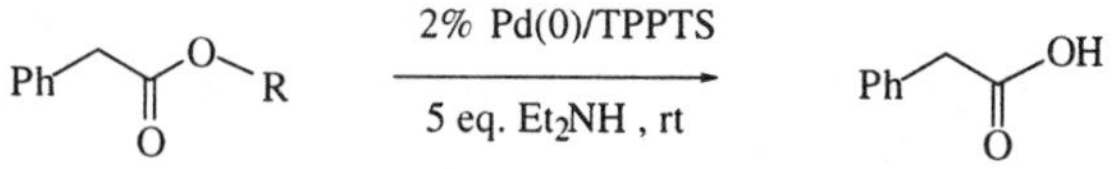

homogeneous:		MeCN/H$_2$O (6/1)	
biphasic:		n-PrCN/H$_2$O (6/1)	

R	Time	Remarks	Yield (%)
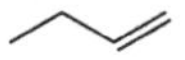 [allyl]	5 min	Homogeneous	100
	10 min	Biphasic	100
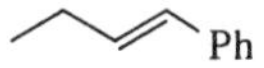 [cinnamyl, Ph]	20 min	Homogeneous	100
	72 h	Biphasic	0
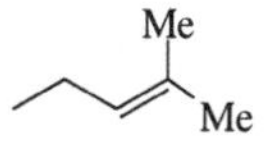 [Me...Me]	1 h 15 min	Homogeneous	100
	72 h	Biphasic	0

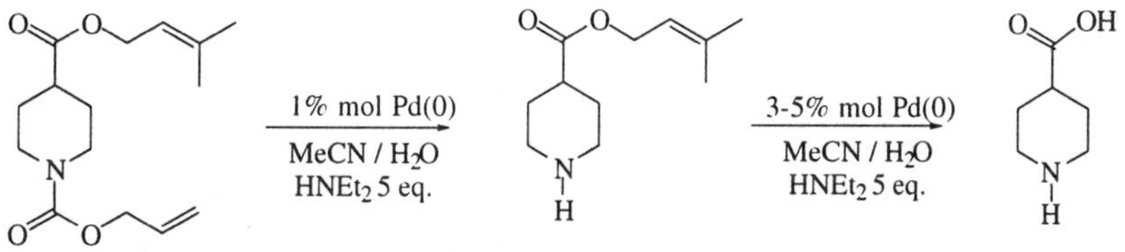

Pd(0) = Pd(OAc)$_2$/TPPTS (1 : 2)

Scheme 3.26.

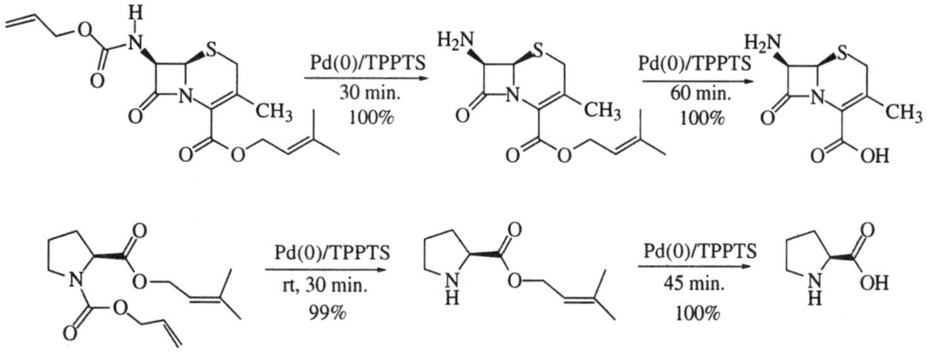

Scheme 3.27.

The same scheme of selective deprotections was achieved on a base-sensitive cephalosporin; with 2.5% water-soluble catalyst, the Alloc moiety was selectively removed to give the dimethylallyl carboxylate within 30 min and then the carboxylic acid was recovered quantitatively using 5% Pd(0).

These conditions of selective deprotection on amino acids were then applied with success for the cleavage of N-allyloxycarbamate of L-proline without affecting the carboxylic acid protected by the dimethylallyl moiety. Moreover, the reaction occurred without any trace of the competitive N-allylation process on the secondary amine (Scheme 3.27).

4.5   *Application to the synthesis of peptides in solution*

The assembly of peptides using conventional methods of organic chemistry is seriously limited by purification, analytical and characterization steps for each intermediate of the sequence. In the early 1960s Merrifield [47] proposed a new strategy of synthesis on a solid support that proved to be very efficient for the rapid preparation of medium-sized peptide chains. However, the final product is sometimes contaminated by impurities that cannot be eliminated, resulting in poor homogeneity of the peptide. Some of these problems have been circumvented by using continuous solution techniques, which may be used with Boc and Fmoc strategies [48]. Having developed a powerful methodology for the chemoselective deprotection of *N*-allyloxycarbonyl-*O*-dimethylallyl-α-aminoesters in the presence of the water-soluble Pd(OAc)$_2$/TPPTS system, we applied these conditions to the synthesis of peptides. Because it was demonstrated that both the zerovalent catalyst and the volatile *N*-allyldiethylamine byproduct were easily removed by aqueous workup and evaporation, we aimed at assembling the peptide by repetition of the sequence of selective cleavage of the terminal allylcarbamate/peptide coupling, avoiding any intermediate purification. This strategy was illustrated through the preparation of tetrapeptides [49]. The classical coupling agent TBTU [50] or the cyclic propylphosphonic anhydride PPA [51] were used and it was possible to eliminate the byproducts by either acid–base or simple aqueous treatments, to provide very clean, crude, coupled compounds.

For instance, *N*-Alloc-L-valine and *N*-Alloc-L-alanine were coupled with the O-protected amino acids in the presence of PPA (1.3 equiv) to give the corresponding dipeptides in 82% and 68% yield, respectively (Scheme 3.28).

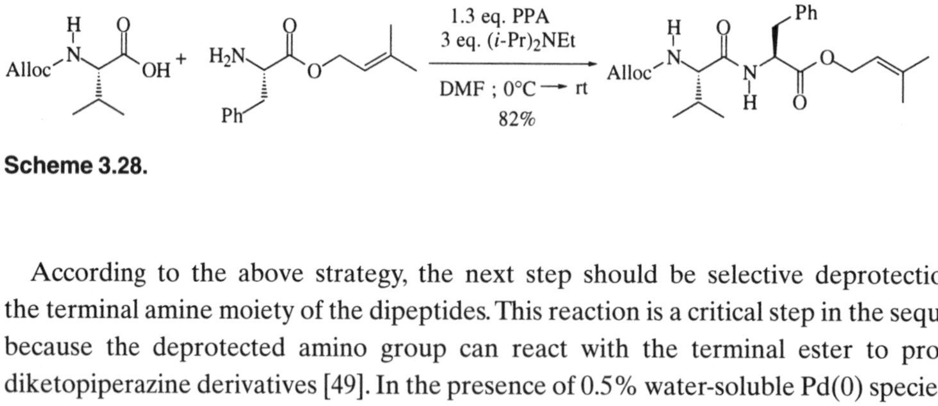

Scheme 3.28.

According to the above strategy, the next step should be selective deprotection of the terminal amine moiety of the dipeptides. This reaction is a critical step in the sequence because the deprotected amino group can react with the terminal ester to produce diketopiperazine derivatives [49]. In the presence of 0.5% water-soluble Pd(0) species, the terminal allylcarbamate was selectively removed without any formation of the undesired cyclized byproduct. Within a very short reaction time, the amino-free dipeptides were obtained in 83% and 88% yield, respectively, and were of very good purity after simple aqueous workup. As a result, the key sequence—peptide coupling/selective deprotection of the terminal amine—could be realized without purification.

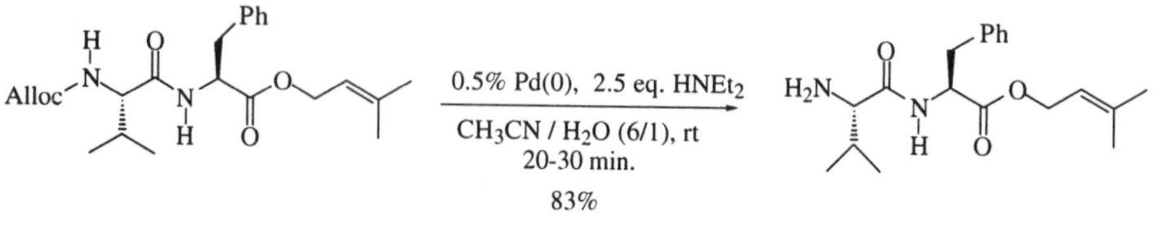

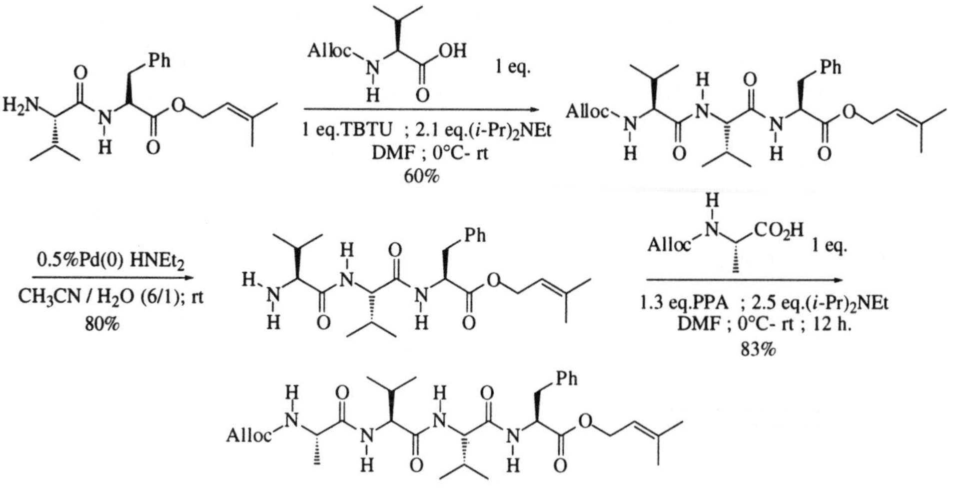

Scheme 3.29.

The synthesis of tetrapeptides was then achieved according to this methodology. An illustrative example is shown in Scheme 3.29: L-valyl-L-phenylalanyl-O-dimethylallyl was coupled with N-allyloxycarbonyl-L-valyl in the presence of TBTU as the coupling reagent. As noted previously, no purification was required after acid–base treatment and extraction, affording pure crude tripeptide that was directly engaged in the deprotection step under aqueous conditions. The terminal amine moiety was regenerated rapidly and selectively in good yields. The fourth amino acid was then assembled using PPA, to produce the corresponding tetrapeptides. As the length of the peptide increased, its water-solubility decreased, enabling better yields to be obtained due to less loss of material during the aqueous washings.

A rapid methodology was thus developed for the preparation of peptides in solution, using the selective removal of an allylcarbamate in the presence of a substituted allylic ester, promoted by the water-soluble Pd(OAc)$_2$/TPPTS catalytic system. This technique proved to be efficient for the synthesis of tetrapeptides, which could be used as building fragments for assembling larger molecules. As no purification was required throughout the elongation, these conditions are suitable for industrial processes because the coupling and deprotection steps could be carried out on a large scale and with short reaction times.

5 Conclusion

Following the development of the valuable Ruhrchemie-Rhône Poulenc process using rhodium water-soluble catalysts in hydroformylation, the important concept of complex-catalysed reactions in aqueous medium (homogeneous and two-phase systems) has been extended to palladium. The water-soluble Pd(0) species, as preformed Pd(TPPMS)$_3$ or prepared *in situ* from Pd(OAc)$_2$ and TPPTS, were excellent catalysts for C—C bond formation and allylic substitution and a valuable tool in deprotective group chemistry in organic–aqueous medium, the organic solvent used being a nitrile. These reaction conditions allowed very easy separation of the catalyst from the reaction product(s), which in most cases immediately could be used for the subsequent sequence without any purification. New selectivities and reactions have also been discovered and the catalyst can be recycled. Some reactions also proceeded under milder conditions compared to the same reactions using anhydrous conditions. The excellent compatibility of palladium complex catalysts with water was not expected, even by experts in catalysis in the past years, and other types of metals could be considered in the future because water is an environmentally friendly solvent.

6 References

1 Wilkinson G. *Comprehensive Organometallic Chemistry*. New York: Pergamon Press, 1982.

2 Wollaston WH. *Philos Trans R Soc London* 1804; **94**: 419; 1805; **95**: 316.

3 Tsuji J. *Palladium Reagents and Catalysts*, New York: Wiley, 1995; Malleron JL, Fiaud JC, Legros JY. *Handbook of Palladium-Catalysed Organic Reactions*. London: Academic Press, 1997.

4 Tsuji J. *Synthesis* 1990: 739–49.

5 Hafner W, Jira R, Sedlmeir J, Smidt J. *Chem Ber* 1962; **95**: 1575–81.

6 Heck RF. *Palladium Reagents in Organic Syntheses*. New York: Academic Press, 1985; Mimoun H, Charpentier R, Mitschler A, Fisher J, Weiss R. *J Am Chem Soc* 1980; **102**: 1047–54; Thompson DJ. In: *Comprehensive Organic Synthesis*. Oxford: Pergamon Press, 1991; Colquhoun HM, Thompson DJ, Twigg MW. *Carbonylation*. New York: Plenum Press, 1991.

7 Tsuji J. *Acc Chem Res* 1973; **6**: 8–15.

8 Heck RF. *Org React* 1982; **27**: 345; Heck RF. In: *Comprehensive Organic Synthesis*. Oxford: Pergamon Press, 1991; De Meijere A, Meyer FE. *Angew Chem Int Ed Engl* 1994; **33**: 2379–411; Negishi E, Coperet C, Ma S, Liou SY, Liu F. *Chem Rev* 1996: 365–93.

9 Sonogashira K. In: *Comprehensive Organic Synthesis*, Vol. 2. Oxford: Pergamon Press, 1991.

10 Suzuki A. *Acc Chem Res* 1982; **15**: 178–84; Suzuki A. *Pure Appl Chem* 1984; **66**: 213–22; Hunt AR, Stewart SK, Whiting A. *Tetrahedron Lett* 1993; **34**: 3599–602; Miyaura N, Suzuki A. *Chem Rev* 1995; **95**: 2457–83.

11 Trost BM, Van Vranken DL. *Chem Rev* 1996; **96**: 395–422.

12 Godleski SA. In: *Comprehensive Organic Synthesis*. New York: Pergamon Press, 1991; Tsuji J. *Acc Chem Res* 1969; **2**: 144–52.

13 Trost BM, Verhoeven TR. *Organopalladium Compounds in Organic Synthesis and in Catalysis in Comprehensive Organometallic Chemistry*, Vol. 8. Oxford: Pergamon Press, 1982: 799; Heck RF. *Palladium Reagents in Organic Synthesis*. New York: Academic Press, 1985.

14 Hermann WA, Kohlpaintner CW. *Angew Chem Int Ed Engl* 1993; **32**: 1524–44; Sinou D. *Bull Soc Chim Fr* 1987: 480–86.

15 Gärtner R, Cornils B, Bexten L, Kupies D. *DE-B 3 235 030*, Ruhrchemie AG, 1982; Bexten L, Cornils B, Kupies D. *DE-B 3 431 643*, Ruhrchemie AG, 1984; Kuntz EG. *Fr-B 2 366237*, Rhône-Poulenc Industries, 1976; Mignani G, Morel D, Colleuille Y. *Tetrahedron Lett* 1985; **26**: 6337–40; Mignani G, Morel D, Colleuille Y. *Tetrahedron Lett* 1986; **27**: 2591–4; Reithel FJ. *Concepts in Biochemistry*. New York: McGraw-Hill, 1967.

16 Beletskaya IP. *New Aspects of Organic Chemistry II*. Tokyo: Kodansha, 1992: 31; Bumagin NA, More PG, Beletskaya IP. *J Organomet Chem* 1989; **371**: 397–401.

17 Casalnuovo AL, Calabrese JC. *J Am Chem Soc* 1990; **112**: 4324–30.

18 For characterization of the catalyst, see: Amatore C, Blart E, Genet JP *et al*. *J Org Chem* 1995; **60**: 6829–39.

19 Genêt JP, Blart E, Savignac M. *Synlett* 1992: 715–17.

20 Safi M, Sinou D. *Tetrahedron Lett* 1991; **32**: 2025–8; Blart E, Genêt JP, Safi M, Savignac M, Sinou D. *Tetrahedron* 1993; **50**: 505–14.

21 Jeffery T. *J Chem Soc Chem Commun* 1984: 1287–9; Jeffery T. *Tetrahedron Lett* 1985; **26**: 2667–70; Larock RC, Baker B. *Tetrahedron Lett* 1988; **29**: 905–8.

22 Karabelas K, Westerlund C, Hallberg A. *J Org Chem* 1985; **50**: 3896–900; Karabelas K, Hallberg A. *J Org Chem* 1986; **51**: 5286–90; Abelman MM, Overman LE. *J Org Chem* 1987; **52**: 4133–5; Grigg R, Loganathan V, Santhakumar V, Sridharan V, Teasdale. *J Org Chem* 1987; **52**: 4130–3.

23 Karabelas K, Hallberg A. *J Org Chem* 1988; **53**: 4909–14; Cabri W, Candiani I. *Acc Chem Res* 1995; **28**: 2–7, and references cited therein.

24 Jeffery T. *Tetrahedron Lett* 1994; **35**: 3051–4.

25 Overman LE, Abelman MM, Kucera DJ, Tran VD, Ricca DJ. *Pure Appl Chem* 1992; **64**: 1813–19.

26 Lemaire-Audoire S, Savignac M, Dupuis C, Genêt J-P. *Tetrahedron Lett* 1996; **37**: 2003–6.

27 Trost BM, Dumas J. *Tetrahedron Lett* 1993; **34**: 19–21.

28 Rigby JM, Hughes RC, Heeg MJ. *J Am Chem Soc* 1995; **117**: 7834–5.

29 Ratovelomanana V, Linstrumelle G. *Tetrahedron Lett* 1981; **22**: 315–18; Crombie L, Horsham MA, Blade RJ. *Tetrahedron Lett* 1987; **28**: 4879–82; Schreiber SL, Kiessling LL. *J Am Chem Soc* 1988; **110**: 631–3.

30 Montalbetti C, Savignac M, Bonnefis F, Genêt J-P. *Tetrahedron Lett* 1995; **36**: 5891–4.

31 Glaser C. *Ber* 1869; **2**: 422; Glaser C. *Ann* 1870; **137**: 154; Eglington G. *Chem Ind* **1956**: 737.

32 Rossi R, Carpita A, Bigelli C. *Tetrahedron Lett* 1985; **26**: 523–6.

33 Cadiot P, Chodkiewicz W. In: Viehe HG, ed. *Chemistry of Acetylenes*, pp. 597–647. New York: Dekker, 1969.

34 Sonogashira K. In: Trost BM, Fleming eds. *Comprehensive Organic Chemistry*. pp. 551–61. New York: Pergamon Press, 1991: 2.5.

35 Genêt JP, Lindquist A, Blart E, Mouries V, Savignac M. *Tetrahedron Lett* 1995; **36**: 1443–6.

36 Galland JC, Savignac M, Genêt JP. *Tetrahedron Lett* 1997; **38**: 8695–8.

37 Oppolzer W. In: *Organometallic Reagents in Organic Synthesis*. New York: Academic Press, 1994: 161.

38 Gomez-Bengoa E, Cuerva JM, Echavarren AM, Martorell G. *Angew Chem Int Ed Engl* 1997; **36**: 767–9.

39 Greene JW, Wut PGM. *Protective Group in Organic Synthesis*. New York: Wiley, 1991.

40 Minami I, Ohashi Y, Shimizu I, Tsuji J. *Tetrahedron Lett* 1985; **26**: 2449–52; Hayakawa Y, Wakabayashi S, Kato H, Noyori R. *J Am Chem Soc* 1990; **112**: 1691–6.

41 Kunz H, Waldmann H. *Angew Chem Int Ed Engl* 1984; **23**: 71–2; Kunz H, Waldmann H, Klinkhammer U. *Helv Chim Acta* 1988; **71**: 1868–74; Kunz H, Unverzagt C. *Angew Chem Int Ed Engl* 1984; **23**: 436–7.

42 Guibé F, Saint M'Leux Y. *Tetrahedron Lett* 1981; **22**: 3591–4; Guibé F, Dangles O, Balavoine G, Loffet A. *Tetrahedron Lett* 1989; **30**: 2641–4; Dangles O, Guibé F, Balavoine G, Lavielle S, Marquet A. *J Org Chem* 1987; **52**: 4984–93; Boullanger P, Descotes G. *Tetrahedron Lett* 1986; **27**: 2599–602.

43 Jeffrey PD, MacCombie SW. *J Org Chem* 1982; **47**: 587–90.

44 Mermouk A, Guibé F, Loffet A. *Tetrahedron Lett* 1992; **33**: 477–80.

45 Dessolin M, Guillerez MG, Thieriet N, Guibé F, Loffet A. *Tetrahedron Lett* 1995; **36**: 5741–4.

46 Lemaire-Audoire S, Blart E, Savignac M *et al. Tetrahedron Lett* 1994; **35**: 8783–6; Bernard JM, Blart E, Genêt JP *et al. French Patent FR2703687*; Lemaire-Audoire S, Savignac M, Genêt JP, Pourcelot G, Bernard, JM. *J Mol Catal* 1997; **116**: 247–58.

47 Merrifield RB. *J Am Chem Soc* 1963; **85**: 2149; Mitchell AR, Kent SBH, Engelhard M, Merrifield RB. *J Org Chem* 1978; **43**: 2845–52.

48 Bodansky M. *Principles of Peptide Synthesis*. Berlin: Springer-Verlag, 1984.

49 Lemaire-Audoire S, Savignac M, Blart E, Bernard JM, Genêt JP. *Tetrahedron Lett* 1997; **38**: 2955–8.

50 Knorr R, Trzeciak A, Bannwarth W, Gillessen D. *Tetrahedron Lett* 1989; **30**: 1927–30.

51 Wissmann H, Kleiner HJ. *Angew Chem Int Ed Engl* 1980; **19**: 123–4.

4 Palladium-catalysed Molecular Queuing Processes: Relay Switches and the Maximization of Molecular Complexity

RONALD GRIGG and VISUVANATHAR SRIDHARAN

Molecular Innovation, Diversity and Automated Synthesis (MIDAS) Centre, School of Chemistry, Leeds University, Leeds LS2 9JT, UK

1 Introduction

Palladium salts and complexes are exceptionally versatile catalysts for the construction of carbon–carbon and carbon–heteroatom bonds [1]. Much recent attention has focused on the Heck reaction [2] (Scheme 4.1) due to developments that have considerably enhanced the scope of this palladium-catalysed vinylation of aryl, heteroaryl, vinyl and benzyl halides. Thus, the Heck reaction has been extended to the synthesis of bridged rings, spiro-cycles and tetrasubstituted carbon centres [3–5]. These latter developments and the ongoing high level of activity have been fostered further by the advent of a range of additives that variously enhance the rate of Heck reactions, control the regioselectivity of the β-hydride elimination step and suppress double bond isomerization in the product. Thus, addition of tetraalkylammonium salts often allows Heck reactions to be carried out at, or near, room temperature in good yield [6], whilst addition of Ag(I) salts [7] or Tl(I) [8] salts can control the direction of β-hydride elimination, suppress double bond isomerization and influence the reaction rate. Thallium(I) additives have also proved useful in natural product synthesis [9]. However, the Heck reaction fails to take advantage of the inherent ability of Pd(0) catalysts to process a wide range of chemically distinctive substrates and suffers from the drawback that, as usually practised, only one C–C bond is made.

The full power of Pd(0) catalysis can be liberated by designing cascade reactions that take advantage of the diverse range of substrates accepted by these catalysts. Cascade reactions may be defined as multireaction 'one-pot' sequences in which the first reaction creates the functionality to trigger the second reaction, and so on.

2 Advantages of cascade reactions

1 High atom efficiency, low waste and clean technology.

2 More efficient use of manufacturing plant.

3 Major increase in molecular complexity. In industrial terms, manufacturers add focused molecular complexity to simpler starting materials. Thus, molecular complexity = added value.

4 The ability to assemble combinations of 2, 3, 4 . . . n different substrates involving both C–C and C–heteroatom bond formation. In this context it is useful to reflect that while enzymes are exquisitely selective catalysts, they only make or break the same type of bond. On the other hand, polymerization methodology can rarely copolymerize more than two types of monomer and control of product molecular weight and stereochemistry present

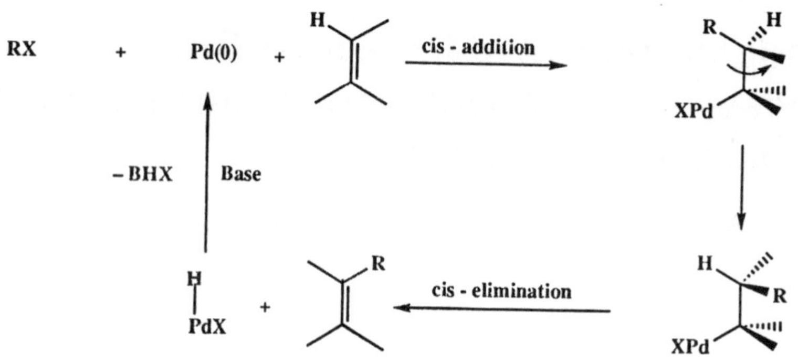

Scheme 4.1.

substantial problems. In contrast, Pd(0)/Pd(II) catalysts are much more versatile and useful in assembling a wide range of substrates and bond types.

5 The ability to design cascades that switch between inter- and intramolecular processes.
6 A high degree of tactical and structural versatility. For example, in designing a cascade that incorporates two reactions A and B, A could precede B, or vice versa.

3   Requirements for cascade design

1 A good understanding of the relative rates of disparate reactions and their sensitivity to temperature and pressure.
2 An ability to engineer the control of chemo-, regio- and stereoselectivity.
3 Mild reaction conditions.

Our initial experiments on palladium-catalysed cascade reaction design focused on ring formation with concomitant incorporation of additional functionality. These studies led to our Pd(0)-catalysed cyclization–anion capture methodology outlined in Scheme 4.2 for a monocyclization with an alkyne-terminating (see below) species.

In Scheme 4.2 the reaction would normally become unproductive with the accumulation of compound **2**. However, anion exchange of X for Y, which produced compound **3** susceptible to reductive elimination, would establish the required catalytic cycle. The deleterious direct capture or 'shunt' pathway **1** → **4** would be expected to be susceptible to ring size and the nature of Y. Initial experiments with Y = hydride ion (Scheme 4.3) [10] established the viability of Scheme 4.2 for various ring sizes.

In most of our work the Pd(0) catalyst is either generated *in situ* from 10 mol.% Pd(OAc)$_2$/20 mol.% PPh$_3$, or 5–10 mol.% of preformed Pd(0) [Pd(PPh$_3$)$_4$, Pd$_2$dba$_3$] is used.

A range of sources of the hydride ion were evaluated and formate salts (HCO$_2$Na, HCO$_2$H/piperidine) were found to be the most generally effective and to suppress the shunt pathway. Further exploration of ring size and anion capture agent Y led to Table 4.1, which illustrates the scope of the cyclization–anion capture methodology.

The Pd(0) catalyst undergoes oxidative addition to the starter species (halide, triflate, acetate, etc.) to generate an organopalladium(II) intermediate. In processes involving the formation of one ring, this intermediate successively engages the terminating species and the anion capture agent Y. In polycyclization processes the initial organopalladium(II) intermediate engages one or more relay species before passing to the terminating phase and then anion capture. This is illustrated for a bis-cyclization of an aryl starter species in

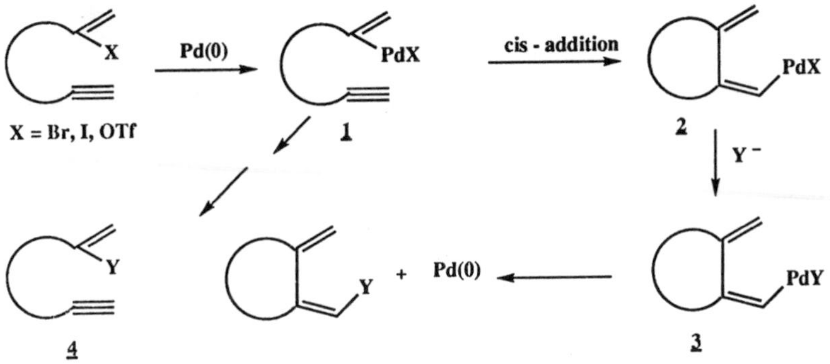

Scheme 4.2.

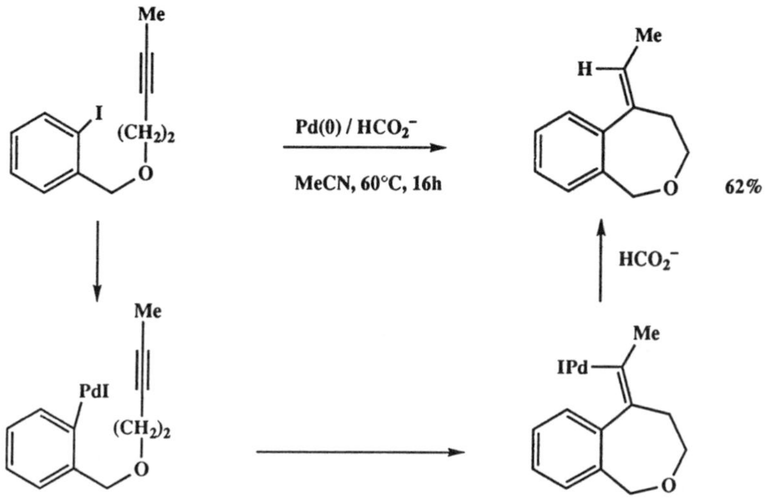

Scheme 4.3.

Table 4.1. Potential combinations for (poly)cyclization–anion capture processes

Starter species	Relay species (R)	Terminating species (T)	Y
Alkyl	Alkene	Alkene	*Anionic* $(H, OAc CN, N_3$ $SO_2Ph, CH(CO_2R)_2)$
Aryl	Alkyne	Alkyne	*Neutral* (amines, MeOH/CO, acrylates)
Vinyl	1,2-Diene	1,2-Diene	*Organometallic* (RM, where
Allyl	1,3-Diene	1,3-Diene	$M = Sn(IV), B, Zn)$
Allenyl			

Scheme 4.4 [11]. Product **5** is obtained as a single diastereomer and the stereochemistry, which is based on a chair-like pretransition state conformer for the second cyclization [12], is provisional. Note that *exo*-cyclization to generate the smallest ring is invariably preferred over *endo*-cyclization.

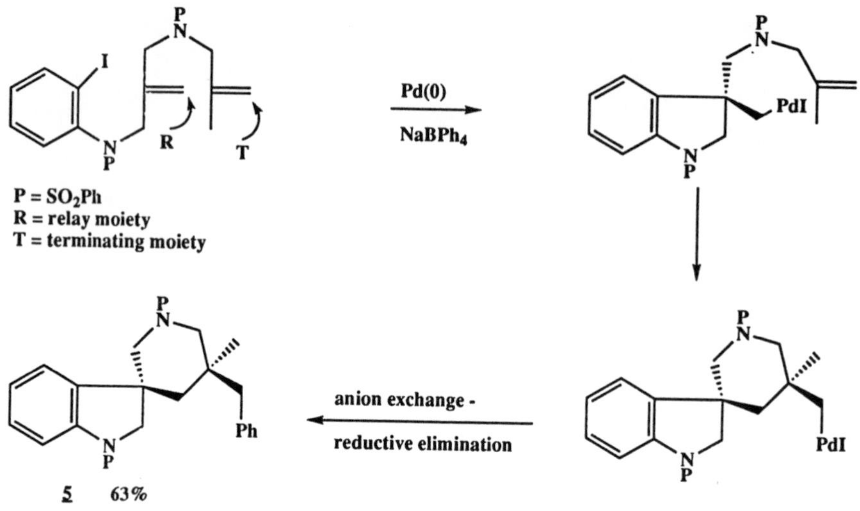

P = SO$_2$Ph
R = relay moiety
T = terminating moiety

Scheme 4.4.

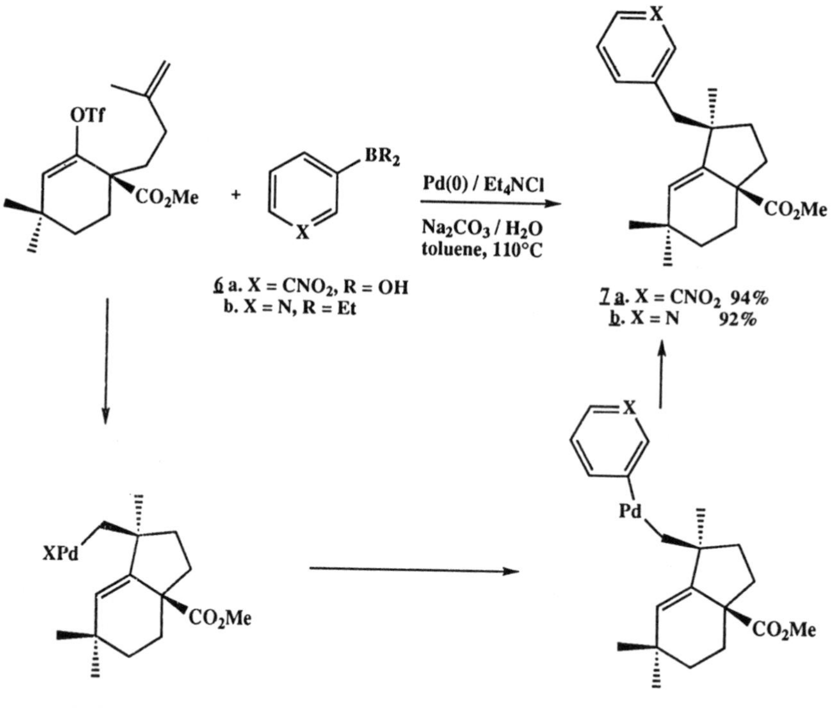

6 a. X = CNO$_2$, R = OH
b. X = N, R = Et

7 a. X = CNO$_2$ 94%
b. X = N   92%

Scheme 4.5.

Table 4.1 is capable of considerable further extension, especially with respect to the anion capture agent Y. Examples that employ all the starter species, terminating species and anion capture agents Y shown in Table 4.1 have been achieved for monocyclization processes but not all possible combinations have been explored. With ring sizes 3–7 the shunt pathway is rarely a problem and additives such as Et$_4$NCl [6] or Tl(I) [8] salts help to suppress this deleterious process. A typical monocyclization process is shown in Scheme 4.5 [11].

Oxidative addition of Pd(0) to the vinyl triflate starter species is followed by stereo-specific 5-*exo*-trig cyclization, anion exchange with compound **6a** or **6b** and reductive elimi-

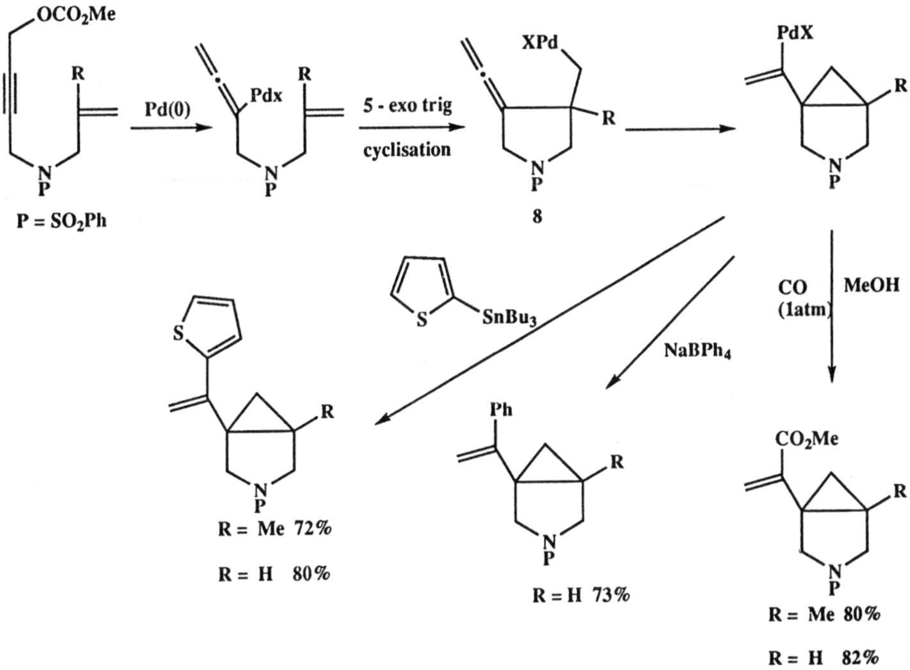

Scheme 4.6.

nation to afford compounds **7a** and **7b** as single diastereomers in excellent yield. Boronic acids, by virtue of their ease of accessibility, stability and diverse structures, form a valuable resource for the cyclization–anion capture methodology.

A typical bis-cyclization involving the formation of three new bonds is shown in Scheme 4.6 [13], which utilizes Tsuji's simple and efficient method [14] for generating allenylpalladium(II) species. Subsequent to the publication of our results, similar observations were published by Oppolzer *et al.* [15]. An interesting rate feature emerges from these cascades. Thus, the alkylpalladium(II) intermediate **8** (R=H) could undergo a β-hydride elimination and hence interrupt the catalytic cycle. That this deleterious process is not observed, is ascribed to a combination of complexation of the palladium by the adjacent allenyl moiety and a rapid 3-*exo*-trig cyclization.

Organotin reagents RSnBu$_3$ and RSnMe$_3$ comprise a valuable source of diversity and added complexity. They are easily accessible and stable and can be interfaced readily with our relay switch reagents (see below).

Tris-cyclization–anion capture involving an organotin anion capture reagent is illustrated in Scheme 4.7 [16]. An alternative mechanism for Scheme 4.7 is that involving a palladium-catalysed [2 + 2 + 2]-cycloaddition to give compound **9b** followed by oxidative addition of Pd(0) to give compound **9a**, an intermediate common to both cascades. Apparently, both mechanisms can operate because mixtures of compounds **9b** and **10** are obtained under certain conditions. However, under the conditions of Scheme 4.7 [16], compound **9b** is *not* converted into compound **10**.

Spiro (Scheme 4.8) and bridged-ring (**11**) → (**12**) macrocycles can be accessed via bis-cyclization processes employing *in situ* generation of vinyl stannanes. The latter are produced by regioselective palladium-catalysed hydrostannylation of alkynes [17].

In Scheme 4.6 the cyclization of alkylpalladium(II) species **8** occurs at the proximal allenyl carbon center, whilst in Scheme 4.7 the vinylpalladium(II) species cyclizes at the

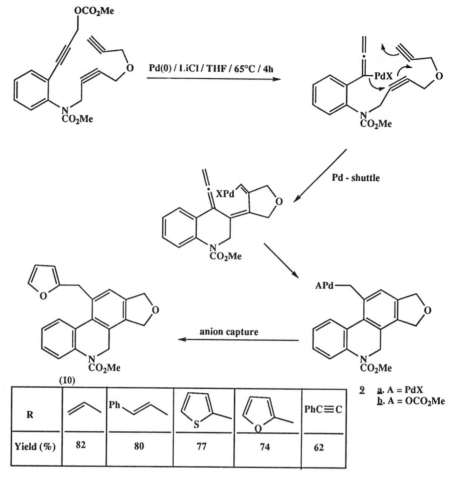

Scheme 4.7.

R	![allyl]	Ph	![thienyl]	![furyl]	PhC≡C
Yield (%)	82	80	77	74	62

center carbon of the allene moiety. The latter site selectivity is the one observed in the absence of any adverse steric factors (*vide infra*) and results in the formation of a π-allyl species. When the allenyl moiety is employed as a terminating species, the regioselectivity of anion capture by the resulting π-allyl Pd(II) species offers scope for increasing the reaction diversity. This is illustrated in Scheme 4.9 [18], where the choice of base allows clean reversal of regiochemistry.

Monitoring of the processes in Scheme 4.9 shows that the reaction employing K_2CO_3 as base initially produces mixtures of both regioisomers but prolonged reaction times result in rearrangement to a single isomer. Thus attack at the least hindered allylic site affords the thermodynamically most stable product, whilst attack at the more substituted allylic site is kinetically favoured.

4   Polycomponent processes and relay switches

Table 4.1 emphasizes the versatility of the general concepts but much detail remains to be explored. However, as initially conceived, our cyclization–anion capture methodology suffers from the constraint that most cascades are two-component processes, i.e. the ring 'zipper' precursor and Y. This shortcoming would be circumvented if polycomponent

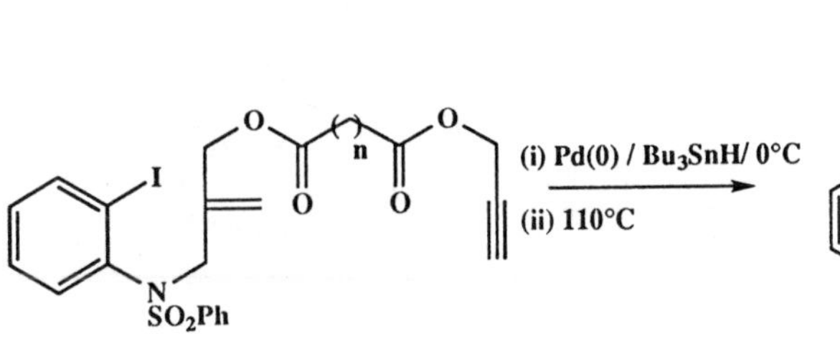

n	macrocyclic ring	Yield (%)[a]
3	12	53(71)
4	13	52(70)
5	14	52(70)
6	15	53(71)
7	16	50(67)
8	17	53(71)

a. Yields in brackets are corrected for the $\alpha -/ \beta$ –stannane ratio

Scheme 4.8.

processes could be achieved by extension of the relay phase with incorporation of both inter- and intramolecular segments. In this context the formation of the ester in Scheme 4.6 would be an example of a three-component process.

Initially we showed that a series of three-component processes can be readily achieved by employing CO (1 atm) in combination with the anionic and MR groups of capture agents Y (Table 4.1). Typical examples are shown in Scheme 4.10 [19]. Thus the relative rates of cyclization, carbon monoxide insertion and capture of Y allow high-yielding three-component processes to be designed where the various components can be considered to be queuing for access to the palladium center. Scheme 4.10 provides the first examples of palladium-catalysed cyclization–carboformylation. The reaction is regiospecific and employs diphenylmethylsilane as the hydride ion source. When sodium hydride is employed as the hydride source, the carbonylation fails to occur and N-phenylsulfonyl-3,3-dimethylindoline is the product. The catalytic hydroformylation of alkenes is a major industrial process that has attracted substantial application in the fine-chemical industry despite the fact that control of regioselectivity is a perennial problem [20].

These polycomponent processes create the flexibility for engineering major increases in molecular complexity and offer rapid entry to advanced intermediates in natural product synthesis. We have applied such three-component processes to the synthesis of indolic spider toxins [21], of which Scheme 4.11 (unpublished) is an example. The precise stage at which double bond migration occurs in Scheme 4.11 has yet to be elucidated.

Substrates such as CO in Schemes 4.10 and 4.11 are termed relay switch reagents because they prolong the relay phase of the cyclization–anion capture methodology whilst offering the potential to switch the cascade between intra- and intermolecular processes (see Scheme 4.13). Allenes are also valuable substrates in this respect and Scheme 4.12 illustrates a three-component polycyclization cascade utilizing allene (1 atm) [16].

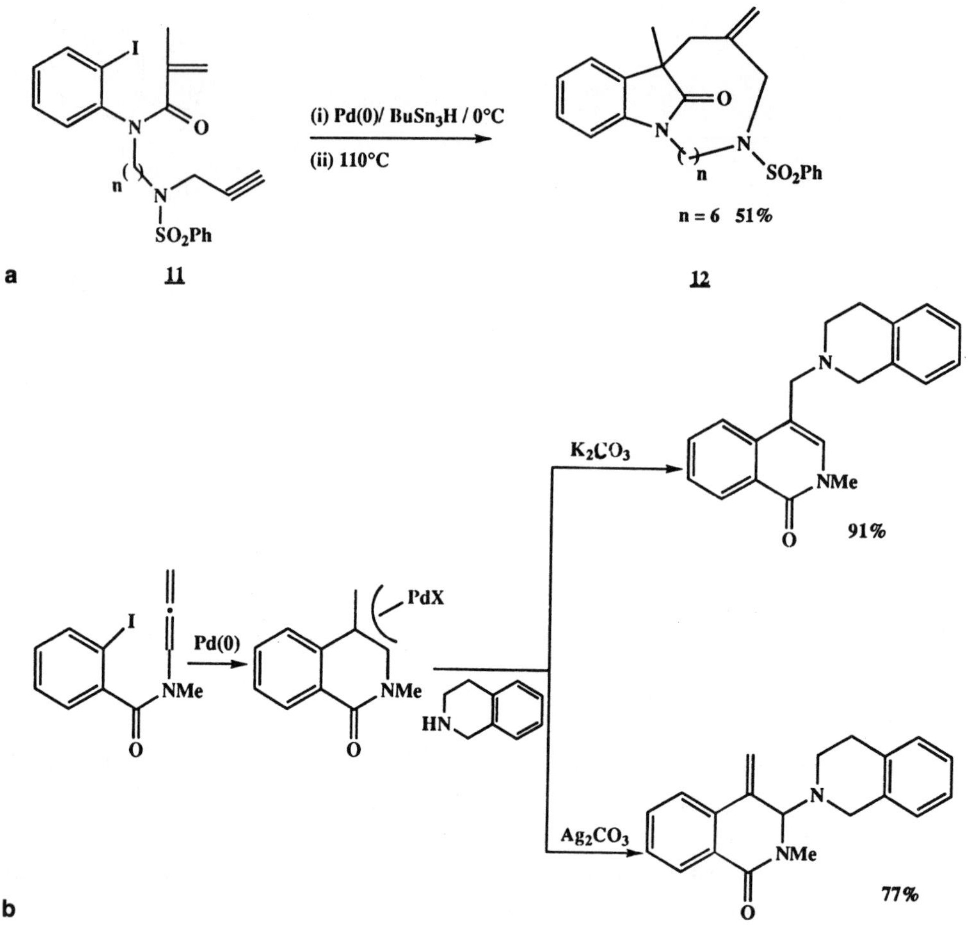

a

(i) Pd(0)/ BuSn₃H / 0°C

(ii) 110°C

11

12

n = 6   51%

b

Pd(0)

K₂CO₃

91%

Ag₂CO₃

77%

Scheme 4.9.

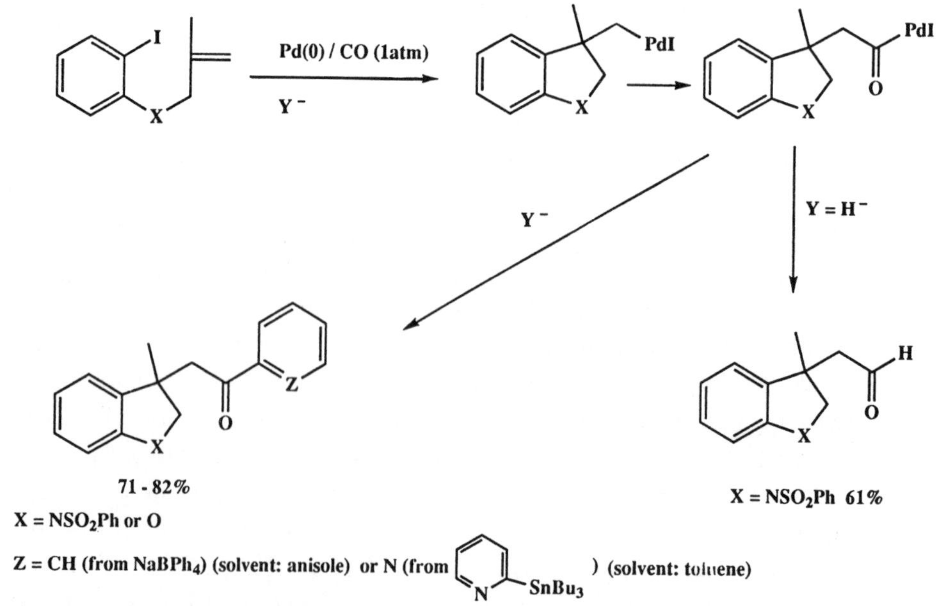

Pd(0) / CO (1atm)

Y⁻

Y⁻

Y = H⁻

71 - 82%

X = NSO₂Ph   61%

X = NSO₂Ph or O

Z = CH (from NaBPh₄) (solvent: anisole) or N (from [pyridine]—SnBu₃) (solvent: toluene)

Scheme 4.10.

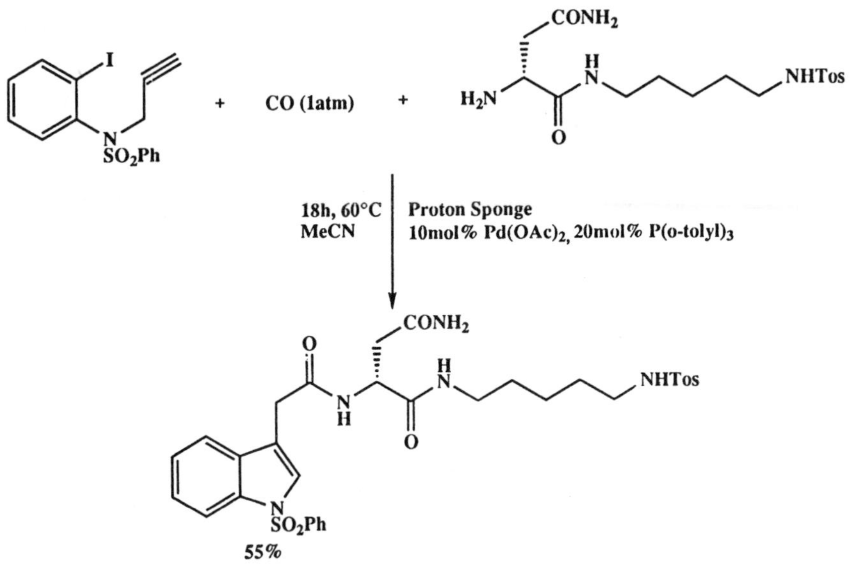

Scheme 4.11.

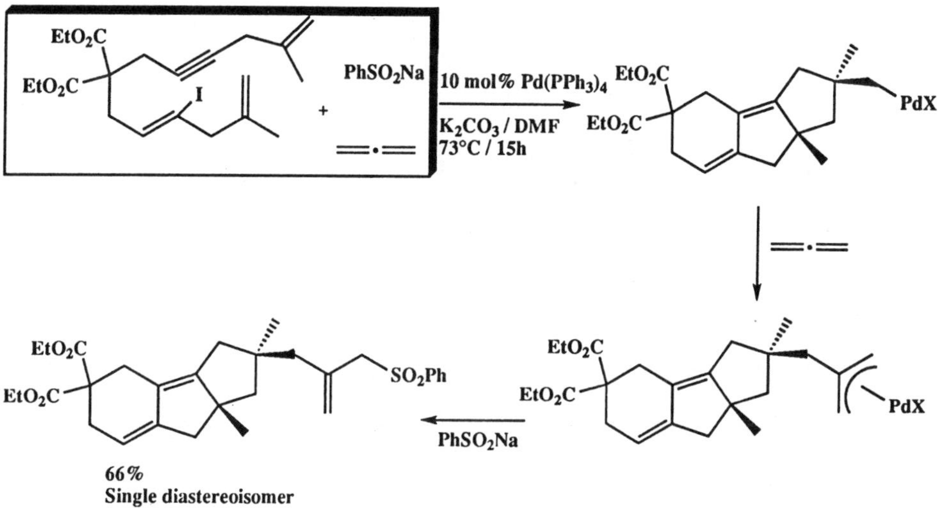

Scheme 4.12.

Once again a well-ordered molecular queue results and the process, which involves the formation of five new bonds and two stereocenters, is highly diastereoselective. The employment of allenes as relay switch components also offers great synthetic flexibility by virtue of the formation of π-allylpalladium(II) intermediates. The rich chemistry of these latter species allows Pd(0) chemistry to be interfaced with other core synthetic reactions such as 1,3-dipolar cycloaddition reactions and Diels–Alder reactions (*vide infra*).

Tactically these polycomponent molecular queuing cascades can be engineered such that the relay switch component is incorporated pre- or post-cyclization. A four-component cascade (CO is utilized twice) involving the former strategy is shown in Scheme 4.13 [22].

In this cascade the initial oxidative addition product from compound **13** and Pd(0) could conceivably undergo a 4-*exo*-trig cyclization. Such cyclizations are known but are

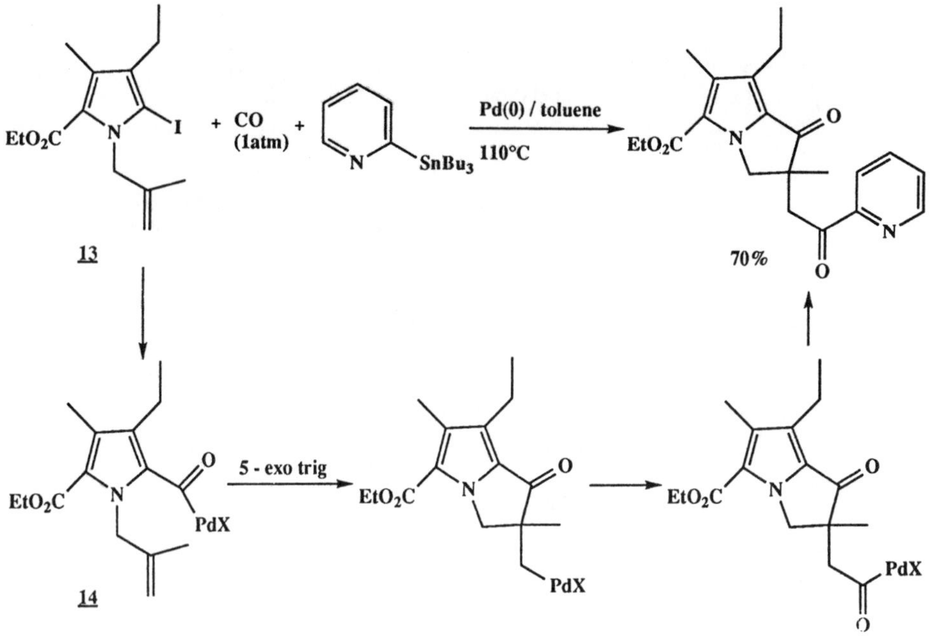

Scheme 4.13.

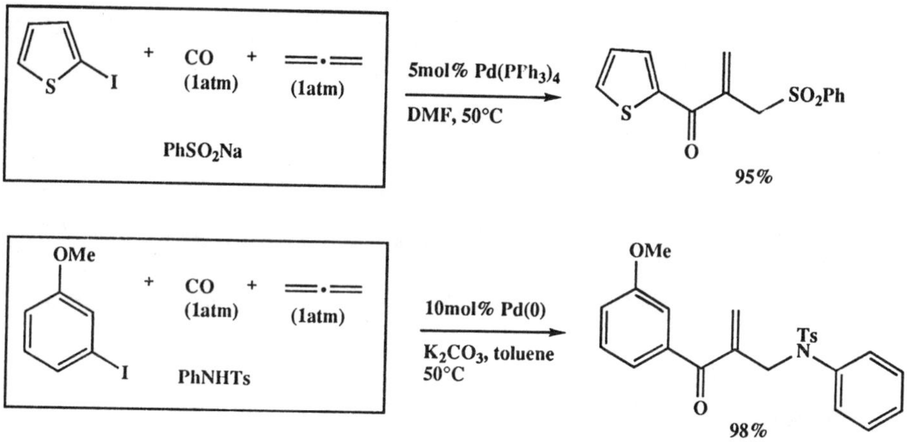

Scheme 4.14.

expected to be slow [11]. Hence, insertion of carbon monoxide occurs first to provide the acylpalladium(II) intermediate **14**. The 5-*exo*-trig cyclization of compound **14** is now fast and is followed by further incorporation of carbon monoxide, with transfer of the 2-pyridyl group from Sn(IV) being kinetically the slowest process. Thus the initial relay phase step is intermolecular and is followed by an intramolecular step. In this cascade four new bonds are formed.

Examples of four-component queuing processes employing allene and CO are shown in Scheme 4.14 [23]. In these cascades there is an orderly molecular queue with incorporation of CO occurring prior to incorporation of allene. The acylpalladium(II) intermediate adds to the center carbon of the allene, generating the corresponding π-allylpalladium(II) species, which is intercepted by the sulfinate anion.

Pentamolecular queuing processes have also been achieved, as illustrated by Schemes 4.15 [24] and 4.16 (unpublished). The strategy employed in Scheme 4.15 is analogous to that employed in Scheme 4.13, in that the initial oxidative addition product of compound **15** and Pd(0) undergoes CO insertion in preference to a 4-*exo*-trig cyclization. Once again orderly queuing processes occur under the reaction conditions shown and result in the

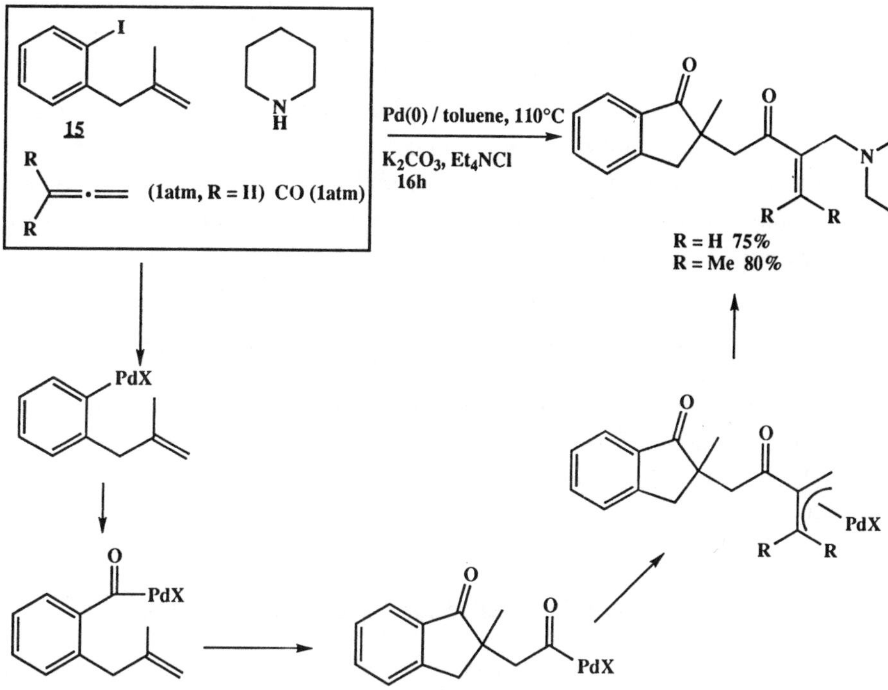

Scheme 4.15.

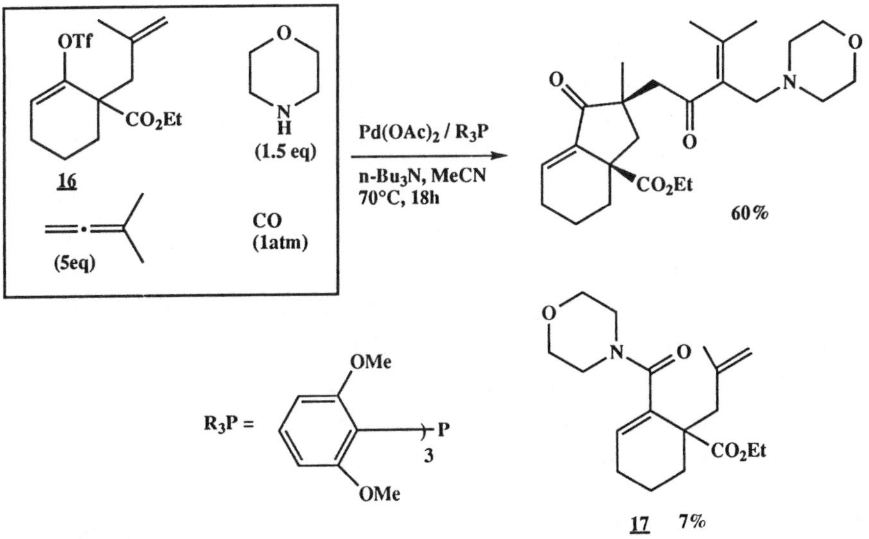

Scheme 4.16.

formation of five new bonds, with regiospecific functionalization of the intermediate π-allylpalladium(II) species by morpholine. Scheme 4.16 illustrates an analogous process for a vinyl triflate. In this case 2,6-dimethoxyphosphine proved to be an effective ligand for palladium and facilitated a highly diastereoselective cascade. A small amount of the direct capture product **17** was also obtained.

Our preliminary studies have shown that palladium catalysts enable the orderly assembly of diverse molecular building blocks via polymolecular queuing processes. These processes exhibit high chemo-, regio- and diastereoselectivity and result in substantial increases in molecular complexity. There is substantial untapped potential for interfacing this cascade methodology with a wide variety of core synthetic reactions.

There is a hierarchy of such tactical combinations with respect to the overall increase in molecular complexity achievable. Cycloaddition processes are particularly valuable partners in tactical combinations because they generate two new C–C or C–heteroatom bonds, they generate up to four stereocenters and asymmetric induction is well developed in many instances. We have begun developing such combinations as cascades or one-pot sequential processes.

5   [2 + 2]-Cycloaddition–palladium-catalysed cyclization [25]

We have achieved a one-pot combination of a range of examples of ketene–imine [2 + 2]-cycloadditions (Staudinger reaction [26]) with palladium-catalysed cyclizations (e.g. Scheme 4.17). This methodology allows ready access to unusual tri- and polycyclic β-lactams.

6   1,3-Dipolar cycloaddition–palladium-catalysed cyclization

We have developed a number of one-pot sequential or cascade protocols that employ several types of 1,3-dipole. Scheme 4.18 illustrates the interfacing of our imine–metallo-azomethine ylide–cycloaddition cascade [27] with a palladium-catalysed carbonylation [28]. Use of a chiral auxiliary on the dipolarophile results in enantiopure products [29] with the LiBr/base combination generating the *E,E*-lithioazomethine ylide. Overall, four new bonds, two rings and three stereocenters are created.

A further example of a cycloaddition preceding palladium chemistry is provided by Scheme 4.19 (R. Grigg, S. Suganthan and J. Virica, unpubl. obs.).

Scheme 4.19 is, strictly speaking, a sequential one-pot process in that the Pd(0) catalyst is added after the cycloaddition is complete. The reaction results in the formation of five bonds and three rings. Moreover, two bonds are formed at each of the contiguous carbon centers a and b.

Processes in which palladium-catalysed cyclization precedes 1,3-dipolar cycloaddition have also been designed. For example, the versatility of π-allyl chemistry allows allylic azides to be generated readily, as shown in Scheme 4.20 [30]. The drawback to implementing this protocol as a cascade is the reaction of phosphine ligands from the catalyst with the acetylenic dipolarophile. Thus a non-phosphine-containing catalyst is required.

Palladium phosphine complexes do not pose problems with norbornadiene as the dipolarophile. Cascade processes are readily achievable in this case and can be engineered

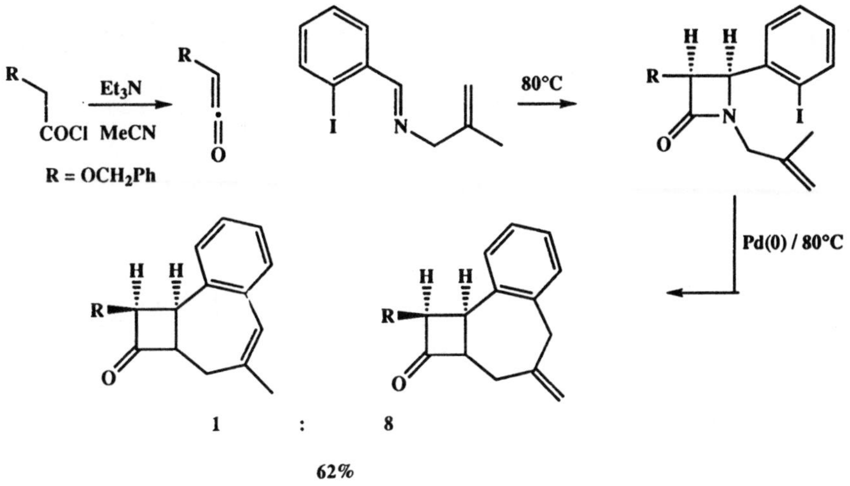

Scheme 4.17.

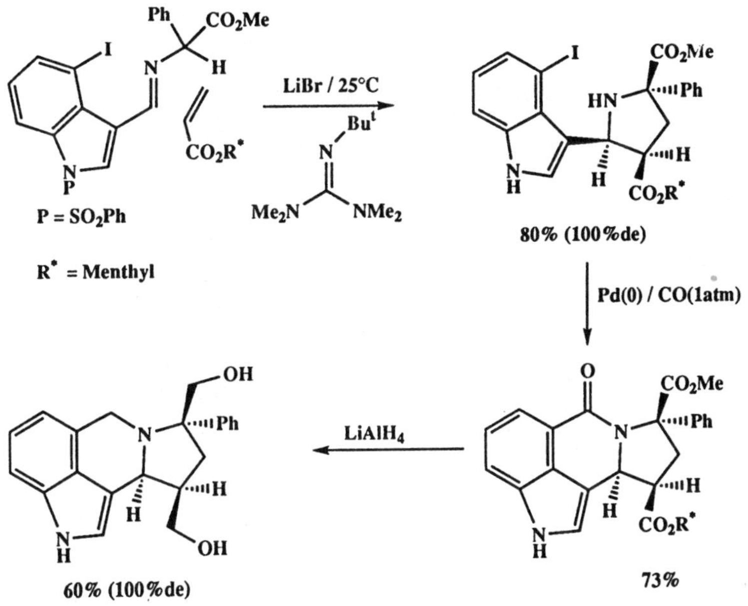

Scheme 4.18.

to terminate with a retro Diels–Alder reaction furnishing unsubstituted triazoles (Scheme 4.21) [30]. In this cascade, two rings and four bonds are formed.

7 Palladium-catalysed cyclization and Diels–Alder reaction

Cascade protocols that provide access to the extensive and rich chemistry of the Diels–Alder reaction are particularly attractive. Our efforts in this area have initially concentrated on cases where a palladium-catalysed cascade precedes the Diels–Alder reaction. A typical example [10] involving cyclization onto an allene is shown in Scheme 4.22. The 6-*exo*-dig cyclization of the arylpalladium(II) intermediate generates a π-allylpalladium(II) species that, in the absence of a suitable nucleophile, undergoes a β-hydride elimination to generate the amino-activated 1,3-diene required for the Diels–Alder reaction.

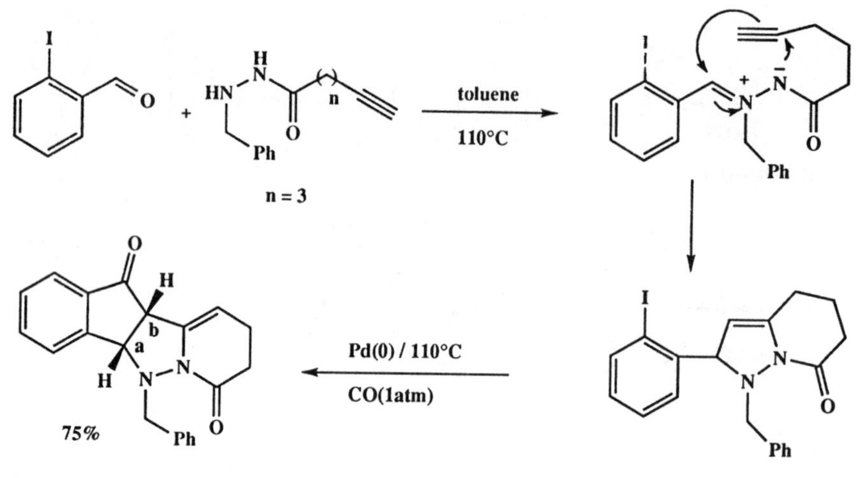

Scheme 4.19.

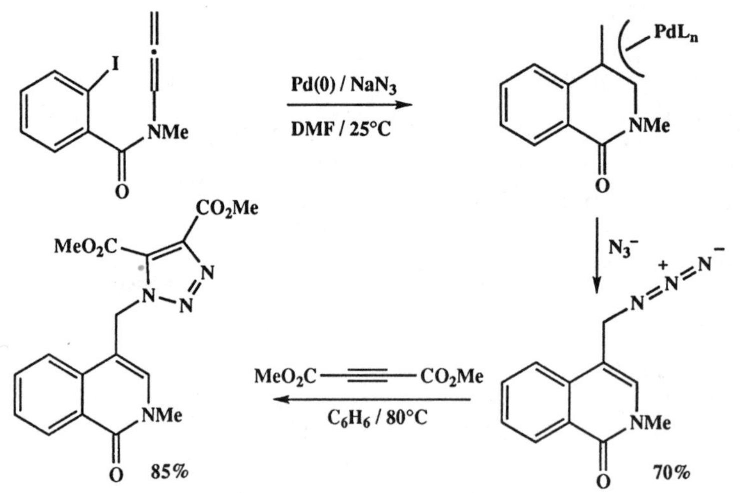

Scheme 4.20.

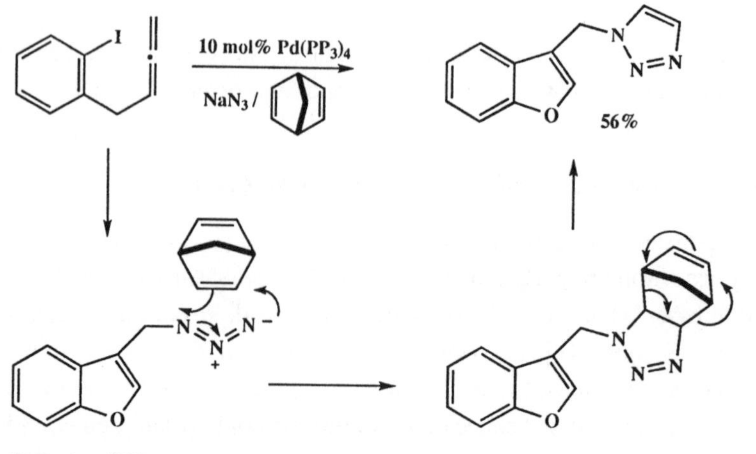

Scheme 4.21.

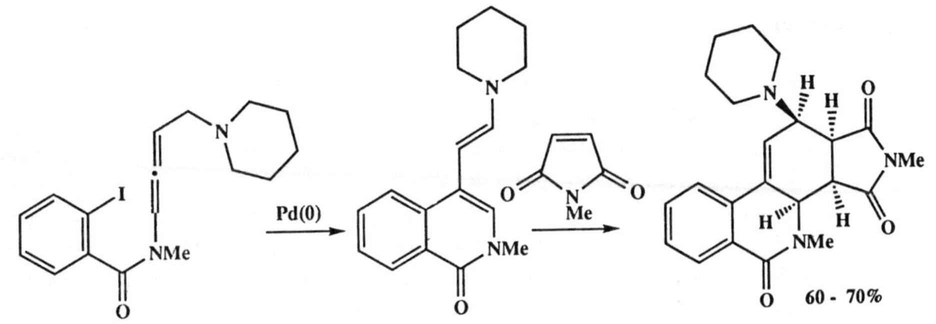

Scheme 4.22.

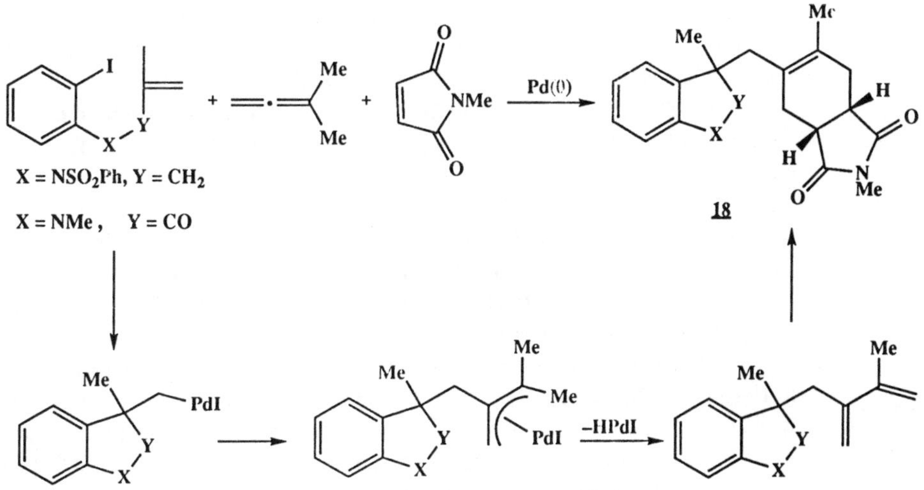

Scheme 4.23.

A series of three-component cascade processes have been developed recently (Scheme 4.23) [31]. These processes also proceed via a π-allylpalladium(II) intermediate and require an allene with a substituent capable of participating in a β-hydride elimination. The products **18a** and **18b** are obtained in 72% yield as 1.7:1 and 1.5:1 mixtures of diastereomers, respectively.

8 Summary

The development of palladium-catalysed cascade processes incorporating a wide and diverse range of anion capture agents and relay switches provides a wealth of novel chemistry for creating heterocycles and carbocycles and maximizing their molecular complexity. The processes are capable of much further development and, by interfacing with other powerful synthetic processes such as cycloaddition reactions, enable the creation of a wide range of structural motifs and the incorporation of diverse bond-forming processes.

9 Acknowledgments

We thank the EPSRC, Leeds University, Organon, Pfizer, Rhône-Poulenc Rorer, Roussel, SmithKline Beecham and Zeneca for support.

10 References

1 Tsuji J. *Palladium Reagents and Catalysts*, Chichester: Wiley, 1995.

2 Soderberg BC. In: Hegedus LS, ed. *Comprehensive Organometallic Chemistry II*, Vol. 12. Amsterdam: Elsevier Science, 1995: 260–87; de Meijere A, Meyer FE. *Angew Chem Int Ed Engl* 1994; **33**: 2379–411; Gibson SE, Middleton RJ. *Contemporary Organic Synthesis*, 1996: 447–71.

3 Grigg R, Sridharan V, Stevenson P, Worakun T. *J Chem Soc Chem Commun* 1986: 1697–9; Grigg R, Sridharan V, Stevenson P, Sukirthalingam S. *Tetrahedron* 1989; **45**: 3557–68; Grigg R, Sridharan V, Stevenson P, Sukirthalingam S, Worakun T. *Tetrahedron* 1990; **46**: 4003–18; Grigg R, Santhakumar S, Sridharan V *et al. Tetrahedron* 1991; **46**: 9703–20.

4 Abelman MM, Oh T, Overman LE. *J Org Chem* 1987; **52**: 4130–33; Abelman MM, Overman LE, Tran VD. *J Am Chem Soc* 1990; **112**: 6959–94; Overman LE. *Pure Appl Chem* 1994; **66**: 1423–30.

5 Negishi EI, Zhang Y, O'Connor B. *Tetrahedron Lett* 1988; **29**: 2915–18; Negishi EI, Nguyen T, O'Connor B. *Heterocycles* 1989; **28**: 55–8; Larock RC, Song H, Baker BE, Gong WH. *Tetrahedron Lett* 1988; **29**: 2919–22; Young WS, Masters JJ, Danishefsky S. *J Am Chem Soc* 1995; **117**: 5228–34; Tietze LF, Schirok H. *Angew Chem Int Ed Engl* 1997; **36**: 1124–5; Rigby JH, Hughes RC, Hee MJ. *J Am Chem Soc* 1995; **117**: 7834–5; Negishi EI, Coperet C, Ma S, Liou SY, Liu F. *Chem Rev* 1996; **96**: 365–93.

6 Jeffery T. *Tetrahedron* 1996; **52**: 10113–30.

7 Karabelas K, Westerlund C, Hallberg A. *J Org Chem* 1985; **50**: 3896–900; Karabelas K, Hallberg A. *J Org Chem* 1986; **51**: 5286–90, 1989; **54**: 1773–6; Abelman MM, Overman LE. *J Am Chem Soc* 1988; **110**: 2328–9; Larock RC, Gong WH. *J Org Chem* 1989; **54**: 2047–50; Nilsson K, Hallberg A. *J Org Chem* 1990; **55**: 2464–70; Jeffrey T. *Tetrahedron Lett* 1992; **33**: 1989–92.

8 Grigg R, Loganathan V, Sukirthalingam S, Sridharan V. *Tetrahedron Lett* 1990; **31**: 6573–6; Grigg R, Loganathan V, Santhakumar V, Sridharan V, Teasdale A. 1991; **32**: 687–90; Carfagna C, Musco A, Sallese G, Santi R, Fiorani G. *J Org Chem* 1991; **56**: 261–3; Cabri W, Candiani I, Bedeschi A, Senti R. *Tetrahedron Lett* 1991; **32**: 1753–6; Jeffrey T. *Tetrahedron Lett* 1992; **33**: 1989–92; Ripa L, Hallberg A. *J Org Chem* 1996; **61**: 7147–55.

9 Chida N, Ohtsuki N, Ogawa S. *Tetrahedron Lett* 1991; **32**: 4525–8; McIntosh MC, Weinreb SM. *J Org Chem* 1993; **58**: 4823–32; Hudlicky T, Olivo HF. *J Am Chem Soc* 1992; **114**: 9694–6.

10 Grigg R, Loganathan V, Sridharan V *et al. Tetrahedron* 1996; **52**: 11479–502.

11 Grigg R, Sansano JM, Santhakumar V *et al. Tetrahedron* 1997; **53**: 11803–26.

12 Burns B, Grigg R, Santhakumar V *et al. Tetrahedron* 1992; **48**: 7297–320.

13 Grigg R, Rasul R, Redpath J, Wilson D. *Tetrahedron Lett* 1996; **37**: 4609–12.

14 Tsuji J, Mandai T. *Angew Chem Int Ed Engl* 1995; **34**: 2589–612.

15 Oppolzer W, Pimm A, Stammen B, Hulme W. *Helv Chim Acta* 1997; **80**: 623–39.

16 Grigg R, Rasul R, Savic V. *Tetrahedron Lett* 1997; **38**: 1825–8.

17 Casaschi A, Grigg R, Sansano JM, Wilson D, Redpath J. *Tetrahedron Lett* 1996; **37**: 4413–16.

18 Grigg R, Sridharan V, Xu L-H. *J Chem Soc Chem Commun* 1995: 1903–4.

19 Grigg R, Sridharan V. *Tetrahedron Lett* 1994; **34**: 7471–4; Brown S, Clarkson S, Grigg R, Sridharan V. *J Chem Soc Chem Commun* 1995: 1135–6; Grigg R, Putnikovic B, Urch C. *Tetrahedron Lett* 1996; **37**: 695–8.

20 For reviews, see: Stille JK. In: Trost BM, Fleming I, Semmelhack MF, eds, *Comprehensive Organic Synthesis*, Vol. 4. Oxford: Pergamon Press, 1991: 913–31; Ojima I, Eguchi M, Yzamarioudaki M. In: Abel EW, Stone FG, Wilkinson G, Hegedus L, eds. *Comprehensive Organometallic Chemistry II*, Vol. 12. Oxford: Pergamon Press, 1995: 27–31.

21 Schäfer A, Benz H, Fielder W *et al.* In: Cordell GA, Brossi A, ed. *The Alkaloids*, Vol. 45. New York: Academic Press, 1994: 1–125.

22 Grigg R, Redpath J, Sridharan V, Wilson D. *Tetrahedron Lett* 1994; **35**: 7661–4.

23 Grigg R, Brown S, Sridharan V, Uttley MD. *Tetrahedron Lett* 1997; **38**: 5031–4.

24 Grigg R, Pratt R. *Tetrahedron Lett* 1997; **38**: 4489–92.

25 Burwood M, Davies B, Diaz, I *et al. Tetrahedron Lett* 1995; **36**: 9053–6.

26 Staudinger H. *Liebigs Ann Chem* 1907; **356**: 51–93.
27 Grigg R, Sridharan V. In: Curran DP, ed. *Advances in Cycloaddition*, Vol. 3. London, JAI Press, 1993: 161–204.
28 Grigg R, Sridharan V, Suganthan S, Bridge AW. *Tetrahedron* 1995; **51**: 295–306.
29 Barr DA, Dorrity MJ, Grigg R *et al. Tetrahedron* 1995; **51**: 273–94.
30 Gardiner M, Grigg R, Vicker N, Sridharan V. *Tetrahedron Lett* 1998; **39**: 435–8.
31 Grigg R, Brown S, Sridharan V, Uttley MD. *Tetrahedron Lett* (In press.)

5 Palladium-catalysed Cross-coupling Reactions: a Personalized Account

ARMIN DE MEIJERE and STEFAN BRÄSE

Institut für Organische Chemie der Georg-August Universität Göttingen, Tammannstrasse 2, 37077 Göttingen, Germany

1 Introduction

Transition-metal-catalysed reactions have gained steadily increasing importance in the last decade. Fine tuning of reaction parameters of known or newly discovered metal-catalysed transformations have not only had an important impact on the synthesis of natural and non-natural biologically active compounds as well as theoretically interesting molecules, but also on process development for valuable intermediates in the pharmaceutical and agrochemical industry as well as on the research towards new materials. Among these processes, palladium-catalysed transformations have played a leading role in developing and understanding totally new reaction types. Two of the most general and widely used palladium-catalysed processes are the arylation and alkenylation of alkenes, generally known as the Heck reaction [1], and the nucleophilic substitution of allylic substrates [2]. This account will focus on new applications of Heck reactions and mechanistically related cross-coupling reactions from the authors' own laboratory, and only briefly point out the context of other groups' most important developments within the full spectrum of palladium-catalysed reactions, including Suzuki [3], Stille [4], Sonogashira–Stephens–Castro [5], Wacker [6], Trost cycloisomerization [7], pallada-ene [8] and related reaction types.

The scope of metal-catalysed reactions often can be broadened by repeating the same reaction type several times or combining it with a different type of metal-catalysed or even a simple thermal reaction in a domino fashion [9]. Such strategies can be subdivided into: multiple cross-coupling reactions, i.e. iterative reactions of the same type with the same reagent; domino reactions (also called 'cascade reactions' [10]), i.e. sequential reactions with several steps of the same mechanistic type but with a changing substrate; and domino reactions (also called cascade reactions [10]), i.e. sequential reactions with several steps of two or more different reaction types. Such sequential processes offer a wide range of possibilities for the efficient construction of highly complex molecules in a single procedural step, frequently with enhanced regio-, diastereo- and even enantioselectivity for the overall transformation.

2 Heck-type reactions: past–present–future

2.1 *The mechanism*

The palladium-catalysed arylation and alkenylation of alkenes, discovered independently by Mizoroki *et al.* in Japan [11] and by Heck *et al.* in the USA [12] around 1970—now generally called the Heck reaction—has become one of the most frequently applied metal-catalysed C—C bond-forming processes in the last 10–15 years and has therefore attracted a great deal of interest [13, 14]. The rapid development of new and vastly improved reaction protocols, the discovery of diastereoselective and—since 1989—even ligand-induced enantioselective couplings [15] has made it possible to apply the Heck reaction in elegant syntheses of various biologically active compounds. Recent work has also led to a deeper insight into the key steps of the widely accepted mechanism of this reaction [16]. A catalytically active coordinatively unsaturated Pd(0) complex is mostly generated *in situ* from a stable palladium complex or salt such as Pd(OAc)$_2$ (step A in Scheme 5.1). After oxidative addition to the 16-electron complex **6** of an aryl or alkenyl halide or sulfonate **1** (C) (see [17] for details) or even an alkyl halide [18], an electrophilic Pd(II) complex **7** is formed. Exchange of one of the ligands with an alkene **2** results in a π-alkene complex **8** (D), which reacts to give a σ-alkyl complex by *syn* addition of the Pd–R bond (E). This complex normally undergoes a *syn* β-hydride elimination (H)—after internal rotation around the C–C bond to place a hydrogen *syn* to the palladium residue—to give the most stable π-alkene complex **10**. However, with certain structural prerequisites fulfilled, such an intermediate **9** can react again in a sequence of coordination and carbopalladation, and this can go on several times before eventually a β-hydride elimination occurs (Section 2.3). Yet another possibility is that complex **9** undergoes rearrangement before any further reaction (Section 2.7). The β-hydride elimination is reversible, and readdition of [HPdX] followed by re-elimination can lead to double bond migration [19] or/and hydrogen scrambling [20]. The new alkene complex **10** can release the alkene product **3** and give the hydridopalla-

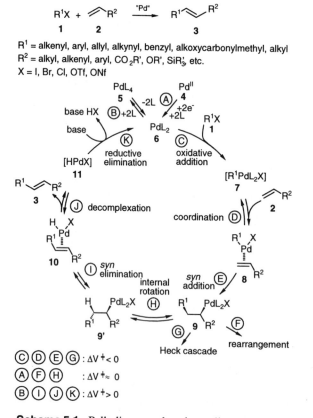

Scheme 5.1. Palladium-catalysed coupling of aryl and alkenyl halides: mechanism and opportunities.

dium species [HPdX] **11**, which eventually undergoes a base-initiated reductive elimination. The catalytically active palladium complex **6** thus reformed can coordinate another alkene and initiate the next catalytic cycle.

2.2　*Rate enhancement under high pressure*

It is generally assumed that the rate-determining step in the catalytic cycle of the Heck reaction usually is the oxidative addition (step C in Scheme 5.1). This is in line with the low reaction rate of aryl and alkenyl chlorides, which less readily undergo oxidative addition than bromides and iodides. Chlorides can be brought to react at elevated temperatures with the use of sufficiently stable catalysts [21] or more nucleophilic trialkylphosphane ligands [22]. It was also found that the application of high pressure (2–10 kbar) enhances the rate of the Heck reaction in general, and thus high pressure can be applied to couple alkenyl and aryl chlorides at ordinary temperatures with better yields (Scheme 5.2) [23].

By simple analogy with other types of reactions it can be foreseen that steps C, D, E and K in the catalytic cycle should be accelerated, whereas steps B, I and J should be retarded by high pressure [24, 25].

A more detailed kinetic study monitoring the Heck reactions under high pressure by on-line FTIR spectroscopy revealed that alkenyl iodides have a lower activation entropy, activation volume and rate coefficient than aryl iodides. Interestingly, the reaction rates of aryl bromides depend more strongly on the pressure than those of aryl iodides, and hence can be forced to react more efficiently by increased pressure rather than elevated temperature (Fig. 5.1) [26].

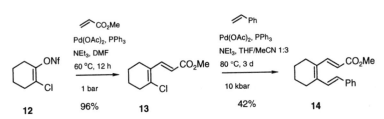

Scheme 5.2. Rate enhancement of the Heck reaction by high pressure.

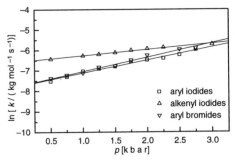

Figure 5.1. Pressure dependence of the overall rate coefficients for the Heck reactions of different organyl halides. (From [26].)

2.3   *Multiple cross-couplings: synthesis of theoretically interesting molecules*

The application of palladium-catalysed cross-coupling reactions on oligohaloalkenes and -arenes can most elegantly and simply lead to highly substituted carbo- and heterocyclic systems [1,13]. By far the shortest and most efficient route to symmetrically substituted dibenzoannelated [2.2]paracyclophanedienes of type **18** is the one that applies a fourfold Heck reaction on 1,2,9,10-tetrabromo[2.2]paracyclophanediene (**15**) with styrene and various substituted styrenes **16**, with subsequent 6π-electrocyclization and aromatization (Scheme 5.3) [27]. It is important to note that good yields in these multiple Heck coupling reactions were only achieved under the modified conditions of Jeffery [28], employing a base like potassium carbonate and a phase-transfer catalyst (a quaternary ammonium salt) rather than the classical Heck conditions (tertiary amine base in the presence of a phosphane) [29].

Under analogous conditions, the readily available 4,5,12,13- and 4,7,12,15-tetrabromo[2.2]paracyclophanes were coupled with styrenes to yield the hydrocarbons **19** and **20** (7–70% isolated), with phenylacetylene under palladium–copper co-catalysis (Yamamoto protocol) [30] to give the tetraalkynylated [2.2]paracyclophane derivative **21** (70%) and with phenylmagnesium bromide to yield 4,7,12,15-tetraphenyl[2.2]paracyclophane (**22**), albeit in low yield (6%) [31]. The tetraalkynyl-substituted compound **21** follows in the footsteps of the hexakisalkynylbenzene derivatives **23** of Moroz *et al.* [32], Vollhardt *et al.* [33], Praefcke *et al.* [34] and Heck *et al.* [29]. Various 1,2- 1,3-, 1,4-di-, 1,2,3-, 1,3,5-tri- and 1,2,4,5-tetrasubstituted benzene derivatives, such as compounds **24**, **25** and **26**, were also prepared by this protocol (Fig. 5.2) [35, 36].

Surprisingly, the twofold alkenylations of *cis*-1,2-dibromoethene, 1,2-dibromocyclo-pentene (**27**) and 1,2-dibromocyclohexene (**28**) proceed with better yields under the classical Heck conditions (see Scheme 5.4) [37]. The resulting (*E*,*Z*,*E*)-hexatrienes **29** and **30** reasonably cleanly undergo 6π-electrocyclizations upon heating to 130–150 °C in an inert solvent such as di-*n*-butyl ether or xylene in the absence of oxygen to give the ring-annelated *cis*-5,6-disubstituted cyclohexadienes **31** and **32**. In the presence of oxygen these products are easily dehydrogenated to the corresponding ring-annelated benzene derivatives.

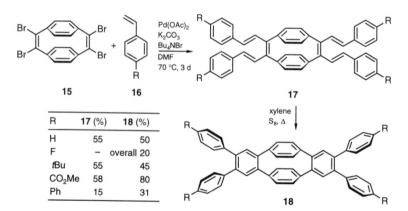

R	17 (%)	18 (%)
H	55	50
F	–	overall 20
tBu	55	45
CO₂Me	58	80
Ph	15	31

Scheme 5.3. 1,2:9,10-Dibenzo[2.2]paracyclophanedienes by fourfold Heck reactions followed by 6π-electrocyclizations.

The central tetrasubstituted and thereby most nucleophilic double bonds in trienes **29** and **30** can be epoxidized selectively with various reagents. The resulting dialkenylepoxides **35** and **36** are set up for a Cope rearrangement. But only the six-membered ring derivatives **36** equilibrate at elevated temperatures (60–80 °C) via Cope rearrangement with the

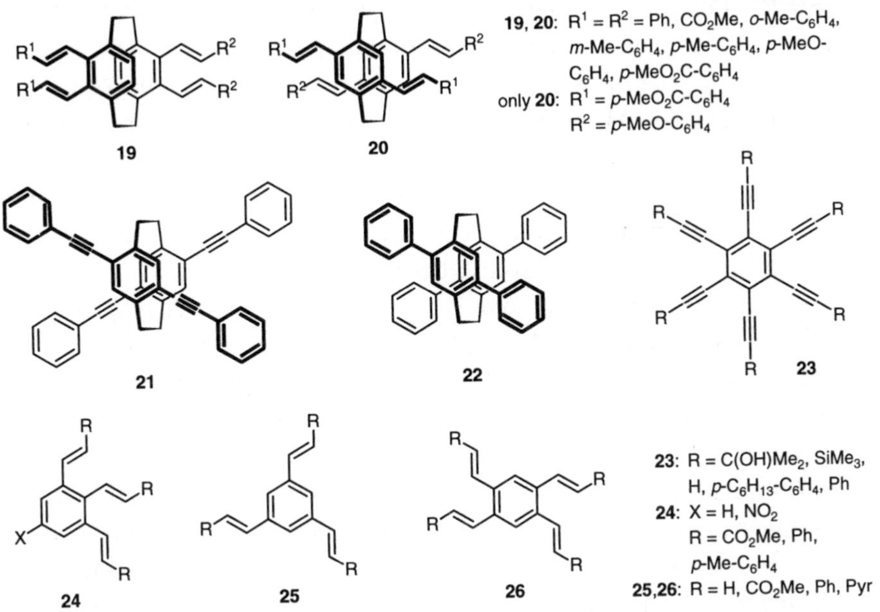

19, 20: $R^1 = R^2 = Ph$, CO_2Me, o-Me-C_6H_4, m-Me-C_6H_4, p-Me-C_6H_4, p-MeO-C_6H_4, p-MeO$_2$C-C_6H_4

only **20:** $R^1 = p$-MeO$_2$C-C_6H_4
$R^2 = p$-MeO-C_6H_4

23: $R = C(OH)Me_2$, $SiMe_3$, H, p-C_6H_{13}-C_6H_4, Ph
24: $X = H$, NO_2
$R = CO_2Me$, Ph, p-Me-C_6H_4
25,26: $R = H$, CO_2Me, Ph, Pyr

Figure 5.2. Highly alkenylated and alkynylated benzene derivatives by palladium-catalysed multifold cross-coupling reactions.

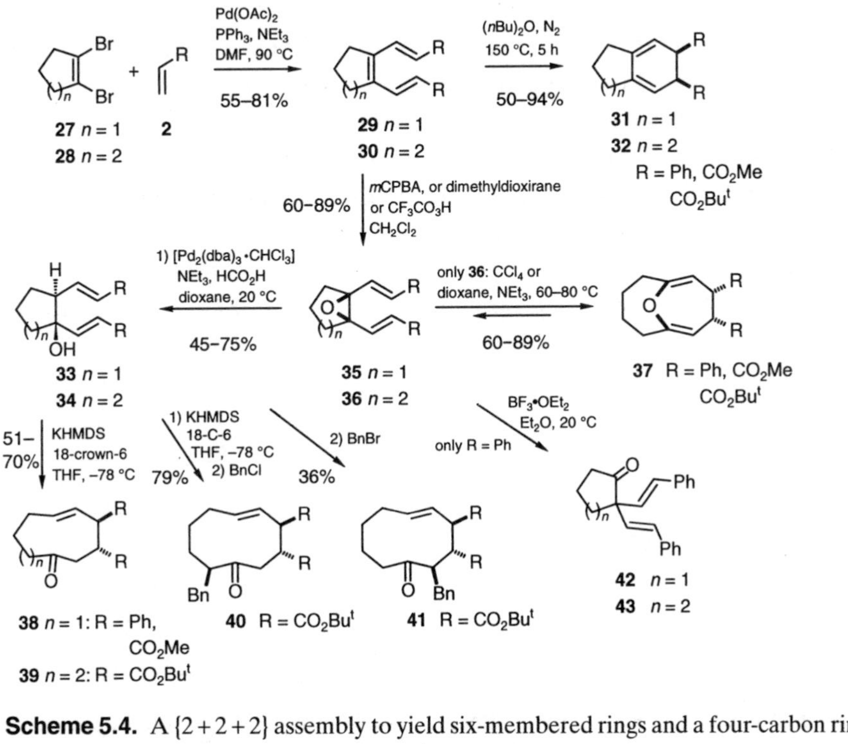

Scheme 5.4. A {2 + 2 + 2} assembly to yield six-membered rings and a four-carbon ring-enlargement methodology.

oxygen-bridged cyclodeca-1,5-dienes **37**, which are strained bridge-head dienes. The five-membered ring derivatives **35** do not rearrange to the corresponding more highly strained oxygen-bridged nonadienes, but undergo an irreversible acid-catalysed 1,2-alkenyl shift to the cyclopentanone derivatives **42** when heated for extended times [37]. The same rearrangement of compounds **35** and **36** (R = Ph) can be brought about with a drop of acid (e.g. $BF_3 \cdot OEt_2$) at room temperature to yield compounds **42** and **43**. On the other hand, both types of epoxides can be ring-opened selectively by palladium-catalysed reduction according to the protocol developed by Tsuji and Shimizu et al. [38] for vinyloxiranes. The resulting alcohols **33** and **34**, upon deprotonation at −78 °C, undergo an oxyanion-accelerated [39] Cope rearrangement to give the trans-3,4-disubstituted trans-cyclonon-5-en-1-ones (**38**) and trans-cyclodec-5-en-1-ones **39**, respectively. The intermediate enolates in this rearrangement can be trapped with other electrophiles such as benzyl halides. Surprisingly, with the less reactive benzyl chloride, only the 8-benzyl derivative **40** (structure and configuration proved by X-ray crystal structure analysis) is obtained in high yield (79%). The more reactive benzyl bromide, however, traps the initially formed enolate to give the 10-benzyl derivative **41** (structure also proved by X-ray structure analysis), albeit in lower yield (36%), and accompanied by the 8-benzyl derivative (31% yield) (P. von Zezschwitz, K. Voigt and A. de Meijere, unpubl. data) [42].

Essentially, an enantioselective variant for the preparation of such highly substituted cyclononenones and cyclodecenones **38–41** can be developed, because the enantioselective desymmetrization of the dialkenylepoxides **35** and **36** by palladium-catalysed reduction can be effected in the presence of chiral ligands. So far, a β-(naphthylphenyl)phosphino-carboxylate ligand derived from tert-butyl myrtenate [40] with [$Pd_2(dba)_3 \cdot CHCl_3$] gave a ligand-induced enantiomeric excess of 15%; with (R)-PROPHOS [41] as a ligand the enantiomeric excess was raised to 36%. An optimization of this procedure is under way [42].

The scope of the twofold coupling to give 1,3,5-hexatrienes could be extended by applying the 1-bromo-2-(trifluoromethanesulfonyloxy)cyclohexene (**45**), which was easily prepared from 2-bromocyclohexanone (**44**) (Scheme 5.5). Favorably, the yield in the twofold Heck coupling of compound **45** with t-butyl acrylate was considerably higher than that obtained from 1,2-dibromocyclohexene (**28**). In addition, a sequence of a Stille coupling with a vinylstannane and a Heck reaction with an alkene can be applied to compound **45** in a one-pot operation to give unsymmetrically 1,6-disubstituted 1,3,5-hexatrienes such as compound **46** [42]. With appropriate nucleophiles, the 1,6-bis(alkoxycarbonyl)-substituted

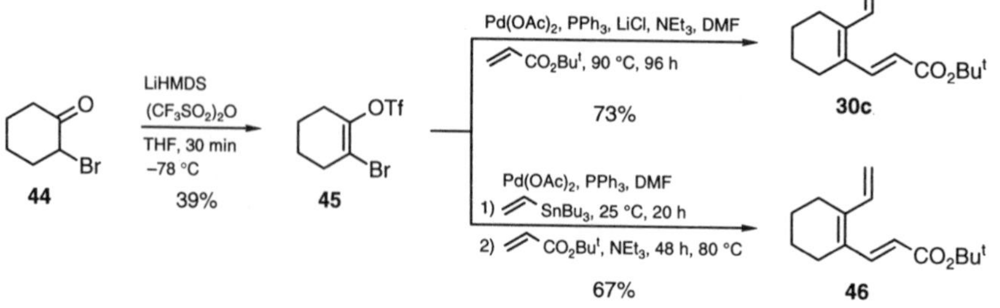

Scheme 5.5. A route to unsymmetrically 1,6-disubstituted 1,3,5-hexatrienes such as compound **46** from 2-bromocyclohexanone (**44**).

hexatrienes **30** undergo a completely diastereoselective domino–Michael reaction, e.g. with ammonia equivalents, to yield protected five-membered ring β-amino acids that are *cis*-pentacin analogs. The analogous domino–Michael reactions to give the corresponding indane derivatives **51** have been carried out with the twofold coupling products **48** of *o*-dibromobenzene with acrylates (Scheme 5.6) [43].

Five-membered ring closure has also been observed under palladium catalysis when *o*-halostyrene derivatives **50** were coupled with alkenes [44]. Apparently an intra-molecular carbopalladation with 5-*exo*-trig ring closure can favorably compete with β-hydride elimination in the intermediate β-(*o*-ethenylphenyl)ethylpalladium halide **54**. This reaction mode for compound **50** is observed especially under Jeffery conditions when R^3 in compound **54** is H, Me or OR. Under the same conditions, however, *o*-dibromobenzene **47** gives very high yields of *o*-dialkenylbenzene derivatives **48** [35, 43]. The *o*-dialkenylbenzenes **48** and **49** can also be cyclized to indane derivatives **52** under Pd(II) catalysis (Wacker-type conditions) [44].

Yet another cyclization was observed for *o*-bromostilbenes **55**, which actually competes with the second coupling step for *o*-dibromobenzene when only one equivalent of styrene is used. *o*-Bromostilbene **55a** and substituted analogs **55b–g**, prepared from *o*-bromobenzaldehydes by Wittig–Horner–Emmons olefination, undergo efficient cyclodimerization to (*E/Z*)-9,10-dibenzylidene-9,10-dihydroanthracenes **56a–g** (Scheme 5.7). The (*Z*)-diastereomers of the parent compound **56a**—characterized by an X-ray crystal structure analysis—and its dimethyl derivative **56b** preferentially crystallize from the crude mixtures, while the (*E*)-diastereomers could never be obtained pure in crystalline form. When heated to over 120 °C in solution, the (*Z*)-form isomerizes to the (*E*)-form and back [45].

Sixfold Heck coupling of hexabromobenzene **57** with styrene and substituted styrenes under conditions of the Jeffery protocol readily occurred [46]. The products obtained in high yields all showed a dominant molecular ion peak for the correct mass in the mass spectrum; however, according to the ^{1}H- and ^{13}C-NMR spectra, as well as an HPLC analysis, they were hard to separate from mixtures of a large number of isomers, apparently

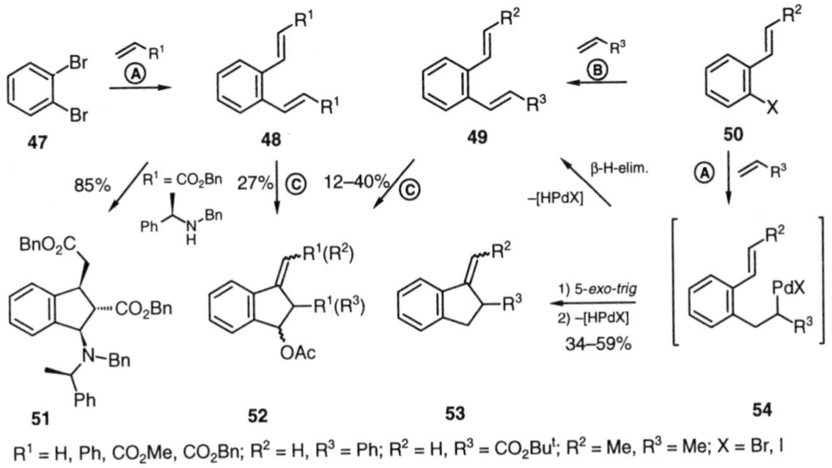

R^1 = H, Ph, CO$_2$Me, CO$_2$Bn; R^2 = H, R^3 = Ph; R^2 = H, R^3 = CO$_2$Bu^t; R^2 = Me, R^3 = Me; X = Br, I

Scheme 5.6. Vicinal dialkenylbenzene derivatives and their follow-up reactions: (A) Pd(OAc)$_2$, K$_2$CO$_3$ (or KHCO$_3$), LiCl, Bu$_4$NBr, DMF (or NMP), 60–100 °C; (B) Pd(OAc)$_2$, PPh$_3$ [or P(*o*-Tol)$_3$], NEt$_3$, MeCN (or NMP), 60–100 °C; (C) Pd(OAc)$_2$, benzoquinone, MnO$_2$, HOAc, 40–50 °C.

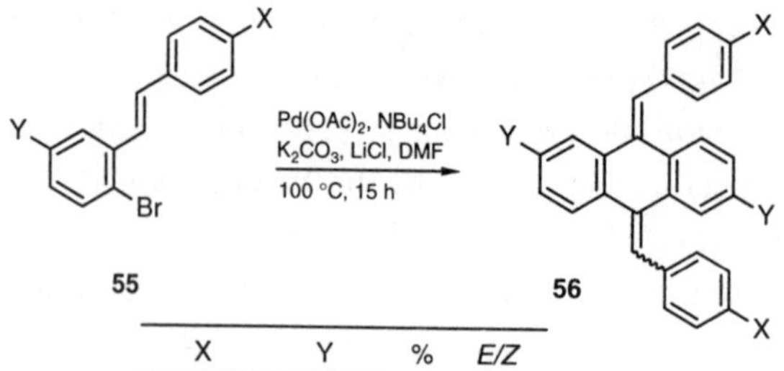

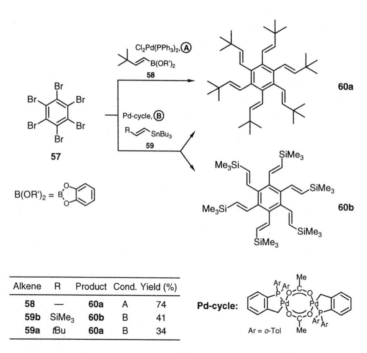

Scheme 5.7. Cyclodimerization of *o*-bromostilbenes to 9,10-dibenzylidene-9,10-dihydroanthracenes.

	X	Y	%	E/Z
a	H	H	80	1:2.9
b	Me	H	75	1:2
c	OMe	H	70	1:2
d	CO₂Me	H	72	1:1
e	NO₂	H	56	1.8:1
f	H	CO₂Et	78	~1:1
g	OMe	CO₂Et	64	~1:1

Alkene	R	Product	Cond.	Yield (%)
58	—	60a	A	74
59b	SiMe₃	60b	B	41
59a	tBu	60a	B	34

Scheme 5.8. Sixfold Suzuki and Stille couplings of hexabromobenzene: (A) [PdCl₂(PPh₃)₂], NaOH, toluene/THF (1:1), 100 °C, 24 h, (B) Pd-cycle, toluene, 100–120 °C, 1–4 days. (From [49].)

formed by additional intramolecular cyclization of intermediates of type **54** [47,48]. Because such σ-alkyl intermediates are not formed in Suzuki- and Stille-type couplings, the sixfold coupling reaction of compound **57** with the alkenylboronate **58** prepared by hydroboration of *tert*-butylacetylene with catecholborane worked marvelously to give the C_6-symmetric hydrocarbon **60a** in analytically pure crystalline form in up to 74% yield (Scheme 5.8) [49]. It turned out that the *t*-butyl groups in compound **58** are essential for the success; other alkenylboronates, except for the adamantyl-substituted one (6% yield), did not give any of the corresponding sixfold coupling products. An interesting self-organizing effect associated with the bulky alkyl groups and van der Waals' attractive interactions

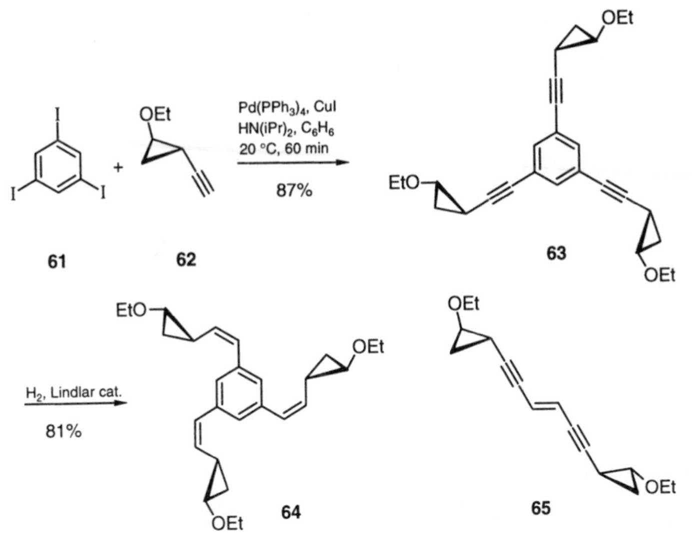

Scheme 5.9. Multifold Sonogashira–Hagihara coupling reactions.

may play an important role in these couplings. Compound **60a** as well as the analogous hexakis(trimethylsilylethenyl)benzene **60b** could be obtained by sixfold Stille coupling of compound **57** with the corresponding alkenylstannanes **59a** and **59b**. Compounds **60a** and **60b** are cup-shaped molecules with six arms rotated about 50° out of the plane of the central ring, all to one side [49].

Multifold coupling reactions of oligohaloarenes with alkynes under palladium–copper co-catalysis (Sonogashira–Hagihara protocol) apparently are no problem, as impressively demonstrated for hexabromobenzene with various alkynes [33] and even butadiynes [50]. The threefold coupling even of the enantiomerically pure ethoxyethynylcyclopropane **62** with 1,3,5-triiodobenzene (**61**) therefore went smoothly and led to the C_3-symmetric product **63** (87% yield) (Scheme 5.9). Partial hydrogenation of the triple bonds furnished the final product, an interesting model compound for chiral discotic liquid crystals [51]. The analogous twofold intermolecular cross-coupling of ethynylcyclopropane **62** with 1,4-diiodobenzene proceeded with 71% yield, while that with cis- and trans-dibromoethene gave modest yields (24 and 19%, respectively); interestingly, however, the cross-coupling of the chlorozinc derivative prepared from compound **62** by lithiation and metal–metal exchange, with the mixture of cis- and trans-1,2-dibromoethene gave a high yield (76%) of only the trans-configurated enediyne **65** with two terminal 2-ethoxycyclopropyl substituents [51]. Terminally substituted vinylcyclopropanes are also accessible by hydroalumination of ethynylcyclopropanes followed by palladium-catalysed cross-coupling with iodoarenes, for example [52].

Even highly strained cyclopropenes, which are known to undergo various transition-metal-catalysed co-oligomerization reactions [53], can be cross-coupled with aryl, alkenyl and alkynyl halides under Pd or Pd–Cu catalysis via zinc and tin derivatives. The 1,2-bis(trimethylstannyl)-3,3-dimethylcyclopropene (**66**) undergoes a twofold Stille coupling with iodobenzene (**67**) to give diphenylcyclopropene **68**, while the chlorozinc derivative **70** could be coupled with haloalkenes such as 1,2-dibromoethene to yield 1-alkenylcyclopropenes such as compound **69**, as well as with haloalkynes to yield 1-alkynylcyclopropenes like compound **71** [54]. Bridge-head bromomagnesium- and

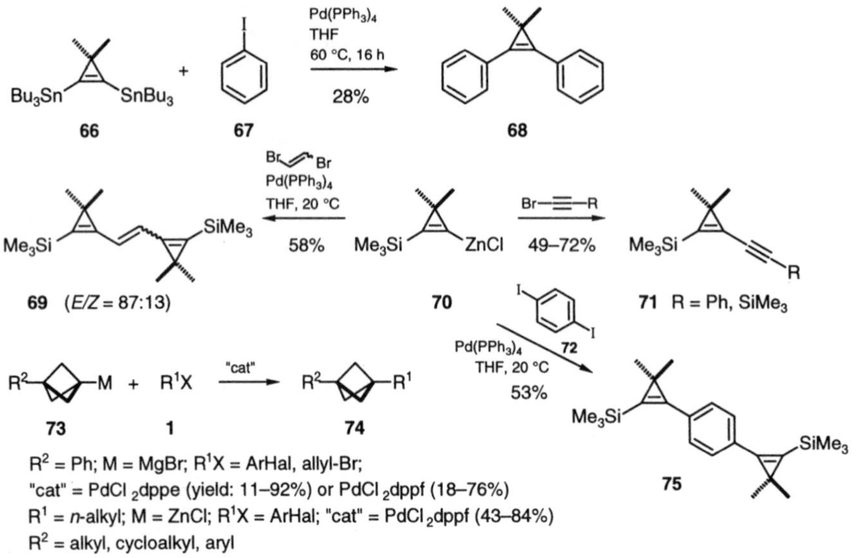

Scheme 5.10. Metallated cyclopropenes and bicyclo[1.1.1]pentanes as coupling partners.

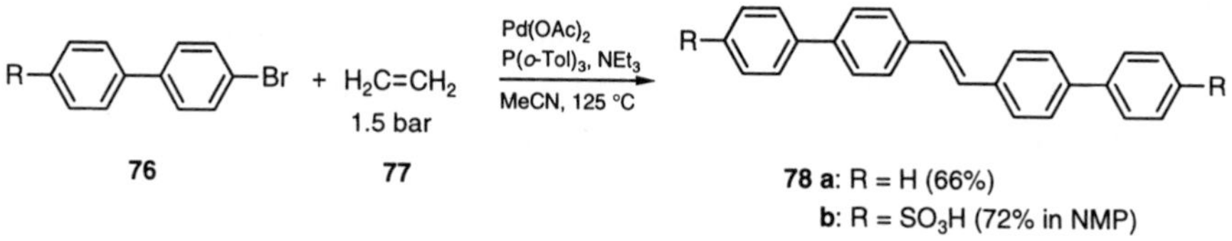

Scheme 5.11. Efficient twofold Heck couplings with ethene.

chlorozincbicyclo[1.1.1]pentane derivatives (**73-MgBr** and **73-ZnCl**, respectively) can be cross-coupled with aryl and alkyl halides under Pd(0) catalysis to give compounds of type **74**, some of which have quite interesting liquid crystalline properties (Scheme 5.10) [55].

As was pointed out by Heck *et al.* in one of their early publications [56], the palladium-catalysed alkenylation of aryl halides can be performed very well with ethene itself. The twofold Heck coupling of ethene (**77**) with 4-bromobiphenyl derivatives **76** can be employed favorably to prepare 4,4′-diarylstilbenes **78a** and **78b**, which are used as laser dyes (Scheme 5.11) [57].

2.4   *Palladacycle intermediates: obstacles and new opportunities*

Quite unusual domino-type multiple coupling reactions can be performed with certain cyclic alkene substrates: the first *syn*-carbopalladation leads to an alkylpalladium intermediate that cannot undergo a *syn*-β-hydride elimination. The σ-alkylpalladium intermediate then continues to react with another alkene, another alkenyl halide or aryl halide. The outcome of such reactions, which have been investigated thoroughly for the strained alkene norbornene by Chiusoli *et al.* [58], can be 2:1, 1:2 or even 1:3 coupling products of the alkene with an alkenyl halide or an aryl halide, respectively. Palladacycles such as compound **79**, which are formed by hydrogen halide elimination from the *syn*-carbopalladation product, and alkyldiarylpalladium(IV) halide intermediates of type **84** play a key role in these reactions. Under traditional Heck conditions, the 1:2 cross-

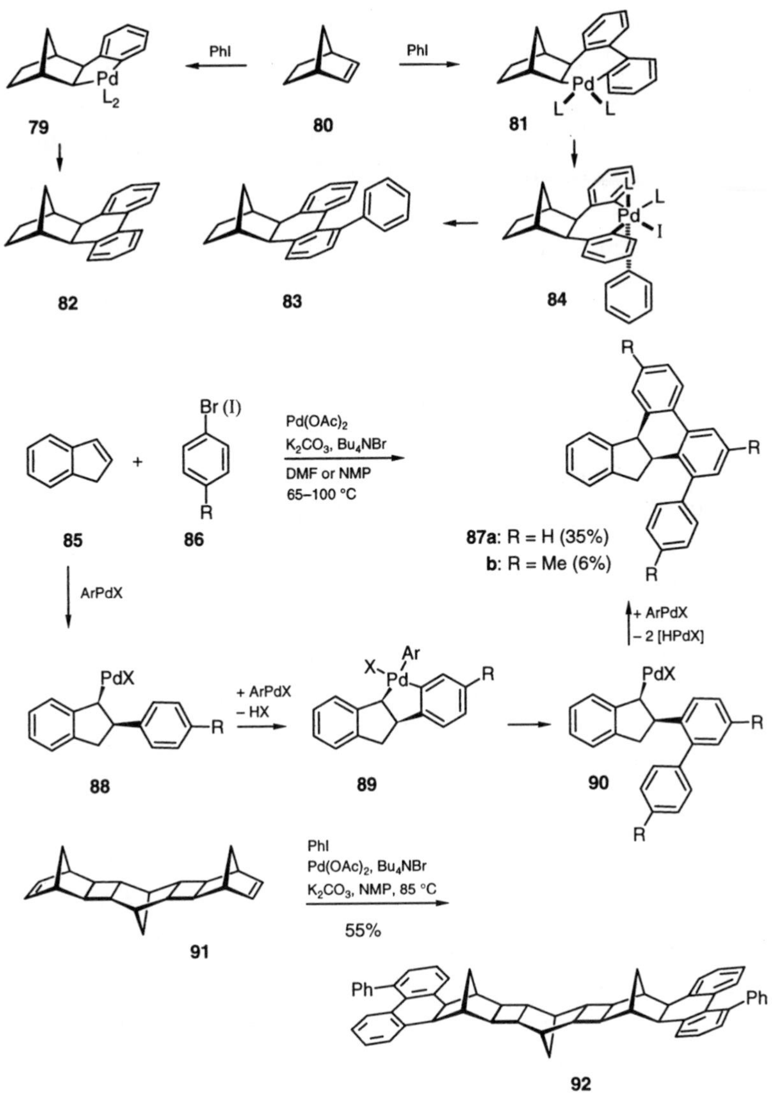

Scheme 5.12. A domino–Heck coupling of norbornene, indene and norbornadiene trimer involving an *ortho*-C–H activation.

coupling-cyclization product **82** is obtained [58], whereas under Jeffery conditions the 1:3 coupling product **83** is formed exclusively (Scheme 5.12) [59, 60].

Indene **85** undergoes *syn*-carbopalladation with arylpalladium halides regioselectively to give compound **88**, which reacts further via the palladacycles **89** and intermediates **90** to yield 1:3 coupling products **87** corresponding to the products of type **83** from norbornene, but differs in its regiochemistry [60]. This methodology can be used to annelate 5- or 8-substituted 9,10-dihydrophenanthrene units to strained cyclic alkenes of the indene and norbornene type. Thus, the extended norbornene-type hydrocarbon **91**, which is obtained by $Ni(COD)_2$-catalysed trimerization of norbornadiene, can be transformed to compound **92** by a 1:6 coupling with iodobenzene performed in a single operation with 55% yield [61].

In essence, these domino-coupling reactions form cyclohexadiene fragments from three two-carbon fragments. The 1:2 coupling of norbornene and iodobenzene discovered by

Chiusoli *et al.* [58] can also be adopted to couple norbornene with β-bromostyrene [62–64]. In an attempt to apply this palladium-catalysed {2 + 2 + 2} assembly for an alternative and more productive access to Hopf's trifoliaphane **95** [65], a 1:2 mixture of [2.2]paracyclophan-1-ene (**93**) and 1-bromo[2.2]paracyclophan-1-ene **94** was treated with palladium acetate under Jeffery conditions. Surprisingly, the expected dihydro derivative of compound **95** was not found at all, and compound **95** was obtained as a minor byproduct (2%) only. The main product was the hydrocarbon **98**, consisting of three [2.2]paracyclophane units linked by a common bicyclo[3.3.0]octene unit (Scheme 5.13) [66].

Apparently, the key intermediate **97** formed via the palladacycle **96**, an alkyldialkenyl-palladium(IV) species, preferentially undergoes a 5-*exo*-trig carbopalladation with subsequent formation of another palladacycle by *ortho*-attack on the neighboring aromatic ring, rather than 6-*endo*-trig carbopalladation to give the precursor to compound **95**. The tribenzoanalog of compound **98**, the interesting $C_{60}H_{38}$ hydrocarbon **99**, was obtained in 52% yield from 9,10-benzo[2.2]paracyclophan-1-ene and 9,10-benzo-1-bromo[2.2]paracyclophan-1-ene (1:2 ratio) under the same conditions [66]. Similar types of C–H activation have been observed by Dyker *et al.* in the Heck-type reactions of *o*-iodo-*tert*-butyl- or *o*-iodomethoxyarenes to give defined polycondensed oligomers [67].

It is interesting to note in this context that aryl–aryl homocouplings were observed as a side reaction under the conditions for the preparation of compounds **83, 87, 92**, etc. In fact, this side reaction has been elaborated further to an efficient biaryl synthesis from aryl halides in the absence of alkenes that does not require the unpleasant triphenylarsine addi-

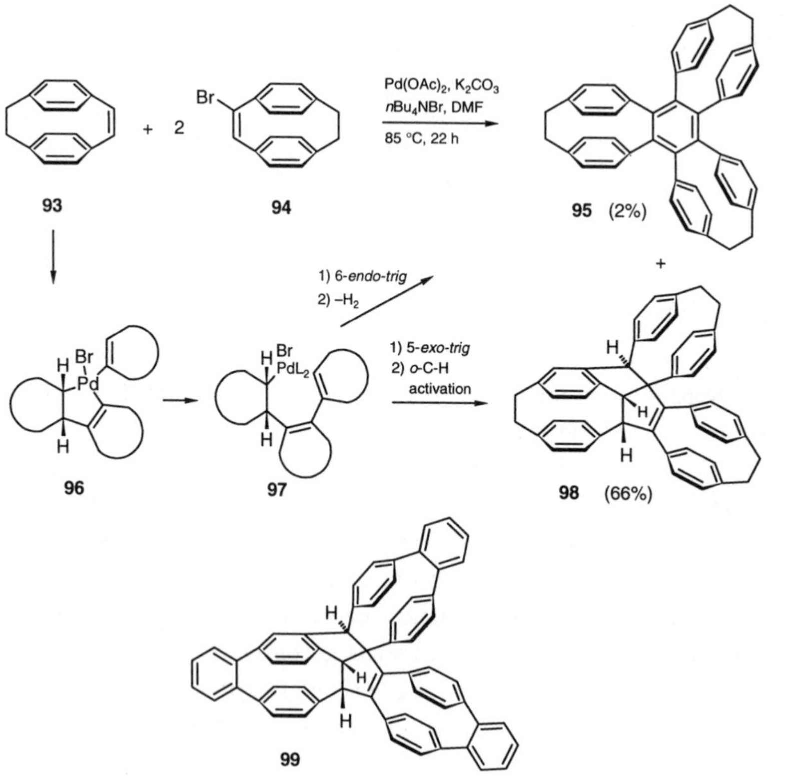

Scheme 5.13. Unusual hydrocarbons formed by 1:2 couplings of strained alkenes with strained alkenyl bromides.

tive as in a previously published procedure (Scheme 5.14) [68]. The advantage of this process, in comparison to the well-established Ullmann reaction in its various forms, is that with only catalytic amounts of palladium acetate used, the workup and purification of the products are easier (M. Weber, K. Albrecht and A. de Meijere, unpubl. data) [61, 69]. [The synthesis of biaryls by palladium-catalysed couplings of arylhalides with other substrates is a well-established process, e.g. with boronic acids (Suzuki protocol), with stannanes (Stille protocol, with silanes or with the corresponding arenes themselves]

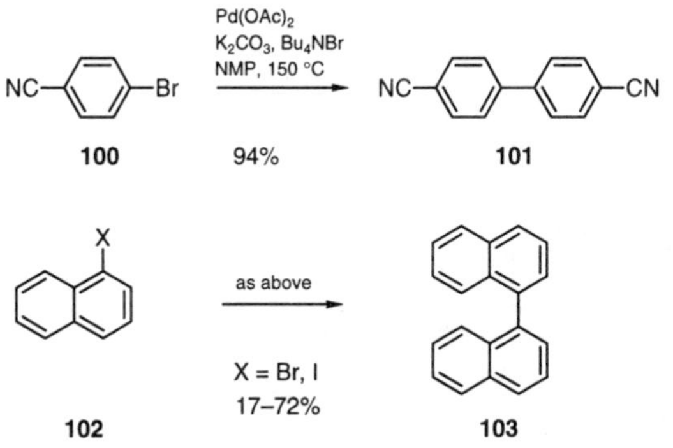

Scheme 5.14. Palladium-catalysed homocoupling of haloarenes to give biaryls.

2.5 *Reaction cascades starting with one or more Heck reactions: facile formation of bicycles, tricycles, tetracycles and more*

Non-aromatic polycyclic systems play an important role as skeletons of many biologically active compounds. Because of their high efficiency, domino-type and other sequential reactions have attracted a great deal of interest in recent years [9]. An in-depth investigation into the scope and limitations of reaction cascades consisting of several inter- or intramolecular Heck-type couplings followed by a purely thermal reaction, such as a 6π-electrocyclization (Scheme 5.15) or a cycloaddition (Scheme 5.16), has unveiled a number of highly efficient processes for the construction of complex oligocyclic skeletons [70–73].

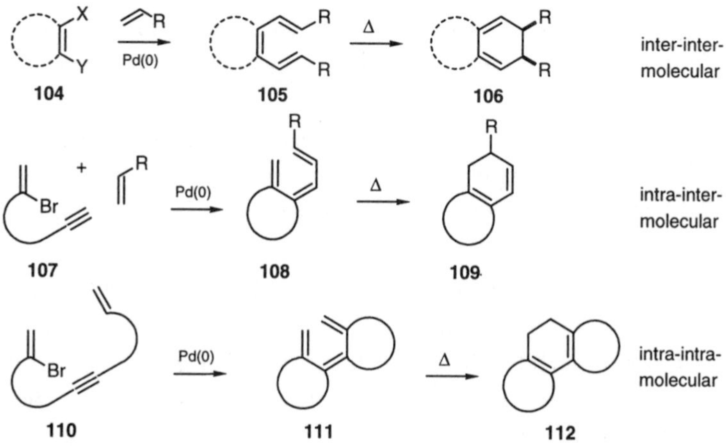

Scheme 5.15. Possible domino reactions consisting of Heck couplings and consecutive 6π-electrocyclizations.

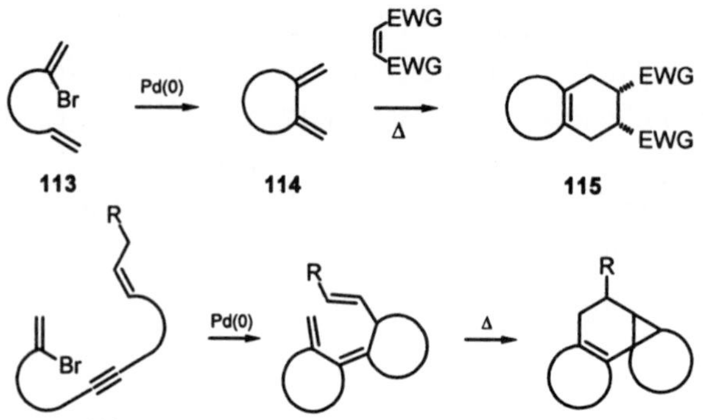

Scheme 5.16. Conceivable domino reactions consisting of one or more Heck couplings and subsequent inter- or intramolecular cycloadditions.

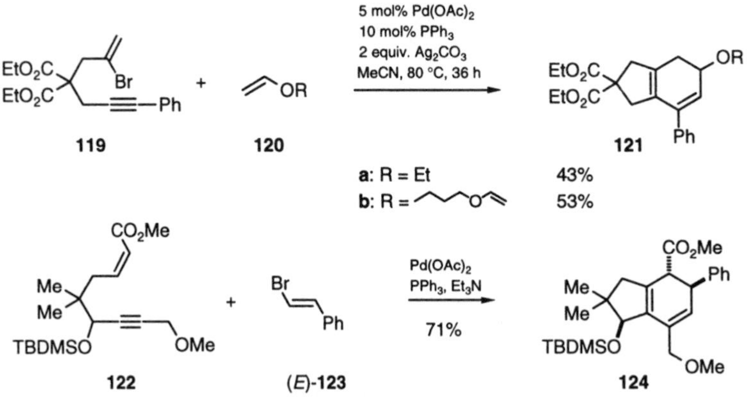

a: R = Et   43%
b: R = ⌒⌒O⌒   53%

71%

Scheme 5.17. Sequences of intra- and intermolecular Heck reactions followed by 6π-electrocyclization.

The combination of two intermolecular Heck reactions on a 1,2-dihalocycloalkene **27** and **28** with a subsequent thermal 6π-electrocyclization of the resulting 1,3,5-hexatrienes **29** and **30** leads to ring-annelated *cis*-5,6-disubstituted cyclohexadienes **31** and **32** (see Scheme 5.4). In these cases, the 6π-electrocyclization requires significantly higher temperatures (50–70 °C higher) than the Heck reaction. The sequence of an intra- and an intermolecular Heck reaction followed by a 6π-electrocyclization gives ring-annelated aromatic compounds from 2-bromoalkeneynes such as **119** and alkynes [74], yet can also yield ring-annelated cyclohexa-1,3-dienes of type **121** particularly well when performed with enol ethers as an alkene component (Scheme 5.17) [70]. The sequence of intra- and intermolecular Heck reactions prior to 6π-electrocyclization can also be reversed, as demonstrated for the coupling of (*E*)-β-bromostyrene ((*E*)-**123**) with the eneyne **122** to yield the bicyclic diene **124** by Trost *et al.* [75].

The fully intramolecular version of this three-step sequential reaction for 2-bromododeca-1,11-diene-6-ynes such as **125** and **127**, 2-bromotrideca-1,12-diene-7-ynes such as **129** and -6-ynes such as **131**, in which an alkenyl bromide starter, an alkynyl relay and an alkenyl terminator are all tethered in a single acyclic precursor molecule, proceeds smoothly to form three new cycles in a tricyclic array with a central cyclohexa-1,3-diene moiety (Scheme 5.18) [76]. In all these cases the 6π-electrocyclization occurs under the conditions of the Heck reaction (60–100 °C), and the yields are good to excellent (up to

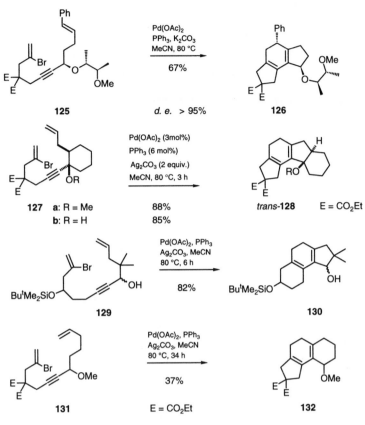

Scheme 5.18. Sequences of two intramolecular Heck-type reactions immediately followed by 6π-electrocyclization.

95%) as long as only five-membered rings are formed in the Heck-type cyclization steps. The overall yields are not as good when one of the Heck-type cyclizations leads to a six-membered ring, especially if this is the second cyclization step, as in the transformation **131 → 132** [77].

An analogous sequence of events occurs in the palladium-catalysed tricyclization of enediynes developed by Trost *et al.* [78], in which the initiating step is an addition of a hydridopalladium species to a triple bond rather than an oxidative addition of an alkenyl bromide to a Pd(0) complex.

A complementary cascade tricyclization of 2-bromoalk-1-enediynes such as **133, 135** and **137** leads to angularly bisannelated benzene derivatives (Scheme 5.19) [72, 77, 79]. This tricyclization gives reasonable yields even of octahydrophenanthrene skeletons such as **138**, but only when the precursor bromoenediyne is terminally substituted with a trialkylsilyl group. Terminally unsubstituted bromoenediynes such as **139a,b** apparently undergo a 5-*exo*-trig instead of a 6-*endo*-trig cyclization in the third step, with subsequent further cyclization and rearrangement to eventually yield bisannelated fulvenes **140a,b** in quite good yields [77].

Nitrogen- and oxygen-containing tricyclic dienes of type **126** can be prepared from appropriately substituted acyclic 2-bromodieneyne precursors without problems. Thus, the diaza-**142** and dioxatricycle **144** were obtained from compounds **141a** and **143** in 55 and >80% yield, respectively (Scheme 5.20) (L. Verhoeven, A. Steinig, L. Bhat and A. de

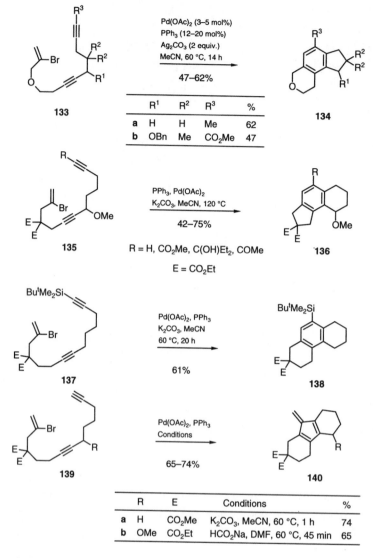

Scheme 5.19. Angularly bisannelated benzene derivatives and fulvenes obtained by palladium-catalysed tricyclization cascades.

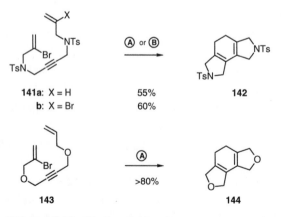

Scheme 5.20. Bis-heterotricyclic skeletons by a Heck–Heck–6π-electrocyclization sequence: (A) palladacycle from $Pd(OAc)_2$ + (o-Tol)$_3$P (see Scheme 5.8), MeCN, K_2CO_3, 80 °C, 48 h; (B) $Pd(OAc)_2$, PPh$_3$, DMF, K_2CO_3, 100 °C, 4 days.

Meijere, unpubl. data). The crude yield of compound **142** is actually much better, and higher yields can be obtained in larger scale runs in which losses upon purification are minimized. Surprisingly, the dibromodieneyne **141b** could also be cyclized to the same tricyclic product **142**. This reaction involves a reduction analogous to the one in the palladium-catalysed homocoupling of aryl halides (see above).

2.5.1 TOWARDS STEROIDS AND UNNATURAL ANALOGS OF STEROIDS

The cascade tricyclization presented above offers itself to construct oligocyclic skeletons of various natural products such as the steroids. Two strategies can be envisaged to approach the tetracyclic system of steroids (I and II in Scheme 5.21). Starting with the five-membered D-ring in a monocyclic precursor with two appropriately functionalized tethers as in compound **145**, the Pd-catalysed tricyclization would be expected to lead to compound **147**, a steroid with a cyclohexadiene B-ring, while a six-membered ring, serving as the later A-ring, with the two appropriate tethers as in compound **148**, would yield the corresponding tetracycle with a diene unit in the C-ring (II in Scheme 5.21).

The feasibility of the second strategy has been demonstrated by Trost *et al.* [78], applying the related enediyne cycloisomerization methodology to the bis-tethered cyclohexane derivative **151**, yet the product **152** is deficient of most of the essential functionalities. Negishi *et al.* [80] have developed an even more efficient assembly of the tetracyclic compound **154** with the basic skeleton of steroids by a zipper-type tetracyclization of the completely open-chain iodotrienediyne **153** (Scheme 5.22) [81].

The first strategy (see Scheme 5.21) appeared to be particularly attractive to approach a steroid with all the essential functionalities, because a precursor like compound **145** should be accessible even in enantiomerically pure form from the Hajos–Wiechert ketone [82] developed in the context of other steroid total syntheses. Several monocyclic bromodieneynes of type **145**, including the diastereomeric mixture *trans/cis*-**155**, were thus assembled. Cyclization of the latter under palladium catalysis, surprisingly, gave the pentacyclic compounds *trans/cis*-**156**, as proved by an X-ray crystal structure analysis for *cis*-**156**, rather than the expected tetracyclic system of type **147** (Z. Z. Song *et al.*, *Agnew. Chem. Int. Ed. Engl.*, 1999; **38**, in press.).

Bromodieneynes **157a** and **157c** with a different type of substitution and combination of substituents, under optimized conditions, eventually gave the steroid-type tetracyclic compounds **158**, **162**, **164** and **165**, albeit in poor yields. It is obvious that the β-hydride elimination in the intermediate **160** formed by the second 6-*exo*-trig cyclization must be relatively

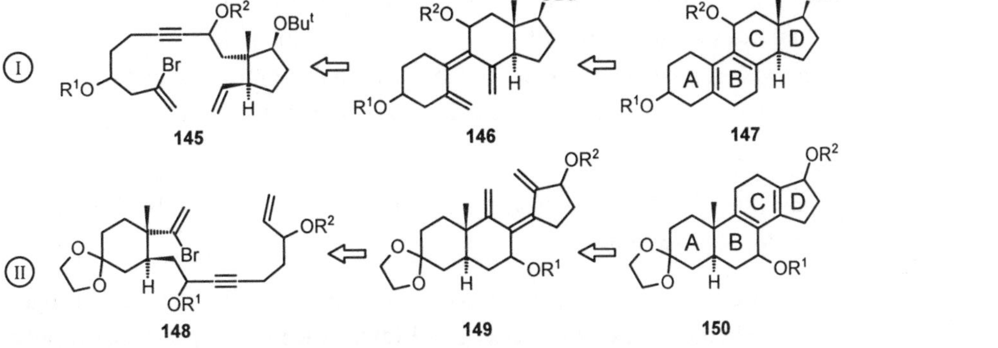

Scheme 5.21. Two new strategies for the construction of the steroid skeleton.

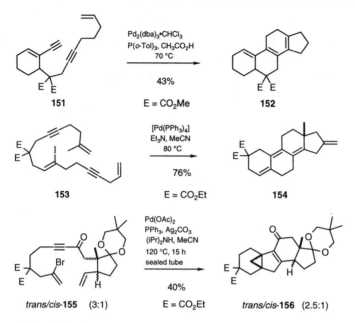

Scheme 5.22. Multiple intramolecular palladium-catalysed cross-coupling in the construction of the steroid skeleton and the [4.3.1]propellane obstacle.

slow, so that the normal tricyclization product **158** was observed as a minor product only from compound **157a** (Scheme 5.23). Two intramolecular carbopalladations successfully compete with the β-hydride elimination in compound **160**, a 5-*exo*-trig to give compound **161b** from compound **160b** and subsequent 3-*exo*-trig cyclization eventually leading to *trans*-**156**, and a 6-*endo*-trig process apparently favored in compound **160c** to give compound **163c**, which can lead to compounds **162**, **164** and **165** (Z. Z. Song *et al.*, 1999).

A vast series of cyclization studies on simple bromodieneyne model compounds with different tether lengths between the bromoene starter unit, the alkyne triple bond relay and the alkene terminator disclosed that the Heck–Heck–6π-electrocyclization cascade is not a feasible process to assemble the decahydrophenanthrene skeleton **169** (Scheme 5.24). The tricyclization works particularly well for angular 5.6.5 tricycles **167** and reasonably well for 6.6.5 systems **171**. When the second intramolecular Heck-type coupling leads to a six-membered or larger ring, the process works less efficiently, as for the 5.6.6 system **175**, or not at all, as for system **177** (Scheme 5.24).

According to these systematic model studies, strategy II above (see Scheme 5.21) should be not only much more feasible than strategy I but also a workable one to yield a fully functionalized steroid analog with a cyclohexadiene unit in the C-ring. It should be noted in this context that it is essential to add a silver salt to the mixture or use silver carbonate as a base to prevent the intermediate hexatriene of type **111** from isomerization [73].

2.5.2   [n.3.1]PROPELLANES FROM ACYCLIC PRECURSORS—FOUR NEW C–C BONDS IN A SINGLE OPERATION

Even though the tricyclization does not work well for angular arrangements of three six-membered rings, the cascade all-intramolecular Heck-type coupling reaction of

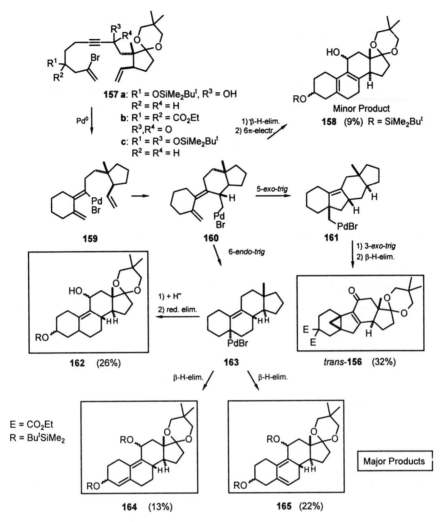

Scheme 5.23. Mechanistic details concerning formation of the steroid skeleton and a pentacyclic analog from bis-tethered cyclopentanone derivatives **157** (for clarity, the substituents have been omitted from presumed intermediates).

corresponding 2-bromotetradeca-1,13-diene-7-ynes **178** and **180a** leads to interesting and potentially useful tetracyclic systems **179** and **181a**, respectively, with a bridging cyclopropane ring joining the A- and B-rings (Scheme 5.25). At least one natural product, the plant growth regulator 3α-hydroxy-9,15-cyclogibberellin A (**182**) [83], contains the same basic skeleton as compound **181a**.

The same tetracyclization is also very efficient for the next higher homolog of compound **180a**, the bromopentadecadieneyne **180b**, in which the first cyclization yields the seven-membered ring in the final product **181b**, and even the eight-membered ring containing tetracycle **181c** could be isolated in 30% yield from the corresponding precursor **180c** [77]. After all, the bridging cyclopropane ring as in compounds **179** and **181** is an equivalent of an angular methyl substituent. A tetracycle of type **181** was obtained as a byproduct (5–10% yield) even from the oxabromodieneyne precursor **183a**, which was set up to give the 6.6.5 tricyclic combination **188** [73]. The tetracycles **185b,c** were obtained in 62 and 71% yield, respectively, from the corresponding acyclic precursors **183b,c** with

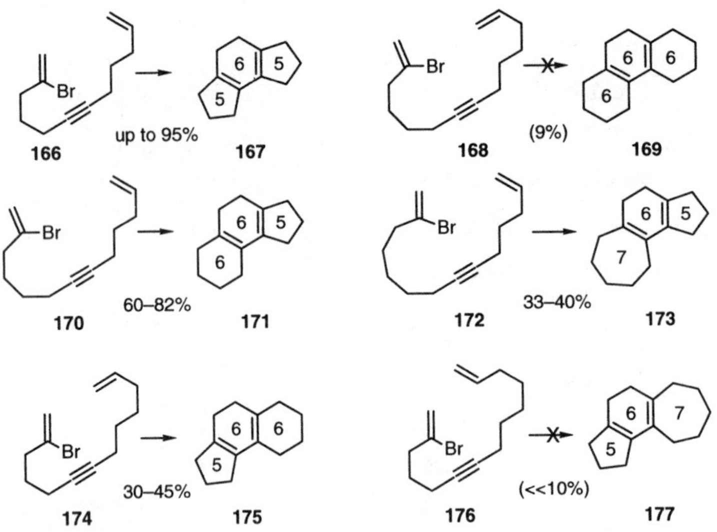

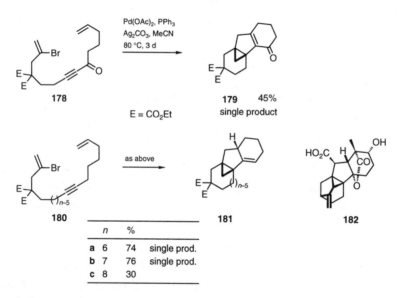

Scheme 5.24. Rules for achievable ring sizes in palladium-catalysed tricyclizations of 2-bromoalka-1, $(n+m)$-diene-$(n+1)$-ynes (substituents of actual model compounds are left off for clarity).

	n	%	
a	6	74	single prod.
b	7	76	single prod.
c	8	30	

Scheme 5.25. Efficient tetracyclizations of 2-bromotetradeca-1,13-diene-7-ynes and homologs to give [n.3.1]propellanes.

substituents $R^1 \neq H$ preventing β-hydride elimination at the intermediate **184** (Scheme 5.26).

Apparently, there are no rules without exceptions. In the above-mentioned cascade Heck-type tetracyclizations, substituents in the acyclic precursors can play a major role and cause the sequential reaction to proceed in an unprecedented direction. The bromodieneyne **189** with a terminal phenyl group apparently sequentially cyclizes to the usual neopentylpalladium intermediate **190**, which, due to the proximity of the phenyl group, undergoes an intramolecular electrophilic substitution [84] to give the pentacycle **191** rather than a system corresponding to compound **181a** by 3-*exo*-trig cyclization and β-hydride elimination (Scheme 5.27). On the other hand, the acyclic precursors **192a,b**, which

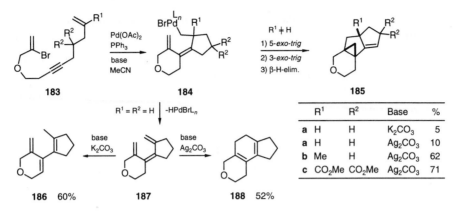

Scheme 5.26. Tetracyclization of 4-oxabromotrideca-1,12-diene-7-ynes.

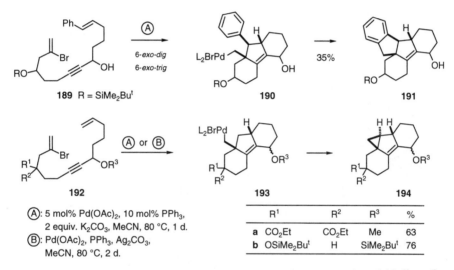

R^1	R^2	R^3	%
a CO$_2$Et	CO$_2$Et	Me	63
b OSiMe$_2$Bu^t	H	SiMe$_2$Bu^t	76

Scheme 5.27. Two other types of tetracyclization of 2-bromotetradeca-1,13-diene-7-ynes caused by a terminal phenyl or a 9-alkoxy substituent.

differ from compounds **178** and **180** only by their 9-methoxy or 9-silyloxy substituent, eventually yield the novel tetracyclic systems **194a,b**, as proved by an X-ray crystal structure analysis for compound **194a**. This sequence must also proceed via the tricyclic intermediates **193**; however, it is unclear whether compound **194** is simply formed by an unprecedented γ-hydride elimination or along a more complicated route [77].

2.6 *Domino Heck–Diels–Alder reactions: efficient combinations*

A variety of fascinating cascade reactions consisting of one or more intra- or intermolecular cross-coupling reactions and some other reaction types have been developed by other groups [9, 85]. Among the ones that achieve the most striking increase in molecular complexity from starting material to product are the molecular zipper reactions by Negishi *et al.* [86], Trost *et al.* [87] and Overman *et al.* [88], and the carbonylation intramolecular cross-coupling cascades by Negishi *et al.* [89] in which up to seven new carbon–carbon bonds are formed in a single operation. Another potentially powerful sequence arises by combining one or two intramolecular Heck-type couplings with an intra- or intermolecular Diels–Alder addition [90]. An all-intramolecular version of such a sequence has been shown to proceed reasonably smoothly for terminally alkoxycarbonyl-substituted 2-bromotrideca-1,11-diene-6-ynes (E/Z)-**195** under palladium catalysis at 130 °C. At 80 °C, the sequential reaction stops after the two consecutive Heck-type cyclizations and subsequent β-hydride elimination to give (E/Z)-**196** (Scheme 5.28); apparently only the (E)-isomer (E)-**195** undergoes the intramolecular Diels–Alder reaction, because (Z)-**196** is observed unchanged in the mixture with compound **197**.

Obviously, systems that can undergo intramolecular Diels–Alder reactions can also be set up by a Trost-type eneyne cycloisomerization. Thus, the dieneynes **198a,b** upon treatment with [Pd$_2$(dba)$_3$·CHCl$_3$] in the presence of acetic acid and triphenylphosphane at 80 °C gave the bis-heterotricycles **201a,b** in good to very good yields (Scheme 5.29) (L. Verhoeven, A. Steinig, L. Bhat and A. de Meijere, unpubl. data). It is particularly noteworthy that the intermediate triene **200** formed via compound **199** undergoes the intramolecular Diels–Alder reaction under the conditions of the eneyne cycloisomerization **198** → **200**, i.e. at 80 °C.

The intermolecular version of this domino Heck–Diels–Alder process previously has been exercised frequently in two steps, especially by Trost *et al.* [91] and Grigg *et al.* [92],

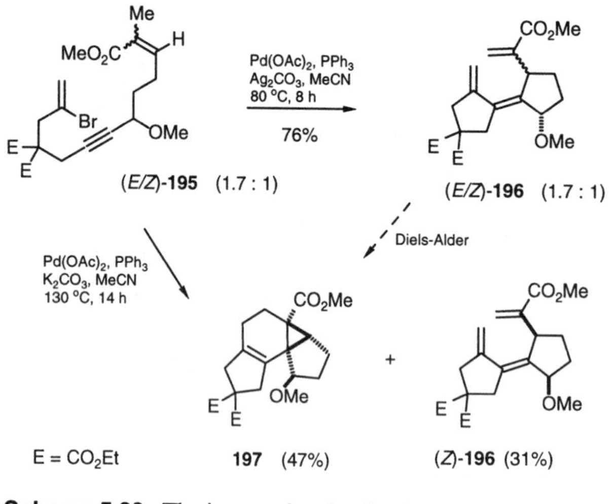

Scheme 5.28. The intramolecular domino Heck–Diels–Alder reaction.

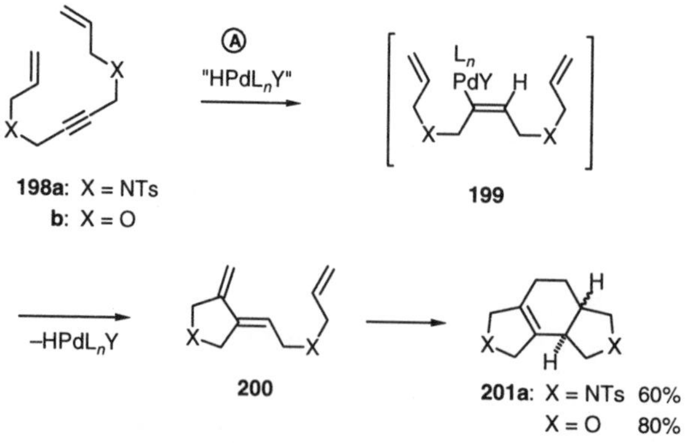

Scheme 5.29. A domino of enyne cycloisomerization and intramolecular [4 + 2] cyclo-addition. Step A: Pd$_2$(dba)$_3$·CHCl$_3$, AcOH (1 equiv), PPh$_3$, C$_6$H$_6$, 80 °C, 100 min.

who have developed the palladium-catalysed eneyne cycloisomerization [7], the bromo-eneyne cyclization–anion capture sequence [93] and the intramolecular Heck coupling of a 2-bromoalka-1,ω-diene to form vicinal dialkylidenecycloalkanes [94].

Because an intramolecular coupling always wins over an intermolecular one by at least a factor of 10^5, such a sequential reaction consisting of an intramolecular Heck coupling or eneyne cycloisomerization and an intermolecular Diels–Alder addition can favorably be performed in a single one-pot operation in the presence of the dienophile, except with dienophiles such as tetracyanoethene (TCNE) and benzoquinones, which are strong oxidants [95,96]. The yields without isolating the intermediate dimethylenecyclopentane derivatives **203** are consistently higher than those obtained in two steps (Table 5.1).

This single-operation domino process was further developed to conveniently prepare heteroanalogous bicyclo[4.3.0]non-1(6)-ene derivatives **207** from appropriate acyclic bromodiene precursors **206** [97].

These examples (Table 5.2) demonstrate that the Thorpe–Ingold effect exerted by the *gem*-diester groups in the model 2-bromohepta-1,6-dienes **202** is not essential for the cyclization to occur. In fact, carbocyclic bicyclo[4.3.0]nonene derivatives without the malonate moiety, such as compound **207g** (Table 5.2), are also formed without problems. The intramolecular Heck coupling has certain advantages over the eneyne cycloisomerization methodology by Trost *et al.* [7]. One advantage is that it also works reasonably well for the formation of six-membered rings, as shown by the formation of compound **207h** from compound **206h** (Table 5.2). The other advantage is its applicability to 2-bromo-1,6-dienes of

Table 5.1. Examples for the intra–intermolecular domino Heck–Diels–Alder reaction to yield bicyclo[4.3.0]non-1(6)-ene skeletons

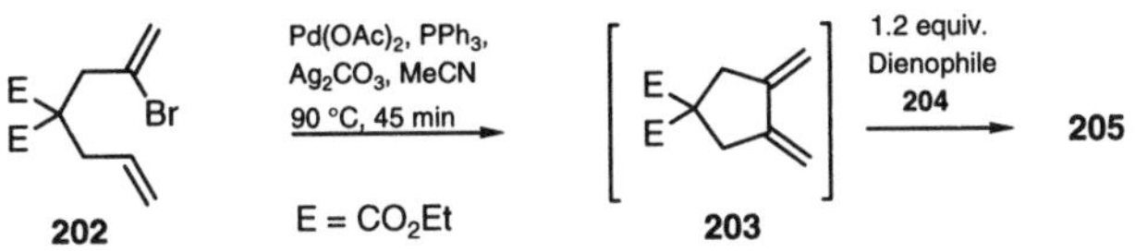

Dienophile	Product	Yield (%)
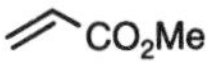 204a	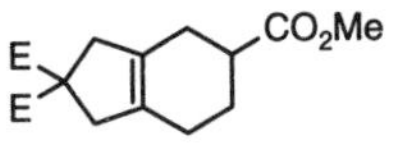 205a	93[a] 60[b]
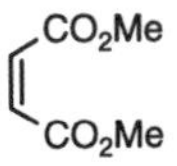 204b	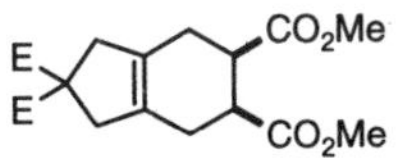 205b	93[a] 67[b]
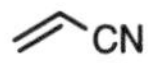 204c	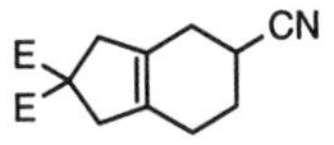 205c	83[a] 70[b]
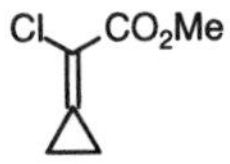 204d	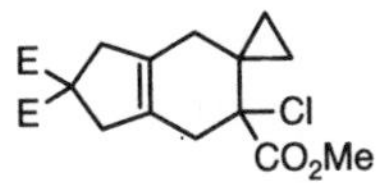 205d	83[a] 61[b]

[a] One-pot procedure.
[b] Overall yield in two-step procedure.

Table 5.2. Examples for the domino Heck–Diels–Alder reaction forming hetera-bicyclo[4.3.0]non-1(6)-enes **207a–f** and carbocyclic analogs

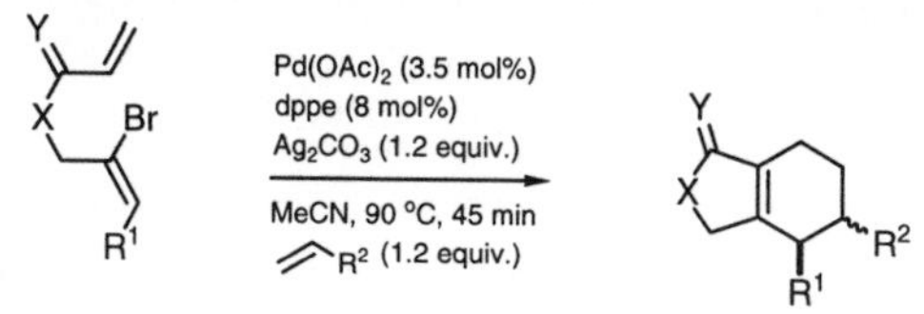

	X	Y	R^1	R^2	%
a	O	H$_2$	Me	CO$_2$Me	55[a]
b	OCH-N	H$_2$	H	CO$_2$Me	62
c	Bn-N	H$_2$	H	CO$_2$Me	46
d	Ac-N	H$_2$	H	CO$_2$tBu	63
e	p-NO$_2$-C$_6$H$_4$SO$_2$-N	H$_2$	H	CO$_2$tBu	72
f	Bn-N	O	Me	CO$_2$Me	51[b]
g	CHOH	H$_2$	H	CO$_2$Me	87[c]
h	CH$_2$CHOSiMe$_2$Bu^t	Me$_2$	H	CO$_2$Me	67[d]

[a] Mixture of regioisomers and diastereomers, with cis-2,3-(quasi-ortho) predominating (67%).
[b] Mixture of regioisomers and diastereomers, with cis-2,3-predominating (71%).
[c] PPh$_3$ used instead of dppe.
[d] Mixture of regioisomers and diastereomers.

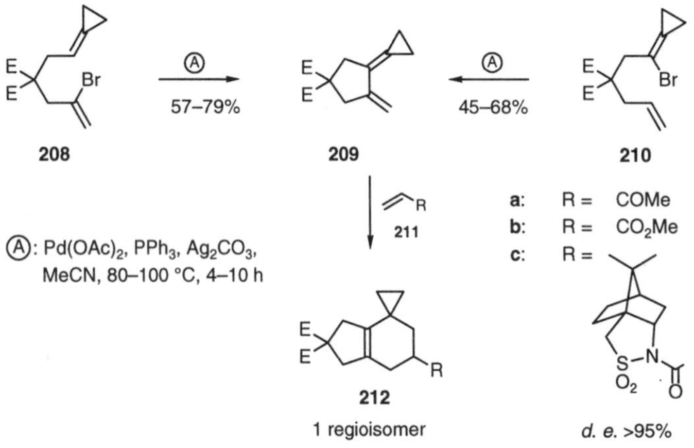

Scheme 5.30. The domino Heck–Diels–Alder reaction leading to (spirocyclopropane)bicyclo[4.3.0]nonenes. d.e. = diastereomer excess.

type **208** and **210** with a methylenecyclopropane terminator and a methylene-cyclopropane starter moiety, respectively. Both compounds cyclize smoothly in the presence of dienophiles **211** to give the Diels–Alder adducts **212** derived from the rather sensitive intermediate diene **209** as single regioisomers without destruction of the three-membered ring as was observed in the attempted cyclization of an eneyne corresponding to compound **208** (K. H. Ang, S. Bräse, A. Stolle and A. de Meijere, unpubl. data) [98].

The domino reaction carried out in the presence of the chiral non-racemic acrylamide **211c** gave the cycloadduct **212c** as a single diastereo- and enantiomer (Scheme 5.30) [99]. These domino cyclizations to give bicyclo[4.3.0]non-1'(6')-ene-2'-spiro-1-cyclopropane

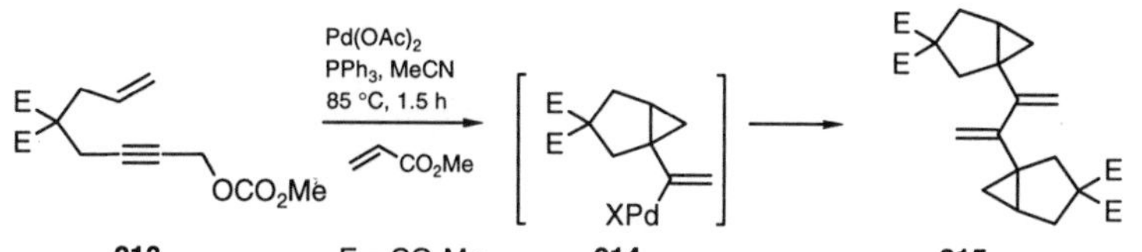

Scheme 5.31. Unprecedented formation of a 1-ethenylbicyclo[3.1.0]hexane dimer.

derivatives in high yields are quite remarkable because methylenecyclopropane derivatives previously have been observed to undergo ring-opening under Heck reaction conditions [100, 101].

In an attempt to prepare the Diels–Alder adduct of the eneallene corresponding to **209** by a domino reaction, the propargyl carbonate **213** [102] was treated with Pd(OAc)$_2$ and PPh$_3$ in the presence of methyl acrylate. But rather than the expected domino reaction product, only the interesting 1-ethenylbicyclo[3.1.0]hexane dimer **215** was isolated (Scheme 5.31), seemingly as a single diastereomer [97, 103]. This product must have been formed via an ethenylpalladium intermediate of type **214**, most probably arising by a 3-*exo*-trig cyclization of an intermediate that could have led to the expected eneallene by β-hydride elimination. Surprisingly, the coupling product arising from compound **214** and methyl acrylate was not observed [104].

2.7   *Dendralenes by intra- and intermolecular Heck-reaction–rearrangement cascades*

Bromodienes with methylenecyclopropane moieties, such as compounds **208** and **210**, cleanly undergo the intramolecular Heck reaction to give allylidenecyclopropanes of type **209** (see above). A different situation arises when the methylenecyclopropane moiety has a tetrasubstituted double bond, as in compounds **218a,b**, which are readily accessible by Pd-catalysed substitution on 1-propenylcyclopropyl tosylate or chloride [105]. The intramolecular Heck reactions of compounds **218a,b** do not proceed by an ($n − 1$)-*exo*-trig cyclization to give a cyclopropylpalladium intermediate, but by an *n-endo*-trig process to yield the cyclopropylcarbinylpalladium species **221**, which rapidly ring-opens to the homoallylpalladium intermediate **222**. Subsequent β-hydride elimination eventually leads to the cross-conjugated trienes **219** [106], so-called [3]dendralenes (Scheme 5.32). Under the conditions of the so-called cycloisomerization developed by Trost *et al.* [7], the 1,6- and 1,7-eneynes **220a** and **220b** gave the [3]dendralenes **219a** ($n = 6$) and **219b** ($n = 7$) in 78 and 100% yield, respectively. In the presence of iodobenzene under Heck conditions, the eneyne **220a** gave the (*Z*)-phenylsubstituted [3]dendralene **216** (45%) together with the phenylsubstituted eneyne **217** (42%).

The same cascade reaction was successful in transforming the enediyne **223** to the cross-conjugated tetraene **224** via the intermediates **225–227** (Scheme 5.33) [107].

In spite of their tetrasubstituted double bonds, these methylenecyclopropane derivatives are outstandingly reactive towards intramolecular Heck coupling, which must be attributed to the relief of strain upon any addition to such a double bond and to the high-lying HOMO of any methylenecyclopropane derivative, which makes them particularly good ligands for transition metals [108]. The extremely strained and highly nucleophilic [109] bicyclopropylidene (**232**), which is readily available from methyl cyclopropanecarboxylate in three efficient steps [110], even surpasses styrene and methyl acrylate in its reac-

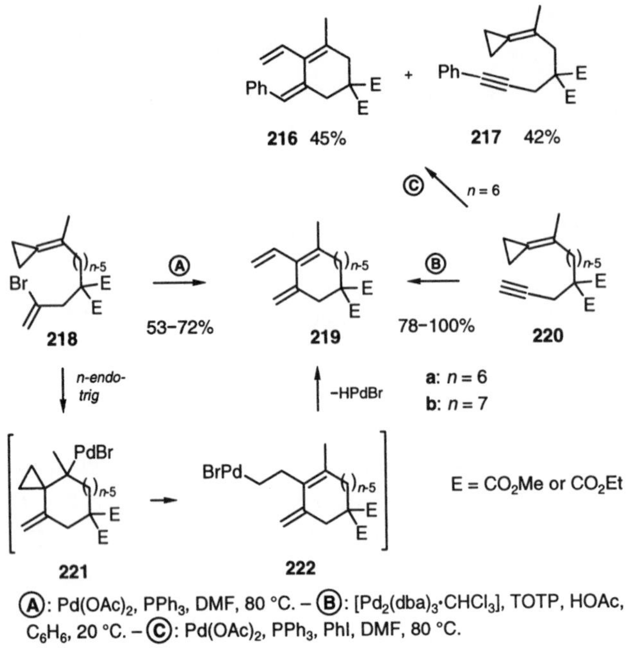

Scheme 5.32. A Heck-reaction–rearrangement cascade forming cross-conjugated trienes, so-called [3]dendralenes.

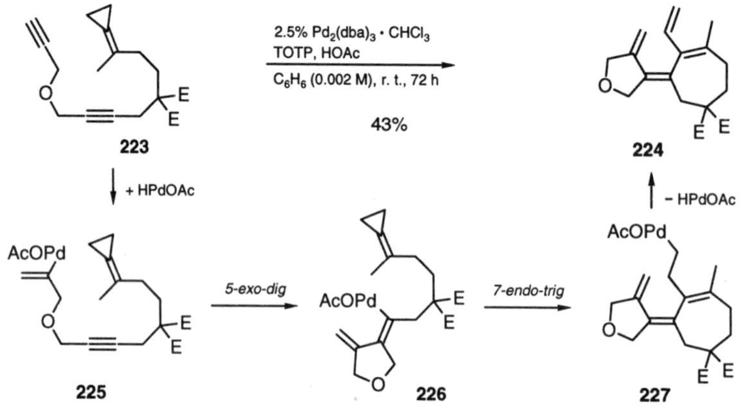

Scheme 5.33. Facile formation of dendralenes: the tetraene case.

tivity towards alkenyl- and arylpalladium intermediates formed under Heck reaction conditions from the corresponding halides or perfluoro-alkanesulfonates. Thus, compound **232** reacts with iodobenzene or iodoethene in the presence of methyl acrylate or dimethyl maleate in an inter–intermolecular domino Heck–Diels–Alder reaction to give the spiro[2.5]octane and (spirocyclopropane)bicyclo[4.4.0]decene derivatives **228** and **229** in 61 and 49% yield, respectively (Scheme 5.34). Surprisingly, the direct Heck coupling product of iodobenzene and methyl acrylate—methyl cinnamate—was formed as a trace byproduct only.

The overall yields from these inter–intermolecular domino Heck–Diels–Alder reactions can be improved, as demonstrated for the reaction of 1,4-diiodobenzene. When performed under 10 kbar pressure—high pressure accelerates both the Heck (see above) and the

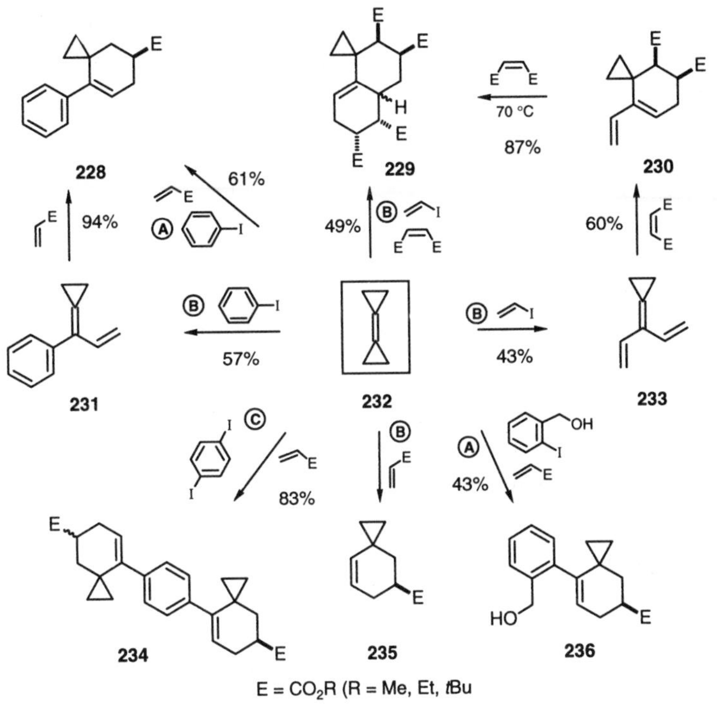

Scheme 5.34. Inter–intermolecular domino Heck–Diels–Alder reactions with bicyclopropylidene: (A) Pd(OAc)$_2$ (5 mol. %), PPh$_3$ (15 mol. %), NEt$_3$, DMF, 80 °C, 20 h–5 days; (B) Pd(OAc)$_2$, PPh$_3$, NEt$_3$, DMF, 80 °C, 4 h; (C) as in (A) but under 10 kbar.

Diels–Alder reaction—this formal five-component reaction of 1,4-diiodobenzene with two molecules each of bicyclopropylidene (**232**) and methyl acrylate gave the product **234** in 83% yield (H. Nüske, S. Bräse, S. I. Kozhushkov and A. de Meijere, unpubl. data). The corresponding domino reaction of compound **232** with *o*-iodobenzyl alcohol and methyl acrylate furnished the axially and centrally chiral adduct **236** as a mixture of two diastereomers in a ratio of 2.5 : 1.

In the absence of an aryl or alkenyl halide, bicyclopropylidene (**232**) undergoes a palladium-catalysed rearrangement to allylidenecyclopropane, which reacts with added acrylate to give the Diels–Alder adduct **235** (H. Nüske, S. Bräse, S. I. Kozhushkov and A. de Meijere, unpubl. data). A different mode of ring opening in bicyclopropylidene (**232**) occurs upon the palladium-catalysed codimerization with α,β-unsaturated esters and strained alkenes, leading to ring-fused vinylcyclopropanes in a formal [3 + 2] cycloaddition [111]. This type of reaction of compound **232** with, for example, diethyl fumarate yields the methylenespiro[2.4]heptane derivative **237**, which is set up for a subsequent rhodium(I)-catalysed formal [5 + 2] cycloaddition of an alkyne such as 2-butyne to give compound **238** with a seven-membered ring annelated to a five-membered ring (Scheme 5.35). The latter sequence of metal-catalysed cocyclizations of bicyclopropylidene (**232**) obviously bears great potential for the assembly of terpene skeletons.

3 Conclusion and future perspectives

Although various types of palladium-catalysed cross-coupling reactions have originally been discovered several decades ago, their application towards real problems in organic

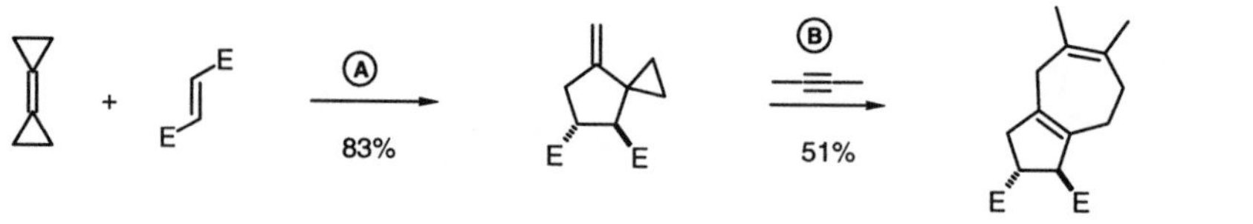

232   E = CO₂Et   **237**   **238**

A: Pd(dba)₂, P(iPr)₂(*t*Bu), toluene, 110°C, 3h. —B: [RhCl(PPH₃)₃], AgOTf, toluene, 110°C, 3h

Scheme 5.35. Palladium(0)-catalysed [3 + 2] cocyclization of bicyclopropylidene with diethyl fumarate and Rhodium(I)-catalysed intermolecular [5 + 2] cycloaddition of the resulting vinylcyclopropane.

synthesis has appeared only in the last decade. Along with drastically improved protocols for practical performance of such reactions came elaborate studies into their scope and limitations.

The current collection of multifold and domino-type transformations, which lead to an impressive increase in molecular complexity in a single operation, offers convincing evidence that the next decades and the next century will see a continuing growth and advancement of this chemistry. Obvious further improvements will be achieved by applying novel low-cost multifunctional building blocks and developing more environmentally friendly reaction conditions, such as ambient temperature for coupling with more active and reusable catalysts. This will undoubtedly lead to an ever-increasing use in industrial processes, especially for the production of high-end chemical products such as pharmaceuticals, as well as modern agrochemicals and eventually highly advanced new materials for the electronics industry.

4   Acknowledgments

The work presented here, including the fruitful collaborations with B. Waegell (Marseille) and P. J. Parsons (formerly Reading, now Brighton), has been supported financially by the Deutsche Forschungsgemeinschaft, the Volkswagen-Stiftung, the Alexander-von-Humboldt-Stiftung, the Fonds der Chemischen Industrie, the Studienstiftung des Deutschen Volkes, the European Community and Bayer, BASF, Degussa, Hoechst and Hüls AG (chemicals).

The authors are grateful to Dr Burkhard Knieriem for his careful reading of the manuscript and to Stefan Beußhausen for his effort in drawing the chemical structures and producing a flawless version at the end. S. B. is indebted to the Fonds der Chemischen Industrie for a graduate student and Liebig fellowship, as well as to NATO-DAAD for a postdoctoral fellowship.

The research covered in this chapter has been conducted by K. Albrecht, K. H. Ang, L. Bhat, S. Bräse, M. Drögemüller, F. Funke, H. Henniges, B. König, S. Kozhushkov, A. Lansky, A. Llebaria, G. McGaffin, M. Messner, F. E. Meyer, H. Nüske, J.-U. Peters, T. Perkovic, P. Prinz, K. Rauch, O. Reiser, K. Rosauer, J. Rümper, U. Schick, S. Schweizer, Z. Z. Song, A. G. Steinig, V. V. Sokolov, B. Stulgies, C. Thies, S. Untiedt, L. Verhoeven, K. Voigt, M. Weber, M. Wrobel and P. v. Zezschwitz. Unpublished work can be found in the respective dissertations as quoted.

A. de M. is grateful to Professor Günther Helmchen for a generous gift of the β-(naphthylphenyl)phosphinocarboxylate ligand.

5 References

1 Bräse S, de Meijere A. In: Stang PJ, Diederich F, eds. *Metal-Catalysed Cross Coupling Reactions.* Weinheim: Wiley–VCH, 1998: 99–166.

2 Trost BM. *Angew Chem* 1989; **101**: 1199–219; *Angew Chem Int Ed Engl* 1989; **28**: 1173–93.

3 Suzuki A. In: Stang PJ, Diederich F, eds. *Metal-Catalysed Cross Coupling Reactions.* Weinheim: Wiley–VCH, 1998: 49–97.

4 Farina V, Krishnamurthy V, Scott WJ. *Org React* 1997; **50**: 1–652; Mitchell TN. In: Stang PJ, Diederich F, eds. *Metal-Catalysed Cross Coupling Reactions.* Weinheim: Wiley–VCH, 1998: 167–202.

5 Sonogashira K. In: Stang PJ, Diederich F, eds. *Metal-Catalysed Cross Coupling Reactions.* Weinheim: Wiley–VCH, 1998: 203–29; Campbell IB. In: Taylor RJK, ed. *Organo-copper Reagents–A Practical Approach.* Oxford: Oxford University Press, 1994: 217–35; Sonogashira K, Tohda Y, Hagihara N. *Tetrahedron Lett* 1975: 4467–70.

6 Heumann A, Jens KJ, Réglier M. In: Karlin KD, ed. *Progress in Inorganic Chemistry*, Vol. 42. Chichester: Wiley, 1994: 483–576.

7 Trost BM. *Acc Chem Res* 1990; **23**: 34–42; *Science* 1991; **254**: 1471–7; *Angew Chem* 1995; **107**: 285–307; *Angew Chem Int Ed Engl* 1995; **21**: 259–81; Trost BM, Krische MJ. *Synlett* 1988: 1–16.

8 Oppolzer W. In: Trost BM, Fleming I, eds. *Comprehensive Organic Synthesis*, Vol. 5. Oxford: Pergamon Press, 1991: 23–61.

9 Tietze LF, Beifuss U. *Angew Chem* 1993: **105**: 137–70; *Angew Chem Int Ed Engl* 1993; **32**: 131–63; Tietze LF. *Chem Rev* 1996; **96**: 115–36.

10 See the discussion of this terminology in ref. 9 and in: Denmark SE, Thoraensen A. *Chem Rev* 1996; **96**: 137–65; see also the titles in: *Chem Rev* 1996: **96**, 4A–5A. Such reactions frequently have been termed tandem reactions. However, real tandem reactions would be processes in which two steps occur simultaneously, and this would of course be extremely rare.

11 Mizoroki T, Mori K, Ozaki A. *Bull Chem Soc Jpn* 1971; **44**: 581.

12 Heck RF, Nolley JP. *J Org Chem* 1972; **37**: 2320–2.

13 Heck RF. *Acc Chem Res* 1979; **12**: 146–51; *Org React* 1982; **27**: 345–90; In: Tsutsui M, Ishii Y, Yaozeng H, eds. *Fundamental Research in Organometallic Chemistry.* New York: Van Nostrand Reinhold, 1982: 447–62; *Palladium Reagents in Organic Synthesis.* London: Academic Press, 1985; Tsuji J. *Organic Synthesis with Palladium Compounds.* Berlin: Springer, 1980; Trost BM, Verhoeven TR. In: Wilkinson G, ed. *Comprehensive Organometallic Chemistry,* Vol. 8. Oxford: Pergamon; 1991: 854–83; Cabri W, Candiani I. *Acc Chem Res* 1995: **28**, 2–7; Negishi EI, Coperet C, Ma S, Liou SY, Liu F. *Chem Rev* 1996; **96**: 365–93; Jeffery T, *Adv Met-Org Chem* 1996; **5**: 153–260.

14 de Meijere A, Meyer FE. *Angew Chem* 1994; **106**: 2473–506; *Angew Chem Int Ed Engl* 1994; **33**: 2379–411.

15 Shibasaki M, Boden CDJ, Kojima A. *Tetrahedron* 1997; **53**: 7371–95, and references cited therein; Hayashi T, Kubo A, Ozawa F. *Pure Appl Chem* 1992; **64**: 421–7, and references cited therein; Ashimori A, Matsuura T, Overman LE. *J Org Chem* 1993; **58**: 6949–51; Loiseleur O, Hayashi M, Schmees N, Pfaltz A. *Synthesis* 1997: 1338–45, and references cited therein; Tietze LF, Schimpf R. *Angew Chem* 1994; **107**: 1138–9; *Angew Chem Int Ed Engl* 1994; **33**: 1089–91.

16 Hii KK, Claridge TDW, Brown JM. *Angew Chem* 1997; **109**: 1033–6; *Angew Chem Int Ed Engl* 1997; **36**: 984–7, and references cited therein.

17 For the mechanism of oxidative addition, see: Amatore C, Pflüger F. *Organometallics* 1990; **9**: 2276–82; Jutand A, Mosleh A. *Organometallics* 1995; **14**: 1810–17; Cianfriglia P, Narducci V, Sterzo CL *et al. Organometallics* 1996; **15**: 5220–30.

18 Bräse S, Waegell B, de Meijere A. *Synthesis* 1998: 148–52.

19 Albéniz AC, Espinet P, Lin Y-S. *J Am Chem Soc* 1996; **118**: 7145–52, and references cited therein.

20 See: Heesing A, Müllers W. *Chem Ber* 1980; **113**: 19–23.

21 Herrmann WA, Brossmer C, Oefele K *et al. Angew Chem* 1995; **107**: 1989–92; *Angew Chem Int Ed Engl* 1995; **34**: 1844–8; Herrmann WA, Elison M, Fischer J, Köcher C, Artus GRJ. *Angew Chem* 1995; **107**: 2602–5; *Angew Chem Int Ed Engl* 1995; **34**: 2371–4; Herrmann WA, Köcher C. *Angew Chem* 1997; **109**: 2257–82; *Angew Chem Int Ed Engl* 1997; **36**: 2162–82.

22 Ben-David Y, Portnoy M, Gozin M, Milstein D. *Organometallics* 1992; **11**: 1995–6.

23 Voigt K, Schick U, Meyer FE, de Meijere A. *Synlett* 1994: 189–90.

24 Hillers S, Sartori S, Reiser O. *J Am Chem Soc* 1996; **118**: 2087–8; Hillers S, Reiser O. *Tetrahedron Lett* 1993; **34**: 5265–8; Hillers S, Reiser O. *Chem Commun* 1996: 2197–8.

25 For a change of product distribution by the application of high pressure in a Heck reaction, see: Sugihara T, Takebashi M, Kaneko C. *Tetrahedron Lett* 1995; **36**: 5547–55.

26 Perkovic T, de Meijere A. To be published.

27 Reiser O, Reichow S, de Meijere A. *Angew Chem* 1987; **99**: 1285–6; *Angew Chem Int Ed Engl* 1987; **26**: 1277–8; Reiser O, König B, Meerholz K *et al. J Am Chem Soc* 1993; **115**: 3511–18.

28 Jeffery T, *Tetrahedron Lett* 1985; **26**: 2667–70; *Tetrahedron* 1996; **52**: 10113–30.

29 For the application of Heck conditions in multiple Heck couplings, see: Tao W, Nesbitt S, Heck RF, *J Org Chem* 1990; **55**: 63–9.

30 Sanechika K, Yamamoto T, Yamamoto A. *Bull Chem Soc Jpn* 1984; **57**: 752–5.

31 König B, de Meijere A. *Synlett* 1997: 1221–32.

32 Moroz AA, Shvatsberg MS, Kotlyarevskii IL. *Bull Acad Sci USSR Div Chem Sci* 1979; **28**: 795–8.

33 Diercks R, Armstrong JC, Boese R, Vollhardt KPC. *Angew Chem* 1986; **98**: 270–1; *Angew Chem Int Ed Engl* 1986; **25**: 268–9; Diercks R, Vollhardt KPC. *J Am Chem Soc* 1986; **108**: 3150–2.

34 Praefcke K, Singer D, Gündogan B, Gutbier K, Langner M. *Ber Bunsenges Phys Chem* 1994; **98**: 118–22; Praefcke K, Kohne B, Singer D. *Angew Chem* 1990; **102**: 200–202; *Angew Chem Int Ed Engl* 1990; **29**: 177–9.

35 Lansky A, Reiser O, de Meijere A. *Synlett* 1990: 405–7.

36 König B, Knieriem B, de Meijere A. *Chem Ber* 1993; **126**: 1643–50.

37 Voigt K, von Zezschwitz P, de Meijere A *et al. J Org Chem* 1999; **64**: 1521–34.

38 Oshima M, Yamazaki H, Shimizu I, Nisar M, Tsuji J. *J Am Chem Soc* 1989; **111**: 6280–7.

39 Evans DA, Golob AM, *J Am Chem Soc* 1975; **97**: 4765–6.

40 Knühl G, Sennhenn P, Helmchen G. *J Chem Soc Chem Commun* 1995: 1845–6.

41 Fryzuk MD, Bosnich B. *J Am Chem Soc* 1978; **100**: 5491–4.

42 von Zezschwitz P, forthcoming *Dissertation*, Universität Göttingen.

43 Voigt K, Lansky A, Noltemeyer M, de Meijere A. *Liebigs Ann Chem* 1996: 899–911.

44 Bräse S, Rümper J, Voigt K *et al. Eur J Org Chem* 1998; **1**: 671–8.

45 de Meijere A, Song ZZ, Lansky A *et al. Eur J Org Chem* 1998: 2289–9.

46 Lansky A. *Dissertation*, Universität Göttingen, 1992.

47 For the parent hexaethenylbenzene, see: Bollinger JM, Burke JJ, Arnett EM. *J Org Chem* 1966; **31**: 1310. Krüger C, Yalpani M. *Chem Ber* 1983; **116**: 3359–65.

48 For a stepwise approach to hexastyrylbenzenes, see: Meier H, Hanold N, Kalbitz H. *Synthesis* 1997: 3276–8.

49 Prinz P, Lansky A, Haumann T *et al. Angew Chem* 1997; **109**: 1343–6; *Angew Chem Int Ed Engl* 1997; **38**: 1289–92.

50 Boese R, Green JR, Mittendorf J, Mohler DL, Vollhardt KPC. *Angew Chem* 1992; **104**: 1643–5; *Angew Chem Int Ed Engl* 1992; **31**: 1643–5.

51 McGaffin G, de Meijere A. *Synthesis* 1994: 583–91.

52 McGaffin G. *Dissertation*, Universität Hamburg, 1993.

53 Binger P, Büch HM. *Topics Curr Chem* 1987; **135**: 77–151; Binger P, Schuchardt U. *Chem Ber* 1981; **114**: 1649–55.

54 Untiedt S, de Meijere A. *Chem Ber* 1994; **127**: 1511–15.

55 Messner M, Kozhushkov SI, de Meijere A. *Eur J Org Chem* (In preparation); de Meijere A, Messner M, Vill V. *Mol Cryst Liq Cryst Sci Technol Sect A* 1994; **257**: 161–8.

56 See: Plevyak JE, Heck RF. *J Org Chem* 1978; **43**: 2454–6; Brenda M, Greiner A, Heitz W. *Makromol Chem* 1990; **191**: 1083–100; Heitz W, Brügging W, Freund L *et al. Makromol Chem* 1988; **189**: 119–27; Beller M, Fischer H, Kühlein K. *Tetrahedron Lett* 1994; **35**: 8773–6.

57 Rümper J, Sokolov VV, Rauch K, de Meijere A. *Chem Ber/Recueil* 1997; **130**: 1193–5.

58 For a review, see: Catellani M, Chiusoli GP. *Gazz Chim Ital* 1993; **123**: 1–7; Catellani M, Ferioli L. *Synthesis* 1996: 769–72; for 2:1 coupling products, see: Catellani M, Chiusoli GP. *J Organomet Chem* 1982; **239**: C35–37; Catellani M, Chiusoli GP, Costa M. *J Organomet Chem* 1995; **500**: 69–80.

59 Albrecht K, Reiser O, Weber M, Knieriem B, de Meijere A. *Tetrahedron* 1994; **50**: 383–401.

60 Reiser O, Weber M, de Meijere A. *Angew Chem* 1989; **101**: 1071–2; *Angew Chem Int Ed Engl* 1989; **28**: 1037–8.

61 Weber M. *Dissertation*, Universität Hamburg, 1992.

62 Albrecht K, Reiser O, Weber M, Knieriem B, de Meijere A. *Tetrahedron* 1994; **50**: 383–401.

63 Albrecht K, de Meijere A. *Chem Ber* 1994; **127**: 2539–41.

64 Albrecht K, Reiser O, Weber M, de Meijere A. *Synlett* 1992: 521–3.

65 Psiorz M, Hopf H. *Angew Chem* 1982; **94**: 639–40; *Angew Chem Int Ed Engl* 1982; **21**: 628–9.

66 Rauch K, Albrecht K, de Meijere A. (To be published.)

67 Dyker G. *Angew Chem* 1994; **106**: 117–19; *Angew Chem Int Ed Engl* 1994; **33**: 103–35; *Chem Ber* 1994; **127**: 739–42; *Angew Chem* 1992; **104**: 1079–81; *Angew Chem Int Ed Engl* 1992; **31**: 1023–5.

68 Brenda M, Knebelkamp A, Greiner A, Heitz W. *Synlett* 1991: 809–10.

69 Albrecht K. *Dissertation*, Universität Göttingen, 1992.

70 Henniges H, Meyer FE, Schick U *et al. Tetrahedron* 1996; **52**: 11545–78.

71 Meyer FE, Brandenburg J, Parsons PJ, de Meijere A. *J Chem Soc Chem Commun* 1992; 390–92.

72 Meyer FE, de Meijere A. *Synlett* 1991: 777–8.

73 Meyer FE, Parsons PJ, de Meijere A. *J Org Chem* 1991; **56**: 6487–8.

74 Parsons PJ, Stefanovic M, Willis P, Meyer FE. *Synlett* 1992: 864–6.

75 Trost BM, Pfrengle W, Urabe H. *J Am Chem Soc* 1992; **114**: 1923.

76 Meyer FE, Henniges H, de Meijere A. *Tetrahedron Lett* 1992; **33**: 8039–42.

77 Schweizer S, *Dissertation*, Universität Göttingen, 1998.

78 Trost BM, Shi Y. *J Am Chem Soc* 1992; **114**: 791–2.

79 For some other examples of palladium-catalysed cyclizations, see: Bouyssi D, Balme G, Faure R, Gore J. *Tetrahedron* 1992; **48**: 10103–14; Larock RC, Yum EK. *J Am Chem Soc* 1991; **113**: 6689–90.

80 Zhang Y, Wu G-Z, Agnel G, Negishi E. *J Am Chem Soc* 1990; **112**: 8590–92.

81 Inter- and intramolecular Pd-catalysed cross-coupling reactions have also been used by other groups in the construction of the steroid skeleton. See: Gauthier V, Cazes B, Goré J. *Tetrahedron Lett* 1991; **32**: 915–18; *Bull Soc Chim Fr* 1996; **133**: 563–79; Deng W, Jensen

MSA, Overman LE, Rucker MV, Vionnet JP. *J Org Chem* 1996; **61**: 6760–61; Tietze LF, Nöbel T, Spescha M. *Angew Chem* 1996; **108**: 2385–6; *Angew Chem Int Ed Engl* 1996; **35**: 2259–61.

82   Micheli RA, Hajos ZG, Cohen N *et al. J Org Chem* 1975; **40**: 675–81.

83   Yamauchi T, Oyama N, Yamane H, Mander LN. *Phytochemistry* 1995; **38**: 1345–8.

84   This type of reaction has previously been regarded as involving *o*-C-H activation, but the term 'Friedel Crafts type' alkylation might be more appropriate in view of the probable mechanism of this process: Brown D, Grigg R, Sridharan V, Tambyrah V. *Tetrahedron Lett* 1995; **36**: 8137–40.

85   Parsons PJ, Penkett CS, Shell AJ. *Chem Rev* 1996; **96**: 195–206; Heumann A, Réglier M. *Tetrahedron* 1996; **52**: 9289–346.

86   Negishi E. *Pure Appl Chem* 1992; **64**: 323–34.

87   Trost BM, Shi Y. *J Am Chem Soc* 1991; **113**: 701–3; **115**: 9421–38.

88   Overman LE, Abelman MM, Kucera DJ, Tran VD, Ricca DJ. *Pure Appl Chem* 1992; **64**: 1813–19.

89   Coperet C, Ma S, Negishi E. *Angew Chem* 1996; **108**: 2255–7; *Angew Chem Int Ed Engl* 1996; **35**: 2125–6.

90   For early examples of inter–intermolecular one-pot domino Heck–Diels–Alder reaction, see: Dieck HA, Heck RF. *J Org Chem Soc* 1975; **40**: 1083–90; Mitsudo T, Fischetti W, Heck RF. *J Org Chem Soc* 1984; **49**: 1640–46.

91   Trost BM, Chung JYL. *J Am Chem Soc* 1985; **107**: 4586–8; Trost BM, Hipskind PA. *Tetrahedron Lett* 1992; **33**: 4541–4; Trost BM, Hipskind PA, Chung JYL, Chan C. *Angew Chem* 1989; **101**: 1559–61; *Angew Chem Int Ed Engl* 1989; **28**: 1502–4.

92   Grigg R, Sridharan V. In: Wilkinson G, ed. *Comprehensive Organometallic Chemistry.* Vol. 12. Oxford: Pergamon Press, 1995: 299–321; Grigg R, Redpath J, Sridharan V, Wilson D. *Tetrahedron Lett* 1994; **35**: 7661–4; Grigg R, Kennewell P, Teasdale A, Sridharan V. *Tetrahedron Lett* 1993; **34**: 153–6; Burns B, Grigg R, Sridharan V, Worakun T. *Tetrahedron Lett* 1988; **29**: 4325–8; Grigg R, Stephenson P, Worakun T. *Tetrahedron* 1988; **44**: 2033–48.

93   Grigg R, Rasul R, Savic V. *Tetrahedron Lett* 1997; **38**: 1825–8; Grigg R, Sansano JM, Santhakumar V *et al. Tetrahedron* 1997; **53**: 11803–26; Grigg R, Rasul R, Redpath J, Wilson D. *Tetrahedron Lett* 1996; **37**: 4609–12; Grigg R, Loganathan V, Sridharan V *et al. Tetrahedron* 1996; **52**: 11479–502; Burns B, Grigg R, Sridharan V, Worakun T. *Tetrahedron Lett* 1988; **29**: 4325–8.

94   Grigg R, Stevenson P, Worakun T. *Tetrahedron* 1988; **44**: 2033–48.

95   Ang KH, Bräse S, Steinig AG *et al. Tetrahedron* 1996; **52**: 11503–28.

96   Meyer FE, Ang KH, Steinig A, de Meijere A. *Synlett* 1994: 191–3.

97   Bhat L, Steinig A, de Meijere A. (To be published.)

98   Stolle A. *Dissertation*, Universität Hamburg, 1992.

99   Bräse S. *Dissertation*, Universität Göttingen, 1995.

100 Fournet G, Balme G, Barieux JJ, Gore J. *Tetrahedron* 1988; **44**: 5821–32.

101 Larock RC, Varaprath S. *J Org Chem* 1984; **49**: 3432–5.

102 Propargyl carbonates have been shown to react with methyl acrylate under palladium catalysis to give ethenylallenes. See: Mandai T, Ogawa M, Yamaoki H *et al. Tetrahedron Lett* 1991; **32**: 3397–8.

103 Steinig A. *Dissertation*, Universität Göttingen, 1997.

104 Products arising from intermediates of type **206** and certain nucleophiles have previously been isolated. See: Grigg R, Sridharan V, Xu L-H. *J Chem Soc Chem Commun* 1995: 1903–4.

105 McGaffin G, Michalski S, Stolle A *et al. Synlett* 1992: 558–60; Stolle A, Ollivier J, Piras PP, Salaün J, de Meijere A. *J Am Chem Soc* 1992; **114**: 4051–67.

106 Bräse S, de Meijere A. *Angew Chem* 1995; **107**: 2741–3; *Angew Chem Int Ed Engl* 1995; **34**: 2545–7.

107 Bräse S, Nüske H, de Meijere A. (To be published.)

108 Foerstner J, Kozhushkov SI, Binger P, Wedemann P, Noltemeyer M, de Meijere A, Butenschön H. *Chem Commun* 1998; 239–40.
109 de Meijere A, Kozhushkov SI, Khlebnikov AF. *Zh Org Khim* 1996; **32**: 1607–26; *Russ J Org Chem* 1996; **32**: 1555–75.
110 de Meijere A, Kozhushkov SI, Späth T, Zefirov NS. *J Org Chem* 1993; **58**: 502–5.
111 Binger P, Wedemann P, Kozhushkov SI, de Meijere A. *Eur J Org Chem* 1998: 113–19.

6 The New Age of Divalent Palladium Catalysis: an Account of Our Own Story

XIYAN LU and SHENGMING MA

Shanghai Institute of Organic Chemistry, Chinese Academy of Sciences, 354 Fenglin Lu, Shanghai 200032, China

1   Introduction

Recently, much attention has been focused on the transition-metal-catalysed coupling of alkynes with alkenes [1], as well as the cyclization of dienes, enynes and diynes [2]. Among them, particularly noteworthy in this regard are the ruthenium-catalysed coupling of alkynes and alkenes [3] and palladium-catalysed cycloisomerization of enynes to monocycles [2a]. Their synthetic applications in complex molecules have been highly successful.

We are interested in the method of assembling the γ-lactone ring by intramolecular enyne coupling: α-alkylidene-γ-butyrolactones would be constructed conveniently from the easily available acyclic allylic 2-alkynoate precursors (Scheme 6.1).

Scheme 6.1.

However, zerovalent palladium or palladium-hydride-catalysed cyclization of unsaturated allylic esters has not been studied, probably due to the possible allylic carbon–oxygen bond cleavage by the Pd(0) catalyst [4] or isomerization of the triple bond in the starting materials catalysed by the [PdH] species [5]. Therefore, it would be a challenge to find a catalyst for the cyclization of an allylic ester of unsaturated acids. In the literature where a divalent palladium complex is the catalytically active species, zerovalent palladium is generally formed during the reaction and then reoxidized to complete the catalytic cycle [6]. In some cases, the divalent palladium complex is used in stoichiometric amounts [7].

In spite of the wide application of palladium-catalysed enyne coupling reactions [2a], the corresponding enyne coupling methodologies suitable for lactone synthesis were limited. In our synthetic application studies directed at a number of bioactive γ-lactone natural products, we developed the facile intramolecular enyne cyclization of allylic 2-alkynoates to construct polysubstituted γ-butyrolactones. The unique feature of this strategy in stereochemical control was gradually recognized and applied in the synthesis of enantiopure targets [8].

2   Halopalladation-initiated cyclization of allylic alkynoates

Kaneda *et al.* reported the bis(benzonitrile)palladium-dihalide-catalysed codimerization of alkynes and allylic halides in which the divalent palladium was regenerated by dehalopalladation [9], thus it occurred to us that it might be possible to develop divalent palladium-catalysed reaction without the participation of the zerovalent palladium

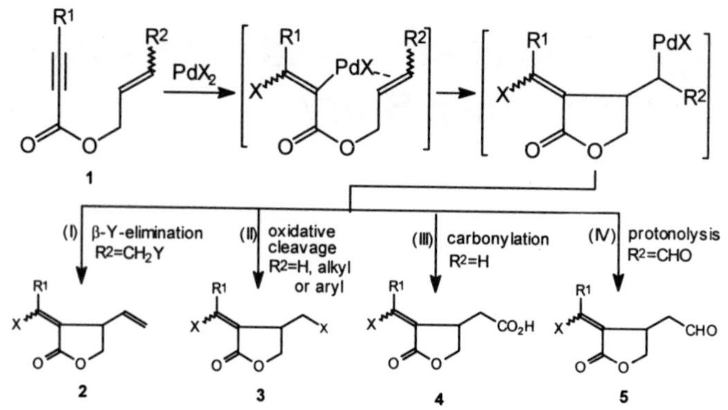

Scheme 6.2.

species. Allylic 2-alkynoates are a group of special enynes with an electron-deficient triple bond and an ester linkage between the double bond and triple bond. Based on the halopalladation elementary reaction of the addition of palladium halide to a triple bond [10], the α-alkylidene-γ–butyrolactone structure could be assembled easily through a halopalladation–carbon–carbon double bond insertion sequence and, in the carefully designed catalytic and substrate systems, β-vinyl-[8, 11], β-halomethyl-[12], β-oxycarbonyl-methyl-[13] and β-formylmethyl-substituted [14] γ-lactones were obtained by different quenching methods of the carbon–palladium bonds, e.g. β-heteroatom elimination, copper-halide-mediated oxidative cleavage, carbonylation and protonolysis of the carbon–palladium bond, respectively (Scheme 6.2).

These divalent palladium-catalysed cyclizations have several advantages over conventional Pd(0)- or PdH-mediated enyne couplings [15]: oxygen-free conditions are no longer required; and the organic-ligand-free catalyst could be recovered more easily. The α-haloalkylidene-γ-butyrolactone products provide great opportunities for further chemical transformations to give sophisticated molecules in a stereoselective manner.

The results of the four types of Pd(II)-catalysed reactions are summarized in Table 6.1.

2.1   *Diastereoselectivity and natural product syntheses*

2.1.1   DIASTEREOSELECTIVITY IN THE PD(II)-CATALYSED CYCLIZATION OF 4′-HETEROATOM-2′-ALKENYL 2-ALKYNOATES

When a substituent is introduced into the 1′-position of the 2′-alkenyl group of the starting 2-alkynoates, significant and interesting results were found in the stereochemistry of the cyclization of different alkynoates [11].

Table 6.2 is a brief summary of the stereochemical control in this cyclization reaction, in which *the substitution pattern of the triple bond determines the β,γ-relative stereochemistry of the lactone product.* Cyclization of *unsubstituted propynoates* mainly afforded *trans*-β,γ-disubstituted lactones (*Z*-allylic double bond in the substrate further increased the selectivity), while cyclization of *3-substituted 2-alkynoates* yielded *cis*-β,γ-disubstituted products (*E*-allylic double bond further increased the selectivity). The bulkiness of the R³ group also increases the diastereoselectivity. Thus, this reaction constitutes a highly efficient route for

Table 6.1. Palladium(II)-catalysed synthesis of γ-lactones from allylic alkynoates

Method	1	Method of quenching C—Pd bond	Reaction conditions[a]	Product	Reference
I	$R^2 = CH_2Y$ (Y = halogen, OAc, OH, OR)	β-Heteroatom elimination	$Pd(OAc)_2$, LiX, HOAc, rt	2	[8], [11]
II	R^2 = H, alkyl, aryl	Oxidative cleavage	PdX_2 CuX_2 LiX, CH_3CN, rt	3	[12]
III	R^2 = H	Carbonylation	$Pd_2(dba)_3 \cdot CHCl_3$, CO, $CuCl_2$, LiCl, HOAc, rt	4	[13]
IV	R^2 = CHO	Protonolysis	$Pd(OAc)_2$ LiX, HOAc, rt	5	[14]

[a] rt = room temperature.

Table 6.2. Factors controlling the diastereoselectivity of cyclization reaction (From [11])

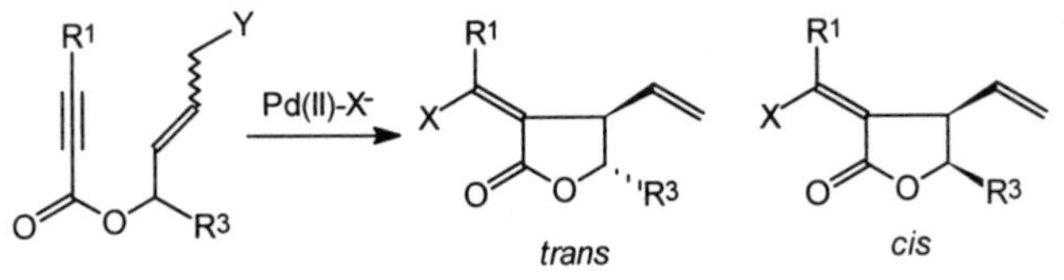

R^1	R^3	Configuration of allylic double bond	β,γ-Stereochemistry
H	Me	E	trans > cis
	Me	Z	trans
	i-Pr	E	trans
	i-Pr	Z	trans
Alkyl	Me	E	cis
	Me	Z	cis > trans
	i-Pr	E	cis
	i-Pr	Z	cis

constructing the γ-butyrolactone structural units with different stereochemistry in a single operation.

2.1.2 DIASTEREOSELECTIVITY IN THE PD(II)–CUCL$_2$-CATALYSED CYCLIZATION OF 2′-ALKENYL 2-ALKYNOATES

The diastereoselectivity in the Pd(II)–LiX–CuX$_2$-catalysed cyclization of 2′-alkenyl 2-alkynoates is similar to that in the Pd(II)–LiX-catalysed cyclization of 4′-heteroatom-2′-alkenyl 2-alkynoates but with different stereochemistry for the formation of exocyclic double bonds, as shown in Table 6.3 [12b,c].

Table 6.3. Factors controlling the diastereoselectivity of cyclization reaction II (From [12b,c])

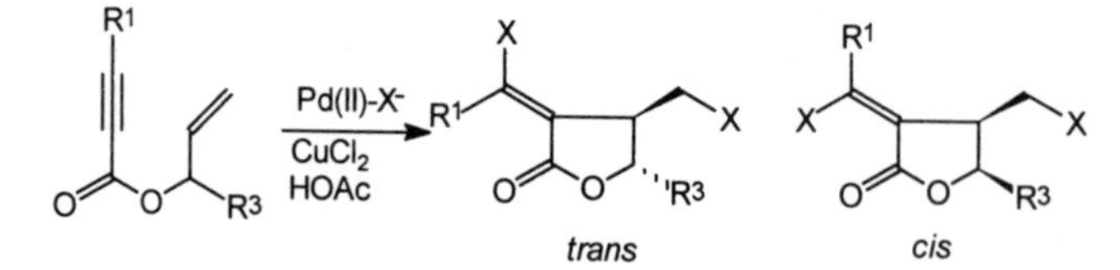

R¹	Configuration of exocyclic double bond	β,γ-Stereochemistry
H	E	trans
Alkyl	Z	cis

The diastereoselectivity is further supported by the cascade cyclization reaction [12b]. A monocyclic product was obtained in the reaction of 1′,5′-hexadien-3′-yl propynoate (**6**). The failure of the second cyclization might be due to the *trans*-configuration of the β, γ-disubstituents in the product. The reaction of 1′,5′-hexadien-3′-yl 3-substituted 2-alkynoates (**8**) under the same conditions yielded the *cis*-fused bicyclic α-(Z)-chloroalkyli-dene-γ-butyrolactone derivatives, implying *cis* diastereochemistry of the β,γ-disubstituents in the first cyclization (Scheme 6.3) [12b].

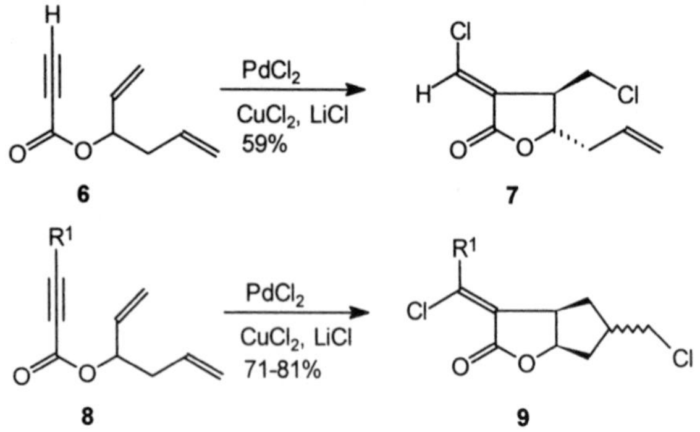

Scheme 6.3.

Similiar diastereoselectivity was observed in the cyclization of *N*-alkenyl 2-alkynamides under similiar conditions [16].

2.2   Natural product syntheses

2.2.1   SYNTHESIS OF (±)-A-FACTOR

With these methodologies, natural products with the α-alkylidene-γ-butyrolactone struc-tural unit are readily synthesized. For example, A-factor, an inducer of the biosynthesis of streptomycin in inactive mutants of *Streptomyces griseus* [17], was efficiently synthesized from cyclization product **11** by two simple transformations: alkaline hydrolysis of the bromomethyl unit gave the β-(hydroxymethyl)-γ-lactonic compound **12**; and the treatment of compound **12** with diethylamine followed by acid hydrolysis transformed the vinylic bromide to ketone to afford (±)-A-factor (**13**) (Scheme 6.4) [12c].

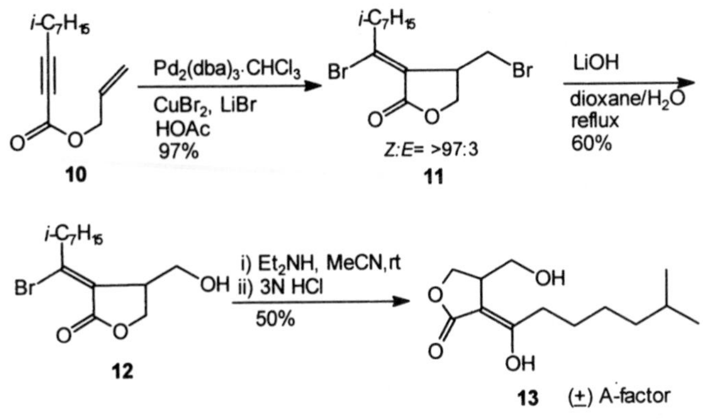

Scheme 6.4.

2.2.2 RELATIVE STEREOCHEMISTRY CONTROL

As stated, the above two reactions (methods I and II) provide a highly efficient route to stereodefined lactones in a single operation, i.e. the β,γ-diastereoselectivity of the lactone product can be controlled by the substituents on the triple bond of the starting 2-alkynoates [11, 12b,c].

For the α,β-disubstituted lactones, the cis-α,β-disubstituted lactones can be obtained easily from hydrogenation of the exocyclic double bond of α-methylene-γ-butyrolactones, as exemplified by the synthesis of ($\pm$)-isohinokinin (Scheme 6.5) [18].

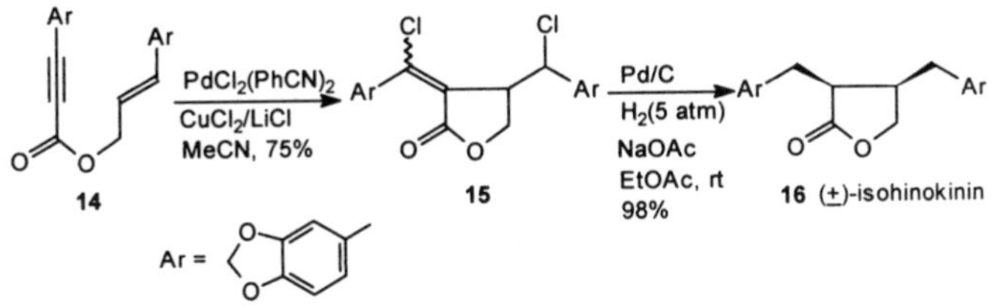

Scheme 6.5.

Lactone **15** was formed easily from the reaction of the starting ester **14** by method II. Hydrogenation of lactone **15** gave ($\pm$)-isohinokinin (**16**), which contains the cis-α,β-disubstituents. The cis-stereochemistry of the product can be deduced from the mechanism of hydrogenation [19].

For the trans-α,β-disubstituted lactones, the exocyclic double bond was first reacted with phenylthiol to give the thermodynamically favored trans-α,β-disubstituted product, followed by further transformation to yield the desired product, which was exemplified by the synthesis of phaseolinic acid (see next section).

2.2.3 ABSOLUTE STEREOCHEMISTRY CONTROL

α,β,γ-Trisubstituted lactones. As we have mentioned already, the most attractive feature of this enyne coupling methodology for the synthesis of lactones is that the stereochemistry of the β,γ-disubstituents can be controlled simply by the substituent pattern of the triple bond of the alkynoates. In addition, due to the well-documented methods for the

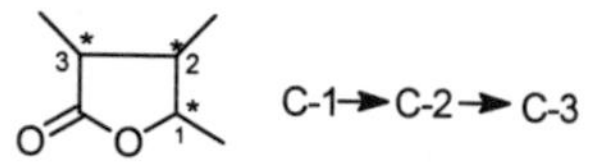

Scheme 6.6.

preparation of enantiopure allylic alcohols [20], it is easy to prepare the optically active starting esters, the chirality of which could be used as the chirality source for further stereocontrol. Thus, the configurations of different carbon atoms were induced in the sequence shown in Scheme 6.6.

Methylenolactocin, a small but richly functionalized and isomerization-prone γ-lactone, was synthesized using compound **17** as the starting material [21]. Intramolecular cyclization of **17** was accomplished under mild conditions (LiBr, Pd(OAc)$_2$, HOAc, room temperature (rt)). The lactone product **18** was obtained in 65% yield with extremely high diastereoselectivity. Having the basic structure and the defined stereochemistry of our target, the vinylic bromide of compound **18** was first reduced using the Zn–Ag couple in almost quantitative yield. The Michael addition reaction with PhSH protected the electron-deficient C=C bond selectively to give compound **20**. Ozonolysis of the C=C bond in compound **20** at −78 °C afforded an unstable aldehyde, which was further oxidized to the carboxylic acid with pyridinium dichromate (PDC) in dimethylformamide (DMF). Finally, (−)-methylenolactocin was obtained by regeneration of the α-methylene unit, which was performed by sodium metaperiodate oxidation to sulfoxide followed by thermolysis (Scheme 6.7) [21].

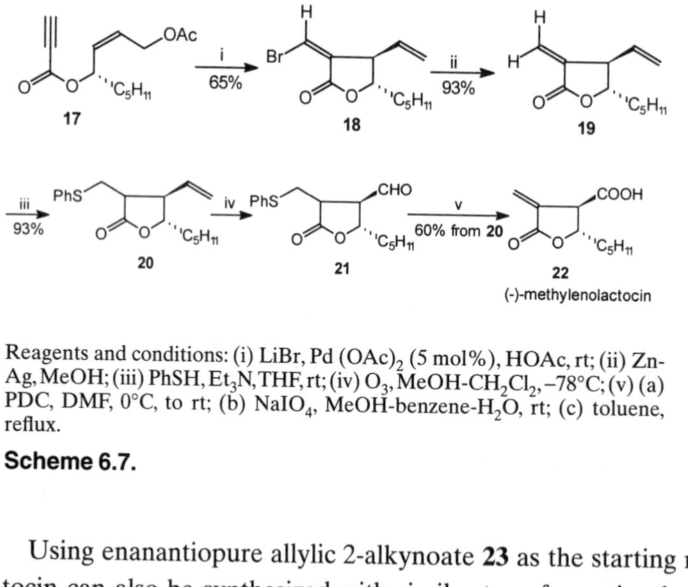

Reagents and conditions: (i) LiBr, Pd (OAc)$_2$ (5 mol%), HOAc, rt; (ii) Zn-Ag, MeOH; (iii) PhSH, Et$_3$N, THF, rt; (iv) O$_3$, MeOH-CH$_2$Cl$_2$, −78°C; (v) (a) PDC, DMF, 0°C, to rt; (b) NaIO$_4$, MeOH-benzene-H$_2$O, rt; (c) toluene, reflux.

Scheme 6.7.

Using enanantiopure allylic 2-alkynoate **23** as the starting material, (−)-methylenolactocin can also be synthesized with similar transformation by method II [22]. The cyclic product **24** was obtained in 90% yield with 71:29 (*trans*:*cis*) diastereoselectivity. After hydrolysis of the sp³ C—Br bond, the desired isomer **25** was separated easily by column chromatography on silica gel from its *cis*-isomer **26** (Scheme 6.8).

It should be pointed out that the relative stereochemistry of β,γ-disubstituents in methylenolactocin is *trans*, thus we used the allylic *unsubstituted* propynoate **17** or **23** as the starting material. The stereochemical control of this methodology can be exemplified further by the synthesis of phaseolinic acid, which contains *cis*-β,γ-substituents.

According to the diastereoselectivity mentioned above, the problem of synthesizing

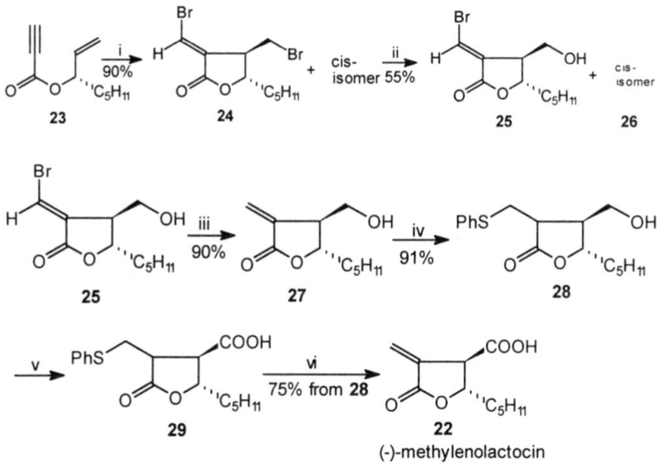

Reagents and conditions: (i) CuBr$_2$, LiBr, Pd(OAc)$_2$ (5 mol%), HOAc, rt; (ii) CaCO$_3$, DMSO-H$_2$O, 100°C; (iii) Zn-Ag, MeOH, 70°C; (iv) PhSH, Et$_3$N, THF rt; (v) (a) PDC, DMF, 0°C, to rt; (vi) (b) NaIO$_4$, MeOH-benzene-H$_2$O, 0°C rt; then reflux in toluene.

Scheme 6.8.

enantiopure phaseolinic acid lies in how to choose a proper substituent on the triple bond in the starting alkynoates. The substituent should induce cis-β,γ-selectivity and be removed easily after the cyclization. It occurred to us that the trimethylsilyl group would meet these two criteria [23].

Thus, allylic 3-trimethylsilylpropynoate **30** was cyclized in the presence of Pd(OAc)$_2$ and LiCl in HOAc. The cyclic product **31**, formed as an *E,Z* mixture referred to the exocyclic double bond, behaves with 100% cis-β,γ-diastereoselectivity. The interesting point is that the single product **33** was obtained from both isomers through desilylation and dechlorination in high yield. As mentioned before, the electron-deficient double bond was protected selectively by reaction with phenylthiol to give the thermodynamically favored *trans*-α,β–*cis*-β,γ-diastereomer The one-pot oxidation with NaIO$_4$–RuCl$_3$.H$_2$O in CCl$_4$–MeCN–H$_2$O of the C=C bond at the β-position to the carboxyl group and of the thioether to sulfone, followed by desulfonation with Na–Hg, afforded phaseolinic acid **35** (Scheme 6.9) [23].

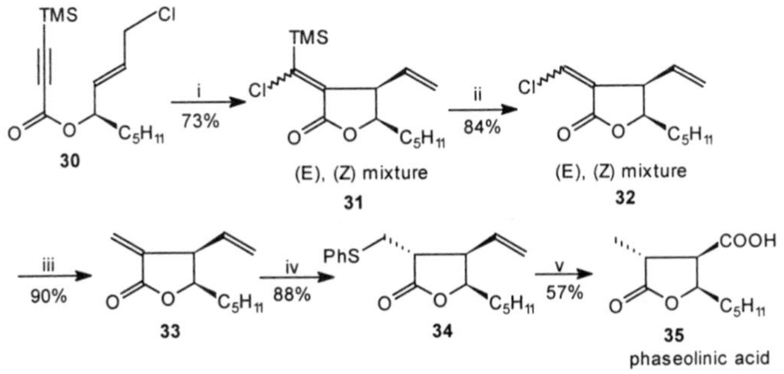

Reagents and conditions: (i) Pd(OAc)$_2$, LiCl HOAc, rt; (ii) TBAF, HOAc, THF, 0°C-rt; (iii) Zn-Ag, MeOH, reflux; (iv) PhSH, Et$_3$N, THF rt; (v) (a) NaIO$_4$, RuCl$_3$ · H$_2$O, CCl$_4$-MeCN-H$_2$O, rt; (b) Na-Hg, NaH$_2$PO$_4$, MeOH, –20°C.

Scheme 6.9.

It should be pointed out that with this method the relative stereochemistry of α,β-substituents is *trans*, which is what we expect in the synthesis. If hydrogenation was used, a *cis*-α,β-disubstituted product would be predicted [18]. Thus, the stereochemistry of α,β-substituents can be controlled conveniently in this way.

From the above examples it is interesting to us that we can completely control the diastereochemistry of α,β,γ-trisubstituted lactones very easily just by choosing different procedures. It is also worth noting that this method permits the synthesis of target molecules in either of the enantiomeric forms, by simply starting with different enantiomers of the allylic alcohols [22].

α,β-Disubstituted lactones. In all α,β,γ-trisubstituted lactones described above, the stereochemistry of the γ-carbon atom derived from the enantiomerically pure allylic alcohols is the source of chirality for the whole molecule. For the synthesis of α,β-disubstituted lactones without a γ-substituent, we need a new strategy to synthesize them.

trans-α,β-Disubstituted lactones [24]. In lactones synthesized from method II, there is a chloromethyl group in the β-position of the lactone ring product **37**. Should the lactone ring be opened and the carboxyl group linked with the chloromethyl group, a new α,β-disubstituted-γ-butyrolactone (**38**) could be obtained with a defined configuration. If we started from a homochiral ester **36** derived from an optically active allylic alcohol, α,β-disubstituted-γ-butyrolactones (**38**) could be obtained in optically active form (Scheme 6.10).

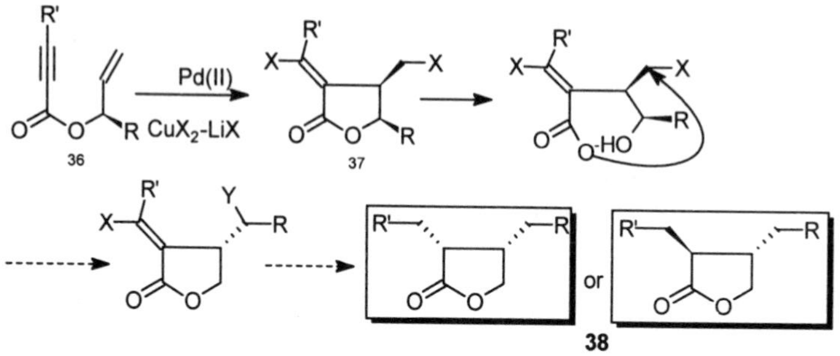

Scheme 6.10.

This strategy has been exemplified by the synthesis of isohomopilopic acid **42** [24]. It is suggested that hydrogenation of the exocyclic double bond will occur first to give the *cis*-α,β-disubstituted product. Then hydrogenolysis opens the lactone ring, followed by lactonization again to form the *trans*-α,β-disubstituted product **41**. By this method, the *trans*-α,β-disubstituted product can be obtained highly stereoselectively (Scheme 6.11).

cis-α,β-Disubstituted lactones. The synthesis of *cis*-α,β-disubstituted lactones can be realized by using method IV. Recently, 4'-oxo-2'-butenyl butynoate **44** was found to cyclize under the catalysis of Pd(OAc)$_2$ in the presence of LiBr and HOAc at room temperature (rt) to yield the lactonic aldehyde **45** (Scheme 6.12) [14].

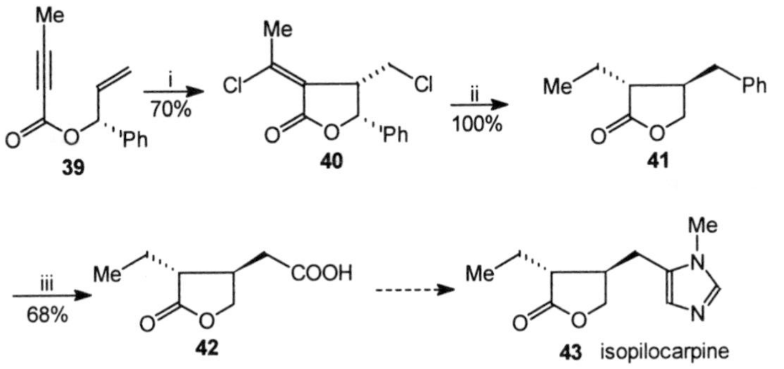

Reagents and conditions: (I) CuCl$_2$, LiCl, Pd(OAc)$_2$, HOAc, rt; (ii) NaOAc, Pd-C, MeOH, H$_2$ (6 atm), rt; (iii) NaIO$_4$, RuCl$_3 \cdot$ H$_2$O, CCl$_4$-MeCN-H$_2$O, 0°C to art.

Scheme 6.11.

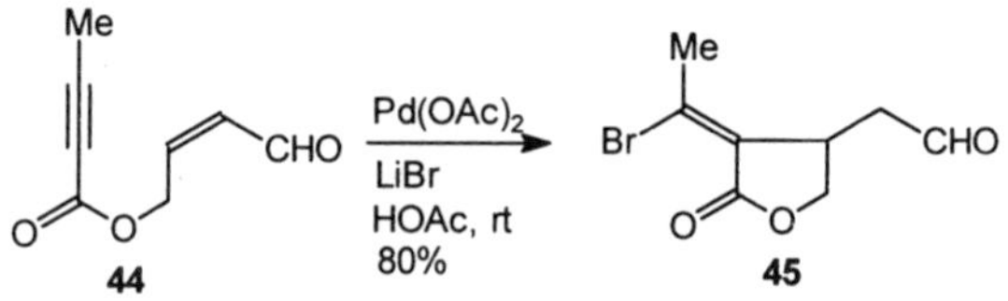

Scheme 6.12.

This cyclization reaction can be regarded as an intramolecular version of the nucleophile–alkyne–α,β-unsaturated carbonyl coupling reaction [25]. The mechanism of this reaction can be speculated as follows: halopalladation of the triple bond gives the vinylic palladium intermediate **46** followed by double bond insertion to form the 2-oxoalkylpalladium intermediate **47**. The key step to form the product **45** and regenerate the Pd(II) catalytic species is protonolysis of the carbon–palladium bond in compound **47** (Scheme 6.13).

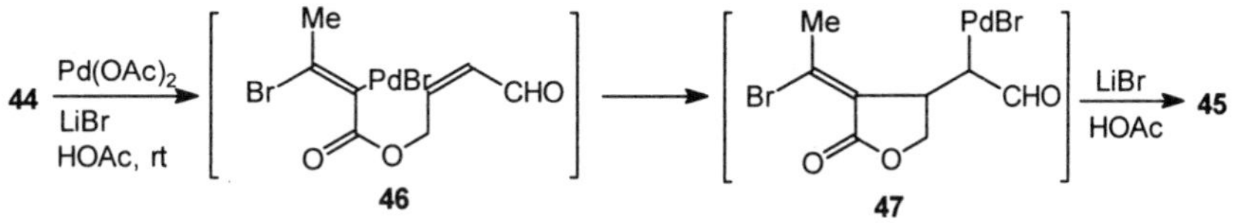

Scheme 6.13.

Compound **45** can be transformed to both enantiomers of lactonic aldehyde **51** by first converting to the diastereomeric acetals with (+)-dimethyl 2,3-*O*-isopropylidene-D-tartrate, separating the diastereomers using column chromatography, hydrogenating and then hydrolyzing. Thus, a formal synthesis of (+)-pilocarpine was developed (Scheme 6.14) [14].

Here, the strategy for synthesizing *cis*- or *trans*-α,β-disubstituted lactones is the same as before. We obtain the enantiopure product by resolution of the diastereomers of the acetals of the lactonic aldehyde.

Using these methodologies, γ-butyrolactones with different stereochemistry can be synthesized, as shown in Scheme 6.15.

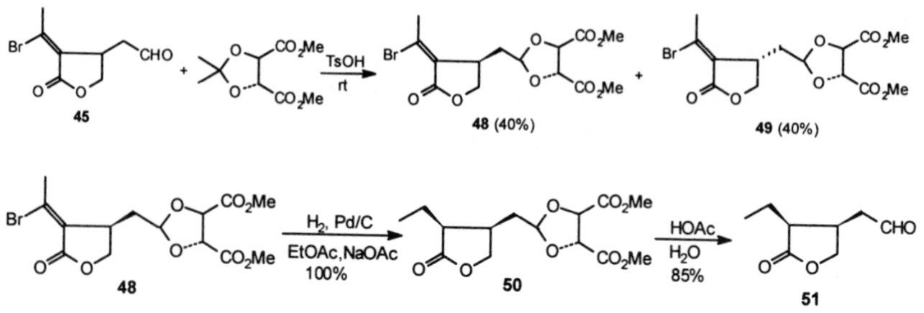

Scheme 6.14.

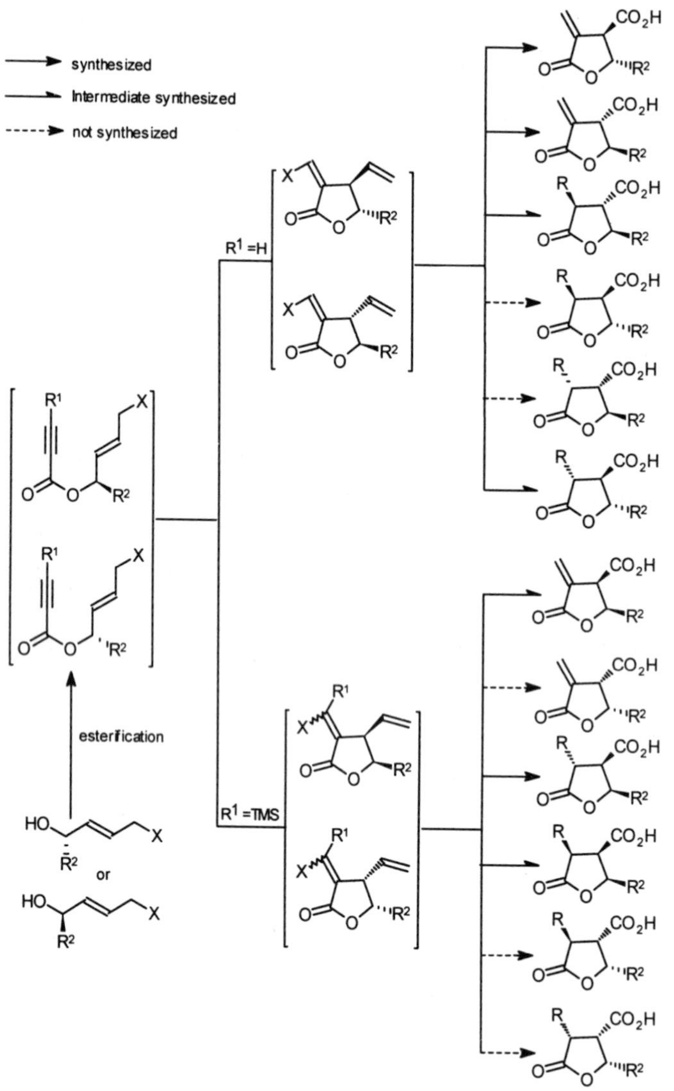

Scheme 6.15.

2.3   Elementary reactions related to the cyclization of enyne esters

After the careful study of PdCl$_2$-catalysed cyclization of allylic alkynoates, we studied in detail the following elementary reactions: halopalladation (for methods I–IV), β-heteroatom elimination (for method I), oxidative cleavage of carbon–palladium bonds (for method II) and protonolysis of carbon–palladium bonds (for method IV).

2.3.1 HALOPALLADATION

Vinylpalladium complexes are versatile intermediates in a number of catalytic reactions such as vinyl couplings [26], carbonylations [27], Heck reactions [28] and enyne cycliza-tions [2a]. In general, the vinylpalladium species is generated by oxidative addition of a vinyl halide or triflate to Pd(0), which requires oxygen-free conditions. Another way of generating a vinylpalladium species is by the addition of a nucleophile to an alkyne–Pd(II) complex [2a, 9, 10c, 29].

The *E*- or *Z*-selectivity of the exocyclic double bond in the cyclization product was mostly controlled by the stereochemical course of the halopalladation [8,10]. While the polarity-directed halopalladation determines the regiospecificity of the reaction to give a five-membered ring product, *cis*- and *trans*-halopalladations lead to *E*- and *Z*-isomers, respectively. Halopalladation of alkynes has been studied extensively [10,30]. The higher mole ratio of LiCl and polar solvents usually favors *trans*-halopalladation [9,10,30]. We found that the stereochemistry was affected mainly by the polarity of the solvent and the concentration of halide, which is in accordance with the results reported [9,30]. In non-polar solvents only low to moderate yields with poor selectivity were observed. Table 6.4 summarizes the influence of solvent and halide concentration in terms of the *E,Z*-selectivity of the reaction. While low halide concentration was not applicable in these reactions, the increase of halide amount from low to high provided facile control of the *E,Z*-selectivity.

The stereochemistry of the exocyclic double bond of this cyclization reaction is much higher than that of the PdCl$_2$-catalysed intermolecular coupling reaction of alkynoic esters with allylic halides [9, 31]. This might be rationalized by the following two reasons: the for-mation of the palladium–enyne complex in the reactants may favor the attack of halide ion from outside of the coordination sphere, which would be expected to give *trans*-halopal-ladation leading to *Z*-selectivity [30, 32]; and in the intramolecular cyclization, the insertion of a carbon–carbon double bond into the carbon–palladium bond is faster than the isomer-ization of the kinetically controlled *trans*-halopalladation product (Scheme 6.16).

Table 6.4. Influence of halide concentration on the stereochemistry of the exocyclic double bond in cyclization products

Reaction type	**1**	LiX (equiv)	Product	Yield (%)	*Z/E*	Reference
I (X=Br)	R^1 = Me	0.5	**2**	68	78:22	[8b]
		2		82	89:11	[8b]
		4		82	>99:5	[8b]
II (X=Br)	R^1 = Me	0	**3**	42	77:23	[12c]
		2		85	90:10	[12c]
		4		95	>97:3	[12c]
		6		93	>97:3	[12c]
		10		91	>97:3	[12c]
III (X=Cl)	R^1 = Me	2	**4**	70	33:67	[13]
		4		66	90:10	[13]
		6		64	>97:3	[13]

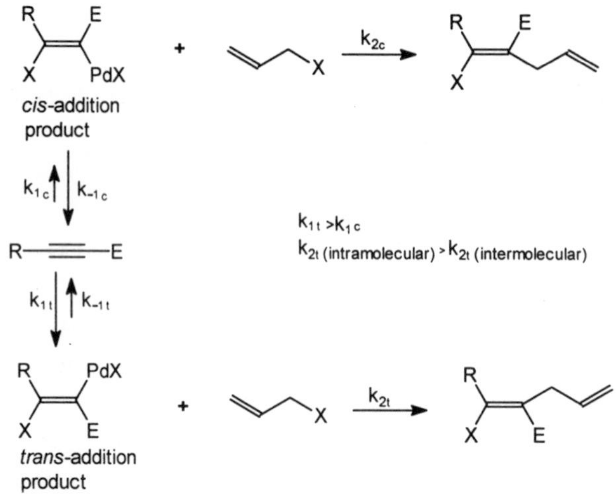

Scheme 6.16.

Table 6.5. Influence of CuX_2 on the stereochemistry of the exocyclic double bond in cyclization products

Reaction type	1	Product	Z or E	Reference
I (without CuX_2)	$R^1 = H$	2	Z	[8b]
	$R^1 = $ alkyl		Z	[8b]
II (with CuX_2)	$R^1 = H$	3	E	[12a]
	$R^1 = $ alkyl		Z	[12a]

Another factor influencing the selectivity of halopalladation was observed in the type II reaction of allylic propynoates. In the presence of copper(II) salt, halopalladation of propynoates uniformly gave E-isomers of the cyclization product, in contrast to the results of other 3-substituted alkynoates (Table 6.5) [12a].

The unusual selectivity of method II may be due to the different behavior of the terminal alkynes and non-terminal alkynes to CuX_2 or $CuCl_2$–$PdCl_2$ complexes [33], the mechanism of which is still to be explored [12d].

2.3.2   β-HETEROATOM ELIMINATION

β-Hydride elimination, known as β-elimination, is the *syn*-elimination of β-hydrogen with Pd as H–Pd–X, which takes place with alkyl palladium complexes to afford palladium hydride and alkene. The palladium hydride species is easily converted to Pd(0) species, especially in the presence of a base [6]. In order to regenerate a divalent palladium species in the catalytic cycle, a heteroatom-substituted methyl group was introduced into compound **1** in method I, in which β-heteroatom elimination was involved in the quenching of the carbon–palladium bond.

Although the elimination of a β-heteroatom with the metal might be a general reaction, relatively few examples are known and the generality of the process is still limited [34–38]. Using the cyclization of 4′-heteroatom-2′-alkenyl 2-alkynoates to afford α-alkylidene-γ-butyrolactone (method I) as a model, we studied the reactivity and stereochemistry of the β-heteroatom elimination.

Reactivity of elimination of Pd and different heteroatoms β to Pd [8c, 39]. The cyclization of 4′-heteroatom-2′-heptenyl propynoate (**52**) occurred smoothly under the catalysis of Pd(II) in the presence of LiCl in HOAc at room temperature (reaction condition: alkynoates (1.0 mmol), Pd(OAc)$_2$ (0.05 mmol), LiCl (4.0 mmol) and HOAc (5 mL) at rt).

When the leaving group in substrate **52** was a chlorine atom (Y = Cl), the cyclization proceeded quickly to give the β-chloride elimination product as the sole cyclic product in good yield without competitive β-hydride elimination. For the acetate substrates (Y = OAc), the cyclization reaction also occurred smoothly to give the β-acetoxy elimination product as the sole product under similar conditions, but the reaction was slower than the corresponding chloro-substituted substrates. When the leaving group was switched to a methoxyl (Y = OMe), the reaction also proceeded quickly to give one cyclic product through β-methoxyl elimination. Much slower reaction was observed in the case of a hydroxy substrate (Y = OH). In addition, besides the β-heteroatom elimination product **53**, the β-hydride elimination product **54** was also isolated (Scheme 6.17). Thus, the ability of β-elimination of different leaving groups at the β-position of palladium follows the order: Cl > OMe > OAc > OH ~ H.

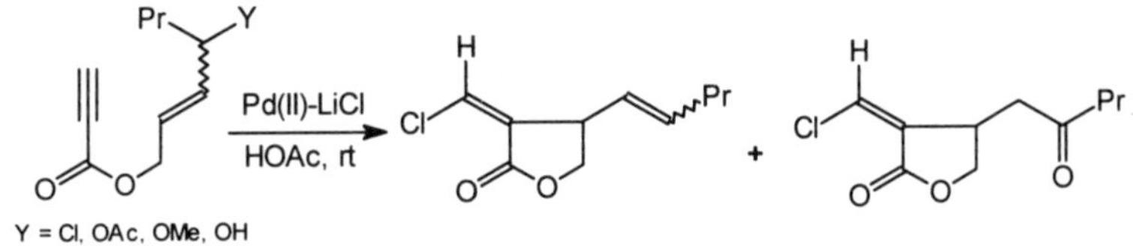

Scheme 6.17.

Stereochemistry of elimination of palladium and the leaving group β to palladium [39]. Although the fact that *syn*-periplanar arrangement of palladium and β-hydride is required for the β-hydride elimination reaction is well documented [40], studies on the stereochemistry of elimination of palladium and a heteroatom on the carbon β to palladium are rare. Daves reported that palladium acetate elimination required an antiperiplanar arrangement of palladium and the acetoxy group [37]. Non-stereospecific elimination of palladium chloride was reported by Herry [32]. The difficulty in studying the stereochemistry of β-heteroatom elimination lies in the lack of a suitable model for the reaction. Even in the cyclization of propynoates (**52**), both the *re*-face and *si*-face of the carbon–carbon double bond in the substrate could coordinate to vinylpalladium, which made the intermediate formed by intramolecular insertion variable, and a random stereochemistry of the elimination product was obtained [39].

In our previous work [11a], we found that diastereoselectivity in the intramolecular carbon–carbon double bond insertion step could be controlled by the substituents at the 3- and 1′-positions of the cyclization precursors. For 2-butynoates (**55**), the substituent propyl group at the 1′-position could control the stereoselectivity in the intramolecular insertion step to give the *cis*-lactone as the sole product. On the basis of such a stereoselectivity, when we choose stereodefined 2-butynoate (**55**) as the model compound to study the stereochemistry of β-heteroatom elimination, a stereodefined cyclic intermediate **57** with a newly formed σ-carbon–palladium bond could be generated. Subsequent elimination of palladium and the leaving group would afford the lactone product with a Z- or E-double bond at the β-position of the lactone ring according to the steric requirement of the elimination (Scheme 6.18).

The results in Table 6.6 show that the elimination of palladium and leaving groups such as chloride, acetoxy and hydroxy all occurred in a highly stereoselectively manner as shown in Scheme 6.18.

Anti-elimination of palladium and the leaving group would afford an E-carbon–carbon double bond of the alkenyl group at the β-position of the lactone, while a Z-carbon–carbon double bond, which is stable under the cyclization conditions [11b], would result from *syn*-elimination. During our study, E-carbon–carbon double bond products were obtained as the sole products in all cases, implying that the elimination required the antiperiplanar

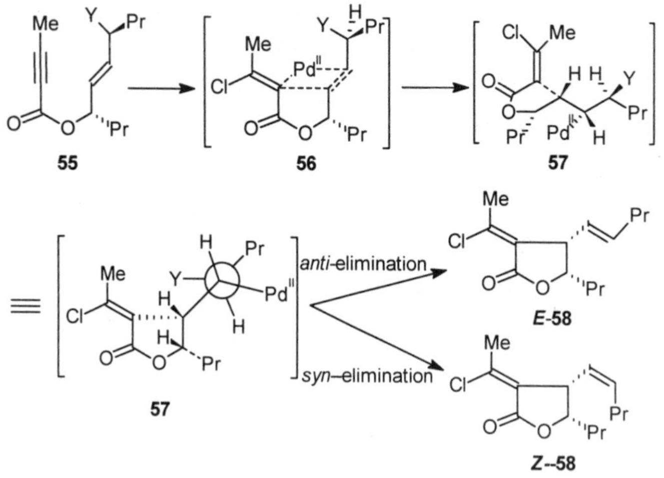

Scheme 6.18.

Table 6.6. Stereochemistry of elimination of palladium and the leaving group β to palladium

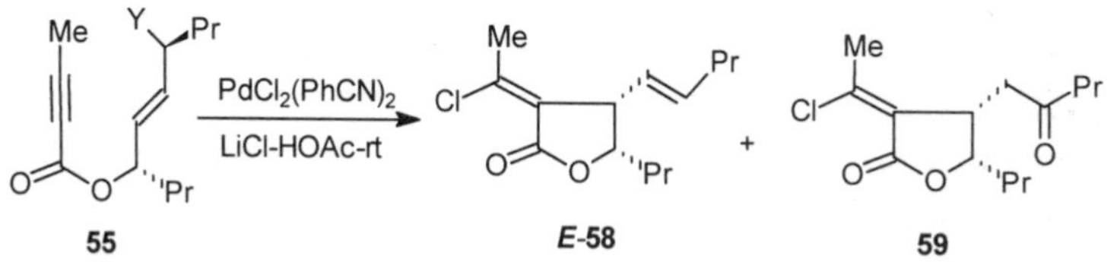

			Isolated yield (%)	
Substrate	Leaving group Y	Time (h)	**58** (*E/Z*)	**59**
55a	OH	60	23 (>97:3)	51
55b	OAc	60	91 (>97:3)	
55c	Cl	60	85 (>97:3)	

Reaction conditions: alkynoates (1.0 mmol), PdCl$_2$(PhCN)$_2$ (0.05 mmol), LiCl (4.0 mmol) and HOAc (5 mL) at rt.

arrangement of the palladium and the leaving group, such as chloride, acetoxy and hydroxyl.

In all cases, high chloride concentration and acid conditions were used in the reaction. Based on the stereochemistry of the β-heteroatom elimination, it is most probable that the β-heteroatom elimination occurred in an E_2-like mechanism. The chloride ion will attack the palladium and then eliminate the protonated leaving group in a *trans* manner (Scheme 6.19). Of course, further detailed study of the mechanism is required.

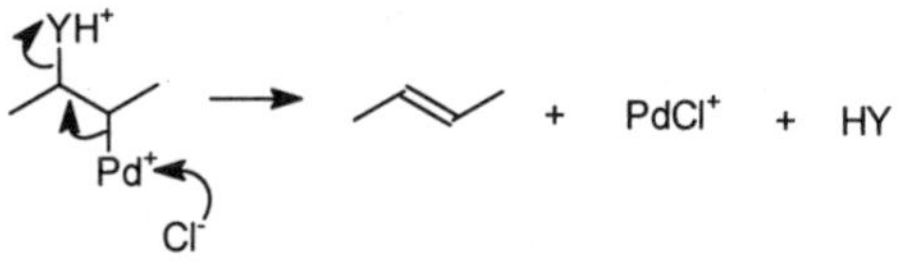

Scheme 6.19.

2.3.3 OXIDATIVE CLEAVAGE OF THE CARBON–PALLADIUM BOND

In method II, oxidative cleavage with $CuCl_2$ was used to quench the carbon–palladium bond. Oxidative cleavage of carbon–palladium bonds takes place in a number of palladium-catalysed reactions in the presence of a nucleophile [41]. The oxidizing agent usually weakens the carbon–palladium bond so that the palladium is turned into a good leaving group [42]. As a result, an organic product is formed by nucleophilic substitution (Scheme 6.20).

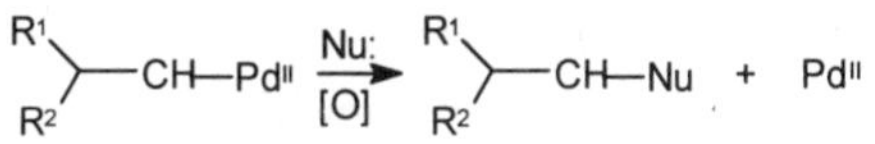

Scheme 6.20.

Although oxidative cleavage of carbon–palladium bonds has been studied exclusively, the detailed mechanism of these reactions is still unknown. In order to provide evidence for the mechanism of the reaction, the stereochemistry of such cleavage has been studied. According to the stereochemical results, the cleavage reaction has been proposed to proceed by reductive elimination [43], by a radical mechanism [44] or by an ionic mechanism involving nucleophilic displacement at the carbon atom [45]. Bäckvall observed that cupric chloride induced cleavage of β-oxoallylpalladium bonds occurred with predominant inversion at the carbon atom in the presence of free chloride ion [45a].

We used method II of the cyclization reaction as a model to study the stereochemistry of oxidative cleavage of the carbon–palladium bond. When 3′-phenyl-2′(Z)-[and 2′(E)-]-propenyl 2-butynoates ((Z)-**60** and (E)-**60**) were used as model compounds, they afforded different diastereoisomers **61** and **62**, respectively, under the same cyclization conditions ($CuCl_2$ (5 equiv), LiCl (2 equiv), $PdCl_2(PhCN)_2$ (0.05 equiv), in MeCN at rt) [12a], implying that the oxidative cleavage of carbon–palladium bonds shows high stereoselectivity in our system (Scheme 6.21).

Because the configuration of the diastereoisomers could not be assigned exactly from the ^{1}H-NMR spectra, they were assigned tentatively according to Bäckvall's mechanism in our early publications [12a,45a]. We were surprised to learn from the X-ray diffraction results that the stereochemistry of the oxidative cleavage in our cyclization reaction is in contrast to Bäckvall's results, i.e. in our system the oxidative cleavage of carbon–palladium

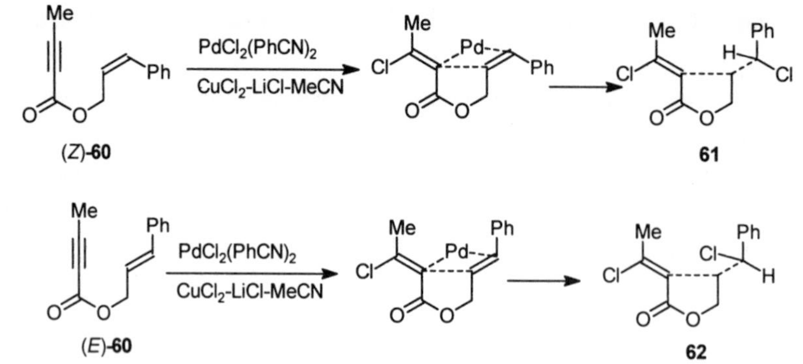

Scheme 6.21.

bonds by $CuCl_2$ occurred with predominant retention at the carbon atom even in an excess of chloride ion [47]. To the best of our knowledge, the only example that oxidative cleavage of the carbon–palladium bond by $CuCl_2$ occurred with retention of configuration is the $CuCl_2$ cleavage of the β-phenethyl–palladium bond [46], in which carbonium character was important, as revealed by the anchimeric assistance. In order to rule out the unusual stereochemistry of the phenyl group, compounds with alkyl (methyl and propyl) groups were tried and these showed the same stereochemical results. It is suggested that oxidative cleavage of the carbon–palladium bond in this case may proceed via a reductive elimination or S_N1-S_Ni type of mechanism [48].

Our model has the advantage that an isolated carbon–palladium bond was used as the model to study the stereochemistry of the oxidative cleavage reaction, as compared to other studies in which a bulky or possible coordination group was present near the palladium atom that may participate and induce different stereochemical results [45].

2.3.4   PROTONOLYSIS OF THE CARBON–PALLADIUM BOND

The carbon–palladium bond, like most second- and third-row transition-metal–carbon bonds, reacts very slowly in hydrolysis reactions. This is in contrast to alkaline- and alkaline-earth-metal–carbon bonds, where hydrolysis is rapid.

Vinylpalladium species usually react with α,β-unsaturated carbonyls to give vinylation products through β-hydride elimination of the insertion intermediate rather than addition products through protonolysis (Scheme 6.22) [28,49]. There are a few reports concerning the quenching of carbon–palladium bonds by acid hydrolysis [50], but the formation of addition products was influenced by many factors [51].

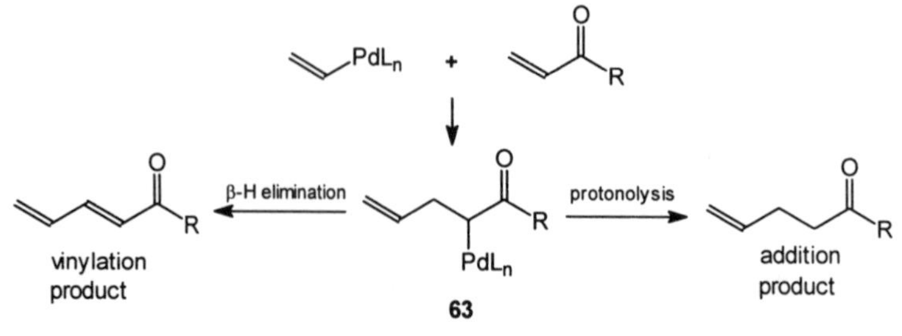

Scheme 6.22.

Inhibition of the normal β-hydride elimination pathway was achieved by addition of excess phosphine [52], but this also inhibited the insertion step when we attempted the reaction of acrolein with the vinylpalladium generated by halopalladation of alkyne. With the idea that an increase in electron density at the palladium center might also diminish *cis*-hydride elimination [52b] and that excess halide ligand might serve this purpose, we found that the addition of the vinylpalladium intermediate to acrolein did occur in the presence of excess LiBr in HOAc [25b]. Thus, a divalent palladium-catalysed halide–alkyne–α,β-unsaturated carbonyl coupling reaction was developed to form the γ,δ-unsaturated carbonyl compounds [25]. In this case, the protonolysis of the (2-oxoalkyl)palladium intermediate effectively recycles the catalytic species. The intramolecular version of this halide–alkyne–α,β-unsaturated carbonyl coupling was developed as method IV, in which the protonolysis was used to quench the carbon–palladium bond and was applied to the synthesis of (+)-homopilopic aldehyde (see Section 2.2.3) [14].

The ease of this process in our case might be due to several factors: the large excess of halide ion may make the β-hydride elimination less feasible through electron donation to palladium and occupancy of the coordination site needed for *cis*-hydride elimination; and intermediate **63**, which is actually a mesomeric palladium enolate, may readily undergo a heterolytic Pd—O fission by nucleophilic attack of halide ion on the palladium center [50i].

2.4 *Comparison between [PdCl]- and [PdH]-catalysed enyne cyclization*

2.4.1 CYCLIZATION OF HOMOALLYLIC ALKYNOATES

We have discussed the cyclization of allylic alkynoates catalysed by $PdCl_2$. Trost developed the elegant palladium-hydride-catalysed enyne cycloisomerization reaction, which is stereoselective and atom economical [2a]. In our early strategy of palladium-catalysed cyclization of allylic alkynoates, low-valent palladium and palladium hydride were avoided on account of the possible allylic carbon–oxygen bond cleavage in the allylic ester substrates [4]. Thus, the cyclization of homoallylic 2-alkynoates with [PdCl] and [PdH] catalytic systems was compared [15].

The results of cyclization of homoallylic 2-alkynoates using the $Pd(OAc)_2$–HOAc catalyst system are shown in Table 6.7.

The cyclizations proceeded well, as expected, and afforded the δ-lactones **65** in good yields, with *Z*-isomers (referring to the α-exocyclic double bond) predominating. For homoallylic 2-heptynoate (R^1 = *n*-Bu, R^2 = H), however, the reaction afforded no identifiable products, which might be due to the side reaction of the carbon–carbon triple bond with [PdH] species [5]. The reaction path can be rationalized via a mechanism similar to one that has appeared in the literature: hydropalladation of the triple bond gave the vinylpalladium intermediate **66**, which after intramolecular olefin insertion and β-hydride elimination afforded the cyclization product **65** (path a, Scheme 6.23).

The stereochemistry of the exocyclic carbon–carbon double bond in compound **65** depends upon the manner of the hydropalladation step. Transition metal hydrides generally add to the carbon–carbon triple bond to give *cis*-addition complexes [53]. The formation of *E,Z*-mixtures in the products may be due to the result of the [PdH]-mediated equilibrium of the products (Scheme 6.24) [5a].

Table 6.7. The Pd(OAc)$_2$–HOAc-catalysed cycloisomerization of homoallylic 2-alkynoates

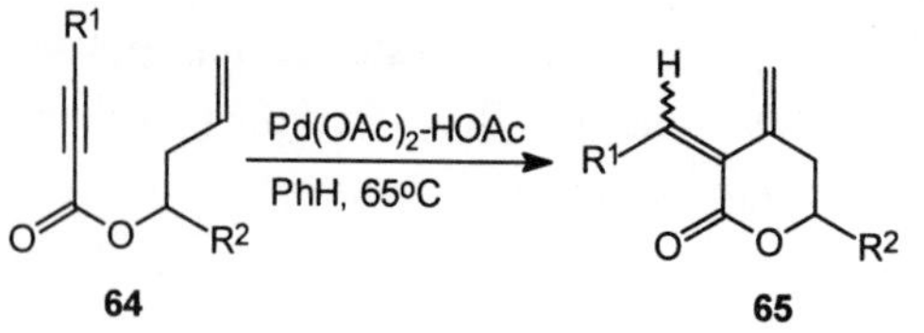

Substrate **64**		Product **65**	
R^1	R^2	Yield (%)	Z/E
Me	H	87	82:18
Ph	H	82	80:20
Me	Me	69	67:33
Ph	Me	75	75:25
n-Bu	H	Complex mixture	

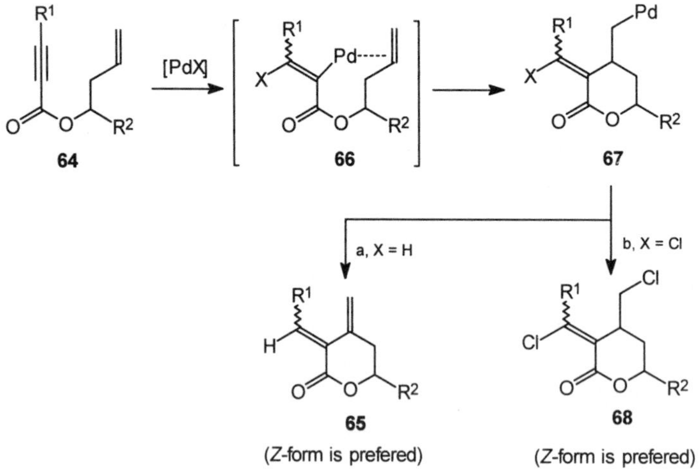

Scheme 6.23.

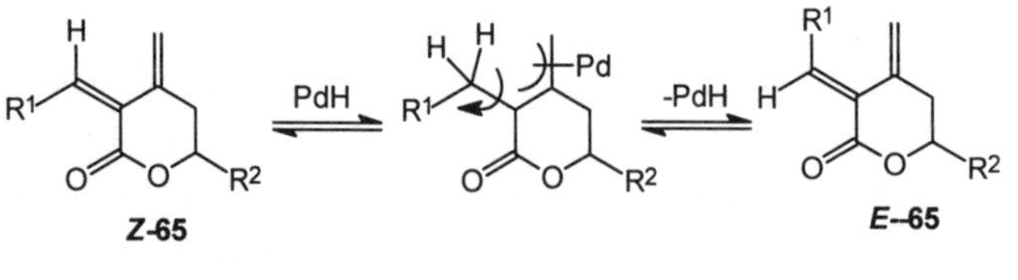

Scheme 6.24.

In the presence of PdCl$_2$, CuCl$_2$ and LiCl in MeCN, the reaction of compound **64** at room temperature yielded product **68** as shown in Table 6.8. In the case of homoallylic 2-butynoates, the reaction afforded product (Z)-**68** only, while in the case of homoallylic 2-propynoates, product (E)-**68** was the sole product. This result is consistent with our early reports [12].

1,3-Stereoinduction in the present reaction was also studied through the introduction of 1′-substituents in the ester group. For both 2-butynoates and 2-propynoates of 1′-

Table 6.8. The PdCl$_2$-catalysed cyclization of homoallylic 2-alkynoates

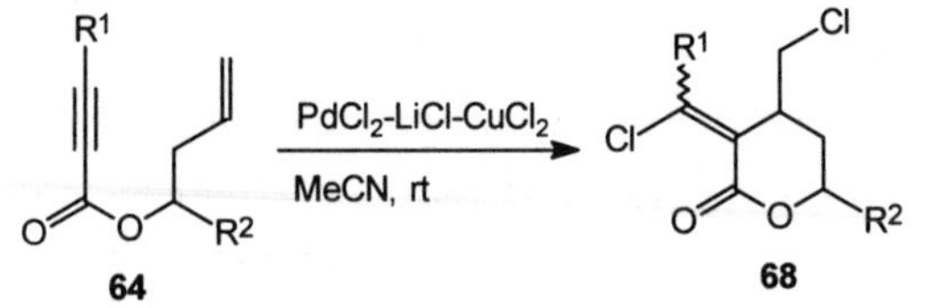

Substrate **64**		Product **68**		
R^1	R^2	Yield (%)	Z/E	*trans/cis*
Me	H		>97:3	
Me	Me	86	>97:3	>97:3
Me	i-Pr	72	>97:3	>97:3
Me	Ph	77	>97:3	>97:3
H	H	67	<3:97	
H	Me	68	<3:97	>97:3
H	i-Pr	76	<3:97	>97:3
H	Ph	77	<3:97	>97:3

substituted homoallylic alcohols, the reaction gave all one diastereomer. In contrast to our previous results on allylic alkynoate cyclization [12b,c], the *cis–trans* selectivity was irrelevant to the substitution of the triple bond; all the substrates studied gave β,δ-*trans* products only.

2.4.2　CYCLIZATION OF ω-ALKEN-2-YNOATES

Under Trost's [PdH] conditions, the cycloisomerization of 1,6- and 1,7-enynes took place in an *exo, exo* manner (5-*exo*-dig and 5-*exo*-trig for the triple bond and the double bond), giving alkylidene–cyclopentane and alkylidene–cyclohexane derivatives, respectively. The same regioselectivity was obtained when an electron-withdrawing group was introduced on the triple bond (path a, Scheme 6.25) [2a].

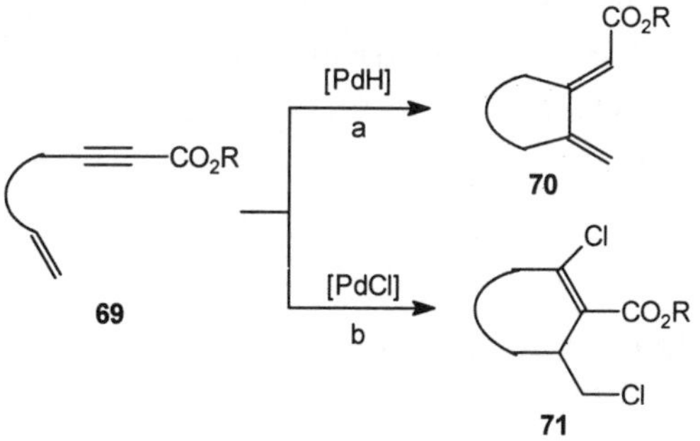

Scheme 6.25.

For the PdCl$_2$-catalysed reactions, the direction of chloropalladation of the bond conforms to their electronic properties. The nucleophilic chloride ion attacks the β-carbon of the 2-alkynoates, forcing the carbon–palladium bond to form at the α-carbon; and a different regiochemistry for cyclization of ω-alken-2-alkynoates was observed (path b) [15].

From the results of the cyclization of compounds **64** and **69** mediated by the two catalytic systems, some conclusions with respect to regioselectivity, stereoselectivity and substrate limitations could be drawn.

Although both hydropalladation and chloropalladation of a carbon–carbon triple bond can give vinylpalladium intermediates, their regioselectivities are different: in contrast to the palladium hydride species whose addition direction is irrelevant to the polarization of the triple bond [5], halopalladation strictly conforms to the electronic requirements. Thus, [PdH]-catalysed cyclization of compound **69** afforded an *exo, exo* product, which stems from the ring strain requirement; [PdCl] on the other hand, only performs Michael-type addition to electron-deficient triple bonds, leading to *endo, exo* cyclization products in spite of the unfavorable factors.

In the cyclization of homoallylic alkynoates, [PdH] and [PdCl] catalytic systems both gave δ-valerolactone derivatives in good yields, but propynoates or alkynoates with longer alkyl chains are not suitable substrates in the [PdH]-catalysed cyclization: with the former, oxidative coupling products between two substrate molecules formed quickly with pre-cipitated metallic palladium; with the latter, the reaction only afforded a complex mixture.

Furthermore, in the reaction preparing bis(alkylidene)cycloalkane or δ-valerolactone derivatives under [PdH] conditions, the yield decreased markedly on elongation of the reaction time. The [PdH]-induced side reactions of the polyunsaturated products may explain these results [10b]. This, compared with the clean reactions with the [PdCl] cat-alytic system, also indicated that chloropalladation is more selective in discriminating dif-ferent types of carbon–carbon multiple bonds in the substrates and products than hydropalladation.

In contrast to the cycloisomerization of simple enynes using [PdH] conditions, poor *Z/E* selectivity was encountered for homoallylic alkynoate substrates. However, when the [PdCl] catalytic system was used, the reaction gave a single geometric isomer for each sub-strate (*Z* for 3-substituted 2-alkynoates and *E* for unsubstituted propynoates). This may also be rationalized on the basis of the chemical selectivity of chloropalladation: unlike [PdH] species, palladium chloride does not react with the α-alkylidene-δ-valerolactone products to induce isomerization. When isomerization was prohibited, the stereochemistry was mainly determined by the step of alkyne chloropalladation: stereospecific chloropal-ladation (*cis* addition for propynoates and *trans* addition for 3-substituted 2-alkynoates) leads to a stereodefined carbon–carbon double bond with high selectivity.

The above results implied that the chloropalladation process of electron-deficient alkynes is much more selective than the corresponding hydropalladation with regard to regioselectivity and stereoselectivity.

3   Divalent palladium-catalysed tandem coupling of nucleophile, alkyne and α,β-unsaturated carbonyls

3.1   *Divalent palladium-catalysed halide–alkyne–α,β-unsaturated carbonyl coupling*

The results that halide ion can change the direction of the reaction of vinylpalladium inter-mediates and α,β-unsaturated carbonyls from vinylation to conjugate addition were applied to the coupling of alkynes and α,β-unsaturated carbonyls to give the conjugate addition products, γ,δ-unsaturated carbonyls. When methyl propynoate (**72**) was treated with acrolein in the presence of $Pd(OAc)_2$ and LiBr in acetic acid, two products **73** and **74** were isolated (Scheme 6.26) [25a].

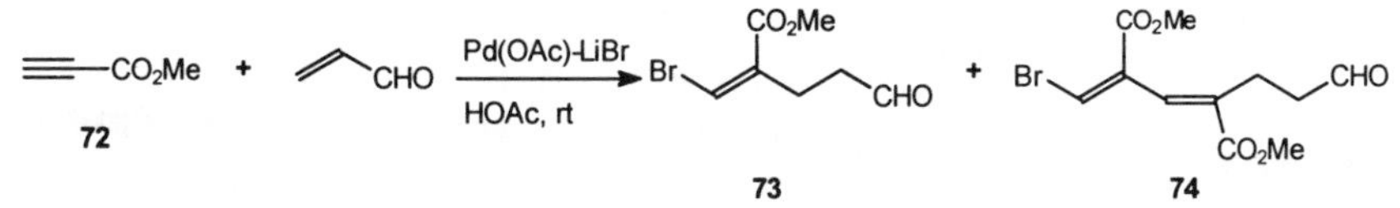

Scheme 6.26.

The formation of these products was rationalized by the mechanism in Scheme 6.27.

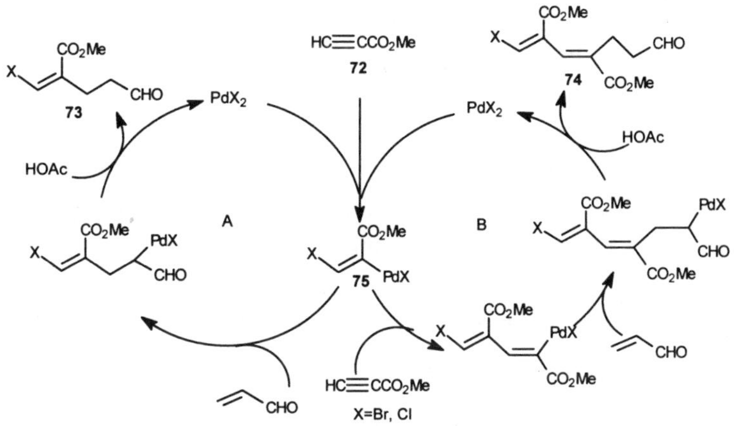

Scheme 6.27.

Vinylpalladium species **75** was first formed by *trans*-halopalladation of methyl propynoate (**72**) in HOAc, followed by acrolein insertion (path A) or subsequent alkyne and acrolein insertion (path B). The competing alkyne insertion may be ascribed to the higher reactivity of alkynes than alkenes. Both reaction paths involved protonolysis of the carbon–palladium bond to regenerate the catalytic Pd(II) species, giving compounds **73** and **74**, respectively.

The reaction selectivity can be controlled by the rate of addition of methyl propynoate (**72**) to give higher yields of compound **73**. Lithium chloride can also be used. Electron-deficient alkynes and phenylacetylene all gave good yields of 1:1 codimerized products with acrolein and methylvinylketone. The regiochemistry was affected by the electronic properties of the substituent on the triple bond, in accordance with the results reported by Kaneda *et al.* [9]; the halide ion always attacks the more electropositive carbon atom. Highest stereoselectivity was observed for the propynoates: only Z-isomers were produced.

3.2 *Divalent palladium-catalysed nucleophile–alkyne–α,β-unsaturated carbonyl coupling*

Oxypalladation of triple bonds is a process analogous to halopalladation [54]. With the hope that oxypalladation of alkynes followed by insertion of α,β-unsaturated carbonyls may lead to 1,5-carbonyl compounds, we found that an intramolecular carboxyl did effect this oxypalladation–acrolein coupling sequence [25b]. Using $Pd(OAc)_2$ as the catalyst, 3-heptynoic acid (**76**) and acrolein (5 equiv) in the presence of excess LiBr (2 equiv) in HOAc gave the cyclized product **77** in 85% yield (path a, Scheme 6.28).

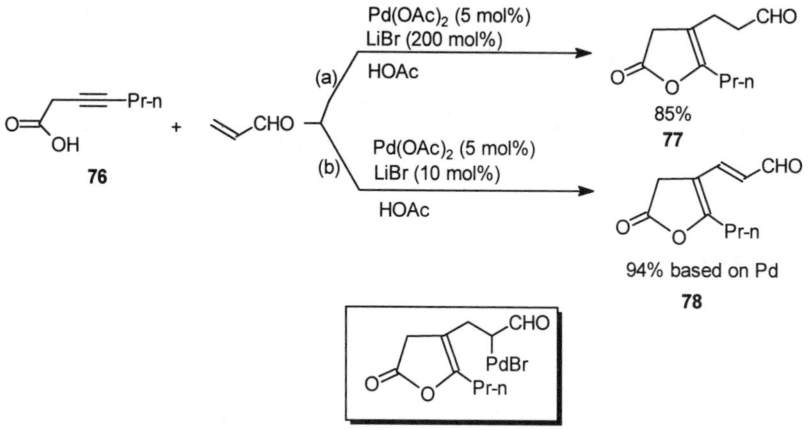

Scheme 6.28.

In the presence of decreased amounts of LiBr, compound **78** produced by β-hydride elimination was obtained in 94% yield (based on Pd), together with precipitated palladium black (path b). This, again, demonstrates the significant role of halide ion in the present reaction.

The scope of the cyclizing coupling reaction was examined as shown in Scheme 6.29.

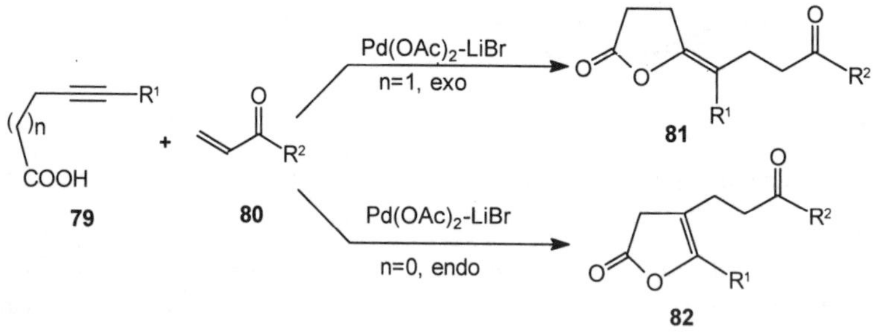

Scheme 6.29.

It was noteworthy that only intramolecular oxypalladation products were obtained, although halopalladation was in competition. 3-Butynoic acid itself did not give the cyclized product.

4   Conclusion

The new enyne coupling methodology for the synthesis of bioactive lactones has the advantages of facile control of the stereochemistry of both α,β- and β,γ-substituents, the simplicity of introducing a stereogenic center into the starting materials and the successful resolution of the lactonic aldehyde through the tartrate acetals. It is still a challenge to develop the asymmetric version of this enyne coupling.

The nucleophile–alkyne–α,β-unsaturated carbonyl coupling reaction permits facile entry to γ,δ-unsaturated carbonyl compounds and lactonic aldehydes (or ketones) from halide and carboxyl nucleophiles, respectively, and may well be extended to other nucleophiles to give carbocyclic or different heterocyclic structures. The good regio- and stereoselectivity, the high catalytic efficiency and the mild conditions ensuring compatibility for sensitive functional groups should make these methods an attractive tool in synthesis.

All the reactions discussed were initiated with a divalent palladium complex followed by a sequence of elementary reactions: nucleopalladation, olefin insertion and quenching of the carbon–palladium bond to afford the final products. Different methods of quenching of carbon–palladium bonds were studied, which were symbolized by direct regeneration of the divalent palladium species to complete the catalytic cycle. These divalent palladium-catalysed reactions have the great advantage that oxygen-free conditions are no longer needed. In addition, all these reactions showed the significant role of halide ions, which stimulated our interest in a mechanistic study.

5 Acknowledgments

We wish to thank our co-workers whose names appear on the cited papers. Financial support of these works by the National Natural Science Foundation of China and the Chinese Academy of Sciences is greatly acknowledged.

6 References

1 Trost BM, Müller TJJ, Martinez J. *J Am Chem Soc* 1995; **117**: 1888; Trost BM, Indolse AF, Müller TJJ, Treptow B. *J Am Chem Soc* 1995; **117**: 615; Trost BM, Flygare JA. *J Org Chem* 1994; **59**: 1078; Darcel C, Bruneau C, Albert M, Dixneuf PH, *Chem Commun* 1996: 919.

2 (a) Trost BM, *Acc Chem Res* 1990; **23**: 34; (b) RajanBabu YV, Nugent WA, Taber DF, Fagen PJ, *J Am Chem Soc* 1988; **110**: 7128; (c) Negishi E. *Pure Appl Chem* 1992; **64**: 323; (d) Vollhardt KPC. *Angew Chem Int Ed Engl* 1984; **23**: 539; (e) Negishi E, Takahashi T. *Synthesis* 1988: 1; (f) Shove, NE. *Chem Rev* 1988; **88**: 1081.

3 Trost BM. *Chem Ber* 1996; **129**: 1313.

4 Yamamoto A. *Organotransition Metal Chemistry.* New York: Wiley, 1986: 233.

5 (a) Ma D, Lin Y, Lu X, Yu Y. *Tetrahedron Lett* 1988; **29**: 1045; (b) Trost BM, Schmidt T. *J Am Chem Soc* 1988; **110**: 2303; (c) Ma D, Yu Y, Lu X. *J Org Chem* 1989; **54**: 1105; (d) Lu X, Ma D. *Pure Appl Chem* 1990; **62**: 723.

6 Tsuji J. *Palladium Reagents and Catalysis: Innovations in Organic Synthesis.* Chichester: J Wiley, 1995: 19.

7 Fugami K, Oshima K, Utimoto K. *Tetrahedron Lett* 1976: 2975; Hosokawa T, Shimo N, Maeda K, Sonoda A, Murahashi S. *Tetrahedron Lett* 1976: 383; Jintoku T, Fujiwara Y, Kawata I, Kawauchi T, Taniguchi H. *J Organomet Chem* 1990; **385**: 297.

8 (a) Ma S, Lu X. *J Chem Soc Chem Commun* 1990: 733; (b) Ma S, Lu X. *J Org Chem* 1991; **56**: 5120; (c) Ma S, Lu X. *J Organomet Chem* 1993; **447**: 305; (d) Lu X, Ma S, Ji J, Zhu G, Jiang H. *Pure Appl Chem* 1994; **66**: 1501.

9 Kaneda K, Uchiyama T, Fujiwara Y, Imanaka T, Teranish S. *J Org Chem* 1979; **44**: 55.

10 (a) Dieth H, Reinheimer H, Moffat J, Maitlis PM. *J Am Chem Soc* 1970; **92**: 2276; (b) Maitlis PM. *The Organic Chemistry of Palladium,* Vol. 1. New York: Academic Press, 1971: 47; (c) Bäckvall JE, Nilsson YIM, Gatt RGP. *Organometallics* 1995; **14**: 4242.

11 (a) Ma S, Zhu G, Lu X. *J Org Chem* 1993; **58**: 3692; (b) Zhu G, Ma S, Lu X. *J Chem Res Suppl* 1993: 366.

12 (a) Ma S, Lu X. *J Org Chem* 1993; **58**: 1245; (b) Ji J, Lu X. *Synlett* 1993: 745; (c) Ji J, Zhang C, Lu X. *J Org Chem* 1995; **60**: 1160; (d) Zhu G, Lu X. *J Organomet Chem* 1996; **508**: 83.

13 Ji J, Lu X. *Tetrahedron* 1994; **50**: 9067.

14 Wang Z, Lu X. *Tetrahedron Lett* 1997; **38**: 5213.

15 Ji J, Wang Z, Lu X. *Organometallics* 1996; **15**: 2821.

16 Jiang H, Ma S, Zhu G, Lu X. *Tetrahedron* 1996; **52**: 10945.

17 Kleiner EM, Pliner SA, Soifer VS *et al. Bioorg Khim* 1976; **2**: 1142.

18 Lu X, Zhu G. *Synlett* 1993: 68.

19 Botterbee JE, Burden RS, Crombie L, Whiting DA. *J Chem Soc (C)* 1969: 2470.

20 Noyori R. *Asymmetric Catalysis in Organic Synthesis.* New York: Wiley, 1994: 310; Ojima I. *Catalytic Asymmetric Synthesis*, New York: VCH, 1993: 132.

21 Zhu G, Lu X. *J Org Chem* 1995; **60**: 1087.

22 Zhu G, Lu X. *Tetrahedron: Asymmetry* 1995; **6**: 885.

23 Zhang Z, Lu X. *Tetrahedron: Asymmetry* 1996; **7**: 1923.

24 Zhu G, Lu X. *Tetrahedron: Asymmetry* 1995; **6**: 1637.

25 (a) Wang Z, Lu X. *Chem Commun* 1996: 535; (b) Wang Z, Lu X. *J Org Chem* 1996; **61**: 2254.

26 Negishi E. *Acc Chem Res* 1982; **15**: 340; Stille JK. *Pure Appl Chem* 1985; **57**: 1771; Miyaura N, Maeeda K, Suzuki A. *J Org Chem* 1982; **47**: 2117.

27 Coloqubour HM, Thompson DJ, Twigg MV. *Carbonylation.* New York: Plenum Press, 1991.

28 Heck RF. *Org React* 1982; **27**: 345.

29 Lambert C, Utimoto K, Nozaki H. *Tetrahedron Lett* 1984; **25**: 5323; Iritani K, Matsubara D, Utimoto K. *Tetrahedron Lett* 1988; **29**: 1799; Liebskind LS, Mitchell D, Foster BS. *J Am Chem Soc* 1987; **109**: 7908.

30 Maitlis PM. *The Organic Chemistry of Palladium*, Vol. 2. New York: Academic Press, 1971: 150; Wiger G, Albelo G, Rettig MF. *J Chem Soc Dalton Trans* 1974: 2242; Wipke WT, Goeke GL. *J Am Chem Soc* 1974; **96**: 4244.

31 Zhang Z, Lu X. (To be published.)

32 Herry PM. *Acc Chem Res* 1973; **6**: 16.

33 Murahashi S, Hosokawa T. *J Am Chem Soc* 1996; **118**: 3990.

34 Shin L, Yu C, Wang K *et al. Organometallics* 1993; **12**: 1018; Steinborn D. *Angew Chem Int Ed Engl* 1992; **31**: 401.

35 Cheng JC, Daves GD Jr. *Organometallics* 1986; **5**: 1753.

36 Hacksell V, Daves GD Jr, *Organometallics* 1983; **2**: 772; Cheng JC, Hacksell V, Daves GD Jr. *J Org Chem* 1986; **51**: 3093.

37 Daves GD Jr. *Acc Chem Res* 1990; **23**: 201.

38 Bäckvall JE, Akermark B, Ljunggren SO. *J Am Chem Soc* 1979; **101**: 2411.

39 Zhu G, Lu X. *Organometallics* 1995; **14**: 4899.

40 Collman JP, Hegedus LS, Norton JR, Finke RG. *Principles and Application of Organotransition Metal Chemistry.* Mill Valley, CA: University Science Books, 1987.

41 Daub GW. *Prog Inorg Chem* 1977; **22**: 409; Bäckvall JE. *Acc Chem Res* 1983; **16**: 335.

42 Wong PK, Stille JK. *J Organomet Chem* 1974; **70**: 121.

43 Heck RF. *Organotransitionmetal Chemistry.* New York: Academic Press, 1974: 110.

44 Kochi JK. *Organometallic Mechanisms and Catalysis.* New York: Academic Press, 1978.

45 (a) Bäckvall JE. *Tetrahedron Lett* 1977: 467; (b) Akermark B, Ljunggreen SO. *J Am Chem Soc* 1979; **101**: 2411; (c) Heumann A, Kaldy S, Tenagllia A, *Tetrahedron* 1994; **50**: 539.

46 Bäckvall JE, Nordberg RE. *J Am Chem Soc* 1980; **102**: 393.

47 Zhu G, Ma S, Lu X, Huang Q. *J Chem Soc Chem Commun* 1995: 271.

48 Wells AP, Kitching W. *J Org Chem* 1992; **57**: 2517.

49 Cabri W, Canadiani I. *Acc Chem Res* 1995; **28**: 2.

50 (a) Friestad GK, Branchaud BP. *Tetrahedron Lett* 1995; **36**: 7047; (b) Denmark SE, Schnute ME. *J Org Chem* 1995; **60**: 1013; (c) Amorese A, Arcadi A, Bernocchi E *et al. Tetrahedron* 1989; **45**: 813; (d) Yamamura K. *J Org Chem* 1978; **43**: 724; (e) Horino H, Arai M, Inoue N. *Bull Chem Soc Jpn* 1974; **47**: 1683; (f) Cacchi S. *Pure Appl Chem* 1996; **68**: 45; (g) Cacchi S. *Pure Appl Chem* 1990; **62**: 713; (h) Copéret C, Sugihara T, Wu G, Shimoyama I, Negishi E. *J Am Chem Soc* 1995; **117**: 3422; (i) Cacchi S, La Torre F, Mistiti D. *Tetrahedron Lett* 1979; **25**: 4591; (j) Yamamura K. *J Chem Soc Chem Commun* 1976: 438.

51 Arcadi A, Cacchi S, Fabrizi G, Marinelli F, Pace P. *Synlett* 1996: 568.

52 (a) Yagyosky G, Mowat W, Shortland A, Wilkinson G. *J Chem Soc Chem Commun* 1970: 1369; (b) Cross RJ. In: Hartley FR, Patai S, eds. *The Chemistry of the Metal–Carbon Bond*, Vol. 2. Chichester: Wiley, 1985: 559.

53 James BP. In: Wilkinson G, Stone FGA, Abel EW, eds. *Comprehensive Organometallic Chemistry*, Vol. 8. Oxford: Pergamon Press, 1982: 285; Siegel S. In: Trost BM, Fleming I, eds. *Comprehensive Organic Synthesis*, Vol. 8. Oxford: Pergamon Press, 1991: 417.

54 Hosokawa T, Murahashi S. *Heterocycles* 1992; **33**: 1079.

7 Chemo-, Regio- and Stereocontrolled Carbon–Carbon Bond Formation using Nickel Catalysts

ANDRE MORTREUX

Laboratoire de Catalyse Hétérogène et Homogène, Groupe de Chimie Organique Appliquée, URA CNRS402, ENSC Lille, USTL, Bât. C7, 59652 Villeneuve d'Ascq Cedex, France

1 Introduction

May 20th 1974 saw my first steps in homogeneous catalysis. At that time, Professor M. Blanchard (my supervisor) and myself decided to look at the homogeneous counterpart of heterogeneous systems for alkyne metathesis ($2R^1C\equiv CR^2 \rightleftharpoons R^1C\equiv CR^1 + R^2C\equiv CR^2$). We had been working since 1970 on the same reaction [1]. From the observation that well-defined catalysts, synthesized using tetrakis π-allyl molybdenum grafted on silica were far superior to those prepared by simple impregnation with molybdate, we imagined that some Mo–O–Si linkage might be responsible for the observed activities and selectivities. This was why we chose to try $Mo(CO)_6$ as a catalytic precursor, together with an excess of a hydroxyl-containing organic compound. The use of *n*-heptanol did not give any metathesis product; however, upon addition of a small amount of resorcinol, the metathesis reaction occurred in a selective way [2]. This discovery was the subject of the last chapter of my PhD thesis, and my research area became mostly devoted to homogeneous catalysis. After a post-doctoral position in Bristol (UK), in Professor F.G.A. Stone's laboratory, where I gained a better insight into organometallic chemistry, I joined my colleague and friend, the late Professor F. Petit, in Lille where he had already started research in this field. Since that time (1976), several research areas, all of them using transition metal complexes in homogeneous catalysis, have been explored by our group.

Among these research areas, the topic I would like to summarize in this paper will be exclusively devoted to nickel-based catalytic systems for C–C bond formation using olefins and dienes as substrates, to show and discuss the determining influence of the ligand environment for the activity and selectivity of these reactions [3].

2 Carbon–carbon bond formation using aminophosphine–nickel catalysts

We were first interested in the use of aminophosphine ligands for C–C bond formation in the codimerization of ethylene with cyclic conjugated dienes (Equation (7.1)).

$$\text{(CH}_2)_n + CH_2=CH_2 \xrightarrow[L]{[Ni]} \text{(CH}_2)_n \tag{7.1}$$

The use of Ni(II)/L/AlR$_2$Cl (L = PBu$_3$) systems has been thoroughly studied for this reaction [4], followed by a systematic study where the organoaluminum co-catalyst, the ligand and the nickel source have been varied [5]. However, this reaction suffered from a lack of selectivity, due to a rapid isomerization of the 1,4-diene into its vinylidene-conjugated isomer. Upon using aminophosphines as ligands (R$_2$P-N<), both the activity

159

Table 7.1. Ethylene–cyclohexadiene codimerization over nickel aminophosphine catalysts[a] (From [6])

Ligand	PPh$_3$	PCy$_3$	P(OPh)$_3$	PPh$_2$NEt$_2$	PPh(NEt$_2$)$_2$	P(NEt$_2$)$_3$
T (°C)	80	80	80	35	35	35
Initial turnover rate (s^{-1})	0.8	1.25	1	0.5	1.2	1.5
R[b]	5.8	0.06	3.3	1.1	0.7	0.1

[a] Ni(COD)$_2$/L/AlEt$_2$Cl (1:1:4); CH$_2$Cl$_2$ (33 ml); cyclohexa-1,3-diene (50 mmol); substrate/Ni (125:1); C$_2$H$_4$ (6 bar, 50 mmol).
[b] 3-Ethylidene-cyclohex-1-ene/3-vinylcyclohex-1-ene at 100% conversion.

Table 7.2. Asymmetric synthesis of 3-vinylcyclohex-1-ene over chiral aminophosphine–nickel catalysts[a] (From [7])

Ligand	P/Ni	T (°C)	ee[b] (%)	Config.
(R)-PhC̊H(Me)N(H)PPh$_2$	1	0	8.3	S
(R)-PhC̊H(Me)N(Me)PPh$_2$	1	0	17	S
	1	−35	21.7	S
	5	−35	32.7	S
	11[c]	−35	38.7	S
	10[c]	−70	46	S

[a] Ni(COD)$_2$ (0.36 mmol); cyclohexa-1,3-diene (144 mmol); C$_2$H$_4$ (12 bar, 144 mmol); CH$_2$Cl$_2$ (33 ml); the reactions were completed within 6 h at 0 °C, with selectivities ranging from 97 to 99% into 3-vinylcyclohex-1-ene.
[b] ee = enantiomeric excess.
[c] Selectivity = 87%.

and the selectivity for the initial product were strongly enhanced. In Table 7.1 some results are reported related to ethylene–cyclohexadiene codimerization into 3-vinylcyclohex-1-ene (Equation (7.2)) [6].

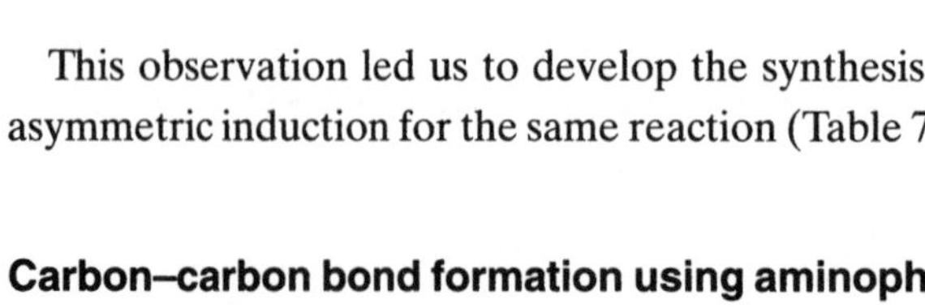

$$\text{(7.2)}$$

This observation led us to develop the synthesis of chiral aminophosphines to induce asymmetric induction for the same reaction (Table 7.2) [7].

3 Carbon–carbon bond formation using aminophosphine–phosphinite–nickel catalysts

3.1 *Ethylene–cyclohexadiene codimerization*

The above results observed with aminophosphines encouraged us to look at the use of bidentate ligands bearing P–N and P–O bonds, namely aminophosphine–phosphinites

(AMPP) of general formulae $Ph_2POCHR^{1*}CHR^{2}NPPh_2$, whose synthesis was undertaken in the early 1980s within a collaboration with Professors G. Peiffer and G. Buono in Marseille [8], according to Scheme 7.1.

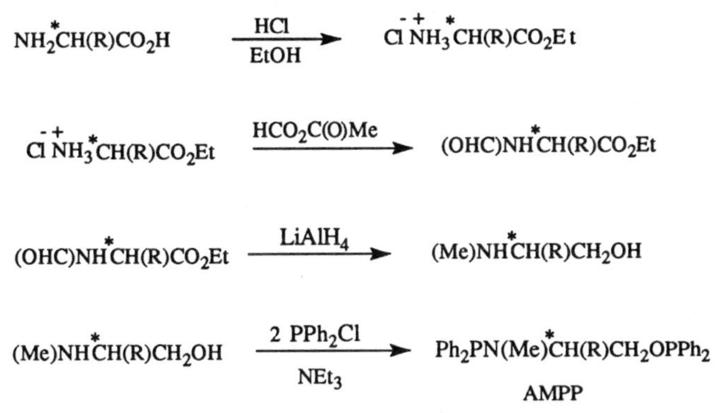

$$NH_2\overset{*}{C}H(R)CO_2H \xrightarrow[\text{EtOH}]{\text{HCl}} Cl^{-}\overset{+}{N}H_3\overset{*}{C}H(R)CO_2Et$$

$$Cl^{-}\overset{+}{N}H_3\overset{*}{C}H(R)CO_2Et \xrightarrow{\text{HCO}_2\text{C(O)Me}} (OHC)NH\overset{*}{C}H(R)CO_2Et$$

$$(OHC)NH\overset{*}{C}H(R)CO_2Et \xrightarrow{\text{LiAlH}_4} (Me)NH\overset{*}{C}H(R)CH_2OH$$

$$(Me)NH\overset{*}{C}H(R)CH_2OH \xrightarrow[\text{NEt}_3]{\text{2 PPh}_2\text{Cl}} Ph_2PN(Me)\overset{*}{C}H(R)CH_2OPPh_2$$

AMPP

Scheme 7.1.

This four-step, straightforward synthesis from readily available amino acids has allowed a large variety of new chiral bidentate ligands to be produced in a cheap way, which is one of the major challenges in the field of catalytic asymmetric synthesis.

The results obtained for ethylene–cyclohexadiene codimerization (Table 7.3) indicate that, as usual, the use of more strained structures enhances the enantioselectivity [9]. Furthermore, the use of tridentate ligands, particularly threo-NOOP synthesized from threonine, which bears two asymmetric centers in the framework, has led to extremely high enantiomeric excesses, which could be improved by lowering the temperature. Among the asymmetric C—C bond catalysed reactions, this system is one of the best known so far for producing chirality at a C_6 ring (93% enantiomeric excess).

However, due to the fact that 1,3-cyclohexadiene is not readily available, we turned our research to the use of a cheaper substrate to produce a stereogenic center at a C_6 ring, namely 1,3-butadiene, whose cyclodimerization into 1,4-vinylcyclohexene has been known for a long time using nickel-based catalysts.

Table 7.3. Nickel–catalysed asymmetric cyclohexa-1,3-diene-ethylene codimerization using aminophosphine-phosphinite $Ph_2PN(Me)\overset{*}{C}H(R^1)\overset{*}{C}H(R^2)OPPh_2$ ligands[a] (From [9])

Ligand precursor	(S)-Alanine	(S)-Valine	(S)-Phenylalanine[b]	(1S,2R)-Ephedrine	(2S,3R)-Threonine[c]
R^1, R^2	Me, H	$CHMe_2$, H	CH_2Ph, H	Me, Ph	$CH(Me)OPPh_2$, H
Enantiomeric excess (%)	15.3	5.8	19.2	8.4	84.9
(Configuration)	(R)	(R)	(R)	(S)	(S)

[a] $Ni(COD)_2$ (0.4 mmol); AMPP (0.4 mmol); toluene (10 ml); $EtAlCl_2$ (1.6 mmol); ethylene (87.5 mmol); 40 °C; reaction time <15 min.
[b] 51.8% enantiomeric excess at −25 °C.
[c] 93% enantiomeric excess at −30 °C.

Table 7.4. Nickel–catalysed asymmetric cyclodimerization of butadiene using aminophosphine–phosphinite $Ph_2PN(Me)\overset{*}{C}HR^1CHR^2OPPh_2$ ligands[a] (From [11])

Ligand	(S)-AlaNOP	(S)-ValNOP	(R)-PheGlyNOP	(1S,2S)-EPHOS	(2R,3R)-ThreoNOOP
R^1, R^2	Me, H	$CHMe_2$, H	Ph, H	Me, Ph	$CH(Me)OPPh_2$, H
VCH/COD[b]	1.7	2.5	1.6	0.7	1.1
Enantiomeric excess (%)	6.4	8.9	7.7	20.7	25.9
(Configuration)	(S)	(S)	(R)	(S)	(R)

[a] $Ni(COD)_2$ (2.5 mmol); ligand (2.5 mmol); butadiene (125 mmol); toluene (10 ml); $T = 40\,^\circ C$; 90% conversion within 24 h.
[b] VCH/COD = 4-vinylcyclohex-1-ene/cycloocta-1,5-diene.

3.2   Butadiene cyclodimerization

This reaction has been studied extensively by Wilke's group since the early 1960s [10]. The use of zerovalent nickel–phosphorus complexes in aprotic solvents gives rise to the production of a mixture of 1,4-vinylcyclohexene, cycloocta-1,5-diene and cis,-1,2-divinylcyclobutane (Equation (7.3)).

$$2 \quad \diagdown\!\!\!\diagup\!\!\!\diagdown \quad \xrightarrow[\text{L}]{Ni^\circ} \quad \text{(ring)} \quad + \quad \text{(ring)} \quad + \quad \text{(ring)} \qquad (7.3)$$

Again, the AMPP ligands have shown some trend for asymmetric induction at the C_6 ring (Table 7.4) [11].

Although the results are far lower than those found in the previous cyclodiene–ethylene codimerization, they led us to discover a new and fruitful catalytic system for the *linear* dimerization of butadiene, which is described below.

4   Carbon—carbon bond formation using aminophosphinite–nickel catalysts

4.1   Linear dimerization of conjugated dienes

During the above studies related to butadiene cyclodimerization, a reaction conducted with EPHOS as ligand (where EPHOS = $Ph_2PN(Me)\overset{*}{C}H(Me)\overset{*}{C}H(Ph)OPPh_2$) gave rise unexpectedly, at a much higher rate than the cyclization, to the production of 1,3,6-octatrienes, which further isomerized into 2,4,6-octatrienes (Equation (7.4)) [12–14].

$$2 \quad \diagdown\!\!\!\diagup\!\!\!\diagdown \quad \xrightarrow[\text{EPHOS-NH}]{Ni^\circ} \quad \diagup\!\!\!\diagdown\!\!\!\diagup\!\!\!\diagdown\!\!\!\diagup \quad \longrightarrow \quad \diagup\!\!\!\diagdown\!\!\!\diagup\!\!\!\diagdown \qquad (7.4)$$

A ^{31}P-NMR analysis of the EPHOS ligand led us to the conclusion that a small part of the starting ephedrine was still only monophosphinylated in this peculiar synthesis, giving the aminophosphinite (AMP) $HN(Me)\overset{*}{C}H(Me)\overset{*}{C}H(Ph)OPPh_2$ in small amounts, together with the expected EPHOS: we came to the conclusion that an extremely active nickel–aminophosphinite species was at the origin of this linear dimerization.

In fact, the nickel-catalysed linear dimerization of butadiene had already been discovered by Heimbach, using phosphite ligands and morpholine as proton donor [15]. This system, however, needs large amounts of morpholine (vs Ni) as well as a much higher temperature (60–80 °C vs 20 °C).

In this new catalytic system, the dimerizing properties of the nickel catalyst are strongly enhanced due to the ability of the N–H bond to act as proton donor, maintained in close proximity within the coordination sphere (Scheme 7.2).

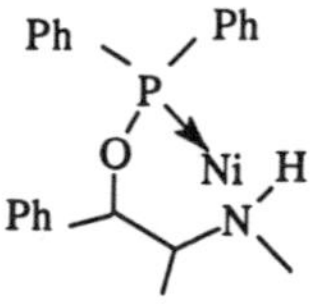

Scheme 7.2.

Table 7.5. Linear dimerization of substituted conjugated dienes over Ni(COD)$_2$–EPHOS-NH catalysts[a] (From [14] and [16])

Diene (Diene/Ni)	T°C	Reaction time (Conversion)	Products
 (50)	40	6h (95%)	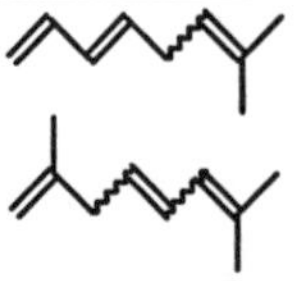 41.4% 10% + cyclodimers
 (100)	40	3h (95%)	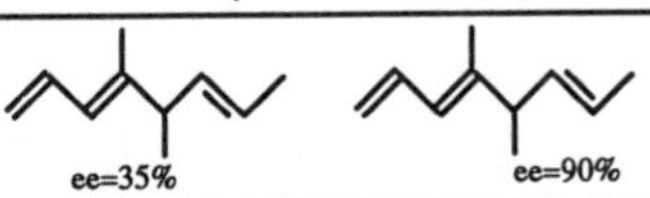 ee=35% ee=90%
 CO$_2$Me (50)	40	0.5h (98%)	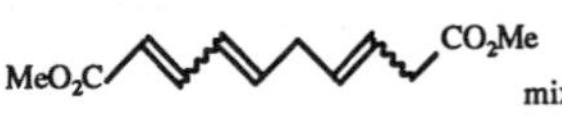 MeO$_2$C CO$_2$Me mixture of isomers
 CO$_2$Me (50)	80	46h (92%)	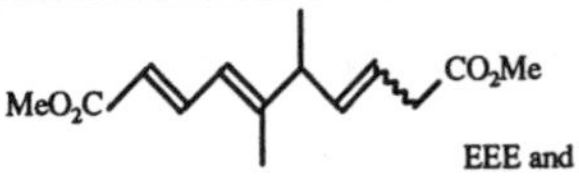 MeO$_2$C CO$_2$Me EEE and EEZ isomers
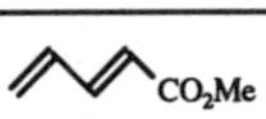 CO$_2$Me (50)  + (50)	20	0.5h (80%)	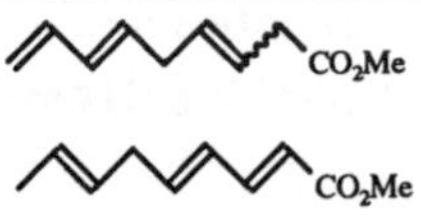 CO$_2$Me 56% CO$_2$Me 43%
 CO$_2$Me (50)  (50)	40	6h (90%)	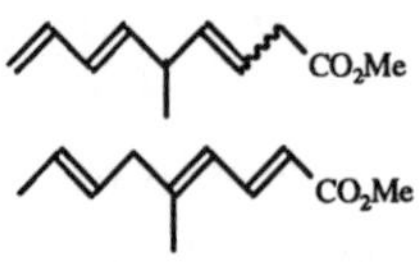 CO$_2$Me 54% CO$_2$Me 31%

[a] Ni(COD)$_2$ (0.4 mmol); toluene (5 ml); n-hexane or n-decane as internal standard (2 g).
[b] ee = enantiomeric excess.

Turnover numbers as high as 5000 h-1 could be reached on this system for butadiene dimerization using $Ni(COD)_2$ (COD = cycloocta-1,5-diene) as the source of Ni(0) and one equivalent of pure HN(Me)ĊH(Me)ĊH(Ph)OPPh$_2$(EPHOS-NH) synthesized selectively from ephedrine and Ph_2PNMe_2.

This unprecedented reactivity allowed the reaction to be extended to substituted conjugated dienes, some of which are summarized in Table 7.5 [14, 16].

The major interest in these reactions arises from the fact that, particularly in the homo-dimerization reactions, the products are obtained regioselectively.

4.2 Consecutive C—C bond formation: olefin-conjugated diene co-oligomerization using nickel–aminophosphinite catalysts

Nickel catalysts are also known to catalyse the oligomerization of olefins and conjugated dienes: we have therefore attempted to apply the Ni0/EPHOS-NH system using butadiene and methylacrylate as coreactants.

The uncatalysed reaction carried out at 80 °C leads to the Diels–Alder product, which decreased to 70% selectivity when $Ni(COD)_2$ was added as catalyst (the cyclotrimer cyclododeca-1, 5, 9-triene was the only byproduct).

Upon addition of one equivalent of EPHOS-NH (Ni/EPHOS-NH = 1:1) using butadiene/nickel and acrylate/nickel ratios of 80:1 and 40:1, respectively, two different products were observed (Equation (7.5)) [17].

$$2 \quad \diagup\!\!\!\!\diagdown\!\!\!\!\diagup \quad + \quad \diagup\!\!\diagdown^{CO_2Me} \quad \longrightarrow \tag{7.5}$$

Labeling experiments using EPHOS-ND followed by analysis (IR, ^{13}C-NMR, MS) of the ester product proved that deuterium was transferred at the C_2 position.

These results can therefore be explained by the same mechanism as depicted for dienic compounds, where the acrylate insertion would occur before the proton transfer leading to the co-oligomers (Scheme 7.3, paths A and B) [17].

5 Allylic substitution reactions using nickel–aminophosphine–phosphinite and nickel–bisaminophosphine (BAMP) catalysts

In the field of C—C bond formation, the nucleophilic substitution of allylic acetates has also been known for a long time, and recent results in this field are mainly concerned with the asymmetric version of this reaction [18]. On the other hand, Ni(diphosphine)$_2$ catalysts have been revealed recently by our group to catalyse this reaction with high turnover frequencies in polar solvents [19]. We have therefore focused our attention on the asymmetric substitution reaction of cyclohexenylacetate by the malonate anion, using AMPP and bisaminophosphine (BAMP) ligands (Equation (7.6)) [20]. Some selected results related to this reaction are reported in Table 7.6.

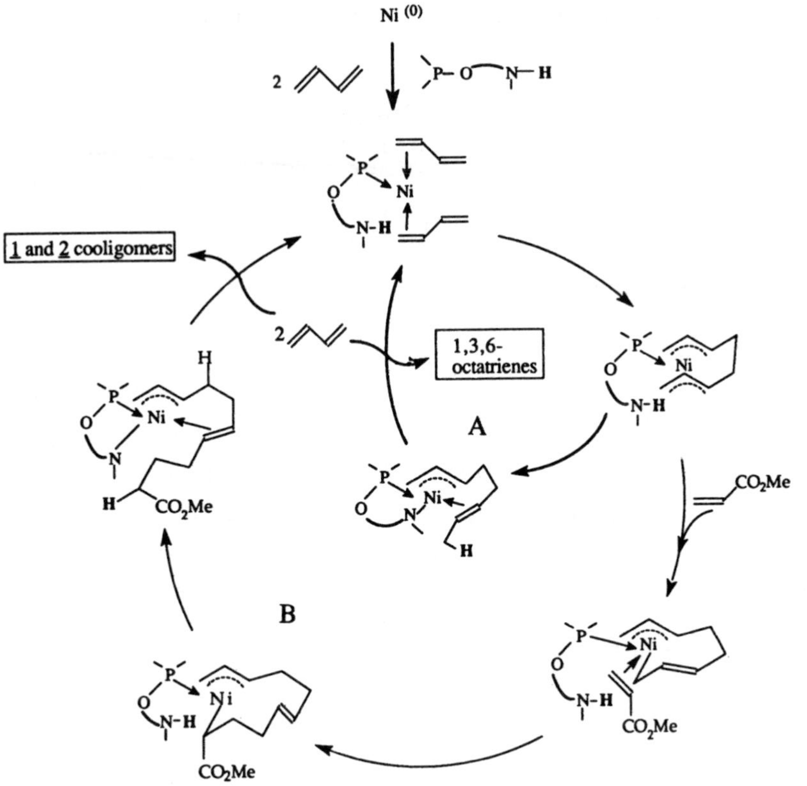

Scheme 7.3.

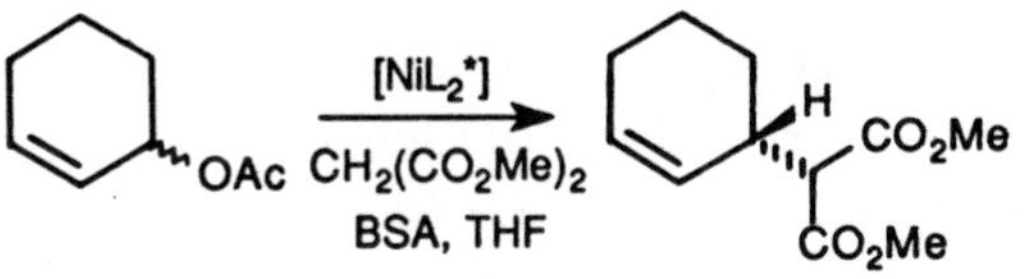

$$(7.6)$$

Of great interest is the trend in activity observed upon changing the ligand: an enhancement of the rate by two orders of magnitude is obtained upon using more basic AMPP bidentates (cyclohexyl vs phenyl groups at phosphorus in ProNOP); and a further tenfold increase is obtained with BAMP ligands.

Meanwhile, this activating effect allows the reaction to occur at reasonable rates at low temperature, which is often a prerequisite for an enhancement of the asymmetric induction. By itself, this spectacular improvement of the reaction rate obtained by changing the electronic and/or steric properties of the ligands provides strong evidence for the usefulness of such variations, especially for cases where new ligands could be synthesized in a minimum of steps.

6   Conclusion

This brief survey on selected C–C bond formation using nickel-based catalysts leads to the conclusion that whatever the reaction, tuning the ligand may lead to strong improvements in both reactivity and selectivity. Particularly interesting are the following points:

Table 7.6. Asymmetric alkylation of 3-acetoxycyclohex-1-ene with dimethylmalonate using nickel-bidentate catalysts[a] (From [20])

$L_2^* =$

(R,R) DIOP **1**

(R) Binap **2**

(S)-Ph,Ph-ProNNP **7**

X=H,H ; R=Ph:(S)-Ph,Ph-ProNOP **3**

X=H,H ; R=Cy:(S)-Cy,Cy-ProNOP **4**

X=O ; R=Ph : (S)-Ph,Ph-oxoProNOP **5**

X=O ; R=Cy : (S)-Cy,Cy-oxoProNOP **6**

Ligand	(R,R)-DIOP (1)	(R)-BINAP (2)	3	4	5	6	7
TOF (h⁻¹)[b]	3.6	17	0.6	110	>1500	10	45
Enantiomeric excess (%)	5	11	29	14	17	40	5
(Configuration)	(R)	(R)	(R)	(R)	(R)	(R)	(R)

[a] Ni(COD)$_2$ (35 mg or 0.13 mmol); Ni/L$_2$/3-acetoxycyclohex-1-ene/dimethylmalonate/BSA (1:2:50:75:75); THF (12.5 ml).
[b] TOF = turnover frequency = mol product mol⁻¹ nickel h⁻¹.

1 The new bifunctional aminophosphinite ligands, which act as both electronic modifier (the phosphinite moiety) and proton-transfer agent (the amino group), provide the opportunity to enhance reaction rates by at least two orders of magnitude compared to classical systems, and lead to regio- and enantioselective reactions when a chiral ligand is used for suitable conjugated dienes.

2 As far as C—C bond formation via allylic substitution is concerned, the above results are very instructive because they prove that upon changing the ligand framework and its electronic properties, nickel catalysts, long neglected compared with their palladium homologs, could be more reactive than the latter.

The above results lead to the conclusion that when one would like to enhance the properties of a particular catalytic system, there are obviously some possibilities in using the concept of 'metal–ligand interaction'.

The problem that still remains is how to rationalize this concept for catalytic reactions. A catalytic reaction generally occurs via successive elementary steps, one of them being the so-called rate-determining step. If this latter step is well defined (it may change according to the nature of the ligand), then using basic principles of organometallic chemistry may help to define the variation to be done on the metal to improve the rate. To achieve this, there is no doubt that the systematic use of a series of ligands whose electronic steric properties are well known is still a good approach. However, the tremendous development of computer science and its application to molecular modeling leads me to think that during the beginning of the 21st century these calculations certainly will be very helpful for the design of the best metal–ligand combination for a given reaction. Up to now, some results have already been published using extended Hückel or *ab initio* calculations for catalytic reactions such as hydroformylation and polymerization [21], or even asymmetric syntheses [22], as a tool to explain or to propose mechanistic insights into such reactions.

Time has therefore arrived when 'theoretical' chemists and 'experimental' chemists find subjects of common interest in the field of homogeneous catalysis. The success of these approaches will depend strongly upon a very close relationship between people from sometimes quite different scientific backgrounds: these challenges hopefully will be very helpful in that sense.

7 Acknowledgments

I would like to dedicate this chapter to the memory of my friend, the late Professor Francis Petit, and to thank all my co-workers who participated in this research, whose names appear on the cited papers.

8 References

1 Mortreux A, Blanchard M. *Bull Soc Chim Fr* 1970; **4**: 1641–3.
2 Mortreux A, Blanchard M. *J Chem Soc Chem Commun* 1974: 786–7.
3 For a review, see: Wilke G. *Angew Chem Int Ed Engl* 1988; **27**: 185.
4 Miller RG, Kesley TJ, Barney AL. *J Am Chem Soc* 1967; **89**: 3756.
5 Adler VB, Beger J, Duschek *et al. J Prakt Chem* 1974; **316**: 952.
6 Peiffer G, Cochet X, Petit F. *Bull Soc Chim Fr* 1979; **7/8**: 415–20.
7 Buono G, Peiffer G, Mortreux A, Petit F. *J Chem Soc Chem Commun* 1980: 937–9.
8 Petit M, Mortreux A, Petit F, Buono G, Peiffer G. *Nouv J Chim* 1983; **7**: 593.
9 Buono G, Siv C, Peiffer G *et al. J Org Chem* 1985; **50**: 1781.
10 Heimbach P, Jolly PW, Wilke G. *Adv Organomet Chem* 1970; **8**: 29.
11 Cros P, Buono G, Peiffer G *et al. N Engl J Chem* 1987; **11**: 573–9.
12 Denis P, Mortreux A, Petit F, Buono G, Peiffer G. *J Org Chem* 1984; **49**: 5275–6.
13 Denis P, Jean A, Croizy JF, Mortreux A, Petit F. *J Am Chem Soc* 1990; **112**: 1292–4.
14 Denis P, Croizy JF, Mortreux A, Petit F. *J Mol Catal* 1991; **68**: 159–75.
15 Heimbach P. *Angew Chem Int Ed Engl* 1973; **12**: 975.
16 El Amrani MA, Mortreux A, Petit F. *Tetrahedron Lett* 1989; **30**: 6515–16.
17 El Amrani MA, Suisse I, Mortreux A, Knouzi N. *J Mol Catal* 1995; **101**: 211–15.
18 Trost BM. *Angew Chem Int Ed Engl* 1989; **28**: 1173.
19 Bricout H, Carpentier JF, Mortreux A. *J Chem Soc Chem Commun* 1995: 1863–4.
20 Bricout H, Carpentier JF, Mortreux A. *Tetrahedron Lett* 1996; **37**: 6105–8.
21 Klein R, Schmid R. In: Cornils B, Herrmann WA, eds. *Applied Homogeneous Catalysis with Organometallic Compounds,* Vol. 2. Weinheim: VCH, 1996: 654–71.
22 Agbossou F, Carpentier JF, Mortreux A, Surpateanu G, Welch AJ. *N J Chem* 1996; **20**: 1047.

8 New Aspects of Group VIII Metal Chemistry of Carbon–Carbon Bond-forming Aromatic Substitution

MARTA CATELLANI

Dipartimento di Chimica Organica e Industriale dell'Università, Viale delle Scienze, 1-43100 Parma, Italy

1 Introduction

Carbon–carbon bond-forming aromatic substitution is a very wide topic of chemistry. This chapter will be limited to noble metals and particularly to palladium chemistry and focused on very recent progress, consisting of multistep alkylation and arylation of aromatic substrates. This progress is based on a series of important advances [1,2] concerning the behavior of metals, ligands, solvents and substrates, which will be treated briefly, in order to give the reader an overview of the relevant developments in the area.

2 Types of aromatic substitution

According to classic systematization, three broad categories of aromatic substitution exist [3]: electrophilic, nucleophilic and homolytic. The first one includes Friedel–Crafts reactions (Equation (8.1)) [4–7], the second Grignard-type reactions (Equation (8.2)) [8–10] and the third homolytic reactions [11–14] (attack by radicals formed by decomposition of peroxides, nitrosoacetanilides, diazonium salts or organometal compounds (Equation (8.3)).

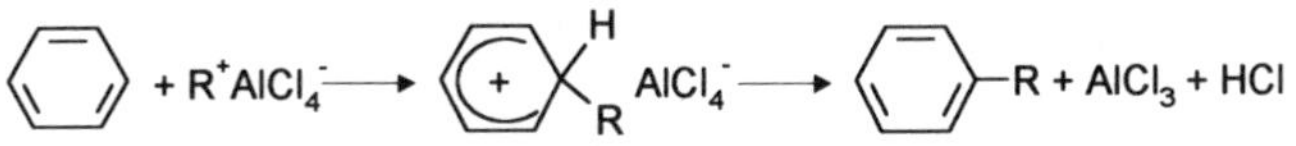

$$
\text{Ar} + R^{+}AlCl_{4}^{-} \longrightarrow \left[\text{Ar}^{+}\!\!<^{H}_{R}\right] AlCl_{4}^{-} \longrightarrow \text{Ar–R} + AlCl_{3} + HCl \tag{8.1}
$$

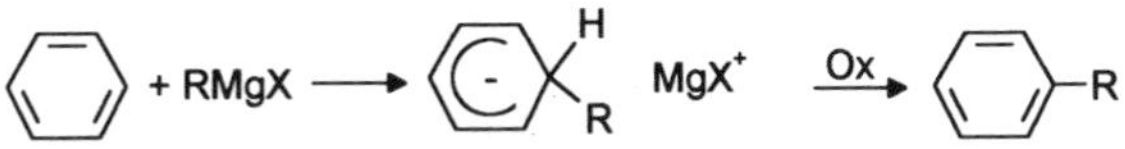

$$
\text{Ar} + RMgX \longrightarrow \left[\text{Ar}^{-}\!\!<^{H}_{R}\right] MgX^{+} \xrightarrow{\ Ox\ } \text{Ar–R} \tag{8.2}
$$

$$
\text{Ar} + R^{\cdot} \longrightarrow \left[\text{Ar}^{\cdot}\!\!<^{H}_{R}\right] \xrightarrow{\ Ox\ } \text{Ar–R} \tag{8.3}
$$

In Equation (8.2), electron-withdrawing substituents on the arene and an oxidant are necessary to cause hydride expulsion. In Equation (8.3), atomic hydrogen can also be expelled by oxidants.

The original picture of aromatic substitution is still valid today in general. A deeper insight has been gained, however, through the study of superacids [15–17], electron-transfer processes [18–20], and *cine* and *tele* substitution [21,22]. These reactions are mostly performed on the laboratory scale except for Friedel–Crafts alkylations and acylations [23], which are industrially important processes. In particular, the former have catalytic character with reagents not containing heteroatoms. Examples are the synthesis of ethylbenzene from benzene and ethylene and that of cumene from benzene and propylene. Friedel–Crafts acylation is a stoichiometric process still used for the synthesis of anthraquinone from phthalic anhydride and benzene [23].

Laboratory and industrial problems are mainly connected with regioselectivity and catalysis in the presence of oxygenated functional groups. The problem of regioselectivity is particularly important because: mixtures of *ortho*, *meta* and *para* isomers are generally formed (mainly *ortho* and *para*) in electrophilic aromatic substitution on arenes substituted with electron-donating substituents and mainly *meta* with electron-withdrawing substituents; there are variable ratios in homolytic substitution, depending on the polar character of the radicals; and products of *ipso-cine-* and *tele-*substitution are formed in nucleophilic aromatic substitution. Thus, more selective processes are needed. Catalysis in the presence of oxygenated functions that inhibit Friedel–Crafts catalysts requires a search for suitable metal salts and complexes.

3   The use of transition metal complexes opens a new era in aromatic substitution

Metal carbonyls, particularly nickel tetracarbonyl, were found to be able to attack aromatic halides, replacing halides with carbonyl functions, initially under drastic conditions [24, 25], and then under mild conditions (Equation (8.4)) [26–30].

$$\langle\ \rangle\!-\!X\ +\ Ni(CO)_4\ +\ MeOH\ \longrightarrow\ \langle\ \rangle\!-\!CO_2Me\ +\ Ni^0\ +\ 3CO\ +\ HX \tag{8.4}$$

The latter circumstance enabled the study of the reaction course and led to the isolation of the relevant complexes of nickel, as well as of other transition metals prepared analogously [31, 32]. The attack of a metal on an aryl halide bond was recognized as an oxidative addition [1,2], the detailed study of which has been carried out recently by electrochemical techniques [33, 34]. The complexes thus obtained were shown to be able to give rise to a variety of reactions with themselves and with other reagents [35, 36]. Some reaction prototypes are given below, omitting non-reactive ligands.

Arylpalladium halides were shown to couple by disproportionation [37, 38]:

$$2\langle\ \rangle\!-\!Pd\!-\!X\ \longrightarrow\ \langle\ \rangle\!-\!Pd\!-\!\langle\ \rangle\ +\ PdX_2\ \longrightarrow\ \langle\ \rangle\!-\!\langle\ \rangle\ +\ 2Pd^0 \tag{8.5}$$

Substitution of the leaving group X by several types of metal-bonded R groups led to a large range of C–C coupled compounds [39–50], for example:

$$\langle\ \rangle\!-\!Pd\!-\!X\ +\ RMgCl\ \longrightarrow\ \langle\ \rangle\!-\!R\ +\ Pd^0\ +\ MgClX \tag{8.6}$$

As can be seen from Equation (6), this type of reactivity gives stoichiometric reactions in the added R-bonded metal while PhArX is regenerated from Pd° and RX. An efficient way to perform coupling reactions is based on the use of arylboronic acids [51–54]:

$$\langle\ \rangle\!-\!Pd\!-\!X\ +\ {HO \atop HO}\!\!>\!\!B\!-\!\langle\ \rangle\ \xrightarrow[-HBO_2,\ -HX]{Na_2CO_3}\ \langle\ \rangle\!-\!\langle\ \rangle\ +\ Pd^0 \tag{8.7}$$

Catalytic reactions can also be carried out by replacing the anion X of the appropriate transition metal (M) complexes with that of active hydrogen compounds such as alkynes [55–57] or HCN [58, 59] in the presence of neutralizing agents:

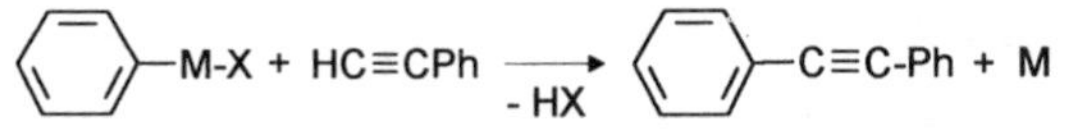

$$\text{Ph}-\text{M-X} + \text{HC}\equiv\text{CPh} \xrightarrow{-\text{HX}} \text{Ph}-\text{C}\equiv\text{C-Ph} + \text{M} \tag{8.8}$$

Not only coupling but also insertion reactions result in useful syntheses. Carbon monoxide insertion is involved in Equation (8.4). The acyl group thus formed is then trapped by methanol (or by water or other nucleophiles) [24–30]. Other trapping agents can be used, however; for example (Equation (8.9), where X = halide and R = alkyl or aryl): dihydrogen [60, 61] or hydrogen donors [62–64]; olefins [65, 66], alkynes and CO [67], CONEt$_2$ [68–71], alkynes [72, 73] and alkyls or aryl donors [74–78].

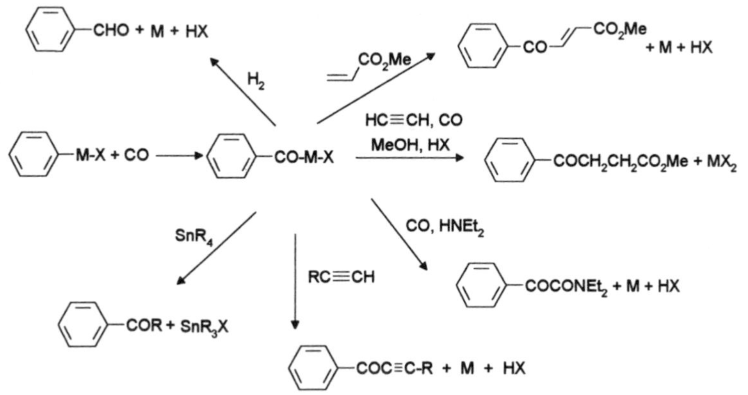

$$\tag{8.9}$$

The reaction of metal-coordinated aryl groups with CO was later extended to olefin, palladium being the metal of choice [79–82]:

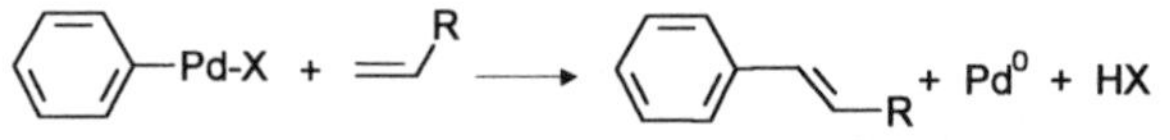

$$\text{Ph}-\text{Pd-X} + \ \overset{R}{=\!\!/} \longrightarrow \text{Ph}-\!\!\!=\!\!\!\diagdown_R + \text{Pd}^0 + \text{HX} \tag{8.10}$$

This is the well-known Heck reaction, which has been applied under a variety of conditions and has made several synthetic developments possible, as we shall see later. The aryl group can also be generated through an indirect reaction consisting of oxidative addition of acyl halides [65, 66, 83, 84] to palladium(0) followed by decarbonylation:

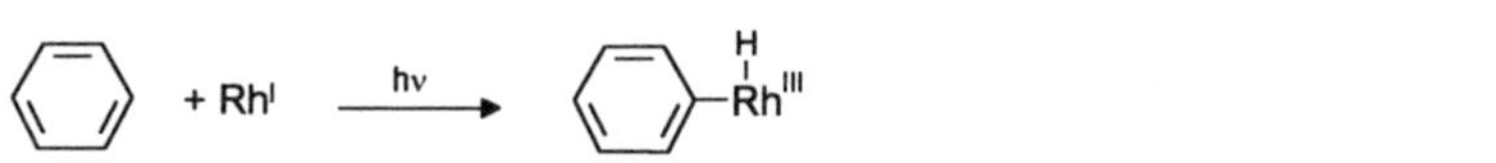

$$\text{Ph}-\text{CO-Pd-X} \xrightarrow{-\text{CO}} \text{Ph}-\text{Pd-X} \xrightarrow{=\!\!\overset{R}{/}} \text{Ph}-\!\!\!=\!\!\!\diagdown_R + \text{Pd}^0 + \text{HX} \tag{8.11}$$

Direct aromatic substitution reactions were achieved through activation of aromatic C–H bonds. To overcome unfavorable thermodynamics for oxidative addition [85–87], high-energy species were generated such as those derived from photochemical activation of rhodium complexes [88–90]:

$$\bigcirc + \text{Rh}^{\text{I}} \xrightarrow{h\nu} \bigcirc\!\!-\!\overset{\text{H}}{\text{Rh}}^{\text{III}} \tag{8.12}$$

or from elimination of a naphthyl ligand from an iron complex with dimethylphosphino-ethane (P–P) [91, 92]:

$$\tag{8.13}$$

$$L_2Ni(Ar)Me + ArX \longrightarrow L_2Ni(Ar)Me^+ Ar^{\cdot-}$$

$$L_2Ni(Ar)Me^+ \longrightarrow L_2Ni^+ + ArMe$$

$$L_2Ni^+ + ArX^{\cdot-} \longrightarrow L_2Ni(Ar)X$$

$$L_2Ni(Ar)X + Me\text{-}m \longrightarrow L_2Ni(Ar)Me + mX$$

Scheme 8.1.

Another instance is offered by the electrophilic activation brought about by ruthenium or platinum (Equation (8.14)) or palladium (Equation (8.15)) complexes [93–97].

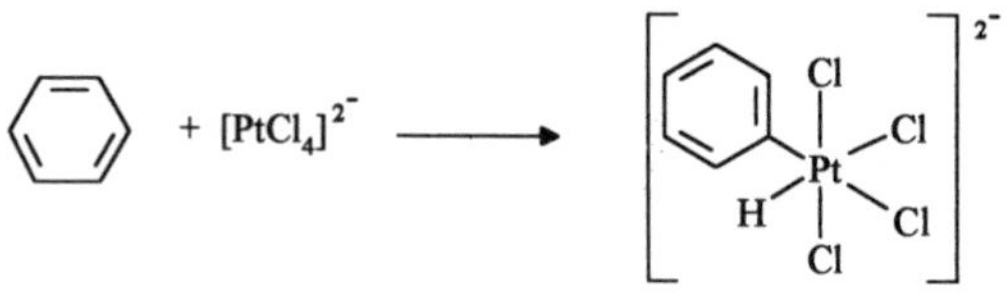

$$(8.14)$$

$$ArH + CO \xrightarrow[K_2S_2O_8,\ TFA]{Pd(OAc)_2} ArCOOH \tag{8.15}$$

The use of the rhodium or osmium carbonyl clusters also proved to be effective for aromatic C–H activation and reaction with several substrates [98]. Nucleophilic aromatic substitution was achieved by first coordinating aromatic compounds to chromium car-bonyls, and then attacking the aromatic ring by nucleophiles [99–104]:

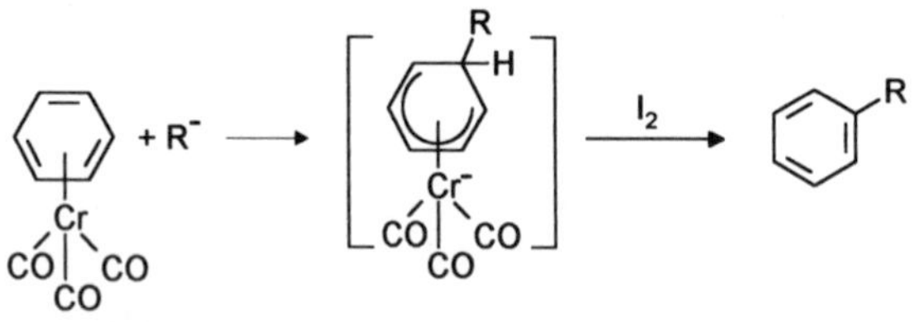

$$(8.16)$$

Electrophilic substitutions on chromium-bonded arenes are also possible. Arene ruthenium and osmium complexes can undergo carbon attack on the arene as well [105].

Homolytic aromatic substitution [12] was achieved by causing radicals generated by decomposition of transition metal complexes to attack aromatic compounds.

Single electron-transfer (SET) mechanisms have also been described by Kochi [106–108], for example the coupling of aryl halides with methylmagnesium halides on nickel passes through intermediates deriving from electron transfer (Equation (8.17) and Scheme 8.1, where m = Li or MgX and L = triethylphosphine).

$$ArBr + Me\text{-}m \xrightarrow{L_2NiBr_2} ArMe + mBr \tag{8.17}$$

4 Development of group VIII metal chemistry of carbon–carbon bond-forming aromatic substitution

The widest scope in aromatic substitution reactions involving catalysis of carbon–carbon bond formation was offered by group VIII metals, particularly palladium. We shall first consider studies aimed at controlling reactions by the use of suitable metals, ligands and solvents, and then consider substrate-controlled reactions.

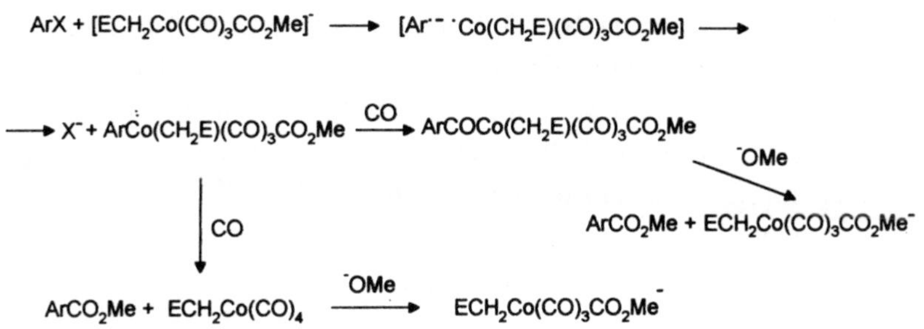

$$ArX + [ECH_2Co(CO)_3CO_2Me]^- \longrightarrow [Ar^{-} \cdot Co(CH_2E)(CO)_3CO_2Me] \longrightarrow$$

$$\longrightarrow X^- + Ar\dot{C}o(CH_2E)(CO)_3CO_2Me \xrightarrow{CO} ArCOCo(CH_2E)(CO)_3CO_2Me$$

$$\xrightarrow{^-OMe} ArCO_2Me + ECH_2Co(CO)_3CO_2Me^-$$

$$\Big\downarrow CO$$

$$ArCO_2Me + ECH_2Co(CO)_4 \xrightarrow{^-OMe} ECH_2Co(CO)_3CO_2Me^-$$

Scheme 8.2.

4.1 *Metals*

The choice of suitable metals is governed by their ability to form aryl–metal bonds either by metathesis of other complexes or by oxidative addition to zero- or low-valent metals. This is not the only requisite, however, because the metal-bonded aryl group must be able to react further, leading eventually to metal elimination. Thus the metal must be sufficiently noble to allow elimination, but its ionization potential must allow the increase in oxidation state when required. Suitable metals are the noble ones of the first and the second series [1,2]. The requirement for kinetic lability of the intermediates involved would further limit the choice to the second series, where palladium usually appears to be able to perform catalytic carbon–carbon bond-forming reactions [109] more efficiently than rhodium and ruthenium. The metal activity, however, is determined also by other properties, such as its oxidation state, number and distribution of d-electrons, coordination number and formal charge, and the correct combination of these properties can result in high catalytic performance. For example, aryl bromides and activated chlorides are not sufficiently reactive to undergo oxidative addition to cobalt(I) but if the latter is present as anion, as in $[NCCH_2–Co(CO)_3CO_2Me]^-$ (from $NCCH_2–Co(CO)_4$ and OMe^-), chloride substitution and carbonylation readily occur through electron transfer according to [110]:

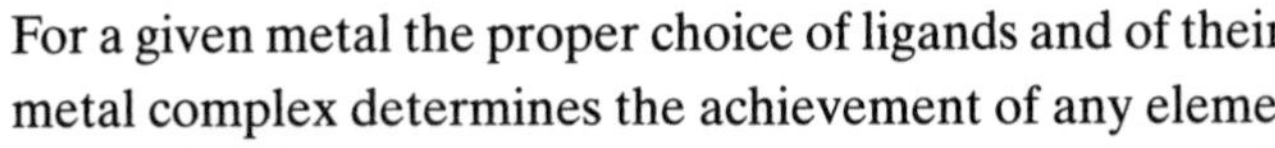

$$\langle\!\langle\rangle\!\rangle\text{--Cl} + CO + MeOH \xrightarrow[\text{base}]{\text{Co cat}} \langle\!\langle\rangle\!\rangle\text{--CO}_2Me + HCl \qquad (8.18)$$

The proposed mechanism is shown in Scheme 8.2 (where E = CN or other electron-withdrawing groups).

4.2 *Ligands*

For a given metal the proper choice of ligands and of their stereochemistry in the resulting metal complex determines the achievement of any elementary step occurring in the coordination sphere of a metal that interacts with an aromatic nucleus [111]. In particular, the effect of ligands such as tertiary phosphines or nitrogen bases on the stabilization of the intermediates resulting from oxidative addition, insertion and reductive elimination was studied with important mechanistic consequences, which also allowed the achievement of efficient organic syntheses. Thus, the previously mentioned process of oxidative addition proved much more complex than initially believed [1, 2] and different mechanisms were shown to be followed for a given substrate in the presence of different ligands, ranging from SN2 to three-center and to electron transfer [106–108]. A significant example of the importance of the ligand adopted is offered by the difficult oxidative addition of aryl chlorides to palladium(0), which was obtained by Milstein (Equation (8.19)) [112, 113] using a chelating ligand (bis(diisopropylphosphino)propane) in the methoxycarbonylation reaction. The ligand has sufficient basicity to allow oxidative addition, bulkiness to prevent carbon monoxide coordination before the formation of the Pd–Cl bond and the ability to dissociate to allow chlorobenzene coordination.

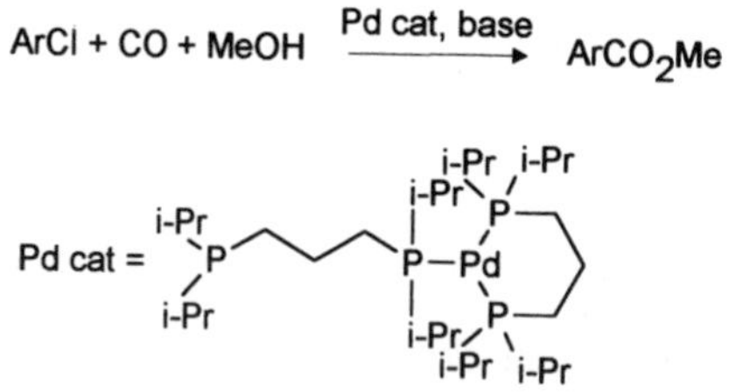

$$ArCl + CO + MeOH \xrightarrow{\text{Pd cat, base}} ArCO_2Me$$

(8.19)

Other ways to activate the Ar–Cl bond, this time involving radicals generated by chemical or photochemical reactions, have been described by Caubère [114, 115] using cobalt carbonyl in association with complex bases. Even the very strong Ar–F bond could be activated by rhodium catalysts [116–118].

The choice of appropriate ligands can cause different pathways to be followed, also in insertion reactions into aryl palladium bonds, where the inserting molecule or group must be *cis* to the arylpalladium bond [119]. The formation of a cationic arylpalladium complex has been shown to enhance the olefin insertion process [120–122]. Substrate or reagent association or anion or ligand dissociation can be at work, however, in connection with different leaving groups [79–81], as exemplified by the use of halides in comparison with triflates (for a systematization of these aspects in the case of the Heck reaction, see the work by Cabri [82]).

The use of chelating ligands that can force the substrate to occupy *cis* or *trans* positions in a complex was shown to be very useful in relation to the stereochemistry required for insertion [116–118]. Herrmann recently showed that the use of a special chelating ligand with P–C coordinating sites, together with promoters such as ammonium salts, could solve the problem of the reactivity of aryl chlorides in a very efficient way [123–125] (Equation (8.20), where R = *o*-tolyl).

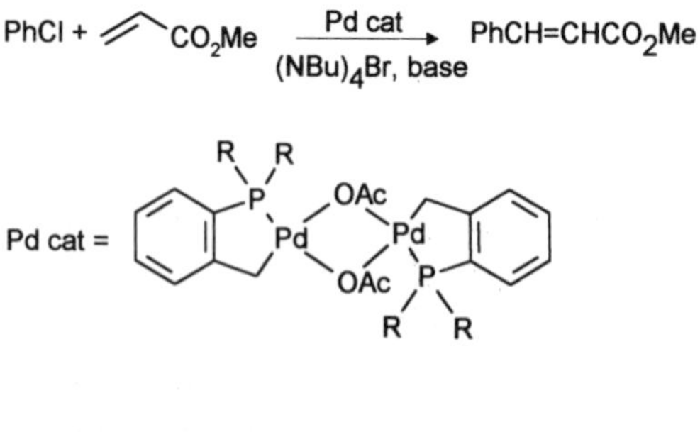

$$PhCl + \text{\textbackslash}CO_2Me \xrightarrow[\text{(NBu)}_4\text{Br, base}]{\text{Pd cat}} PhCH=CHCO_2Me$$

(8.20)

4.3   *Solvents and additives*

The use of suitable solvents also led to substantial progress in conducting aromatic substitution reactions. Solvents can be regarded both as media, able to influence the reaction course by their polarity and basicity, and as weak ligands [126], which can dissociate to accommodate reagents and substrates in the coordination sphere of the metal or associate to stabilize transition states or intermediate complexes. In their role as ligands they can enable a metal to act catalytically more efficiently than in the presence of stabilizing ligands. Thus Heck-type reactions can be performed in dimethylformamide (DMF) in the presence of palladium acetate without using tertiary phosphines. Efficient procedures, including the use of neutralizing agents such as potassium carbonate and additives such as phase-transfer agents (ammonium salts), were described by Jeffery [127, 128]. Apart from solvents, several types of additives have been discovered that can help anion dissociation, e.g. silver and thallium salts [129–132]. Phase-transfer techniques with double organic and aqueous layers were also described for carbonylation reactions [133].

The use of soluble catalysts in an aqueous medium has led to remarkable developments in the area of palladium-catalysed aromatic coupling reactions [134, 135]. Thus, Beletskaya succeeded in obtaining several coupling and insertion reactions, e.g.:

$$Ar_4BNa \ + \ 4Ar'X \ \xrightarrow[\text{base}]{\text{Pd(OAc)}_2} \ 4\,Ar\text{-}Ar' \tag{8.21}$$

4.4   Substrates

Substrate-controlled reactions are based on the observation that the reacting molecules and groups themselves can influence the properties of metal–carbon bonds in such a way as to induce preferential reactions with other substrates. Aromatics can thus be functionalized chemoselectively through the formation of one or more C–C bonds, as initially shown with nickel complexes bonded to a dissociable ligand such as carbon monoxide [39, 136]:

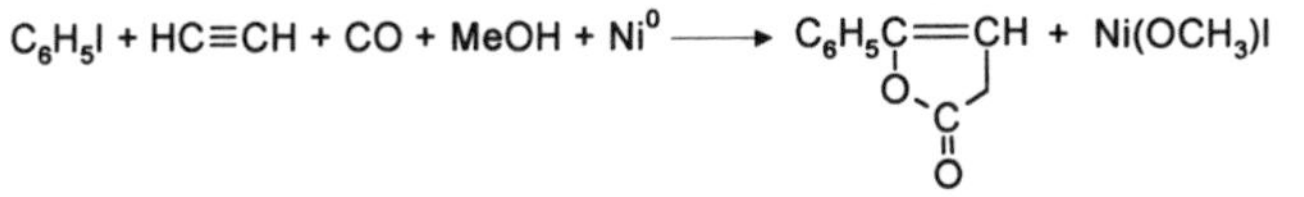

$$\tag{8.22}$$

The different substrates are in competition with each other and the relative predominance of a reaction step is governed by the nature of the complex involved, by the reagents and by the substrate itself. The reaction pathway is shown in Scheme 8.3.

When hydrogen elimination is rather difficult for steric reasons [137], as with *cis, exo* norbornylmetal complexes, e.g.:

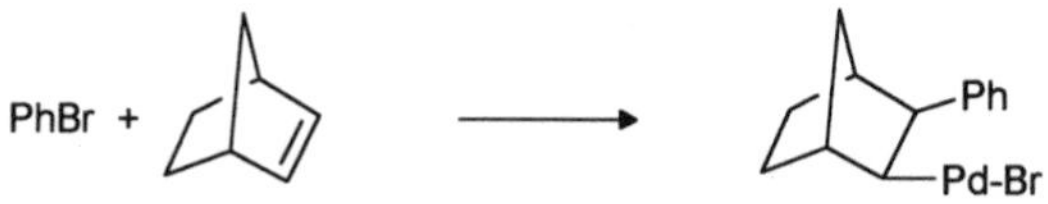

$$\tag{8.23}$$

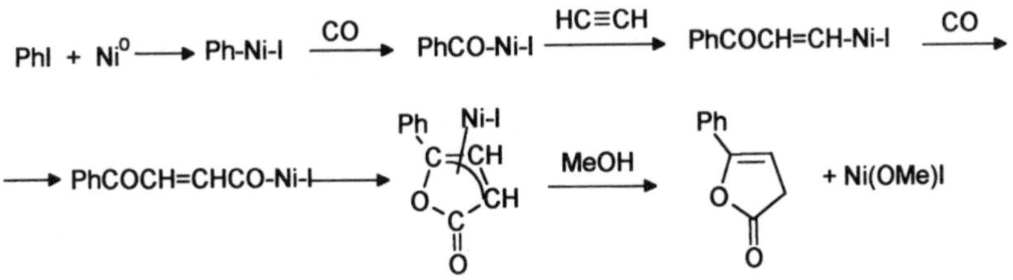

Scheme 8.3.

then coupling or other insertions followed by elimination can occur. For example, termination could be effected by coupling [138–142]:

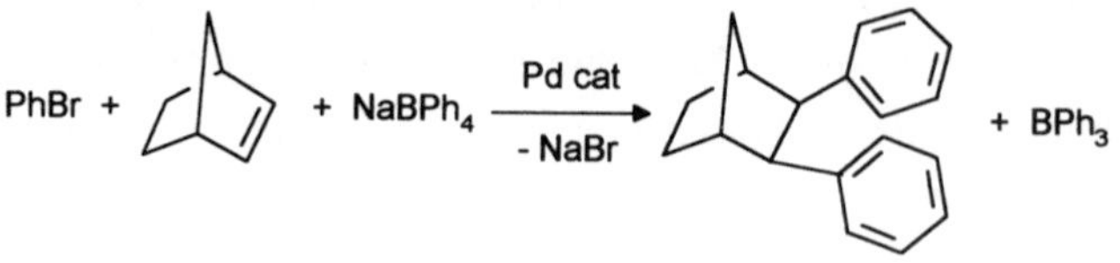

$$PhBr + \quad \text{[norbornene]} \quad + NaBPh_4 \quad \xrightarrow[- NaBr]{Pd\ cat} \quad \text{[product]} \quad + BPh_3 \qquad (8.24)$$

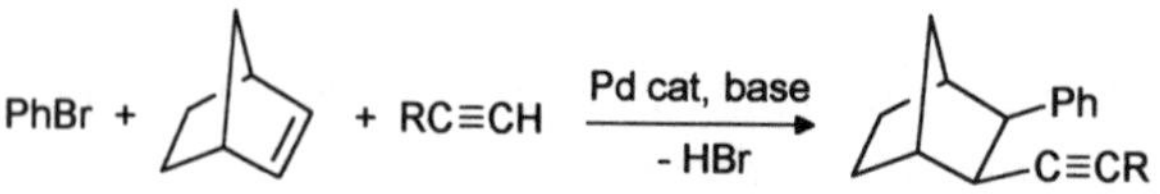

$$PhBr + \quad \text{[norbornene]} \quad + RC{\equiv}CH \quad \xrightarrow[- HBr]{Pd\ cat,\ base} \quad \text{[product Ph, C}{\equiv}CR] \qquad (8.25)$$

In the absence of steric limitations, one of the most common termination steps is hydrogen elimination, as exemplified by the Heck reaction [79–82] (Equation (8.10)).

A consequential development of multistep processes is the synthesis of cyclic from acyclic compounds. An early example was offered by the reaction of o-bromostyrene with norbornene [143]:

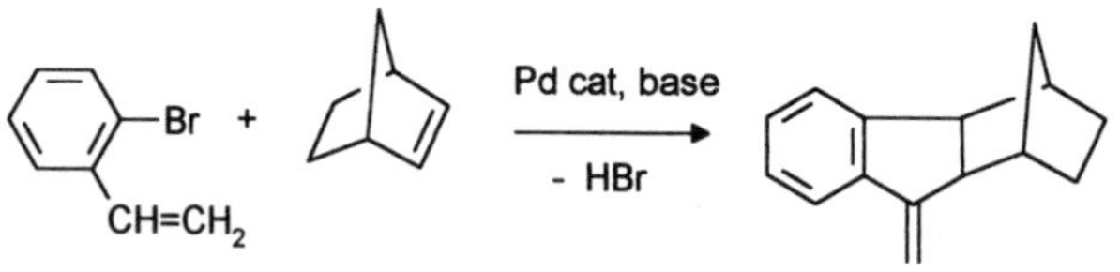

$$\text{[o-bromostyrene]} + \quad \text{[norbornene]} \quad \xrightarrow[- HBr]{Pd\ cat,\ base} \quad \text{[product]} \qquad (8.26)$$

The insertion of carbon monoxide into the C–Pd bond, formed in its turn by double bond insertion into the arylpalladium bond, is another possibility and in this case the resulting acyl group can be trapped by nucleophiles such as carboxylates [144]:

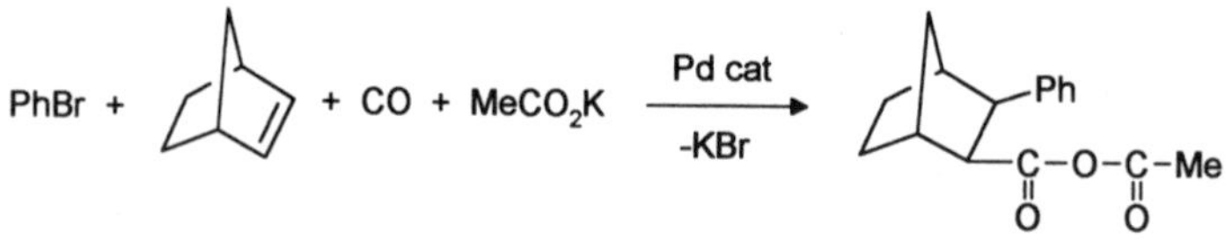

$$PhBr + \quad \text{[norbornene]} \quad + CO + MeCO_2K \quad \xrightarrow[-KBr]{Pd\ cat} \quad \text{[product Ph, C-O-C-Me]} \qquad (8.27)$$

The use of o-iodophenol also allowed an efficient synthesis of coumarin using norbornadiene in place of norbornene and exploiting the spontaneous retro-Diels–Alder reaction, which liberates cyclopentadiene [145, 146]:

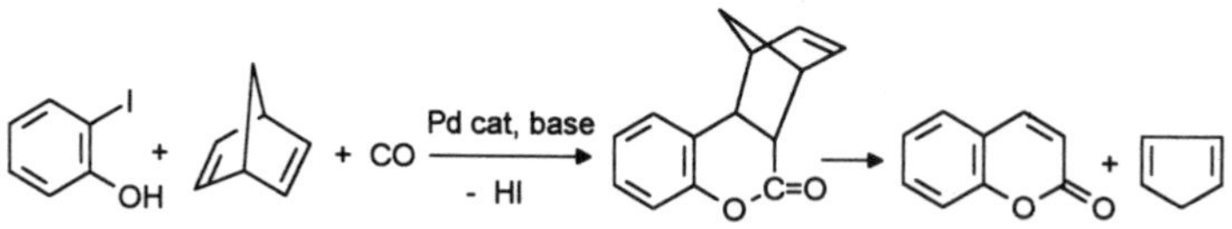

$$\text{[o-iodophenol]} + \quad \text{[norbornadiene]} \quad + CO \quad \xrightarrow[- HI]{Pd\ cat,\ base} \quad \text{[intermediate]} \rightarrow \text{[coumarin]} + \text{[cyclopentadiene]} \qquad (8.28)$$

Highly stereoselective reactions were obtained from intramolecular Heck-type or carbonylative ring closure [147–162] using o-iodo-substituted aromatics. Enantioselective reactions have also been obtained [163] both from intermolecular [164–166] and intramolecular [167, 168] Heck reactions.

The concept controlling the reactions shown above is based on the use of molecules able to insert into palladium–carbon bonds in sequence until the last palladium–carbon bond formed undergoes reductive elimination.

A variant that has been widely applied consists of the use of o-substituted aryl halides containing more than one function able to react in sequence after the reaction of the first one with the arylpalladium bond formed by oxidative addition. A description of these highly stereoselective reactions would be outside the scope of the present work which is restricted to aromatic substitution and we limit ourselves to an example (Equation (8.29)) [169] out of a selection of many others [170–172] that also often require prevention of unwanted secondary reactions such as isomerization or H-elimination [173].

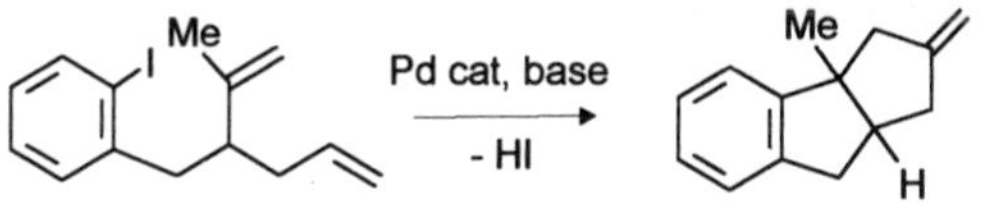

$$
\text{(8.29)}
$$

So far we have seen that palladium–aryl bonds can react with functions bonded to the aromatic substrate to form rings. It is also possible, however, that an aryl-bonded chain, containing palladium or other group VIII metals, reacts with the aromatic nucleus, e.g. [174]:

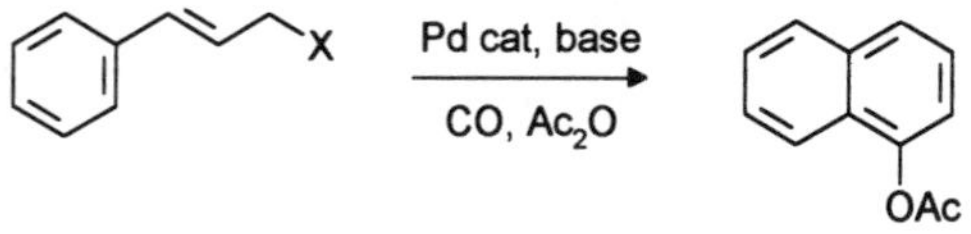

$$
\text{(8.30)}
$$

Insertion of two molecules of alkynes into bromobenzene can end up with ring closure [175–177]:

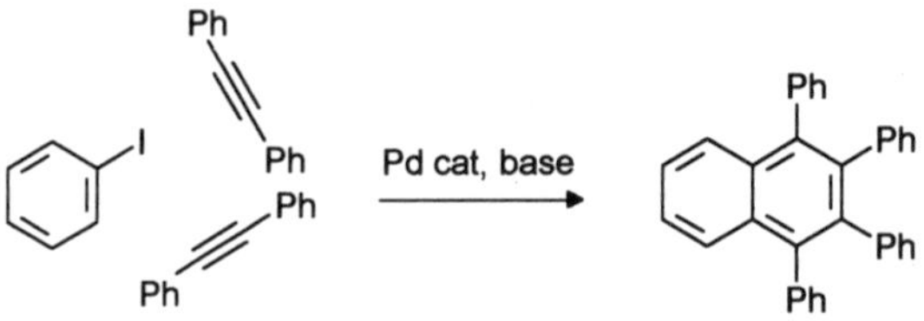

$$
\text{(8.31)}
$$

Intramolecular coupling between an aryl halide and another aryl group bonded to the former was also obtained [178, 179] using palladium(0) as catalyst:

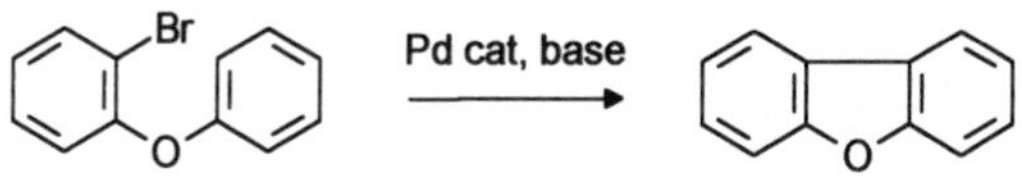

$$
\text{(8.32)}
$$

Application of this technique to a variety of substrates led to many complex molecules containing condensed rings [180].

Other transition metals can be used to obtain cyclization on an unactivated aromatic carbon. Thus, with $RhCl_3$, diphenylmethylencyclopropane opens up the cyclopropane ring, inserts 3-butenoic acid and alkylates the aromatic ring through the primary CH_2 (Equation (8.33)) [181] to form a condensed seven-membered ring.

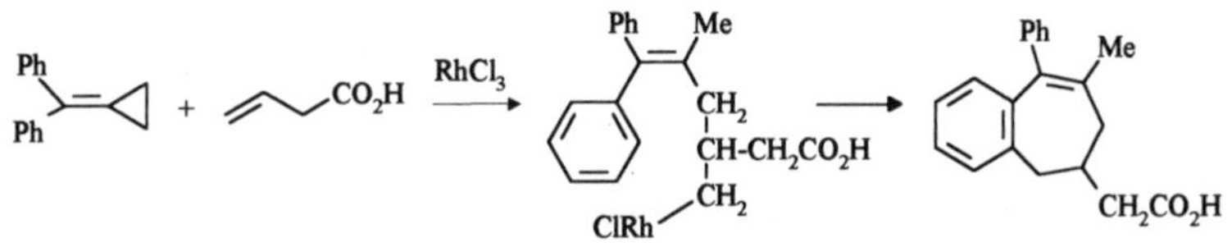

$$
\text{(8.33)}
$$

Equations (8.30)–(8.33) probably involve intermediate metallacycles, although more complex patterns, implying intramolecular cyclization to spirocycles and concomitant η^3-allylpalladium complex formation, are at work in the case of Equation (8.31) [175–177]. For related literature concerning alkyne insertion into an arylpalladium bond, see [182].

The type of ring closure involving palladacycles will be encountered again in Sections 6 and 7.

It has also to be observed that transition metal complexes can also catalyse homolytic reactions such as ring closure [183, 184], e.g.:

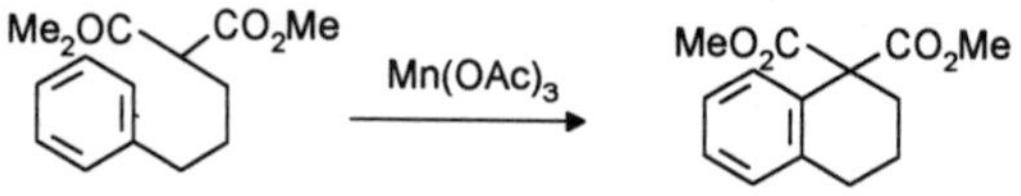

$$(8.34)$$

5   Aromatic substitution directed by chelating substrates

The concept of controlling regioselectivity through chelation has been known for many years (for a general outlook on this subject, see [185]). Its application to the orientation of alkyl attack on the aromatic nucleus was described as far back as 1974, when Murahashi reported the following alkylation reaction of a chelated palladium complex [186]:

$$(8.35)$$

Another interesting example was described by Holton [187]:

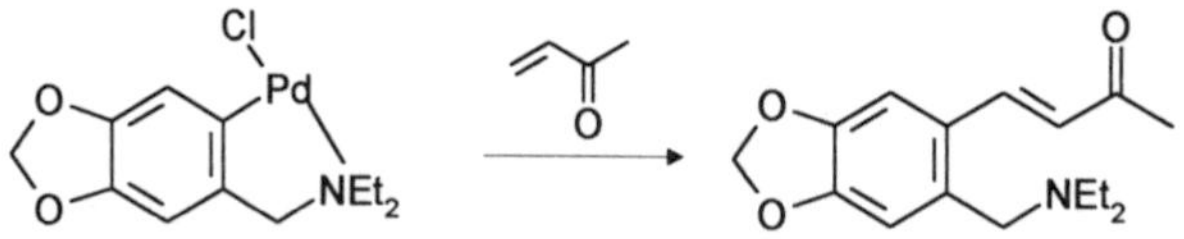

$$(8.36)$$

ortho-Alkylation of acetanilide through metallation by palladium and attack by methyl iodide was obtained by Tremont [188]:

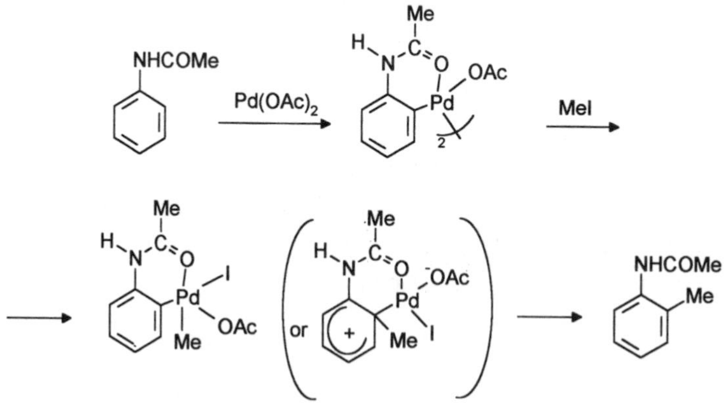

$$(8.37)$$

It is worth noting that for this reaction (Equation (8.37)) the authors postulated either a palladium(IV) intermediate (not yet known) or an arenonium ion bonded to palladium(II), as described by van Koten [189–191] (see also later).

Phenolic oxygen in the presence of Cs_2CO_3 has also been shown to be a good directing group for palladium-catalysed mono- and biarylation of unactivated aromatic C–H bonds [192], e.g.:

$$(8.38)$$

Reactions involving attack at an aromatic position *ortho* to a chelating group were also described with other metals [193, 194]:

$$(8.39)$$

In the context of chelation-directed aromatic substitution, it is appropriate to mention another important result by Milstein [195], who succeeded in activating an aryl-bonded CH_2 and in cleaving an aromatic to aliphatic C–C bond by taking advantage of the steric effect exerted by a bisphosphine substrate on a $PhCH_2$–Rh bond:

$$(8.40)$$

6 Alkylaromatic and aromatic metallacycles

Although many ring-forming reactions involving aromatic substitution probably occur via metallacycles [37, 196–200], isolation of the relevant intermediates is not common. Among group VIII metals, only a few examples of nickel(II), palladium(II) and platinum(II) metallacycles of this type have been reported. Several types of ligands were used to stabilize these and other metallacycles obtained from different group VIII metals. Thus, nickel complex **1** (Fig. 8.1) was isolated with $Cy_2PCH_2CH_2PCy_2$ as ligand in the case of a metallacycle containing a benzylic methylene [201] and with PMe_3 as ligand [202, 203] in the case of geminal methyl groups in the same benzylic position (see also [201] for a vinylaromatic

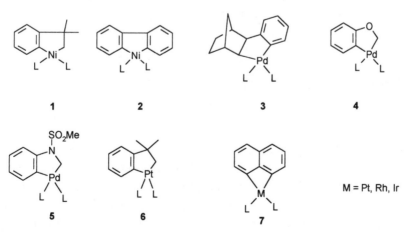

Figure 8.1.

nickel complex and [204] for a recent review mentioning other still unpublished alkylaromatic complexes). Nickel complex **2** was stabilized with PEt$_3$ [205]; 1,10-phenanthroline or other monodentate nitrogen compounds [206–208] or PPh$_3$ [209, 210] were used as ligands for palladium complex **3** (*cis, exo*) and for the analogous complex obtained from norbornadiene; triphenylphosphine or chelating phosphines, isonitriles, bipyridyl or phenanthroline stabilize palladium complexes **4** and **5** [211, 212]; for platinum complex **6** several ligands comprising 1,5-cyclooctadiene, 2,2′-bipyridyl, 2,2′-bipyrimidyl, 1,10-phenanthroline and its substituted derivatives and chelating phosphorus ligands are effective [213]; finally, the unusual four-membered metal complex **7** was obtained with Pt (PEt$_3$ as ligand), Rh and Ir (Cp*, PPh$_3$ as ligands) [214].

Some of these complexes were functionalized at the aromatic carbon by insertion of unsaturated molecules, e.g. CS$_2$ [203]:

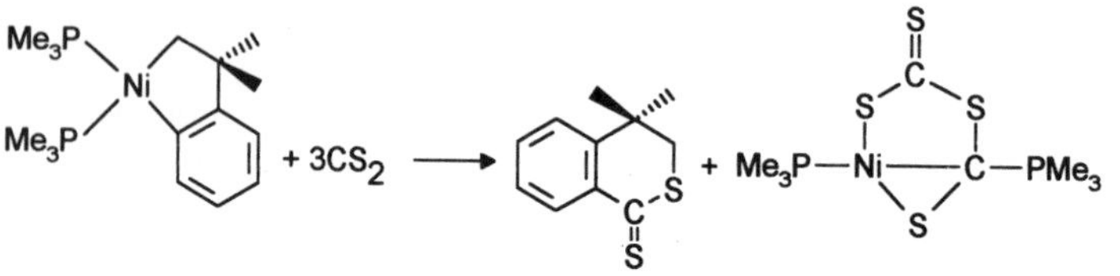

$$(8.41)$$

Ring contraction was described for complex **3** (Equation (8.42)) [210, 215–217]. Enlargement by alkyne insertion was described for most of the compounds reported in Figure 8.1. An example refers to the same palladacycle **3** (Equation (8.43)) [208, 210].

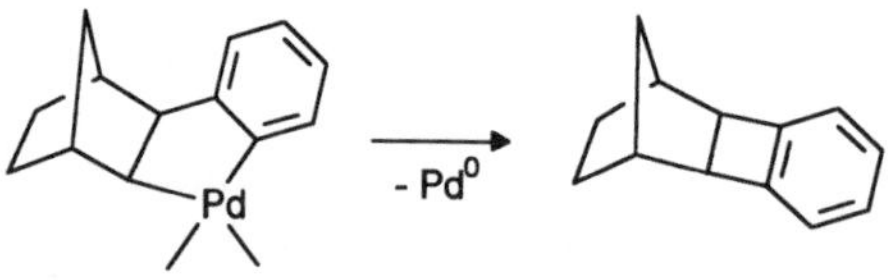

$$(8.42)$$

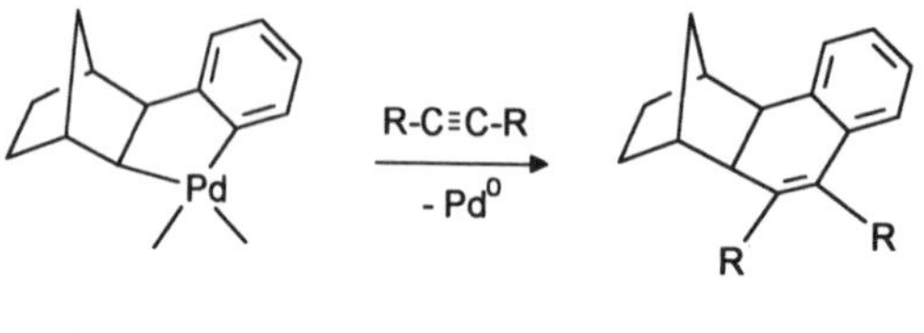

$$(8.43)$$

Of the complexes reported above, the one that has the best chances of acting catalytically is complex **3** because it is itself produced by norbornene insertion into an aryl—palladium bond that can be regenerated catalytically [209, 210, 218]:

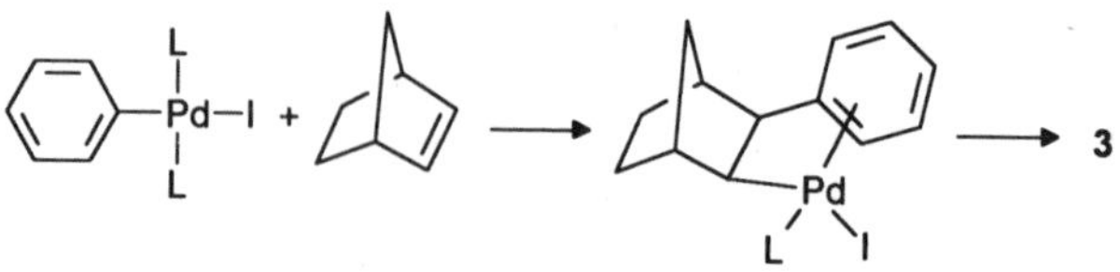

$$(8.44)$$

The ring closure was shown to be an electrophilic attack on the aromatic ring [207]. An η^2-coordination of the aryl group to palladium was observed on a triphenylphosphine-substituted complex [122, 219]. This feature is probably a requisite for ring closure.

With regard to palladium(IV) alkylaromatic metallacycles, the only examples of well-characterized complexes are those derived from oxidative addition of alkyl halides to complex **3** [218, 220, 221], e.g.:

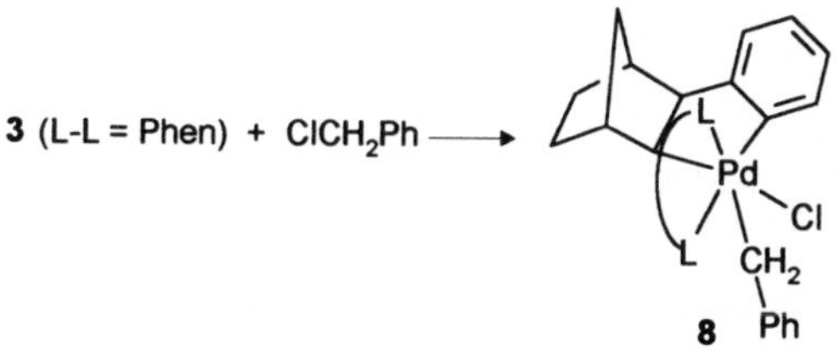

$$\textbf{3} \text{ (L-L = Phen)} + ClCH_2Ph \longrightarrow \qquad \qquad (8.45)$$

Other complexes such as **4** and **5** will probably undergo similar types of oxidative addition but at the moment this is a matter of hypothesis (for palladium(IV) chemistry in general, see [222–228]).

7 Palladium(II) and palladium(IV) metallacycles and their use for a new type of aromatic substitution via metallacycles

Some years ago it was shown that an unactivated aromatic C–H bond could be alkylated or arylated in an intermolecular reaction. Thus according to the process shown in Equation (8.46) ring closure of the phenylnorbornylpalladium halide readily occurs at room or lower temperature, leading to an alkylaromatic type **3** palladacycle. The latter is attacked by reactive species such as methyl [206], benzyl [221] and allyl halides [220] to give palladium(IV) intermediates that can be isolated in the presence of suitable ligands as shown in Equation (8.45). Palladium(IV) metallacycles spontaneously undergo reductive elimination by selective migration of the R group to the aromatic ring. The crystal structure of the palladium(II) complex resulting from compound **8** was reported in [221].

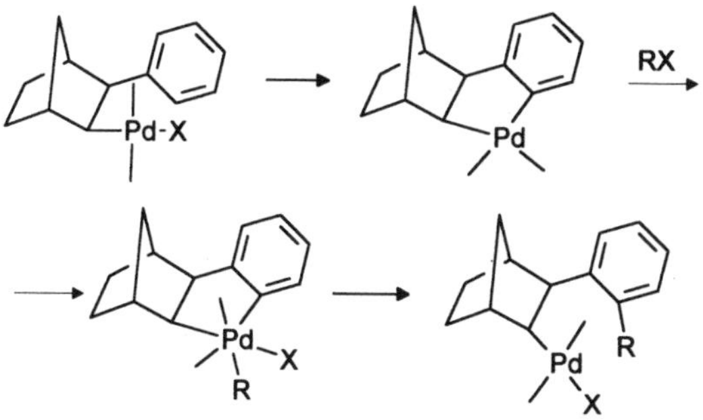

$$ \qquad \qquad (8.46) $$

It is worth noting that platinum(II) undergoes a similar oxidative addition to give platinum(IV), as described by van Koten [189–191], and can reversibly evolve towards an arenonium ion:

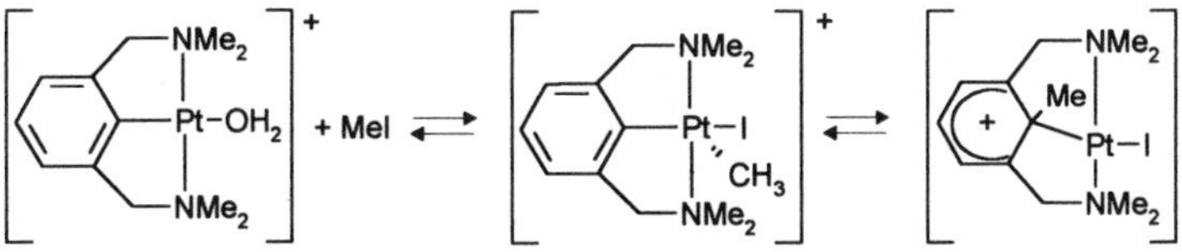

$$ \qquad \qquad (8.47) $$

When R in Equation (8.46) was an aryl group, special conditions were required (*t*-BuOK, anisole, 105 °C). Methanotriphenylene derivatives could be isolated from catalytic reactions (Equation (8.48), where Y = substituent) [229].

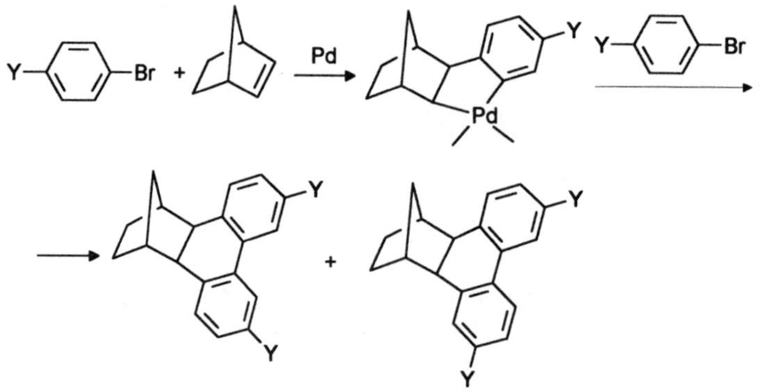

$$(8.48)$$

The process clearly involved two species in which the aryl halide attacked first the alkyl or the aryl site of the metallacycle (Fig. 8.2).

Although it is not yet clear with aromatic halides whether the reaction passes through a palladium(IV) intermediate that could not be detected or by concerted attack on the palladium–aryl bond, the result shows that it is possible to arylate catalytically an unactivated aryl–carbon bond through a metallacycle. It is worth noting that the intermolecular activation of an aromatic C–H bond shown in Equation (8.49) has not been reported.

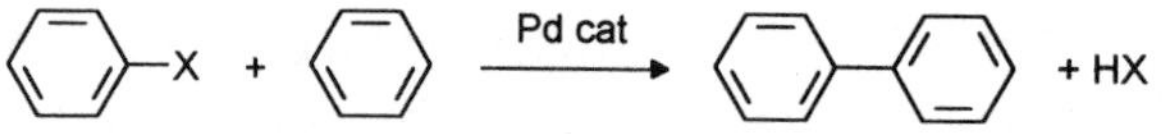

$$(8.49)$$

From this point on, two topics of aromatic substitution via metallacycles were the object of further investigation: alkylation and arylation.

Alkylation was carried out stoichiometrically at 20 °C in DMF as solvent in the absence of other ligands. It proceeded nicely without stopping at the level of the first migration of the R group on the aromatic ring, forming instead a new palladacycle with the aryl moiety onto which a second migration of the R group occurred selectively [230]:

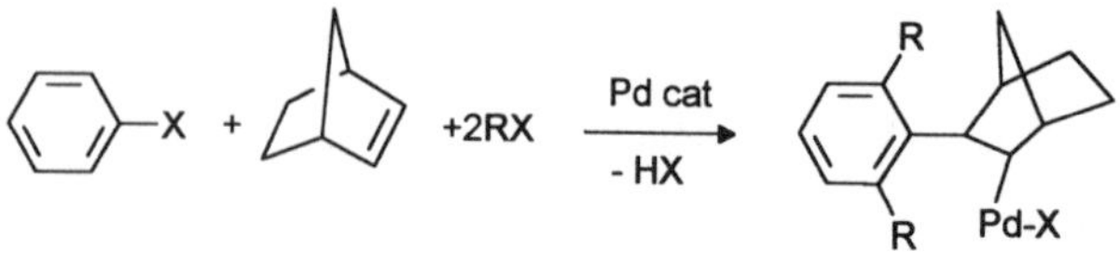

$$(8.50)$$

The single steps were identified by isolating the intermediate complexes involved. The resulting picture of the reaction is represented in Scheme 8.4.

A surprising fact was observed at this point. The final complex of Scheme 8.4 was not stable and spontaneously expelled the same norbornene that had been readily inserted in the initial step of the reaction sequence shown. Apparently, this behavior must be the consequence of the strong steric effect exerted by the two alkyl groups. Interestingly the

Figure 8.2. Proposed palladium species from reaction of the alkylaromatic metallacycle (eqn 8.46) with YC_6H_4X.

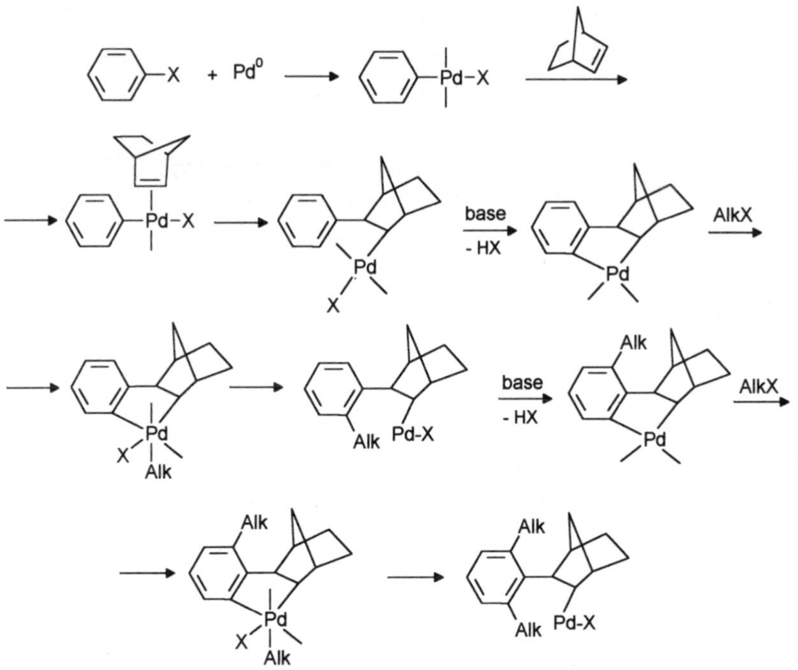

Scheme 8.4.

resulting complex is thus obtained through a unique way for preparing sterically hindered palladium compounds [230]:

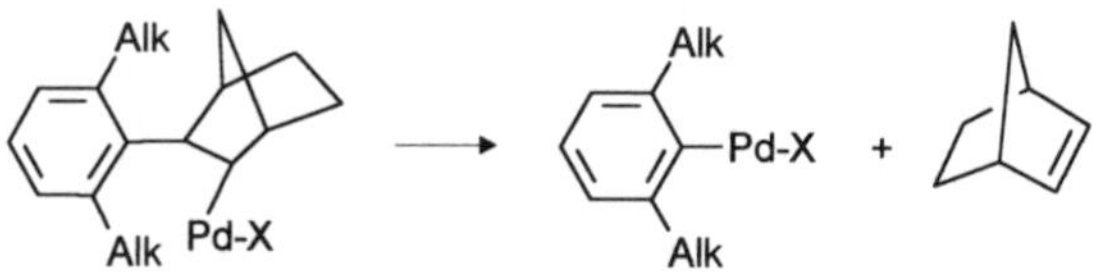

$$(8.51)$$

As shown by the combination of Scheme 8.4 and Equation (8.51) norbornene turns out to play the role of a scaffold that helps to build up the dialkylated molecule and is then removed.

At this point we reasoned that the use of a Heck-type insertion reaction would cleave the aryl–palladium bond, reducing palladium to the zero oxidation state and liberating HX (to be neutralized by an added base) [231]. This reaction indeed occurs with a variety of terminal olefins also bearing functional groups. In fact, while the alkyl groups exert a strong steric effect on the adjacent bulky norbornyl, a terminal olefin can be inserted without difficulty:

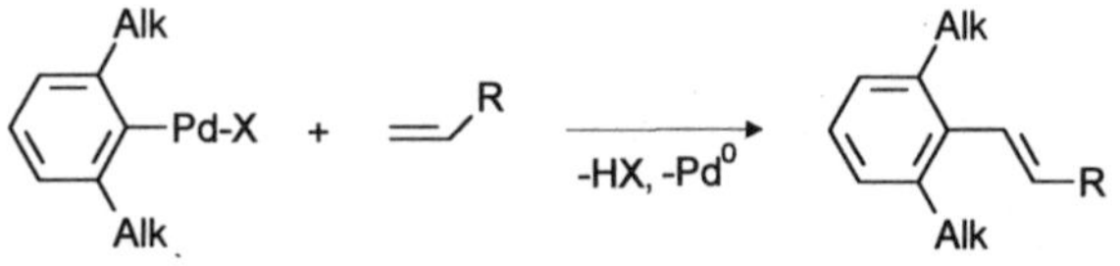

$$(8.52)$$

Equation (8.52) restores palladium(0) at the end of the sequence shown in Scheme 8.4 and in Equation (8.51) thus allowing a catalytic cycle that indeed has been achieved [231]. It has to be observed that in Equation (8.52) the situation is the opposite to that at the beginning of the cycle, where norbornene was definitely preferred to terminal olefins.

Through the new method it is possible to prepare a variety of aromatic compounds with extreme regiospecificity. In addition, it is possible to vary all the building blocks (aromatic halides, aliphatic halides and olefins) introducing functional groups.

It is worth noting, however, that the conditions for optimization must be chosen accurately for each combination of building blocks. In fact, the reaction is based on a delicate balance of factors that govern the competition for insertion and for oxidative addition at the levels of palladium(0) and palladium(II). Scheme 8.5 shows the choice made by the latter in the presence of couples of different substrates.

The problems connected with aromatic arylation are much more complex. As mentioned at the beginning of this section, arylation to form hexahydromethanotriphenylene had been observed initially to occur in part at the aryl site of the alkylaromatic palladacycle using a palladium/triphenylphosphine/*t*-BuOK system in anisole. When the same reaction was carried out in DMF in the presence of potassium carbonate as a base, aryl migration occurred essentially at the norbornyl site [232–234]. However, another instance of aromatic arylation was reported by de Meijere [235, 236] (Equation (8.53)), who described a reaction that, at variance with the other case reported above, occurs in DMF in the absence of phosphine ligands.

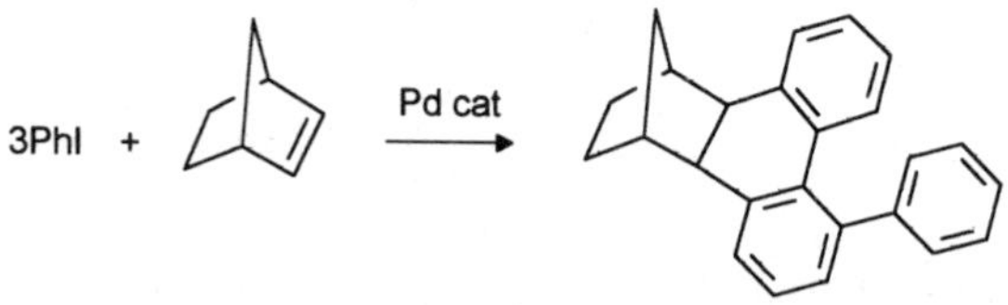

$$3PhI \ + \quad \xrightarrow{\text{Pd cat}} \qquad\qquad (8.53)$$

The mechanism put forward for the synthesis of hexahydromethanotriphenylenes seemed inadequate to interpret the formation of an additional aryl–carbon bond, so a mechanism based on coordinated aryne formation [204] was proposed as in Scheme 8.6 [236].

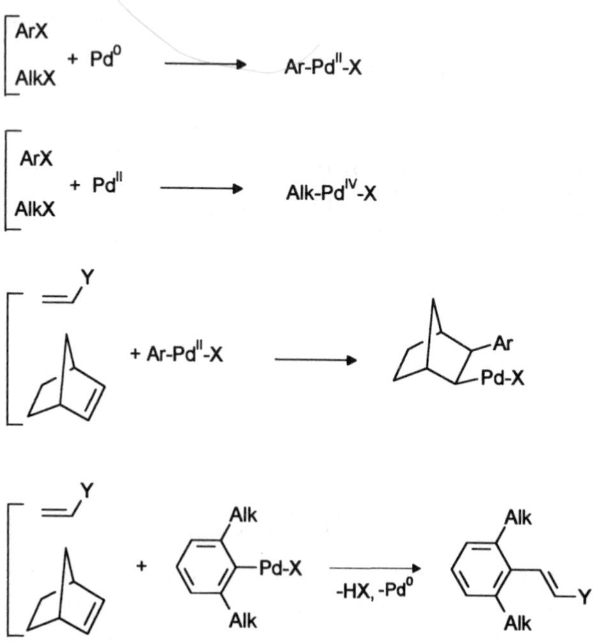

Scheme 8.5.

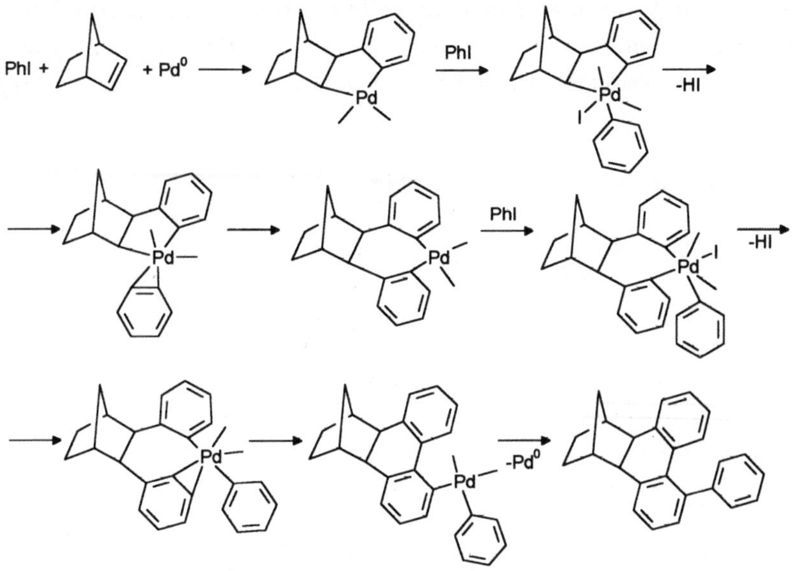

Scheme 8.6.

This interpretation seemed to us highly improbable because of the difficulty in forming an aryne under the mild reaction conditions adopted [204] and because of the absence of products (with appropriate substituent) resulting from aryl attack on either of the two carbon atoms of the aryne. We therefore wondered whether our previous mechanism, shown in Scheme 8.4 for alkylation, could be at work also in this case. Let us first consider the initial metallacycle and its evolution:

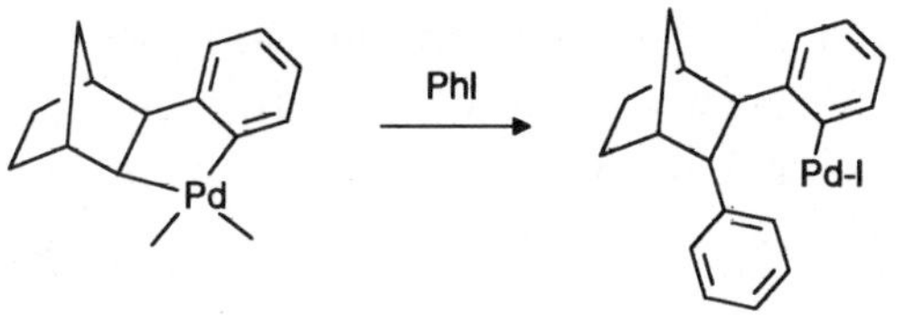

$$(8.54)$$

The arylpalladium–arene coupling to hexehydromethanotriphenylene is not particularly favored and the complex thus formed undergoes another norbornene insertion. This has been proved [232] by the isolation of the corresponding benzocyclobutene product:

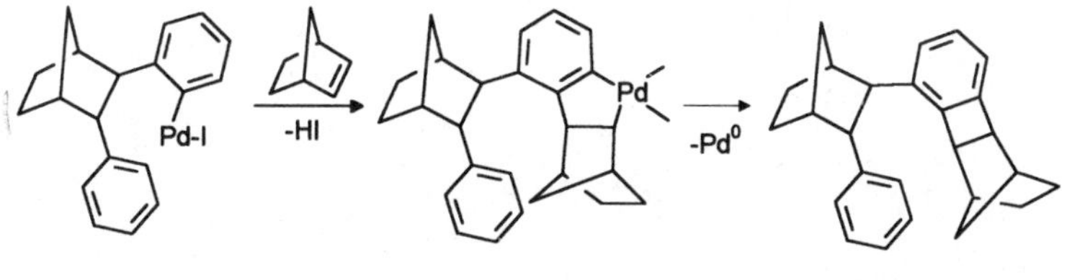

$$(8.55)$$

If now the two palladacycles of Equations (8.54) and (8.55) are compared, it can be observed that the situation is quite similar, the only difference being that an alkyl group *ortho* to the aromatic to aliphatic C–C bond of the palladacycle is present in the latter. One therefore would expect that in the case of the palladacycle shown in Equation (8.55) the reaction with an aryl iodide led to arylation at the norbornyl site rather than at the aromatic site. However, we observed that this kind of palladacycle favors aryl attack on the aromatic site. Evidence for this behavior was gained by carrying out a stoichiometric reaction using *o*-tolylnorbornyl palladium chloride. The latter reacted with iodobenzene to form substantial amounts of the product deriving from phenylation at the aromatic site. Norbornene expulsion and treatment with NaBH₄ affords 3-methyl-1, 1′-biphenyl in 60% yield [234]:

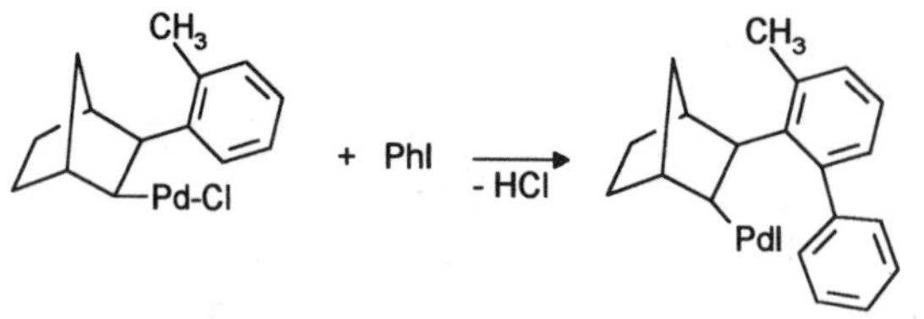

(8.56)

This fact can explain the formation of arylhexahydromethanotriphenylenes satisfactorily, the *ortho* substituent clearly causing preferential palladacycle opening and functionalization at the aryl site:

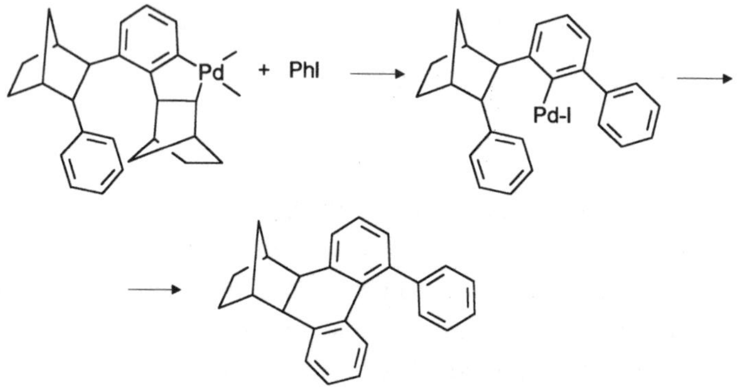

(8.57)

Further support in favor of the proposed mechanism was gained by comparing conditions for the formation of hexahydromethanotriphenylene and phenylhexahydromethano-triphenylene [237]. The former was obtained selectively by causing the initial palladacycle to react with iodobenzene in the absence of norbornene, while the latter could be obtained only in the presence of norbornene, in agreement with the proposed mechanism (Equation (8.58)).

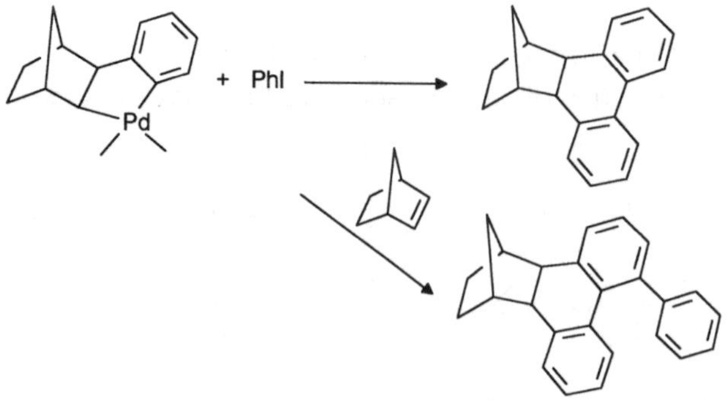

(8.58)

As shown above, there are conditions where arylation of aryl sites takes place substantially. A systematic study will certainly uncover general conditions to effect this type of arylation, which, when properly selectivized, could offer ready access to a host of compounds containing aryl–aryl bonds. It is worth noting that the palladium intermediate formed by norbornene expulsion in a catalytic reaction involving iodobenzene and norbornene (potassium acetate as a base in DMF at 105 °C) has been trapped by added olefins such as methyl acrylate or styrene [237], e.g.:

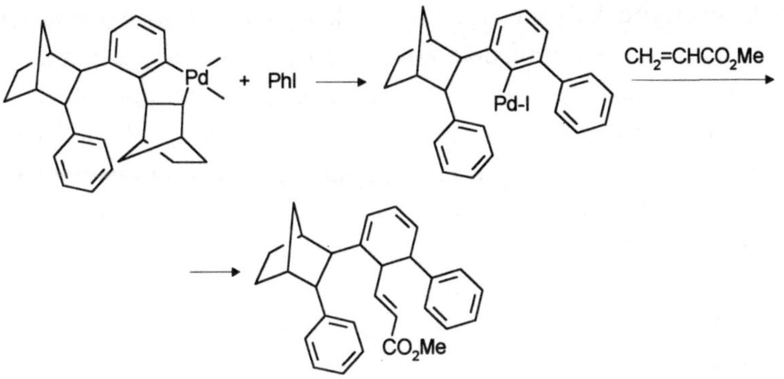

(8.59)

This is another instance of the extremely versatile behavior of the reactions via metallacycles with more than two components.

A point that has not been settled so far is the nature of the intermediate involved in arylation (both at the aliphatic or the aromatic site). Is this a palladium(IV) complex, analogous to the isolated alkyl complex? So far we have not been able to find evidence for any other mechanism. Possible alternatives have been put forward [211, 212, 238, 239].

It may well be, however, that palladium(IV) is actually formed but the tendency towards reductive elimination is so high that this species cannot be detected. It is proper to mention at this point other recent studies concerning the reactivity of metallacycles and mechanistic interpretations. Dyker [240–242] has shown that the palladacycle formed *in situ* from *o*-iodo-*tert*-butylbenzene can react further with another molecule of *o*-iodo-*tert*-butylbenzene to form arylation products according to a mechanism that is in line with that proposed by us for arylation.

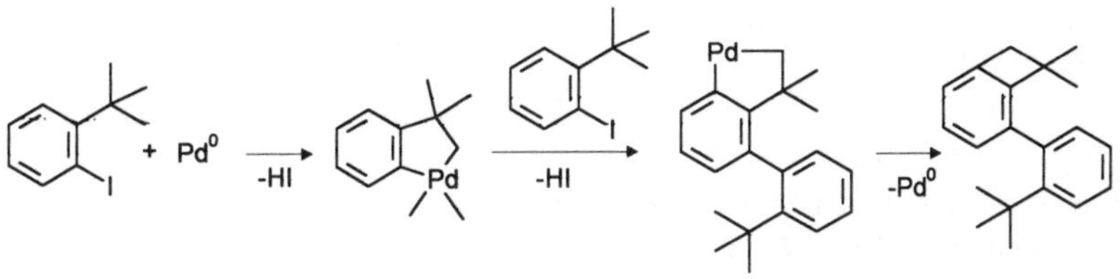

(8.60)

The metallacycle formed from *o*-iodoanisole (isolated by Echavarren [212]) also gives rise to further arylation possibly through palladium (IV).

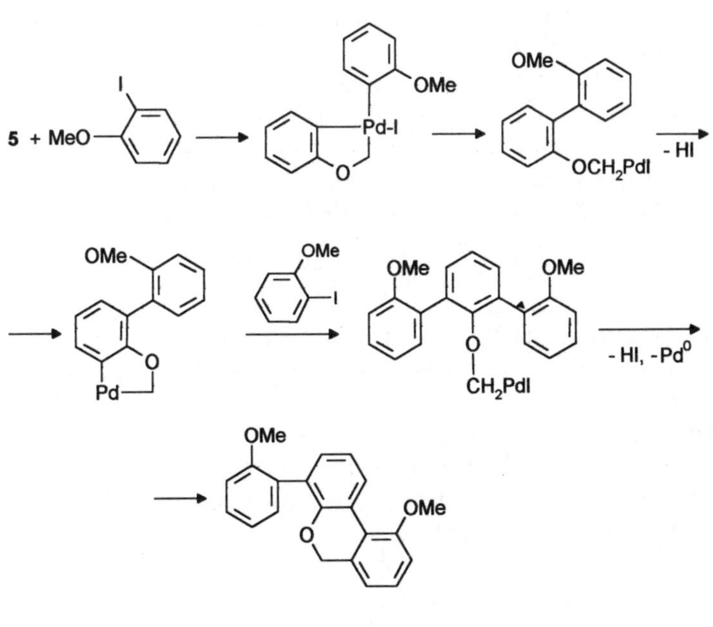

(8.61)

It is possible that more than one arylation mechanism is at work. However, there is not yet any clear evidence that arylation does not proceed via the formation of palladium(IV).

In conclusion, on the basis of the results achieved so far it is possible to envisage important developments connected with varying the nature of the building blocks and with the type of palladium complexes involved. Aromatic functionalization can be expected to be improved remarkably in connection to selectivity and catalytic efficiency. The rules governing aromatic substitution and the reactivity of palladium (and possibly of other metals) in high oxidation state will be the object of intense research. The area of aromatic and aliphatic C–H activation may benefit from a better knowledge of the chemistry and stereochemistry of palladium(IV) complexes and it cannot be excluded that in the future the chemistry of organic halides is replaced by that of the parent hydrocarbons.

8 References

1 For general information, see: Collman JP, Hegedus LS, Norton JR, Finke RG. *Principles and Applications of Organotransition Metal Chemistry.* Mill Valley, CA: University Science Books, 1987.
2 Halpern J. *Acc Chem Res* 1970; **3**: 386–92.
3 *Rodd's Chemistry of Carbon Compounds*, 2nd Edn, Vol. IIIA. Amsterdam: Elsevier, 1971: 5–129.
4 Olah GA, Krishnamurti R, Surya Prakash GK. In: Trost BM, Fleming I, eds. *Comprehensive Organic Synthesis*, Vol. III. Oxford: Pergamon Press, 1991: 293–339.
5 Heaney H. In: Trost BM, Fleming I, eds. *Comprehensive Organic Synthesis*, Vol. II. Oxford: Pergamon Press, 1991: 733–68.
6 Wynberg H. In: Trost BM, Fleming I, eds. *Comprehensive Organic Synthesis*, Vol. II. Oxford: Pergamon Press, 1991: 769–75.
7 Meth-Cohn O, Stanforth SP. In: Trost BM, Fleming I, eds. *Comprehensive Organic Synthesis*, Vol. II. Oxford: Pergamon Press, 1991: 777–94.
8 Paradisi C. In: Trost BM, Fleming I, eds. *Comprehensive Organic Synthesis*, Vol. IV. Oxford: Pergamon Press, 1991: 423–50.
9 Kessar SV. In: Trost BM, Fleming I, eds. *Comprehensive Organic Synthesis*, Vol. IV. Oxford: Pergamon Press, 1991: 483–515.
10 Bunnett JF. *Q Rev* 1958; **12**: 1–16.
11 Perkins MJ. In: Kochi JK, ed. *Free Radicals*, Vol. II. New York: Wiley, 1973: 231–71.
12 Kochi JK. *Free Radicals*, Vols I and II. New York: Wiley, 1973.
13 Norris RK. In: Trost BM, Fleming I, eds. *Comprehensive Organic Synthesis*, Vol. IV. Oxford: Pergamon Press, 1991: 451–82.
14 Araneo S, Arrigoni R, Biørsvik H *et al. Tetrahedron Lett* 1996; **37**: 7425–8.
15 Olah GA, Surya Prakash GK, Sommer J. *Superacids.* New York: Wiley, 1985.
16 Olah GA, Arpad M. In: *Hydrocarbon Chemistry.* New York: Wiley, 1995: 276–7.
17 Olah GA, Farooq O, Morteza S, Farnia F, Olah JA. *J Am Chem Soc* 1988; **110**: 2560–5.
18 Eberson L. *Electron Transfer Reactions in Organic Chemistry.* Berlin: Springer, 1987.
19 Bunnett JF. *Acc Chem Res* 1978; **11**: 413–20.
20 Speiser B. *Angew Chem Int Ed Engl* 1996; **35**: 2471–4.
21 Pietra F. *Q Rev* 1969; **23**: 504–21.
22 Novi M, Guanti G, Dell'Erba C, Calabrò D, Petrillo G. *Tetrahedron* 1980; **36**: 1879–83.
23 *Kirck Othmer Encyclopedia of Chemical Technology*, 2nd Edn, Vol. 6. New York: Interscience, 1951: 883–92.
24 Prichard WW. *J Am Chem Soc* 1956; **78**: 6137–9.
25 Dieterle H, Eschenbach W. *German Patent 537610,* 1927.
26 Weil TA, Cassar L, Foà M. In: Wender I, Pino P, eds. *Organic Syntheses via Metal Carbonyls*, Vol. I. New York: Wiley, 1977: 517–44.

27 Bauld NL. *Tetrahedron Lett* 1963: 1841–5.

28 Corey EJ, Hegedus LS. *J Am Chem Soc* 1969; **91**: 1233–4.

29 Cassar L, Foà M. *J Organomet Chem* 1973; **51**: 381–93.

30 Schanberg A, Bartoletti I, Heck RF. *J Org Chem* 1974; **39**: 3318–26.

31 Heck RF. *Organotransition Metal Chemistry*. New York: Academic Press, 1974.

32 Fitton P, Rick EA. *J Organomet Chem* 1971; **28**: 287–91.

33 Amatore C, Jutand A, M'Barki MA. *Organometallics* 1992; **11**: 3009–13.

34 Amatore C, Carré E, Jutand A *et al. Chem Eur J* 1996; **2**: 957–66, and references therein.

35 Tsuji J. *Organic Synthesis with Palladium Compounds.* Berlin: Springer, 1980.

36 Davies SC Jr. *Organotransition Metal Chemistry: Applications to Organic Synthesis*. Oxford: Pergamon Press, 1982.

37 Ryabov AD. *Chem Rev* 1990; **90**: 403–24.

38 Clark FRS, Norman ROC, Thomas CB. *J Chem Soc Perkin Trans 1* 1975: 121–5.

39 Jolly PW. In: Wilkinson G, Stone FGA, Abel EW, eds. *Comprehensive Organometallic Chemistry*, Vol. VIII. New York: Pergamon Press, 1982: 713–72.

40 Farina V. In: Abel EW, Stone FGA, Wilkinson G, eds. *Comprehensive Organometallic Chemistry II*, Vol. XII. Oxford: Pergamon Press, 1995: 161–240.

41 Milstein D, Stille JK. *J Am Chem Soc* 1979; **101**: 4992–8.

42 Aoki S, Fujimura T, Nakamura E, Kuwajima I. *J Am Chem Soc* 1988; **110**: 3296–8.

43 Bringmann G, Walter R, Weirich R. *Angew Chem Int Ed Engl* 1990; **29**: 977–91.

44 Carfagna C, Musco A, Sallese G, Santi R, Fiorani T. *J Org Chem* 1991; **56**: 261–3.

45 Mori M, Kaneta N, Shibasaki M. *J Org Chem* 1991; **56**: 3486–93.

46 Ritter K. *Synthesis* 1993: 735–62.

47 Hagashi T, Niizuma S, Kamikawa T, Sazuki N, Uozumi Y. *J Am Chem Soc* 1995; **117**: 9101–2.

48 Yamada I, Yamazaki N, Yamaguchi M, Yamagishi T. *J Mol Catal A* 1997; **130**: L13–15.

49 Kang S-K, Yamaguchi T, Ho PS, Kim WY, Yoon S-K. *Tetrahedron Lett* 1997; **38**: 1947–50, and references therein.

50 Shirakawa E, Yoshida H, Takaya H. *Tetrahedron Lett* 1997; **38**: 3759–62.

51 Suzuki A. *Pure Appl Chem* 1985; **57**: 1749–58.

52 Suzuki A. *Pure Appl Chem* 1994; **66**: 213–22.

53 Miyaura N, Suzuki A. *Chem Rev* 1995; **95**: 2457–83.

54 Indolese A. *Tetrahedron Lett* 1997; **38**: 3513–16.

55 Cassar L. *J Organomet Chem* 1973; **54**: C57–8.

56 Cassar L. *J Organomet Chem* 1975; **93**: 253–7.

57 Catellani M, Chiusoli GP, Salerno G, Dallatomasina F. *J Organomet Chem* 1978; **146**: C19–22.

58 Cassar L, Ferrara S, Foà M. *Adv Chem Ser* 1974; **132**: 252–73.

59 Cassar L, Foà M, Montanari F, Marinelli GP. *J Organomet Chem* 1979; **173**: 335–9.

60 Schoenberg A, Heck RF. *J Am Chem Soc* 1974; **96**: 7761–4.

61 Ben-David Y, Portnoy M, Milstein D. *J Chem Soc Chem Commun* 1989: 1816–17.

62 Baillargeon VP, Stille JK. *J Am Chem Soc* 1986; **108**: 452–61.

63 Pri-Bar I, Buchman O. *J Org Chem* 1984; **49**: 4009–11.

64 Boukherroub R, Chatgilialoglu C, Manuel G. *Organometallics* 1996; **15**: 1508–10.

65 Biavati A, Chiusoli GP, Costa M, Terenghi G. *Transition Met Chem* 1979; **4**: 398–9.

66 Chiusoli GP. *Pure Appl Chem* 1980; **52**: 635–48.

67 Chiusoli GP, Merzoni S, Mondelli G. *Tetrahedron Lett* 1964: 2777–81.

68 Ozawa F, Yamamoto A. *Chem Lett* 1982; **6**: 865–8.

69 Ozawa F, Soyama H, Yamamoto T, Yamamoto A. *Tetrahedron Lett* 1982; **23**: 3383–6.

70 Kobayashi T, Tanaka M. *J Organomet Chem* 1982; **233**: C64–6.

71 Chen JT, Sen A. *J Am Chem Soc* 1984; **106**: 1506–7.

72 Kobayashi T, Tanaka M. *J Chem Soc Chem Commun* 1981: 333–4.

73 Bumagin NA, Bumagin IG, Kashin AN, Beletskaya IP. *Dokl Akad Nauk SSSR* 1981; **261**: 1141–4.

74 Echavarren AM, Stille JK. *J Am Chem Soc* 1988; **110**: 1557–65.

75 Davies SG, Pyatt D, Thomson C. *J Organomet Chem* 1990; **387**: 381–90.

76 Kikukawa K, Idemoto T, Katayama A *et al. J Chem Soc Perkin Trans 1* 1987: 1511–20.

77 Kang S-K, Liu K-H, Ho P-S, Kim NY. *Synthesis* 1997: 874–6.

78 Brunet JJ, Chauvin R. *Chem Soc Rev* 1995; **25**: 89–95.

79 Heck RF. *Palladium Reagents in Organic Syntheses*. London: Academic Press, 1985.

80 de Meijere A, Meyer FE. *Angew Chem Int Ed Engl* 1994; **33**: 2379–411.

81 Rix FC, Brookhart M, White PS. *J Am Chem Soc* 1996; **118**: 2436–48.

82 Cabri W, Candiani I. *Acc Chem Res* 1995; **28**: 2–7.

83 Blaser H-U, Spencer A. *J Organomet Chem* 1982; **233**: 267–74.

84 Spencer A. *J Organomet Chem* 1983; **247**: 117–22.

85 Halpern J. *Inorg Chim Acta* 1985; **100**: 41–8.

86 Crabtree RH. *Chem Rev* 1985; **85**: 245–69.

87 Bergman RG. *Science* 1984; **223**: 902–8.

88 Kunin AJ, Eisenberg R. *Organometallics* 1988; **7**: 2124–9.

89 Jones WD, Feher FJ. *Acc Chem Res* 1989; **22**: 91–100.

90 Tanaka M, Sakakura T. *Pure Appl Chem* 1990; **62**: 1147–50.

91 Ittel SD, Tolman CA, English AD, Jesson JP. *J Am Chem Soc* 1976; **98**: 6073–5.

92 Ittel SD, Tolman CA, English AD, Jesson JP. *Adv Chem Ser* 1979; **173**: 67–80.

93 Garnett JL. *Catal Rev* 1971; **5**: 229–67.

94 Moritani I, Fujiwara Y. *Synthesis* 1973: 524–33.

95 Shul'pin GB, Shilov AE, Kitaigorodskii AN, Zeile-Krevor JV. *J Organomet Chem* 1980; **201**: 319–25.

96 Jintoku T, Fujiwara Y, Kawata I, Kawauchi T, Taniguchi H. *J Organomet Chem* 1990; **385**: 297–306.

97 Fujiwara Y, Takaki K, Taniguchi Y. *Synlett* 1996: 591–9.

98 For a review, see: Süss-Fink G, Meister G. *Adv Organomet Chem* 1993; **35**: 41–134.

99 Semmelhack MF. In: Trost BM, Fleming I, eds. *Comprehensive Organic Synthesis*, Vol. IV. Oxford: Pergamon Press, 1991: 517–49.

100 Semmelhack MF. In: Abel EW, Stone FGA, Wilkinson G, eds. *Comprehensive Organometallic Chemistry II*, Vol. XII. Oxford: Pergamon Press, 1995: 979–1070.

101 Ceccon A, Gambaro A, Gottardi F, Manoli F, Venzo A. *J Organomet Chem* 1989; **363**: 91–102.

102 Kundig EP, Ripa A, Liu R, Bernardinelli G. *J Org Chem* 1994; **59**: 4773–83.

103 Semmelhack MF, Schmaltz HG. *Tetrahedron Lett* 1996; **37**: 3089–92, and references therein.

104 Pearson AJ, Gontcharov AV, Zhu PY. *Tetrahedron* 1997; **53**: 3849–62.

105 LeBozec H, Touchard D, Dixneuf PH. *Adv Organomet Chem* 1989; **29**: 163–247.

106 Kochi JK. *Organometallic Mechanisms and Catalysis*. New York: Academic Press, 1978.

107 Kochi JK. *Acta Chem Scand* 1990; **44**: 409–32, and references therein.

108 Amatore C, Carré E, Jutand A *et al. Organometallics* 1995; **14**: 5605–14.

109 Trost BM, Verhoeven TR. In: Wilkinson G, Stone FGA, Abel EW, eds. *Comprehensive Organometallic Chemistry*, Vol. VIII. Oxford: Pergamon Press, 1982: 799–938.

110 Foà M, Francalanci F, Bencini E, Gardano A. *J Organomet Chem* 1985; **285**: 293–303.

111 Braterman PS, ed. *Reactions of Coordinated Ligands*. New York: Plenum Press, 1985.

112 Ben-David Y, Portnoy M, Milstein D. *J Am Chem Soc* 1989; **111**: 8742–4.

113 Milstein D. *Acc Chem Res* 1988; **21**: 428–34.

114 Brunet JJ, Sidot C, Caubère P. *J Org Chem* 1983; **48**: 1166–71.

115 Brunet JJ, Sidot C, Caubère P. *Tetrahedron Lett* 1981: 1013–16.

116 Aizenberg M, Milstein D. *J Am Chem Soc* 1995; **117**: 8674–5.

117 Aizenberg M, Milstein D. *Science* 1994; **265**: 359–61.

118 Edelbach BL, Jones WD. *J Am Chem Soc* 1997; **119**: 7734–42.

119 Anderson GK, Cross RJ. *J Chem Soc Dalton Trans* 1979: 1246–50.

120 Kawataka F, Shimizu I, Yamamoto A. *Bull Chem Soc Jpn* 1995; **68**: 654–60.

121 Portnoy M, Ben-David Y, Milstein D. *Organometallics* 1993; **12**: 4734–5.

122 Li C-S, Jou D-C, Cheng C-H. *Organometallics* 1993; **12**: 3945–54.

123 Herrmann WA, Brossmer C, Öfele K *et al. Angew Chem Int Ed Engl* 1995; **34**: 1844–8.

124 Beller M, Fischer H, Herrmann WA, Öfele K, Brossmer C. *Angew Chem Int Ed Engl* 1995; **34**: 1848–9.

125 Herrmann WA, Brossmer C, Reisinger C-P, Riermeier TH, Öfele K, Beller M. *Chem Eur J* 1997; **3**: 1357–64.

126 Hartley FR. *Chem Rev* 1981; **81**: 79–90.

127 Jeffery T. *Adv Met Org Chem* 1996; **5**: 153–266.

128 Jeffery T. *Tetrahedron Lett* 1985; **26**: 2667–70.

129 Abelmann MM, Oh T, Overmann LE. *J Org Chem* 1987; **52**: 4130–33.

130 Larock RG, Gong WH. *J Org Chem* 1989; **54**: 2047–50.

131 Sonesson C, Larhed M, Nyqvist C, Hallberg A. *J Org Chem* 1996; **61**: 4756–63.

132 Ripa L, Hallberg A. *J Org Chem* 1997; **62**: 595–602.

133 Alper H. *Adv Organomet Chem* 1981; **19**: 183–211.

134 Beletskaya IP. *Pure Appl Chem* 1997; **69**: 471–6.

135 Joó F, Kathó A. *J Mol Cat A* 1997; **116**: 3–26.

136 Chiusoli GP, Cassar L. In: Wender I, Pino P, eds. *Organic Syntheses via Metal Carbonyls*, Vol. II. New York: Wiley, 1997: 297–319.

137 Sicher J. *Angew Chem Int Ed Engl* 1972; **11**: 200–14.

138 Kosugi M, Tamura H, Sano H, Migita T. *Chem Lett* 1987: 193–4.

139 Kosugi M, Tamura H, Sano H, Migita T. *Tetrahedron* 1989; **45**: 961–7.

140 Catellani M, Chiusoli GP, Concari S. *Tetrahedron* 1989; **45**: 5263–8.

141 Catellani M, Chiusoli GP. *Tetrahedron Lett* 1982; **23**: 4517–20.

142 Catellani M, Chiusoli GP, Mari A. *J Organomet Chem* 1984; **275**: 129–38.

143 Catellani M, Chiusoli GP, Sgarabotto P. *J Organomet Chem* 1982; **240**: 311–19.

144 Catellani M, Chiusoli GP, Peloso C. *Tetrahedron Lett* 1983; **24**: 813–16.

145 An Z, Catellani M, Chiusoli GP. *J Organomet Chem* 1989; **371**: C51–2.

146 An Z, Catellani M, Chiusoli GP. *Gazz Chim Ital* 1990; **120**: 383–5.

147 Soderberg BC. In: Abel EW, Stone FGA, Wilkinson G, eds. *Comprehensive Organometallic Chemistry II*, Vol. XII. Oxford: Pergamon Press, 1995: 241–97.

148 Heck RF. *Org React* 1982; **27**: 345–90.

149 Grigg R, Stevenson P, Worakun T. *Tetrahedron* 1988; **44**: 2033–48.

150 Catellani M, Chiusoli GP, Fagnola MC, Solari G. *Tetrahedron Lett* 1994; **35**: 5919–22.

151 Catellani M, Chiusoli GP, Marzolini G, Rossi E. *J Organomet Chem* 1996; **525**: 65–9.

152 Brown JM, Pérez-Torrente JJ, Alcock N, Clase HJ. *Organometallics* 1995; **14**: 207–13.

153 Tietze LF, Burkhardt O, Henrich M. *Liebigs Ann Recueil* 1997: 1407–13.

154 Catellani M, Chiusoli GP, Cugini F, Lasagni B, Neri Mari M. *Inorg Chim Acta* 1998; **270**: 123–9.

155 Tour JM, Negishi EI. *J Am Chem Soc* 1985; **107**: 8289–91.

156 Brennan CM, Johnson CD, McDonnell PD. *J Chem Soc Perkin Trans 2* 1989: 957–61.

157 An Z, Catellani M, Chiusoli GP. *J Organomet Chem* 1990; **397**: C31–2.

158 Kalinin VN, Shostakovsky MV, Ponomaryov AB. *Tetrahedron Lett* 1990; **31**: 4073–6.

159 Torii S, Okumoto H, Xu LH. *Tetrahedron Lett* 1991; **32**: 237–9.

160 Ciattini PG, Morera E, Ortar G. *Tetrahedron Lett* 1991; **32**: 6449–52.

161 Brocato E, Castagnoli C, Catellani M, Chiusoli GP. *Tetrahedron Lett* 1992; **33**: 7433–6.

162 Catellani M, Chiusoli GP, Fagnola MC, Solari G. *Tetrahedron Lett* 1994; **35**: 5923–6.

163 Heumann A, Réglier M. *Tetrahedron* 1996; **28**: 9289–346.

164 Shibasaki M. *Adv Met Org Chem* 1996; **5**: 119–52.

165 Loiseleur O, Meier P, Pfaltz A. *Angew Chem Int Ed Engl* 1996; **35**: 200–2.

166 Ozawa F, Kubo A, Hagashi T. *J Am Chem Soc* 1991; **113**: 1417–19.

167 Overman LE, Poon DJ. *Angew Chem Int Ed Engl* 1997; **36**: 518–21, and references therein.

168 Shibasaki M, Boden CDJ, Kojima A. *Tetrahedron* 1997; **53**: 7371–95.

169 Abelman MM, Overman LE. *J Am Chem Soc* 1988; **110**: 2328–9.

170 Grigg R, Sridharan V. In: Abel EW, Stone FGA, Wilkinson G, eds. *Comprehensive Organometallic Chemistry II*, Vol. XII. Oxford: Pergamon Press, 1995: 299–321.

171 Copéret C, Ma S, Negishi EI. *Angew Chem Int Ed Engl* 1996; **35**: 2125–6.

172 González JJ, García N, Gómez-Lor B, Echavarren AM. *J Org Chem* 1997; **62**: 1286–91.

173 Grotjahn DB, Zhang X. *J Mol Catal A* 1997; **116**: 99–107.

174 Koyasu Y, Matsuzaka M, Hiroe Y, Uchida Y, Hidai M. *J Chem Soc Chem Commun* 1987: 575–6.

175 Wu G, Rheingold AL, Geib SJ, Heck RF. *Organometallics* 1987; **6**: 1941–6.

176 Tao W, Silverberg LJ, Rheingold AL, Heck RF. *Organometallics* 1989; **8**: 2550–9.

177 Pfeffer M, Sutter J-P, Rotteveel MA, De Cian A, Fischer J. *Tetrahedron* 1992; **48**: 2427–40.

178 Ames DE, Opelko A. *Synthesis* 1983: 234–5.

179 Ames DE, Opelko A. *Tetrahedron* 1984; **40**: 919–25.

180 See, for example: Dyker G, Körning J, Bubenitschek P, Jones PG. *J Chem Res(s)* 1997: 132–3, and references therein.

181 Chiusoli GP, Costa M, Melli L. *J Organomet Chem* 1988; **358**: 495–505.

182 Hosokawa T, Calvo C, Lee HB, Maitlis PM. *J Am Chem Soc* 1973; **95**: 4914–23.

183 Snider BB. *Chem Rev* 1996; **96**: 339–63.

184 Citterio A, Fancelli D, Finzi C, Pesce L, Santi R. *J Org Chem* 1989; **54**: 2713–18.

185 Hoveyda AH, Evans DA, Fu GC. *Chem Rev* 1993; **93**: 1307–70.

186 Murahashi S-I, Tanaba Y, Yamamura M, Moritani I. *Tetrahedron Lett* 1974: 3749–52.

187 Holton RA. *Tetrahedron Lett* 1977: 355–8.

188 Tremont SJ, Rahman HU. *J Am Chem Soc* 1984; **106**: 5759–60.

189 Canty AJ, van Koten G. *Acc Chem Res* 1995; **28**: 406–13.

190 Grove DM, van Koten G, Louwen JN *et al. J Am Chem Soc* 1982; **104**: 6609–16.

191 Terheijden J, van Koten G, Vinke IC, Spek AL. *J Am Chem Soc* 1985; **107**: 2891–8.

192 Satoh T, Kawamura Y, Miura M, Nomura M. *Angew Chem Int Ed Engl* 1997; **36**: 1740–42, and references therein.

193 Murai S, Chatani N, Kakiuchi F. *Pure Appl Chem* 1997; **69**: 589–94.

194 Chatani N, Yutaka I, Kakiuchi F, Murai S. *J Org Chem* 1997; **62**: 2604–10, and references therein.

195 Gozin M, Weisman A, Ben-David Y, Milstein D. *Nature (London)* 1993; **364**: 699–701.

196 Parshall GW. *Acc Chem Res* 1975; **8**: 113–17.

197 Bruce MI. *Angew Chem Int Ed Engl* 1997; **16**: 73–87.

198 Ingrosso G. In: Braterman PS, ed. *Reactions of Coordinated Ligands*, Vol. I. New York: Plenum Press, 1986: 639–77.

199 Evans DW, Baker GR, Newkome GR. *Coord Chem Rev* 1989; **93**: 155–89.

200 Catellani M, Chiusoli GP, Costa M. *J Organomet Chem* 1995; **500**: 69–80.

201 Bennett MA, Hambley TW, Roberts NK, Robertson GB. *Organometallics* 1985; **4**: 1992–2000.

202 Carmona E, Palma P, Paneque M, Poveda ML, Gutierrez Puebla E, Monge A. *J Am Chem Soc* 1986; **108**: 6424–5.

203 Càmpora J, Gutierrez E, Monge A *et al. Organometallics* 1994; **13**: 1728–45.

204 Bennett MA, Wenger E. *Chem Ber* 1997; **130**: 1029–42.

205 Eisch JJ, Piotrowski AM, Han KI, Kruger C, Tsay YH. *Organometallics* 1985; **4**: 224–31.

206 Catellani M, Chiusoli GP. *J Organomet Chem* 1988; **346**: C27–30.

207 Catellani M, Chiusoli GP. *J Organomet Chem* 1992; **425**: 151–4.

208 Catellani M, Marmiroli B, Fagnola MC, Acquotti D. *J Organomet Chem* 1996; **597**: 157–62.

209 Catellani M, Chiusoli GP. *J Organomet Chem* 1992; **437**: 369–73.

210 Liu C-H, Li C-S, Cheng C-H. *Organometallics* 1994; **13**: 18–20.

211 Cardenas DJ, Mateo C, Echavarren AM. *Angew Chem Int Ed Engl* 1994; **33**: 2445–6.

212 Mateo C, Cárdenas DJ, Fernández-Rivas C, Echavarren AM. *Chem Eur J* 1996; **2**: 1596–606.

213 Griffiths DC, Young GB. *Organometallics* 1989; **8**: 875–86.

214 Tinga MAGM, Schat G, Akkerman OS, Bickelhaupt F, Smeets WJJ, Spek AL. *Chem Ber* 1994; **127**: 1851–6.

215 Catellani M, Chiusoli GP, Ricotti S. *J Organomet Chem* 1985; **296**: C11–15.

216 Catellani M, Chiusoli GP, Ricotti S, Sabini F. *Gazz Chim Ital* 1985; **115**: 685–9.

217 Catellani M, Ferioli L. *Synthesis* 1996: 769–72.

218 Catellani M, Chiusoli GP. *Gazz Chim Ital* 1993; **123**: 1–7.

219 Li C-S, Cheng C-H, Liao F-L, Wang S-L. *J Chem Soc Chem Commun* 1991: 710–12.

220 Catellani M, Mann BE. *J Organomet Chem* 1990; **390**: 251–5.

221 Bocelli G, Catellani M, Ghelli S. *J Organomet Chem* 1993; **458**: C12–15.

222 Canty AJ. In: Abel EW, Stone FGA, Wilkinson G, eds. *Comprehensive Organometallic Chemistry II*, Vol. IX. Oxford: Pergamon Press, 1995: 225–90.

223 Canty AJ. *Acc Chem Res* 1992; **25**: 83–90, and references therein.

224 Uson R, Fornies J, Navarro RJ. *J Organomet Chem* 1975; **96**: 307–12.

225 Milstein D, Stille JK. *J Am Chem Soc* 1979; **101**: 4981–91.

226 Byers PK, Canty AJ, Skelton BW, White AH. *J Chem Soc Chem Commun* 1986: 1722–4.

227 de Graaf W, Boersma J, Smeets WJJ, Spek AL, van Koten G. *Organometallics* 1989; **8**: 2907–17.

228 van Asselt R, Rijnberg E, Elsevier C. *Organometallics* 1994; **13**: 706–20.

229 Catellani M, Chiusoli GP. *J Organomet Chem* 1985; **286**: C13–16.

230 Catellani M, Fagnola MC. *Angew Chem Int Ed Engl* 1994; **33**: 2421–2.

231 Catellani M, Frignani F, Rangoni A. *Angew Chem Int Ed Engl* 1997; **36**: 119–22.

232 Catellani M, Chiusoli GP, Castagnoli C. *J Organomet Chem* 1991; **407**: C30–3.

233 Catellani M, Fagnola MC. *Gazz Chim Ital* 1992: 481–3.

234 Catellani M, Motti E. (To be published.)

235 Reiser O, Weber M, de Meijere A. *Angew Chem Int Ed Engl* 1989; **28**: 1037–8.

236 Albrecht K, Reiser O, Weber M, Knieriem B, de Meijere A. *Tetrahedron* 1994; **50**: 383–401.

237 Catellani M, Paterlini L, Farioli M. *New J Chem* 1998; **122**: 759–61.

238 Ozawa F, Fujimori M, Yamamoto T, Yamamoto A. *Organometallics* 1986; **5**: 2144–9.

239 Dyker G. *Chem Ber* 1994; **127**: 739–42.

240 Dyker G. *Angew Chem Int Ed Engl* 1992; **31**: 1023–5.

241 Dyker G. *Angew Chem Int Ed Engl* 1994; **33**: 103–5.

242 Dyker G, Körning J, Nereuz F *et al. Pure Appl Chem* 1996: 323–6.

9 Highly Efficient Method for the Addition of Unactivated Carbon–Hydrogen Bonds to Carbon–Carbon Multiple Bonds with the Aid of Ruthenium Complexes as Catalysts

FUMITOSHI KAKIUCHI and SHINJI MURAI

Department of Applied Chemistry, Faculty of Engineering, Osaka University, Suita, Osaka 565-0871, Japan

1 Introduction

Over the last three decades, direct use of C–H bonds in organic synthesis has been the focus of intensive research interest because this methodology is believed to become one of the most powerful synthetic methods. A large number of reports and over 50 review articles have appeared in the literature with respect to stoichiometric reactions [1, 2]. On the contrary, with respect to the catalytic use of otherwise unreactive C–H bonds in organic synthesis, only a limited number of examples have been reported. Almost all of these examples suffer from the low efficiencies (low chemical yields, high catalyst loading, the need to use one of the reactants in large excess and the need for extra energy input by photo-irradiation, etc.), the low selectivities and narrow applicability [3].

Herein, we wish to describe our recent results of the transition-metal-catalysed addition of carbon–hydrogen bonds to carbon–carbon multiple bonds.

2 Ruthenium-catalysed aromatic ketone C–H/olefin coupling

Recently, we reported a new transition-metal-catalysed reaction that enabled the addition of aromatic C–H bonds to olefins [4, 5]. The new Ru-catalysed reaction could bring about the addition of C–H bonds in a broad range of substrates, from aromatic ketones to double bonds of olefins. A C–C bond could be formed from a C–H bond in a single step with a high efficiency that had never been attained hitherto. A representative example is given in Equation (9.1).

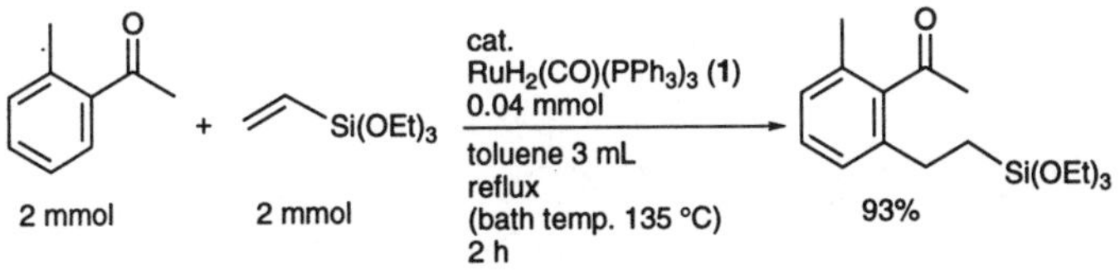

$$(9.1)$$

The reaction gave a 1:1 adduct in almost quantitative yield based on both starting materials. Only the C–H bond *ortho* to the carbonyl group had reacted. Various transition metal complexes were examined for their catalytic activities. Among those examined, only ruthenium complexes exhibited catalytic activity: $RuH_2(CO)(PPh_3)_3$ and $Ru(CO)_2(PPh_3)_3$ were found to be most effective, while $RuH_2(PPh_3)_4$ and $Ru(CO)_3(PPh_3)_2$ were moderate and $RuHCl(CO)(PPh_3)_3$, $RuCl_2(PPh_3)_3$, $RuCl_2(CO)_2(PPh_3)_2$ and $Ru_3(CO)_{12}$ were inactive. Interestingly, the ruthenium complex RuCl(OAc)(CO)(PPh$_3$)$_2$, which was known to undergo cyclometallation of acetophenone [6] with cleavage of the *ortho* C–H bond, did not show any catalytic activity in the present reaction. Some other transition metal complexes, such as RhCl(PPh$_3$)$_3$, RhH(PPh$_3$)$_4$, IrCl(CO)(PPh$_3$)$_2$ and [IrH$_2$(acetone)$_2$(PPh$_3$)$_2$]BF$_4$, showed no catalytic activity.

Vinylsilanes and allylsilanes gave coupling products with *o*-methylacetophenone in excellent yields (Equations (9.2) and (9.3)). For less reactive olefins such as allyltrimethylsilane, the use of five equivalents of olefins and higher catalyst loading resulted in high yields of the coupling products. Triethoxyvinylsilane seemed to be the olefin of choice for the examination of reactivities for different ketones (*vide infra*).

(9.2)

2 mmol   2 mmol   1 0.04 mmol
          4 mmol   toluene 3 mL
                    reflux
                    (bath temp. 135 °C)
                    4 h   72%
                    2 h   97%

(9.3)

2 mmol   10 mmol   1 0.12 mmol
                     toluene 3 mL
                     reflux
                     (bath temp. 135 °C)
                     4 h   quant.

The results obtained for ethylene and *tert*-butylethylene are given in Equations (9.4) and (9.5). The reaction with ethylene took place smoothly to give the corresponding coupling product (Equation (9.4)). In the case of *tert*-butylethylene the 1:1 coupling product was formed quantitatively (Equation (9.5)). It should be noted that these products are of the type that are not easily obtainable by conventional synthetic methods. Acylation of an alkylbenzene would give a mixture of *ortho*- and *para*-isomers with the latter predominating. Alkylation of acylbenzenes as an alternative synthetic route would be even more difficult because the electron deficient aromatic ring is usually unreactive to known alkylating reagents.

(9.4)

2 mmol   7 kg/cm^2 1 0.04 mmol
          (14 mmol)   toluene 3 mL
                         in an autoclave, 135 °C
                         24 h   quant.

(9.5)

2 mmol   2 mmol   1 0.04 mmol
          10 mmol   2 h   53%
                    8 h   99%

Various aromatic ketones have been subjected to the catalytic reaction using triethoxyvinylsilane as the olefin. Acetophenone has two *ortho* hydrogens and indeed 1:1 and 1:2 coupling products were obtained. By employing an excess of the olefin and prolonged reaction times, the 1:2 adduct was produced almost exclusively (Equation (9.6)). A 1:2 adduct was no longer formed even in the presence of an excess amount of the olefin in the case of *tert*-butylphenyl ketone ((Equation (9.7)). The α-tetralone showed quite high reactivity (Equation (9.8))

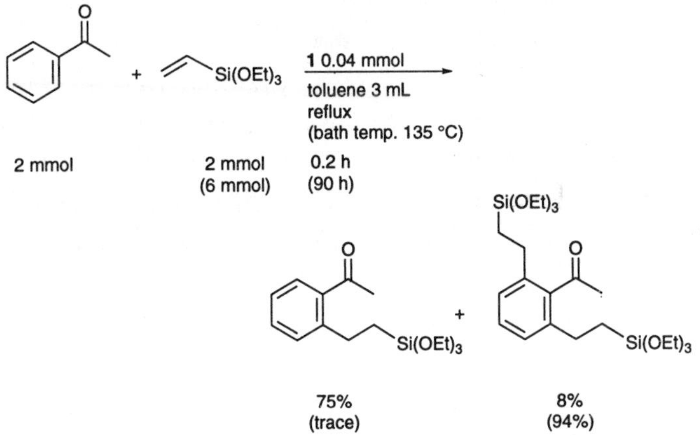

(9.6)

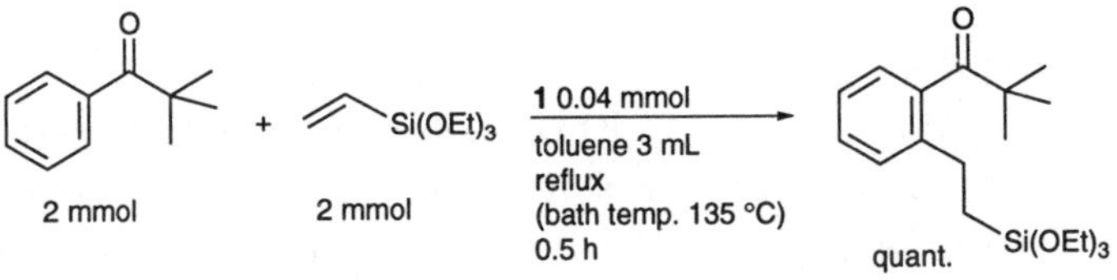

(9.7)

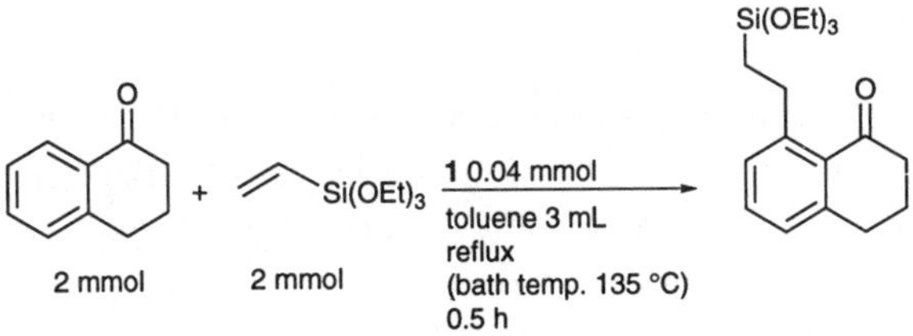

(9.8)

The catalytic reaction can be applied also to naphthalenes (Equation (9.9)). The result of Equation (9.9) is worth noting because only one of the two different C–H bonds adjacent to the carbonyl group reacted.

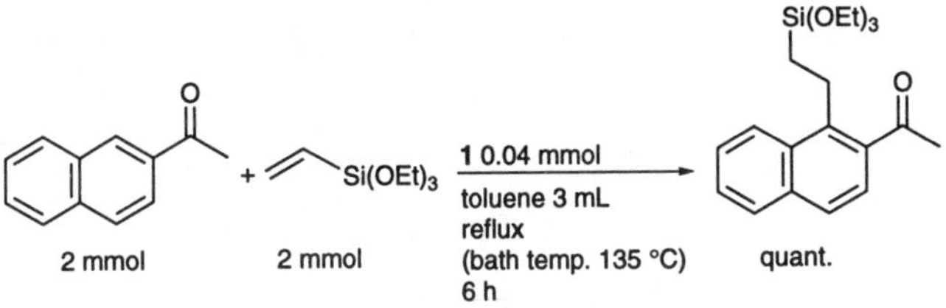

(9.9)

Five-membered heteroaromatic ketones also reacted with triethoxyvinylsilane (Equations (9.10) and (9.11)). The reaction of 1-methyl-2-acetylpyrrole gave the corresponding coupling product in almost quantitative yield, although a large amount of vinylsilane was necessary to obtain a high yield (Equation (9.10)). Complete regioselectivity was also observed in the reaction of 3-acetylthiophene (Equation (9.11)).

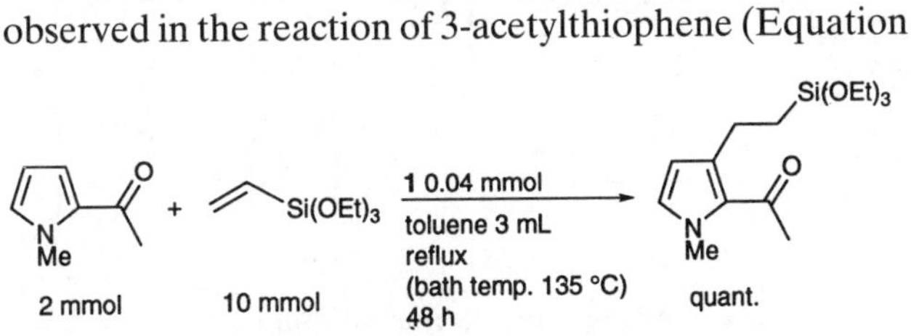

(9.10)

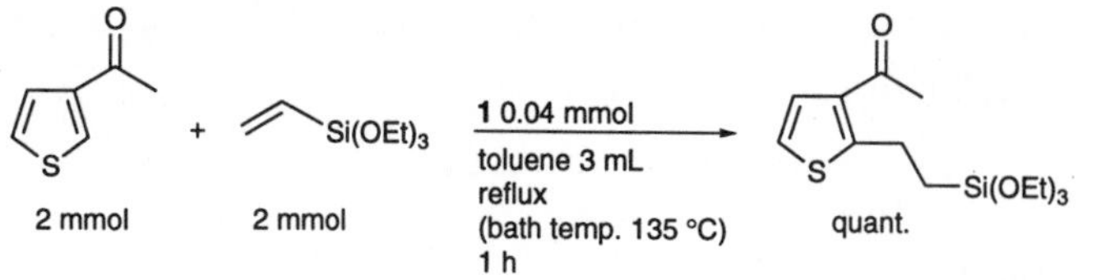

(9.11)

This new catalytic reaction achieves the efficiency, selectivity and generality to be of general practicability in organic synthesis. Moreover, the catalytic reaction is operationally simple and proceeds cleanly: simple mixing of the reactants and the catalyst in toluene and heating them under vigorous reflux, then simple bulb-to-bulb distillation of the reaction mixture usually allows isolation of the analytically pure product. That the reaction can be run equally well on a larger scale (10 g) of reactant has been confirmed for the reaction of Equation (9.8).

3   Suggested mechanism

The mechanism of the new Ru-catalysed C–H/olefin coupling reaction is intriguing and important, but it is far from clear at the present time. We suggest just the possibilities here. The course of the catalytic reaction is outlined in Fig. 9.1, with important alternative routes and with simplified structures. The reaction begins with the coordination of a carbonyl group to the Ru atom in the catalyst, bringing the Ru atom close to the *ortho* C–H bond.

After coordination of an olefin to compound **5**, insertion of the olefin into the Ru–H bond gives compound **6**, or alternatively insertion into the Ru–C bond gives compound **7**. The reductive elimination from compound **6** or **7** leads to compound **8**, from which decom-

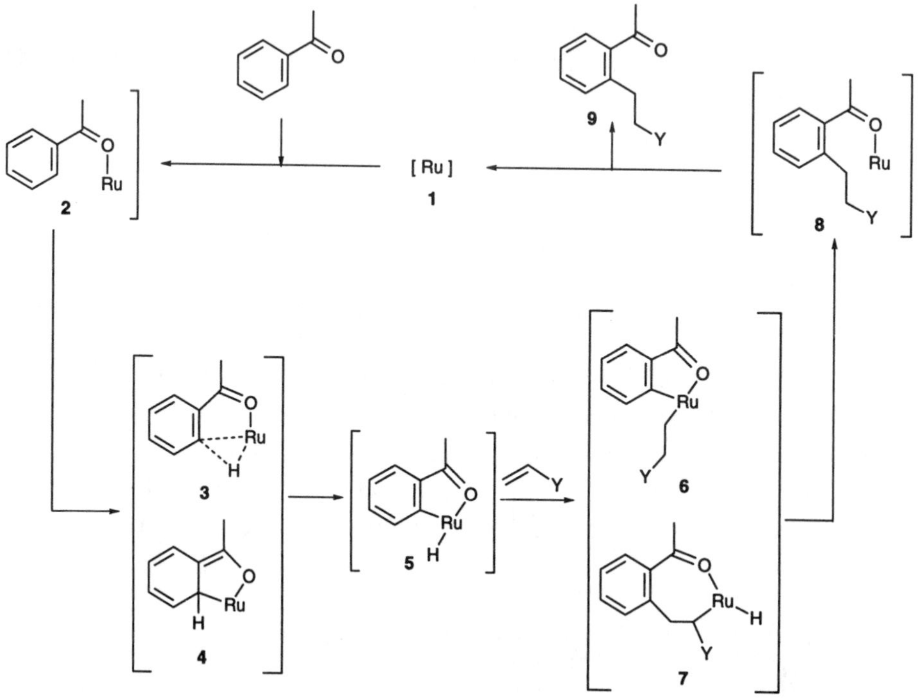

Figure 9.1. An outline of the catalytic cycle with alternative routes.

plexation of the coupling product **9** regenerates the catalyst **1**. Before the reductive elimination, the chelating carbonyl group can also dissociate (this alternative is not shown in Figure 9.1).

4 Manipulation of other types of C–H bonds

The present discovery should open up a myriad of fascinating opportunities in synthetic organic chemistry. Indeed, we have already found aromatic C–H/olefin coupling, aromatic C–H/acetylene coupling and olefinic C–H/olefin coupling, as shown below.

We found that acetylenes as well as olefins participate in the catalytic reaction [7]. We were pleased to see that internal acetylenes did participate in the present coupling reaction, even though the internal acetylenes are rather sterically crowded. Because α-tetralone was a highly reactive aromatic ketone in the C–H/olefin coupling reaction, substituted acetylenes were subjected to the catalytic reaction using α-tetralone. The selected examples of the results of catalytic addition of the C–H bond in α-tetralone to acetylenes are given in Table 9.1.

Symmetrically substituted diphenylacetylene (run 1) gave 1:1 coupling products in good yields. The reactivity of diphenylacetylene was the highest among those examined so far and one-third (i.e. 0.04 mmol) of the catalyst was enough in this case (run 1). Virtually complete regio- and stereoselectivities were attained when 1-trimethylsilyl-1-propyne was used as the acetylene (run 2). Only the *E*-isomer was formed in 83% yield with the regiochemistry shown in run 2 of Table 9.1. This result indicates that the addition of a C–H bond to acetylene proceeds in *cis* fashion. In the case of other trimethylsilylacetylenes, the regioselectivities were exclusive (runs 2–4), although mixtures of stereoisomers were obtained.

Table 9.1. Addition of α-tetralone to acetylenes catalysed by $RuH_2(CO)(PPh_3)_3$[a]

Run	Acetylene	Time/h	Product	Yield[b]/%
1[c]	Ph–≡–Ph	1	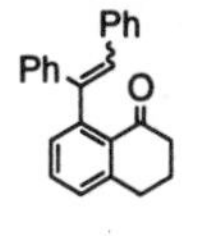	85% (9/1)[d]
2	Me–≡–SiMe₃	3	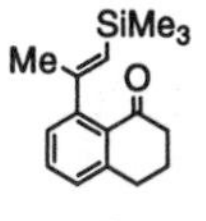	83% (only *E*-isomer)
3	Hex–≡–SiMe₃	2	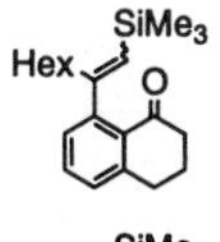	quantitative (*E/Z* = 11/1)
4	Ph–≡–SiMe₃	1	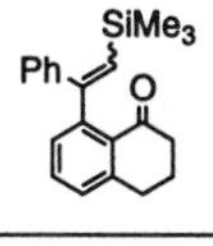	76% (*E/Z* = 2/1)

[a] Reaction conditions: α-tetralone (2 mmol), acetylene (4 mmol), $RuH_2(CO)(PPh_3)_3$ (0.12 mmol), toluene (3 mL), 135 °C (oil bath temperature).
[b] Isolated yield.
[c] Catalyst: 0.04 mmol.
[d] The stereochemistry could not be determined.

Substituted vinylsilanes are known to be highly versatile synthetic building blocks [8]. The present catalytic reaction (e.g. runs 3 and 4) provides an entirely new entry to vinyl-silanes, i.e., addition of C–H bonds to silylacetylenes, and seems promising as a synthetic reaction. A further example, shown in Equation (9.12), illustrates that a heteroaromatic ketone also undergoes the desired coupling reaction with a silylacetylene.

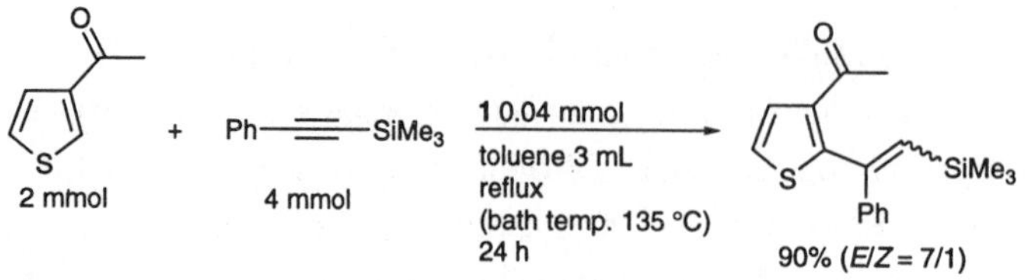

(9.12)

It was found that olefinic C–H bonds can be added to different olefins [9]. We first studied the reaction of 1-acetylcyclohexene with triethoxyvinylsilane in the presence of $RuH_2(CO)(PPh_3)_3$ as the catalyst. The reaction gave the expected product but in only moderate yield (Equation (9.13)). Interestingly, however, the reaction of 1-pivaloylcyclo-hexene gave an almost quantitative yield of the coupling product in a short reaction period (Equation (9.14)).

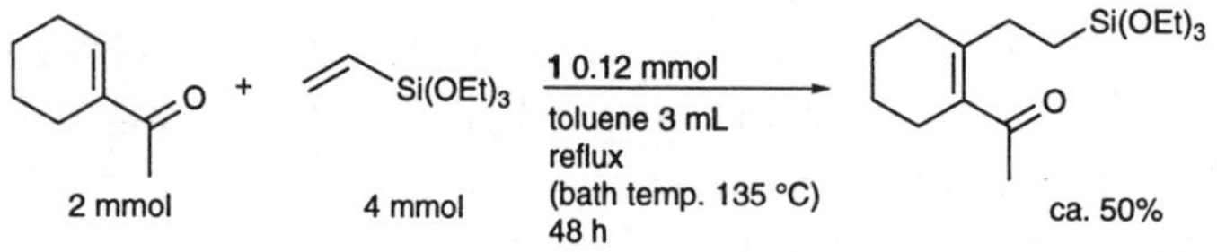

(9.13)

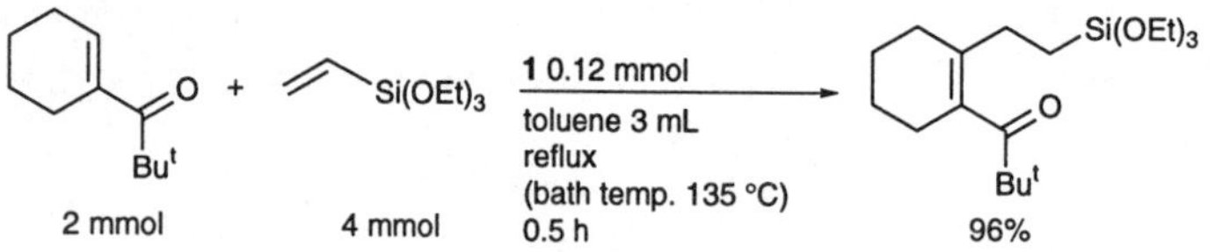

(9.14)

Next the catalytic reactions of heterocyclic enones were examined. The corresponding coupling products were obtained from heterocyclic enones and triethoxyvinylsilane in good to excellent yields (Equation (9.15)–(9.17)). As shown in Equations (9.15)–(9.17), the oxygen substitution can be either at the α- or β-position of the enone moiety or at both positions.

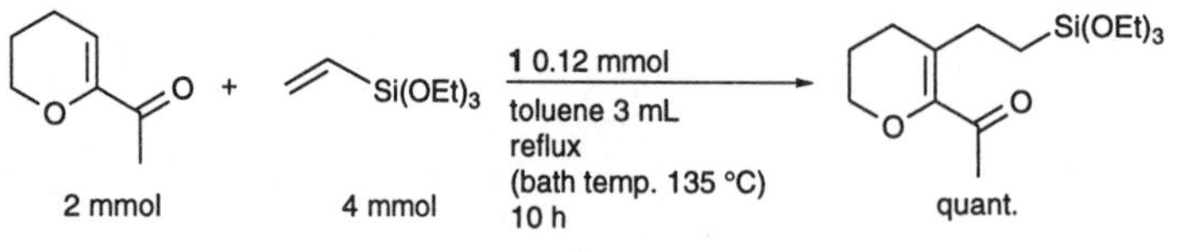

(9.15)

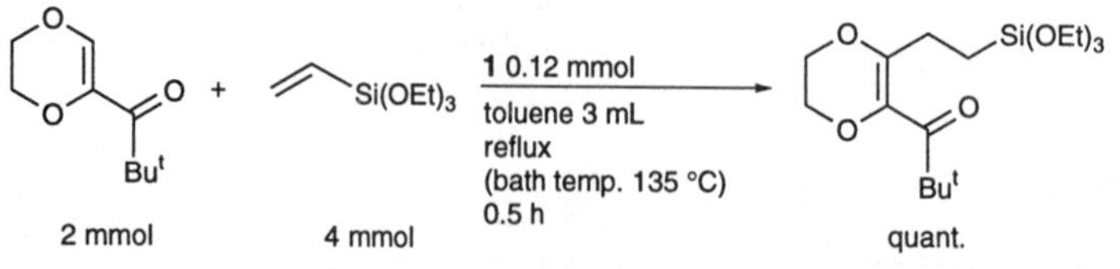

(9.16)

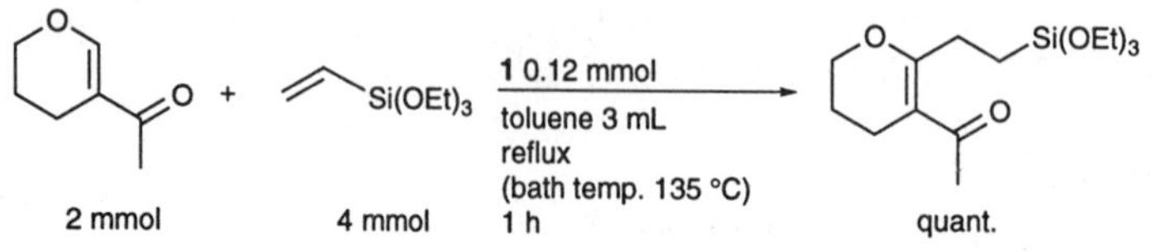

(9.17)

Table 9.2. Products of the catalytic reaction of α,β-enone with olefins[a]

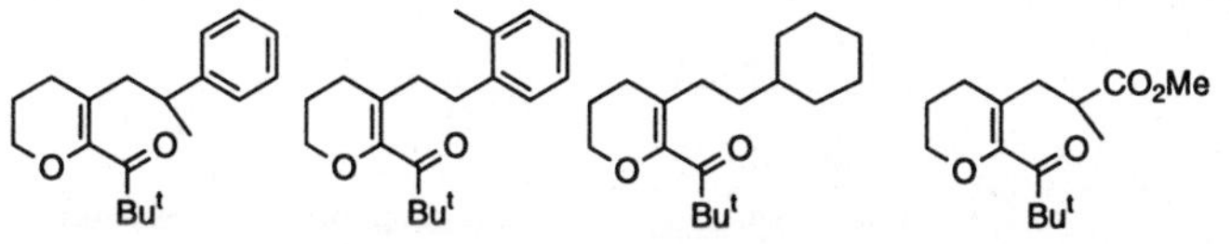

(96%, 15 h)   (96%, 1 h)   (98%, 12 h)[b]   (30%, 48 h)[b]

[a] Reaction conditions: α,β-enone (2 mmol), olefin (4 mmol), $RuH_2(CO)(PPh_3)_3$ (**1**) (0.12 mmol), toluene (3 mL), 135 °C (oil bath temperature).
[b] Olefin: 10 mmol.

The new olefinic C—H/olefin coupling can be applied to various terminal olefins as shown in Table 9.2. From α-methylstyrene, o-methylstyrene, vinylcyclohexane and methyl methacrylate, the coupling products were obtained in moderate to excellent yields. It is interesting to note that α-methylstyrene and methyl methacrylate are the olefins that do not undergo coupling with aromatic ketones [4,5].

In addition, we have found that the new catalytic reaction can be extended to aromatic esters [10]. We carried out the reaction of methyl benzoate with triethoxyvinylsilane in the presence of a catalytic amount of $RuH_2(CO)(PPh_3)_3$ (6 mol.%), however the desired coupling reaction did not take place. We were pleased to observe, however, that the reaction of methyl 2-trifluoromethylbenzoate with triethoxyvinylsilane occurred smoothly to give the 1:1 coupling product in 97% gas chromatographic yield under the reaction conditions depicted in Equation (9.18).

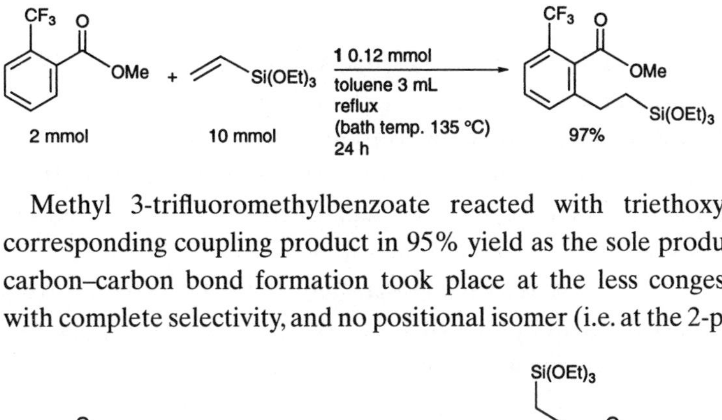

$$(9.18)$$

Methyl 3-trifluoromethylbenzoate reacted with triethoxyvinylsilane to give the corresponding coupling product in 95% yield as the sole product (Equation (9.19)). The carbon–carbon bond formation took place at the less congested position (6-position) with complete selectivity, and no positional isomer (i.e. at the 2-position) was detected.

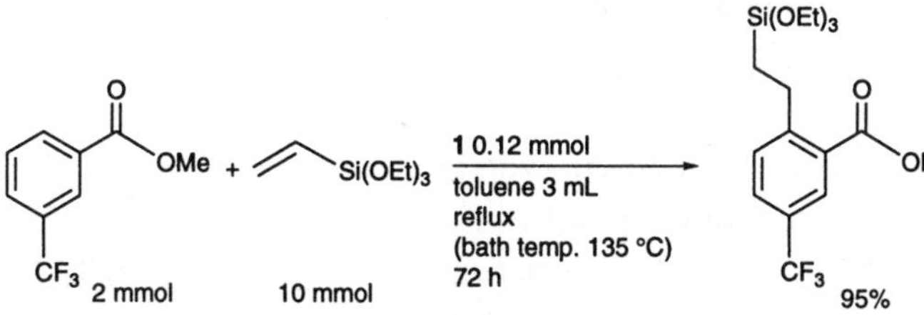

$$(9.19)$$

Similarly, the reaction of ethyl 3-fluorobenzoate with triethoxyvinylsilane gave the coupling products (42% yield) predominantly, along with a small amount (1% yield) of the regioisomer (Equation (9.20)). In this case the carbon–carbon bond is formed at the 2-position, in contrast to that observed for methyl 3-trifluoromethylbenzoate (Equation (9.19)). The site selectivity of these reactions may be attributable to both steric factors and the electronic properties of the substituents [11]. The unshared electron pairs on the fluorine in ethyl 3-fluorobenzoate may have a weak interaction with the ruthenium, directing it to the adjacent position.

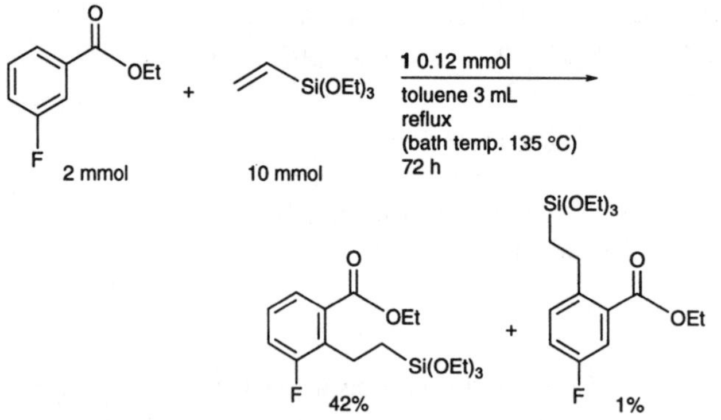

(9.20)

The electron-withdrawing substituent is not essential for attaining this C—H/olefin coupling reaction. For example, the reaction of a six-membered lactone with triethoxyvinylsilane afforded the corresponding coupling product in moderate yield (Equation (9.21)).

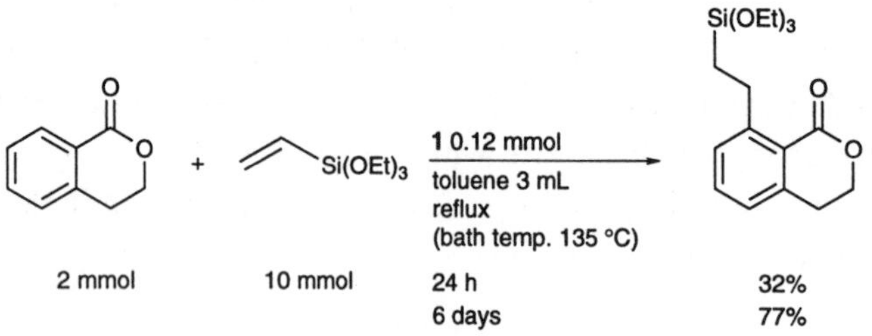

(9.21)

During the course of the present work, Trost and co-workers reported a similar catalytic reaction of α,β-unsaturated esters [12].

Coordination of sp^2 nitrogen atoms to transition metals, leading to cleavage of C—H bonds by the metals, is a well-documented phenomenon in stoichiometric reactions. We thought that imines should be good candidates for the chelation-assisted C—H/olefin coupling reaction, and this is indeed so.

Reaction of *N*-(2-methylbenzylidene)-*tert*-butylamine with triethoxyvinylsilane in the presence of $RuH_2(CO)(PPh_3)_3$ as the catalyst gave the corresponding 1:1 coupling product and the dehydrogenated coupling product in 26% and 8% yields, respectively (Equation (9.22)) [13]. This result shows that aromatic aldimines are also applicable to the ruthenium catalysed C—H/olefin coupling. To improve the efficiency of this new catalytic reaction, the reaction parameters were examined. Complexes $Ru(CO)_2(PPh_3)_3$, $Ru(CO)_3(PPh_3)_2$ and $Ru(OAc)(H)(CO)(PPh_3)_2$ showed similar catalytic activities to that of $RuH_2(CO)(PPh_3)_3$. It is noteworthy that $Ru(OAc)(H)(CO)(PPh_3)_2$ was totally ineffective in the case of aromatic ketones. The use of $Ru_3(CO)_{12}$ greatly increased the yields of the coupling product and the dehydrogenated coupling product (81% and 10% yields, respectively) (Equation (9.22)).

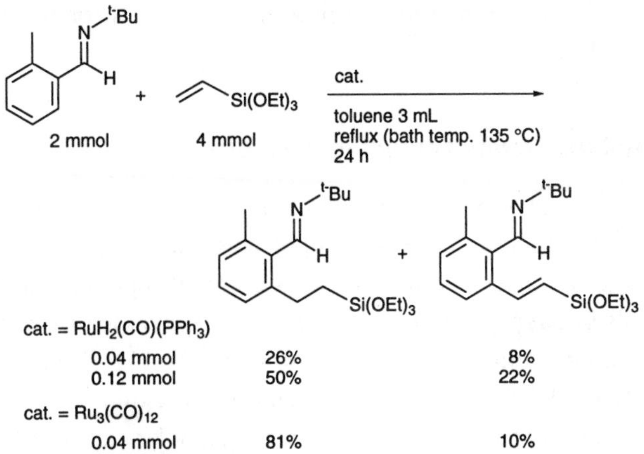

(9.22)

Selected results obtained for various types of aromatic imines with triethoxyvinylsilane are listed in Table 9.3. Reaction of an imine having a trifluoromethyl group at the *ortho* position afforded exclusively the corresponding coupling product in 75% yield. Fluorine-substituted imine also reacted with triethoxyvinylsilane to give the coupling product and dehydrogenative coupling product in 79% and 8% yields, respectively. It is interesting to note that even though a carbon–fluorine bond is prone to react with transition metals, the fluorine substituent remained in the product. As shown, both electron-releasing (Me) (Equation (9.22)) and electron-withdrawing (CF_3 and F) (Table 9.1) groups are applicable in this coupling reaction. The reaction of a ketimine derived from acetophenone also proceeded smoothly to afford the 1:1 product in 78% yield as a mixture of *syn* and *anti* isomers. Interestingly, no 1:2 coupling product was formed even in the presence of an excess amount of triethoxyvinylsilane, which is in contrast to the $RuH_2(CO)(PPh_3)_3$-catalysed coupling reaction of acetophenone with triethoxyvinylsilane, which yields a mixture of the corresponding 1:1 and 1:2 coupling products. Steric congestion around the

Table 9.3. Ruthenium-catalysed addition of a carbon–hydrogen bond in aromatic imines to triethoxyvinylsilane[a]

Imine	Product and Yield[b]	
	75%	
	79%	8%
	78% (*anti*:*syn* = 62:38)	

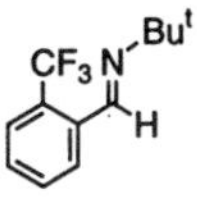

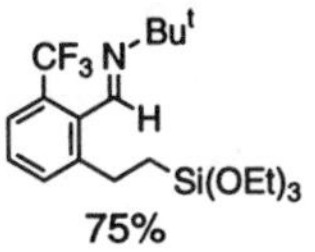

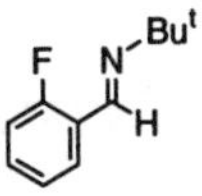

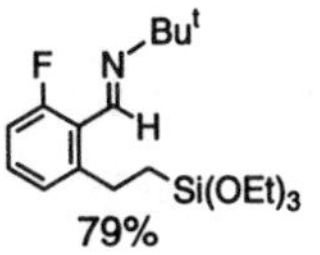

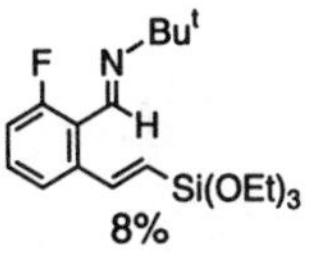

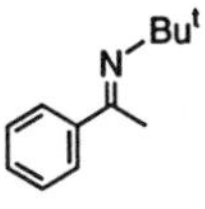

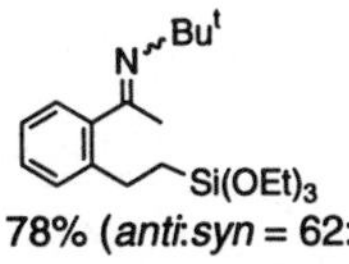

[a] Reaction conditions: imine (2 mmol), triethoxyvinylsilane (4 mmol), $Ru_3(CO)_{12}$ (0.04 mmol), toluene (3 mL), 135 °C (oil bath temperature), 24 h.
[b] Gas chromatographic yields.

imino group in the starting ketimine and the product probably strongly suppresses the second C–H/olefin coupling step.

5   Transition-metal-catalysed related coupling reactions

After our initial report in 1993 on the ruthenium-catalysed addition of C–H bonds to olefins [4], related catalytic reactions have been reported from several research groups [12, 17–21]. In 1994, Kim and co-workers found that a rhodium–phosphine-complex-catalysed addition of C–H bonds in phenylpyridine derivatives to olefins took place to give *ortho*-alkylated products in good to high yields (Equation (9.23)) [17]. Trost and co-workers reported that the $RuH_2(CO)(PPh_3)_3$-catalysed addition of olefinic C–H bonds at the β-position in α,β-unsaturated carbonyl compounds to olefins took place with high regio- and stereoselectivities (Equation (9.24)) [12]. In 1996, Woodgate and co-workers applied the $RuH_2(CO)(PPh_3)_3$-catalysed addition of C–H bonds in aromatic ketones to olefins for the functionalization of aromatic diterpenoids (Equation (9.25)) [18]. The silylation of an aromatic ring in aromatic imines with hexaorganodisilanes was performed by Tanaka and co-workers (Equation (9.26)) [19]. Very recently, Grigg and co-workers found that the acetylpyridines also reacted with olefins in the presence of $RuH_2(CO)(PPh_3)_3$ as catalyst to give the corresponding coupling products in good yields (Equation (9.27)) [20]. From Weber's group, a large number of examples of the application of the C–H/olefin coupling reaction to polymer preparations were reported. They developed a new entry to the copolymerization of aromatic ketones with dienes by using $RuH_2(CO)(PPh_3)_3$ as the catalyst (Equation (9.28)) [21].

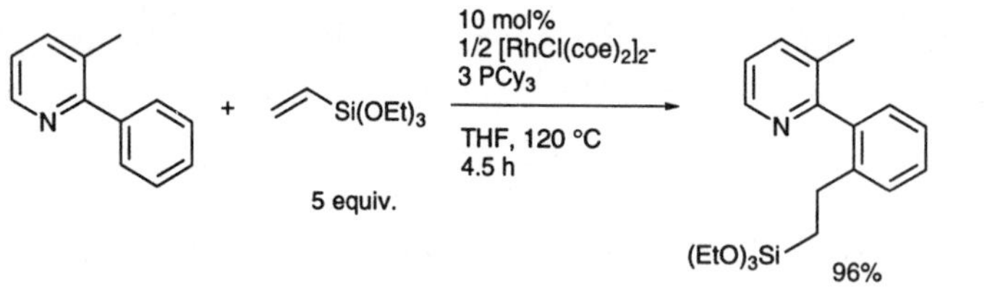

(9.23)

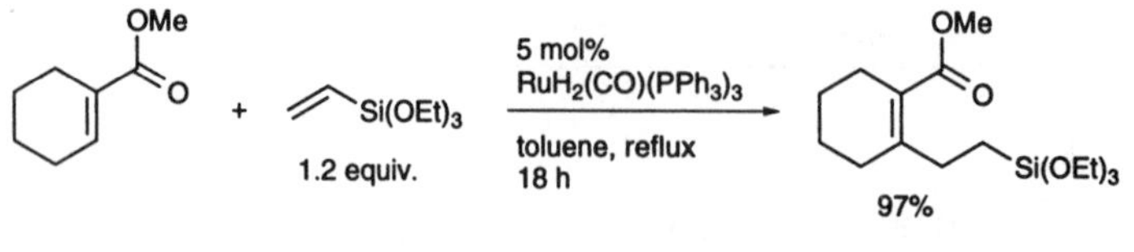

(9.24)

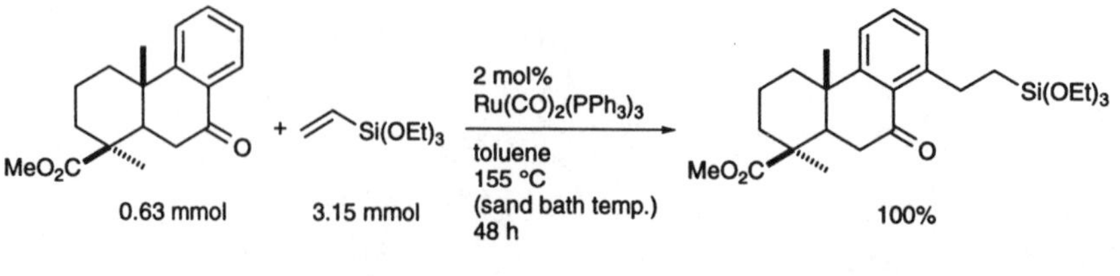

(9.25)

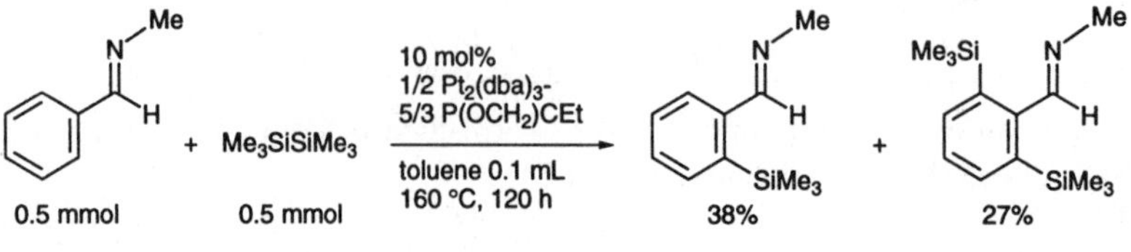

(9.26)

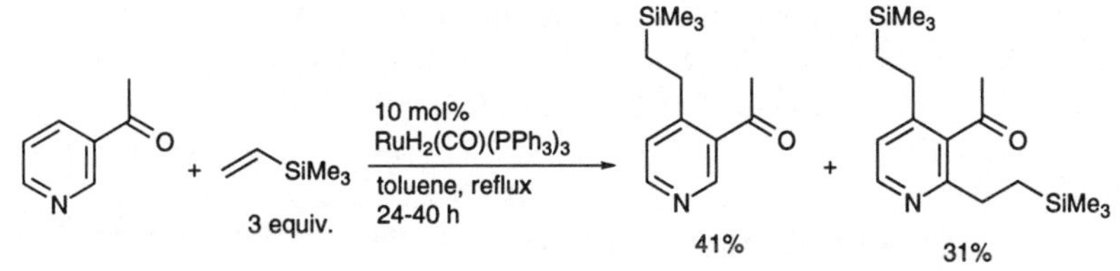

$$(9.27)$$

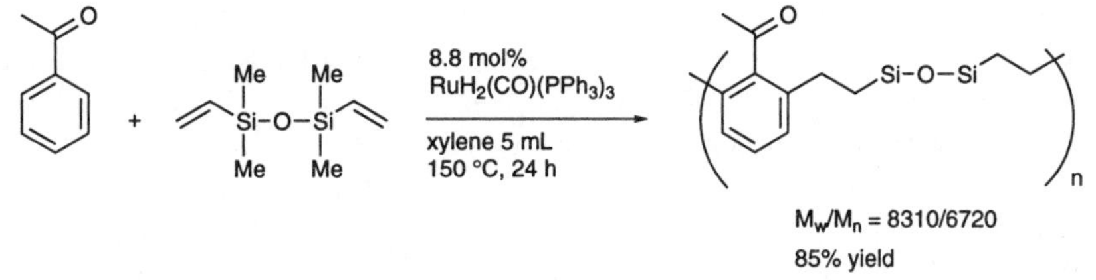

$M_w/M_n = 8310/6720$
85% yield

$$(9.28)$$

As briefly surveyed above, transition-metal-catalysed reactions involving a cleavage of otherwise unreactive C–H bonds have already become useful tools in organic synthesis.

6 Conclusion

In the next few years we anticipate that our efforts will be concentrated on extending the scope of the new ruthenium-catalysed C–H/olefin and C–H/acetylene couplings, primarily from the synthetic point of view. Also, attempts to find new and different catalytic reactions by making use of the already rich chemistry of C–H bond cleavage [14, 15] will be made. For this purpose, we will rely on the following working hypothesis: the cleavage of C–H bonds is a facile step, so for the construction of new catalytic cycles it is more important to devise a system to make the last steps, such as reductive elimination, easier.

7 Acknowledgments

This work would have never been developed without the contribution of our colleague, Professor Naoto Chatani. The authors also wish to express their sincere thanks to his talented co-workers: Mrs Shinya Sekine, Yasuo Tanaka, Asayuki Kamatani, Motohiro Sonoda, Masakazu Yamauchi, Taisuke Sato, Yoshinari Santo, Yoshimi Yamamoto, Naoaki Fujii, Takuya Tsujimoto and Ms Airi Yamada. Without their intellectual and experimental efforts, this review would have never appeared.

8 References

1 For representative reviews, see: Parshall GW. *Acc Chem Soc* 1970; **3**: 139; Webster DE. *Adv Organomet Chem* 1977; **15**: 147; Komiya S, Yamamoto A, Yamamoto T. *Yuki Gosei Kagaku Kyokaishi*, 1980; **38**: 633; *Chem Abstr* 1980; **93**: 203386; Yamazaki H, Hong P. *J Mol Catal* 1983; **21**: 133; Bergman RG. *Science* 1984; **223**: 902; Shilov AE. *Activation of Saturated Hydrocarbons by Transition Metal Complexes*. Dordrecht: Reidel, 1984, Chapter 5; Green MLH, O'Hare D. *Pure Appl Chem* 1987; **57**: 1897; Crabtree RH. *Chem Rev* 1985; **85**: 245; Tanaka M. *Yuki Gosei Kagaku Kyokaishi* 1988; **46**: 832; *Chem Abstr* 1989; **110**: 113908; Tanaka M. *CHEMTECH* 1989; **19**: 59; Jones WD, Feher FJ. *Acc Chem Res* 1989; **22**: 91; Hill CL. *Activation and Functionalization of Alkanes*. New York: Wiley, 1989, pp. 79–149; Davies JA, Watson PL, Liebman JF, Greenberg A. *Selective Hydrocarbon Activation*. New York: VCH, 1990, Chapter 5; Bergman RG. *Adv Chem Ser* 1992; **230**: 211.

2 For reviews on *ortho*-metallation, see: Dehand J, Pfeffer M. *Coord Chem Rev* 1976; **18**: 327; Bruce MI. *Angew Chem Int Ed Engl* 1977; **16**: 73; Omae I. *Chem Rev* 1979; **79**: 287; Ryabov AD. *Synthesis* 1985: 233; Newkome GR, Puckett WE, Gupta VK, Kiefer GE. *Chem Rev* 1986; **86**: 451; Ryabov AD. *Chem Rev* 1990; **90**: 403; Pfeffer M. *Pure Appl Chem* 1992; **64**: 335.

3 Lin Y, Ma D, Lu X. *Tetrahedron Lett* 1987; **28**: 3249; Diamond SE, Szalkiewicz A, Mares F. *J Am Chem Soc* 1979; **101**: 409; Hong P, Yamazaki H. *Chem Lett* 1979: 1335. Hong P, Cho B-R, Yamazaki H. *Chem Lett* 1980: 507. Yamazaki H, Hong P. *J Mol Catal* 1983; **21**: 133; Hong P, Yamazaki H. *J Mol Catal* 1984; **26**: 297; Lewis LN, Smith JF. *J Am Chem Soc* 1986; **108**: 2728; Sakakura T, Sodeyama T, Tanaka M. *Chem Lett* 1988: 683; Sasaki K, Sakakura T, Tokunaga Y, Wada K, Tanaka M. *Chem Lett* 1988: 685; Tokunaga Y, Sakakura T, Tanaka M. *J Mol Catal* 1989; **56**: 305; Moore EJ, Pretzer WR, O'Connell TJ, *et al. J Am Chem Soc* 1992; **11**: 5888; Fisher BJ, Eisenberg R. *Organometallics* 1983; **2**: 764; Kunin AJ, Eisenberg R. *J Am Chem Soc* 1986; **108**: 535; Sakakura T, Tanaka M. *J Chem Soc Chem Commun* 1987: 758; Sakakura T, Tanaka M. *Chem Lett* 1987: 249; Sakakura T, Hayashi T, Tanaka M. *Chem Lett* 1987: 859; Sakakura T, Tanaka M. *Chem Lett* 1987: 1113; Sakakura T, Sasaki K, Tokunaga Y, Wada K, Tanaka M. *Chem Lett* 1988: 155; Gordon EM, Eisenberg R. *J Mol Catal* 1988; **45**: 57; Kunin AJ, Eisenberg R. *Organometallics* 1988; **7**: 2124; Sakakura T, Sodeyama T, Sasaki K, Wada K, Tanaka M. *J Am Chem Soc* 1990; **112**: 7221; Boese WT, Goldman AS. *J Am Chem Soc* 1992; **114**: 350; Tanaka M, Sakakura T, Tokunaga Y, Sodeyama T. *Chem Lett* 1987: 2373; Jones WD, Foster GP, Putinas JM. *J Am Chem Soc* 1987; **109**: 5047; Jones WD, Hessell ET. *Organometallics* 1990; **9**: 718; Jones WD, Kosar WP. *J Am Chem Soc* 1986; **108**, 5640; Hsu GC, Kosar WP, Jones WD. *Organometallics* 1994; **13**: 385; Ishikawa M, Okazaki S, Naka A, Sakamoto H. *Organometallics* 1992; **11**: 4135; Uchimaru Y, Sayed AMME, Tanaka M. *Organometallics* 1993; **12**: 2065; Ishikawa M, Naka A, Ohshita J. *Organometallics* 1993; **12**: 4987; Sakakura T, Tokunaga Y, Sodeyama T, Tanaka M. *Chem Lett* 1987: 2375.

4 Murai S, Kakiuchi F, Sekine S *et al. Nature (London)* 1993; **366**: 529; Kakiuchi F, Sekine S, Tanaka Y *et al. Bull Chem Soc Jpn* 1995; **68**: 62.

5 Murai S, Kakiuchi F, Sekine S *et al. Pure Appl Chem* 1994; **66**: 1527; Murai S. *J Synth Org Chem Soc Jpn* 1994; **52**: 992.

6 McGuiggan MF, Pignolet LH. *Inorg Chem* 1982; **21**: 2523.

7 Kakiuchi F, Yamamoto Y, Chatani N, Murai S. *Chem Lett* 1995: 681.

8 Colvin EW. *Silicon in Organic Synthesis*. London: Butterworths, 1981: Chapt. 7. Weber WP. *Silicon Reagents for Organic Synthesis*. Berlin: Springer-Verlag, 1983: Chapt. 7.

9 Kakiuchi F, Tanaka Y, Sato T, Chatani N, Murai S. *Chem Lett* 1995: 679.

10 Sonoda M, Kakiuchi F, Kamatani A, Chatani N, Murai S. *Chem Lett* 1996: 109.

11 Sonoda M, Kakiuchi F, Chatani N, Murai S. *J Organomet Chem* 1995; **504**: 151.

12 Trost BM, Imi K, Davies IW. *J Am Chem Soc* 1995; **117**: 5371.

13 Kakiuchi F, Yamauchi M, Chatani N, Murai S. *Chem Lett* 1996: 111.

14 Moore EJ, Pretzer WR, O'Connell JT *et al. J Am Chem Soc* 1992; **114**: 5888. We found that the reaction of 1,2-disubstituted imidazoles with CO and olefins in the presence of $Ru_3(CO)_{12}$ resulted in carbonylation of a C–H bond at the 4-position on the imidazole ring to give acylation products. Chatani N, Fukuyama T, Kakiuchi F, Murai S. *J Am Chem Soc* 1996; **118**: 493.

15 Jones WD, Kosar WP. *J Am Chem Soc* 1986; **108**: 5640; Hsu GW, Kosar WP, Jones WD. *Organometallics* 1994; **13**: 385.

16 Jones has shown that the slow step in the catalytic formation of indole from 2,6-xylyl isocyanide is not the step of C–H bond cleavage but the last step involving reductive elimination (see ref. 15).

17 Lim Y-G, Kang J-B, Kim YH. *J Chem Soc Chem Commun* 1994: 2267; 1996: 585; *J Chem Soc Perkin Trans 1* 1996: 585.

18 Harris PWR, Woodgate PD. *J Organomet Chem* 1996; **506**: 339; 1997; **530**: 211.

19 Williams NA, Uchimura U, Tanaka M. *J Chem Soc Chem Commun* 1995: 1129.

20 Grigg R, Savic V. *Tetrahedron Lett* 1997; **38**: 5737.

21 Guo H, Weber WP. *Polym Bull (Berlin)* 1994; **32**: 525; Guo H, Tapsak MA, Weber WP. *Macromolecules* 1995; **28**: 4714; Guo H, Wang G, Tapsak MA, Weber WP. *Macromolecules* 1995; **28**: 5696; Wang G, Guo H, Weber WP. *J Organomet Chem* 1996; **521**: 351; Londergan TM, Weber WP. *Macromol Rapid Commun* 1997; **18**: 207.

10 New Palladium Catalysts for Heck Reactions of Unreactive Aryl Halides

MANFRED T. REETZ

Max-Planck-Institut für Kohlenforschung, Kaiser-Wilhelm-Platz 1, 45470 Mülheim an der Ruhr, Germany

1 Introduction

The Heck reaction constitutes a versatile method for C–C bond formation in which the vinylic hydrogen of an olefin is replaced by vinyl, aryl or benzyl moieties [1]. The latter components are generally introduced by reaction of the corresponding halides, although triflates, hypervalent iodo derivatives and diazonium salts of aryl compounds may also be used. Because acids (HX) are generated in stoichiometric amounts, bases such as amines, sodium acetate (NaOAc) or sodium carbonate (Na_2CO_3) are required. Traditionally, Pd(0) catalysts such as $Pd[P(C_6H_5)_3]_4$ or *in situ* catalysts of the type $Pd(OAc)_2/[P(C_6H_5)_3]_n$ are employed. Quite early, Heck reported that the *in situ* catalyst derived from tris-*o*-tolylphosphine (P(*o*-tolyl)) is more reactive than the analog based on triphenylphosphine $P(C_6H_5)_3$. However, P(*o*-tolyl)$_3$ is 40–50 times as expensive as the traditional $P(C_6H_5)_3$. It has been shown that in such *in situ* processes Pd(II) is reduced to Pd(0) by the phosphine [1].

Although the question of the mechanism of the Heck reaction has not been answered unambiguously, most authors prefer the standard catalyst cycle in which the reaction is initiated by oxidative insertion of a Pd(0) species into the aryl halide bond (Scheme 10.1) [1].

A wide variety of compounds have been subjected to the Heck reaction, making it one of the standard reactions of organic chemistry. Furthermore, asymmetric versions based on the use of chiral ligands also have attracted a great deal of attention recently [1, 2].

It is therefore surprising that the Heck reaction has not been applied industrially to any significant extent [3]. One reason has to do with the fact that as a consequence of HX neutralization stoichiometric amounts of salts are formed, which need to be disposed of. Perhaps more unfortunate is the well-known decrease in reactivity in the order ArI > ArBr >> ArCl, which imposes real restrictions with respect to industrial viability [1, 3]. The temperature necessary for the reaction to occur increases from 80–90 °C for the iodides to 120–130 °C for the bromides and to >140 °C for the chlorides. These are only approximate values because substituents in the aryl halide may strongly influence reactivity. For example, electron-withdrawing groups activate the substrates. Whereas academic laboratories most often use the relatively reactive aryl iodides under fairly mild conditions, the costs of these compounds prohibit their industrial use. Aryl bromides are considerably cheaper, but the most easily accessible substrates industrially are the chlorides. This dilemma is all the more unfortunate in view of the instability of catalysts and/or ligands at temperatures above 130 °C. Thus, the model reaction involving styrene (**1**) and chlorobenzene (**2**) fails using any of the traditional catalysts such as $Pd[P(C_6H_5)_3]_4$. Another drawback of the Heck reaction is the extreme difficulty in recycling the catalyst, which traditionally needs to be used in fairly large amounts (3–5 mol. %).

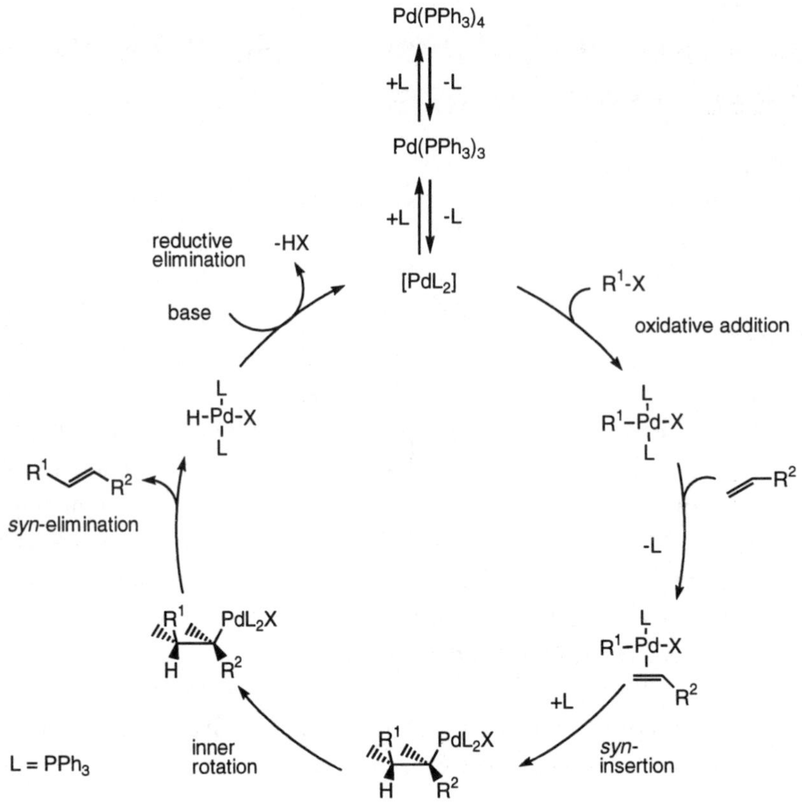

Scheme 10.1.

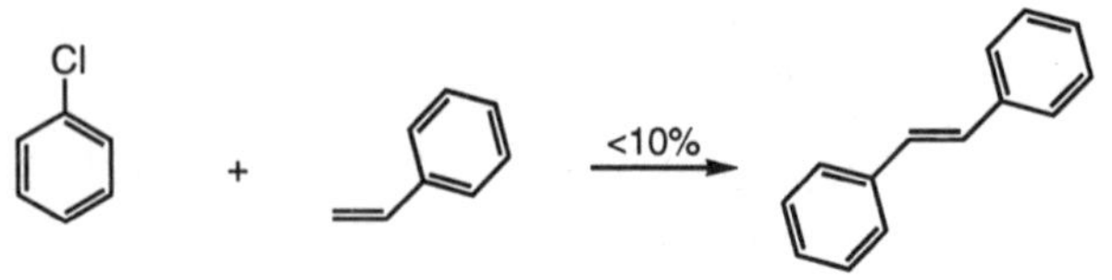

$$(10.1)$$

The above-mentioned reactivity scale correlates with the increase in carbon–halogen bond energy in the sequence C–I ($65\,\text{kcal mol}^{-1}$) < C–Br ($81\,\text{kcal mol}^{-1}$) < C–Cl ($96\,\text{kcal mol}^{-1}$) [3]. Indeed, it was long believed that the first step of the catalytic cycle, namely oxidative insertion of Pd(0) into the carbon–halogen bond, is the slow step and therefore responsible for the failure of aryl chlorides to react. However, it was shown later that this need not be the case [1]. Actually, oxidative insertion into aryl chlorides may occur at temperatures as low as 60 °C!

Of the many early attempts to solve the problem of sluggishly reacting aryl halides [4], the Milstein ligand system using the highly basic diphosphine bis[(diisopropyl)phosphino]butane (dippb) deserves special attention [5]. Accordingly, the phenylation of styrene (**2**) by chlorobenzene (**1**) with the formation of stilbene (**3**) proceeds to the extent of 80% at 140 °C, provided that 1 mol.% of the dippb-based catalyst **4a** is used in the presence of NaOAc as the base. Interestingly, the propano-bridged analog **4b** shows no activity, indicating that in the olefin insertion step a four-coordinate intermediate, formed via chelate opening, is required.

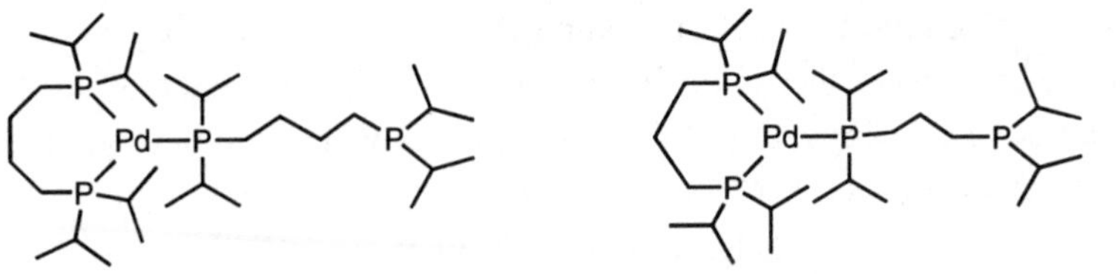

(10.2)

4a   4b

Unfortunately, the process has not been generalized to include electron-poor olefins such as acrylates, the reason being that the highly basic phosphines are not only expensive but also cause oligomerization of such substrates.

More recently, Herrmann has described the use of palladacycles (6), prepared from tri(o-tolyl)phosphine (5), as highly active and stable catalysts or precatalysts for Heck reactions of aryl bromides and activated aryl chlorides [6]. Temperatures of up to 150°C are tolerated. The mechanism has not been elucidated, but an oxidation switch of Pd(II) ⇌ Pd(IV) instead of the usual Pd(0) ⇌ Pd(II) has been suggested for these intriguing Heck processes. In the case of fairly reactive aryl halides, turnover numbers (TON) of more than 100000 are typical. In the case of less reactive substrates such as bromobenzene (7), more catalyst is required, resulting in a TON of 48. Chlorobenzene (1) is essentially inert. Activated aryl chlorides such as p-chlorobenzaldehyde are also unreactive, unless an excess of tetrabutylammonium chloride is added (Jeffery conditions) [1e].

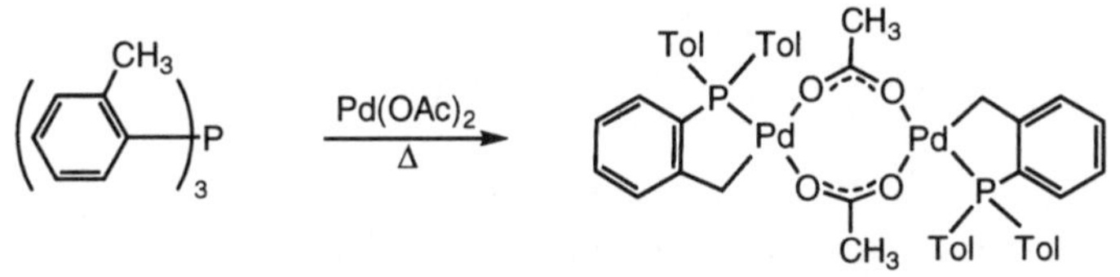

(10.3)

5   6

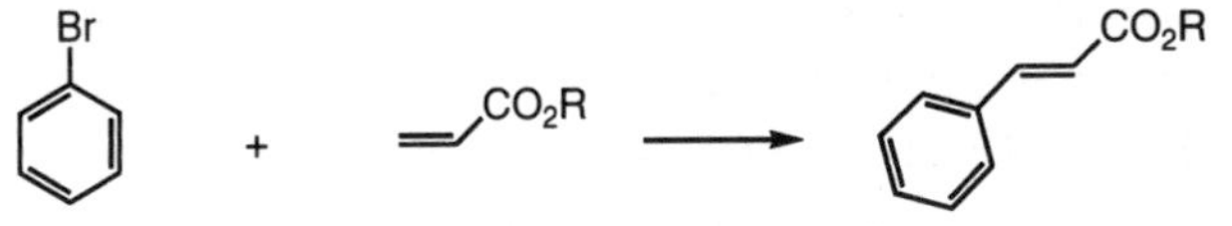

(10.4)

7   8 a R = CH$_3$ 9
         b R = n-C$_4$H$_9$
         c R = CH$_2$CH(C$_2$H$_5$)(CH$_2$)$_3$CH$_3$

Subsequently, Milstein reported the use of a different palladacycle 11 as a highly active catalyst for Heck reactions [7]. Although mainly aryl iodides were employed, in two cases aryl bromides were shown to react with acrylate 8a. For example, the reaction of bromobenzene (7) with acrylate 8a proceeds with a conversion of 79% after 63h at 140°C, the TON being 132900. Unfortunately, chlorobenzene (1) turned out to be almost inactive. The authors suggest that the classical Pd(0) cycle is not involved.

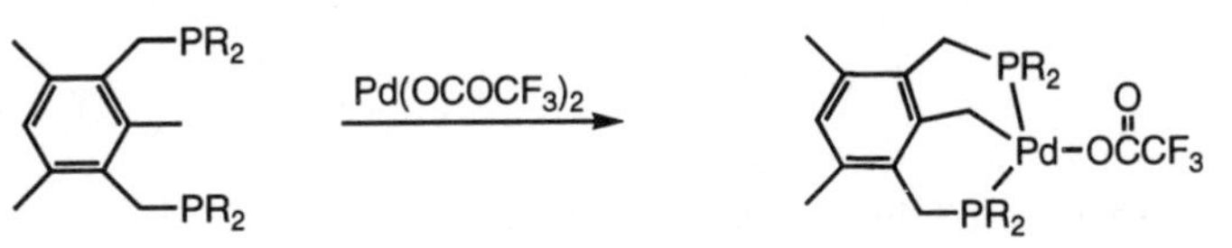

(10.5)

10   11

Although phosphine-free systems using palladium salts (PdX_2) in aqueous medium have been shown to be successful in the case of active aryl halides and (water-soluble) olefins [8], this interesting methodology does not solve the problem of non-activated aryl bromides or chlorides. Heterogeneous Pd catalysts such as palladium black [9], highly dispersed Pd on MgO [10], ethylsilyldiphenylphosphine–palladium supported on montmorillonite [11] or other heterogenized forms [12] have been used to effect Heck reactions. However, these methods do not solve the longstanding problem of unreactive aryl halides.

2 Carbon–chlorine activation of vinyl chlorides

Because vinyl chlorides were known to be inert in Heck reactions using traditional Pd catalysts [1], we decided to test 1,4-bis(dicyclohexylphosphino)butane ($Cy_2P(CH_2)_4PCy_2$ = dcypb) as a possible ligand in the potential reaction of 1-chloro-2-methylprop-1-ene (**12**) with styrene (**2**) [13]. This ligand is related to the dippb used by Milstein in Heck arylations [5] (cf. compound **4a**), but it has the advantage of simple synthesis via hydrogenation of the corresponding tetraphenyl derivative. Unfortunately, only 5–20% of the desired Heck coupling product **13** was observed.

$$(10.6)$$

12   **2**   **13**

Speculating that the problem could be due to the formation of a catalytically non-active π-allyl palladium species following Heck-type C—C bond formation, we repeated the reaction in the presence of sodium dimethylmalonate (**14**) as a carbon nucleophile in the hope of inducing a tandem Heck–allyl substitution process (2 mol.% $Pd(OAc)_2$; 5 mol.% dcypb; DMF; 150 °C, 18 h) [13]. To our surprise none of the expected tandem reaction product was observed, the major products being 2-methylallylmalonic acid dimethyl ester (**15**, 55%) and its decarboxylated form (**16**, 20%). Because styrene (**2**) is not involved, the reaction was repeated in the absence of this olefin at a lower temperature (2 mol.% Pd(OAc); 5 mol.% dcypb; DMF; 120 °C, 18 h). Again the methallylated products **15** (70%) and **16** (12%) were formed. Other ligands are less effective, the yields of compounds **15/16** being lower, as in the case of PPh_3 (8%/3%), 1,2-bis(diphenylphosphino)ethane (4%/5%) and 1,4-bis(diphenylphosphino)butane (32%/26%). In further optimization it was discovered that the best catalyst system is the dimethylpalladium complex of the ligand, (dcypb) $PdMe_2$ (2 mol.%), in the presence of additional dcypb (3 mol.%) at 120 °C/2 h. This led to the products **15** (76%) and **16** (1.4%) [13].

$$(10.7)$$

12   **14**   **15 (76%)**   **16 (1.4%)**

A plausible mechanism for these novel processes involves palladium-catalysed isomerization of the otherwise non-reactive vinyl chloride (**12**) to methallyl chloride, followed by classical palladium-catalysed allylic substitution in which the malonate participates as the carbon nucleophile [13]. Separate NMR experiments revealed that the reaction is initiated by thermolytic decomposition of (dcypb)PdMe$_2$ with the formation of (dcypb)Pd(0) (or its dimer). It is likely that the reactive Pd(0) catalyst then coordinates to the double bond of compound **12** and induces CH activation in one of the methyl groups, the first step in the isomerization. Methallyl chloride may not actually be set free, because palladium can remain coordinated to the π-allyl system, chloride being expelled with intermediate formation of [Pd{η^3-(CH$_2$CMe=CH$_2$)}(dcypb)]$^+$. Indeed, upon heating (dcypb)PdMe$_2$ in the presence of compound **12** as the sole reaction partner, this cationic π-allyl palladium species was identified as the major product by NMR spectroscopy; its BF$_4^-$ salt was synthesized by an independent route and its structure was proved by X-ray structural analysis (Fig. 10.1). Scheme 10.2 shows a possible catalytic cycle that is in accord with all of the present data, including the observation that other ligands are less effective.

Regioselectivity and therefore the possibility of two different products becomes relevant in the case of unsymmetrically substituted vinyl chlorides. In order to test whether the reaction is regioselective, substrates **17** were subjected to the isomerization–substitution reaction. In all cases complete regioselectivity was observed, in that CH activation occurs solely at the methyl groups with the formation of products **18** [13].

$$(10.8)$$

17
a *E*,R = cyclo-C$_6$H$_{11}$
b *Z*,R = cyclo-C$_6$H$_{11}$

18
75%
65%

Functionalized substrates such as **19** were also shown to react regioselectively. A limitation of the method has to do with steric factors. Specifically, substrates bearing groups larger than methyl do not react analogously [13].

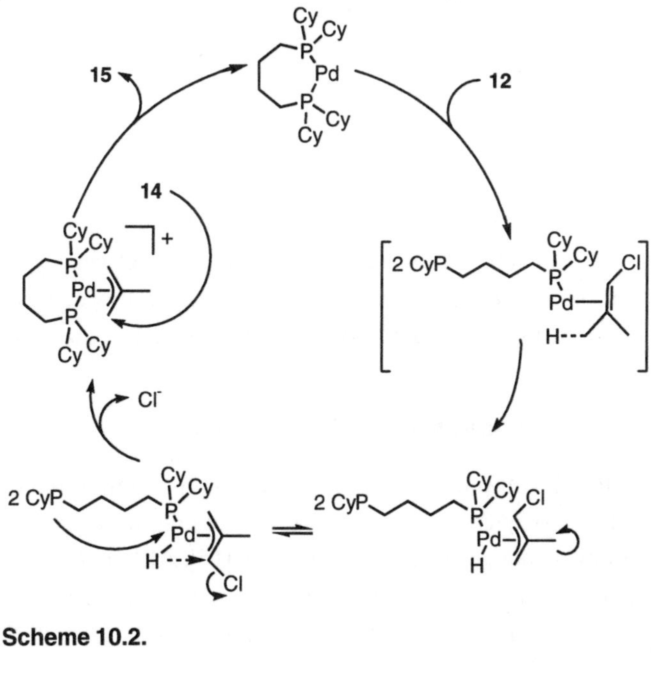

Scheme 10.2.

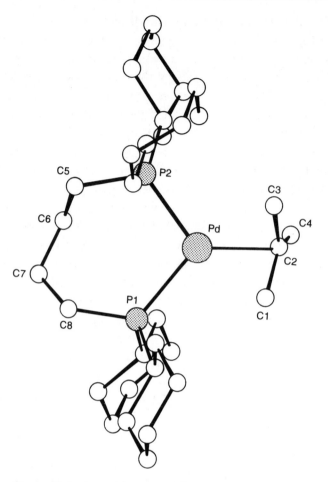

Figure 10.1. Crystal structure of [Pd{μ³-(CH₂CMe=CH₂)}(dcypb)]⁺.

$$ (10.9) $$

19   **20**

Although the process of C–C bond formation described here does not involve vinylation, it does constitute an interesting way to activate unreactive vinyl chlorides. The concept of isomerization of an otherwise unreactive substrate followed by reaction may be of interest in other cases using palladium or other transition metal catalysts.

3   Palladium-containing dendrimers: re-usable catalysts

The preparation of metal-containing dendrimers with defined inner and outer structural elements provides access to macromolecular materials having special properties and functions [14] that are, in principle, also useful in catalysis [15]. Although a number of metal-containing dendrimers have been described, they were rarely prepared for use as catalysts. As recently delineated by van Koten *et al.*, such catalysts can be considered to be at the interface between homogeneous and heterogeneous catalysis [16]. In the most favorable case they combine the advantages of both classes of catalysts, namely the existence of a specific number of structurally defined, catalytically active metal centers as in traditional homogeneous catalysis, as well as the possibility of simple separation and re-use of the catalyst as in heterogeneous catalysis. However, it was not possible to recycle the catalyst in this case. Thus, the potential advantage of dendritic catalysts remained unclear.

We recently devised a simple synthesis of dendritic diphosphines based on commercially available dendritic amines such as compound **21** [17]. Double phosphinomethylation of each of the 16 primary amino end-groups affords DAB-dendr-[N(CH$_2$PPh$_2$)$_2$]$_{16}$ (**22**), in almost quantitative yield.

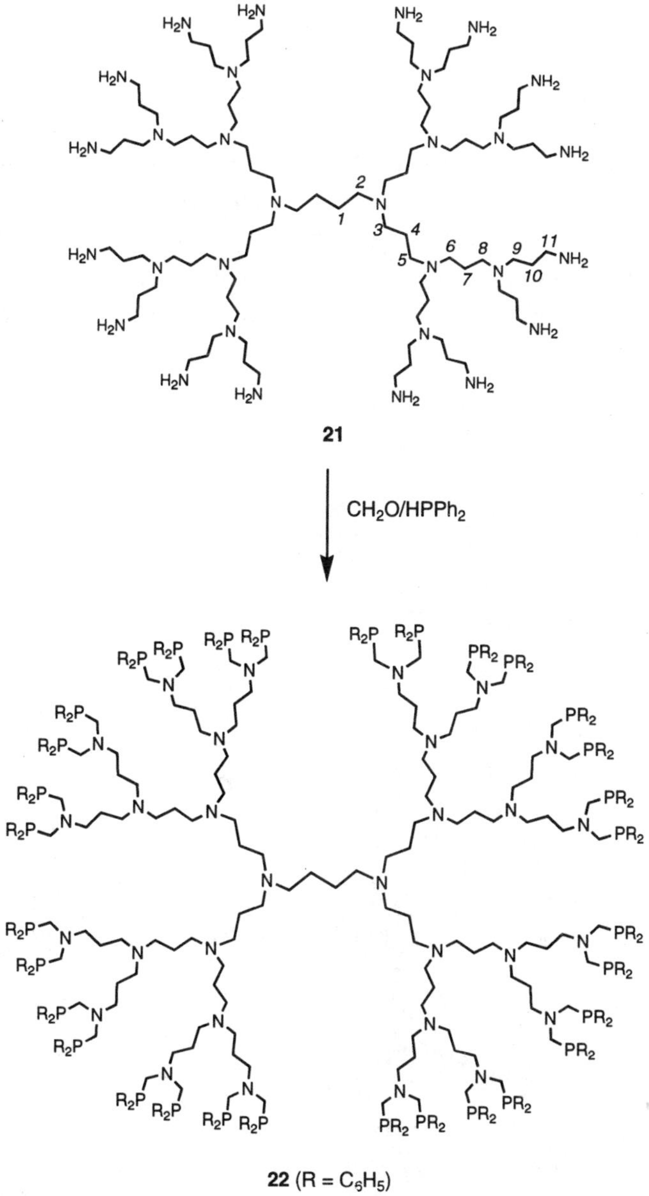

21

$$CH_2O/HPPh_2 \qquad (10.10)$$

22 (R = C$_6$H$_5$)

Dendrimer **22** with 16 bidentate ligands on the outer surface provides the possibility of specific complexation with a variety of transition metals [17]. Loading can be complete or partial on an optional basis. One possibility is compound **23**, prepared by reacting dendri-mer **22** with [Pd(CH$_3$)$_2$(tmeda)]. Partial loading, e.g. 50%, was also achieved. For purposes of comparison, the parent 'monomer' **24** was also prepared and characterized by X-ray analysis [18]. Figure 10.2 shows a chair geometry in which nitrogen is not coordinated to palladium.

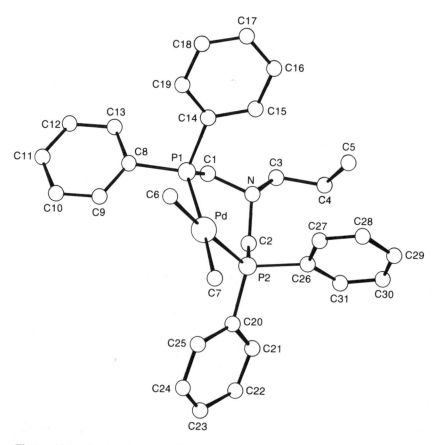

(10.11)

23   **24**

In a simple test reaction, bromobenzene (**7**) and styrene (**2**) were dissolved in dimethyl-formamide (DMF) and heated for 3 h at 130 °C in the presence of sodium acetate as the base and complex **23** (0.125 mol.% corresponding to 2 mol.% 'Pd monomer') as catalyst. The insoluble inorganic components were filtered from the reaction mixture and the clear solution was diluted with diethyl ether. Addition of this relatively unpolar solvent caused essentially complete precipitation (>98%) of a Pd-containing dendrimer, the precise struc-

Figure 10.2. Crystal structure of Pd complex **24**.

ture of which has not been determined. Analysis of the reaction mixture by gas chromatography showed the presence of stilbene (**3**, 89%) and 1,1-diphenylethylene (11%), conversion being 85–90%. The isolated Pd-containing dendrimer was used in the same reaction once more and displayed the same catalytic activity (91% stilbene and 9% 1,1-diphenylethylene, conversion 77%). The catalytic activity of the precatalyst with only 50% Pd loading was similar.

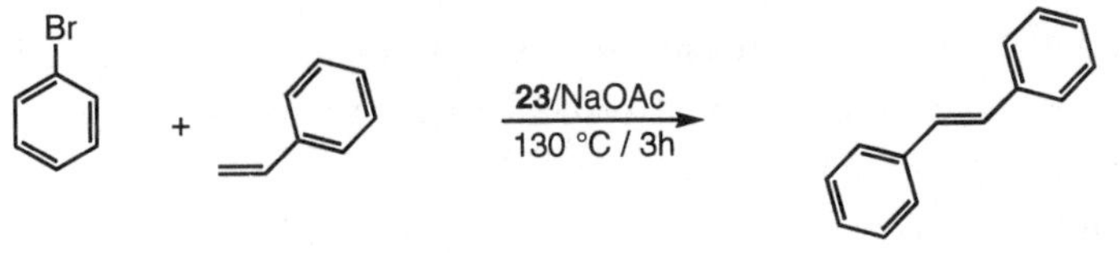

(10.12)

Thus, this is the first metal-containing dendrimer that combines the desired features regarding application in catalysis, namely efficient catalysis on the basis of structurally well-defined metal centers and the possibility of recyclization and re-use (Scheme 10.3) [17]. The reason why recyclization of the dendritic catalyst is so simple has to do with the nanoscopic properties of the material. The 'monomer' **24** has completely different solubility properties and cannot be separated easily from the reaction mixture.

In the few previously known cases of catalysis using dendritic metal complexes, catalytic activity was shown to be slightly or significantly *lower* than that of the corresponding monomeric parent compounds [15,16,19]. In contrast, we observed in our system a significantly *higher* activity [17]. Typically, a TON of 50 for compound **23** versus a TON of 16 for the parent compound **24** was observed. This is attributable to the higher thermal stability of the dendritic complexes. In the case of compound **23**, practically no undesired formation of elemental palladium was observed. In contrast, the Pd complexes of the parent compound **24** undergo partial decomposition with the formation of Pd precipitates. Why is the dendritic Pd compound **23** thermally more stable than the corresponding monomeric complex **24**? Perhaps the 16 Pd complexes on the surface of dendrimer **23** behave as though they are immobilized on a solid support, thereby preventing easy decomposition.

Although it was subsequently found that catalyst **23** cannot be used in the case of less reactive aryl halides such as chlorobenzene (**1**), the catalyst system is remarkable in that it

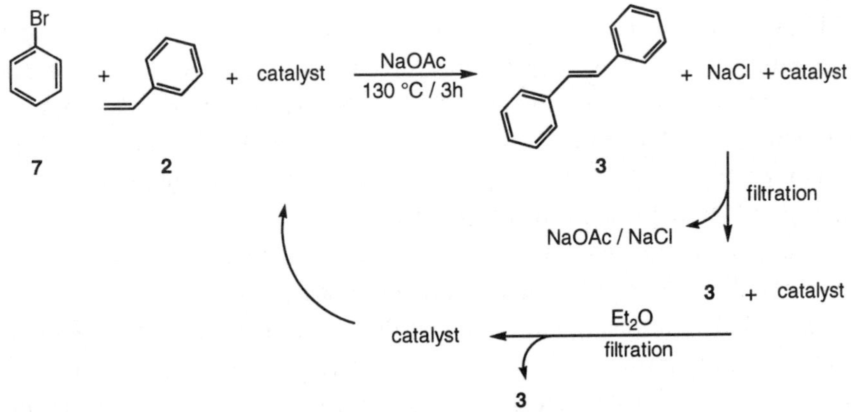

Scheme 10.3..

illustrates for the first time the principle that dendritic catalysts have advantages over the analogous 'monomeric' parent compounds. Thus, research in the area of dendritic metal catalysts may be rewarding after all.

4   A serendipitous discovery: PdX$_2$·6 Ph$_4$PX as highly active catalysts

Upon attempting to extend the use of the dendritic catalyst **23** to include aryl chlorides, the model reaction **1 + 2 → 3** was attempted at 150 °C. Unfortunately, after 10 h only small amounts of compound **3** were observed, together with catalyst decomposition [18]. Small amounts of a phosphorus-containing decomposition product were detected by ^{31}P-NMR spectroscopy and tentatively assigned to some kind of phosphonium salt Ph$_4$PX. Later it was shown that this assignment was in fact wrong. Nevertheless, we decided to test the mixture of a phosphonium salt Ph$_4$PX and a Pd salt PdX$_2$ as a possible catalyst for Heck reactions. This turned out to be the beginning of a remarkable development! Following some exploratory experiments using the test reaction **1 + 2 → 3**, we soon discovered that the soluble form of PdCl$_2$, namely the acetonitrile adduct Pd(CH$_3$CN)$_2$Cl$_2$, in combination with Ph$_4$PCl constitutes a highly active catalyst system in Heck reactions of otherwise unreactive aryl chlorides and bromides. Careful optimization and analysis of the reaction of chlorobenzene (**1**) with styrene (**2**) with the formation of Heck products **3**, **25** and **26** showed that if the ratio of PdX$_2$ to Ph$_4$PX is adjusted to 1:6, the best results are observed. In these reactions sodium acetate and *N,N*-dimethylformamide (DMF) or *N*-methylpyrrolidinone (NMP) are well suited as the base and solvent, respectively (Table 10.1) [20].

$$(10.13)$$

1	**2**	**3**	**25**	**26**

We were also surprised to observe that the use of small amounts of *N,N*-dimethylglycine (DMG) as an additive leads to a pronounced improvement, namely an increase in regioselectivity [20]. Upon employing 6 equiv. DMG relative to Pd and only 2 mol.% of the Pd catalyst, the proportion of undesired regioisomer **26** is significantly reduced (Table 10.1). Although no chemically engineered optimization was carried out, simple exploratory experiments showed that a further decrease in the amount of catalyst is certainly possible. Using 0.5 mol.% of the catalyst, a TON of 130 is reached. The same reaction using bromobenzene (**7**) and styrene (**2**) in NMP in the presence of as little as 0.05 mol.% of the catalyst [Pd(CH$_3$CN)$_2$Cl$_2$]·6 Ph$_4$PCl also proceeded smoothly (24 h; 140 °C; 77% Heck products **3**, **25** and **26** in a ratio of 90:1:9), and in this case the TON is 1300.

Another unusual observation pertains to the influence of the nature of the halide in the phosphonium salt Ph$_4$PX. Whereas the bromide is somewhat less effective than the chloride, the iodide turned out to be practically ineffective (<16% conversion after 8 h). This order is highly unusual. Although detailed mechanistic studies remain to be carried out, we currently suspect that part or all of the reaction proceeds via the phosphonium salt. Indeed, according to Yamamoto, Ph$_4$PX can function as a stoichiometric arylating agent in Pd-

Table 10.1. Heck reaction of chlorobenzene (**1**) with styrene (**2**) at 150 °C[a]

Catalyst	Pd (mol.%)	Solvent	t (h)	Additive[b] (mol.%)	Conversion (%)	Side-products (%)	Selectivity (**3**:**25**:**26**)
[Pd(CH$_3$CN)$_2$Cl$_2$]	2	NMP	12	—	6	6	—
[Pd(CH$_3$CN)$_2$Cl$_2$]· 6Ph$_4$PCl	2	DMF	8	—	89	11	86:1:13
[Pd(CH$_3$CN)$_2$Cl$_2$]· 6Ph$_4$PCl	2	NMP	12	—	79	<1	84:1:15
[Pd(CH$_3$CN)$_2$Cl$_2$]· 6Ph$_4$PCl	2	NMP	12	12	96	<1	96:1:3
[Pd(CH$_3$CN)$_2$Cl$_2$]· 6Ph$_4$PCl	2	DMF	12	12	95	12	96:1:3
[Pd(CH$_3$)$_2$(TMEDA)· 6Ph$_4$PCl	2	NMP	12	9	85	<1	97:1:2
PdCl$_2$(Ph$_3$P)$_2$· 4Ph$_4$PCl	2	DMF	6	—	77	7	85:1:14
[(Ph$_3$P)Pd(Ph)(µ-Cl)]$_2$	2	NMP	12	—	14	8	83:2:15
[(Ph$_3$P)Pd(Ph)(µ-Cl)]$_2$·6Ph$_4$PCl	2	NMP	12	—	86	<1	85:1:14
[(Ph$_3$P)Pd(Ph)(µ-Cl)]$_2$·6Ph$_4$PCl	2	NMP	12	12	80	<1	97:1:2
Pd(OAc)$_2$·6Ph$_4$PCl	2	NMP	12	12	79	<2	97:1:2

[a] Sodium acetate (2 equiv) as base.
[b] N,N-Dimethylglycine.

catalysed Heck reactions, albeit in moderate yield [21]. Triphenylphosphine is thereby generated, which in our case becomes arylated back to the phosphonium salt, a process that has precedence in the literature [22].

Ph$_4$PI + =—CO$_2$CH$_3$ $\xrightarrow[\text{140 °C / NEt}_3]{\text{10 mol% Pd(PPh}_5)_4}$ [styryl CO$_2$CH$_3$] + PPh$_3$ (10.14)

27   **8a**   **9a** (32%)   **28**

Nevertheless, other aspects of the PdX$_2$·6 PH$_4$PX-catalysed Heck reactions are currently unclear. For example, it is uncertain whether PdX$_2$·6 Ph$_4$PX is the actual catalyst in a Pd(II) ⇌ Pd(IV) redox system, or whether the mixture is simply a precatalyst that undergoes reduction to Pd(0) species under the reaction conditions. Although this question remains unanswered, we were able to obtain some structural data on a compound (or intermediate) that is formed upon mixing PdCl$_2$ and Ph$_4$PCl in acetonitrile. The Pd(II) complex [PdCl$_4^{2-}$]$_2$[Ph$_4$P$^+$] was isolated in high yield [18]. It is a yellowish, crystalline compound that is extremely stable to reduction. For example, upon heating a mixture of the complex and ethanol as a potential reducing agent at 130 °C, no sign of Pd precipitation in the form of powder or mirror was observed! Crystals of the complex, grown from acetonitrile, were analysed by X-ray crystallography and the crystal structure shows the $PdCl_4^{2-}$ component as a square planar entity (Fig. 10.3) (M.T. Reetz, R. Goddard and G. Lohmer, unpubl. data).

This compound is in fact a highly effective catalyst (or precatalyst) for the Heck reaction of chlorobenzene (**1**), provided that four parts of additional Ph_4PCl are used [18]. These observations are useful, but they do not answer the question as to why the phosphonium salt stabilizes palladium so effectively under the rather drastic conditions under which all other known Pd complexes begin to decompose with the formation of black Pd precipitates. Perhaps anionic complexes [23] are formed via halide addition of Ph_4PCl to such possible intermediates as $[(Ph_3P)Pd(Ph)X]$ [24] or the more stable dimeric form $[(Ph_3P)Pd(Ph)(\mu\text{-}Cl)]_2$. Indeed, the latter compound, which we were able to prepare easily as a relatively insoluble and difficult to characterize complex via reaction of $Pd(OAC)_2$ with Ph_4PCl in the presence of ethanol as the reducing agent, is a highly active and thermally stable catalyst or precatalyst in the presence of excess Ph_4PCl [20]. Accordingly, at 60–70 °C the dimeric precursor **29** converts into a soluble and presumably anionic form. Possibly, anions of the type **30** or **31** are involved.

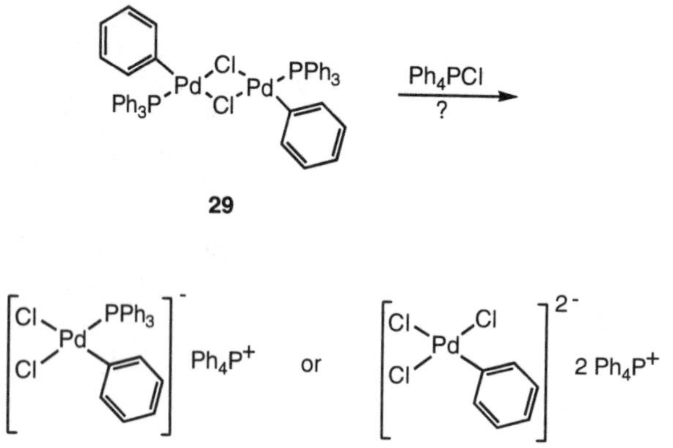

$$(10.15)$$

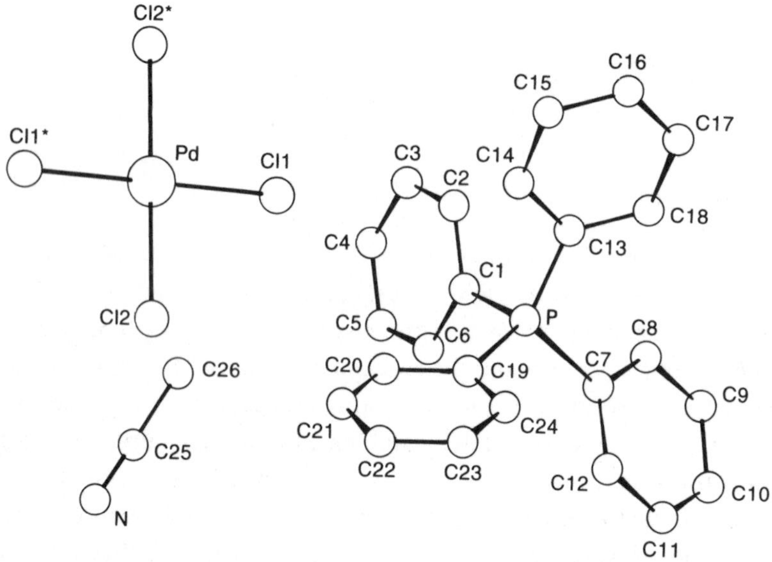

Figure 10.3. Crystal structure of $[PdCl_4^{2-}]\,2\,[Ph_4P^+]$ as CH_3CN solvate.

Synthetically and mechanistically significant is also the observation that the use of the spiro-type phosphonium salt **32a,b** under the usual conditions of the test reaction leads to less than 4% conversion [20]. This special phosphonium salt is particularly stable and does not undergo ring-opening arylation.

(10.16)

32 a X = Cl
   b X = I

If phosphonium species are in fact the arylating agents, then the use of stoichiometric amounts of substituted aryl halides and catalytic amounts of $[Pd(CH_3CN)_2Cl_2]\cdot6\ Ph_4PCl$ should afford scrambling products, the amount of which is expected to depend upon the ratio of ArX to Ph_4PCl. Indeed, this was observed [20]. For example, upon reacting p-chlorotoluene (**33a**) with styrene (**2**) in the presence of $[Pd(CH_3CN)_2Cl_2]\cdot6\ Ph_4PCl$ (2 mol.%; NMP; 8 h; 150 °C), not only the normal products **34a**, **35a** and **36a** were formed in a ratio of 90:<1:10, but also the scrambling products **3**, **25** and **26** in a ratio of about 85:1:14 (conversion with respect to product **33a**: 66%; 30% normal products, 27% scrambling products). In the presence of DMG (12 mol.%) the reaction proceeds similarly, but with increased regioselectivity (**34a**:**35a**:**36a** = 97:<1:3). The use of $[Pd(CH_3CN)_2Cl_2]\cdot6$-$(p$-tolyl$)_4PCl$ as the catalyst (2 mol.%; NaOAc; NMP; 19 h; 150 °C) affords by nature solely the desired products **34a**, **35a** and **36a** in a ratio of 90:<1:10 (58% conversion; 4% side-products).

(10.17)

33 a R = CH$_3$ **34**   **35**   **36**
   b R = CHO

Of course, it would be desirable to suppress undesired scrambling without having to prepare new phosphonium salts. Indeed, it is simply necessary to reduce the amount of catalyst $[Pd(CH_3CN)_2Cl_2]\cdot6\ Ph_4PCl$. For example, the reaction of p-chlorobenzaldehyde (**33b**) with styrene (**2**) in the presence of 0.1 mol.% $[Pd(CH_3CN)_2Cl_2]\cdot6$-$Ph_4PCl\cdot6$ DMG (NaOAc; NMP; 48 h 150 °C) affords the desired products **34b**, **35b** and **36b** in a ratio of 98:<1:2, conversion being 98%! (TON = 950). Only 1% of undesired scrambling products **3**, **25** and **26** is formed under these conditions [20].

The method can be applied to other substrates such as acrylates **8** [20]. For example, bromobenzene (**7**) reacts smoothly with acrylate **8b** (0.01 mol.% $[Pd(CH_3CN)_2Cl_2]\cdot6$-Ph_4PCl; NMP; 130 °C; >97% yield; TON 9800). Reaction of the normally very sluggishly reacting p-bromoanisol (**37**) with the acrylate **8a** leads to 71% of the desired product **38** (TON 1300). Here again, only 1% scrambling product and no 1,1-regioisomers are formed.

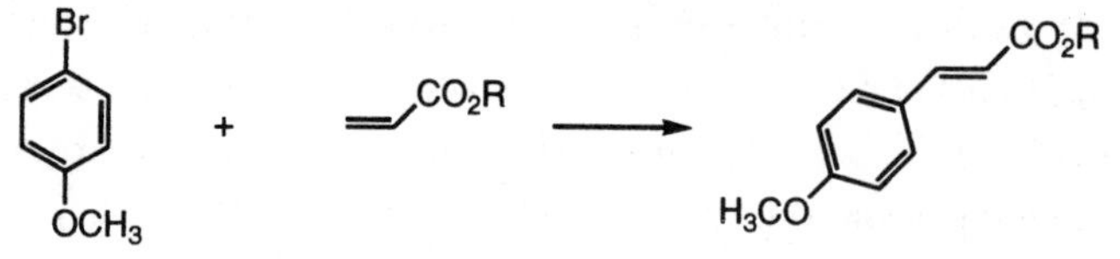

$$(10.18)$$

37   **8 c R = CH$_2$CH(C$_2$H$_5$)(CH$_2$)$_3$CH$_3$**   **38**

Industrially relevant substrates also react smoothly. For example, 2-bromo-6-methoxynaphthalene (**39**) was treated with ethylene in the presence of 0.05 mol.% [Pd(CH$_3$CN)$_2$Cl$_2$]·6 Ph$_4$PCl in a steel autoclave (NaOAc; *N,N*-dimethylacetamide as solvent; 24 h; 140 °C; 20 bar). The desired product **40** was obtained in a yield of 89% (conversion 97%; TON = 1800) [20].

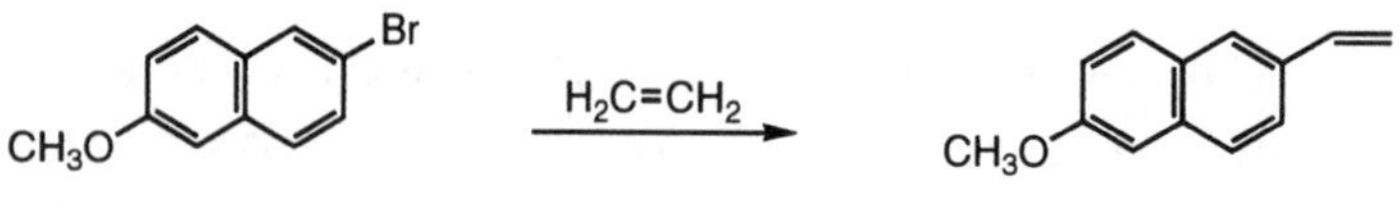

$$(10.19)$$

39   **40 (89%)**

In summary, the combination of PdX$_2$ and Ph$_4$PX in the presence of DMG represents the currently most active and selective catalyst system for Heck reactions of normally unreactive aryl halides [20, 25]. Even the special palladacycles, recently described as being particularly active [6], fail in reactions of such substrates as chlorobenzene (**1**). Indeed, in our hands the coupling of chlorobenzene (**1**) with styrene (**2**) using the usual palladacycle *trans*-di(μ-acetato)-bis[*o*-(di-*o*-tolylphosphino)benzyl]dipalladium **6** with or without DMG as an additive proceeded to less than 2% (TON = 1) [20]. The turnover number in the reaction of bromobenzene (**7**) with acrylate **8b** is also significantly higher in our catalyst system (TON 9800 versus 48 for the palladacycle). The same pertains to the reaction of *p*-bromoanisol (**37**) (TON 1300 versus 87) [20].

It was of interest to see if Suzuki reactions are also catalysed by the PdX$_2$·2 Ph$_4$PX system. Indeed, preliminary studies are promising. For example, the following two reactions of the boronic acid ester **41** proceed very smoothly (M.T. Reetz and E. Westermann, unpubl. data).

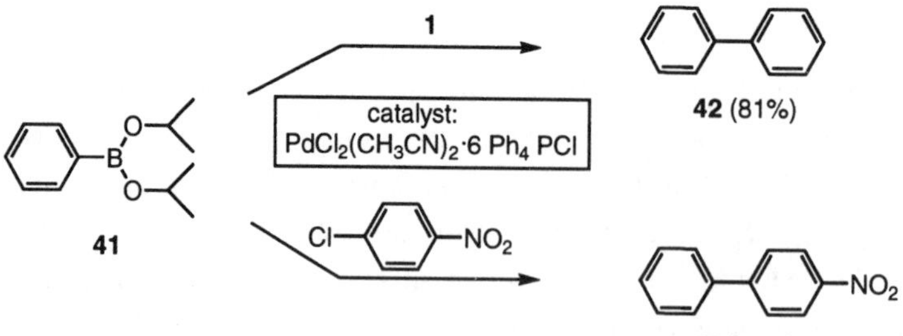

$$(10.20)$$

5   Colloidal solutions of Pd clusters: efficient catalysts?

Nanostructured transition metal clusters are of great potential interest as catalysts for organic transformations, as electrocatalysts in fuel cells and as advanced materials with special electronic and/or magnetic properties [26]. They are generally prepared by chemi-

Anode:	Met_{bulk} $\longrightarrow$	Met^{n+} + ne^-
Cathode:	Met^{n+} + ne^- + stabilizer $\longrightarrow$	Met_{coll}/stabilizer
Sum:	Met_{bulk} + stabilizer $\longrightarrow$	Met_{coll}/stabilizer

Scheme 10.4.

cal reduction of the corresponding transition metal salts by such reagents as hydrazine, $NaBH_4$, $R_3BH^-N^+R_4$, HCO_2Na or $LiAlH_4$ in the presence of stabilizers such as special ligands [27], polymers [28] or surfactants [29] (e.g. ammonium salts R_4NX). This area has been reviewed [26, 30], and only a few relevant aspects are treated here. In spite of many successful methods of preparation some general problems remain, including ease of isolation, purity and size selectivity [31]. We have recently developed electrochemical methods for the preparation of R_4NX-stabilized transition metal clusters, including those containing palladium as the metal [32]. The synthetic advantages include ease of performance, high purity, simple isolation and size selectivity in the range 1–6 nm. The latter can be controlled by proper choice of the current density, solvent polarity, temperature and other parameters [32] (M.T. Reetz, M. Winter and R. Breinbauer, unpubl. data).

In the case of palladium, the metal source is a sacrificial anode composed of a Pd sheet [32]. The ammonium salt R_4NX serves as electrolyte and stabilizer. A combined TEM/STM study showed that the surfactant R_4NX forms a monomolecular layer around the metal cluster, thereby preventing undesired aggregation with the formation of metal powder [33]. Following electrolysis, the addition of water induces the clean precipitation of the R_4NX-stabilized cluster, which can be redispersed in organic solvents (Scheme 10.4) [30].

A different way to stabilize Pd clusters is to use polar solvents such as propylene carbonate (PC) [34]. Interestingly, these PC-stabilized Pd clusters, accessible either electrochemically or simply by heating $Pd(OAc)_2$ in this polar solvent, are stable at temperatures of up to 140–150 °C, i.e. there is no Pd powder formation [34].

Although preformed transition metal clusters prepared by other methods had been used in colloidal form as hydrogenation catalysts [26], their application as catalysts for C–C bond-forming reactions was unknown. We therefore tested them in Heck and Suzuki coupling reactions [35] and in [3 + 2] cycloaddition reactions of methylene cyclopropane [36]. In the case of R_4NX-stabilized Pd clusters and Pd/Ni bimetallic clusters, colloidal solutions turned out to be active catalysts in Heck reactions of activated aryl halides [35]. Hoechst has reported related results [37]. However, the problem of unreactive aryl chlorides and bromides was not solved using colloidal solutions of R_4NX-stabilized Pd clusters. In contrast, preliminary results regarding the immobilization of such preformed clusters in hydrophobic sol–gel materials (M.T. Reetz and M. Dugal, unpubl. data), as previously shown for enzymes [38], appear promising. More work is necessary in this interesting area.

Along similar lines, Pd clusters stabilized by propylene carbonate were used as catalysts in the Heck reaction of chlorobenzene (**1**) with styrene (**3**) [34]. Using 8–10 nm sized Pd clusters, about 30–40% of the desired product **3** was observed. It remains to be seen if smaller Pd clusters (e.g. 2 nm) lead to better results.

Most recently, we discovered an extremely simple and highly reactive phoshine-free catalyst system for Heck reactions of aryl bromides in which colloidal palladium may possibly be involved [39]. Accordingly, aryl bromides such as **7**, p-bromoanisole or p-bromotoluene react smoothly with olefins such as **2** or **8**, the catalyst being $PdCl_2(PhCN)_2$ or $Pd(OAc)_2$ in the presence of DMG. This is the most active catalyst system for aryl bromides known so far. Indeed, it also works in the preparation of tri-substituted olefins (M.T. Reetz and E. Westermann, unpubl. data).

6   Summary and outlook

This progress report revolves around two basic problems associated with the Heck reaction: recyclization and efficient re-use of Pd-catalysts; and catalyst development for the case of traditionally unreactive aryl halides. Progress has been made in both areas, so that industrial viability is closer than ever, but further improvements are necessary. Palladium-containing dendritic catalysts offer the advantage of homogeneous catalysts (defined number of catalytic centers) and of heterogeneous catalysts (simple separation from the products via precipitation). Surprisingly, an added benefit is possible, namely increased thermal stability and therefore better catalytic performance. We have demonstrated these features in the case of Heck reactions catalysed by dendritic Pd catalysts. However, the inherent problem of unreactive aryl chlorides has not been solved on the basis of dendritic catalysts. Rather, the combination of simple Pd salts such as $PdCl_2$ or $Pd(OAc)_2$ and tetraarylphosphonium salts (Ar_4PX) allows for surprisingly efficient Heck coupling reactions involving aryl bromides and chlorides. Such catalysts, e.g. $PdCl_2(CH_3CN)_2 \cdot 6\text{-}Ph_4PCl_2$, also catalyse the Suzuki reaction of aryl chlorides and bromides. Nevertheless, further improvements are necessary, especially with respect to catalyst recyclization. The question whether these simple catalysts also induce carbonylation or amination reactions also needs to be addressed. With respect to Heck reactions of aryl bromides, the most active (and cheapest) catalyst system is $PdCl_2(PhCN)_2$ or $Pd(OAc)_2$ in the presence of *N,N*-dimethylglycine as an additive.

Finally, the entrapment of R_4NX-stabilized Pd clusters in sol–gel materials, if performed in such a way that *individual* clusters become encapsulated in the mesoporous matrix, may well turn out to be a promising method for the creation of active heterogeneous catalysts for Heck couplings and other C—C bond-forming reactions.

7   Acknowledgment

Parts of this work were supported by the Katalyse-Verbund des Landes Nordrhein-Westfalen.

8   References

1   Reviews of Heck reactions: (a) Heck RF. *Palladium Reagents in Organic Syntheses*. London: Academic Press, 1985; (b) Tsuji J. *Palladium Reagents and Catalysts: Innovations in Organic Synthesis*. Chichester: Wiley, 1995; (c) de Meijere A, Meyer FE. *Angew Chem* 1994; **106**: 2473–506; *Angew Chem Int Ed Engl* 1994; **33**: 2379; (d) Cabri W, Candiani I. *Acc Chem Res* 1995; **28**: 2–7; (e) Jeffery T. *Adv Met-Org Chem* 1996; **5**: 153–260.

2   See, for example: Loiseleur O, Meier P, Pfaltz A. *Angew Chem* 1996; **108**: 218–20; *Angew Chem Int Ed Engl* 1996; **35**: 200; and references therein.

3   Cornils B, Herrmann WA. *Applied Homogeneous Catalysis with Organometallic Compounds*. Weinheim: VCH, 1996.

4   Grushin VV, Alper H. *Chem Rev* 1994; **94**: 1047–62.

5   Ben-David Y, Portnoy M, Gozin M, Milstein D. *Organometallics* 1992; **11**: 1995–6.

6 Herrmann WA, Broßmer C, Öfele K *et al. Angew Chem* 1995; **107**: 1989–92; *Angew Chem Int Ed Engl* 1995; **34**: 1844; Herrmann WA, Brossmer C, Reisinger C-P *et al. Chem–Eur J* 1997; **3**: 1357–64.

7 Ohff M, Ohff A, van der Boom ME, Milstein D. *J Am Chem Soc* 1997; **119**: 11687–8.

8 Bumagin NA, Bykov VV, Beletskaya IP. *Russ J Org Chem* 1995; **31**: 439–44; Bumagin NA, Bykov VV, Sukhomlinova LI, Tolstaya TP, Beletskaya IP. *J Organomet Chem* 1995; **486**: 259–62.

9 Augustine RL, O'Leary ST. *J Mol Catal A* 1995; **95**: 277–85.

10 Kaneda K, Higuchi M, Imanaka T. *J Mol Catal* 1990; **63**: L33–6.

11 Choudary BM, Sarma RM, Rao KK. *Tetrahedron* 1992; **48**: 719–26.

12 See, for example: Andersson CM, Karabelas K, Hallberg A. *J Org Chem* 1985; **50**: 3891–5; Terasawa M, Kaneda K, Imanaka T, Teranishi S. *J Organomet Chem* 1978; **162**: 403–14.

13 Reetz MT, Wanninger K, Hermes M. *Chem Commun (Cambridge)* 1997: 535–6.

14 Newkome GR, Moorefield CN, Vögtle F. *Dendritic Molecules: Concepts—Syntheses—Perspectives.* Weinheim: VCH, 1996; Tomalia DA. *Sci Am* 1995; **272**: 62–6.

15 Matthews OA, Shipway AN, Stoddart JF. *Prog Polym Sci* 1998; **23**: 1–56.

16 van Koten *et al.* describe a silane-dendrimer having 12 arylnickel(II)-bearing end-groups, a compound that catalyses the addition of CCl_4 to methacrylic acid methylester in a Kharasch reaction; however, separation from the product and recyclability of the catalyst are not mentioned: Knapen JWJ, van der Made AW, de Wilde JC *et al. Nature (London)* 1994; **372**: 659–63.

17 Reetz MT, Lohmer G, Schwickardi R. *Angew Chem* 1997; **109**: 1559–62; *Angew Chem Int Ed Engl* 1997; **36**: 1526.

18 Lohmer G. *Dissertation*, Universität Bochum, 1997.

19 Seebach *et al.* have prepared a dendrimer with six TADDOL end-groups, the corresponding Ti alkoxides functioning as catalysts (20 mol.%) in the addition of Et_2Zn to benzaldehyde; separation of the metal-free ligand following aqueous workup was accomplished by column chromatography: Seebach D, Marti RE, Hintermann T. *Helv Chim Acta* 1996; **79**: 1710–40.

20 Reetz MT, Lohmer G, Schwickardi R. *Angew Chem* 1998; **110**: 492–5; *Angew Chem Int Ed Engl* 1998; **37**: 481–3.

21 Sakamoto M, Shimizu I, Yamamoto A. *Chem Lett* 1995: 1101–2.

22 Hirusawa Y, Oku M, Yamamoto K. *Bull Chem Soc Jpn* 1957; **30**: 667–70; Horner L, Mummenthey G, Moser H, Beck P. *Chem Ber* 1966; **99**: 2782–8; Ziegler CB Jr, Heck RF. *J Org Chem* 1978; **43**: 2941–6; Migita T, Nagai T, Kiuchi K, Kosugi M. *Bull Chem Soc Jpn* 1983; **56**: 2869–70; Cassar L, Foà M. *J Organomet Chem* 1974; **74**: 75–8.

23 Various anionic Pd species have been postulated as possible intermediates in Heck reactions: Scott WJ, Stille JK. *J Am Chem Soc* 1986; **108**: 3033–40; Amatore C, Jutand A, M'Barki MA. *Organometallics* 1992; **11**: 3009–13; Negishi E, Takahashi T, Akiyoshi K. *J Chem Soc Chem Commun* 1986: 1338–9.

24 This compound as well as soluble derivatives were first prepared by Anderson via the reaction of $[(PR_3)PdCl(\mu\text{-}Cl)]_2$ with Ar_2Hg in yields of 20–40%: Anderson GK. *Organometallics* 1983; **2**: 665–8; López G, Ruiz J, García G *et al. J Organomet Chem* 1990; **393**: C53–5. The reaction of $Pd(OAc)_2$ with 2 equivalents of Ph_4PCl at 25–60 °C in DMF affords only soluble adducts of the type $(Ph_4P)_2[Pd(OAc)_2Cl_2]$, whereas the same reaction in the presence of stoichiometric amounts of ethanol results in the formation of an insoluble light yellow solid material. The elemental analysis is in accord with $(PH_3P)Pd(Ph)(Cl)]_n$.

25 Reetz MT, Lohmer G, Schwickardi R. *German Patent DE-A 19712388.0*, 1997.

26 Schmid G, ed. *Clusters and Colloids.* Weinheim: VCH, 1994.

27 Schmid G, Morun B, Malm J-O. *Angew Chem* 1989; **101**: 772–3; *Angew Chem Int Ed Engl* 1989; **28**: 778; Vargaftik MN, Zagorodnikov VP, Stolarov IP *et al. J Mol Catal* 1989; **53**: 315–48.

28 Bradley JS, Millar JM, Hill EW. *J Am Chem Soc* 1991; **113**: 4016–17; Porta F, Ragaini F, Cenini S, Scari G. *Gazz Chim Ital* 1992; **122**: 361–3.

29 Kiwi J, Grätzel M. *J Am Chem Soc* 1979; **101**: 7214–17; Sasson Y, Zoran A, Blum J. *J Mol Catal* 1981; **11**: 293–300; Boutonnet M, Kizling J, Stenius P, Maire G. *Colloids Surf* 1982; **5**: 209–25;

Toshima N, Takahashi T, Hirai H. *Chem Lett* 1985: 1245–8; Boutonnet M, Kizling J, Touroude R, Maire G, Stenius P. *Appl Catal* 1986; **20**: 163–77; Meguro K, Torizuka M, Esumi K. *Bull Chem Soc Jpn* 1988; **61**: 341–5; Wiesner J, Wokaun A, Hoffmann H. *Prog Coll Polym Sci* 1988; **76**: 271–7; Satoh N, Kimura K. *Bull Chem Soc Jpn* 1989; **62**: 1758–63; Bönnemann H, Brijoux W, Brinkmann R *et al. Angew Chem* 1991; **103**: 1344–6; *Angew Chem Int Ed Engl* 1991; **30**: 1312; Toshima N, Takahashi T. *Bull Chem Soc Jpn* 1992; **65**: 400–9.

30 Reetz MT, Helbig W, Quaiser SA. In: Fürstner A, ed. *Active Metals: Preparation, Characterization, Applications.* Weinheim: VCH, 1996: 279–97.

31 Bradley JS. In: Schmid G, ed. *Clusters and Colloids.* Weinheim: VCH, 1994: Chap. 6.

32 Reetz MT, Helbig W. *J Am Chem Soc* 1994; **116**: 7401–2; Reetz MT, Helbig W, Quaiser SA. *Chem Mater* 1995; **7**: 2227–8; Reetz MT, Quaiser SA. *Angew Chem* 1995; **107**: 2461–3; *Angew Chem Int Ed Engl* 1995; **34**: 2240–1; Reetz MT, Quaiser SA, Winter M *et al. Angew Chem* 1996; **108**: 2228–30; *Angew Chem Int Ed Engl* 1996; **35**: 2092–4.

33 Reetz MT, Helbig W, Quaiser SA *et al. Science* 1995; **267**: 367–9.

34 Reetz MT, Lohmer G. *Chem Commun (Cambridge)* 1996: 1921–2.

35 Reetz MT, Breinbauer R, Wanninger K. *Tetrahedron Lett* 1996; **37**: 4499–502.

36 Reetz MT, Breinbauer R, Wedemann P, Binger P. *Tetrahedron* 1998; **54**: 1233–40.

37 Beller M, Fischer H, Kühlein K, Reisinger C-P, Herrmann WA. *J Organomet Chem* 1996; **520**: 257–9.

38 Reetz MT, Zonta A, Simpelkamp J. *Angew Chem* 1995; **107**: 373–6; *Angew Chem Int Ed Engl* 1995; **34**: 301–3; Reetz MT, Zonta A, Simpelkamp J. *Biotechnol Bioeng* 1996; **49**: 527–34; Reetz MT. *Adv Mater* 1997; **9**: 943–54.

39 Reetz MT, Westermann E, Lohmer R, Lohmer G. *Tetrahedron Lett* 1998; **39**: 8449–52.

11 Selective Allylation Reactions

AKIRA YANAGISAWA and HISASHI YAMAMOTO

Graduate School of Engineering, Nagoya University, CREST, Japan Science and Technology Corporation (JST), Chikusa, Nagoya 464-8603, Japan

1 Introduction

Allylation reactions are an attractive route for the formation of carbon–carbon bonds because the allylated products can be transformed into organic molecules possessing a variety of functional groups [1]. Numerous methods for introducing an allyl group have been developed and applied successfully to organic synthesis. However, some important problems remain regarding the control of regio- and stereochemistry, as shown in Scheme 11.1 [2]. In general, γ-substituted allylmetals react with electrophiles selectively at the γ-position, and thus the regioreversed α-adduct is seldom obtained as a major product [1a]. The key problem concerning the stereochemistry of allylmetals is their facile *trans–cis* isomerization. Even when a stereochemically pure allylic chloride is used as a starting material for α-selective allylation, a nearly 1:1 mixture of *trans* (*E*)- and *cis*(*Z*)-isomers is formed as the α-product. While remarkable progress has been made recently in the field of asymmetric synthesis, the enantioselective allylation of carbonyl compounds remains a challenging problem (Scheme 11.1). Although several studies on the reaction using a stoichiometric amount of chiral Lewis acids have been reported, there have been only a few reports of catalytic versions [1b, 1c, 3]. We describe here our recent methods for the selective allylation of electrophiles with allylic metal compounds using transition metal catalysts. First, we present regioselective cross-coupling reactions between allylic Grignard reagents and allylic phosphates, which proceed in the presence of a catalytic amount of Fe, Ni, or Cu salt. Next, we discuss the preparation of stereochemically homogeneous allylic metal compounds from allylic halides and highly reactive Rieke metals at low temperature. The allylic barium reagent thus obtained indicates high α-selectivity and stereospecificity in the reaction with carbonyl compounds. Finally, we describe the enantioselective addition of allylic stannanes to aldehydes catalysed by a BINAP·Ag(I) complex.

2 Transition-metal-catalysed regioselective cross-coupling reactions of allylic phosphates with Grignard reagents

2.1 *Superiority of phosphate ester as a leaving group for S_N2' selective alkylation of allylic alcohol derivatives with organocopper reagents*

Substitution at allylic carbon is an important process in organic synthesis. Over the past 20 years, organocopper reagents leading to S_N2 or S_N2' coupling products have been studied intensively [4]. Organocuprates generally show γ- and *anti*-selectivities in reactions with allylic carboxylates and allylic sulfonates. In contrast, a preference for γ- and *syn*-substitution is observed with allylic carbamate, allyloxybenzothiazoles and allylic ammonium salts. Although these methods are quite useful and broadly used in syntheses of

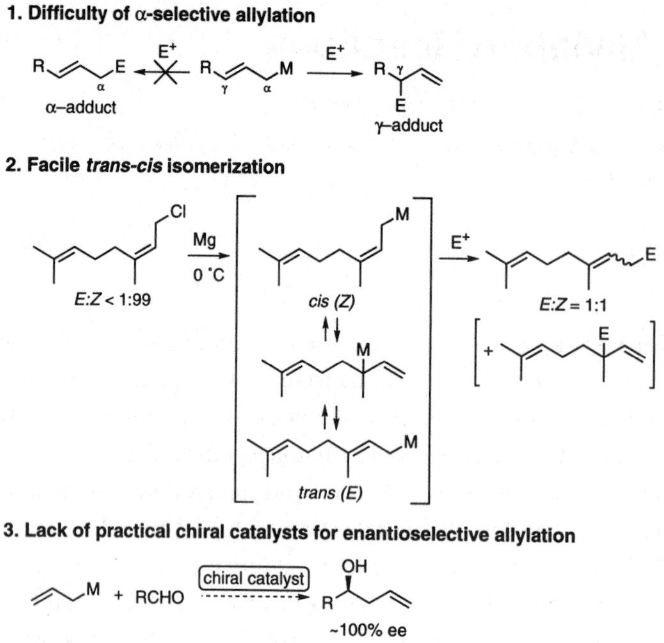

1. Difficulty of α-selective allylation

α–adduct

γ–adduct

2. Facile *trans-cis* isomerization

E:Z < 1:99

cis (Z)

E:Z = 1:1

trans (E)

3. Lack of practical chiral catalysts for enantioselective allylation

chiral catalyst

~100% ee

Scheme 11.1.

various valuable organic molecules, each of the known procedures still has some draw-backs and limitations. We have been interested in using allylic phosphates for these cross-coupling reactions and have anticipated that the phosphate leaving group should exhibit reactivity and selectivity markedly different from those of the allylic alcohol derivatives described above, because the phosphate ester can coordinate strongly with metal. Thus, an extensive study was made of the effect of leaving groups on the regio- and stereo-selectivities of the reaction between Grignard reagent and allylic alcohol derivatives in the presence of a copper(I) salt (Equation (11.1)).

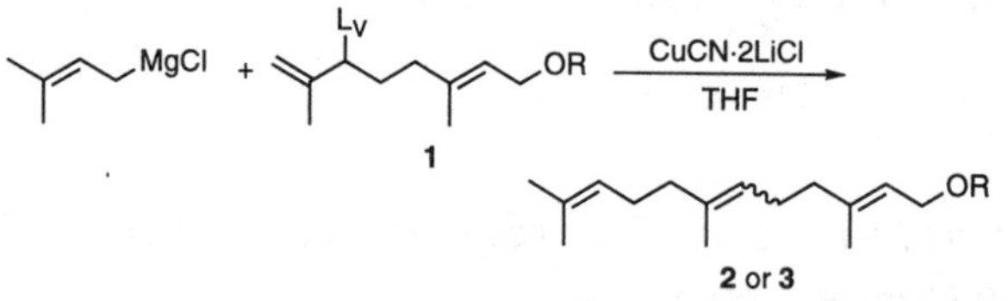

$$\tag{11.1}$$

1

2 or 3

Some of the results of secondary allylic alcohol derivatives **1** are summarized in Table 11.1 [5]. Treatment of the allylic chloride **1a** in THF with prenyl Grignard reagent in the presence of CuCN·2LiCl at 0 °C for 1 h affords the α-S_N2' coupling product **2** in 90% yield with an E/Z ratio of 46:54 without contamination by other regioisomers (entry 1). The phosphate ester is much more convenient and general than a variety of other derivatives. It also solves the problem of stereoselectivity (E/Z = 96:4, entries 3–5). It should be noted that other ester derivatives such as the mesylate derivative **1g** do not provide high stereoselectivity (entry 8). With primary allylic phosphate **4**, S_N2' coupling is still dominant (Equation (11.2)).

Table 11.1. Effect of leaving group of allylic alcohol derivatives **1** on the S_N2/S_N2' regio- and E/Z stereoselectivities[a]

Entry	Substrate	Leaving group	R	Conditions	Product	Yield (%)[b]	$S_N2/$ S_N2'[c]	E/Z[c]
1	**1a**	Cl	Ac	0 °C, 1 h	2	90	<1 : 99	46 : 54
2	**1a**	Cl	Ac	−100 °C, 1 h	2	74	<1 : 99	85 : 15
3	**1b**	$(EtO)_2PO_2$	$Si(t\text{-Bu})Me_2$	−78 °C, 1 h	3	94	<1 : 99	96 : 4
4	**1c**	$(PrO)_2PO_2$	$Si(t\text{-Bu})Me_2$	−78 °C, 1 h	3	96	<1 : 99	96 : 4
5	**1d**	$(c\text{-}C_6H_{11}O)_2PO_2$	$Si(t\text{-Bu})Me_2$	−60 °C, 1 h	3	58	<1 : 99	96 : 4
6	**1e**	$(PhO)_2PO_2$	$Si(t\text{-Bu})Me_2$	−78 °C, 1 h	3	91	<1 : 99	74 : 26
7	**1f**	$(Me_2N)_2PO_2$	$Si(t\text{-Bu})Me_2$	−20 °C, 1.5 h	3	47	<1 : 99	59 : 41
8	**1g**	MsO	$Si(t\text{-Bu})Me_2$	−78 °C, 1 h	3	68	<1 : 99	55 : 45

[a] The reaction was carried out using the allylic alcohol derivative **1** (1 equiv), prenyl Grignard reagent (3 equiv) and CuCN·2LiCl (3 equiv) in THF.

[b] Isolated yield.

[c] Determined by GC analysis. For entries 3–8, the ratios were determined after conversion to the corresponding alcohols.

$$(11.2)$$

However, no remarkable differences in S_N2/S_N2' selectivities are observed among the allylic phosphates (R = Ph, Et, i-Pr) for prenylation [5b].

The present displacement reaction using allylic phosphates is stereospecific, and the transfer of chirality from the secondary alcohol center to a newly formed carbon atom is predictable. The stereochemical result for prenylation of (*R*)-allylic phosphate **7** (94% enantiomeric excess, ee) with Grignard reagent and CuCN·2LiCl is shown in Equation (11.3). The prenylated product **8** is homogeneous and shows 94% ee with the *S*-configuration [5]. Thus, in this acyclic system, the enantioselectivity of the reaction is nearly quantitative and a complete 1,3-chirality transfer occurs.

$$(11.3)$$

2.2 Transition-metal (Fe, Ni, Cu)-catalysed substitution reactions of allylic diphenylphosphates with Grignard reagents

The transition-metal-catalysed substitution reaction of alkyl halides with Grignard reagents is generally described as the Kharasch reaction [6]. In the cross-coupling reaction of allylic substrates, regioselectivity has been studied actively with a variety of leaving groups, but to a lesser extent with phosphate leaving groups [4]. We have developed two catalytic processes: the S_N2-selective coupling reaction using Ni or Fe catalyst [7, 8] and the

S_N2' -selective coupling reaction using a catalytic amount of CuCN·2LiCl (Scheme 11.2) [7, 9]. Based on the results of the S_N2'-selective alkylation reaction of allylic phosphates using a stoichiometric amount of CuCN·2LiCl described above, we examined which transition metal catalysts were most suitable for the regioselective coupling of allylic phosphates with Grignard reagents. Treatment of (E)-2-decenyl 1-diphenylphosphate with 2 equiv of n-butylmagnesium chloride in the presence of various metal catalysts (5–10 mol.%) in THF gives a mixture of S_N2 and S_N2' coupling products. Some results are summarized in Table 11.2. Among the metal catalysts tested, iron, nickel and copper compounds show remarkable catalytic activities and the coupling products are obtained in high yield at low temperature (entries 5 and 7–13) [7]. In addition, nearly exclusive S_N2-selectivities are

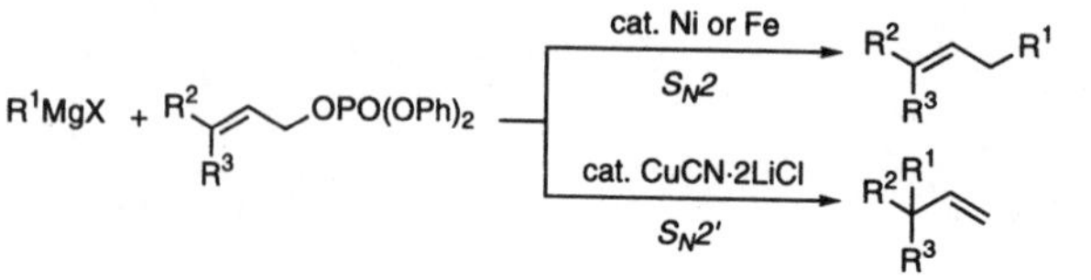

Scheme 11.2.

Table 11.2. Cross-coupling reaction of (E)-2-decenyl 1-diphenylphosphate with ^{n}BuMgCl in the presence of various metal catalysts[a]

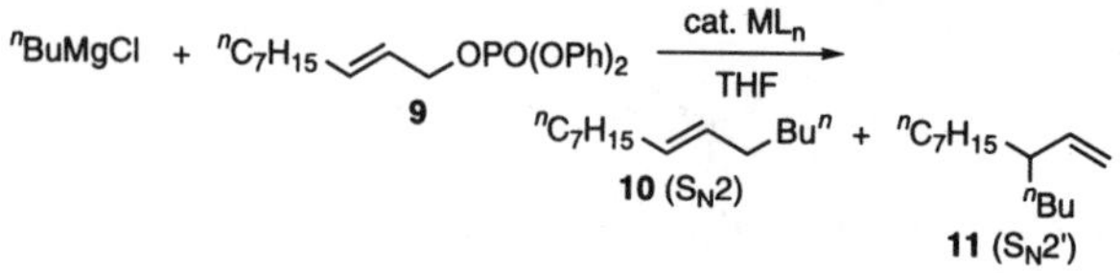

Entry	ML_n	Conditions	Yield (%)[b]	S_N2/S_N2'[c]
1	—	0 °C, 6 h	32	91:9
2	Ti(O^{i}Pr)$_4$	−23 °C, 5 h	10	89:11
3	CrCl$_2$	−23 °C, 6 h	48	81:19
4	MnCl$_2$	−30 °C, 5 h	20	93:7
5	Fe(acac)$_3$[d]	−76 °C, 1 h	94	99:1
6	CoCl$_2$	−73 °C, 1 h	64	81:19
7	NiBr$_2$	−73 °C, 2 h	93	>99:1
8	Ni(acac)$_2$[d]	−73 °C, 1 h	83	>99:1
9	CuCN·2LiCl	−76 °C, 1 h	98	1:99
10	CuBr$_2$	−78 °C, 1 h	88	20:80
11	Li$_2$CuCl$_4$	−78 °C, 1 h	74	51:49
12	CuOTf	−75 °C, 1 h	62	81:19
13	CuSCN	−43 °C, 1 h	94	91:9
14	AgNO$_3$	−23 °C, 5 h	55	96:4
15	CeCl$_3$	−20 °C, 3 h	7	97:3

[a] The reaction was carried out using butylmagnesium chloride (2 equiv), (E)-2-decenyl 1-diphenylphosphate (**9**, 1 equiv) and metal catalyst (0.05–0.1 equiv) in THF.
[b] Isolated yield.
[c] Determined by GC analysis.
[d] acac = acetylacetonato.

observed using Fe [8] and Ni catalysts (entries 5, 7, and 8), while the regioselectivity of the reaction by copper catalysts changes from 1:99 to 91:9 depending on the ligands (entries 9–13). The highest S_N2'-regioselectivity ($S_N2/S_N2' = 1:99$) is seen for CuCN·2LiCl, which is effective for the stoichiometric S_N2'-selective cross-coupling reaction between Grignard reagents and allylic phosphates. Unless these metal catalysts are present, the reaction proceeds slowly even above 0 °C with moderate S_N2 selectivity (entry 1).

Table 11.3 shows the results of the cross-coupling reaction between various Grignard reagents and primary allylic diphenylphosphates using nickel, iron and copper catalysts. In the reaction using Ni(acac)$_2$ catalyst, the S_N2 coupling products **13a**, **13b**, **13d–13g** and **13i** are obtained selectively in moderate yields [7]. However, use of methyl and *iso*-propyl Grignard reagents results in low yields because of a competitive homocoupling reaction or reduction (entries 1 and 4). Catalyst ($^nBuC\equiv C)_2$ Ni is effective for obtaining a higher yield than Ni(acac)$_2$ in the reaction with geranyl diphenylphosphate (entry 16). The *E/Z* ratio of the S_N2 coupling product **13g**, derived from (*Z*)-2-decenyl 1-diphenylphosphate, is 81:19 because of rapid isomerization of the π-allylnickel intermediate (entry 13). These difficulties can be overcome by a Fe(acac)$_3$ catalyst [8]. For example, reaction of methyl Grignard reagent selectively affords the S_N2 coupling product **13a** in high yield without contamination by homocoupling products (entry 2). In the butylation of (*Z*)-2-decenyl 1-diphenylphosphate, the double bond geometry of the product **13g** is mostly retained (entry 14).

In contrast, all of the reactions using a CuCN·2LiCl catalyst proceed smoothly to afford the S_N2' coupling products **14a–14i** in high yields [7, 9]. Nearly exclusive S_N2' selectivities are obtained in the reaction of alkyl Grignard reagents with (*E*)-2-decenyl 1-diphenylphosphate (entries 3, 5 and 6). Even *tert*-butylmagnesium chloride shows a high S_N2' selectivity despite steric hindrance (entry 6). A higher reaction temperature (0 °C) and/or a less polar solvent (4:1 toluene–ether) are required to obtain high regioselectivities for *tert*-butyl, benzyl, vinyl and phenyl Grignard reagents (entries 6, 8, 10 and 12). (*Z*)-2-Decenyl 1-diphenylphosphate gives a result ($S_N2/S_N2' < 1:99$) similar to the corresponding *E*-isomer in the reaction with *n*-butyl Grignard reagent (entry 15). The presence of two alkyl substituents or a conjugated phenyl group at the C-3 position of the allylic diphenylphosphate has no effect on the reaction course (entries 17 and 19).

This method can be applied to regiocontrolled allyl–allyl coupling, which is important for selective 1,5-diene synthesis [10]. Some results of the copper-catalysed reaction between (*E*)-2-decenyl or (*E*)-2-hexenyl 1-diphenylphosphate and γ-substituted allyl Grignard reagents are listed in Table 11.4. These reactions have characteristics [7, 9]:

1 In the presence of the copper catalyst, allylic magnesium reagents react selectively at the less-substituted allylic terminus (α-position) with allylic phosphates. The α-S_N2' selectivity in the reaction of cinnamylmagnesium chloride is improved using a less Polar solvent (3:2 toluene–ether, entries 3 and 4).

2 Use of γ-disubstituted allyl Grignard reagents results in higher α- and S_N2' selectivities (entries 5 and 6). In particular, geranylmagnesium chloride affords the α-S_N2' coupling product **17e** almost exclusively (entry 6). In contrast, an alkyl substituent at the β-position reduces both α- and S_N2'-selectivities (entry 7).

3 No remarkable *E, Z*-stereoselectivities are observed for the α-S_N2' coupling products **17** other than cinnamylmagnesium chloride due to rapid isomerization of the γ-substituted allyl Grignard reagents.

Table 11.3. Transition-metal-catalysed cross-coupling reaction between alkyl Grignard reagents and primary allylic diphenylphosphates[a]

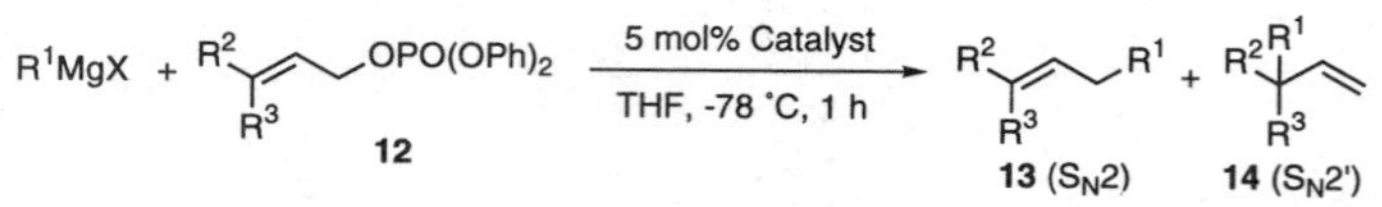

$$R^1MgX + \underset{\textbf{12}}{R^2 \overset{R^3}{\diagup}\!\!\diagdown\!\!/\text{OPO(OPh)}_2} \xrightarrow[\text{THF, }-78\,^{\circ}\text{C, 1 h}]{5\text{ mol\% Catalyst}} \underset{\textbf{13 (S}_N\text{2)}}{R^2\diagdown\!\!\diagup^{R^1}_{R^3}} + \underset{\textbf{14 (S}_N\text{2')}}{R^2\diagup^{R^1}_{R^3}\!\!\diagdown}$$

Entry	R¹MgX	R²	R³	Catalyst[b]	Products	Yield (%)[c]	S_N2/S_N2'[d]
1	MeMgI	$^nC_7H_{15}$	H	Ni	13a + 14a	26[e]	94:6
2				Fe		87	97:3
3				Cu		87	2:98
4	^iPrMgBr	$^nC_7H_{15}$	H	Ni	13b + 14b	53[f]	>99:1
5				Cu		83	2:98
6	^tBuMgCl	$^nC_7H_{15}$	H	Cu[g,h]	13c + 14c	80	2:98
7	$PhCH_2MgBr$	$^nC_7H_{15}$	H	Ni[i]	13d + 14d	75	>99:1
8				Cu[h]		98	8:92
9	$H_2C{=}CHMgBr$	$^nC_7H_{15}$	H	Ni	13e + 14e	78	>99:1
10				Cu[h]		81	33:67
11	PhMgBr	$^nC_7H_{15}$	H	Ni	13f + 14f	65	92:8
12				Cu[g,h]		99	12:88
13	^nBuMgCl	H	$^nC_7H_{15}$	Ni[i]	13g + 14g	81[j]	99:1
14				Fe		93[k]	98:2
15				Cu		91	<1:99
16	^nBuMgCl	$Me_2C{=}CH(CH_2)_2$	Me	$(^nBuC{\equiv}C)_2Ni$[i]	13h + 14h	38	99:1
17				Cu		97	3:97
18	^nBuMgCl	Ph	H	Ni	13i + 14i	84	>99:1
19				Cu[g,h]		79	6:94

[a] Unless specified otherwise, the reaction was carried out using Grignard reagent (2 equiv.), allylic diphenylphosphate (**12**, 1 equiv) and metal catalyst (0.05 equiv) in THF at −78 °C for 1 h.

[b] **Ni** = Ni(acac)₂; **Cu** = CuCN·2LiCl; **Fe** = Fe(acac)₃.

[c] Isolated yield.

[d] Determined by GC analysis.

[e] Homocoupling products of (E)-2-decenyl 1-diphenylphosphate were obtained in 64% combined yield.

[f] Reduced products of the allylic phosphate were formed in 28% combined yield.

[g] A 4:1 mixture of toluene and ether was used as solvent.

[h] The reaction was performed at 0 °C for 1 h.

[i] The reaction was performed at −45 °C for 2 h.

[j] The E/Z ratio of **13g** was 81:19.

[k] The E/Z ratio of **13g** was 6:94.

Table 11.4. Cross-coupling of (E)-2-decenyl or (E)-2-hexenyl 1-diphenylphosphate with γ-substituted allyl Grignard reagents[a]

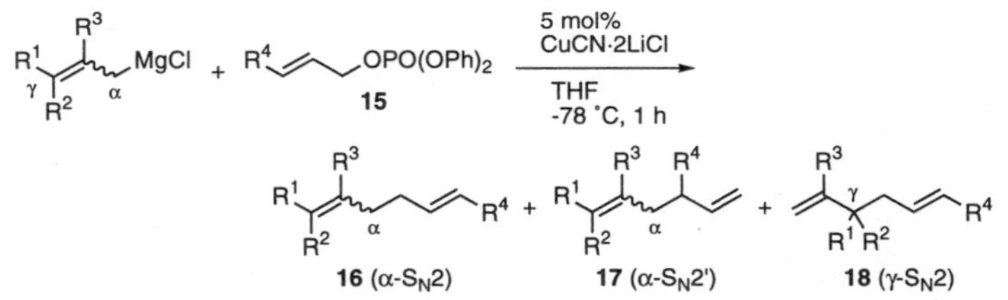

Entry	R^1	R^2	R^3	R^4	Products	Yield (%)[b]	α/γ[c]	S_N2/S_N2'[c]	16/17/18[c,d]
1	Me	H	H	nC_7H_{15}	**16a + 17a + 18a**	90[e]	79:21	27:73	6:73:21
2	nC_3H_7	H	H	nC_7H_{15}	**16b + 17b + 18b**	96[f]	76:24	29:71	5:71:24
3	Ph	H	H	nC_7H_{15}	**16c + 17c + 18c**	97[g]	30:70	76:24	6:24:70
4	Ph	H	H	nC_7H_{15}	**16c + 17c + 18c**	89[g,h]	86:14	19:81	5:81:14
5	Me	Me	H	nC_7H_{15}	**16d + 17d + 18d**	95	96:4	6:94	2:94:4
6	$Me_2C=CH(CH_2)_2$	Me	H	nC_3H_7	**16e + 17e + 18e**	90[i]	>99:1	2:98	2:98:0
7	C_2H_5	H	Me	nC_7H_{15}	**16f + 17f + 18f**	99[j]	66:34	40:60	6:60:34

[a] Unless specified otherwise, the reaction was carried out using allylic Grignard reagent (1.2 equiv), allylic diphenylphosphate (**15**, 1 equiv) and CuCN·2LiCl (0.05 equiv.) in THF at −78 °C for 1 h.

[b] Isolated yield.

[c] Determined by GC analysis.

[d] The corresponding γ-S_N2' product was not obtained in all experiments.

[e] The E/Z ratio of **17a** was 55:45.

[f] The E/Z ratio of **17b** was 63:37.

[g] The E/Z ratio of **17c** was >99:1.

[h] The reaction was performed in a 3:2 mixture of toluene and ether at 0 °C for 1 h.

[i] The E/Z ratio of **17e** was 73:27.

[j] The E-isomer of **17f** was produced as the major product.

3 Preparation of stereochemically homogeneous allylic metal compounds

3.1 *Direct insertion method using reactive Li, Mg and Ba*

Allylic alkali and alkaline-earth metal compounds are popular allylating reagents that exhibit high reactivity toward various functional groups of organic molecules [1]. However, these allylic organometallics readily isomerize between the *E*- and *Z*-isomers probably due to metallotropic 1,3-rearrangements [11]. The facility of this isomerization makes it difficult to control the regio- and *E/Z* stereochemistry of the subsequent reaction. If the stereo-randomization of an allylic metal is due to rapid isomerization through metallotropic rearrangements that are temperature dependent, a stereochemically pure allylic metal should be generated from the corresponding allylic halide by its reaction with reactive metal below the isomerization temperature (Scheme 11.3). The *in situ*-generated allylic metal reagent can be used for the subsequent stereoselective allylation.

Metal biphenylide is the most suitable reactive metal for generating an allylic alkali metal reagent at low temperature (Equation (11.4)) [12]. In contrast, reactive alkaline-earth metals can be prepared according to Rieke's method, in which the corresponding anhydrous metal halide is reduced by lithium naphthalenide or biphenylide at room temperature [13]. Allylmagnesium [14] and barium [15] reagents can be prepared readily by treatment of allyl chloride with these reactive metals (Equations (11.5) and (11.6)).

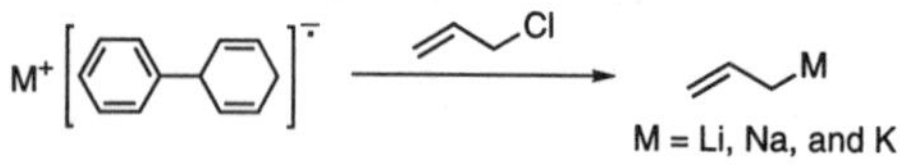

$$M^+ \left[\langle \bigcirc - \bigcirc \rangle \right]^{\bar{\cdot}} \xrightarrow{\qquad} \text{\hspace{1cm}} M \qquad (11.4)$$
$$M = \text{Li, Na, and K}$$

$$\text{MgCl}_2 + \text{Li}^+ \left[\bigcirc\bigcirc \right]^{\bar{\cdot}} \longrightarrow \text{Mg}^* \xrightarrow{\qquad} \text{\hspace{1cm}} \text{MgCl} \qquad (11.5)$$

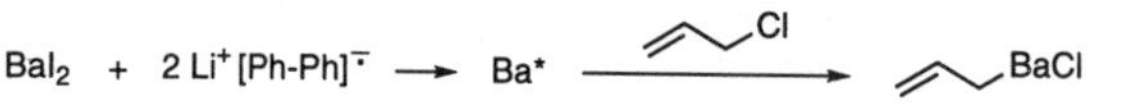

$$\text{BaI}_2 + 2\,\text{Li}^+ [\text{Ph-Ph}]^{\bar{\cdot}} \longrightarrow \text{Ba}^* \xrightarrow{\qquad} \text{\hspace{1cm}} \text{BaCl} \qquad (11.6)$$

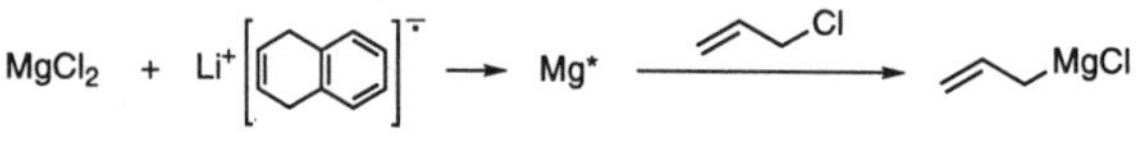

The following method is used to determine the isomerization temperature of allylic metals: a γ-substituted allylic chloride is treated with a reactive metal at $-95, -75, -60, -30$ and $0\,°C$. The mixture is then stirred for 30 min at each temperature and quenched with methanol to

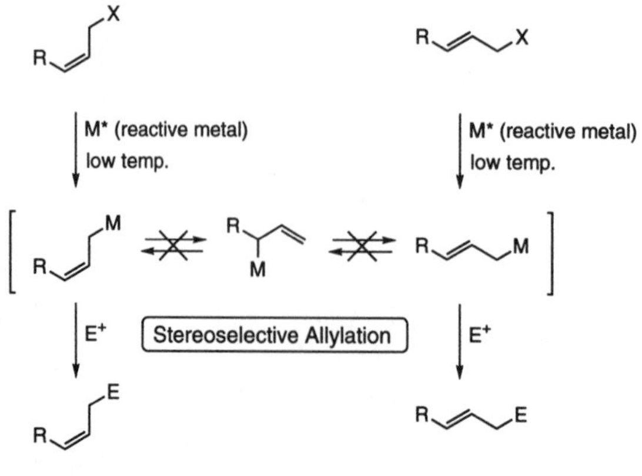

Scheme 11.3.

give a mixture of E- and Z-hydrocarbons. The rate of isomerization is calculated by measuring the E/Z ratio of this hydrocarbon mixture (Equation (11.7)).

$$
\underset{\substack{R^2}}{R^1}\diagdown Cl \xrightarrow[\substack{-95 \sim 0\,°C}]{\substack{1)\ M^*/THF \\ 2)\ MeOH}} \left[\underset{R^2}{R^1}\diagup\diagdown H \right] + \underset{R^2}{R^1}\diagup H
$$

M* = Li, Mg, and Ba   E/Z ratio

(11.7)

The results with geranyl and neryl metal compounds are shown in Fig. 11.1, the implications of which are apparent. There are two experimental variables (the temperature of the system and the choice of metal) and three consequences (the E/Z ratio of the olefin **20** produced, the yield (%) and the α/γ ratio of the protonation products **20** and **21**). Although there is no remarkable E/Z selectivity obtained by protonation of magnesium derivatives above −60 °C, very high stereoretention is observed below −95 °C [16]. In contrast, the double-bond geometry of allylic barium compounds is retained even at −50 °C, which is higher than the temperature for the corresponding lithium compounds [15b]. Thus, the barium reagents are superior for stereoselectivity. In addition, the combined yields of the derived olefins **20** and **21** are sufficiently high for practical purposes. In contrast to the γ-disubstituted allylmetals, significant enhancement of the rate of isomerization is observed for γ-monosubstituted allylmetals. For example, rapid stereoisomerization of (E)- and (Z)-2-decenylmagnesium compounds is found even at −100 °C [16].

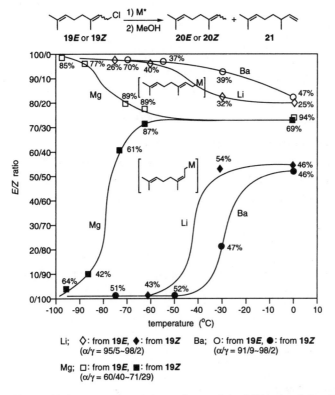

Figure 11.1. Temperature dependence of the E/Z ratio of the allylic metals (Li, Mg and Ba) derived from geranyl chloride (**19E**; E/Z > 99:1) and neryl chloride (**19Z**, E/Z < 1:99). Numbers refer to combined yields of the products **20** and **21**.

3.2   *Reactions of stereochemically homogeneous allylic metals*

The versatility of stereochemically homogeneous allylic metals in organic synthesis is note-worthy, as is their complementary relationship to other key functional groups. Stereoche-mically pure allylic silanes can be prepared from the corresponding magnesium or barium derivatives. Silylation of geranylmagnesium chloride (**22**), generated from activated Rieke-Mg and geranyl chloride (**19E**) at −95 °C, with chlorotrimethylsilane affords geranyl trimethylsilane (**23**) with an *E/Z* ratio of >99:1 (Equation (11.8)) [16]. The *cis*-isomer can also be obtained from neryl chloride (**19Z**) by a similar experimental procedure.

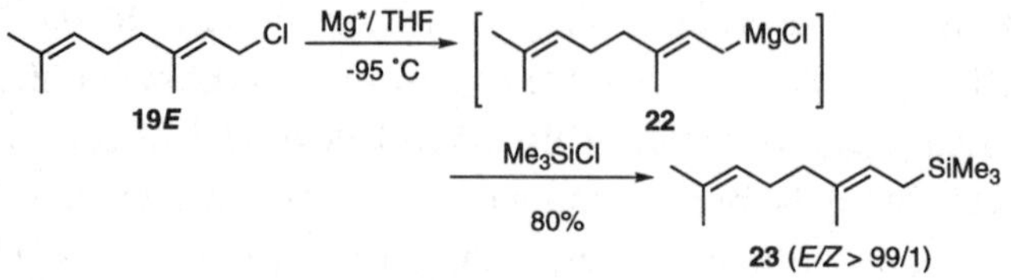

$$\tag{11.8}$$

The α/γ regioselectivity in the reaction of allylic organometallics with carbonyl com-pounds depends on the choice of metal. An allylic magnesium reagent predominantly gives a γ-product, whereas allylation with the lithium reagent is less selective [1a]. In marked contrast, however, the barium reagent furnishes remarkable α-selectivity and stereoselectivity [15]. For example, (*E*)-2-decenylbarium reagent **25** reacts with ben-zaldehyde at −78 °C with an α/γ ratio of 97:3 and retention of configuration of the starting chloride (**24**, *E/Z* > 99:1, Equation (11.9)). Highly α,α′-selective and stereocontrolled homocoupling and cross-coupling reactions of allylic halides can also be attained using barium reagents [17].

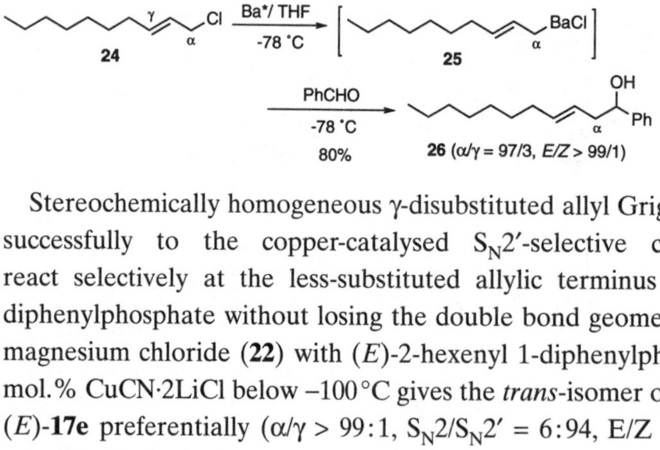

$$\tag{11.9}$$

Stereochemically homogeneous γ-disubstituted allyl Grignard reagents can be applied successfully to the copper-catalysed S_N2'-selective cross-coupling reaction and react selectively at the less-substituted allylic terminus (α-position) with an allylic diphenylphosphate without losing the double bond geometry [7]. Treatment of geranyl-magnesium chloride (**22**) with (*E*)-2-hexenyl 1-diphenylphosphate in the presence of 5 mol.% CuCN·2LiCl below −100 °C gives the *trans*-isomer of the α-S_N2' coupling product (*E*)-**17e** preferentially (α/γ > 99:1, S_N2/S_N2' = 6:94, *E/Z* = 95:5, 30% yield, Equation (11.10)). Similarly, the *cis*-isomer (*Z*)-**17e** is selectively obtained from neryl chloride (**19Z**, Equation 11.11). This is the first example of the generation of stereoretained allylic copper reagents.

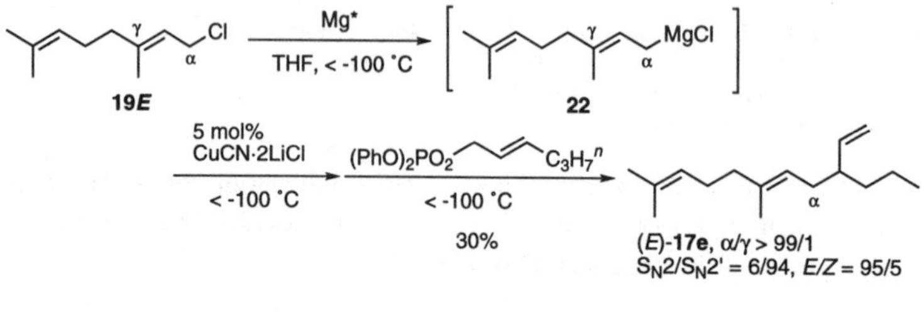

$$\tag{11.10}$$

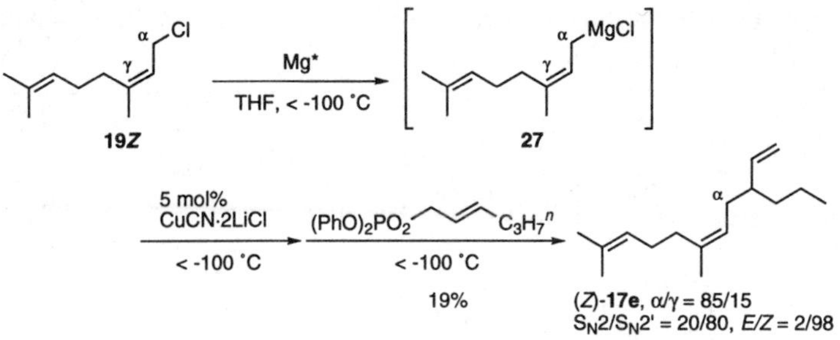

(11.11)

4 Enantioselective addition of allylic stannanes to aldehydes catalysed by BINAP·Ag(I) complex

Enantioselective addition of an allyl group to carbonyl compounds to provide optically active secondary homoallylic alcohols is a valuable synthetic method because the products are readily transformed into β-hydroxycarbonyl compounds and various other chiral compounds [1b,1c,3]. Although numerous important works on the reaction using a stoichiometric amount of chiral Lewis acids have been reported, there are few methods available for a catalytic process including a chiral (acyloxy)borane (CAB) complex [18] or a binaphthol-derived chiral titanium complex [19] as a catalyst. Recently, we found that a BINAP·silver(I) complex also catalyses the asymmetric allylation of aldehydes with allylic stannanes, and high γ-, anti- and enantioselectivities are obtained by this method (Equation (11.12)).

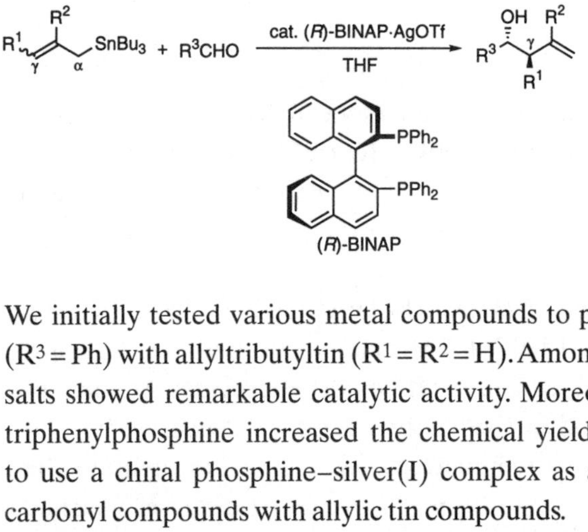

(11.12)

We initially tested various metal compounds to promote the allylation of benzaldehyde (R^3 = Ph) with allyltributyltin ($R^1 = R^2$ = H). Among the metal catalysts examined, silver(I) salts showed remarkable catalytic activity. Moreover, addition of a catalytic amount of triphenylphosphine increased the chemical yield. This rate enhancement prompted us to use a chiral phosphine–silver(I) complex as a catalyst for asymmetric allylation of carbonyl compounds with allylic tin compounds.

The chiral phosphine–silver(I) catalyst can be prepared simply by stirring an equimolar mixture of chiral phosphine and silver(I) compound in tetrahydrofuran (THF) at room temperature. Table 11.5 shows the results obtained by the reaction of benzaldehyde with allyltributyltin under the influence of 5 mol.% of various chiral phosphine–silver(I) complexes in THF at −20 °C; BINAP·silver(I) triflate complex gives the highest yield and enantiomeric excess (entry 4) [20].

Table 11.6 summarizes selected data of the allylation of a variety of aldehydes. Characteristic features are as follows: (1) all of the reactions furnish high yields and remarkable enantioselectivities, not only with aromatic aldehydes but also with α, β-unsaturated alde-

Table 11.5. Allylation reaction of benzaldehyde with allyltributyltin in the presence of various chiral phosphine·silver(I) complexes[a]

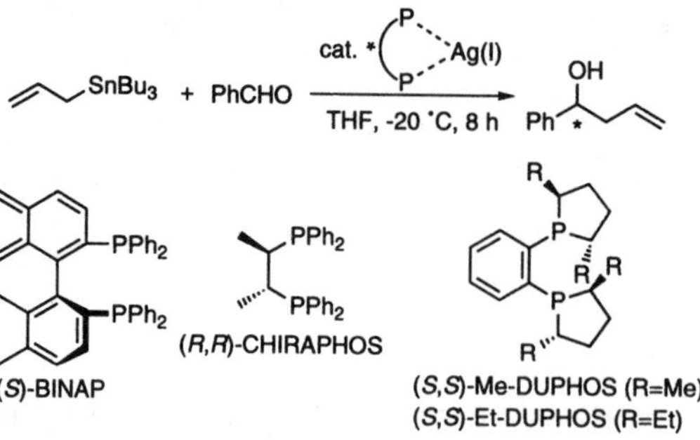

Entry	Chiral phosphine·Ag(I) complex	Yield (%)[b]	% ee[c] (config.)
1	(S)-BINAP·AgOCOCF$_3$	47	40 (S)
2	(S)-BINAP·AgClO$_4$	1	26 (S)
3	(S)-BINAP·AgNO$_3$	26	53 (S)
4	(S)-BINAP·AgOTf	88	96 (S)
5	(R,R)-CHIRAPHOS·AgOTf	97	2 (R)
6	(S,S)-Me-DUPHOS·AgOTf	4	48 (R)
7	(S,S)-Et-DUPHOS·AgOTf	13	3 (R)

[a] Unless specified otherwise, the reaction was carried out using chiral phosphine·Ag(I) complex (0.05 equiv.), allyltributyltin (1 equiv.) and benzaldehyde (1 equiv.) in THF at −20 °C for 8 h.
[b] Isolated yield.
[c] Determined by HPLC analysis (Chiralcel OD-H, Daicel Chemical Industries, Ltd.).

hydes (entries 5 and 6), with the exception of an aliphatic aldehyde, which gives a slightly lower chemical yield and enantioselectivity (entry 7); (2) in the reaction with α, β-unsaturated aldehydes, the 1,2-addition reaction takes place exclusively (entries 5 and 6); and (3) an electron-withdrawing substituent at the *para*-position of benzaldehyde increases the rate of allylation (cf. entries 1–3).

Enantioselective addition of methallyltributylstannane (**28**) to aldehydes can also be achieved using this method [20,21]. For example, the reaction of benzaldehyde with 5 mol.% (R)-BINAP·silver(I) triflate in THF at −20 °C provides the corresponding optically active homoallylic alcohol **29** in 75% yield with 92% ee (Equation (11.13)). In general, the reactivity of the methallyltin compound is lower than that of allyltributyltin. However, the use of an increased amount (up to 20 mol.%) of catalyst results in a satisfactory yield.

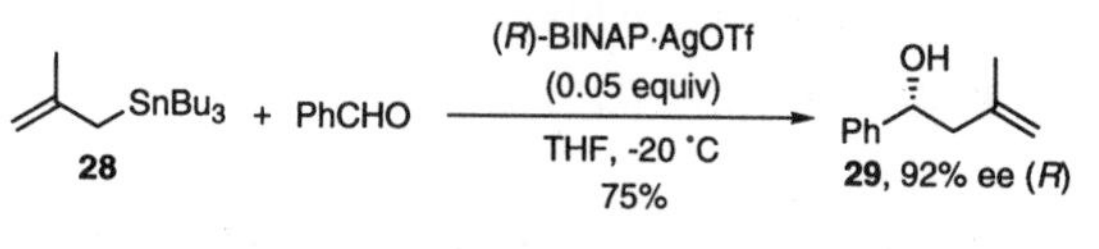

$$(11.13)$$

Condensation of γ-substituted allylmetals with aldehydes is a fascinating subject with respect to regioselectivity (α/γ) and stereoselectivity (*E/Z* or *anti/syn*). Addition of (*E*)-crotyltributyltin (**30E**, *E/Z* = 95:5) to benzaldehyde in the presence of 20 mol.% (R)-BINAP·AgOTf in THF at −20 °C and room temperature exclusively gives the γ-adducts **31** and **32** with an *anti/syn* ratio of 85:15 [21]. The *anti*-isomer **31** indicates 94% ee with a 1R,

Table 11.6. Asymmetric allylation reactions of aldehydes catalysed by BINAP·AgOTf complex[a]

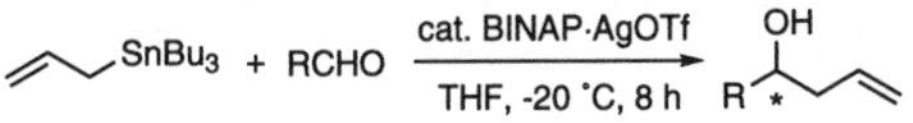

Entry	Aldehyde	Yield (%)[b]	% ee[c] (config.)
1	PhCHO	88	96 (S)
2	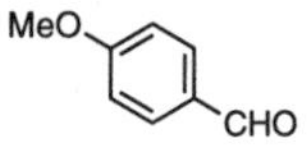 MeO–C₆H₄–CHO	59	97
3	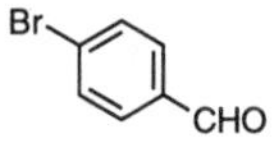 Br–C₆H₄–CHO	95	96
4[d]	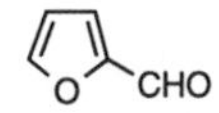 furyl–CHO	94	93
5[e]	(E)-n-C₃H₇CH=CHCHO	72	93[f]
6[g]	(E)-PhCH=CHCHO	83	88 (S)
7[e]	PhCH₂CH₂CHO	47	88

[a] Unless specified otherwise, the reaction was carried out using (S)-BINAP·AgOTf (0.05 equiv), allyltributyltin (1 equiv) and aldehyde (1 equiv) in THF at −20 °C for 8 h.
[b] Isolated yield.
[c] Determined by HPLC analysis (Chiralcel OD-H, AD or OJ, Daicel Chemical Industries, Ltd.).
[d] 4 equiv of allyltributyltin and 0.2 equiv of (S)-BINAP·AgOTf were used.
[e] The reaction was started using 2 equiv of allyltributyltin and 0.1 equiv of (R)-BINAP·AgOTf, and 0.1 equiv of the catalyst was added after 4 h.
[f] Determined by HPLC analysis (Chiralcel AD) of the benzoate ester of the product.
[g] 3 equiv of allyltributyltin and 0.15 equiv of (S)-BINAP·AgOTf were used.

2R configuration (Equation (11.14)). Use of (Z)-crotyltributyltin (**30Z**, E/Z = 2:98) or a nearly 1:1 mixture of the (E)- and (Z)-crotyltributyltin also results in a similar *anti/syn* ratio and enantioselectivity (Equation (11.14)).

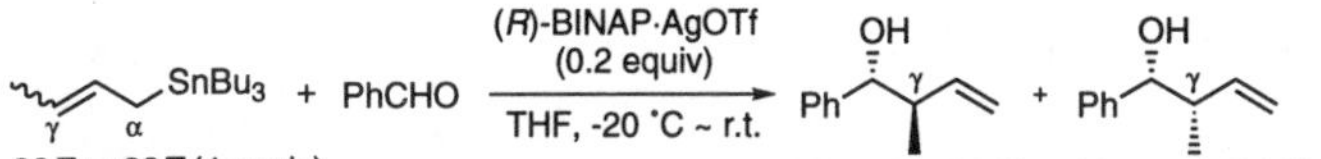

(11.14)

E/Z ratio of crotyltin	Yield, %	anti (% ee)/syn (% ee)
95/5	56	85 (94)/15 (64)
2/98	72	85 (91)/15 (50)
53/47	45	85 (94)/15 (57)

The reaction of aldehydes with 2,4-pentadienylstannanes is also catalysed by BINAP·silver(I) complex, and the corresponding γ-pentadienylated optically active alcohols are obtained with high enantioselectivity [22]. When benzaldehyde is reacted with 1 equiv of pentadienyltributyltin (**33**, E/Z = 97:3) and 0.1 equiv of (S)-BINAP·AgOTf at −20 °C for 8 h, the γ-product **34** is obtained in 61 % yield with 90 % ee (Equation (11.15)). Pentadienyltrimethyltin offers a chemical yield and enantioselectivity comparable to those of pentadienyltributyltin. Ketones are inert under the standard reaction conditions.

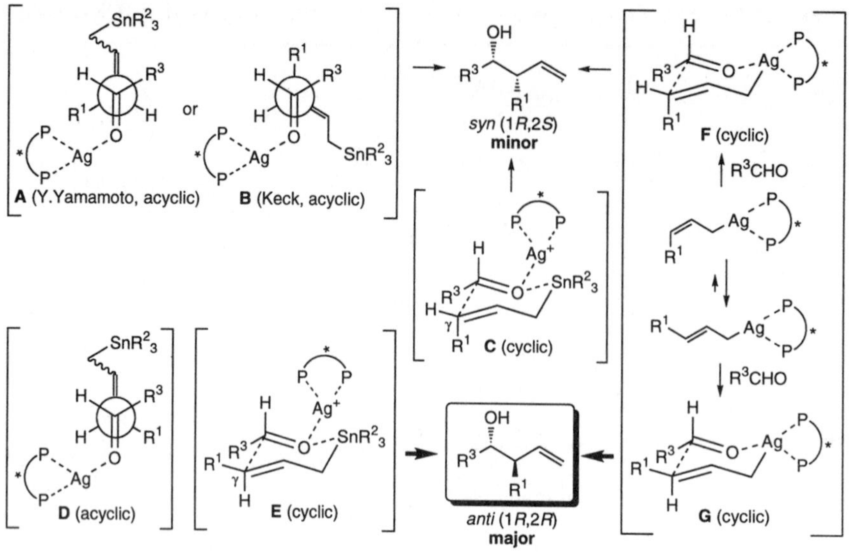

Scheme 11.4.

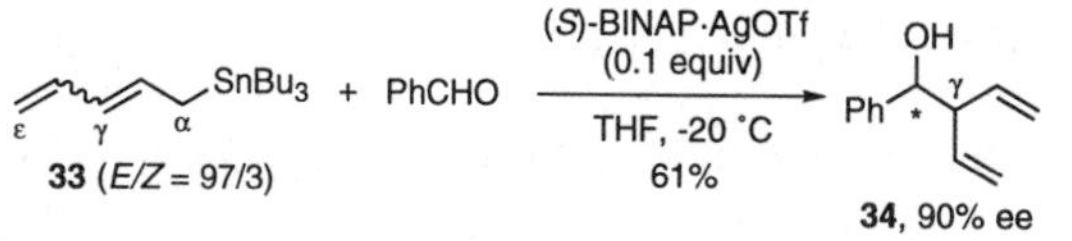

(11.15)

It is not clear why *anti*-selectivity is obtained for the crotyl addition reaction (Equation (11.14)) regardless of the double bond geometry of the tin compound **30**, although some transition state models are conceivable (Scheme 11.4). (*E*)- and (*Z*)-Crotyltributyltin (**30E** and **30Z**) have been reported to react with aldehydes in CH_2Cl_2 with γ- and *syn* (*erythro*)-selectivities in the presence of 2 equiv of $BF_3 \cdot OEt_2$ [23]. For the *syn*-selective reaction of crotylstannane, Yamamoto (Tohoku University) proposed an acyclic antiperiplanar transition state structure **A** [23]. Subsequently, Keck suggested a *syn*-synclinal alternative **B** to explain the higher *syn*-selectivity obtained with the *E*-stannane [24]. If the BINAP·Ag(I) complex acts as a Lewis acid in the *anti*-selective allylation, the reaction might proceed via an acyclic antiperiplanar **D**, which seems to have the least steric interaction between BINAP·Ag(I) and the stannyl methylene carbon and/or the R^1 group of the allylic stannane. A cyclic transition state **E** is also a possible model for a Lewis acid mechanism, because pentadienyltin compounds react selectively at the γ-carbon as shown in Equation (11.15). Nishigaichi and Takuwa proposed a similar cyclic model for $ZnCl_2$-promoted *anti*-selective γ-allylation of aldehydes with γ-substituted allylstannanes [25]. In contrast, a cyclic transition state model **G** containing a BINAP-coordinated silver atom instead of a trialkylstannyl group is a probable alternative leading to the *anti*-product when transmetallation to an allylic silver occurs and *E/Z* isomerization of the silver compound is sufficiently rapid. The corresponding *syn*-homoallylic alcohol should be obtained from the (*Z*)-allylic silver via a cyclic transition state model **F**.

5   Summary and conclusions

We have described here methods for the selective allylation of electrophiles with allylic metal compounds using transition metal catalysts. The main features of the present processes are: (1) S_N2- and S_N2'-selective cross-coupling reactions between Grignard reagents and allylic phosphates are achievable under the influence of Fe, Ni, or Cu catalyst; (2) stereochemically homogeneous allylic metals can be prepared easily by treatment of the corresponding allylic halides with reactive Rieke metals at low temperature and successfully applied to the copper-catalysed stereoretentive α-S_N2' cross-coupling or α-selective allylation of carbonyl compounds with allylic barium reagents; and (3) highly γ-, *anti*- and enantioselective addition of allylic stannanes to aldehydes occurs in the presence of BINAP·Ag(I) catalyst. These reactions represent a new class of allylations catalysed by transition metal compounds and can be performed on a substantial scale using ordinary laboratory equipment, because no complex preparation of catalysts is required. The remarkable regio- and stereoselectivities of the allylation reactions provide unprecedented routes to 1,5-dienes and homoallylic alcohols and are widely applicable to organic synthesis.

6 Acknowledgments

The authors wish to thank their talented colleagues whose names appear in the references. This work was supported in part by the Ministry of Education, Science, Sports and Culture of the Japanese Government.

7 References

1 Reviews: (a) Courtois G, Miginiac L. *J Organomet Chem* 1974; **69**: 1–44; (b) Roush WR. In: Trost BM, Fleming I, Heathcock CH, eds. *Comprehensive Organic Synthesis*, Vol. II. Oxford: Pergamon Press, 1991: 1–53; (c) Yamamoto Y, Asao N. *Chem Rev* 1993; **93**: 2207–93; (d) Hoppe D. In: Helmchen G, Hoffmann RW, Mulzer J, Schaumann E, eds. *Houben-Weyl: Methods of Organic Chemistry*, Vol. E 21b. Stuttgart: Georg Thieme Verlag, 1995: 1357–409.
2 Reviews: Yamamoto Y, Maruyama K. *Heterocycles* 1982; **18**: 357; Hoffmann RW. *Angew Chem Int Ed Engl* 1982; **21**: 555–66.
3 Bach T. *Angew Chem Int Ed Engl* 1994; **33**: 417; Hoveyda AH, Morken JP. *Angew Chem Int Ed Engl* 1996; **35**: 1262.
4 Reviews: Magid RM. *Tetrahedron* 1980; **36**: 1901; Carruthers W. In: Wilkinson G, Stone FGA, Abel EW, eds. *Comprehensive Organometallic Chemistry*, Vol. VII. Oxford: Pergamon Press, 1982: 685–722; Lipshutz BH. *Synlett* 1990: 119–28; Klunder JM, Posner GH. In: Trost BM, Fleming I, Pattenden G, eds. *Comprehensive Organic Synthesis*, Vol. III. Oxford: Pergamon Press, 1991: 220–23; Lipshutz BH, Sengupta S. *Org React* 1992; **41**: 135–631; Yamamoto Y. In: Helmchen G, Hoffmann RW, Mulzer J, Schaumann E, eds. *Houben-Weyl: Methods of Organic Chemistry*, Vol. E 21b. Stuttgart: Georg Thieme Verlag, 1995: 2011–40; Lipshutz BH. In: Abel EW, Stone FGA, Wilkinson G, eds. *Comprehensive Organometallic Chemistry II*, Vol. XII. Oxford: Pergamon Press, 1995: 59–130.
5 (a) Yanagisawa A, Noritake Y, Nomura N, Yamamoto H. *Synlett* 1991: 251; (b) Yanagisawa A, Nomura N, Noritake Y, Yamamoto H. *Synthesis* 1991: 1130.
6 Kharasch MS, Reinmuth O. *Grignard Reactions of Nonmetallic Substances*. Englewood Cliffs, NJ: Prentice-Hall, 1954: 122; Elsom LF, Hunt JD, Mckillop A. *Organomet Chem Rev (A)* 1972; **8**: 135; Felkin H, Swierczewski G. *Tetrahedron* 1975; **31**: 2735; Kochi JK. *Organometallic Mechanisms and Catalysis*. New York: Academic Press, 1978: 372.
7 Yanagisawa A, Nomura N, Yamamoto H. *Tetrahedron* 1994; **50**: 6017.
8 Yanagisawa A, Nomura N, Yamamoto H. *Synlett* 1991: 513.
9 Yanagisawa A, Nomura N, Yamamoto H. *Synlett* 1993: 689.
10 Billington DC. In: Trost BM, Fleming I, Pattenden G, eds. *Comprehensive Organic Synthesis*, Vol. III. Oxford: Pergamon Press, 1991: 413–34.

11 Wardell JL. In: Wilkinson G, Stone FGA, Abel EW, eds. *Comprehensive Organometallic Chemistry*, Vol. I. Oxford: Pergamon Press, 1982: 97–106; Hutchinson DA, Beck KR, Benkeser RA, Grutzner JB. *J Am Chem Soc* 1973; **95**: 7075.

12 Holy NL. *Chem Rev* 1974; **74**: 243; Cohen T, Bhupathy M. *Acc Chem Res* 1989; **22**: 152.

13 Rieke RD, Sell MS, Klein WR *et al.* In: Fürstner A, ed. *Active Metals. Preparation, Characterization, Applications*. Weinheim, Germany: VCH, 1996: 1–59; Rieke RD, Hanson MV. *Tetrahedron* 1997; **53**: 1925.

14 Burns TP, Rieke RD. *J Org Chem* 1987; **52**: 3674.

15 (a) Yanagisawa A, Habaue S, Yamamoto H. *J Am Chem Soc* 1991; **113**: 8955; (b) Yanagisawa A, Habaue S, Yasue K, Yamamoto H. *J Am Chem Soc* 1994; **116**: 6130.

16 Yanagisawa A, Habaue S, Yamamoto H. *J Am Chem Soc* 1991; **113**: 5893.

17 Yanagisawa A, Hibino H, Habaue S, Hisada Y, Yamamoto H. *J Org Chem* 1992; **57**: 6386; Corey EJ, Shieh WC. *Tetrahedron Lett* 1992; **33**: 6435; Corey EJ, Noe MC, Shieh WC. *Tetrahedron Lett* 1993; **34**: 5995; Yanagisawa A, Hibino H, Habaue S *et al. Bull Chem Soc Jpn* 1995; **68**: 1263.

18 Furuta K, Mouri M, Yamamoto H. *Synlett* 1991: 561; Ishihara K, Mouri M, Gao Q *et al. J Am Chem Soc* 1993; **115**: 11490.

19 Aoki S, Mikami K, Terada M, Nakai T. *Tetrahedron* 1993; **49**: 1783; Costa AL, Piazza MG, Tagliavini E, Trombini C, Umani-Ronchi A. *J Am Chem Soc* 1993; **115**: 7001; Keck GE, Tarbet KH, Geraci LS. *J Am Chem Soc* 1993; **115**: 8467.

20 Yanagisawa A, Nakashima H, Ishiba A, Yamamoto H. *J Am Chem Soc* 1996; **118**: 4723.

21 Yanagisawa A, Ishiba A, Nakashima H, Yamamoto H. *Synlett* 1997: 88.

22 Yanagisawa A, Nakatsuka Y, Nakashima H, Yamamoto H. *Synlett* 1997: 933.

23 Yamamoto Y, Yatagai H, Naruta Y, Maruyama K. *J Am Chem Soc* 1980; **102**: 7107; Yamamoto Y, Yatagai H, Ishihara Y, Maeda N, Maruyama K. *Tetrahedron* 1984; **40**: 2239.

24 Keck GE, Savin KA, Cressman ENK, Abbott DE. *J Org Chem* 1994; **59**: 7889.

25 Nishigaichi Y, Takuwa A. *Chem Lett* 1994: 1429.

12 Enantioselective Catalysis of Allylic Substitutions with Palladium Complexes of Phosphinooxazolines

GÜNTER HELMCHEN, HENNING STEINHAGEN and STEFFEN KUDIS

Organisch-Chemisches Institut, Universität Heidelberg, D-69120 Heidelberg, Germany

1 Introduction

Over the last approximately 30 years, organometallic chemistry has been an inexhaustible source of new methods of ever-increasing value to organic synthesis both in the laboratory and in large-scale industrial plants using transition metal catalysts. So far, catalytic processes are not well understood and their mechanistic investigation will certainly be one of the major research areas of 21st century chemistry. A particular challenge is the development of enantioselective syntheses under control of chiral transition metal catalysts for the production of organic intermediates. Successful developments have been achieved mainly for C–H (catalytic hydrogenation) and C–O bond formations (epoxidation and dihydroxylation of alkenes); the metals most useful in these areas are rhodium ruthenium and osmium manganese, respectively. In comparison, C–C bond-forming catalytic asymmetric syntheses are not sufficiently developed, although numerous exciting reports are available [1].

Palladium is perhaps the element most generally useful for the formation of C–C bonds in catalytic reactions [2] and is most often applied in cross-coupling reactions (Heck, Stille and Suzuki reactions) for the connection of sp² centers, e.g. diene synthesis. It was shown by Shibasaki, Overman and others that the Heck reaction with cyclic alkenes can be channeled towards chiral products with high enantioselectivity [3]. These are very recent, important achievements. The traditional battle fields of Pd-catalysed asymmetric synthesis are C–C and C–N bond-forming substitutions at allylic compounds via π-allylpalladium complexes:

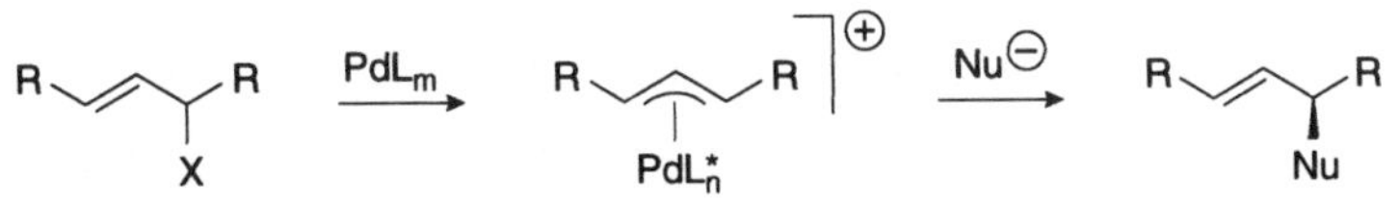

The first π-allylpalladium complex was reported in 1959 by Smidt and Hafner [4], famous for their invention of the Wacker process (1956), which was the starting point of modern Pd chemistry. Carbon–Carbon bond-forming substitution reactions with π-allylpalladium complexes were discovered in 1965 by Tsuji [5], then at Toray Industries, and have been forcefully developed by Trost and co-workers since 1973 [6]. Crucial improvements in this chemistry are due to industrial chemists at Toray Industries and Union Carbide, who demonstrated in 1970 that the precious metal can be employed in catalytic amounts and that phosphines accelerate the reaction [7].

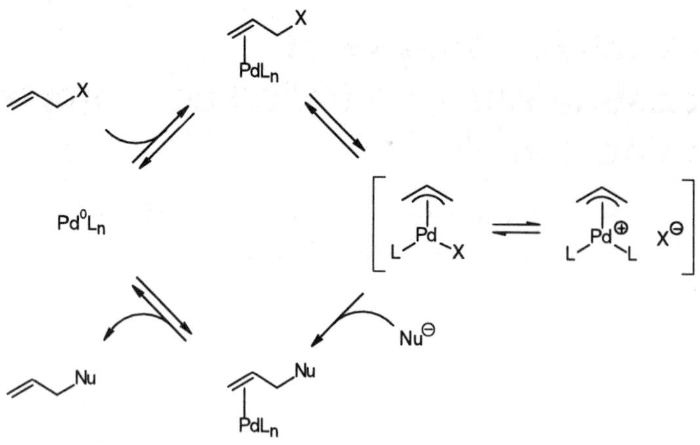

Scheme 12.1.

The first attempt to achieve enantioselectivity in a stoichiometric allylic substitution was reported by Trost and Dietsche in 1973 [8]. The enantioselectivity achieved then was low by today's standards, and applications in organic synthesis required the development of a catalytic reaction. There was some activity towards this goal but for a long time progress was slow. In the present decade dramatic improvements have been achieved [9] and we give here an account of our work and related recent contributions of others in this area.

The catalytic cycle (Scheme 12.1) of a Pd-catalysed substitution first involves co-ordination of a Pd(0) species to the double bond of an allylic system and then expulsion of the leaving group X to give a π-allylpalladium intermediate, which, depending on ligands L and counter ion X, can be a neutral or, presumed to be much more reactive, a cationic complex. A soft carbanion attacks the terminal carbon directly with inversion of configuration rather than via Pd with retention. The substitution is usually irreversible and turnover determining. The resulting Pd(0) olefin complex dissociates to yield the product and regenerate the catalyst.

It is helpful for the understanding of substitutions at π-allyl complexes to consider the fundamental types of substrates leading to chiral products given in Scheme 12.2. In reactions of type I, a new stereogenic center is created in the nucleophile. The starting materials are achiral. In reactions of type II, chiral racemic allylic derivatives are used as starting materials. The three subclasses are distinguished according to the symmetry and modes of isomerization of stereoisomeric π-allyl intermediates via the well-established π–σ–π mechanism [10].

In this article we will deal particularly with reactions of type IIa, which are more closely depicted in Scheme 12.3. Such a reaction begins with oxidative addition of a Pd(0) fragment to a chiral, *racemic* allylic substrate to yield the complex of a symmetric allylic cation. With an achiral auxiliary ligand the intermediary π-complex would be an achiral *meso* structure with enantiotopic electrophilic carbon atoms. Attack of a nucleophile would yield enantiomers in 1:1 ratio. In the presence of a chiral ligand L*, terminal carbon atoms of the allyl unit are diastereotopic and hence enantiomers must be produced in unequal amounts.

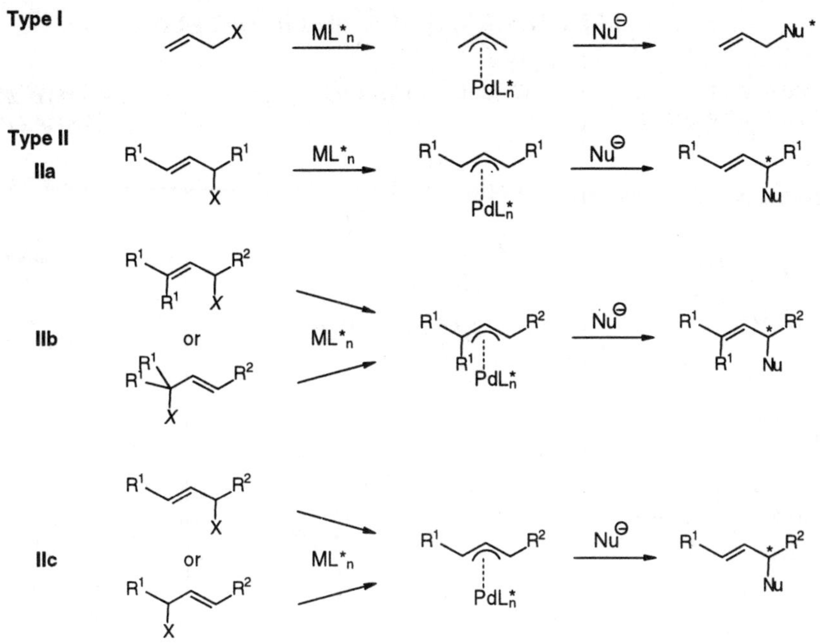

Scheme 12.2.

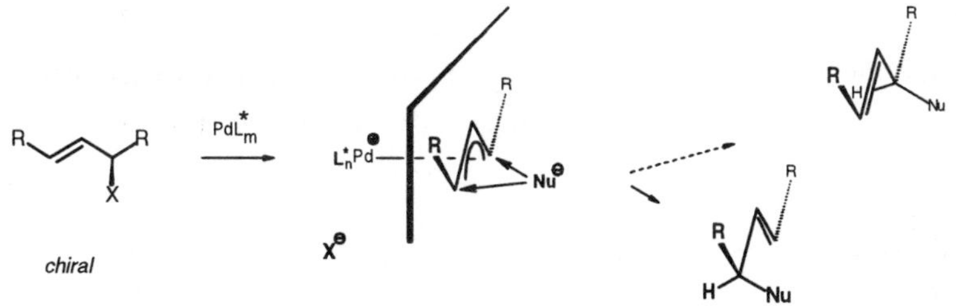

Scheme 12.3.

One quite obvious problem associated with this reaction is the long distance between the chiral information provided by L* and the reaction path of the nucleophile. It was therefore believed for some time that the (now) traditional C_2-symmetric chelate ligands are not suited for differentiation of these two carbon atoms; indeed, with diphosphines that gave excellent results in hydrogenations, i.e. CHIRAPHOS, BINAP, etc., results were not satisfactory, particularly with cyclic allylic substrates. However, over the last few years it was clearly demonstrated by Pfaltz with bisoxazolines [11] and Trost with diphosphines [12] that very high degrees of enantioselection are possible if a proper combination of substrate and ligand is chosen. These and various additional approaches for effectively transmitting chiral information are described in Scheme 12.4.

The combination of a C_2-symmetric and a mirror symmetric object leads to an asymmetric object, as is apparent for the bisoxazoline system in Scheme 12.4: with respect to the coordination plane, the substituents R and R' occur in a *cis* and in a *trans* relationship, respectively. Interaction of the *cis*-R/R' groups leads to distortion, in particular to weakening of the adjacent Pd—C bond, as was demonstrated by X-ray crystal structures [11c]. Note that this effect occurs on the wings of the allylic system.

Transmittance of Chiral Information

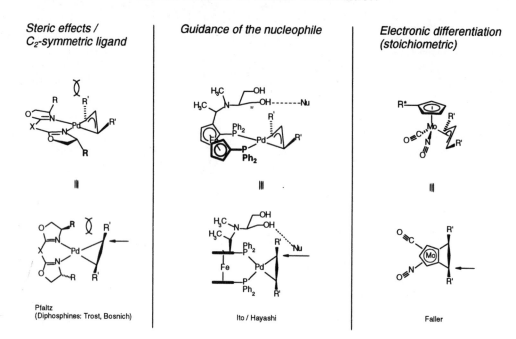

Steric effects / *C₂-symmetric ligand*	*Guidance of the nucleophile*	*Electronic differentiation* *(stoichiometric)*

Pfaltz
(Diphosphines: Trost, Bosnich)

Ito / Hayashi

Faller

Scheme 12.4.

In the modular Trost ligands [12] of C_2-diphosphines with a particularly large bite angle of *ca.* 110°, the allylic moiety sits in a chiral pocket defined by the phenyl rings at phosphorus.

X–Y = CO–O, CO–NH
O–CO, NH–CO

One may express the distinction between the Pfaltz bisoxazolines and the Trost ligands on the basis of the approved convex–concave concept [13]: the bisoxazolines provide for a convex and the Trost diphosphines for a concave environment at the coordination sites occupied by the π-allyl unit. This observation immediately explains why the range of applications of the bisoxazolines is very narrow.

Guidance of the nucleophile by hydrogen bonding is a concept successfully realized by the Ito–Hayashi team of Kyoto with their well-known phosphinoferrocenes that are particularly effective in allylic aminations [14].

Yet another concept is *electronic differentiation*, realized by Faller for a stoichiometric substitution at the molybdenum complex displayed in Scheme 12.4 [15]. Here we have a nitrosyl and a carbonyl ligand of almost identical size and yet the reaction occurs exclu-

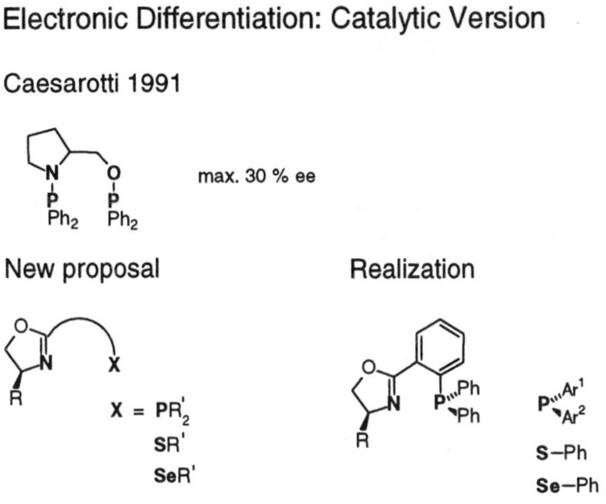

Electronic Differentiation: Catalytic Version

Caesarotti 1991

max. 30 % ee

New proposal   Realization

X = PR$_2'$
   SR'
   SeR'

S—Ph
Se—Ph

Scheme 12.5.

sively *cis* to the better π-acceptor NO. A catalytic version of electronic differentation was apparently first probed by Caesarotti with the ligand PRONOP, with two slightly (by bonding to O or N) differentiated P atoms [16]. A fairly low level of enantioselectivity was achieved. We felt that a more pronounced difference in electronic as well as steric properties was required and therefore chose combinations with a hard (N) and a soft (P, S or Se) donor (Scheme 12.5).

Realization of this proposal made use of the proven stability, variability and usefulness of the oxazoline moiety [17], which was first used in catalysis by Brunner [18]. Aryl groups were preferred as substituents at the P atom because triarylphosphines are normally stable to air. The same concept was independently pursued by the groups of Pfaltz [19] and Williams [20]. Using different P—N chelate ligands, the groups of Brown and Togni have also carried out allylic substitutions with interesting results concerning mechanistic aspects [21].

2   Preparation of oxazolines containing an additional soft donor center

The oxazoline moiety is available from amino alcohols, which in turn can be prepared from the chiral pool of natural amino acids. There are many convenient routes from amino alcohols to oxazolines, which Gant and Meyers have reviewed very recently in *Tetrahedron* [22]. Most people prefer one-step procedures as described in Scheme 12.6.

A convenient method is the one-pot condensation of the amino alcohol with a carboxylic acid under conditions (triphenylphosphine/CCl$_4$/base) developed by Vorbrüggen [23]. Yields are typically 40–50%. Higher yields are achieved with a three-step procedure involving formation of an *N*-acylamino alcohol, activation of the OH group by tosylation and ring closure with base [24]. The groups of Pfaltz [25] and Williams [26] prefer ZnCl$_2$-promoted condensation of the amino alcohol with a nitrile developed by Witte and Seeliger [27]. This method was improved recently by Pfaltz *et al.* [25].

Introduction of phosphorus is described in Scheme 12.7. The most convenient way is *nucleophilic* substitution of fluorine with a diarylphosphide, which proceeds with 70–90% yield. (This approach to obtain phosphinooxazolines was developed independently by our

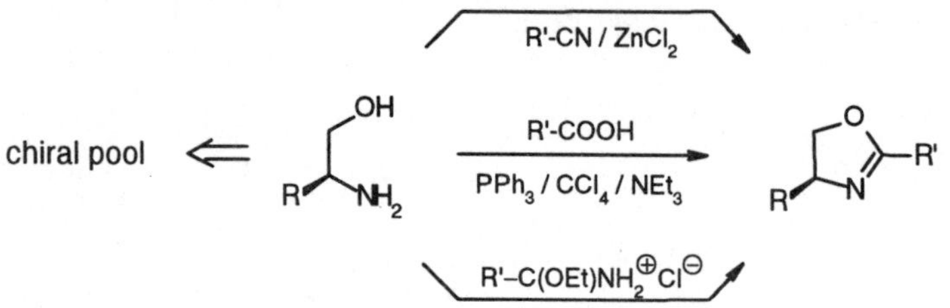

Scheme 12.6.

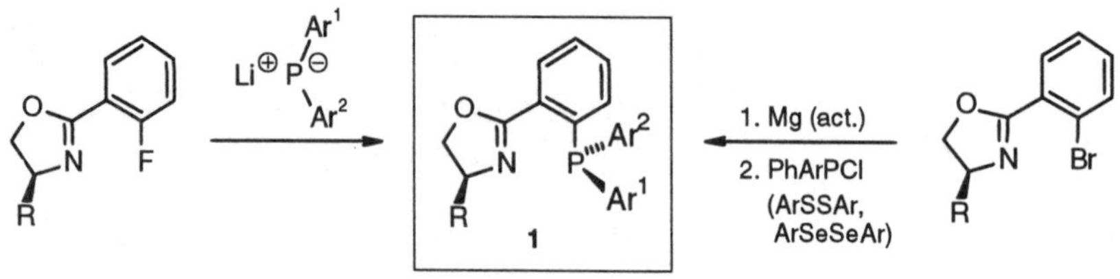

	R	Ar¹	Ar²
1a	*i*Pr	2-biphenylyl	phenyl
epi-1a	*i*Pr	phenyl	2-biphenylyl
1b	*i*Pr	2-biphenylyl	3,5-(CF$_3$)$_2$-C$_6$H$_3$
epi-1b	*i*Pr	3,5-(CF$_3$)$_2$-C$_6$H$_3$	2-biphenylyl
1c	phenyl	2-biphenylyl	phenyl
epi-1c	phenyl	phenyl	2-biphenylyl

Scheme 12.7.

group and Williams' group [20b].) In the case of stereogenic phosphorus with, for example, Ph and 1-naphthyl or 2-biphenylyl substituents, 7:3 mixtures of diastereomers are formed, which can be separated easily by flash chromatography or crystallization. *Electrophilic* phosphorus and also sulfur and selenium compounds can be reacted with the Grignard compounds obtained from the bromo derivatives and activated magnesium. Yields for the reactions of these Grignard compound with halophosphines are only *ca.* 50%; however, when compounds with stereogenic phosphorus are formed, diastereoselectivities of >85:15 can be achieved.

3   Allylic substitutions with acyclic substrates and mechanistic aspects

In the early phase of exploring ligands with two different donor atoms it was important to establish the best combination of donor atoms, i.e. to evaluate which of the soft donor atoms, P, S or Se, would be the best choice, given the nitrogen of the oxazoline, which is indispensable as a carrier of the chiral information. For this evaluation the standard test system in the area, the reaction of 1,3-diphenylallyl acetate with dimethyl malonate, was chosen (Table 12.1) and ligands derived from valine were employed. In addition to the P, S and Se derivatives, the monodentate phenyloxazoline was also tested. The reaction times and yields indicate that the combination of donor atoms N with P is by far the most effective one. The monodentate ligand is inactive. Typically, the most reactive system is also the most selective. It is also significant that with all ligands the product with (*S*)-configuration is formed preferentially.

Table 12.1.

Ligand	Time (h)	Enantioselectivity (% ee)[a]	Yield (%)
	1	98 (*S*)	98
	96	79 (*S*)	49
	72	95 (*S*)	50–84
	120	2 (*S*)	3

[a] ee = enantiomeric excess.

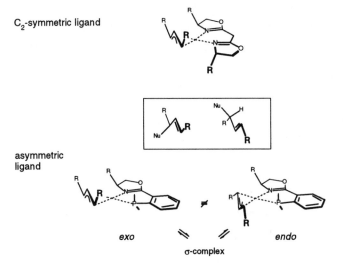

Scheme 12.8.

How can one rationalize the preferred steric course of the reactions: (*S*)-configuration? Finding an answer is much more difficult for asymmetric than for C_2-symmetric ligands. In the case of the π-allyl complex of the C_2-symmetric bisoxazoline (cf. Scheme 12.8) it has to be established which allylic terminus is attacked by the nucleophile. The experimental decision between the *two* possibilities is given by the configuration of the preferred product. As explained already in the introduction (cf. Scheme 12.4), preferential attack occurs at the terminus with R′ and R in *cis* disposition. This is plausible, because the adjacent C—Pd bond is the longer and therefore the weaker bond. Note that the description of the complex is simplified by describing the ligand by the C_2-conformation of the undisturbed ligand. However, as will be explained later, the ligand is highly distorted in the complex [28].

With an asymmetric ligand the problem is more difficult, because according to NMR studies there are two diastereomeric π-allyl complexes present, designated here as *exo* and *endo* isomers. Interconversion of these isomers proceeds via σ-allyl complexes, by π–σ–π interconversion [29], rather than simple rotation. The products can be formed via four pathways and the preferred product, again the enantiomer on the left-hand side of the box in Scheme 12.8, can arise by reaction at the C *trans* to the P atom of the *exo* or at the *cis* to the P atom of the *endo* isomer.

A postulate brought forward in 1985 by Bosnich [30] was initially applied to reach a decision between these possibilities: the reaction traverses an early transition state (Scheme 12.9). Bosnich argued that the energy content of a π-allyl complex is higher than that of the product and therefore, according to the Hammond postulate, the transition state should be similar to the π-allyl complex. Furthermore, in the case of a fast equilibrium between intermediates, the reaction mainly proceeds via the transition state of lower energy according to the Curtin–Hammett principle. It follows that the more abundant isomer is the more reactive. Which of these isomers is the more stable? We were able to solve this problem in 1994 [31]. Earlier, there were predictions without experimental foundation, even in print, that the more stable isomer would be the *endo* isomer. The experimental determination first required preparation of the π-allyl complexes.

A variety of π-allyl complexes of the phosphinooxazolines were prepared via established methods [32] (cf. Scheme 12.10). The standard method involves first reaction of an allylic alcohol with lithium tetrachloropalladate under acidic and reducing conditions to

Bosnich's Hypothesis on the Transition State of Allylic Substitutions

● The reaction traverses an early transition state because of the Hammond Postulate

● The more abundant diastereomer is the more reactive

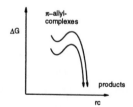

exo   endo

Hammond Postulate

Scheme 12.9.

Preparation of π-Allylpalladium Complexes

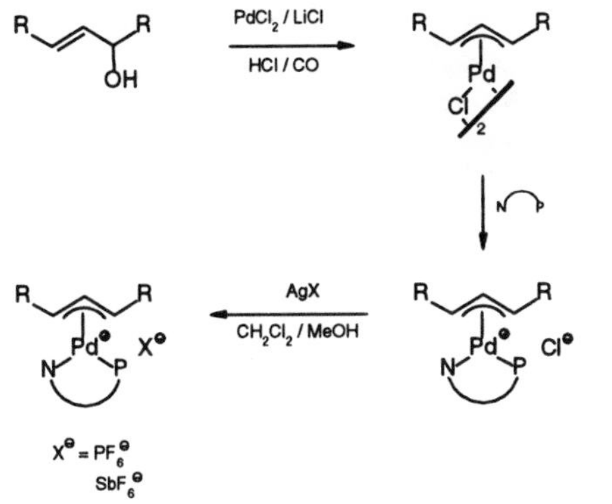

Scheme 12.10.

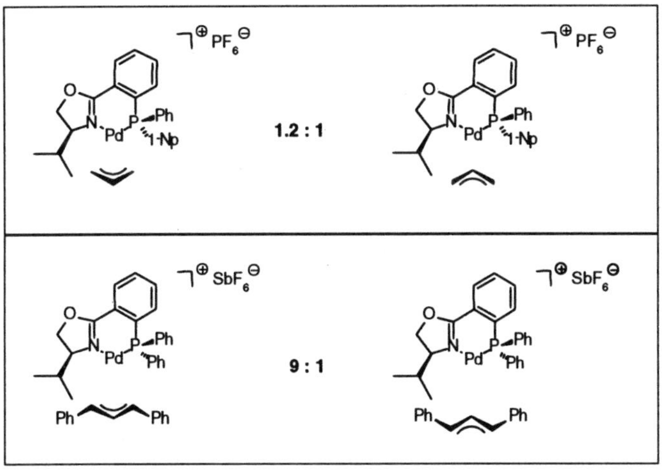

Scheme 12.11.

give the air-stable chloropalladium π-allyl complex, which is simply treated with the chiral ligand to give the chloro or chloride complex; this is usually transformed by treatment with a silver salt into the PF_6^- or SbF_6^- salt, which often yields crystals suitable for X-ray analysis.

The results of the analyses of the complexes were quite remarkable. The crystal of the complex of the unsubstituted allyl system contained both the *exo* and the *endo* complex, only distinguished by the location of C-2 of the allyl unit. Nuclear magnetic resonance revealed a *ca.* 1.2 : 1 isomer ratio for the complex in solution ($CDCl_3$) (Scheme 12.11). The 1,3-diphenylallyl complex in solution displayed a 9 : 1 ratio of isomers, but surprisingly it was the *exo* isomer that was found in the crystal and this was also the predominant isomer in solution, as could be determined easily by dissolving a crystal at −78°C, where no isomerization occurs, and warming up to yield the 9 : 1 equilibrium ratio of *exo* and *endo* isomers. In conjunction with Bosnich's argument of a lower activation energy for the more stable isomer, now known to be the *exo* isomer, we infer from the known configuration of the product of allylic substitution that the nucleophile preferentially attacks the carbon *trans* to the P atom (Scheme 12.12).

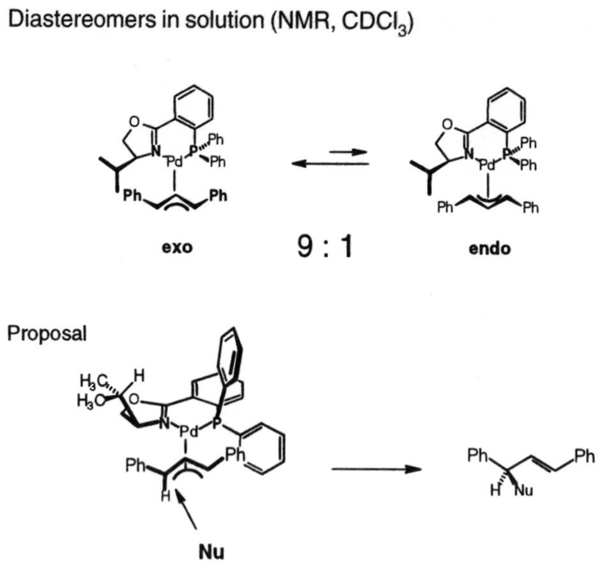

Diastereomers in solution (NMR, CDCl₃)

exo   **9 : 1**   **endo**

Proposal

Nu

Scheme 12.12.

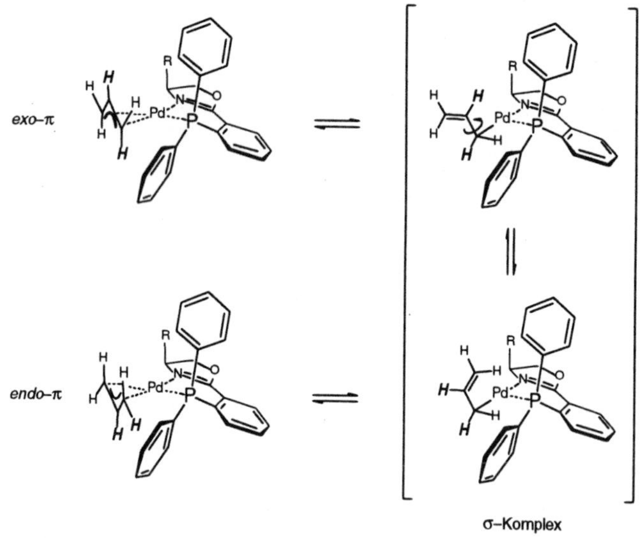

exo–π

endo–π

σ–Komplex

Scheme 12.13.

We have tried hard to verify this hypothesis by treating the isomer mixture in an NMR tube under non-equilibrating conditions with nucleophiles, but unfortunately with all the nucleophiles probed equilibration was faster than substitution.

However, our hypothesis is further strengthened by the observation that according to ¹³C-NMR the positive charge on the carbon *trans* to the P atom is higher than on the carbon *trans* to the N atom. Furthermore, as previously pointed out, interconversion of *exo* and *endo* diastereomers proceeds via the σ-complex. We have closely studied these interconversions by NMR [31]. For the case of the unsubstituted allylic system it was found that the Pd–C bond *trans* to P, not *trans* to N, is opened to give the σ-complex displayed in Scheme 12.13. From the σ-complex one can obtain the *endo*-π-complex by rotation either around the Pd–C or the C–C bond. Exchange experiments show that only C–C rotation occurs.

According to these experiments the Pd—C bond *trans* to the P atom is clearly the kinetically more labile bond. This is corroborated further by the crystal structures (Fig. 12.1, first row) of the allyl and 1,3-diphenylallyl complexes. The C—Pd bond *trans* to the P atom, 224 and 226 pm, is distinctly longer than the C—Pd bond trans to the N atom, 212 and 214 pm. No doubt the C—Pd bond *trans* to the P atom is the weaker bond.

The crystal structures also explained why for the 1,3-diphenylallyl complex the *exo* is preferred to the *endo* isomer. The front view in the second row of Figure 12.1 allows us to recognize the molecules; however, the side view in the third row with the coordination plane perpendicularly arranged to the screen and P hiding N is more informative with respect to several important general points:

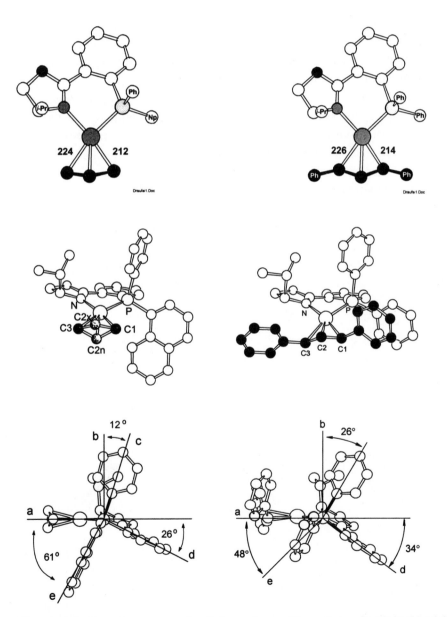

Figure 12.1. Crystal structures of π-allyl complexes with an allyl- and a 1,3-diphenylallyl group; Pd—C bond lengths are given in pm. (From [31].)

1 The conformation of the 'inner' chelate cycle PdNCCCP is bent, not planar as previously given schematically. The reason is very trivial, similar to the case of cyclohexane: a planar ring would have a bond angle at Pd near 120°; the required 90° is reached by bending.

2 A consequence of bending is conformational non-equivalence of the substituents at the P atom: one is in an axial and the other in an equatorial position. Characteristically, ring planes of the aryl groups are nearly perpendicular to each other, with the axial group pointing its edge and the equatorial its face to the metal.

3 A further important observation is that the substituent of the oxazoline ring, isopropyl in this case, occupies an axial position and only the equatorial H can interact with the allylic moiety. The dominating interaction is the one with the equatorial aryl group at the P atom. Minimization of this interaction is the reason for preference of the *exo* over the *endo* diastereomer.

The assumption of an early transition state according to Bosnich is not necessarily realistic. In fact, arguments in favor of a late transition state have been presented for allylic substitutions carried out with QUINAP as chiral ligand [21] and a late transition state is also favored by quantum chemical calculations [33]. On the other hand, the example in Scheme 12.14 illustrates (cf. Scheme 12.15 for the full catalytic cycle with all intermediates) that the reaction path of the substitution step fundamentally involves merely a rotation of the allylic moiety by 30°. Thus, a late transition state structurally should closely resemble the first formed olefin complex. With the hope of gaining detailed insight into the structure of this complex, we studied the reaction course by modern two-dimensional NMR spectroscopic methods [34].

Initially, the stoichiometric reaction between the π-allyl complexes **4** of the ligand **1** (R = *i*Pr, R′ = Ph; 10:1 mixture of **4x** and **4n**) and sodium dimethyl malonate as nucleophile was examined. In the experiment, a sample containing the π-allyl complexes **4** and the nucleophile was prepared at low temperature and warmed up to room temperature

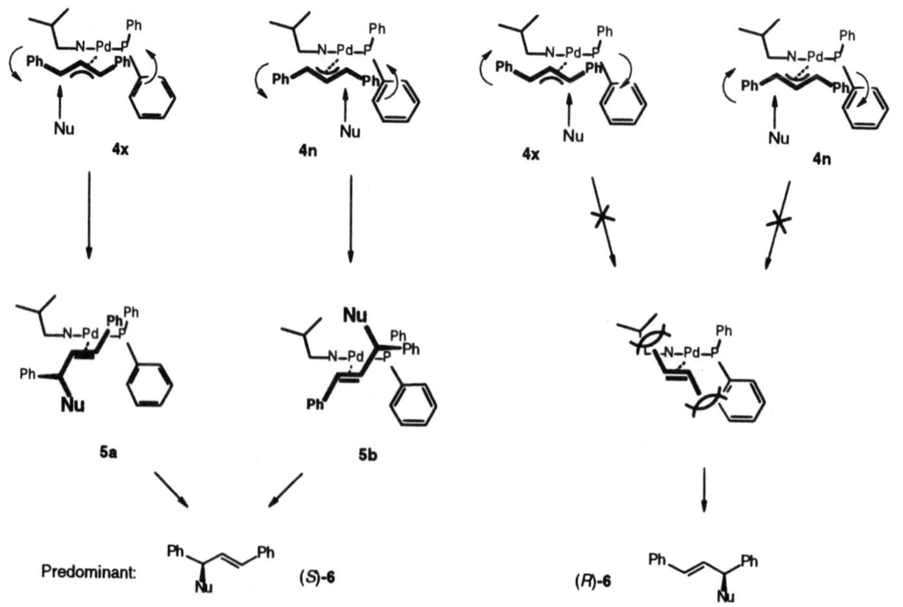

Scheme 12.14.

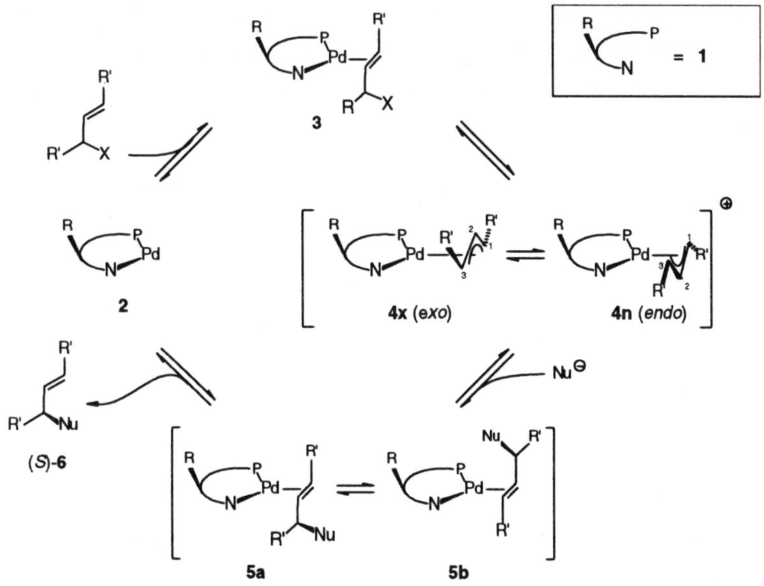

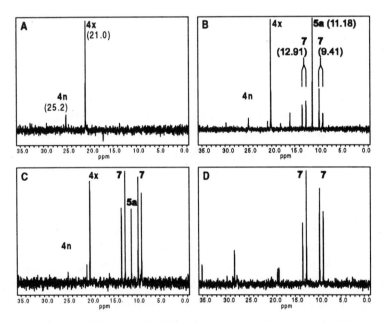

Scheme 12.15.

Figure 12.2. The $^{31}P\{^1H\}$ NMR spectra recorded at various reaction times ($-60\,°C$): (**A**) equilibrating π-allyl complexes **4x** and **4n**; (**B–D**) reaction mixtures 370 s, 500 s and 1 h after the addition of sodium dimethyl malonate.

inside the NMR probe head. The progress of the reaction was then monitored by ^{31}P-NMR spectroscopy (Fig. 12.2).

During the course of the reaction **4x** and **4n** are always in rapid equilibrium. As the first new species a compound with a singlet at $\delta = 11.18$ ppm appears, whose concentration reaches a maximum already after 90 s before it is consumed within a few minutes. This compound is the Pd(0) alkene complex **5a**. Consumption of complex **5a** is accompanied by the appearance of a new species with a characteristic AB spin system ($\delta_A^k = 9.41$, $\delta_B^k = 12.91$, $^2J_{PP} = 128\,\text{Hz}$), which is the main component after a reaction time of $c\,1\,\text{h}$. We assign structure **7** to this long-lived intermediate. The precise geometry of this complex cannot yet be determined.

Only traces of the metal-free product (*S*)-**6**, which is formed from complex **7**, were detected in the reaction mixture when the reaction was stopped at a conversion of $c\,50\%$. The concentration of product (*S*)-**6** only increased when conversion exceeded 50%. Therefore, complex **7** is a stable byproduct of the stoichiometric reaction. For determination of the constitution of the mechanistically meaningful transient species **5a**, the reaction was carried out with $^{13}C_3$-labeled $\text{Na(CHCOOCH}_3)_2$ at -20 to $-30\,^\circ\text{C}$ and stopped after $2\,\text{min}$ by cooling to $-78\,^\circ\text{C}$. A sample prepared in this way contained $c\,75\%$ of the phosphorus in the form of the Pd(0) alkene complex **5a** and was stable for several weeks at $-78\,^\circ\text{C}$ in an inert atmosphere.

Assignment of the resonances of all NMR-active nuclei was possible by the use of $^{13}C_3$-labeled malonate with a large set of two-dimensional NMR experiments (^{1}H,^{1}H-COSY, -TOCSY, -NOESY, -ROESY, ^{13}C,^{1}H-HSQC, -HMBC, -HMQC-TOCSY and ^{31}P,^{1}H-HMBC). By quantitative analysis of NOE and ROE data, information on distances of H nuclei could be obtained, which is only in accordance with conformer **5a** and not with conformer **5b** (Scheme 12.15).

Statements concerning the mechanism of the allylic substitution require the following two plausible assumptions: the attack of malonate at complex **4x** under the formation of complex **5a** proceeds via a 'least motion' reaction path, i.e. a rotation of 30° from conformer **4x** to **5a** is the main process; and rotation of the alkene fragment relative to the N–Pd–P plane in complex **5a** is sufficiently slow so that equilibration between conformers **5a** and **5b** is slow compared with the rate of their formation.

The latter assumption is supported by experiments in which the reaction was carried out in a temperature gradient ($-78\,^\circ\text{C}$ to room temperature) and monitored by ^{1}H-NMR spectroscopy. In this experiment broadening of the resonances of complex **5a**, which would be expected for a dynamic exchange process, was not found. Accepting this assumption, the configuration of the more reactive allyl complex **4x** is conserved in the Pd(0) alkene complex **5a**. Considering the known absolute configuration of the product, an attack of the nucleophile *trans* to phosphorus at the *exo* π-allyl complex **4x** can be derived. This result is in accordance with earlier interpretations [21, 31], but here is based on a precisely characterized Pd(0) olefin complex in the Pd-complex-catalysed allylic substitution.

4   Slim substrates: big problems and their solutions

The front view in Figure 12.1 shows quite clearly that the chiral ligand mainly provides interactions at its wings. It appears likely that allylic systems with big substituents, such as phenyl, should display high *exo/endo* ratios and enantioselectivity, but narrow systems with small substituents or cyclic compounds might give low selectivity. This is exactly what was found. In Scheme 12.6 substrates are ordered according to their 'broadness' and it is quite remarkable how closely the enantiomeric excess (ee) values parallel the steric extension (the isopropyl of case is taken from [19]). The importance of this parameter is underlined further by NMR data of the corresponding π-complexes: ratios of 1.8:1, 4:1 and 9:1 for the cyclohexenyl, the 1,3-dimethyl- and the 1,3-diphenylallyl derivative (CDCl$_3$ solution), respectively. The cause of enantioselectivity, though, is a kinetic phenomenon, i.e. a function of differing reaction rates at the allylic termini in *exo* and *endo* complexes. Recent results indicate a significant difference of relative reaction rates in acyclic and cyclic substrates with respect to *exo* and *endo* isomers.

The rather clear relationship between the size of the π-allyl moiety and enantioselectivity was very satisfactory, because it was in excellent agreement with our mechanistic assumptions. However, the production of a racemic product from the cyclic substrate (cf. Scheme 12.16) was somewhat unsatisfactory from a preparative point of view. Clearly, a ligand was required that would reach into the narrow area directly above or below the allylic sp^2 centers. As such ligands, the biphenylyl derivatives **1a–1c** (Scheme 12.3) were conceived [35]. We were able to obtain a high-resolution X-ray crystal structure of the complex [Pd(η^3-C$_6$H$_9$)(**1a**)]SbF$_6$ derived from ligand **1a** (R = *i*-Pr); indeed, in the crystal, conformer α of the cyclohexenyl π-allylpalladium complex is found, in which the phenyl of the 2-biphenylyl group is located directly above the allylic moiety as described in Scheme 12.17.

Despite the fact that conformer α is favored in the solid state, enantioselectivities

	% ee
cyclohexenyl–X	0
H$_3$C—CH=CH—C(CH$_3$)(X)	56
H$_3$C—CH$_2$—CH=CH—C(CH$_3$)(X)	74
H$_3$C—CH(CH$_3$)—CH=CH—C(CH$_3$)(X)	94
Ph—CH=CH—CH(Ph)(X)	98.5

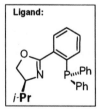

Ligand:

Scheme 12.16.

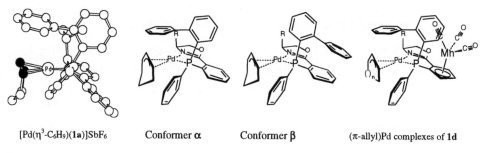

[Pd(η^3-C$_6$H$_9$)(**1a**)]SbF$_6$ Conformer α   Conformer β   (π-allyl)Pd complexes of **1d**

Scheme 12.17.

Table 12.2. Enantioselectivity of allylic substitution remerions with cyclic substrates

Ligand	Method[a]	Five-membered ring (% ee)[b]	Six-membered ring (% ee)	Seven-membered ring (% ee)
1a	(i)	56	51	83
1b	(i)	63	53	83
1c	(ii)	64	72	85
1d	(iii)	95	93	>99

[a] Method (i): 1.5 equiv LiCH(COOCH$_3$)$_2$, dioxane, room temperature; method (ii): 2.5 equiv CH$_2$(COOCH$_3$)$_2$, BSA method, methylene chloride, 0 °C; (iii): 1.5 equiv NaCH(COOCH$_3$)$_2$, dimethylformamide, −50 °C to 0 °C.
[b] ee = enantiomeric excess.

resulting with ligand **1a** (cf. Scheme 12.7) were not satisfactory. Distinct dependence on ring size of the substrate and, to a certain extent, on reaction conditions are apparent from the data given in Table 12.2. In order to enhance electronic effects, the ligand **1b** with electron-withdrawing CF$_3$ groups was prepared. With this ligand improved enantioselectivity was obtained with methylene chloride as solvent.

Nevertheless, results were still not satisfactory. A hint towards improvement was gained by an NMR analysis of the complex [Pd(η^3-C$_6$H$_9$)(**1a**)]SbF$_6$, which indicated the existence of several conformers in solution, including the unfavorable conformer β with the crucial phenyl group rotated away from the allylic moiety. In order to destabilize conformers of this type, the cymantrene-based ligand **1d** was conceived and could be prepared in a reasonably straightforward way [36]. This ligand induces excellent catalytic activity and displays long shelf-life. Conformers analogous to β are apparently destabilized by interaction with the manganese tricarbonyl group. High enantioselectivity with this new ligand was indeed obtained.

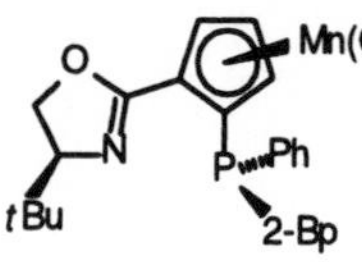

Prior to the development of the new ligand **1d**, high enantioselectivities with cyclic substrates were achieved with salts of the easily available β-phosphinocarboxylic acid (**8**) as chiral ligand [37]: enantiomeric excesses of 85, 98 and >99% for the five-, six- and seven-membered ring derivatives, respectively [with $LiCH(COO-t-Bu)_2$ as nucleophile]. Products with *S*-configuration are formed with complexes of compound **8**. So far we have not been able to obtain crystals or usable NMR spectra of a Pd complex of compound **8**. However, the corresponding *N,N*-dimethylamide, which is a ligand with properties very similar to compound **8**, allowed a clean complex to be prepared and its solution structure to be determined by NMR methods (H. Steinhagen, B. Wiese, Ch. Mürmann and G. Helmchen, unpubl. data.): the amide does not give rise to a chelate complex; rather, it acts as a monodentate phosphine to give complexes of the type $(\pi\text{-allyl})PdL_2$ with two molecules of the ligand.

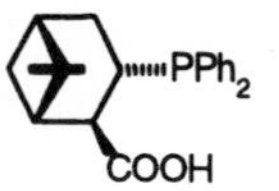

5   Scope and applications

How about other nucleophiles? We and others have investigated, in addition to malonates, a variety of different nucleophiles: amines and *n*-acylamides [38], nitro compounds [39] and *p*-toluenesulfinate [40]. As a rule, these nucleophiles are less reactive than malonates; however, enantioselectivities are very similar, which is in accord with mechanistic proposals discussed previously.

In 1979 one of us developed a method for enantiomer resolution of chiral amines that was based on the formation of diastereomeric amides by heating the racemic amine with enantiomerically pure 3-phenylbutyrolactone [41]. At that time there was no convenient way to prepare this compound. It was gratifying that the standard example of allylic alkylation opened the straightforward access described in Scheme 12.18.

Cyclic allylic derivatives, in particular the cyclopentanes as described in the previous section, are starting materials for many applications. However, we were initially faced with two problems preventing applications: first, enantiomerically pure compounds were needed; secondly, applications required a prohibitively large amount of precious material: for a 1 mol batch of *c* 150 g of product, an amount of 1 mol.% of catalyst means *c* 5 g of the chiral ligand. Our solution to these problems is shown in Scheme 12.19. The first problem could be solved very easily by using a more reactive starting material: the chloride instead of the acetate. Cyclopentenyl chloride is available by simply treating cyclopentadiene with

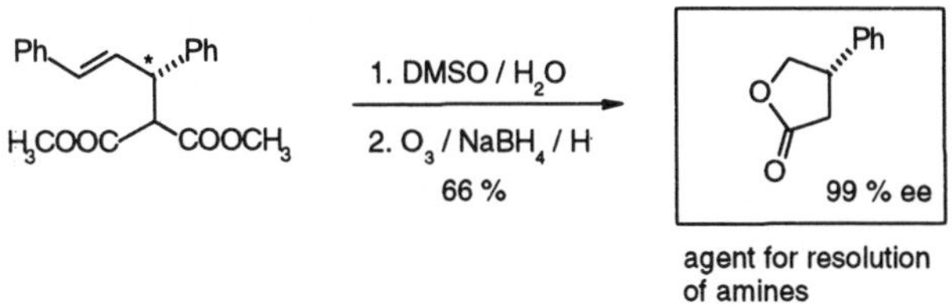

Scheme 12.18.

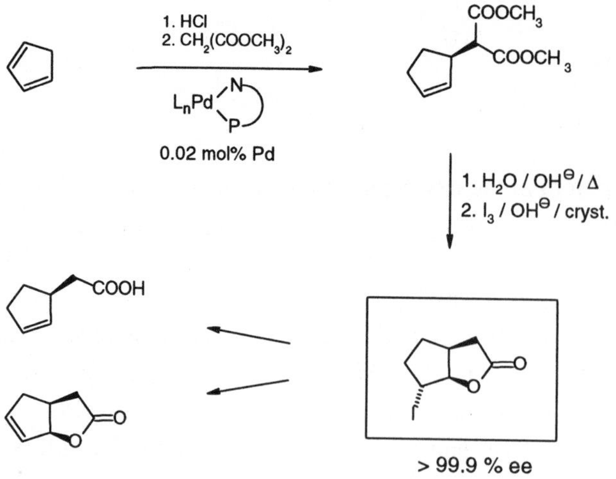

Scheme 12.19.

HCl gas. High reactivity of this compound allowed the amount of catalyst to be reduced from the customary 1–3 mol.% to a really satisfactory 0.02 mol.%, which corresponds to a turnover number of 5000 per *c* 3 h. Enantiomerically pure material could be obtained by saponification, decarboxylation and reaction with iodine to give the iodolactone, which is obtained enantiomerically pure with remarkable ease by recrystallization.

In our report [42] on the syntheses of these synthons we had to rely on ligand **1c**, which provided for a cyclopentenylmalonate yield of typically 60% ee and gave the enantiomerically pure iodolactone in *c* 30% yield. Fortunately, with the new ligand **1d**, under similar conditions (0.08 mol.% of catalyst), methyl cyclopentenylmalonate was formed with 95% ee and the resultant enantiomerically pure iodolactone was obtained in excellent 82% overall yield on a 100-g scale. The iodolactone can be transformed into a variety of useful compounds, e.g. cyclopentenyl acetic acid, which has been used for the synthesis of chaulmoogric acid [43], which in turn has been used in the treatment of leprosy.

6   Conclusions and outlook

The studies described above suggest that by logical development, on the basis of mechanistic arguments, highly effective chiral catalysts may be designed. However, one must always bear in mind that practicality and the possibility to vary easily the structure of the ligand (ligand tuning) are of prime importance in order to enable conjectures and concepts to be tested on a reasonable time scale. The phosphinooxazolines possess this property and the other desirable features listed in Scheme 12.20. Furthermore, they gave excellent results not only in allylic substitutions but also in the hydrosilylation [44] and transfer reduction [45] of ketones, the Heck reaction [46] the Diels–Alder reaction [47] and Aza-Claisen rearrangement [48].

One must, however, not believe that today a truly logical development on first principles is possible in catalysis. The element of change and discovery is still of enormous importance and probably will be so long into the 21st century. For example, for the π-allylpalladium complexes of the PHOX ligands there is an essential piece of knowledge certainly amiss: the relative rate of the reaction of *endo* and *exo* isomers with a nucleophile. Present-day

Requirements for a Good Ligand/Catalyst

☐ Accessibility

☐ Tunabiliy ⟨ electronic / steric

☐ Convenience of handling /
   Stability to air and moisture
   (Crystallinity)

☐ High reactivity / turnover

   Allylic substitutions (Pd)
   Heck reaction (Pd)
   Transfer reduction of ketones (Ru)
   Hydrosilylation of ketones (Rh)
   Diels-Alder reaction (Cu)

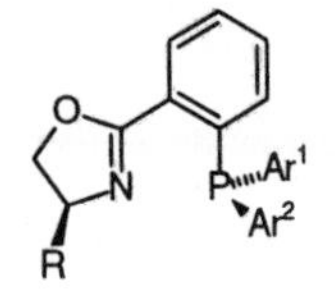

PHOX

Scheme 12.20.

methods, at least those available to the authors, do not allow this crucial parameter to be determined. Nevertheless, excellent results are being obtained, suggesting that the hypotheses given above are essentially correct.

7 Acknowledgments

This work was supported by the Deutsche Forschungsgemeinschaft (SFB 247) and the Fonds der Chemischen Industrie.

8 References

1 Noyori R. *Asymmetric Catalysis in Organic Synthesis.* New York: Wiley, 1994; Ojima I, ed. *Catalytic Asymmetric Synthesis.* Weinheim: VCH, 1993.

2 Tsuji J. *Palladium Reagents and Catalysts.* New York: Wiley, 1995; Heck RF. *Palladium Reagents in Organic Syntheses.* London: Academic Press, 1985; Malleron J-L, Fiaud J-C, Legros J-Y. *Handbook of Palladium-Catalysed Organic Reactions.* London: Academic Press, 1997.

3 de Meijere A, Meyer FE. *Angew Chem* 1994; **106**: 2473; *Angew Chem Int Ed Engl* 1994; **33**: 2379.

4 Smidt J, Hafner W. *Angew Chem* 1959; **71**: 284.

5 Tsuji J, Takahashi H, Morikawa M. *Tetrahedron Lett* 1965; **6**: 4387–8.

6 Trost BM, Fullerton TJ. *J Am Chem Soc* 1973; **95**: 292–4.

7 Hata G, Takahashi K, Miyake A. *Chem Commun* 1970: 1392–3; Atkins KE, Walker WE, Manyik RM. *Tetrahedron Lett* 1970; **11**: 3821–4.

8 Trost BM, Dietsche TJ. *J Am Chem Soc* 1973; **95**: 8200.

9 Frost CG, Howarth J, Williams JMJ. *Tetrahedron Asymm* 1992; **3**: 1089–122; Hayashi T. In: *Catalytic Asymmetric Synthesis.* Ojima I, ed. Weinheim: VCH, 1993: 325–65; Lübbers T, Metz P. In: Helmchen G, Hoffmann RW, Mulzer J, Schaumann E, eds. *Houben-Weyl E21, Stereoselective Synthesis.* 1995: 2371–3, 5643–76; Trost BM, Van Vranken DL. *Chem Rev* 1996; **96**: 395–422.

10 Consiglio G, Waymouth RM. *Chem Rev* 1989; **89**: 257.

11 (a) Müller D, Umbricht G, Weber B, Pfaltz A. *Helv Chim Acta* 1991; **74**: 722–40; (b) Leutenegger U, Umbricht G, Fahrni C, von Matt P, Pfaltz A. *Tetrahedron* 1992; **48**: 2143; (c) von Matt P, Lloyd-Jones GC, Minidis ABE *et al. Helv Chim Acta* 1995; **78**: 265–84.

12 Trost BM, Van Vranken DL, Bingel C. *J Am Chem Soc* 1992; **114**: 9327–43.

13 Helmchen G, Schmierer R. *Angew Chem Int Ed Engl* 1981; **20**: 205–6.

14 Hayashi T, Yamamoto A, Ito Y *et al. J Am Chem Soc* 1989; **111**: 6301.

15 Adams RD, Chodosh DF, Faller JW, Rosan AM. *J Am Chem Soc* 1979; **101**: 2570; Chao K-H, Faller JW. *J Am Chem Soc* 1983; **105**: 3893; Faller JW, Chao K-H, Murray HH. *Organometallics* 1984; **3**: 1231–40.

16 Caesarotti E, Demartin F, Grassi M, Prati L. *J Chem Soc Dalton Trans* 1991: 2073–82.

17 Sprinz J, Helmchen G. *Tetrahedron Lett* 1993; **34**: 1769–72.

18 Brunner H, Obermann U. *Chem Ber* 1989; **122**: 499.

19 von Matt P, Pfaltz A. *Angew Chem Int Ed Engl* 1993; **32**: 566–7.

20 (a) Dawson GJ, Frost CG, Williams JMJ, Coote SJ. *Tetrahedron Lett* 1993; **34**: 3149–50; (b) Coote SJ, Dawson GJ, Frost CG, Williams JMJ. *Synlett* 1993: 509–10.

21 Brown JM, Hulmes DI, Guiry PI. *Tetrahedron* 1994; **50**: 4493; Togni A, Burckhardt U, Gramlich V, Pregosin PS, Salzmann R. *J Am Chem Soc* 1996; **118**: 1031.

22 Gant TG, Meyers AI. *Tetrahedron* 1994; **50**: 2297–360.

23 Vorbrüggen H, Krolikiewics K. *Tetrahedron* 1993; **49**: 9353–72, and earlier work cited therein.

24 Peer M, de Jong JC, Kiefer M *et al. Tetrahedron* 1996; **52**: 7547–83.

25 Koch G, Lloyd-Jones GC, Loiseleur O *et al. Recl Trav Chim Pays-Bas* 1995; **114**: 206–10.

26 Allen JV, Coote SJ, Dawson GJ *et al. J Chem Soc Perkin Trans 1* 1994: 2065, and literature cited therein.

27 Witte H, Seeliger W. *Liebigs Ann Chem* 1974: 996–1009.

28 Pfaltz A. *Acc Chem Res* 1993; **26**: 339–45.

29 Faller JW, Thomsen ME, Mattia MJ. *J Am Chem Soc* 1971; **93**: 2642.

30 Bosnich B, Mackenzie PB. *Pure Appl Chem* 1982; **54**: 189.

31 Sprinz J, Kiefer M, Helmchen G *et al. Tetrahedron Lett* 1994; **35**: 1523–6.

32 Auburn PR, Bosnich B, Mackenzie PB. *J Am Chem Soc* 1985; **107**: 2033–46; Trost BM, Strege PE, Weber L, Fullerton TF, Dietsche TJ. *J Am Chem Soc* 1973; **95**: 8200–1.

33 Plöchl PE, Togni A. *Organometallics* 1996; **15**: 4125–32.

34 Steinhagen H, Reggelin M, Helmchen G. *Angew Chem* 1997; **109**: 2199–202; *Angew Chem Int Ed Engl* 1997; **36**: 2108–10.

35 Sennhenn P, Gabler B, Helmchen G. *Tetrahedron Lett* 1994; **35**: 8595–8.

36 Corresponding ligands with a ferrocene or a benzenechromium tricarbonyl moiety were also prepared and tested in catalytic allylic alkylations. A report by S. Kudis and G. Helmchen on preparations and catalysis data will soon be published.

37 Knühl G, Sennhenn P, Helmchen G. *J Chem Soc Chem Commun* 1995: 1845–6.

38 von Matt P, Loiseleur O, Koch G *et al. Tetrahedron Asymm* 1994; **5**: 573–84.

39 Rieck H, Helmchen G. *Angew Chem Int Ed Engl* 1995; **34**: 2687–9.

40 Eichelmann H, Gais H-J. *Tetrahedron Asymm* 1995; **6**: 643–6.

41 Helmchen G, Nill G. *Angew Chem* 1979; **91**: 66; *Angew Chem Int Ed Engl* 1979; **18**: 65.

42 Sennhenn P, Gabler B, Helmchen G. *Tetrahedron Lett* 1994; **37**: 8595.

43 Mislow K, Steinberg IV. *J Am Chem Soc* 1955; **57**: 3807.

44 Langer Th, Janssen J, Helmchen G. *Tetrahedron Asymm* 1997; **7**: 1599–602; Newman LM, Williams JMJ, McCague R, Potter GA. *Tetrahedron Asymm* 1997; **7**: 1597–8.

45 Langer Th, Helmchen G. *Tetrahedron Lett* 1996; **37**: 1381–4.

46 Loiseleur O, Meier P, Pfaltz A. *Angew Chem* 1996; **108**: 218–20; *Angew Chem Int Ed Engl* 1996; **35**: 200–2.

47 Sagasser I, Helmchen G. *Tetrahedron Lett* 1998; **39**: 261–4.

48 Vozumi Y, Kato K, Hayashi T. *Tetrahedron Asymm* 1998; **9**: 1065–72.

13 New Approaches to the Synthesis of Heterocycles by Transition Metal Catalysis

BASSAM EL ALI* and HOWARD ALPER†

*Chemistry Department, King Fahd University of Petroleum & Minerals, Dhahran 31261, Saudi Arabia, † Department of Chemistry, University of Ottawa, 10 Marie Curie, Ottawa, Ontario K1N 6N5, Canada

1 Introduction

Homogeneous catalysis provides new approaches for the use of transition metal complexes in organic synthesis [1–4]. An area where significant progress has been made is the synthesis of heterocycles using catalysis [5, 6]. Carbonylation reactions catalysed by transition metal complexes have been used extensively for diverse applications [1–6]. The synthesis of lactones and lactams, either by the cyclocarbonylation of unsaturated alcohols, amines and other suitable substrates [7, 8], or by the ring expansion–carbonylation of heterocycles, are two simple and efficient carbonylation processes. The cycloaddition of three- and four-membered ring heterocycles with heterocumulenes has been of considerable interest because of the potential biological activity of the products [11–21]. Palladium(0) complexes are effective catalysts for this class of reactions [12].

In this review, we will discuss new approaches to the synthesis of heterocycles by transition-metal-catalysed cyclocarbonylation reactions by the carbonylative ring expansion of heterocyclic compounds and by the cycloaddition of heterocycles with heterocumulenes.

2 Catalytic synthesis of lactones and lactams via carbonylation reactions

2.1 Four-membered rings

Three-membered ring heterocycles undergo carbonylation and ring expansion in the presence of different metal catalysts. When styrene oxide was treated with a catalytic amount of $RhCl(CO)(PPh_3)_2$ under CO, the β-lactone **1** was obtained in up to 67% yield (Equation (13.1)) [18]. All aliphatic epoxides examined in this study afforded β-lactones in poor yields. The authors attempted to explain the reactivity of styrene oxide by invoking a cationic intermediate generated by coordination of the epoxide to rhodium in a Lewis acid-type manner [18].

$$\text{Ph} \underset{O}{\triangle} + CO \xrightarrow{\text{RhCl(CO)(PPh}_3)_2} \quad \mathbf{1} \tag{13.1}$$

Drent and co-workers prepared the β-lactone **2** using $Co_2(CO)_8$ in the presence of a hydroxy-substituted pyridine ligand. Under these conditions, propylene oxide was transformed to the corresponding β-lactone **2** in 93% conversion and 90% selectivity (Equation (13.2)) [19].

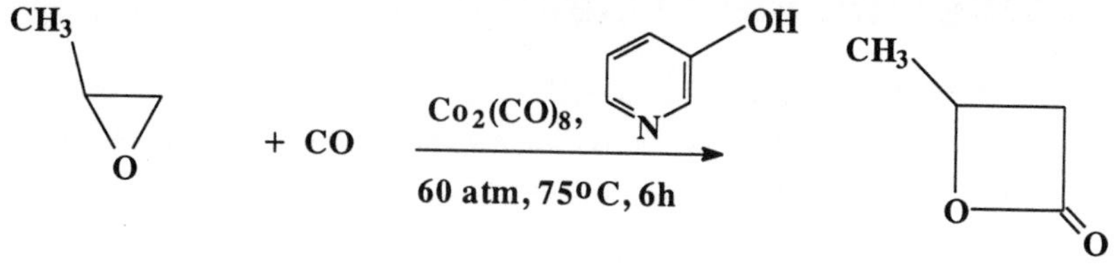

$$\text{(13.2)}$$

The transition-metal-catalysed ring expansion of aziridines is a useful method for the synthesis of β-lactams. Depending on the nature of the substituents on the aziridine ring, carbon monoxide insertion usually proceeds in a regiospecific manner into one of the two aziridine C—N bonds (Equations (13.3) and (13.4)).

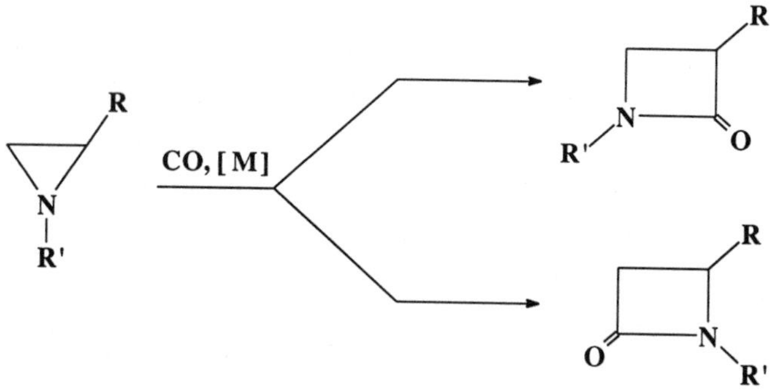

$$\text{(13.3)}$$

$$\text{(13.4)}$$

The complex $[\text{Rh(CO)}_2\text{Cl}]_2$ catalyses the carbonylation of 2-arylaziridines, affording the β-lactams **3** in good to quantitative yields (Equation (13.5)) [22]. Because the carbonylation is regiospecific at the aryl-substituted C—N bond, it is conceivable that the aryl group directs the metal insertion via temporary coordination of the arene to rhodium.

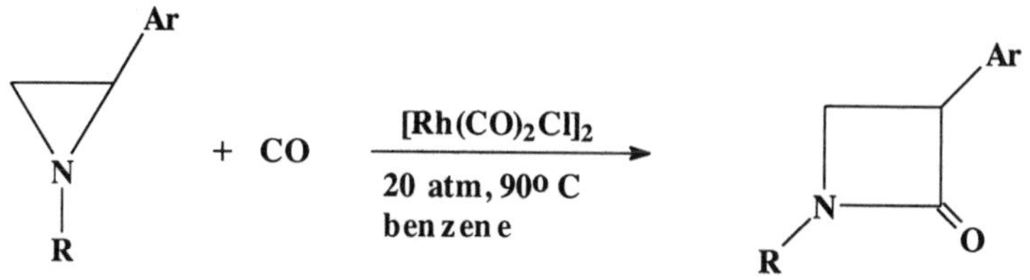

$$\text{(13.5)}$$

R = t-Bu, 1-adamantyl.
Ar = Ph, p-PhC$_6$H$_4$, p-BrC$_6$H$_4$.

The carbonylation of *cis*-2,3 disubstituted aziridines gave β-lactams **4** with retention of configuration (Equation (13.6)) [23].

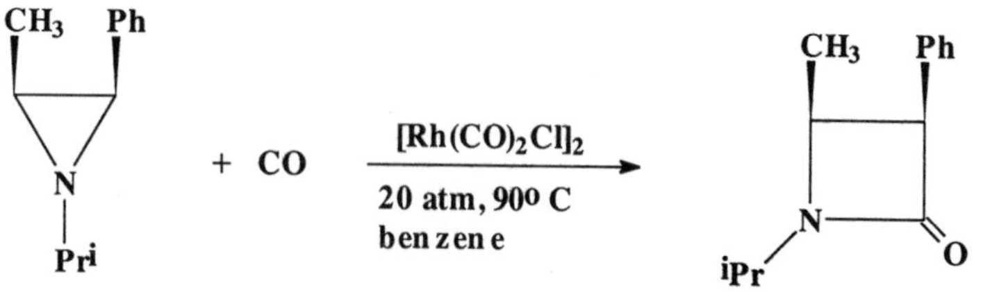

$$\text{(13.6)}$$

The asymmetric carbonylative ring expansion of aziridines in the presence of chiral ligands is a useful process not only for the synthesis of chiral β-lactams but also for the preparation of enantiomerically enriched aziridines. The most effective ligands in this reaction were *d*- and *l*-menthol (Equations (13.7) and (13.8)) [23].

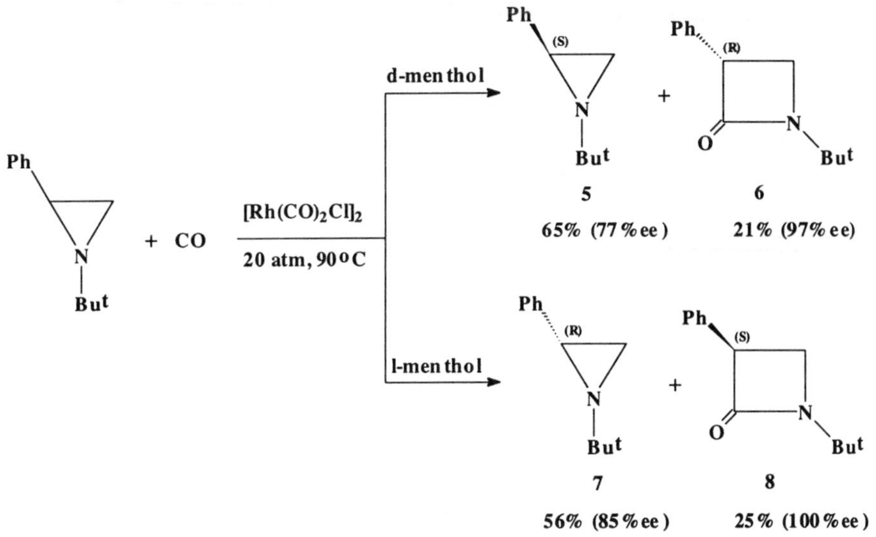

$$
\text{Ph} \overset{\triangle}{\underset{\text{But}}{\text{N}}} + \text{ CO} \quad \xrightarrow[\text{20 atm, 90°C}]{[\text{Rh(CO)}_2\text{Cl}]_2}
$$

d-menthol

(S) Ph — **5** 65% (77 % ee) + (R) Ph — **6** 21% (97 % ee)

(13.7)

l-menthol

(R) Ph — **7** 56% (85 % ee) + (S) Ph — **8** 25% (100 % ee)

(13.8)

The catalytic carbonylation of 2-alkylaziridines readily occurs in the presence of $Co_2(CO)_8$ or $NaCo(CO)_4$ as the catalyst, affording β-lactams with CO insertion into the least substituted ring C—N bond (Equation (13.9)) [24]. These reactions often proceed in high yields and always with inversion of configuration (Table 13.1).

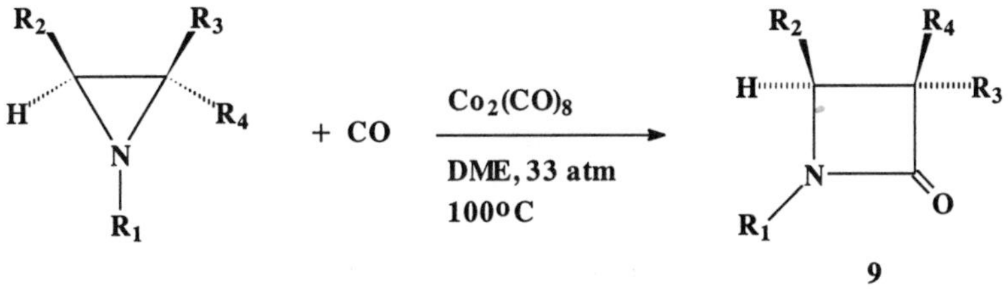

$$
\underset{R_1}{\overset{R_2 \quad R_3}{\underset{\text{N}}{\triangle}}}_{R_4} + \text{ CO} \quad \xrightarrow[\substack{\text{DME, 33 atm} \\ \text{100°C}}]{\text{Co}_2(\text{CO})_8} \quad \mathbf{9}
$$

(13.9)

Table 13.1. Cobalt-catalysed carbonylation of monocyclic aziridines[a]

R¹	R²	R³	R⁴	Isolated yield of compound **9** (%)
PhCH₂CH₂	Et	H	H	94
PhCH₂	Et	H	H	64
PhCH₂CH₂	Me	H	Me	95
PhCH₂CH₂	Me	Me	H	95
t-Bu	Ph	Ph	H	94
i-Pr	Me	Ph	H	94

[a] Reaction conditions: $Co_2(CO)_8$ (0.05 mmol), substrate (0.6 mmol), DME (10 ml), 500 psi CO, 100 °C, 24 h.

A possible mechanism for the reaction is outlined in Scheme 13.1. The reaction of $Co_2(CO)_8$ with most N-, O- and P-containing compounds is known to induce the cleavage of the Co—Co bond, giving an ionic species with $Co(CO)_4^-$ as the anionic component. Thus, aziridine can react with $Co_2(CO)_8$ to generate $Co(CO)_4^-$, which is the active catalyst for the reaction. Nucleophilic ring opening of the aziridine by $Co(CO)_4^-$ would occur at the less substituted carbon with inversion of configuration to form compound **10**. Insertion of CO into the C—Co bond of compound **10** should proceed with retention of configuration to form compound **11**. Ring closure of the acyl complex **11** gives the β-lactam and regenerates the active species.

A novel feature of the $Co_2(CO)_8$-catalysed reaction is the conversion of bicyclic aziridines into highly strained *trans* bicyclic β-lactams. Carbonylation of *cis*-bicyclic aziridines affords the β-lactams **12** in low to good yields using tetrahydrofuran (THF) as the solvent (Equation (13.10)) [24].

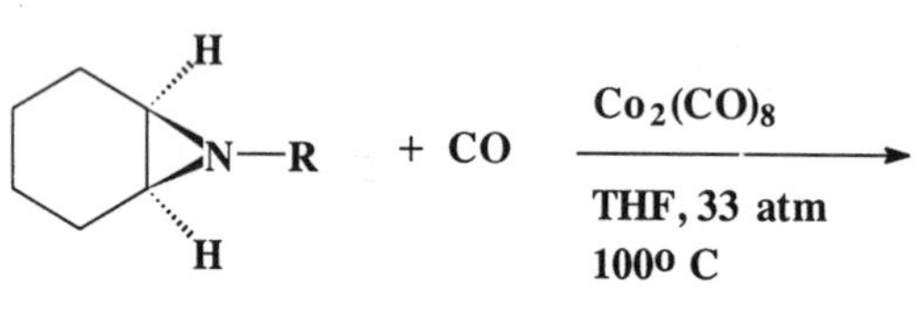

(13.10)

12 (44-88%)

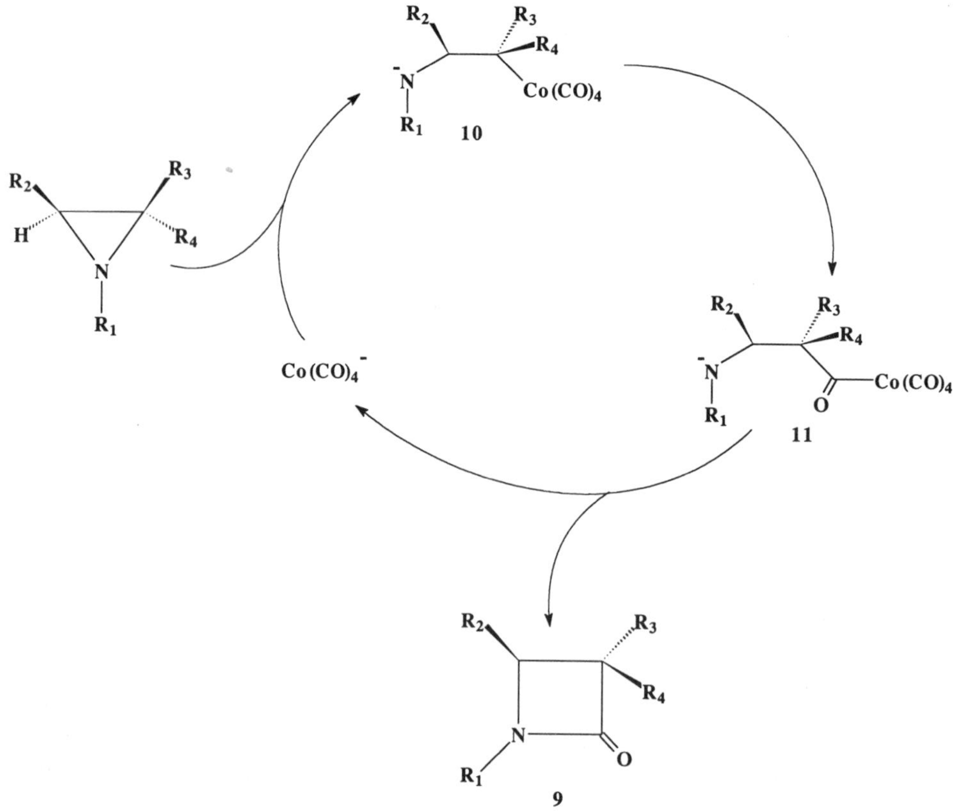

Scheme 13.1.

β-Lactams **13** can be prepared by the reaction of aziridines with CO in the presence of $Pd(PPh_3)_4$ under both homogeneous [25] and phase-transfer catalysis conditions (Equation (13.11)). It was proposed that the reaction involves dimerization followed by carbonylation, both processes being catalysed by palladium.

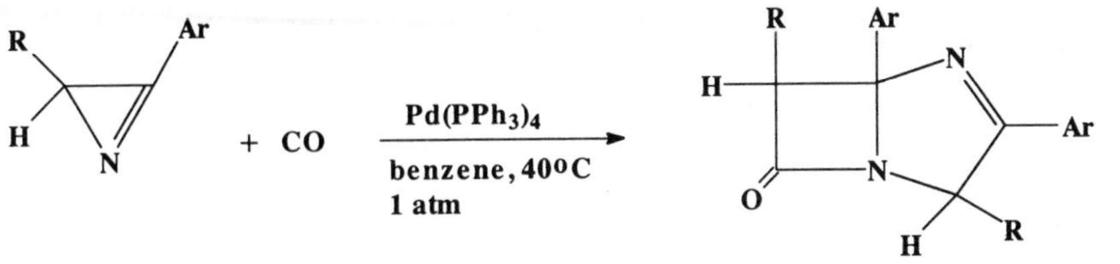

$$ (13.11) $$

R = H, Me.
Ar = Ph, p-MeC$_6$H$_4$, p-BrC$_6$H$_4$, p-ClC$_6$H$_4$.

13 (25-63 %)

The reaction of methyleneaziridines with CO in the presence of $Pd(PPh_3)_4$ or $Pd(OAc)_2$/PPh_3 occurs regiospecifically at the C(sp^2)–N bond, forming α-methylene-β-lactams **15** in fair to good yield (Equation (13.12)) [27]. This reaction may proceed via the generation of a vinylpalladium complex (**14**), followed by CO insertion and subsequent reductive elimination.

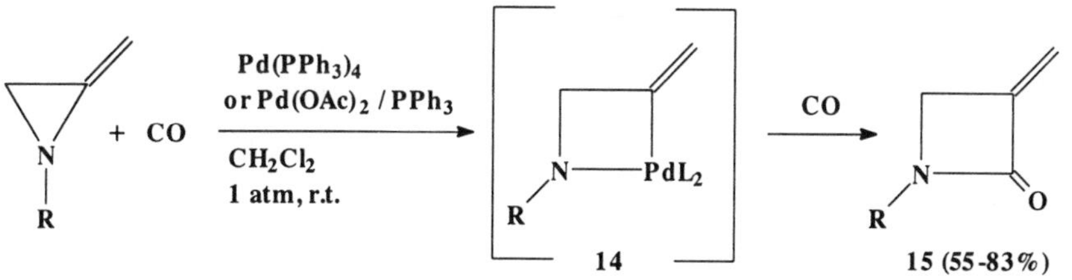

$$ (13.12) $$

14   15 (55-83%)

A different strategy for the synthesis of β-lactams involves palladium-catalysed reaction of an allylic phosphonate with an imine and CO (Equation (13.13)). This reaction can occur in high stereoselectivity, subject to the nature of the imine [28].

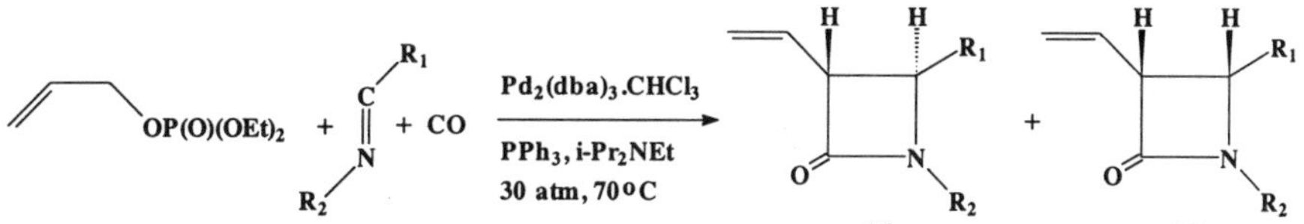

$$ (13.13) $$

16   17

Bicyclic β-lactams **18** were formed by carbonylative coupling and cyclization of 2-aryl-1,3-thiazines with allylphosphates, using a catalytic quantity of palladium(0) and triphenylphosphine in the presence of N,N-diisopropylethylamine as a base. These reactions are stereospecific, with the aryl and vinyl groups on the β-lactam ring being *cis* to each other (Equation (13.14)) [29].

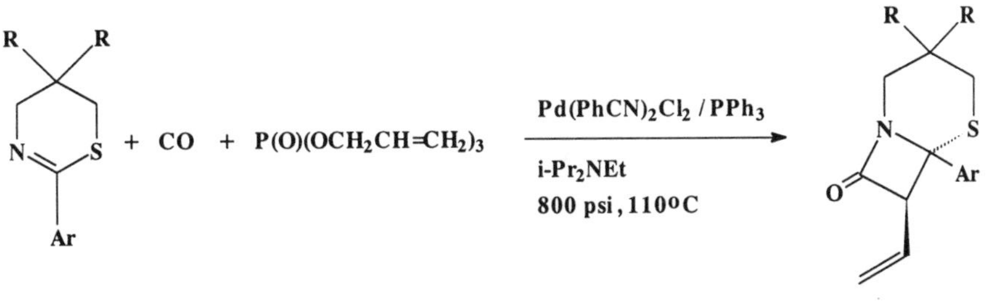

$$ (13.14) $$

18 (32-84%)

2.2   Five-membered rings

The carbonylation and ring expansion reaction can also be performed under phase-transfer catalysis (PTC) conditions. Styrene and β-methylstyrene oxide react with a catalytic amount of $Co_2(CO)_8$ and CO in the presence of MeI in basic solution, with cetyltrimethylammonium bromide (CTAB) as the phase-transfer catalyst to form compound **19**. Two molecules of CO are incorporated in this reaction (Equation (13.15)) [30].

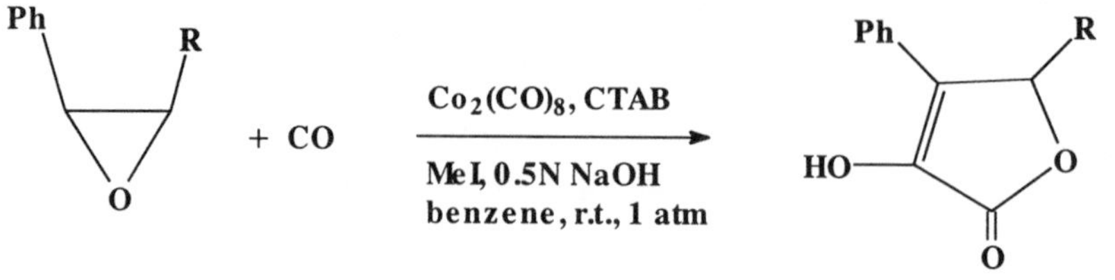

$$(13.15)$$

19   (R = H, 65 %
      R = Me, 34 %)

When 2-aryl-3-(hydroxymethyl) oxiranes were used under PTC conditions (with TDA-1 as the phase-transfer agent), the highly functionalized triple carbonylation product **20** was formed as the major product of the reaction (Equation (13.16)) [31]. The incorporation of three molecules of CO in the final product was demonstrated by [13]CO-labeling experiments. Note that acylcobalt tetracarbonyl, not $Co(CO)_4^-$, is the key catalytic species when $Co_2(CO)_8$ is used with methyl iodide.

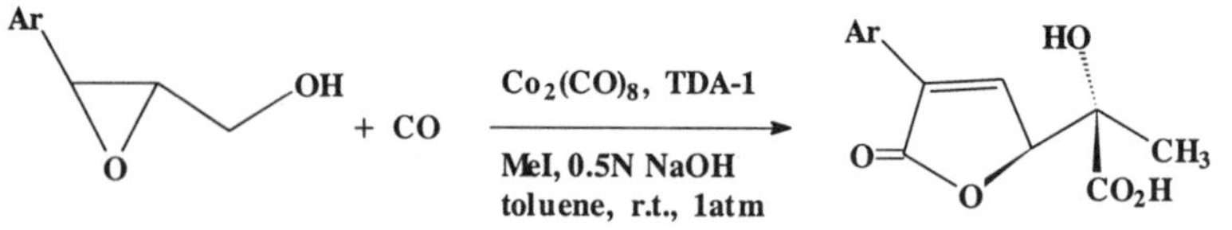

$$(13.16)$$

20 (42-55%)

2(5H)-Furanones, or butenolides, which occur in some biologically active natural products [32], have been prepared via cyclocarbonylation reactions. For example, the intramolecular cyclocarbonylation of terminal alkynols was effected using catalytic quantities of Pd(dba)$_2$ and 1,4-bis(diphenylphosphino(butane)) (dppb), affording 2-(5H)-furanone (**21**) in good yields (65–80%) at 20 atm of CO and 150 °C (Equation (13.17)) [33].

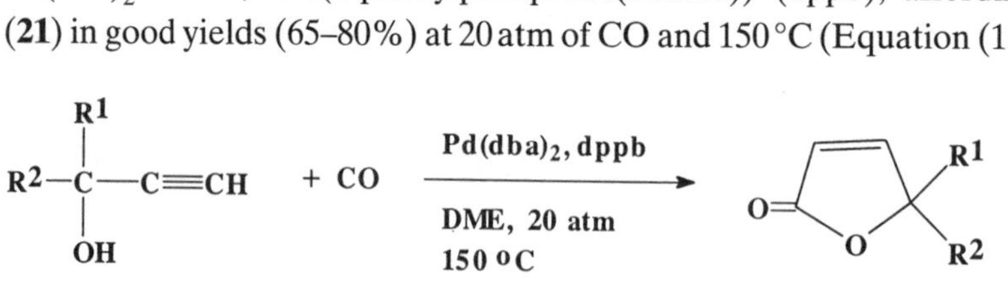

$$(13.17)$$

21

Modification of the procedure for the reaction in Equation (13.17), using a mixture of CO and H$_2$, resulted in extension of the reaction to internal alkynols with alkyl, phenyl and vinyl units attached to one acetylene carbon atom (Equation (13.18)) [34]. This reaction requires catalytic quantities of Pd$_2$(dba)$_3$. CHCl$_3$ and dppb in dichloromethane, CO (600 psi) and H$_2$ (200 psi) at 95 °C. Hydrogen is essential for this reaction and 2(5H)-furanones (**22**) were formed in good to excellent yields (67–97%) (Table 13.2).

Table 13.2. Palladium-catalysed cyclocarbonylation of alkynols to 2(5H)-furanone[a]

Alkynol	Furanone **22**	Isolated yield (%)
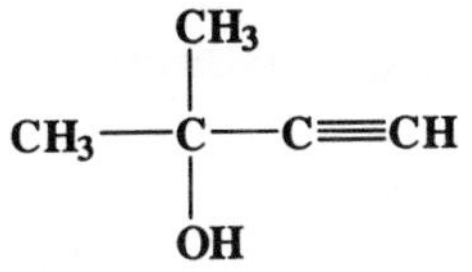	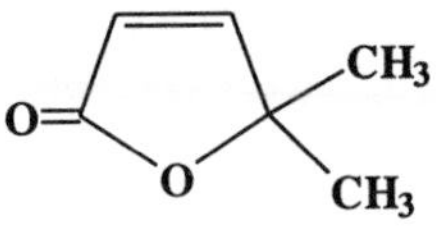	98
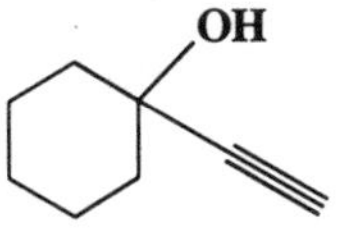	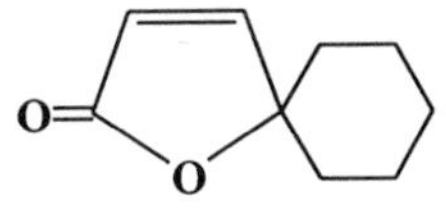	92
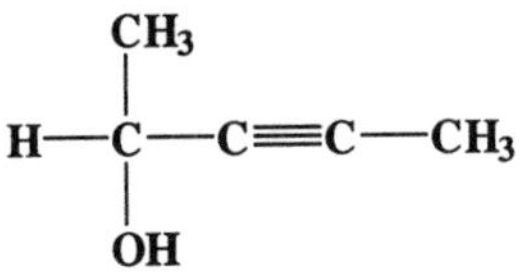	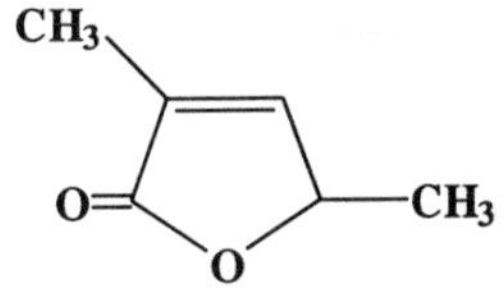	98
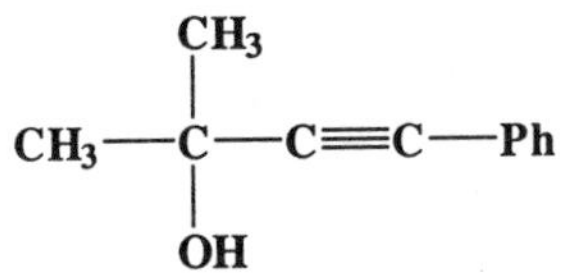	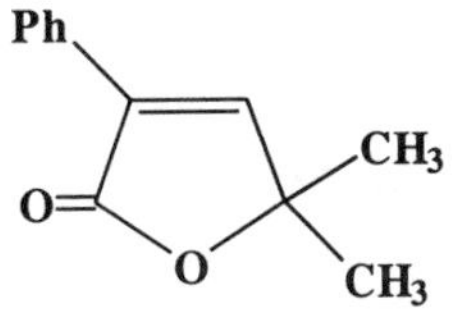	97
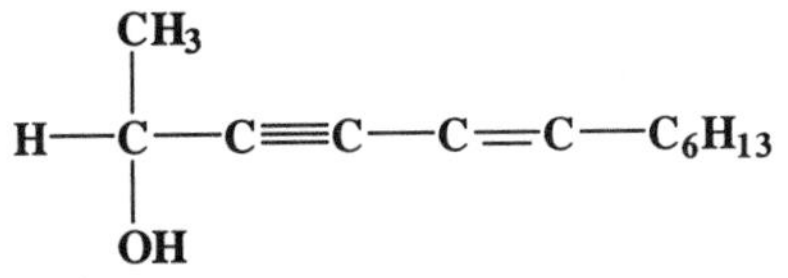	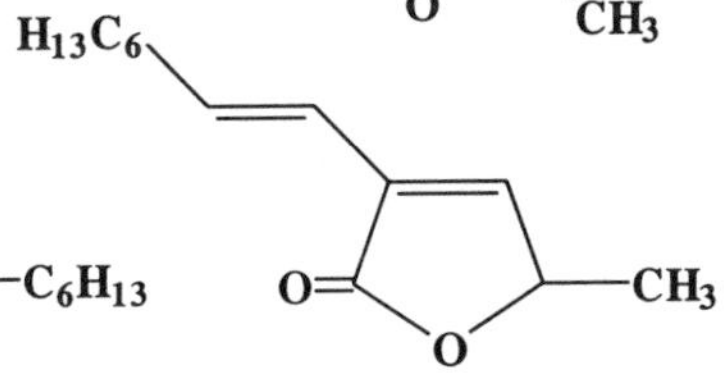	85

[a] Reaction conditions: alkynol (1 mmol), catalyst (0.04 mmol), dppb (0.08 mmol), dry CH_2Cl_2 (10 ml), CO/H_2 (600/200 psi), 95 °C, 36 h.

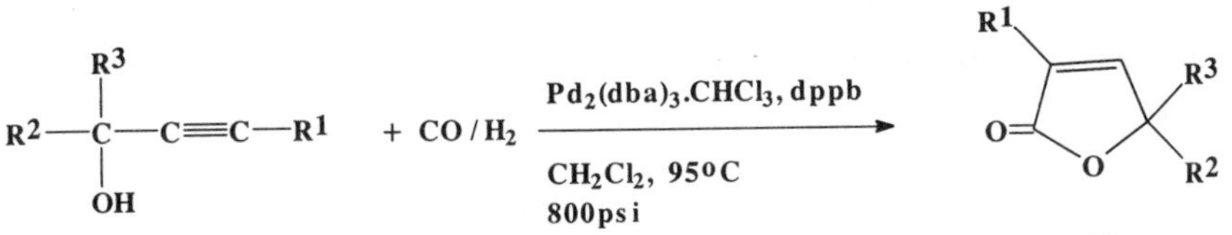

$$(13.18)$$

The cyclocarbonylation reaction is believed to proceed via a palladium intermediate (compound **23**, Scheme 13.2) [35], which is formed by initial insertion of Pd(0) into the C—O bond of the alkynol followed by rearrangement. Insertion of carbon monoxide and subsequent reductive elimination may lead to the 2,3-dienioc acid **24** [36]. Trace quantities of acid present in the solvent can result in cyclization of compound **24** to give a 2(5H)-furanone (**22**) [36].

Palladium(0)-catalysed carbonylation and intramolecular coupling of hydroxyvinyl triflates leads to the formation of α,β-butenolides (**25**) in good yields (60–95%) (Equation (13.19)) [37].

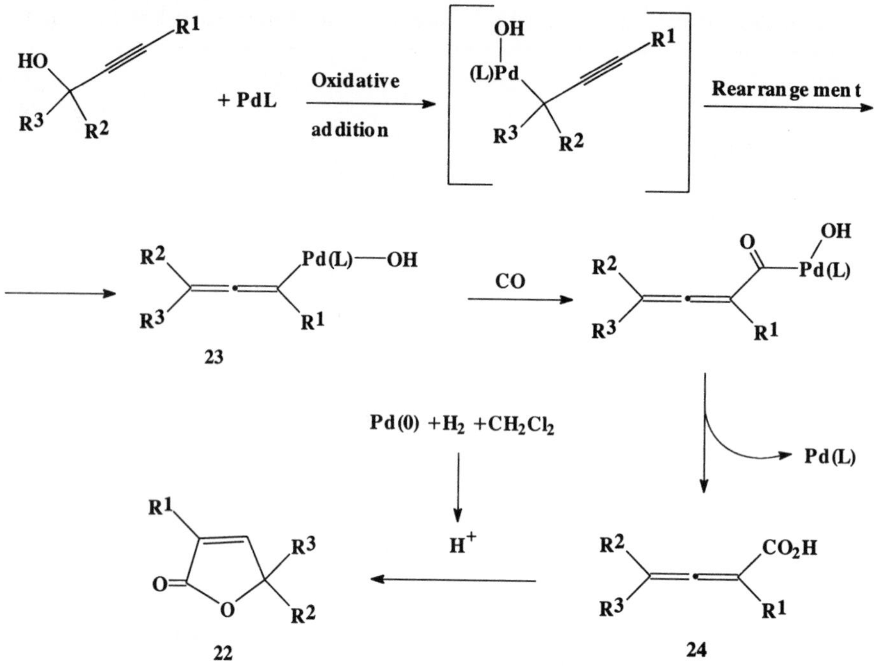

Scheme 13.2.

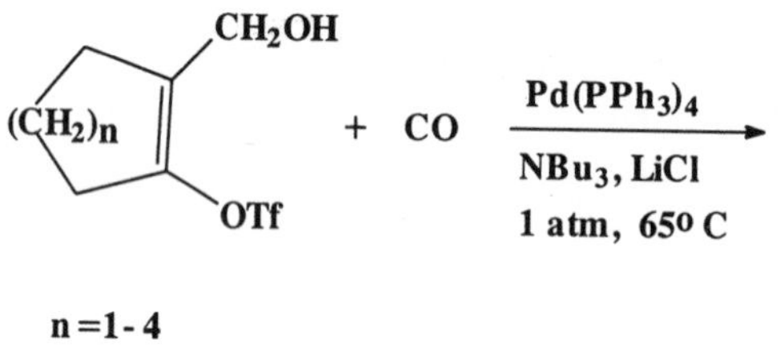

(13.19)

n = 1 - 4

Recently, novel tricyclic lactones were synthesized by rhodium-catalysed carbonylation of 2-alkynylbenzaldehydes under water–gas shift reaction conditions (Equation (13.20)) [38].

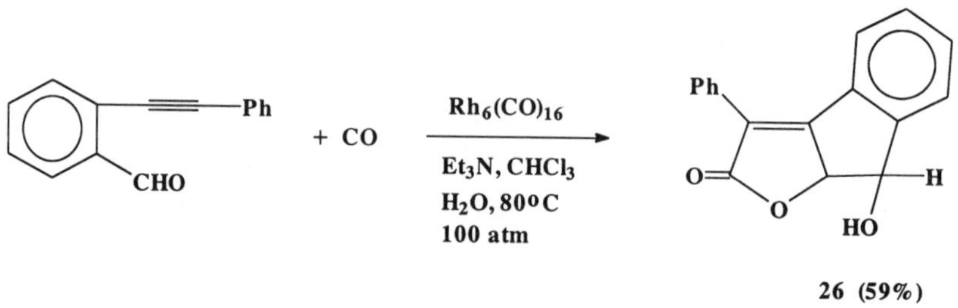

(13.20)

26 (59%)

Negishi and co-workers [8, 39] have reported the palladium-catalysed carbonylation of a mixture of internal alkynes and aryliodides and the acylpalladation of internal alkynes. 2-Butenolides (**27**) are formed in reasonable to excellent yields (49–99%) in the presence of water and triethylamine (Equation (13.21)).

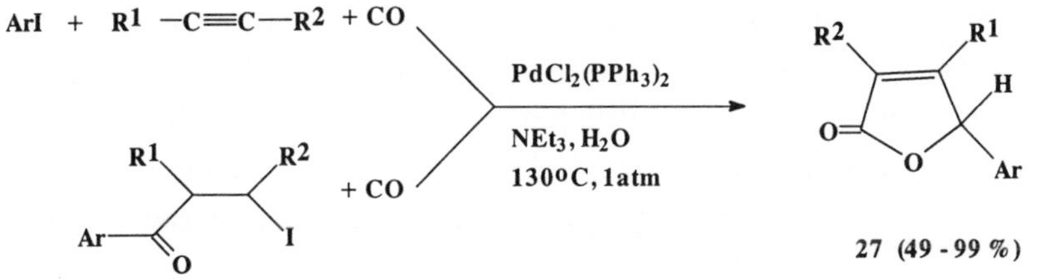

$$\text{ArI} + \text{R}^1-\text{C}\equiv\text{C}-\text{R}^2 + \text{CO} \quad\quad (13.21)$$

27 (49 - 99 %)

The regio- and stereospecific intramolecular cyclocarbonylation of 3-aryl-l-propynes with iodoarenes and CO, catalysed by Pd(OAc)$_2$ and PPh$_3$, leads to the formation of (E)-3-arylidenebutenolides 28 (Equation (13.22)) [40].

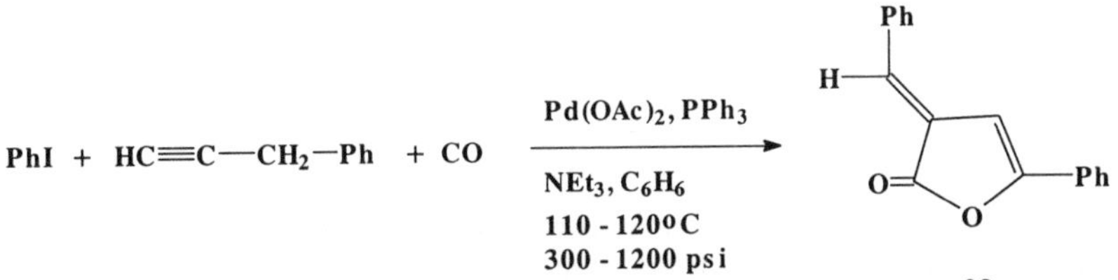

$$\text{PhI} + \text{HC}\equiv\text{C}-\text{CH}_2-\text{Ph} + \text{CO} \quad\xrightarrow[\substack{\text{NEt}_3, \text{C}_6\text{H}_6 \\ 110-120^\circ\text{C} \\ 300-1200\,\text{psi}}]{\text{Pd(OAc)}_2, \text{PPh}_3}\quad (13.22)$$

28

γ-Butyrolactones (29) can be prepared by the carbonylation of oxetanes using mixed metal carbonyls. A 1:1 mixture of Co$_2$(CO)$_8$ and Ru$_3$(CO)$_{12}$ was employed under 60 atm of CO at 165–240 °C. A wide variety of oxetanes were reactive under these conditions, affording the corresponding lactones in good yields. It was also found that the carbonylation reaction proceeds with retention of configuration (Equation (13.23)) [41].

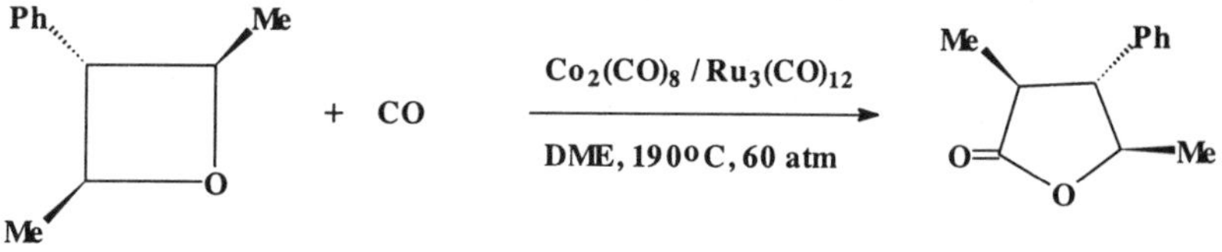

$$+ \text{CO} \quad\xrightarrow[\text{DME}, 190^\circ\text{C}, 60\,\text{atm}]{\text{Co}_2(\text{CO})_8 / \text{Ru}_3(\text{CO})_{12}}\quad (13.23)$$

29 (63 %)

The intramolecular carbonylation of unsaturated alcohols in the presence of catalytic amounts of PdCl$_2$ and CuCl$_2$ along with concentrated hydrochloric acid and oxygen in dry THF at 1 atm of CO leads to the formation of γ-butyrolactones 30 (R = H, n = 1 and R = C$_3$H$_7$, n = 0) (Equation (13.24)) [42].

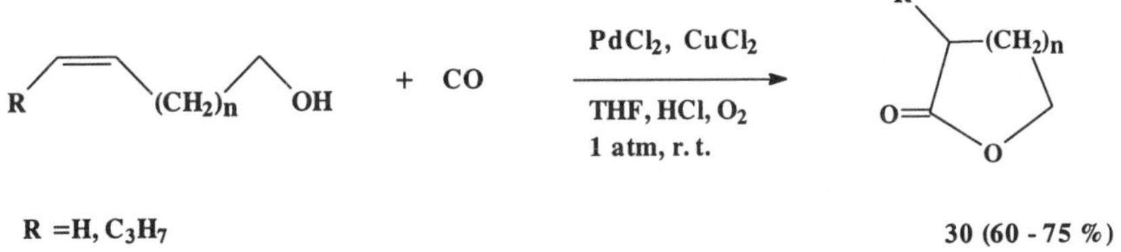

$$+ \text{CO} \quad\xrightarrow[\substack{\text{THF, HCl, O}_2 \\ 1\,\text{atm, r.t.}}]{\text{PdCl}_2, \text{CuCl}_2}\quad (13.24)$$

R = H, C$_3$H$_7$

n = 0, 1

30 (60 - 75 %)

Secondary and tertiary allylic alcohols undergo cyclocarbonylation in the presence of Pd(dba)$_2$ and dppb at 40 atm of CO and 190 °C to form compound 31 in good yields (Equation (13.25)) [33].

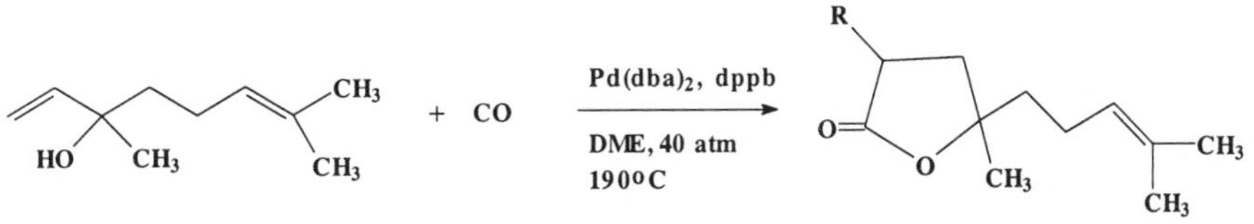

$$+ \text{CO} \quad\xrightarrow[\substack{\text{DME, 40 atm} \\ 190^\circ\text{C}}]{\text{Pd(dba)}_2, \text{dppb}}\quad (13.25)$$

31 (80 %)

Recently, β,γ-substituted allylic alcohols were cyclocarbonylated (1:1 CO/H$_2$) in the presence of catalytic quantities of Pd(OAc)$_2$ and dppb to give α,β-substituted-γ-butyrolactones (**32**) in 42–84% isolated yields. The complete stereoselectivity observed in some cases is a significant feature of the lactonization reaction, with (E)-allylic alcohols affording *trans*-disubstituted lactones (Equation (13.26)) [43]. The presence of hydrogen is essential for this reaction because no lactone was formed in its absence.

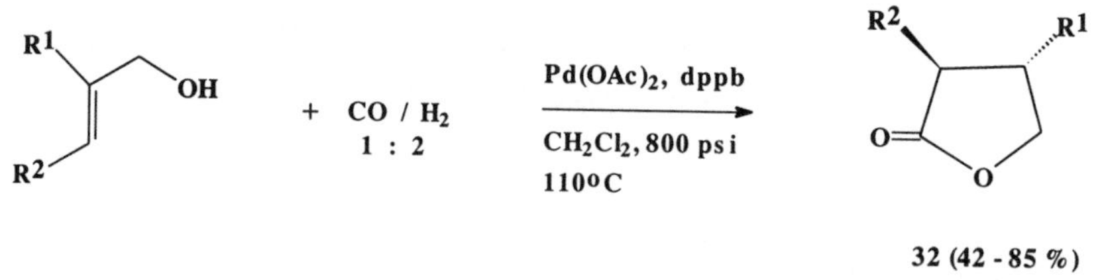

$$(13.26)$$

32 (42 - 85 %)

An efficient method for the synthesis of vinyl-γ-butyrolactones by palladium(0)-catalysed decarboxylative carbonylation has been reported. 4-Vinyl-1,3-dioxan-2-ones (cyclic carbonates) react with CO at room temperature in the presence of Pd(PPh$_3$)$_4$ to give compound **33** in good yields (Equation (13.27)) [44].

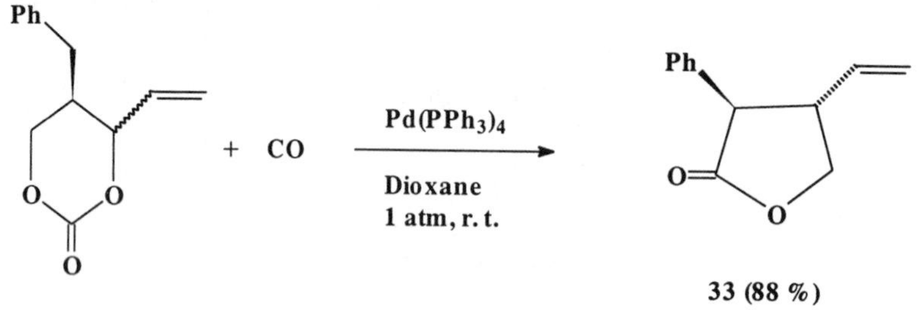

$$(13.27)$$

33 (88 %)

The reaction might be rationalized by involving an equilibrium between alkoxypalladium complexes **34** and **35**. In this equilibrium, inversion of configuration of the ring carbon correlates with inversion of geometry of the double bond. The intermediate **35** might isomerize to the relatively stable isomer **38** via the *cis*-oxopalladacycloheptenes **36** and **37**.

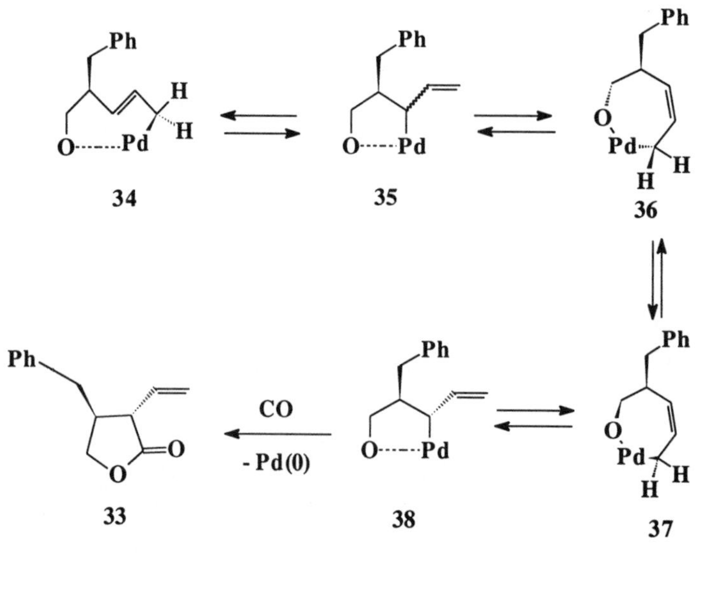

34   **35**   **36**

33   **38**   **37**

Palladium(II)-catalysed dicarbonylation (1 atm of CO) of 3-buten-1-ols in the presence of propylene oxide and ethyl orthoacetate in methanol–dichloromethane afforded γ-butyrolactone carboxylic esters (39) in good yields (Equation (13.28)) [45].

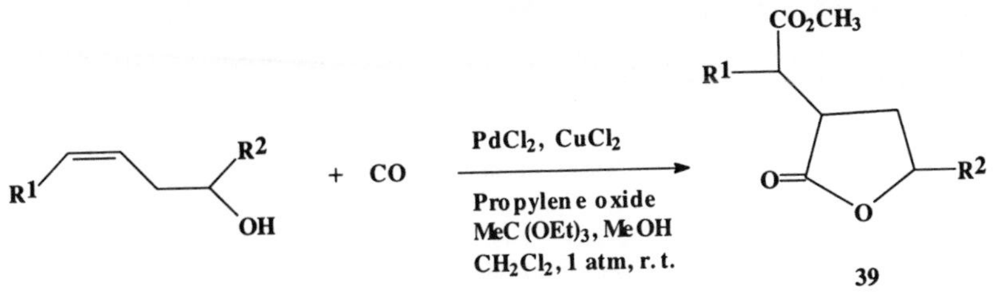

$$(13.28)$$

Because the unsaturated group of the allylic alcohol in Equation (13.24) is prochiral, repetition of the carbonylation reaction in the presence of an appropriate chiral ligand, L*, leads to the formation of optically active lactone 40 (Equation (13.29)) [46].

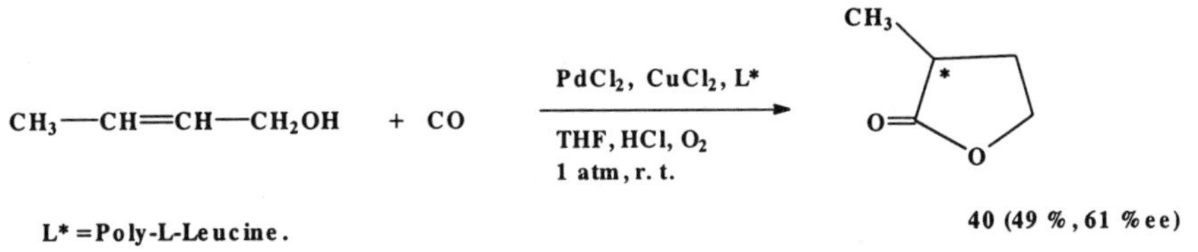

$$(13.29)$$

L* = Poly-L-Leucine.

40 (49 %, 61 % ee)

The palladium(0)-catalysed asymmetric cyclocarbonylation of allyl alcohols to afford γ-butyrolactones gave more fruitful results. For example, 2,3-diphenyl-3-buten-2-ol was subjected to carbonylation in the presence of Pd$_2$(dba)$_3$·CHCl$_3$ and (2S,4S)-N-(tert-butoxycarbonyl)-4-(diphenylphosphino)-2-[(diphenylphosphino)methyl] pyrrolidine ((−)-bppm) in CH$_2$Cl$_2$ at 100 °C. A 1:1 mixture of syn (41) and anti (42) diastereoisomers was formed in 81% and 69% enantiomeric excess (ee), respectively (Equation (13.30)) [47]. The presence of hydrogen is required in order to obtain lactones in good chemical yields.

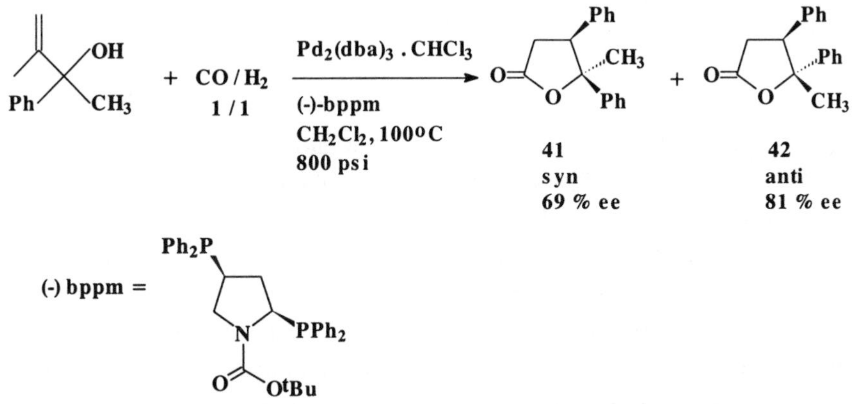

$$(13.30)$$

The cyclocarbonylation reaction of 2-triflyloxyacetophenone derivatives in the presence of Pd(OAc)$_2$ and 1,3-bis(diphenylphosphino)propane (dppp) and NEt$_3$ affords five-membered fused lactones (43) in good yields (Equation (13.31)) [48].

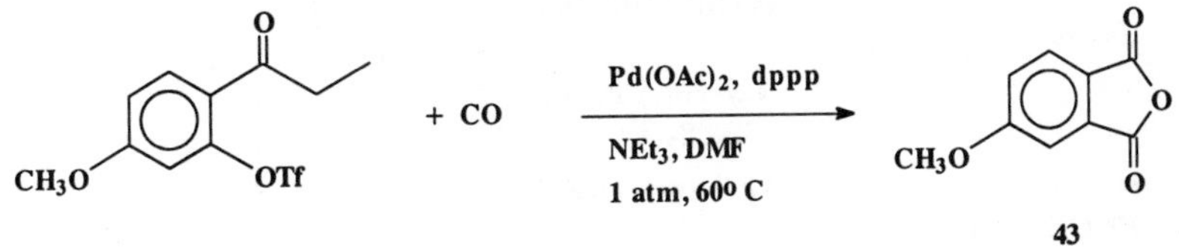

$$(13.31)$$

Recently, the synthesis of bicyclic lactones has been reported by the intramolecular cyclocarbonylation of 2-allylphenol [49]. For instance, 2-allylphenol reacts with CO/H_2 (1:1), a catalytic amount of $trans[Pd(PCy_3)(H)(H_2O)]^+BF_4^-$ and dppb at 120 °C in CH_2Cl_2 to form the five-membered ring lactone **44** as the major product (Equation (13.32)). Six- and seven-membered ring lactones were obtained as byproducts of the reaction [49].

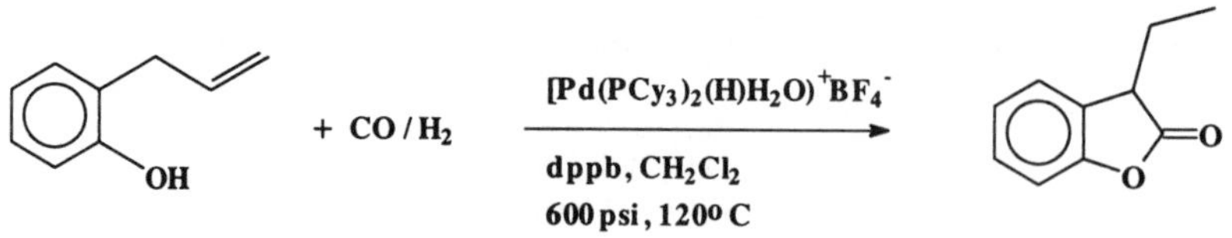

$$\text{[structure]} + CO/H_2 \xrightarrow[\substack{\text{dppb, CH}_2\text{Cl}_2 \\ 600\,\text{psi},\,120^0\,\text{C}}]{[Pd(PCy_3)_2(H)H_2O]^+BF_4^-} \text{[structure]} \qquad (13.32)$$

44 (76%)

The palladium-catalysed oxidative carbonylation ($CO/O_2/4:1$) of 1-substituted prop-2-ynylamines in methanol afforded γ-lactams (e.g. compound **45**) (Equation (13.33)) [50].

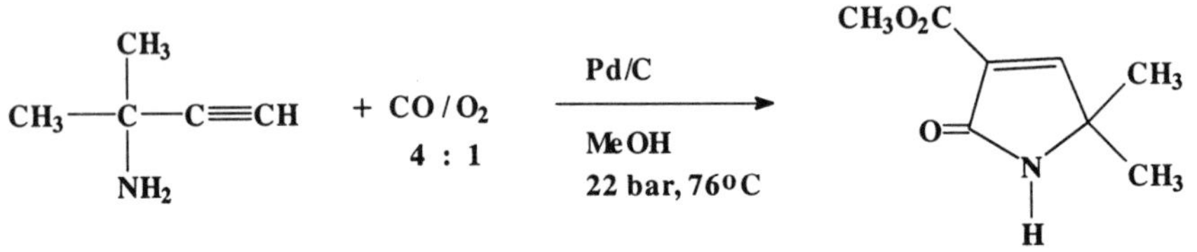

$$CH_3-\underset{\underset{NH_2}{|}}{\overset{\overset{CH_3}{|}}{C}}-C{\equiv}CH + \underset{4\,:\,1}{CO/O_2} \xrightarrow[\substack{\text{MeOH} \\ 22\,\text{bar},\,76^0\text{C}}]{Pd/C} \text{[structure]} \qquad (13.33)$$

45 (45%)

The intramolecular cyclocarbonylation of 3-iodo-2-heptylamines in the presence of $PdCl_2(PPh_3)_2$ at 1 atm of CO resulted in the formation of lactams (e.g. compound **46**) in good yields (Equation (13.34)) [51].

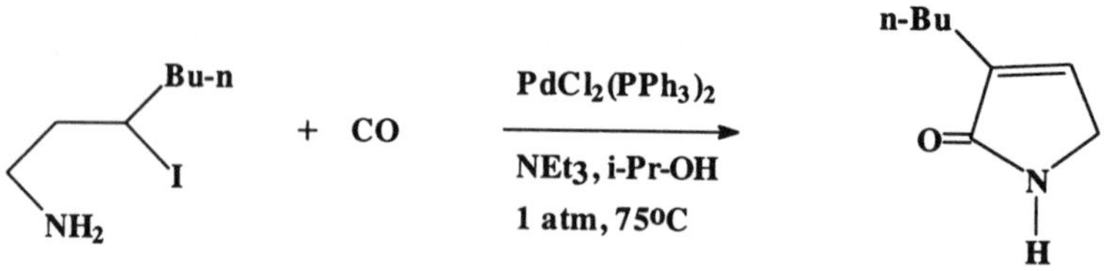

$$\text{[structure]} + CO \xrightarrow[\substack{\text{NEt}_3,\,\text{i-Pr-OH} \\ 1\,\text{atm},\,75^0\text{C}}]{PdCl_2(PPh_3)_2} \text{[structure]} \qquad (13.34)$$

46 (85%)

The carbonylation of azetidines is feasible and can be effected under relatively mild conditions. The regioselectivity of the carbonylation reaction depended on the substituents at the 2-position on the ring. For example, cobalt-carbonyl-catalysed carbonylation of 2-alkylazetidines led to 5-alkylpyrrolidinones **47**, while 2-arylazetidines form 3-arylpyrrolidinones **48** (Scheme 13.3) [52].

The cyclocarbonylation of allylamines can be catalysed, under relatively mild conditions, by rhodium complexes in the presence of sodium borohydride and isopropanol [53]. Pyrrolidinones **49** were isolated as the only products of the reaction when N-allyl or N-(methylallyl)amines were used as reactants (Equation (13.35)).

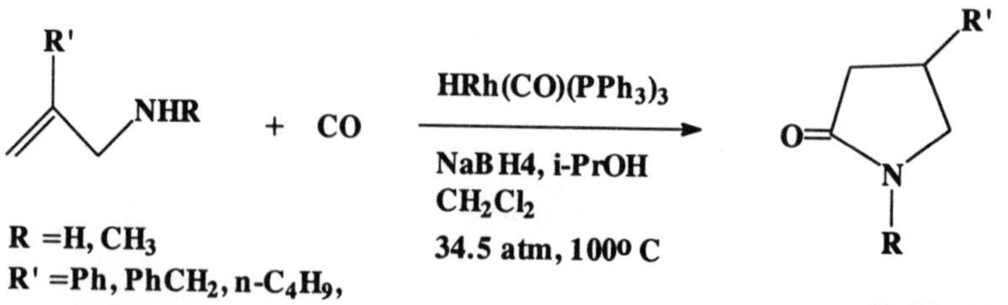

$$\text{[structure]} + CO \xrightarrow[\substack{\text{NaBH4, i-PrOH} \\ \text{CH}_2\text{Cl}_2 \\ 34.5\,\text{atm},\,100^0\,\text{C}}]{HRh(CO)(PPh_3)_3} \text{[structure]} \qquad (13.35)$$

R = H, CH_3
R' = Ph, $PhCH_2$, n-C_4H_9,
   C_8H_{15}, $PhCH_2CH_2$

49 (51-92%)

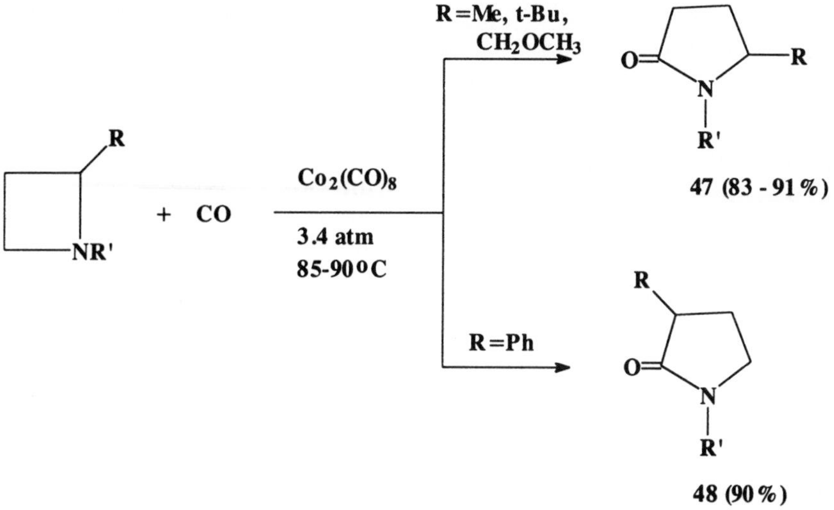

R=Me, t-Bu,
CH$_2$OCH$_3$

47 (83 - 91 %)

+ CO

Co$_2$(CO)$_8$

3.4 atm
85-90°C

R=Ph

48 (90 %)

Scheme 13.3.

Takahashi and co-workers described the rhodium-catalysed cyclocarbonylation of 2-alkynylanilines under water–gas shift reaction conditions. Five-membered ring lactams **50** and **51** were formed in good yields at 100 atm of CO (Equation (13.36)) [54].

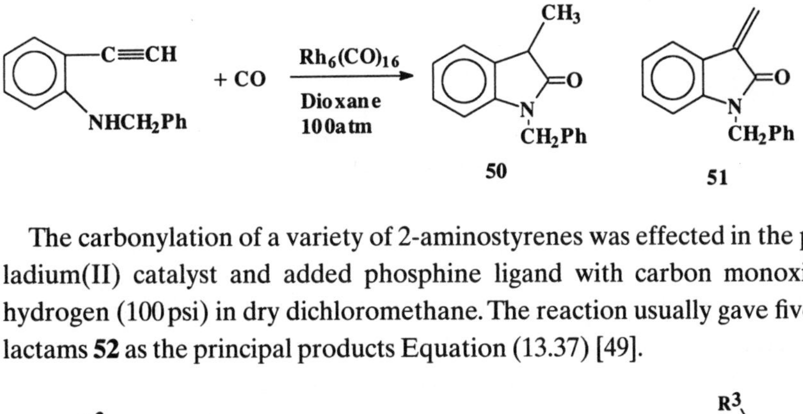

-C≡CH

+ CO

Rh$_6$(CO)$_{16}$

Dioxane
100 atm

NHCH$_2$Ph

CH$_3$

50

51

(13.36)

The carbonylation of a variety of 2-aminostyrenes was effected in the presence of a palladium(II) catalyst and added phosphine ligand with carbon monoxide (500 psi) and hydrogen (100 psi) in dry dichloromethane. The reaction usually gave five-membered ring lactams **52** as the principal products Equation (13.37) [49].

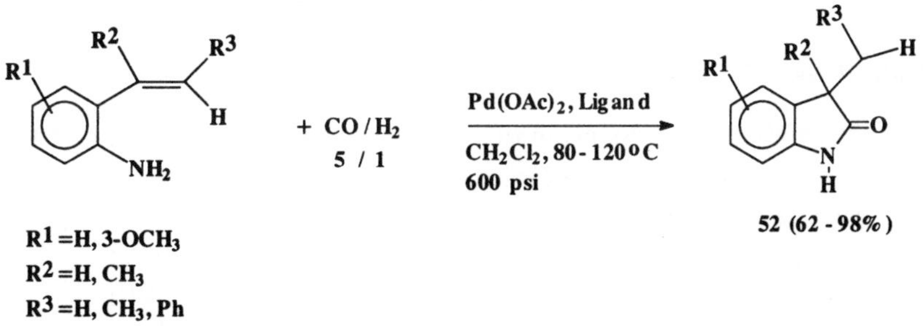

R^2 R^3

R^1

H

+ CO/H$_2$
5 / 1

Pd(OAc)$_2$, Ligand

CH$_2$Cl$_2$, 80-120°C
600 psi

NH$_2$

R^3

R^2 H

R^1

52 (62 - 98%)

(13.37)

R^1 =H, 3-OCH$_3$
R^2 =H, CH$_3$
R^3 =H, CH$_3$, Ph

2.3 Six-membered rings

Chiusoli and co-workers have reported a new palladium(0)-catalysed synthesis of some unusual 3,4-disubstituted coumarins **53** from 3-alkenoates of *ortho*-iodophenol, phenylacetylene and carbon monoxide at 80 °C in the presence of potassium butyrate (Equation (13.38)) [55].

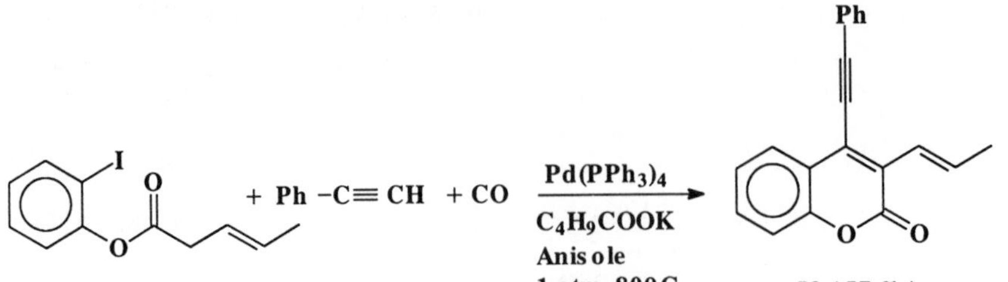

$$\text{(13.38)}$$

$$\textbf{53 (57 \%)}$$

Palladium(0) in combination with thallium acetate (TlOAc), promotes the catalytic cycloaddition of aryl or heteroaryl iodides with norbornene and carbon monoxide at 80 °C to give six-membered ring lactones (e.g. compound **54**) in moderate yields (Equation (13.39)). *o*-Iodoanilines react similarly [56].

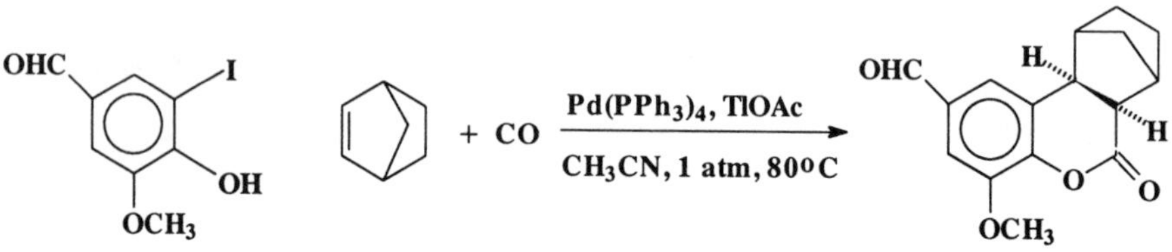

$$\text{(13.39)}$$

$$\textbf{54 (65 \%)}$$

The cyclocarbonylation of 2-aminostyrenes, described above, when catalysed by $Pd(OAc)_2$ and dppb in the presence of a 5:1 mixture of CO/H_2 results in the formation of lactams (e.g. compound **55**) in excellent yields (Equation (13.40)) [49].

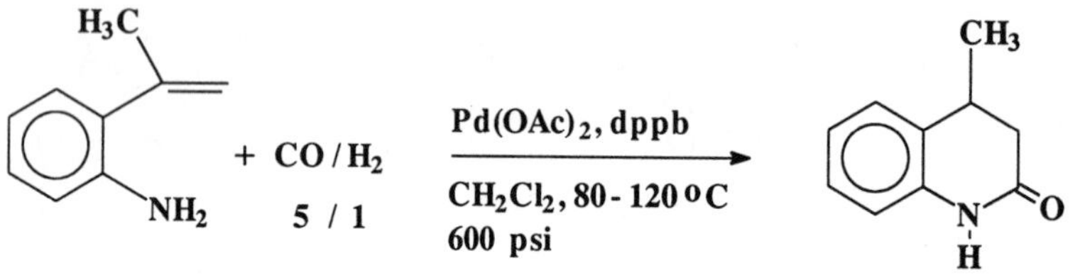

$$\text{(13.40)}$$

$$\textbf{55 (90\%)}$$

The asymmetric cyclocarbonylation of 2-(1-methylvinyl) anilines using a catalyst system consisting of $Pd(OAc)_2$ and (−)-DIOP as the chiral ligand gave 3,4-dihydro-4-methyl-2 (1H)-quinolin-2-ones (**56**) in up to 54% ee (Table 13.3) [57].

2.4   Seven-membered rings

The carbonylative ring expansion of vinylazetidines catalysed by $Co_2(CO)_8$ at 3.4 atm of CO and 85–90 °C afforded azepinones **56** with the vinyl side chain incorporated into the ring (Equation (13.41)) [52].

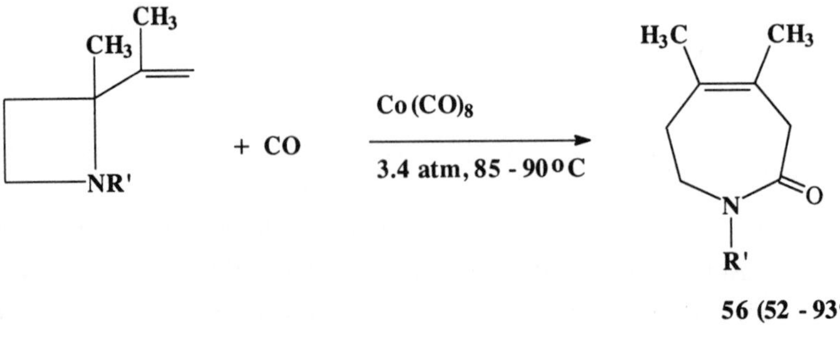

$$\text{(13.41)}$$

$$\textbf{56 (52 - 93\%)}$$

Table 13.3. Asymmetric cyclocarbonylation of substituted 2-(1-methylvinyl) anilines catalysed by Pd(OAc)$_2$ and (−)-DIOP[a]

Substrate	Product **56**	Isolated yield (%)	% ee[b]
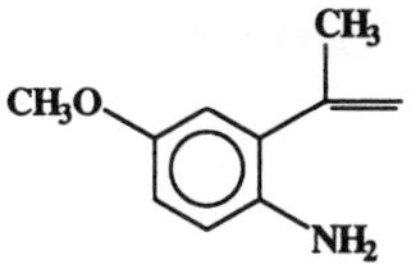	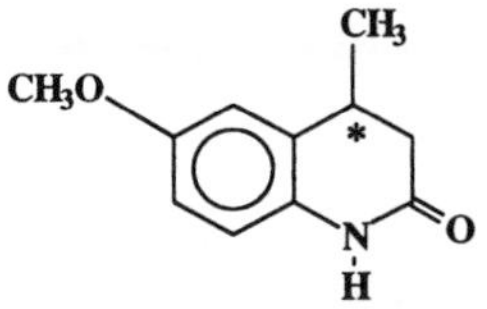	97	20
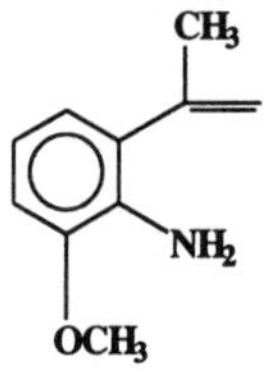	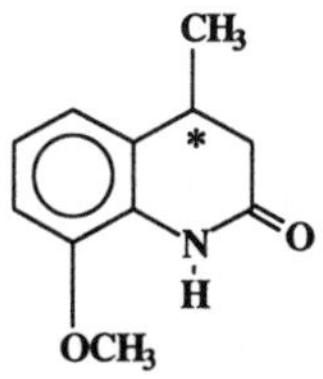	99	54
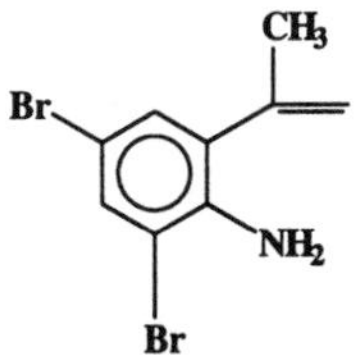	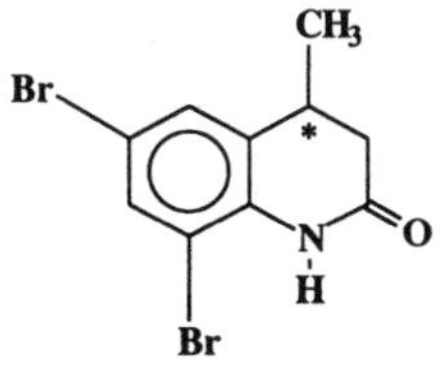	48	43
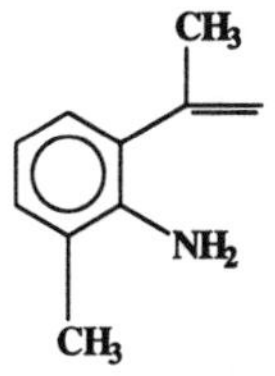	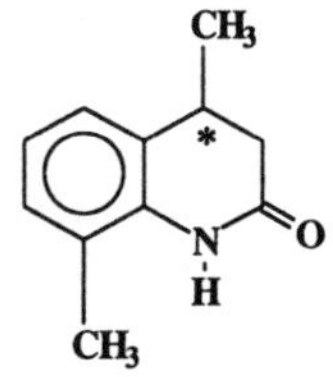	98	32

[a] Reaction conditions: substrate/Pd(OAc)$_2$/(−)-DIOP (1:0.01:0.02), CH$_2$Cl$_2$(5 ml), CO(500 psi), H$_2$(100 psi), 48 h.
[b] Determined by ^{1}H-NMR using Eu(hfc)$_3$ or Eu(dcm)$_3$.

Seven-membered ring lactones can be prepared via palladium(0)-catalysed cycloaddition of heteroaryl iodides with norbornene and carbon monoxide (Equation (13.42)) [56].

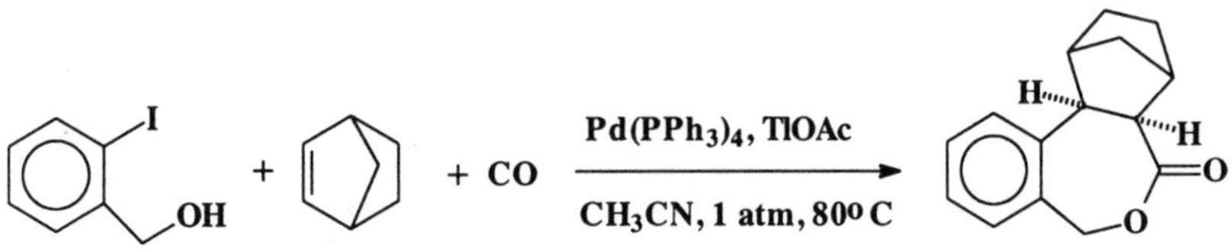

$$(13.42)$$

58 (65 %)

The cyclocarbonylation of 2-allylphenols catalysed by Pd(OAc)$_2$ or *trans*-[Pd(PCy$_3$)$_2$(H)(H$_2$O)]$^+$BF$_4^-$ using dppb and a 1:1 mixture of CO/H$_2$ in toluene afforded seven-membered ring lactones (**59**) in excellent yields and selectivities (Equation (13.43)) [49]. This is an attractive synthesis of bicyclic and polycyclic lactones of this ring size.

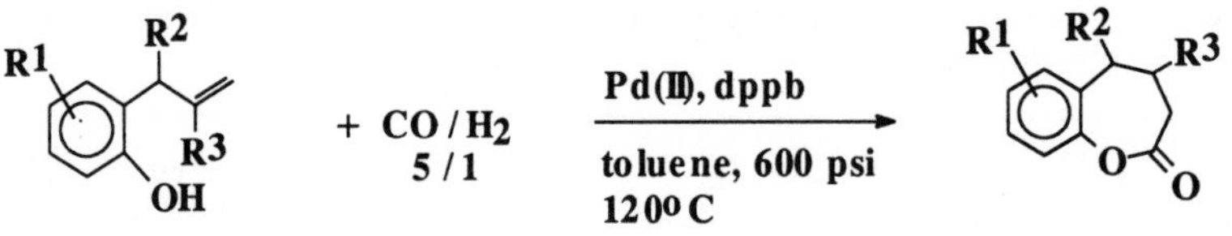

(13.43)

59 (89 - 98%)

R^1 =H, 6-CH_3
R^2 =H, CH_3
R^3 =H, CH_3

A possible mechanism for the formation of the seven-membered ring lactone **59** ($R^1 = R^2 = R^3 = H$) is outlined in Scheme 13.4. Oxidative addition of palladium(0) (generated from Pd(II) on exposure to CO and H_2) to the OH bond of 2-allylphenol (($R^1 = R^2 = R^3 =$ H), and coordination of the olefinic unit to Pd, would give compound **60** in which dppb is coordinated to the metal center in a monodentate fashion. Intramolecular hydropalladation and coordination of carbon monoxide would form compound **61**. Carbonyl insertion (via ligand migration) would form compound **62**, with dppb bound to Pd in a bidentate fashion. The lactone **59** would arise by subsequent reductive elimination.

2-Allylanilines were carbonylated under similar experimental conditions to give benzazepinones (**63**) (Equation (13.44)) [49].

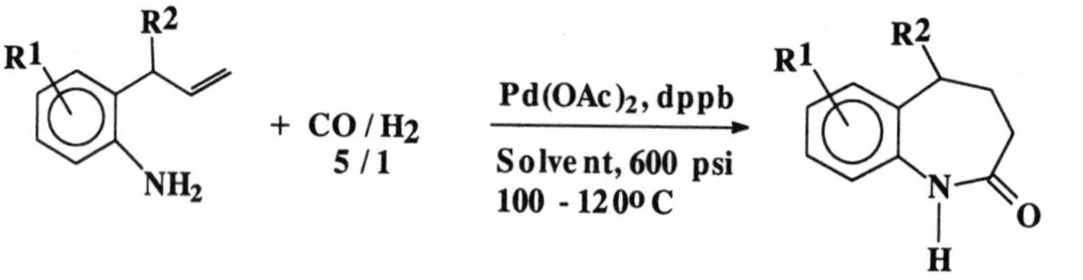

(13.44)

63 (75 - 92%)

R^1 =H, 4-CH_3, 4-OCH_3
R^2 =H, CH_3

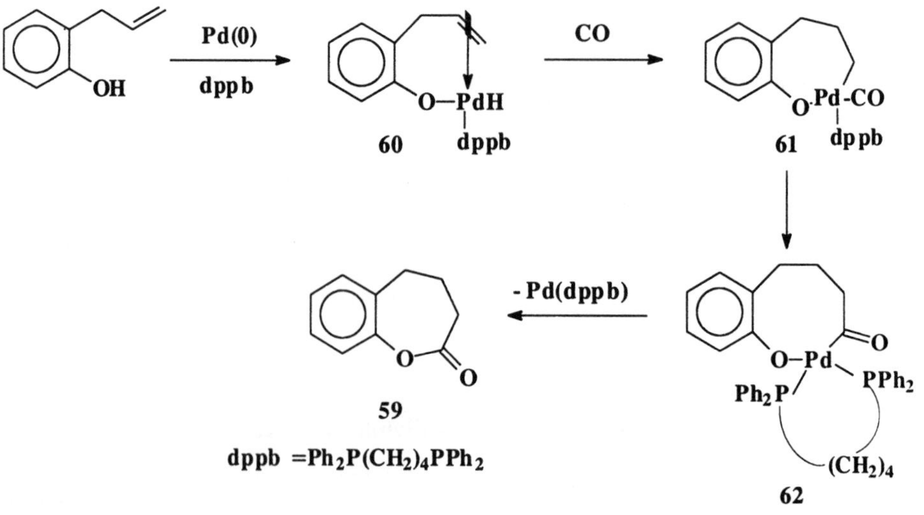

Scheme 13.4.

3 Carbonylative synthesis of other heterocycles

The carbonylation of diaziridines in the presence of an excess of cobalt carbonyl affords 1,3-azetidin-2-ones (**58**). This reaction is only applicable to 3,3-disubstituted diaziridines. The carbonylation of 3-monosubstituted diaziridines was catalysed by Pd(dba)$_2$ at 1 atm of CO and 120°C, leading to 1,3-diazetidin-2-ones (**64**) (R^1 = H) in reasonable yields (39–66%) (Equation (13.45)) [58].

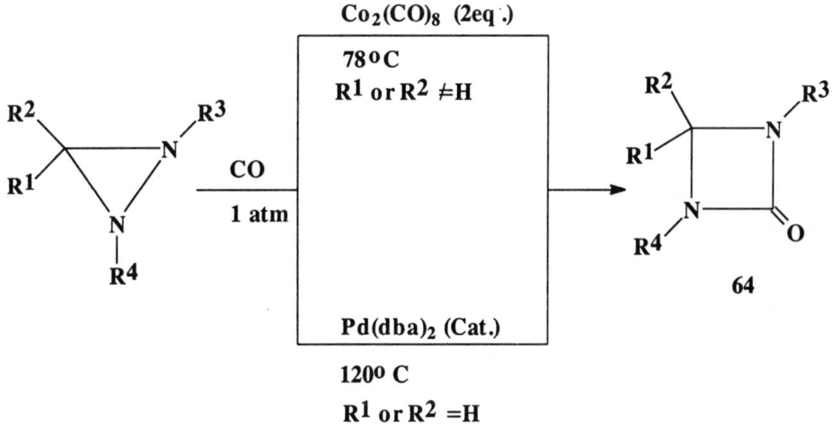

$$\text{(13.45)}$$

Co$_2$(CO)$_8$ (2eq.)

78°C
R^1 or R^2 ≠H

CO

1 atm

Pd(dba)$_2$ (Cat.)

120°C

R^1 or R^2 =H

64

The carbonylative ring expansion of 2-alkylthietanes was realized by using the mixed catalytic system of Co$_2$(CO)$_8$/Ru$_3$(CO)$_{12}$(1:1). Insertion of CO occurred into the less substituted C–S bond in accord with the results of azetidines and oxetanes (Equation (13.46)) [57].

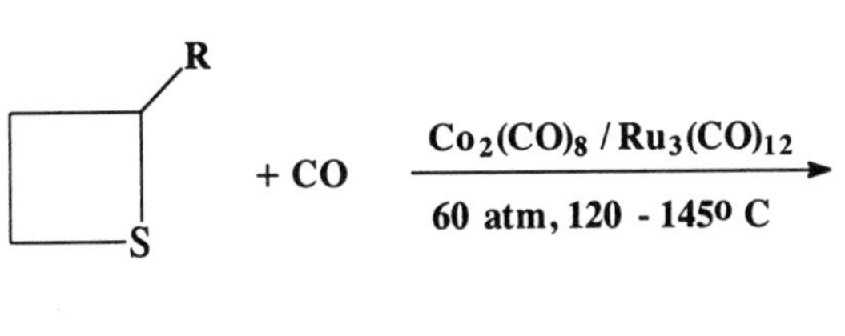

$$\text{(13.46)}$$

+ CO

Co$_2$(CO)$_8$ / Ru$_3$(CO)$_{12}$

60 atm, 120 - 145°C

65 (87-100%)

R =alkyl

The carbonylation of isoxazolidines proceeded with insertion of CO into the N–O bond, affording tetrahydro-1,3-oxazin-2-ones (**66**) in the presence of [Rh(COD)Cl]$_2$ as the catalyst for the reaction (Equation (13.47) [60]. However, when iridium trichloride was used as the catalyst, carbonylation was followed by hydrogen transfer from another molecule of starting material. The heterocycle **67** was formed as the major product of the reaction but in yields below 50% (Equation (13.48)).

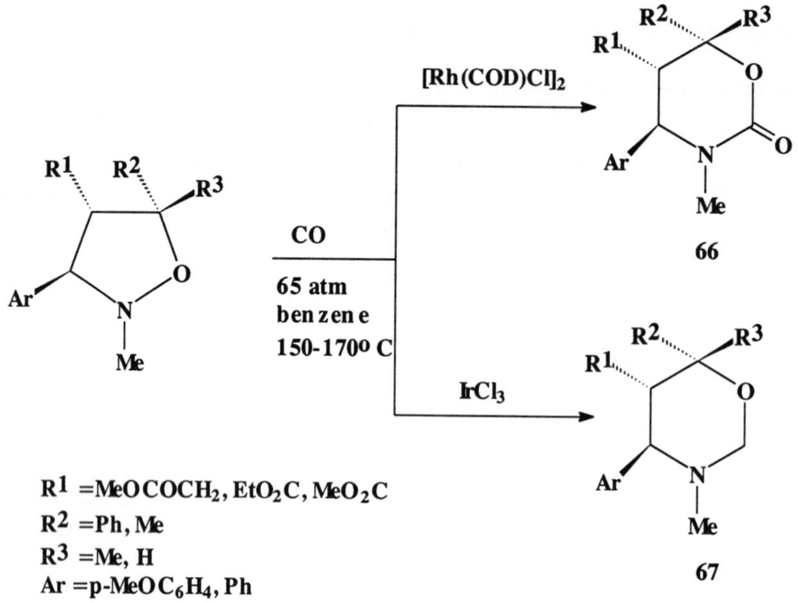

[Rh(COD)Cl]$_2$
CO
65 atm
benzene
150-170° C
IrCl$_3$
66
67
R1 =MeOCOCH$_2$, EtO$_2$C, MeO$_2$C
R2 =Ph, Me
R3 =Me, H
Ar =p-MeOC$_6$H$_4$, Ph

(13.47)

(13.48)

The reaction of thiazolidines with CO in the presence of an Rh catalyst proceeded in an unusual and unexpected manner, affording thiazolidinones (**69**) in good yields (Equation (13.49)) [61]. It was demonstrated that the reaction involves sequential CO insertion, elimination of ketene from the resulting thiazinone (**68**) and finally another carbonylation to give compound **69** [61].

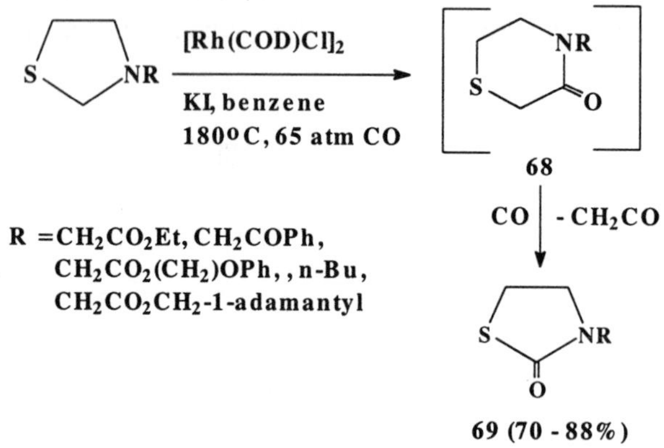

[Rh(COD)Cl]$_2$
KI, benzene
180°C, 65 atm CO
68
CO
- CH$_2$CO
R =CH$_2$CO$_2$Et, CH$_2$COPh,
CH$_2$CO$_2$(CH$_2$)OPh, , n-Bu,
CH$_2$CO$_2$CH$_2$-1-adamantyl
69 (70 - 88%)

(13.49)

The carbonylative ring expansion reaction of 3,6-dehydro-2H-1,2-oxazines proceeded in the presence of $Co_2(CO)_8$ to form 4,7-dihyro-1,3-oxazepin-2(3H)-ones (**70**) in moderate yields (Equation (13.50) [62].

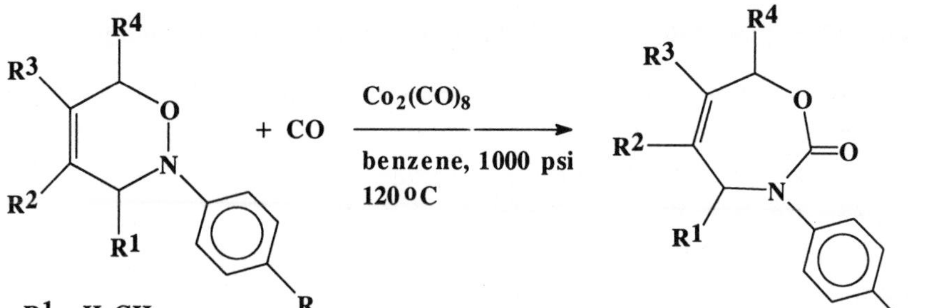

$$(13.50)$$

R¹ = H, CH₃
R² = H, CH₃, OCH₃
R³ = H, CH₃, OCH₃, Ph
R⁴ = H, CH₃
R = H, CH₃, OCH₃, COOCH₃, CH₂-CH=CH₂

Finally, when o-iodophenols were treated with allenes in the presence of a base, 3-methylene-2,3-dihydro-4H-1-benzopyrans (**71**) were formed in good yields (Table 13.4) (Equation (13.51) [63].

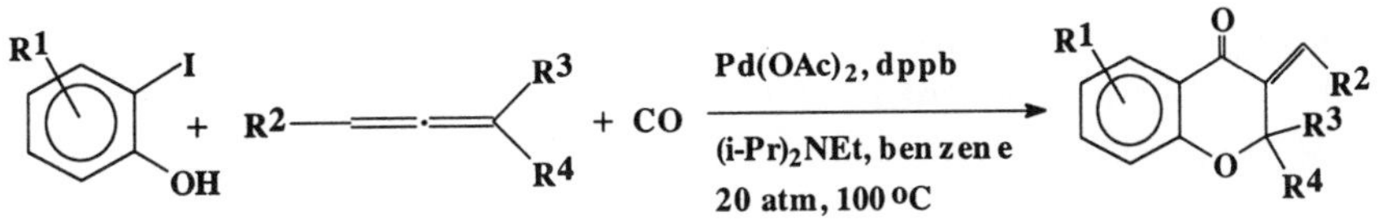

$$(13.51)$$

Table 13.4. Palladium-catalysed carbonylation of o-iodophenols with allenes[a]

Iodophenol	Allene	Product **71**	Yield (%)[b]
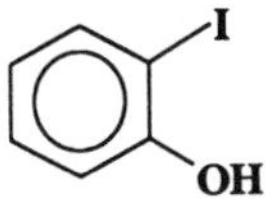	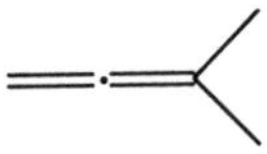	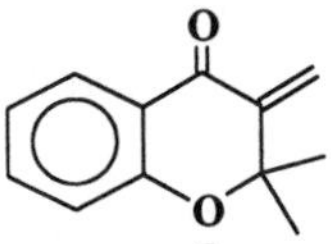	77
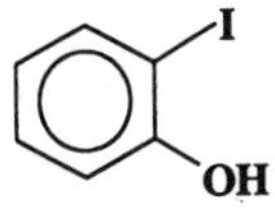	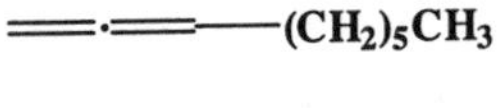	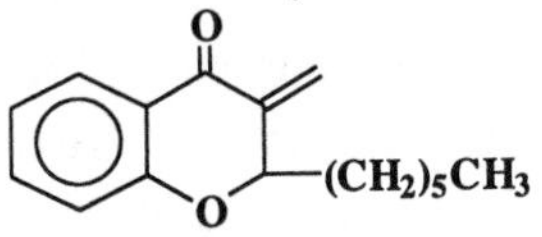	74
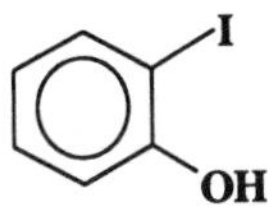	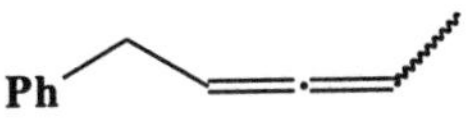	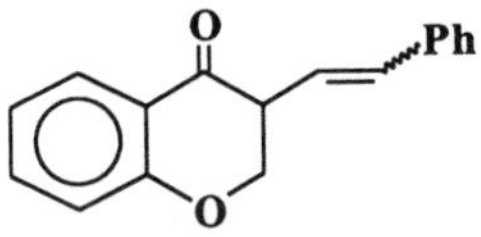	91
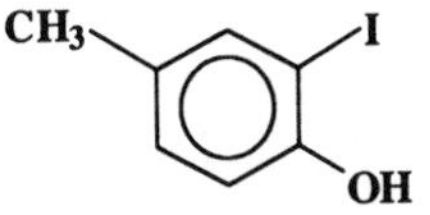	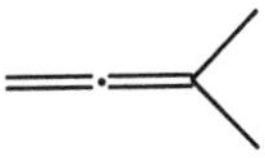	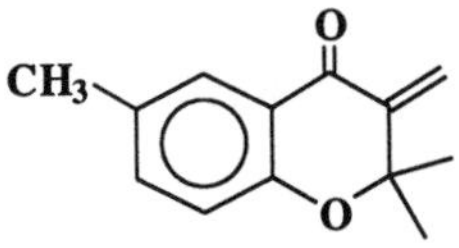	73
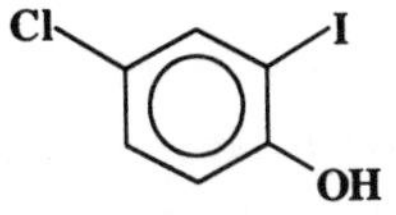	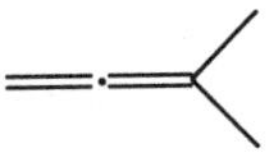	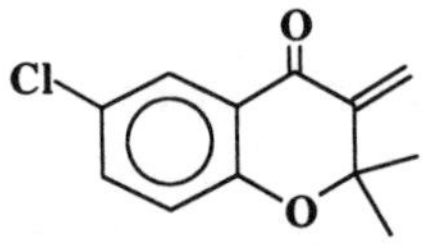	23

[a] Reaction conditions: [iodophenol]/[allene]/[Pd]/[dppb]/[(i-Pr)₂NEt] (1.5:3.0:0.075:2.0), benzene (5.0 ml), 20 atm CO, 100 °C, 20 h.
[b] Isolated yield based on iodophenol.

4   Metal-catalysed reactions of heterocumulenes with heterocycles

The cycloaddition reaction of three- and four-membered ring heterocycles with hetero-cumulenes is an efficient method to prepare five- and six-membered ring heterocycles, respectively [64,65]. Reaction of aziridines with carbodiimides in toluene using bis(benzonitrile) palladium dichloride ($PdCl_2(PhCN)_2$) as the catalyst afforded imida-zolidineimines (**72**) (Equation (13.52)). The use of tetrakis(triphenylphosphine) palladium(0) or bis (dibenzylideneacetone) palladium(0), with or without 2 equiv of benzonitrile (relative to palladium), resulted in complete recovery of the aziridine [66].

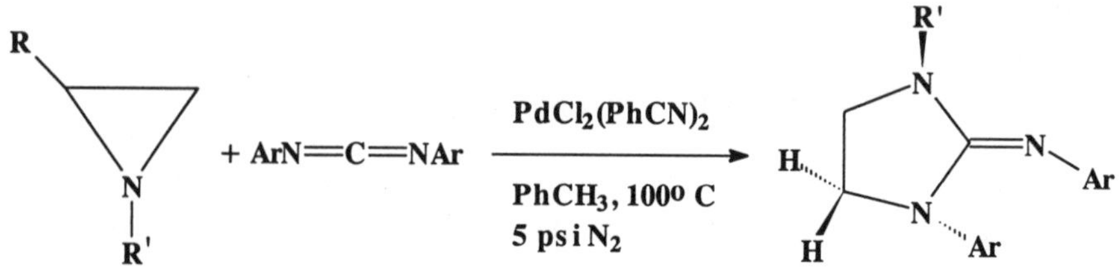

$$ (13.52) $$

72 (40 - 95 %)

R = Ph, p-PhC$_6$H$_4$, p-BrC$_6$H$_4$
R' = C(CH$_3$)$_3$, 1- adamantyl
Ar = Ph, p-tolyl

It was anticipated that replacement of the heterocumulene carbon of carbodiimides (Equation (13.52)) by a sulfur atom (i.e. sulfur diimide) in the palladium-catalysed reaction would result in the formation of a thiodiazolidinethione (**73**) (Scheme 13.5). The latter compound was not obtained; rather, a unique cyclization to form imidazolidinethione occurred instead. Reaction of a labeled aziridine (labeled with a ^{13}C atom as one of the ring carbons) with a sulfur diimide in the presence of catalytic quantities of $PdCl_2(PhCN)_2$ resulted in incorporation of the label at the 2- and 5-positions in the imidazolidinethione (**73**) (Scheme 13.5) [67].

Treatment of 1-(1-adamantyl)-2-phenylaziridine with phenyl isothiocyanate under the same conditions gave thiazolidinimine (**75**) in 85% yield, and no imidazolidinethione was detected in the reaction (Equation (13.53)) [67]. This result is analogous to the organoantimony halide-based cycloaddition of aziridines with phenyl isothiocyanate [14].

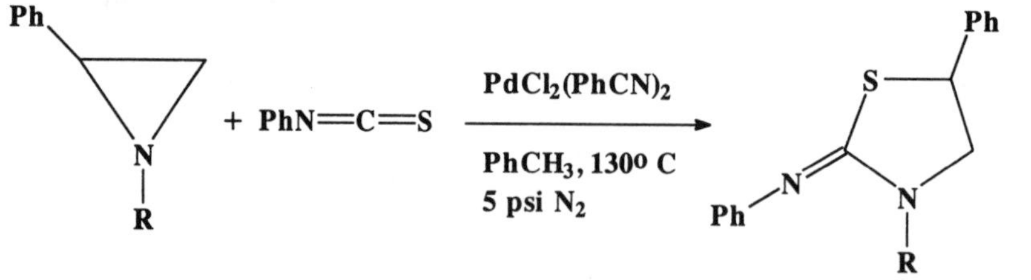

$$ (13.53) $$

R =1- adamantyl

75 (85%)

The cycloaddition of aziridines and heterocumulenes is both stereo- and enantiospecific [68]. Treatment of cis-1-n-butyl-2-carboalkoxy-3-methylaziridines, which contain an ester group, with carbodiimides, isothiocyanates or arylisocyanates in the presence of a catalytic amount of $(PhCN)_2PdCl_2$ exclusively affords the cis-product in good yields (Equation (13.54)) [68].

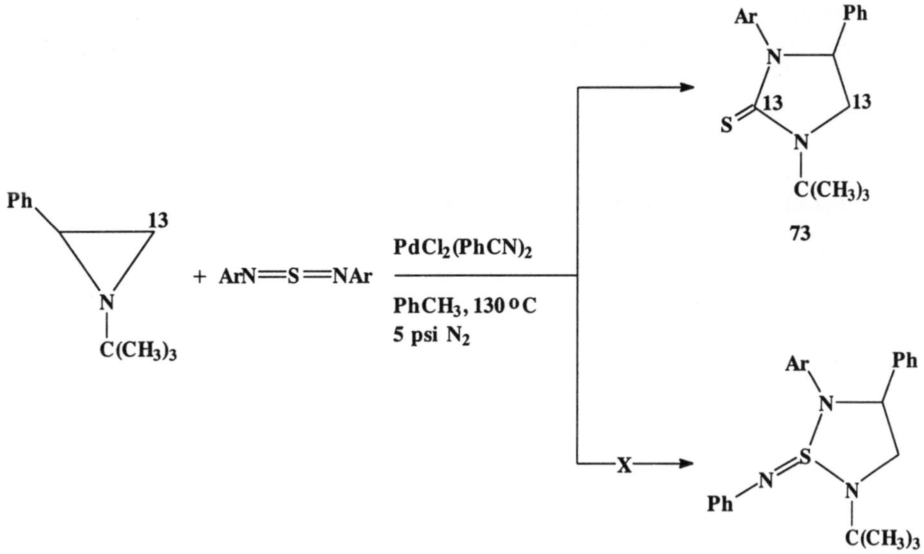

Scheme 13.5.

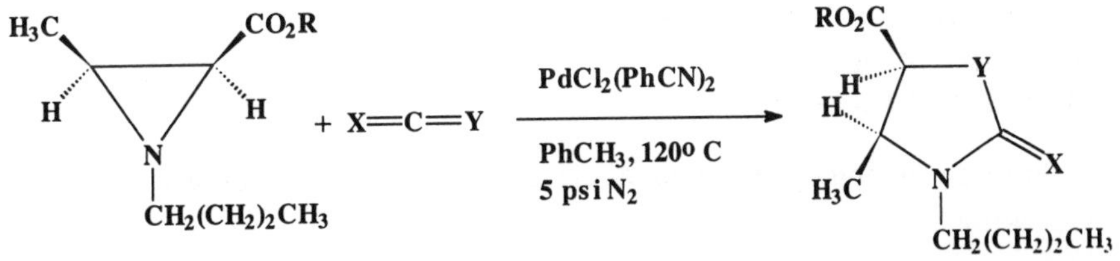

$$\begin{aligned}&R = CH_3, C_2H_5\\&X = PhN, O, p\text{-}ClC_6H_4N\\&Y = PhN, p\text{-}ClC_6H_4N, S\end{aligned} \qquad 76\ (60\text{-}86\%)$$

In order to determine the enantioselectivity of the cycloaddition reaction, several optically pure aziridines were treated with carbodiimides, isothiocyanates or aryl isocyanates under the same conditions as in Equation (13.54). The reaction proceeds with retention of configuration (Equation (13.55)) [68].

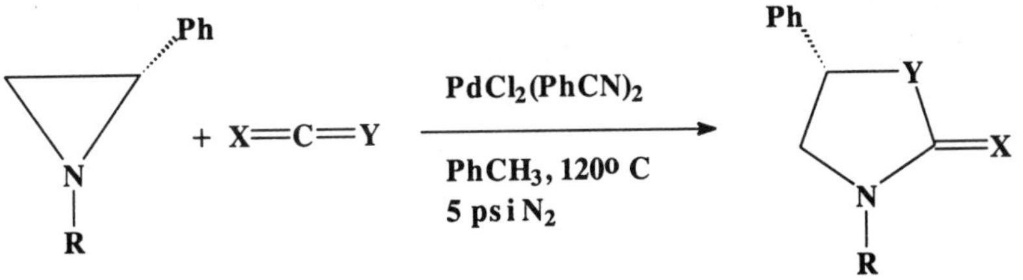

Azetidines undergo reaction with heterocumulenes in the same manner as aziridines, e.g. with carbodiimides, tetrahydropyrimidin-2-imines (**78**) were isolated in high yields (Table 13.5). The reaction is both regio- and stereospecific. The cycloaddition occurs with retention of configuration of the carbon centers bearing the substituent groups (Equation (13.56)) [69].

Table 13.5. Reaction of azetidines with carbodiimides catalysed by $PdCl_2(PhCN)_2$[a]

R^1	R^2	Ar	Isolated yield of compound **78** (%)
$C(CH_3)_3$	CH_3	Ph	92
$C(CH_3)_3$	$PhCH_2$	Ph	94
$C(CH_3)_3$	$PhCH_2$	p-Tolyl	64
C_6H_{11}	CH_3	p-ClC_6H_4	88
$C(CH_3)_3$	$PhCH_2$	p-ClC_6H_4	97

[a] Reaction conditions: azetidine (1.0 mmol), carbodiimide (1.0 mmol), $(PhCN)_2PdCl_2$ (0.1 mmol), $PhCH_3$ (3.0 ml), 130 °C, 5 psi N_2, 48 h.

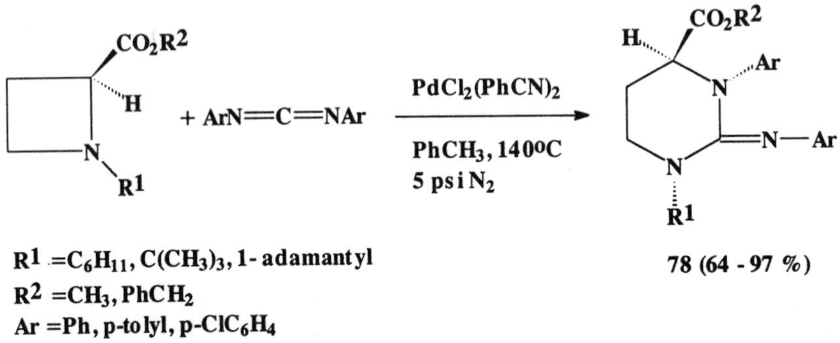

$$R^1 = C_6H_{11}, C(CH_3)_3, 1\text{-adamantyl}$$
$$R^2 = CH_3, PhCH_2$$
$$Ar = Ph, p\text{-tolyl}, p\text{-}ClC_6H_4$$

78 (64 - 97 %)

$$\tag{13.56}$$

Similarly, isothiocyanates gave tetrahydro-1,3-thiazin-2-imines (**79**) in excellent yields (Equation (13.57)) (Table 13.6) [70].

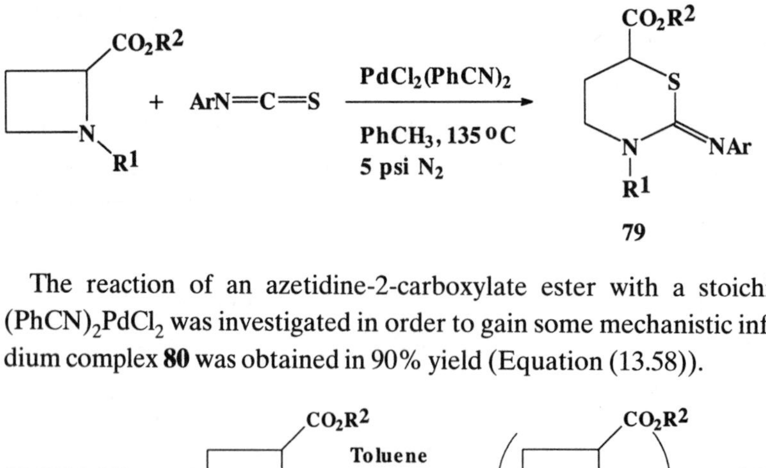

$$\tag{13.57}$$

79

The reaction of an azetidine-2-carboxylate ester with a stoichiometric amount of $(PhCN)_2PdCl_2$ was investigated in order to gain some mechanistic information. The palladium complex **80** was obtained in 90% yield (Equation (13.58)).

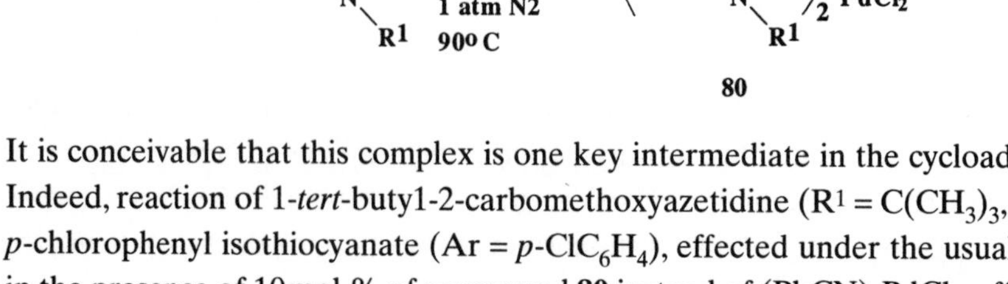

$$\tag{13.58}$$

80

It is conceivable that this complex is one key intermediate in the cycloaddition reaction. Indeed, reaction of 1-*tert*-butyl-2-carbomethoxyazetidine ($R^1 = C(CH_3)_3$, $R^2 = CH_3$) with p-chlorophenyl isothiocyanate (Ar = p-ClC_6H_4), effected under the usual conditions but in the presence of 10 mol.% of compound **80** instead of $(PhCN)_2PdCl_2$, afforded the same cycloaddition product **79** in almost the same yields (Equation (13.59)) [70].

Table 13.6. Reaction of azetidines with isothiocyanates catalysed by $(PhCN)_2PdCl_2$[a]

R^1	R^2	Ar	Isolated yield of compound **79** (%)
$C(CH_3)_3$	$PhCH_2$	Ph	82
$C(CH_3)_3$	CH_3	$p\text{-}ClC_6H_4$	90
C_6H_{11}	CH_3	$p\text{-}NO_2C_6H_4$	85
$C(CH_3)_3$	CH_3	$p\text{-}NO_2C_6H_4$	92
$C(CH_3)_3$	CH_3	$p\text{-}NO_2C_6H_4$	92
1-Adamantyl	CH_3	$p\text{-}NO_2C_6H_4$	89

[a] Reaction conditions: azetidine (1.0 mmol), isothiocyanate (1.0 mmol), $(PhCN)_2PdCl_2$ (0.1 mmol), $PhCH_3$ (2.0 ml), 135 °C, 48 °C, 5 psi N_2.

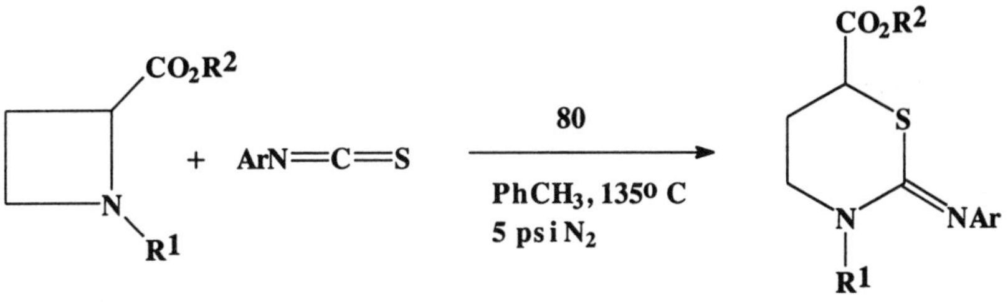

$$R^1 = C(CH_3)_3, R^2 = CH_3$$
$$Ar = p\text{-}ClC_6H_4$$

79 (89 %)

A possible mechanism for the palladium-catalysed cycloaddition reaction is presented in Scheme 13.6. Reaction of $(PhCN)_2PdCl_2$ with 2 molar equiv. of the azetidine affords the palladium–azetidine N-donor ligand complex **80**. Reaction of the latter with the aryl isothiocyanate may give compound **81**, in which there is *p*-complexation of one of the double bonds of the aryl isothiocyanate to palladium [71]. Subsequent cycloaddition of the azetidine to the uncomplexed double bond of the aryl isothiocyanate ligand, possibly via a four-membered transition-state compound **82**, may give compound **83** [67,69]. Decomplexation of compound **82** with additional azetidine would afford the corresponding terahydro-1,3-thiazin-2-imines **79** and regenerate compound **80**.

Another class of heterocumulenes that react with aziridines are ketenimines. The hard Lewis acid $LiClO_4$ proved to be superior to the soft $[PhCN]_2PdCl_2$, affording higher yields of iminopyrrolidines (**84**) under mild conditions (Equation (13.60)) [72].

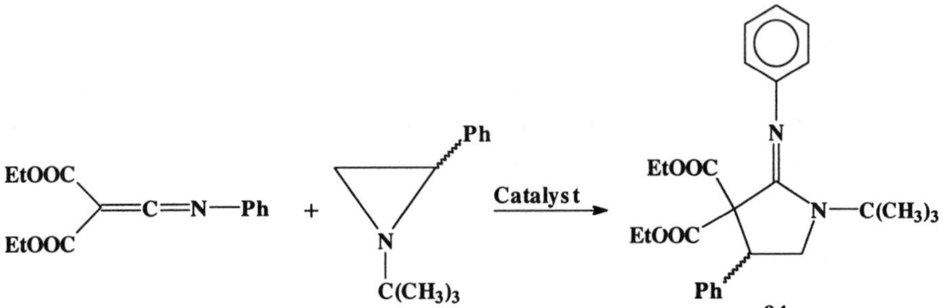

Catalyst	T,°C	time, h	Yield, %
$(PhCN)_2PdCl_2$ (10% mol)	75	18	60
$LiClO_4$ (5% mol)	r.t.	5	76

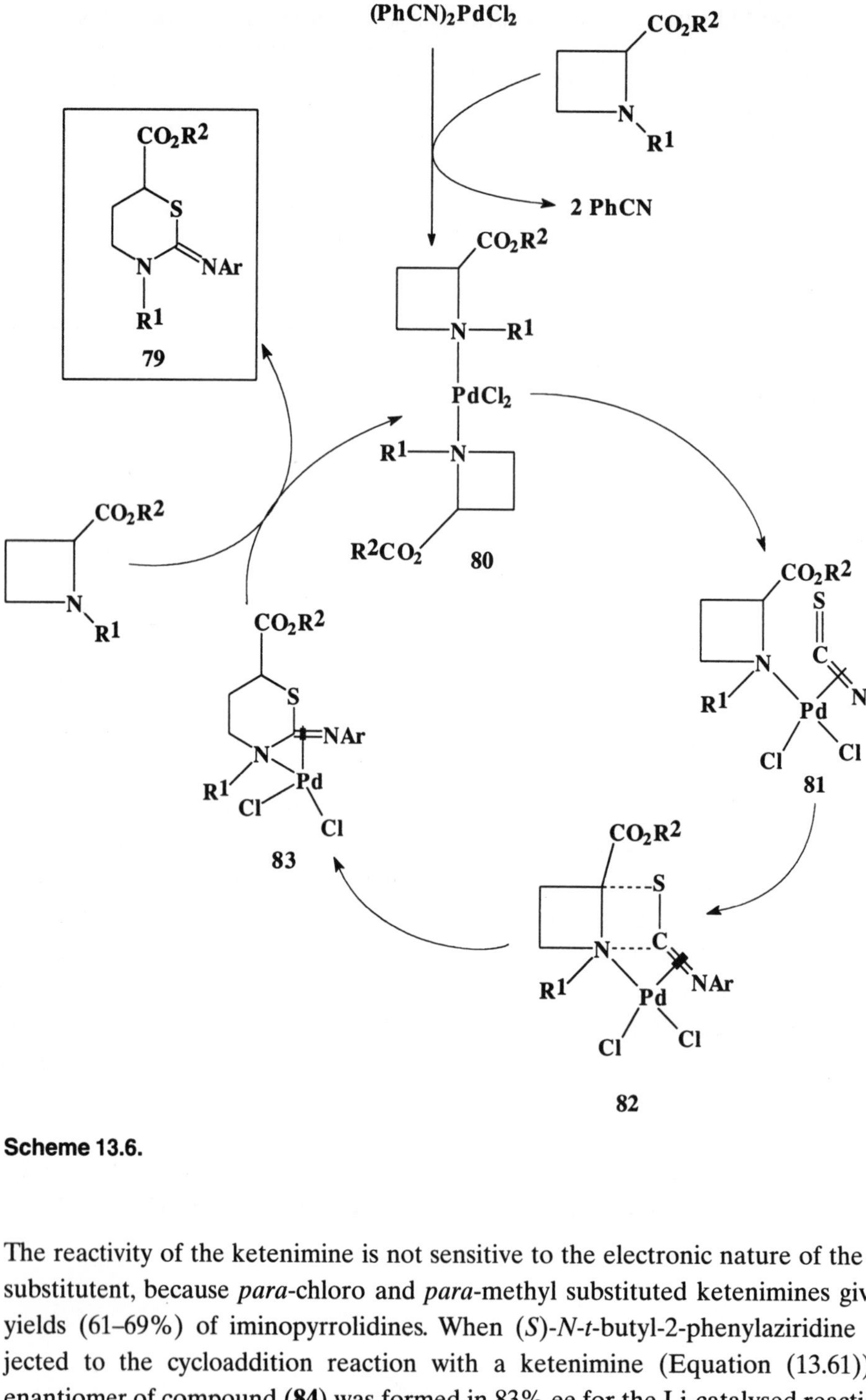

Scheme 13.6.

The reactivity of the ketenimine is not sensitive to the electronic nature of the aromatic substitutent, because *para*-chloro and *para*-methyl substituted ketenimines give similar yields (61–69%) of iminopyrrolidines. When (*S*)-*N*-*t*-butyl-2-phenylaziridine was subjected to the cycloaddition reaction with a ketenimine (Equation (13.61)), the *R*-enantiomer of compound (**84**) was formed in 83% ee for the Li-catalysed reaction and in 93% ee for the palladium-catalysed reaction.

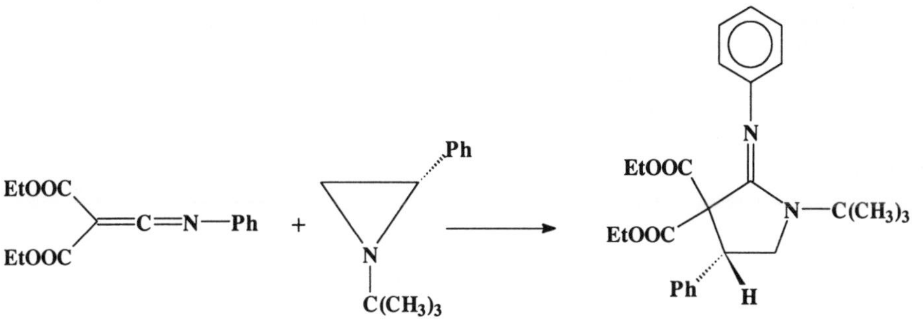

$$\begin{array}{c}\text{EtOOC}\\ \text{EtOOC}\end{array}\!\!=\!\!C\!=\!N\!-\!Ph \quad + \quad \text{(S)} \quad \longrightarrow \quad \text{(R) - 84} \qquad (13.61)$$

The cycloaddition reaction of oxiranes with heterocumulenes is also an efficient method for the synthesis of five-membered ring heterocycles [64]. Trost and Sudhakar reported the effective conversion of vinyloxiranes to *cis*-oxazolidin-2-ones (**85**) catalysed by $Pd_2(dba)_3 \cdot CHCl_3$ and triisopropylphosphite (Equation (13.62)), and the stereochemical features were rationalized on the basis of steric interactions between the **R** group on the vinyloxirane and triisopropylphosphite ligands [73].

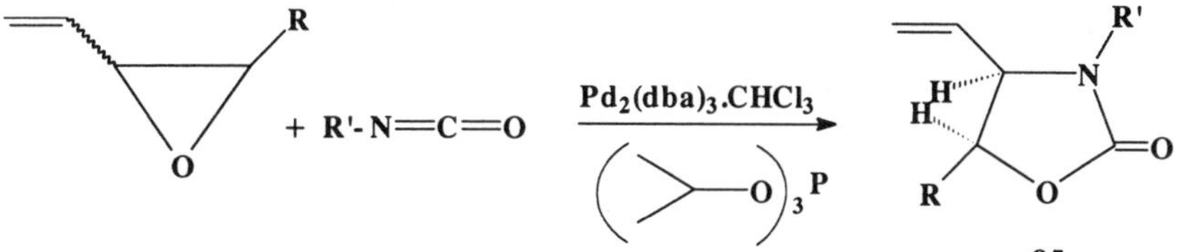

$$+ \quad R'\text{-}N\!=\!C\!=\!O \quad \xrightarrow[\left(\raisebox{-2pt}{$\displaystyle \triangleright$}\!-\!O\right)_3 P]{Pd_2(dba)_3 \cdot CHCl_3} \quad \textbf{85} \qquad (13.62)$$

Reaction of the same oxiranes with carbodiimides or isocyanates, with $Pd(PPh_3)_4/PPh_3$ as the catalyst system, gave oxazolidinoneimines and oxazolidinones, respectively, in 90–98% yields (Equation (13.63)) [74].

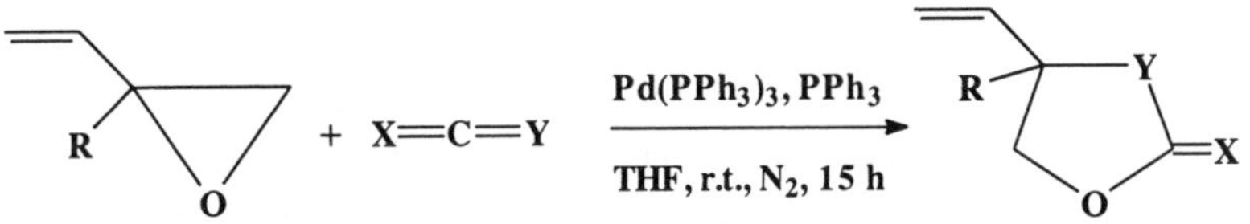

$$+ \quad X\!=\!C\!=\!Y \quad \xrightarrow[\text{THF, r.t., } N_2, \text{ 15 h}]{Pd(PPh_3)_3, PPh_3} \quad \textbf{86 (90-98\%)} \qquad (13.63)$$

R =H, CH₃
X =O, C₆H₅N, p-ClC₆H₄N, p-CH₃C₆H₄N
Y =C₅H₆N, p-CH₃C₆H₄N, p-ClC₆H₄N, p-BrC₆H₄N

A high degree of asymmetric induction was achieved using carbodiimides as the hetero-cumulenes, $Pd_2(dba)_3 \cdot CHCl_3$ as the catalyst and either (S)-TolBINAP or (R)-BINAP as the chiral ligand (Equation (13.64)) (Table 13.7) [74].

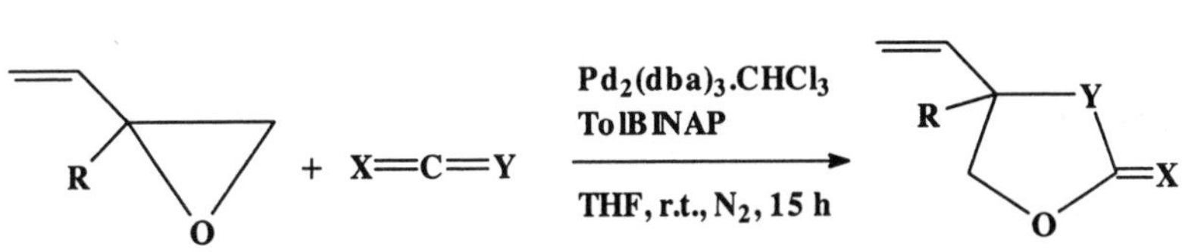

$$+ \quad X\!=\!C\!=\!Y \quad \xrightarrow[\text{THF, r.t., } N_2, \text{ 15 h}]{\substack{Pd_2(dba)_3 \cdot CHCl_3 \\ \text{TolBINAP}}} \quad \textbf{87} \qquad (13.64)$$

The reaction temperature does not affect significantly the enantioselectivity of the reaction. High stereoselectivity is still maintained at a lower reaction temperature (5–22 °C),

Table 13.7. Asymmetric cycloaddition of 2-vinyloxirane with heterocumulenes in the presence of $Pd_2(dba)_3.CHCl_3$/TolBINAP as the catalyst[a]

Oxirane	Heterocumulene			Heterocycle	
R	X	Y	Ligand[b]	yield (%)[c]	ee (%)[d]
H	C_6H_4N	C_6H_4N	A	98	93
H	p-ClC_6H_4N	p-ClC_6H_4N	A	95	94
H	p-$CH_3C_6H_4N$	p-$CH_3C_6H_4N$	A	98	93
H	p-BrC_6H_4N	p-BrC_6H_4N	A	84	94
H	o-$CH_3C_6H_4N$	o-$CH_3C_6H_4N$	A	98	84
H	o-$CH_3C_6H_4N$	o-$CH_3C_6H_4N$	B	98	88
CH_3	C_6H_4N	C_6H_4N	B	87	91
CH_3	p-ClC_6H_4N	p-ClC_6H_4N	A	98	69
CH_3	p-$CH_3OC_6H_4N$	p-$CH_3OC_6H_4N$	A	98	75
H	O	p-ClC_6H_4N	B	94	43
H	O	C_6H_4N	A	99	49

[a] Reaction conditions: oxirane (1 mmol), heterocumulene (1 mmol), $Pd_2(dba)_3.CHCl_3$ (0.03 mmol), ligand (0.06 mmol), THF (5 ml), room temperature, 15 h, N_2 atmosphere.
[b] A = (S)-TolBINAP; B = (R)-TolBINAP.
[c] Isolated yield.
[d] Determined by HPLC analysis using chiral OD with 15% i-PrOH in n-hexane.

which is in contrast to the results reported previously for the preparation of vinyloxazolidin-2-one [75]. The optimum temperature for achieving the asymmetric cycloaddition of 2-vinyloxirane with heterocumulene in the presence of Pd(0) and chiral phosphine ligands is in the range 10–22°C, although excellent results are also obtained at high temperature [74].

4-Vinyl-1,3-oxazolidin-2-ones (**87**; Table 13.7, X = O) were obtained from the reaction of 2-vinyloxirane with iosocyanates. The % ee of these products is appreciably lower than those derived from carbodiimides, suggesting that TolBINAP has less influence in the steric interaction in the enantiodetermination step. The results from Table 13.7 indicate that cycloaddition reactions of 2-vinyloxirane and carbodiimides are influenced not only by the ligands but also by the structure of the reaction partners. In reaction with 2-vinyloxirane, carbodiimides provide greater steric interaction between the substituent of the nitrogen nucleophile and the substituent on the chiral phosphine ligands, in comparison with reaction using isocyanates. Consequently, higher enantiomeric excesses were realized in using carbodiimides than isocyanates.

5   Conclusion

In this account we have described the role of transition metal chemistry in the synthesis of various heterocyclic compounds. Recent advances in carbonylative cyclization catalysed by transition metal complexes demonstrate the utility of these processes in synthetic organic chemistry. A large variety of N-, O- and S-containing heterocyclic compounds undergo carbonylative ring expansion. These reactions afford different substituted heterocycles, some of which are difficult or impossible to obtain by other methods. The intramolecular and intermolecular cyclocarbonylation is also a valuable method for the synthesis of bicyclic, tricyclic and pentacyclic compounds containing five-, six- or seven-membered ring lactones and lactams. Furthermore, the cycloaddition reaction of three- and four-membered ring heterocycles with heterocumulenes is a useful method for the synthesis of five- and six-membered ring heterocycles. The presence of an appropriate chiral ligand resulted in the asymmetric synthesis of various heterocycles in high enantiomeric excess. The cycloaddition reaction displays excellent regio- and stereochemical control, making this a method of value for the construction of heterocycles with stereochemically defined substituent groups.

6 Acknowledgments

We are grateful to the University of Ottawa and the King Fahd University of Petroleum and Minerals for support during the realization of this work.

7 References

1 Cornils B, Herrmann WA. *Applied Homogeneous Catalysis with Organometallic Compounds.* New York: VCH, 1996.

2 Parshall GW, Ittel SD. *Homogeneous Catalysis.* New York: Wiley–Interscience, 1993.

3 Master C. *Homogeneous Transition-Metal Catalysis.* New York: Wiley–Interscience, 1993.

4 Colquhoun HM, Thompson DJ, Twigg MV. *Carbonylation.* New York: Plenum Press, 1991.

5 Hosokawa T, Murahashi SI. *Hetercycles* 1992; **33**: 2.

6 Khumtaveeporn K, Alper H. *Acc Chem Res* 1995; **28**: 414.

7 El Ali B, Alper H. (1998) In: Bolm C, Beller M, eds. *Transition Metals for Fine Chemicals and Organic Synthesis.* p. 49. Weinheim, Germany: VCH (in press).

8 Negishi E, Coperet C, Ma S, Lion SY, Lin F. *Chem Rev* 1996; **96**: 635.

9 Alper H. *J Organomet Chem* 1986; **300**: 1.

10 Alper H. *Pure Appl Chem* 1988; **60**: 35.

11 Shibata I, Baba A, Iwasaki H, Matsuda H. *J Org Chem* 1986; **51**: 2177.

12 Trost BM, Sudhakar AR. *J Am Chem Soc* 1987; **109**: 3792.

13 Fujiwara M, Baba A, Matsuda H. *J Heterocycl Chem* 1988; **25**: 135.

14 Nomura R, Nakano T, Nisho Y *et al. Chem Ber* 1989; **122**: 2409.

15 Fujiwara M, Imada M, Baba A, Matsuda H. *J Org Chem* 1988; **53**: 5974.

16 Aumann R, Frohlich K, Ring H. *Angew Chem Int Ed Engl* 1974; **13**: 275.

17 Aumann R, Ring H. *Angew Chem Int Ed Engl* 1977; **16**: 50.

18 Kamiya Y, Kawato K, Ohta H. *Chem Lett* 1980: 1549.

19 Drent E, Kragtwijk E. *European Patent Application EP 577,206* 1994; *Chem Abstr* 1994; **120**: 191517c.

20 Chamchaang W, Pinhas AR. *J Chem Soc Chem Commun* 1988: 70.

21 Chamchaang W, Pinhas AR. *J Org Chem* 1990; **55**: 2943.

22 Alper H, Urso F, Smith DJH. *J Am Chem Soc* 1983; **105**: 6735.

23 Calet S, Urso F, Alper H. *J Am Chem Soc* 1989; **111**: 931.

24 Piotti ME, Alper H. *J Am Chem Soc* 1996; **118**: 111.

25 Alper H, Perera CP, Ahmed FR. *J Am Chem Soc* 1989; **103**: 1289.

26 Alper H, Mahatantila CP. *Heterocycles* 1983; **20**: 2025.

27 Alper H, Hamel N. *Tetrahedron Lett* 1987; **28**: 3237.

28 Tanaka H, Abdul Hai AKM, Sadadane M, Okumoto H, Torii S. *J Org Chem* 1994; **59**: 3040.

29 Zhou Z, Alper H. *J Org Chem* 1996; **61**: 1256.

30   Alper H, Arzoumanian H, Petrignani JF, Maldonado MS. *J Chem Soc Chem Commun* 1985: 340.

31   Alper H, Eisentat A, Satyanarayana N. *J Am Chem Soc* 1990; **112**: 7060.

32   Nagao Y, Dai W, Ochiai M, Shiro M. *J Org Chem* 1989; **54**: 5211; Knight DW. *Org Synth* 1994; **1**: 287.

33   El Ali B, Alper H. *J Org Chem* 1991; **56**: 5357.

34   Yu WY, Alper H. *J Org Chem* 1997; **62**: 5684.

35   Osakada K, Chiba I, Nakamura Y, Yamamoto T, Yamamoto A. *J Chem Soc Chem Commun* 1986: 1589; Murahashi SI, Imada I, Taniguchi Y, Higashimura S. *J Org Chem* 1993; **58**: 1538.

36   Matsuhita K, Komori T, Oi S, Inoue Y. *Tetrahedron Lett* 1994; **35**: 5889.

37   Crisp GT, Meyer AG. *J Org Chem* 1992; **57**: 6972.

38   Sugioka T, Zhang SW, Morii N, Joh T, Takahashi S. *Chem Lett* 1996: 249.

39   Coperet C, Sugihara T, Wu G, Shimoyama I, Negishi EI. *J Am Chem Soc* 1995; **117**: 3422.

40   Huang Y, Alper H. *J Org Chem* 1991; **56**: 4534.

41   Wang MD, Calet S, Alper H. *J Org Chem* 1989; **54**: 20.

42   Alper H, Leonard D. *J Chem Soc Chem Commun* 1985: 511; *Tetrahedron Lett* 1985; **6**: 5639.

43   Brunner M, Alper H. *J Org Chem* 1997; **62**: 7565.

44   Tamaru Y, Hojo M, Yoshida Z. *J Org Chem* 1991; **56**: 1099.

45   Bando T, Tanaka S, Fugami K, Yoshida Z, Tamaru Y. *Bull Chem Soc Jpn* 1992; **65**: 97.

46   Alper H, Hamel N. *J Chem Soc Chem Commun* 1990: 135.

47   Yu WY, Bensimon C, Alper H. *Chem Eur J* 1997; **3**: 417.

48   Giattini PG, Mastropietro G, Morera E, Ortar G. *Tetrahedron Lett* 1993; **34**: 3763.

49   El Ali B, Okuro K, Vasapollo G, Alper H. *J Am Chem Soc* 1996; **118**: 4264.

50   Coperet C, Sugihara T, Negishi E. *Tetrahedron Lett* 1995; **36**: 1771.

51   Coperet C, Ma S, Sugihara T, Negishi E. *Tetrahedron* 1990; **35**: 11529.

52   Roberto D, Alper H. *J Am Chem Soc* 1989; **111**: 7539.

53   Zhou JQ, Alper H. *J Org Chem* 1992; **57**: 3328.

54   Hirao K, Morii N, Joh T, Takahashi S. *Tetrahedron Lett* 1995; **36**: 6243.

55   Catellani M, Chiusoli GP, Fagnola MF, Salari G. *Tetrahedron Lett* 1994; **35**: 5923.

56   Grigg R, Khalil H, Levett P, Virica J, Sridharan V. *Tetrahedron Lett* 1994; **35**: 3197.

57   Okuro K, Kai H, Alper H. *Tetrahedron Asymm* 1997; **8**: 1.

58   Alper H, Delledonne D, Kameyama M, Roberto D. *Organometallics* 1990; **9**: 762.

59   Wang MD, Calet S, Alper H. *J Org Chem* 1989; **54**: 20.

60   Khumtaveeporn K, Alper H. *J Org Chem* 1995; **60**: 8142.

61   Khumtaveeporn K, Alper H. *J Am Chem Soc* 1994; **116**: 5662.

62   Okuro K, Dang T, Khumtaveeporn K, Alper H. *Tetrahedron Lett* 1996; **37**: 2713.

63   Okuro K, Alper H. *J Org Chem* 1997; **62**: 1566.

64   Katrizky AR, Rees CW. *Comprehensive Heterocyclic Chemistry.* Oxford: Pergamon Press, 1984: Parts 1 and 4B.

65   Shibata I, Toyota M, Baba A, Matsuda H. *J Org Chem* 1990; **55**: 2487; Karikomi M, Yamazaki T, Toda T. *Chem Lett* 1993: 1965.

66   Baeg JO, Alper H. *J Org Chem* 1992; **57**: 157.

67   Baeg JO, Alper H. *J Am Chem Soc* 1994; **116**: 1220.

68   Baeg JO, Bensimon C, Alper H. *J Am Chem Soc* 1995; **117**: 4700.

69   Baeg JO, Bensimon C, Alper H. *J Org Chem* 1995; **60**: 253.

70   Baeg JO, Alper H. *J Org Chem* 1995; **60**: 3092.

71   Hoberg H, Korff J. *J Organomet Chem* 1978; **15**: C20.

72   Maas H, Bensimon C, Alper H. *J Org Chem* 1998; **63**: 17.

73   Trost BM, Sudhakar AR. *J Am Chem Soc* 1988; **110**: 7933.

74   Larksarp C, Alper H. *J Am Chem Soc* 1997; **119**: 3709.

75   Hayashi T, Yamamoto A, Ito Y. *Tetrahedron Lett* 1988; **29**: 99.

14 Catalytic Enantioselective Cyclopropanation, Carbon–Hydrogen Insertion and Ylide Reactions

MICHAEL P. DOYLE and DAVID C. FORBES

Department of Chemistry, University of Arizona, Tucson, Arizona 85723, USA

1 Introduction

The search for asymmetric induction in homogeneous catalytic reactions had its origins in the development of chiral ligands for copper catalysts in metal carbene transformations [1, 2]. In 1966 Nozaki, Moriuti, Takaya and Noyori reported the first example of an asymmetric cyclopropanation reaction [3] and, although the enantiomeric excess was less than 10%, this example set in motion extensive efforts that evolved the field of asymmetric catalysis to its present state [4]. This first chiral catalyst was an imine derived from salicylaldehyde and 1-phenethylamine (**1**), whose impact was to be felt not only in the effective structural

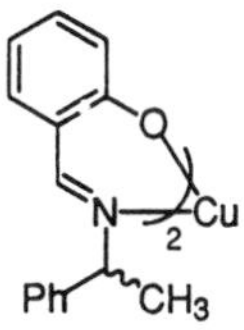

A = CH₃, CH₂Ph

R = —

B = C₄H₉, C₈H₁₇

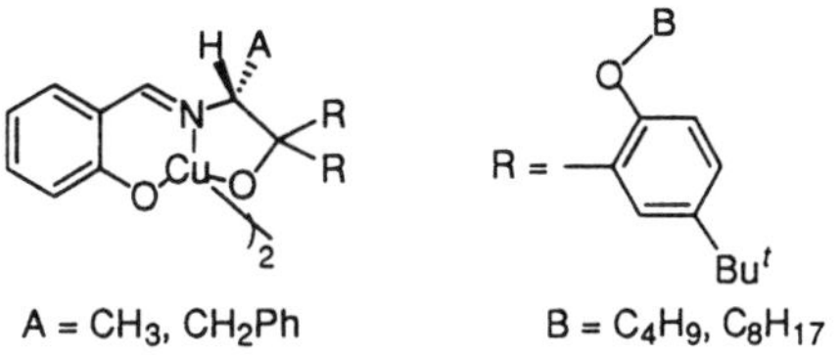

modifications made by Aratani (**2**) [5] but also in salen complexes subsequently developed by Jacobsen [6] and Fukuda and Katsuki [7] for epoxidation and cyclopropanation (**3**), respectively.

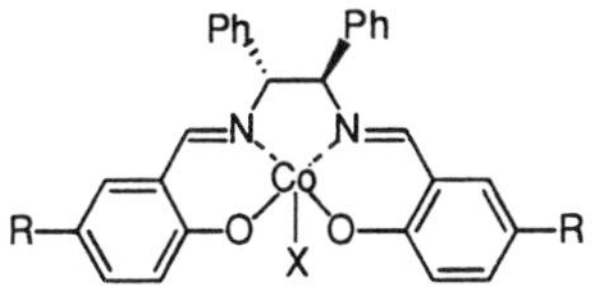

3

(R = H, ᵗBu)

A second major advance in the development of chiral catalysts for asymmetric cyclopropanation reactions was made by Pfaltz and co-workers with the introduction in 1986 of semicorrin ligands (**4**) for copper [8]. Having the design of what came to be known as C_2-symmetric ligands [9], they were soon evolved to bis-oxazoline ligands (**5**) by Masamune *et al.* [10], Evans *et al.* [11] and Pfaltz *et al.* [12], and soon applied broadly in catalytic asymmetric transformations [13].

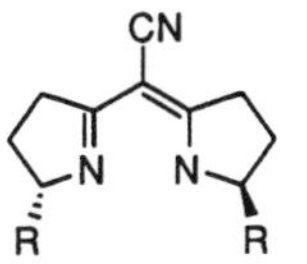

4

R = CMe₂OH,
CH₂OSiMe₂ᵗBu

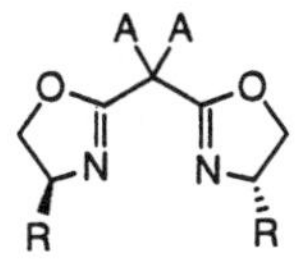

5

R = Ph, ᵗBu, CMe₂OH
A = H, CH₃

The success of these catalysts has been demonstrated in a limited number of intermolecular cyclopropanation reactions with diazoacetates, including those with styrene (**6**), isobutylene (**7**) and 2,5-dimethyl-2,4-hexadiene (**8**). Diastereocontrol is generally low in

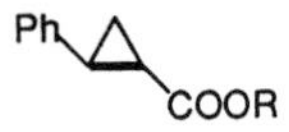

6

up to 99%
ee [11]

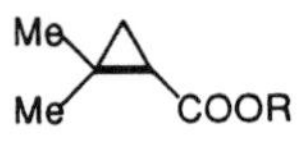

7

(precursor to
cilastatin)
up to >99%
ee [11]

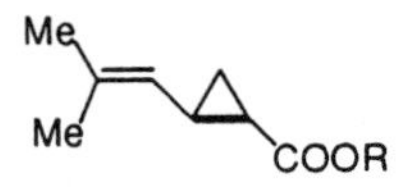

8

(precursor to
pyrethroids)
up to 94%
ee [14]

copper-catalysed reactions, but Nishiyama's chiral ruthenium(II) pybox catalyst (**9**) provides high diastereocontrol and high enantiocontrol in reactions with monosubstituted alkenes [15]. There is opportunity for improvement because at this time there is no single

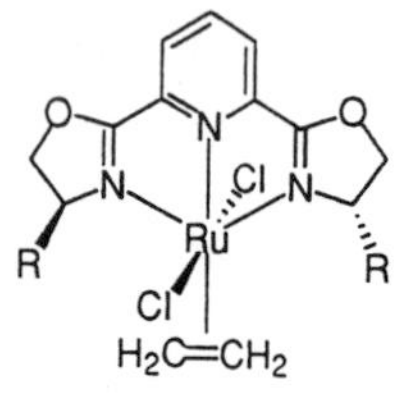

R = ^{i}Pr, Ph, ^{s}Bu, Bn

9

catalyst or ligand that is optimum for all alkenes or diazo compounds. Only reactions leading to compounds **6–8** and related cyclopropanes have been optimized with diazoacetates, and low enantiocontrol characterizes reactions performed with diazomalonates or diazoacetoacetates. However, vinyldiazoacetates have proven to be highly selective towards alkene cyclopropanation when catalysed by homochiral dirhodium(II) prolinate (**10**) in pentane [16], especially when applied to the synthesis of the antidepressant sertraline (**11**) [17].

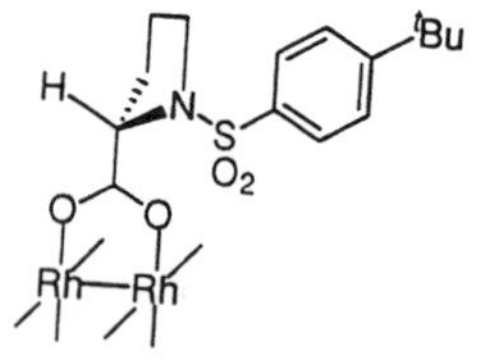

10

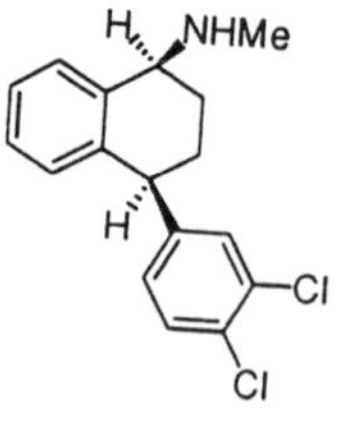

11

2   Chiral dirhodium(II) carboxamidates

Since the initial report of its catalytic activity by Teyssie and co-workers in 1973 [18], dirhodium(II) tetraacetate was known to be an exceptional catalyst for metal carbene transformation, surpassing all others in critical comparisons [19–23]. Despite its relatively high cost, turnover numbers and turnover rates are high, and this catalyst can be recovered and re-used. However, it is the less reactive, more selective carboxamide-ligated dirhodium(II) that has proven to be so highly versatile for highly selective catalytic intramolecular metal carbene reactions.

Constructed by ligand displacement from rhodium(II) acetate using a novel procedure in which acetic acid is trapped by sodium carbonate (Equation (14.1)) in a Soxhlet extraction apparatus [24], dirhodium(II) carboxamidates are formed in high yield with nearly

$$Rh_2(OAc)_4 + 4LH \rightleftharpoons Rh_2L_4 + 4HOAc \qquad (14.1)$$

$$NaOAc + NaHCO_3 \xleftarrow{Na_2CO_3}$$

exclusive selectivity for the (*cis*-2,2)-geometry. Accordingly, surrounding each rhodium of the dirhodium core are two nitrogens and two oxygens, and the two nitrogens (or oxygens) are adjacent to each other [25]. Figure 14.1 depicts the X-ray structure of a representative chiral dirhodium(II) carboxamidate as its nitrile complex [26]. The composite set of chiral carboxamidate ligands that have been developed for dirhodium(II) catalysts are those of oxopyrrolidines (**12**), oxooxazolidines (**13**), oxoimidazolidines (**14**) and 2-oxoazetidines (**15**) [25–28]. Each has its own particular advantage in enantioselective metal carbene transformations, and each has been characterized structurally.

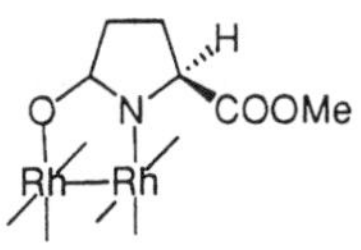

12

Rh$_2$(5*S*-MEPY)$_4$

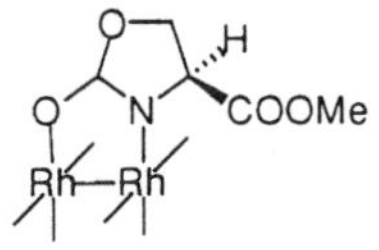

13

Rh$_2$(4*S*-MEOX)$_4$

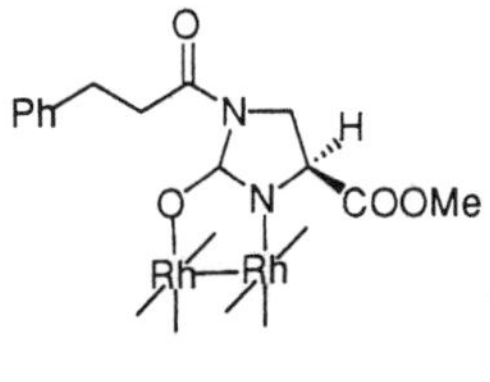

14

Rh$_2$(4*S*-MPPIM)$_4$

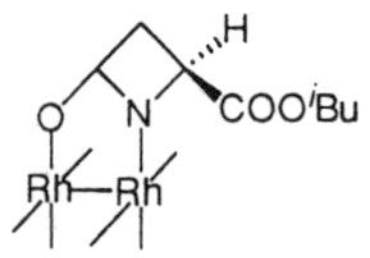

15

Rh$_2$(4*S*-IBAZ)$_4$

The close proximity of the carboxylate group to the carbene center greatly enhances enantioselection and, because only two quadrants of a partitioned volume surrounding rhodium are occupied, there is relatively open access to the carbene center for the reacting substrate.

3   Intramolecular cyclopropanation reactions

Allylic diazoacetates undergo diazo decomposition catalysed by chiral dirhodium(II) carboxamidates to produce bicyclic lactones from intramolecular cyclopropanation reactions in high yield [25, 26]. Enantioselectivities for this transformation (Equation (14.2)) generally exceed 90% enantiomeric excess (ee) with the Rh$_2$(MEPY)$_4$ catalysts [29],

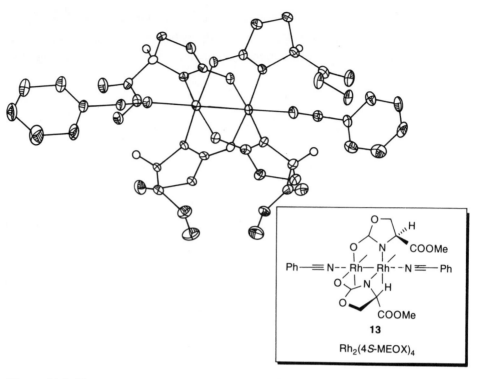

Figure 14.1. X-ray structure of dirhodium(II)tetrakis[methyl 2-oxooxazolidine-4(*S*)-carboxylate] as its benzonitrile complex.

(14.2)

> 90% ee

although with *trans*-disubstituted allylic diazoacetates [30] and methallyl diazoacetates [31], $Rh_2(MPPIM)_4$ catalysts provide the structural framework required to achieve ⩾90% ee. Representative examples of compounds formed by this methodology are compounds **16–20** [29–33]. Several have been used for the construction of peptide isosteres [34, 35], and compound **19**, which was formed in 96% yield from farnesyl diazoacetate, is a direct

16
93% ee

17
94% ee

18
95% ee

19
96% ee

20
98% ee

precursor to presqualene alcohol [33]. As seen by compound **20**, *N*-allylic diazoacetamides also undergo intramolecular cyclopropanation with high levels of enantiocontrol [32]. Turnover numbers for some of these reactions have been up to 1000, and even higher numbers can be anticipated in larger scale reactions.

Homoallylic diazoacetates and diazoacetamides also undergo intramolecular cyclopropanation with high enantioselectivity when catalysed by $Rh_2(MEPY)_4$ catalysts, although % ee values are generally 10–15% lower than those from reactions with their allylic counterparts [29]. Similarly, allylic α-diazopropanoates exhibit lower enantiocontrol than do the corresponding diazoacetates in intramolecular cyclopropanation [36], and diazoketones are relatively unresponsive to enantiocontrol with chiral dirhodium(II) catalysts [37]. By comparison, neither copper(I) catalysts with chiral bis-oxazoline ligands (**5**) nor the chiral ruthenium-pybox catalyst **9** exhibit enantiocontrol competitive with chiral dirhodium(II) carboxamidate catalysts [31] for reactions with diazoacetates or diazoacetamides, although the Pfaltz semicorrin-ligated copper(I) catalysts show high enantiocontrol for intramolecular cyclopropanation of some diazoketones.

4 Macrocyclization in intramolecular cyclopropanation

There is an unexpectedly high preference for macrocycle formation in intramolecular cyclopropanation reactions of selected diazoacetates catalysed by chiral bis-oxazoline ligated copper(I) (**21**). As shown by the example of Scheme 14.1, this preference is opposite to that of $Rh_2(5S\text{-}MEPY)_4$ [38] when there is a choice between allylic cyclopropanation and macrocyclization.

The more reactive catalysts are more suitable for macrocyclization. Ring sizes up to 20 have been achieved [39] and, curiously, the level of enantiocontrol does not change with ring size so long as $21/CuPF_6$ is employed; a significant increase in enantioselectivity accompanies the use of the chiral dirhodium(II) carboxamidate $Rh_2(5S\text{-}MEPY)_4$. The

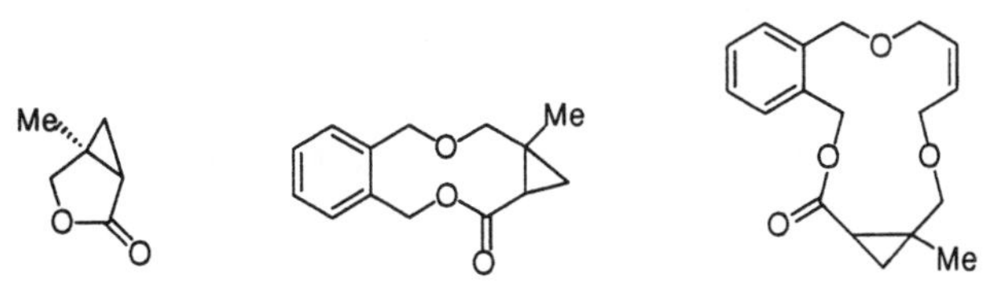

	25	**26**	**27**
$21/CuPF_6$	87% ee	90% ee	90% ee
$Rh_2(5S\text{-}MEPY)_4$	7% ee	47% ee	36% ee

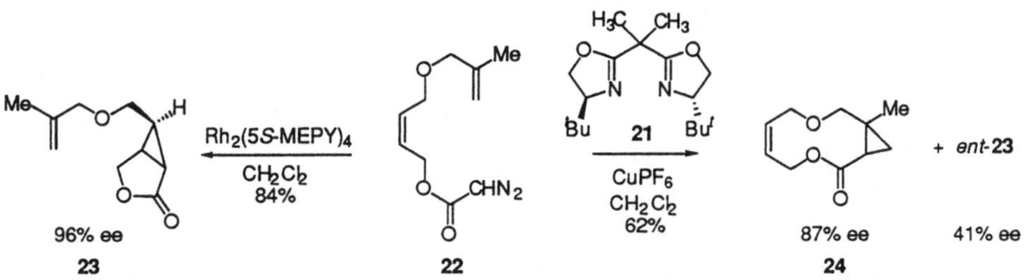

Scheme 14.1.

reason for this selectivity appears to lie in the ability of the metal carbene to form intramolecular π-complexes that collapse to the cyclopropane product. High dilution procedures are not required here.

5   Enantiomer differentiation in intramolecular cyclopropanation

Treatment of racemic 2-cyclohexenyl diazoacetate with $Rh_2(4S\text{-}MEOX)_4$ resulted in the formation of tricyclic lactone **28** in 40% yield and 94% ee (Equation (14.3)) together with the oxidation product 2-cyclohexenone [40]. Use of $Rh_2(4R\text{-}MEOX)_4$ produced the same result, except with the formation of compound **28** having the opposite configuration. In this

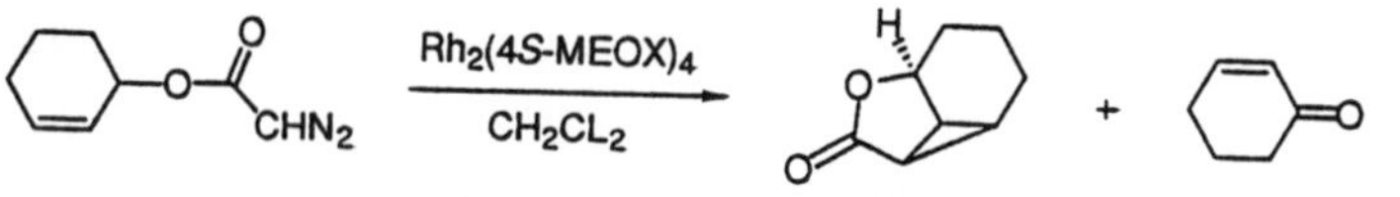

$$
\begin{array}{c}
\text{(14.3)}
\end{array}
$$

28
94%ee

and related examples, the chiral catalyst selects one enantiomer of the racemic mixture for intramolecular cyclopropanation while the other enantiomer undergoes hydride abstraction, which occurs with the loss of ketene [41]. This selectivity, which we call *enantiomer differentiation*, appears to be general for cyclic allyl diazoacetates but not for acyclic systems.

Diastereoselective match/mismatch has also been described [42]. For example, treatment of diazoacetate **29** with $Rh_2(5S\text{-}MEPY)_4$ results in intramolecular cyclopropanation with a diastereoselectivity for the formation of ***endo*-30** of greater than 20:1 (Equation

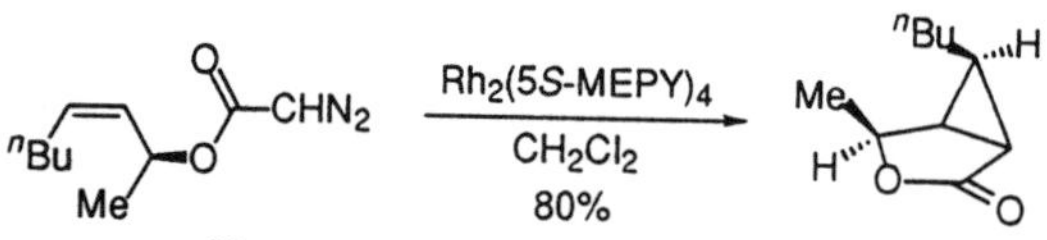

29

***endo*-30**

$$
\text{(14.4)}
$$

(14.4)). Using $Rh_2(5R\text{-}MEPY)_4$, on the other hand, the *endo/exo* ratio for 30 was only 1.0 : 1.5, and only a 30% yield of this product was obtained.

6   Carbon–hydrogen insertion reactions: lactones

The metal carbene transformation that has the highest potential synthetic utility is the insertion of a carbene moiety into a C–H bond, ordinarily to form a five-membered ring product in intramolecular reactions [1,19–22]. Initial efforts by McKervey *et al.* [43] and Hashimoto with Ikegami and co-workers [44] showed that homochiral dirhodium(II) carboxylates were effective for highly enantiocontrolled syntheses leading to products as diverse as compounds **31** and **32**. Although copper(I) catalysts are normally less effective for C–H insertion relative to those of rhodium(II), the process leading to compound **33** occurred readily with CuOTf/**21** [46] but not with dirhodium(II) catalysts. With chiral

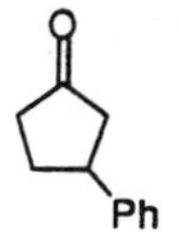

31 [44]   **32** [45]   **33** [46]
76% ee   79% ee   48% ee

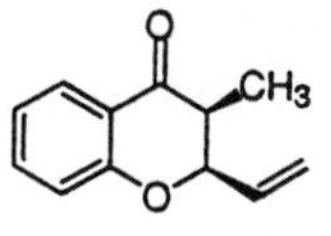

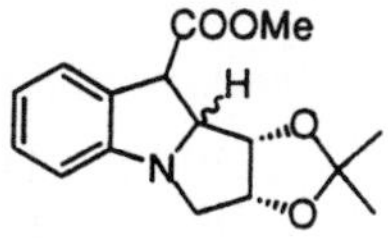

dirhodium(II) carboxamidates such as $Rh_2(5S\text{-MEPY})_4$, however, even higher levels of enantiocontrol could be achieved.

Doyle and co-workers have demonstrated exceptionally high enantiocontrol in the formation of β-alkoxy-γ-butyrolactones [47], as well as their β-alkyl and β-aryl counterparts [48], with results such as those shown for compounds **34–36**. This methodology, generally with $Rh_2(4R\text{-MPPIM})_4$ catalyst, has been employed for the synthesis of a vast array of

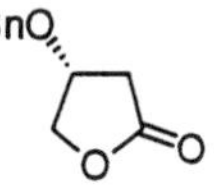

34 [47]   **35** [48]   **36** [48]
89% ee   96% ee   95% ee

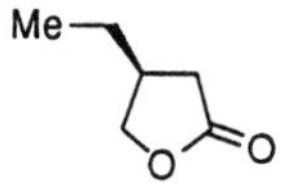

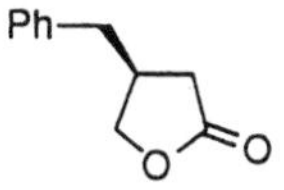

lignin lactones [48] in an efficient procedure that begins with the cinnamic acid. Regioselectivity for the formation of the five-membered ring lactones is ⩾95%.

Intramolecular C—H insertion with cycloalkyl diazoacetates occurs with high enantiocontrol and diastereocontrol in reactions catalysed by $Rh_2(4S\text{-MACIM})_4$ (**37**). Comparable enantioselectivities, but much lower diastereoselectivities, were observed with $Rh_2(5S\text{-MEPY})_4$ and $Rh_2(4S\text{-MEOX})_4$ [49], suggesting that the N-acetyl group is responsible for the high level of diastereocontrol observed in reactions of compound **38** with $Rh_2(4S\text{-MACIM})_4$ (Equation (14.5)). Similar selectivities were observed from reactions

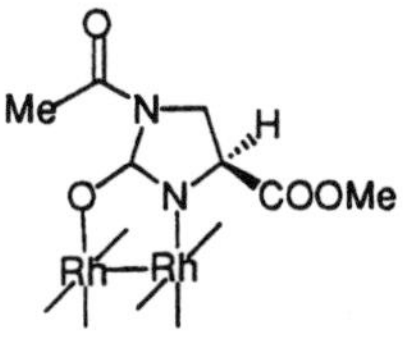

$Rh_2(4S\text{-MACIM})_4$
37

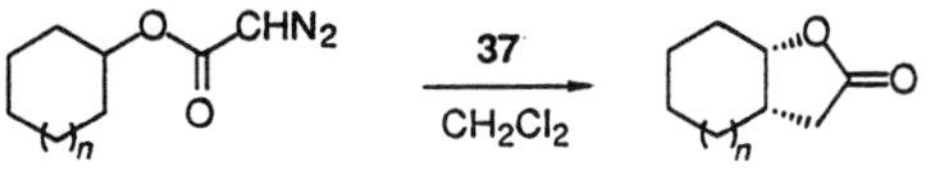

38: $n = 1,2,3$

$\xrightarrow[CH_2Cl_2]{37}$

cis:trans = 99:1
96-97% ee

(14.5)

with 4-substituted cycloalkyl diazoacetates (**39, 40**) and even 2-adamantyl diazoacetate (**41**) in reactions catalysed by $Rh_2(4S\text{-MEOX})_4$ [26].

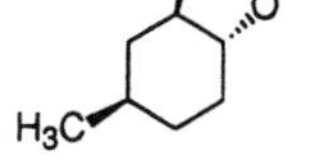

39   **40**   **41**
95% ee   98% ee   98% ee

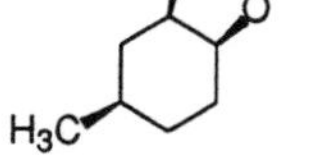

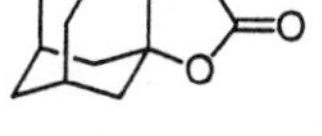

Lower enantiocontrol occurs in C–H insertion reactions of tertiary alkyl diazoacetates than in the reactions of their secondary counterparts [50]. With 1-methylcyclohexyl diazoacetate, insertion occurred selectively into the secondary C–H position (Equation (14.6)) to provide bicyclic lactone **42** in 90% ee. However, applications to acyclic diazoacetates provided variable selectivities.

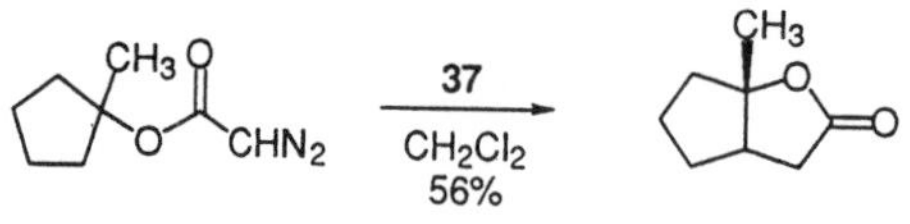

$$\text{(diagram)} \qquad (14.6)$$

42
90% ee

7   Synthesis of 2-deoxyxylolactone

Control of *cis/trans* selectivity is the basis for the successful enantiocontrolled synthesis of 2-deoxyxylolactone from inexpensive 1,3-dichloro-2-propanol [51]. Use of $Rh_2(5R\text{-}MEPY)_4$ on compound **43** (Equation (14.7)) produced the dibenzyl ether of 2-deoxy-D-xylolactone in 94% ee and with a diastereomeric ratio of 93:7. Only 0.1 mol.% of catalyst

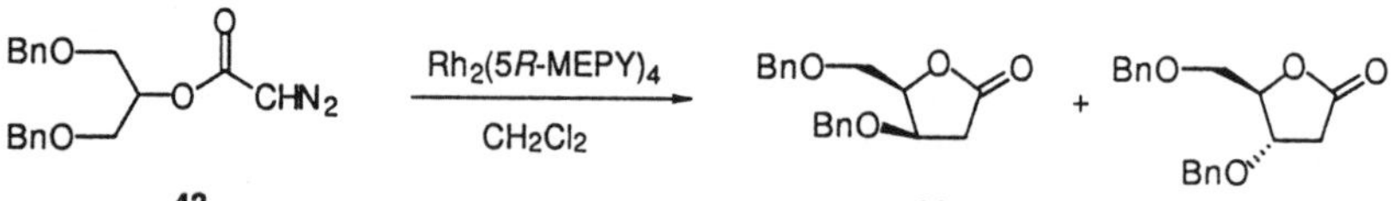

$$\text{(diagram)} \qquad (14.7)$$

43   **44**   **45**
                          93(94% ee)   7(45% ee)

was required and, following chromatographic separation of compound **44** from compound **45**, hydrogenolysis produced pure 2-deoxy-D-xylolactone with 94% ee.

An alternative approach eliminates the need for diastereomer separation. This benzal acetal of 1,3-dihydroxy-2-propyl diazoacetate (**46**) undergoes C–H insertion catalysed by $Rh_2(5S\text{-}MEPY)_4$ to afford only that product from insertion into the axial C–H bond (Equation (14.8)). Hydrogenolysis of compound **47** produced pure 2-deoxy-L-xylolactone with 94% ee.

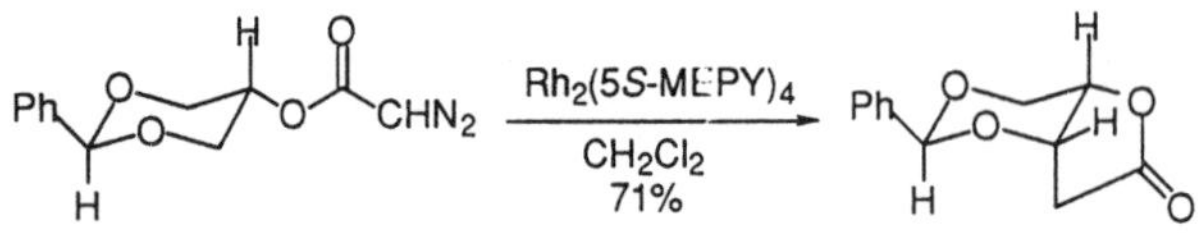

$$\text{(diagram)} \qquad (14.8)$$

46   **47**: 94% ee

The aliphatic counterpart of compound **43** also undergoes C–H insertion to form the analogous five-membered ring lactone in high yield (Equation (14.9)). In this case, use of

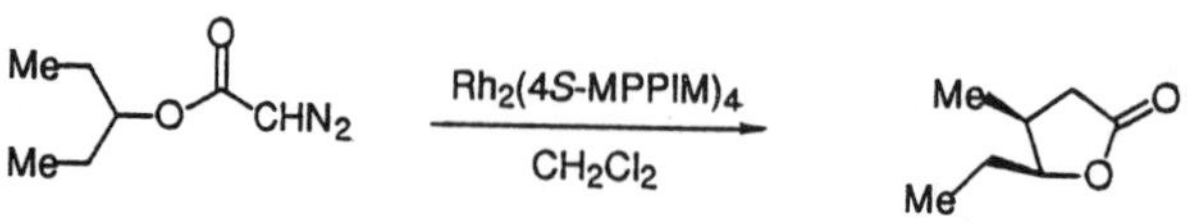

$$\text{(diagram)} \qquad (14.9)$$

48   **49**
                          97:3 dr
                          98% ee

Rh$_2$(4S-MPPIM)$_4$ gave even higher stereocontrol than Rh$_2$(5S-MEPY)$_4$ [30]. Other examples have been provided [52].

8 Carbon–hydrogen insertion reactions: lactams

Highly enantioselective synthesis of β-lactams is possible with the C–H insertion methodology [53]. Cyclic and bicyclic secondary amines converted to diazoacetamides undergo intramolecular C–H insertion onto the nitrogen-activated C–H bond (e.g. Equation

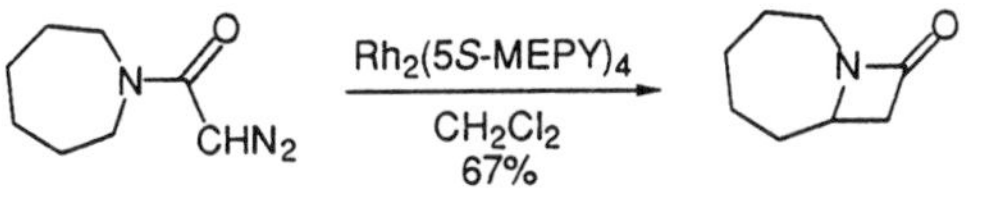

$$\text{(14.10)}$$

50   **51**: 97% ee

(14.10)). However, although compound **50** undergoes this reaction with comparative ease, the next lower homolog does not yield C–H insertion products.

When there is an option to form the five-membered ring insertion product, this process is preferred (e.g. Equation (14.11)), except when the substituent beta to nitrogen is an

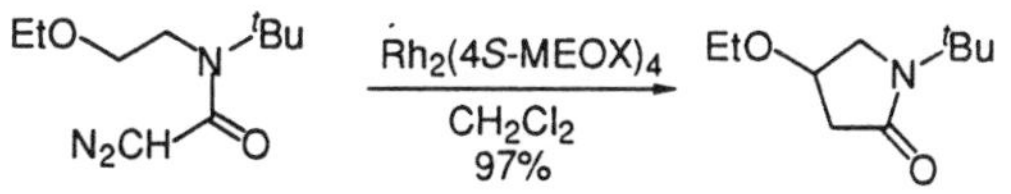

$$\text{(14.11)}$$

52: 78% ee

electron-withdrawing substituent [54]. There are many fewer examples here of effective C–H insertion reactions than with lactone formation.

9 Enantiomer differentiation in intramolecular carbon–hydrogen insertion reactions

Section 5 in this chapter described enantiomer differentiation in intramolecular cyclopropanation reactions whereby one enantiomer matched with the appropriate catalyst underwent cyclopropanation, while the other enantiomer was subject to hydride abstraction. In intramolecular C–H insertion reactions enantiomer differentiation occurs with diastereoselectivity and regioselectivity. As examples, enantiomerically pure compound **53** gave compound **54** predominantly with the *R*-configured catalysts such as Rh$_2$(4*R*-MPPIM)$_4$, and gave compound **55** with *S*-configured catalysts [55]. Similarly, enantio-

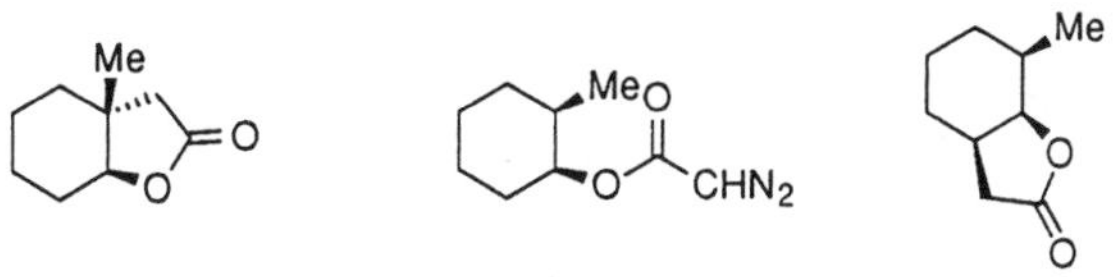

55   **53**   **54**

merically pure compound **56** gave compound **57** as the major product with the *S*-configured catalysts and compound **58** with the *R*-configured catalysts. In some cases the

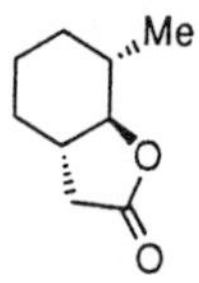

58 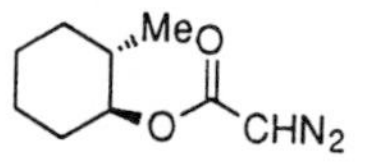   **56** 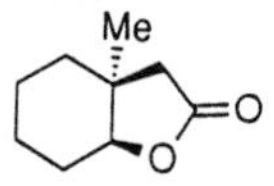   **57**

selectivity for formation of the preferred product was 98:2. With acyclic diazoacetates, enantiomer differentiation takes the form of regiocontrol [55]. Use of the racemic mixtures gives predictable results based on those from the enantiomerically pure diazoacetates [56].

Proline-derived diazoacetamides undergo C–H insertion into the pendant alkyl group to form pyrrolizidine derivatives [57]. For example, compound **59** produces compound **60** in 94% yield with a diastereomeric ratio of 98:2 with Rh$_2$(4S-MACIM)$_4$ as the catalyst (Equation (14.12)). Reduction of compound **60** produced the pyrrolizidine base

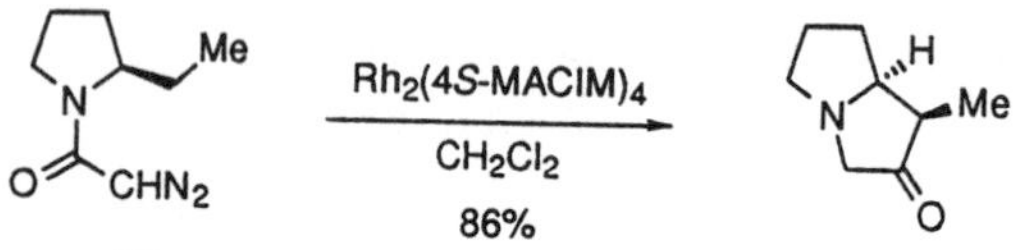

$$(14.12)$$

59   **60**

heliotridane. Noteworthy is the observation that compound **60** is the thermodynamically less stable isomer.

10   Ylide generation and subsequent reaction

One feature unique in ylide transformations is the reversible association between the electrophilic metal carbene and a Lewis base. This association is crucial because it is here where chemoselective processes bifurcate. Interestingly, given the complexity in the sequence of events necessary for ylide-type reactions to occur, high levels of chemo-, diastereo- and enantiocontrol can be achieved. As shown in Equation (14.13), the catalytic diazo decomposition of ethyl diazoacetate in the presence of *trans*-cinnamyl methyl ether afforded the [2,3]-sigmatropic rearrangement adduct in 15:85 dr with exceptionally high enantiocontrol [58]. The degree of diastereocontrol and chemoselectivity (90:10 for compounds **61/62**) with Rh$_2$(4S-MEOX)$_4$ is significantly altered, and diastereocontrol is reversed with the use of achiral dirhodium(II) catalysts (83:17 with Rh$_2$(OAc)$_4$). The effect of catalyst on product distribution and selectivity is compelling evidence for the presence of a metal-associated ylide intermediate in these ylide-derived transformations.

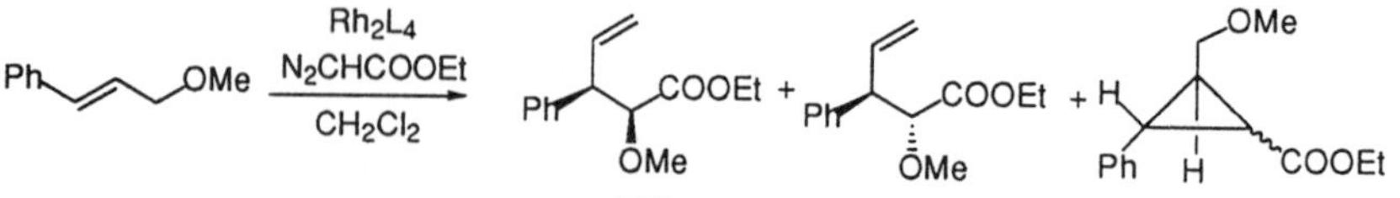

$$(14.13)$$

61E   **61T**   **62**

	61E	61T
Rh$_2$(OAc)$_4$	83	17
Rh$_2$(4S-MEOX)$_4$ (13)	15 (94% ee)	85 (98% ee)
Rh$_2$(4R-MEOX)$_4$	15 (94% ee)	85 (98% ee)

Jacobsen *et al.* have reported recently [59] on the formation of chiral non-racemic aziridines (67% ee), and were the first to provide evidence that ylide-generated intermediates exist as metal-associated intermediates rather than the free ylide in the metal-catalysed decomposition of diazocarbonyl derivatives in the presence of aryl-substituted imines. Even more compelling evidence of metal association would be those processes that undergo ylide-type transformations where the free ylide itself is achiral so that any level of stereoinduction would have to originate from the chiral catalyst. Through the use of allyl halides as precursors to [2,3]-sigmatropic rearrangements with ethyl diazoacetate in the presence of $Cu(CH_3CN)_4PF_6$/**21** (Equation (14.14)), iodoester **63** was obtained in 62% isolated yield and with 69% ee [58].

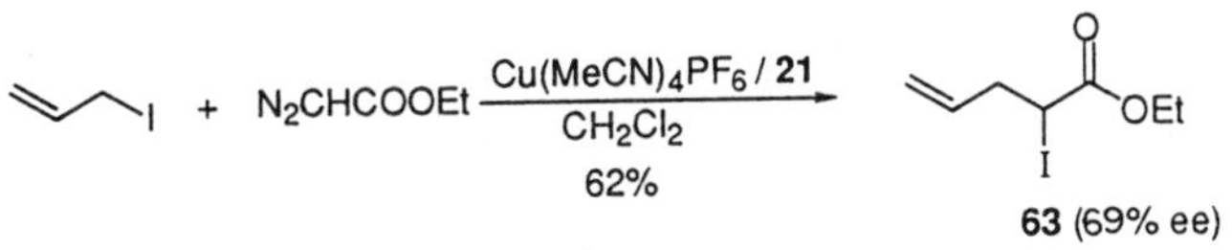

$$(14.14)$$

As initially reported by McKervey *et al.* in 1992 [60], the use of chiral catalysts in the asymmetric diazo decomposition to afford chiral non-racemic [2,3]-sigmatropic rearrangement products serves as a powerful synthetic tool to afford novel heterocycles. Success in achieving enantiocontrol with oxonium ylides has been reported to occur also from macrocyclic templates (Equation (14.15)) [58]. Using a $Cu(CH_3CN)_4PF_6$/**21** complex, diazoester **64** formed a 13-membered ring oxonium ylide intermediate that underwent a [2,3]-sigmatropic rearrangement to afford macrocycle **65** in 35% isolated yield. Analysis of

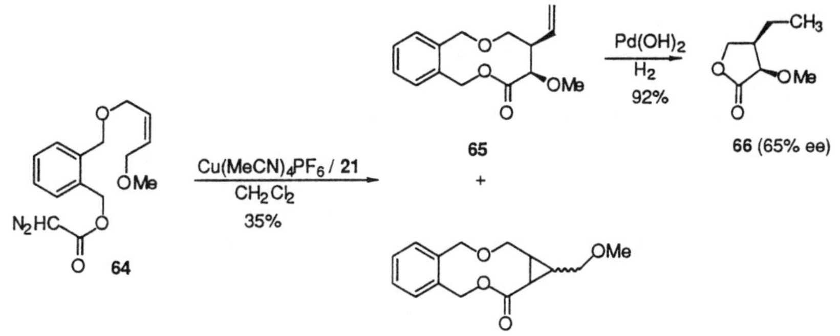

$$(14.15)$$

lactone **66**, obtained by hydrogenolysis of compound **65**, established both the level of enantiocontrol (65% ee) and diastereocontrol (a single diastereomer).

Oxonium ylides that do not utilize a pendant allyl group have been shown to undergo [1,2]-insertion processes: the Stevens rearrangement. High levels of enantiocontrol have been achieved with selected substrates, as shown in Equation (14.16) [61]. Using $Rh_2(4S\text{-}MPPIM)_4$ as the source for asymmetric induction, diazo decomposition of diazoacetate **67** gave bicyclic product **68** in high yield and in 81% ee.

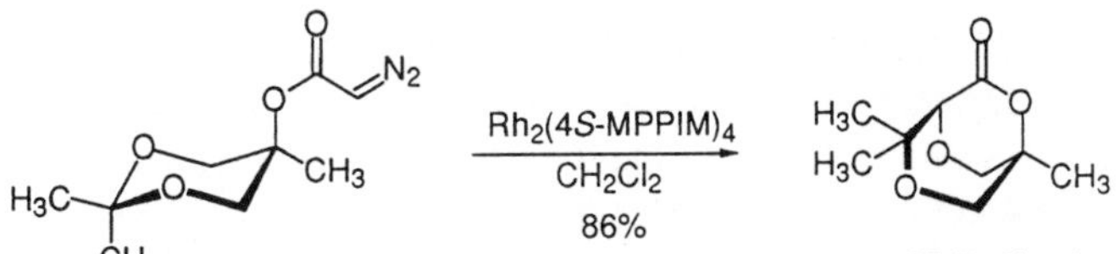

$$(14.16)$$

11 Summary

Chiral dirhodium(II)-carboxamidate-catalysed diazo decomposition of diazoacetates and diazoamides provides access to complex cyclic products with exceptionally high levels of enantiocontrol. These catalysts have demonstrable influence on cyclopropanation, carbon–hydrogen insertion and ylide transformations, and associated regiocontrol, chemoselectivity and diastereocontrol are characteristic for these processes.

12 Acknowledgments

We are grateful to the National Science Foundation and to the National Institutes of Health for their support of the research conducted in our laboratories.

13 References

1 Doyle MP, McKervey MA, Ye T. *Modern Catalytic Methods for Organic Synthesis with Diazo Compounds*. New York: Wiley, 1997.
2 Noyori R. *Science* 1990; **248**: 1194–9.
3 Nozaki H, Moriuti S, Takaya H, Noyori, R. *Tetrahedron Lett* 1966: 5239–42.
4 Ojima I, ed. *Catalytic Asymmetric Synthesis*. New York: VCH, 1993.
5 Aratani T. *Pure Appl Chem* 1985; **57**: 1839–44.
6 Jacobsen EN. In: Wilkinson G, Stone FGA, Abel EW, Hegedus LS, eds. *Comprehensive Organometallic Chemistry II*. New York: Pergamon, 1995.
7 Fukuda T, Katsuki T. *Synlett* 1995: 825–6.
8 Fritschi H, Leutenegger U, Pfaltz A. *Angew Chem Int Ed Engl* 1986; **25**: 1005–6.
9 Pfaltz A. *Acc Chem Res* 1993; **26**: 339–45.
10 Lowenthal RE, Abiko A, Masamune S. *Tetrahedron Lett* 1990; **31**: 6005–8.
11 Evans DA, Woerpel KA, Hinman MM, Faul MM. *J Am Chem Soc* 1991; **113**: 726–8.
12 Müller D, Umbricht G, Weber B, Pfaltz A. *Helv Chim Acta* 1991; **74**: 232–40.
13 Denmark SE, Nakajima N, Nicaise OJC, Faucher AM, Edwards JP. *J Org Chem* 1995; **60**: 4884–92.
14 Lowenthal RE, Masamune S. *Tetrahedron Lett* 1991; **32**: 7373–6.
15 Nishiyama H, Itoh Y, Sugawara Y *et al. Bull Chem Soc Jpn* 1995; **68**: 1247–62.
16 Davies HML, Hutcheson DK. *Tetrahedron Lett* 1993; **34**: 7243–6.
17 Corey EJ, Grant TG. *Tetrahedron Lett* 1994; **35**: 5373–6.
18 Paulissenen R, Reimlinger H, Hayez E, Hubert AJ, Teyssie P. *Tetrahedron Lett* 1973: 2233–6.
19 Maas G. *Top Curr Chem* 1987; **137**: 76–253.
20 Doyle MP. *Chem Rev* 1986; **86**: 919–39.
21 Padwa A, Krumpe KE. *Tetrahedron* 1992; **48**: 5385–453.
22 Ye T, McKervey MA. *Chem Rev* 1994; **94**: 1091–160.
23 Nefedov OM, Shapiro EA, Dyatkin AB. In: Patai S, ed. *Supplement B: The Chemistry of Acid Derivatives*, Chapt. 25. New York: Wiley, 1992.
24 Doyle MP, Winchester WR, Protopopova MN, Kazala AP, Westram LF. *Org Syn* 1996; **73**: 13–24.
25 Doyle MP, Winchester WR, Hoorn JAA *et al. J Am Chem Soc* 1993; **115**: 9968–78.
26 Doyle MP, Dyatkin AB, Protopopova MN *et al. Recl Trav Chim, Pays-Bas* 1995; **114**: 163–70.
27 Doyle MP, Zhon QL, Raab CE *et al. Inorg Chem* 1996; **35**: 6064–73.
28 Doyle MP, Zhou QL, Simonsen SH, Lynch V. *Synlett* 1996: 697–8.
29 Doyle MP, Austin RE, Bailey AS *et al. J Am Chem Soc* 1995; **117**: 5763–75.
30 Doyle MP, Zhou QL, Dyatkin AB, Ruppar DA. *Tetrahedron Lett* 1995; **36**: 7579–82.
31 Doyle MP, Peterson CS, Zhou QL, Nishiyama H. *J Chem Soc, Chem Commun* 1997: 211–12.
32 Doyle MP, Kalinin AV. *J Org Chem* 1996; **61**: 2179–84.
33 Rogers DH, Yi EC, Poulter CD. *J Org Chem* 1995; **60**: 941–5.

34 Martin SF, Austin RE, Oalmann CJ *et al. J Med Chem* 1992; **35**: 1710–21.

35 Martin SF, Oalmann CJ, Liras S. *Tetrahedron* 1993; **49**: 3521–32.

36 Doyle MP, Zhou QL. *Tetrahedron: Asymm* 1995; **6**: 2157–60.

37 Doyle MP, Eismont MY, Zhou QL. *Russ Chem Bull* 1997; **46**: 955–8.

38 Doyle MP, Peterson CS, Parker DL Jr. *Angew Chem Int Ed Engl* 1996; **35**: 1334–6.

39 Doyle MP, Peterson CS, Protopopova MN *et al. J Am Chem Soc* 1997; **119**: 8826–37.

40 Doyle MP, Dyatkin AB, Kalinin AV *et al. J Am Chem Soc* 1995; **117**: 11021–2.

41 Doyle MP, Dyatkin AB, Autry CL. *J Chem Soc, Perkin Trans 1* 1995: 619–22.

42 Martin SF, Spallar MR, Liras S, Hartmann B. *J Am Chem Soc* 1994; **116**: 4493–4.

43 Kennedy M, McKervey MA, Maguire AR, Roos GHP. *J Chem Soc, Chem Commun* 1990: 361–2.

44 Hashimoto S, Watanabe N, Sato T, Shiro M, Ikegami S. *Tetrahedron Lett* 1993; **34**: 5109–12.

45 McKervey MA, Ye T. *J Chem Soc, Chem Commun* 1992: 823–4.

46 Lim HJ, Sulikowski GA. *J Org Chem* 1995; **60**: 2326–7.

47 Doyle MP, van Oeveren A, Westrum LJ, Protopopova MN, Clayton TW Jr. *J Am Chem Soc* 1991; **113**: 8982–4.

48 Bode JW, Doyle MP, Protopopova MN, Zhou QL. *J Org Chem* 1997; **61**: 9146–55.

49 Doyle MP, Dyatkin AB, Roos GHP *et al. J Am Chem Soc* 1994; **116**: 4507–8.

50 Doyle MP, Zhou QL, Raab CE, Roos GHP. *Tetrahedron Lett* 1995; **36**: 4745–8.

51 Doyle MP, Dyatkin AB, Tedrow JS. *Tetrahedron Lett* 1994; **35**: 3853–6.

52 Müller P, Polleux P. *Helv Chim Acta* 1994; **77**: 645–54.

53 Doyle MP, Kalinin AV. *Synlett* 1995: 1075–6.

54 Doyle MP, Protopopova MN, Winchester WR, Daniel KL. *Tetrahedron Lett* 1992; **33**: 7819–22.

55 Doyle MP, Kalinin AV, Ene DG. *J Am Chem Soc* 1996; **118**: 8837–46.

56 Doyle MP, Kalinin AV. *Russ Chem Bull* 1995; **44**: 1729–34.

57 Doyle MP, Kalinin AV. *Tetrahedron Lett* 1996; **37**: 1371–4.

58 Doyle MP, Forbes DC, Vasbinder MM, Peterson CS. *J Am Chem Soc* 1998; **120**: 7653–4.

59 Hansen KB, Finney NS, Jacobsen EN. *Angew Chem Int Ed Engl* 1995; **34**: 676–8.

60 McCarthy N, McKervey MA, Ye T *et al. Tetrahedron Lett* 1992; **33**: 5983–6.

61 Doyle MP, Ene DG, Forbes DC, Tedrow JS. *Tetrahedron Lett* 1997; **38**: 4367–70.

15 Copper(I)-promoted Asymmetric Transformations: a Longstanding Challenge

A. ALEXAKIS

Université de Genève, Département de Chimie Organique 30, quai Eznert Anserment, Genève CH-1211, Switzerland

1 Introduction

Organocopper reagents emerged as useful synthetic tools in the early 1960s and since then have enjoyed continuing popularity among organic chemists [1]. There are hundreds of total syntheses that include a copper(I)-promoted step. During all these years a variety of organocopper reagents have been described, improving the chemical yields, the reactivity, the chemoselectivity and, above all, the thermal stability and ease of handling of these reagents.

There are several synthetic transformations that are best carried out with this class of organometallic:

1 conjugate addition;
2 carbometallation of alkynes;
3 S_N2 substitution;
4 S_N2' substitution;
5 ring opening of epoxides;
6 addition in position 4 of pyridinium salts.

By far the most popular reaction is conjugate addition, which can be performed either with a stoichiometric organocopper reagent or under copper(I) catalysis, usually with a Grignard reagent as the main organometallic.

From the beginning, chemists realized that many of these reactions created a new stereogenic center, therefore it was not surprising to find very early papers dealing with asymmetric organocopper synthesis [2]. Most of these attempts were quite disappointing as far as enantiomeric excess (ee) was concerned. In fact, one of the main problems in handling this task is the lack of knowledge of the mechanistic insights of copper-mediated reactions, as well as the true reactive species. The aggregation state of organocopper reagents is usually believed to be a square planar dimer for R_2CuLi [3], although recent cryoscopic measurements on $R_2Cu(CN)Li_2$ favor a monomeric species [4].

In this review, we shall focus first on conjugate addition, as this reaction is by far the most investigated one.

2 Conjugate addition

Conjugate addition is among the basic carbon–carbon bond-forming reactions [5]. It is best performed with a copper(I) species as promoter, in stoichiometric or catalytic amount. It should be pointed out that the stoichiometric reagent (R_2CuLi) is most often prepared with an organolithium reagent as the primary organometallic, whereas for catalytic purposes organomagnesium (Grignard reagents RMgX) or organozinc reagents (RZnX or R_2Zn) are usually used. The importance of this fact will become apparent later.

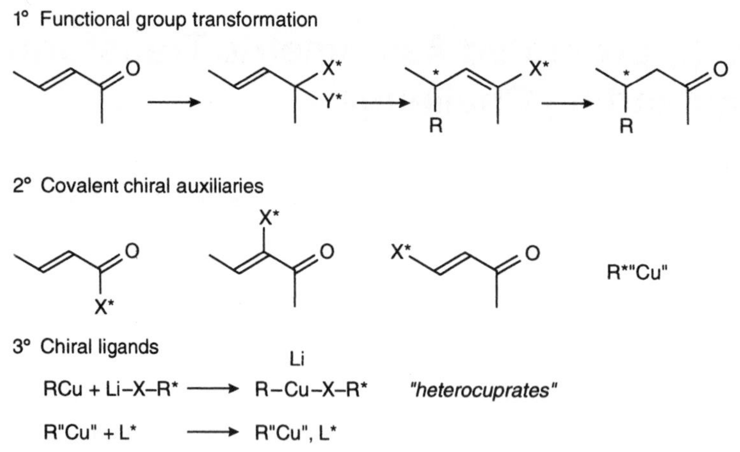

1° Functional group transformation

2° Covalent chiral auxiliaries

3° Chiral ligands

RCu + Li–X–R* ⟶ R–Cu–X–R*   "heterocuprates"

R"Cu" + L* ⟶ R"Cu", L*

Scheme 15.1.

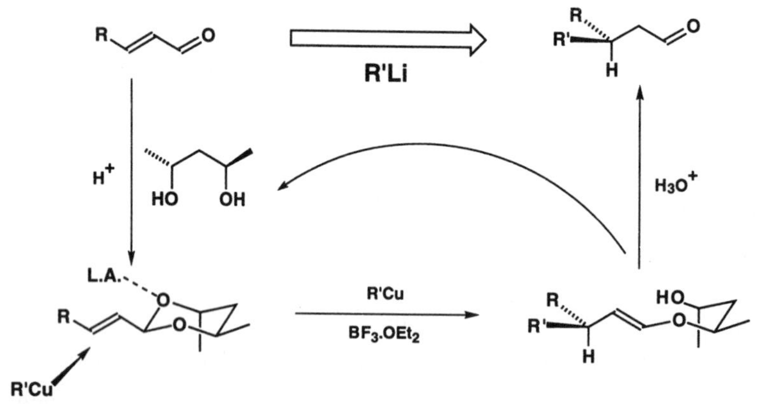

Scheme 15.2.

In conjugate addition, the substrate usually has a double (or triple) bond to which an electron-withdrawing group (most often a carbonyl group) is attached. Conceptually, there are several ways to tackle this problem in an asymmetric sense (Scheme 15.1).

2.1 *With a covalent chiral auxiliary*

The covalent chiral auxiliary approach requires, by definition, a stoichiometric amount of the chiral auxiliary. In this case, either a stoichiometric or catalytic copper reagent may be used. One approach is to transform the sp^2 carbon of the carbonyl into another functionality with an sp^3 carbon. The reaction is now a γ-allylic substitution. Chiral acetals are a typical example of this class of transformations [6], resulting in the formal asymmetric conjugate addition of organolithiums to α,β-ethylenic aldehydes in high enantiomeric excess (85–95%) (Scheme 15.2).

However, the most successful results were obtained on a true conjugate addition with a covalent chiral auxiliary. The reaction is again a diastereoselective one, and the major diastereomer may often be purified by chromatography or crystallization, thus ultimately affording a pure enantiomer. The chiral covalent auxiliary may be attached at different

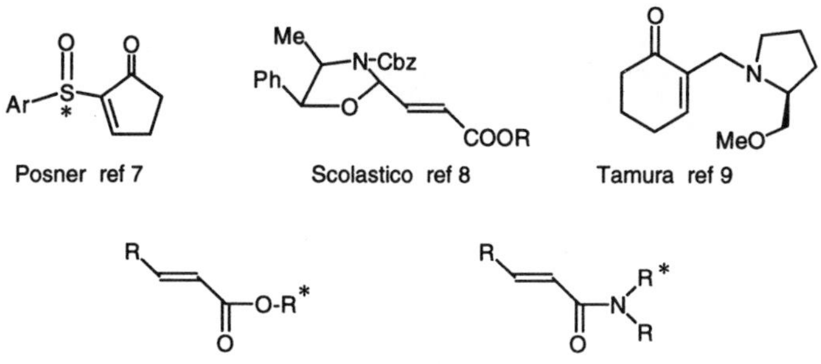

Scheme 15.3.

positions of the substrate. There are several examples in the literature describing such an approach [2] (Scheme 15.3).

Chiral unsaturated esters or amides of various chiral alcohols or amines allow a very large array of chiral auxiliaries [2]. It should be noted that these substrates were also used in other reactions, the most typical being the asymmetric Diels–Alder reaction. Several of the most popular auxiliaries take advantage of the camphor framework (Scheme 15.4). The same enantiomeric form of camphor is modified to allow both enantiomers of the conjugate adduct to be obtained. Although the reactivity of such substrates is usually rather low, good chemical yields may be obtained using various additives to increase reactivity (Me_3SiCl [10], BF_3 [11], etc.). However, care should be taken because some of the additives may strongly affect the stereoselectivity, or even reverse it [10a,12]. It is also possible to take advantage of this reversal of stereoselectivity in order to obtain either enantiomer. Some representative examples are listed in Scheme 15.4.

The enantiomeric excesses obtained via this route are close to 100%, making this approach the most efficient for this type of substrate. As a consequence, the number of articles dealing with this approach have declined considerably over the last few years. In contrast, recent interest has focused on the chiral ligand approach, which, in principle, should provide a more general solution to the problem.

2.2 The chiral ligand approach

This approach uses an external chiral moiety, based on the more general concept of transition metal chemistry. The amount of chiral ligand should therefore reflect the amount of copper metal used. The main consequence is that all studies dealing with stoichiometric reagents (R_2CuLi) are not applicable, as such, to the catalytic version where the main organometallic is an Mg or Zn species.

Again, two different routes have been explored. The discovery of 'heterocuprates' by Posner *et al.* in 1973 [18] immediately gave a strong impetus to the heterocuprate approach.

2.2.1 THE HETEROCUPRATE APPROACH

In such mixed reagents the chirality is brought about by a chiral alcoholate, amide or thiolate (Scheme 15.5). Only the R group is transferred to the enone. When R–M is R–Li, a stoichiometric amount of an easily available chiral auxiliary (with good recovery of the chiral auxiliary) is used. However, a catalytic version has also been studied, where a Grig-

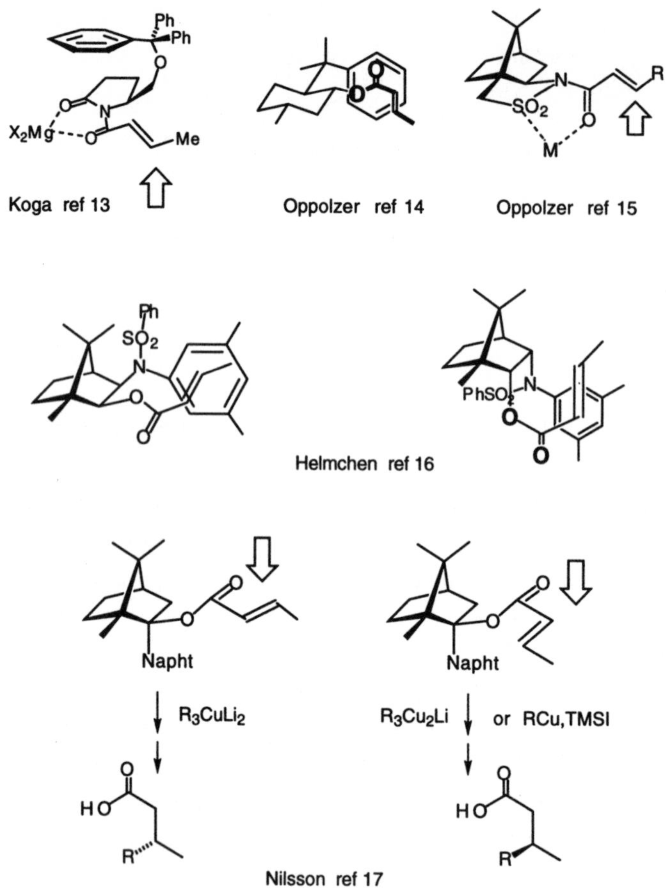

Koga ref 13   Oppolzer ref 14   Oppolzer ref 15

Helmchen ref 16

Napht   Napht

R₃CuLi₂ (as image)

Scheme 15.4.

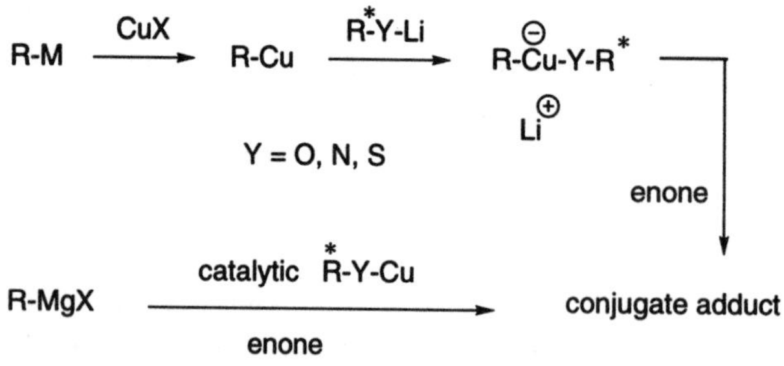

Scheme 15.5.

nard reagent is catalysed by a small amount of copper chiral alcoholate, amide or thiolate, Several dozens of chiral alcohols, amines and thiols have been tested [2]. Some of the most representative examples are listed in Scheme 15.6.

The first attempts with heterocuprates were not very successful, due to the facile disproportionation of these species into two homocuprates according to:

$$2\ R(\overset{*}{RO})CuLi \ \rightleftharpoons \ R_2CuLi \ + \ (\overset{*}{RO})_2CuLi$$

The way to stabilize the mixed heterocuprate species was to introduce additional heteroatoms for better coordination of the metals (see Scheme 15.6). Remarkable results were

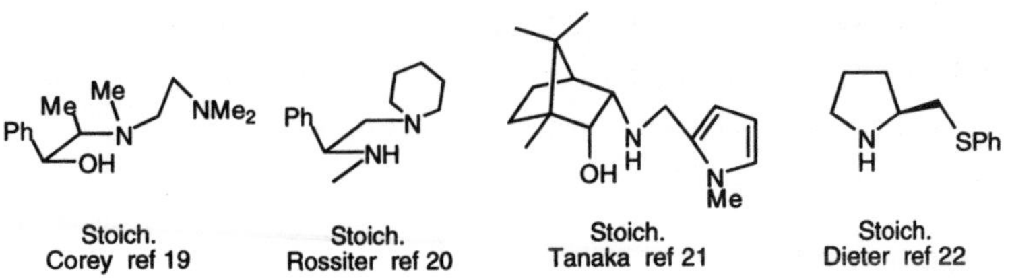

Stoich.
Corey ref 19

Stoich.
Rossiter ref 20

Stoich.
Tanaka ref 21

Stoich.
Dieter ref 22

Scheme 15.6.

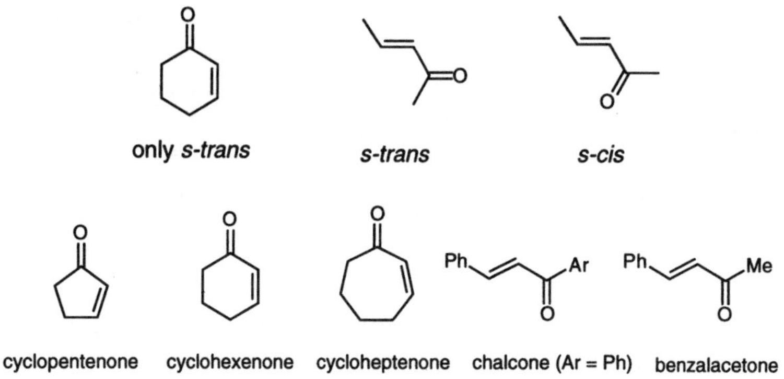

only *s-trans*

s-trans

s-cis

cyclopentenone   cyclohexenone   cycloheptenone   chalcone (Ar = Ph)   benzalacetone

Scheme 15.7.

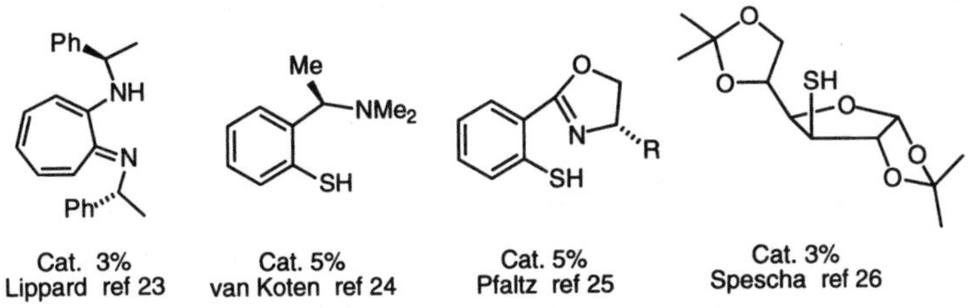

Cat. 3%
Lippard ref 23

Cat. 5%
van Koten ref 24

Cat. 5%
Pfaltz ref 25

Cat. 3%
Spescha ref 26

Scheme 15.8.

achieved in this way by several groups. It has to be noted that the quality of the R—Li used is essential for reproducible results. Old bottles usually contain a variable amount of R—O—Li, which strongly affects the enantioselection [19]. Most of the work done in this field concerns cyclic enones, the most representative being cyclohexenone. The enantioselection usually decreases with the flatness of the enone: cyclopentenone is the worst, while cycloheptenone is the best [20]. Even a large 15-membered ring gives excellent results for the asymmetric synthesis of Muscone, a valuable fragrance [21]. In contrast, few examples deal with acyclic enones, which may occur either in the *S-cis* or the *S-trans* conformation, thus adding one more variable (Scheme 15.7).

Catalytic systems have been developed more recently, although the enantiomeric excesses do not yet match those obtained in the stoichiometric cases. They are all based on the use of Grignard reagents as primary organometallics. One may note that, except for the first report by Lippard [23], all other systems are based on copper *thiolates* (see Scheme 15.8). In these species, the copper atom seems to be tightly bound to the sulfur atom, avoiding the formation of a copper halide species with the halogen of the Grignard reagent.

Again strong differences were noted between cyclic and acyclic enones, the experimental conditions being essential for optimization of the enantiomeric excess. For example,

HMPA is often used as cosolvent [23,25], or a simultaneous addition of the Grignard reagent and the enone [24]. In addition, the nature of the solvent may affect the enantio-selectivity, and cases of total inversion are not uncommon.

2.2.2   THE LIGAND APPROACH

More attention is currently given to the non-covalently bound ligands. As the copper atom needs to complete its coordination sphere, pure coordinating ligands may be used by analogy with all other transition metal catalysts. Again, the first reports were successful using stoichiometric quantities of chiral auxiliary and the efforts are presently directed towards their catalytic use. Early reports with (–)-sparteine [27] or chiral solvents [28] and lithium diorganocuprate or triorganozincates gave disappointingly low enantiomeric excesses. The first successful example was described by Leyendecker and Lancher with a ligand derived from hydroxyproline [29]. Enantiomeric excesses up to 90% were obtained with acyclic enones (Scheme 15.9).

In 1991 we introduced a new class of ligands based on trivalent phosphorus derivatives [30]. In this case the phosphorus atom is itself a stereogenic center. These ligands are easily available from cheap ephedrine or its derivatives. Ligand A* (Scheme 15.10) was very efficient with primary lithium diorganocuprates and cyclic enones, giving enantiomeric excesses in the 70–95% range [31]. With acyclic enones low enantioselection was observed. It was found that two equivalents of this ligand were needed as well as four equivalents of LiBr.

A catalytic (5–10%) use of the chiral copper complex cannot be accomplished with organolithium reagents, necessitating the use of Grignard reagents. However, changing the metal counter ion from lithium to magnesium has a detrimental effect on the enantioselectivity. The experimental conditions were found to be completely different from those found with stoichiometric lithium diorganocuprate and a full study had to be undertaken again. The ephedrine-derived phosphorus ligands gave, at best, 40–60% ee with Grignard reagents and the catalytic (5–10%) chiral copper complex [32].

More recently, Tomioka *et al.* developed a new phosphorus ligand where the chirality is located on the carbon framework of proline [33a] (see Scheme 15.8). Lithium

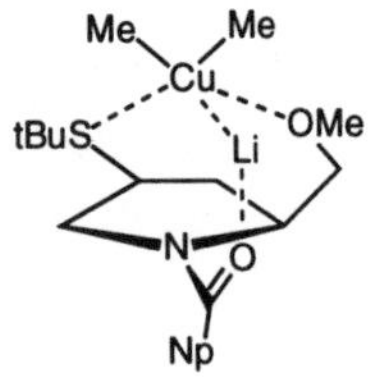

Scheme 15.9.

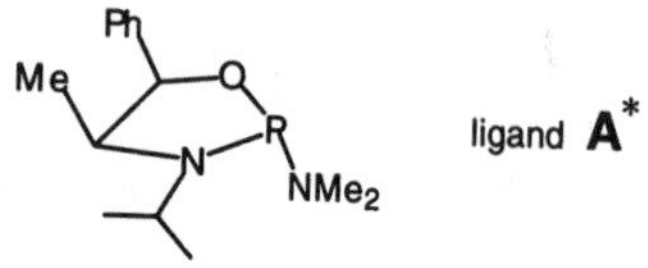

ligand **A***

Scheme 15.10.

diorganocuprate gave excellent enantiomeric excesses with acyclic enones whereas lithium cyanocuprates were better suited for cyclic enones. One of the most amazing observations was the complete reversal of enantioselectivity when stoichiometric *magnesium* cuprates were used with cyclohexenone and cycloheptenone, again with excellent enantiomeric excesses [33b]. This result allowed these authors to perform this reaction with catalytic amounts of copper iodide (8%) and chiral ligand (32%). Enantiomeric excesses range from 70 to 92% but only with cyclic enones [33c].

Comparable results (10% ligand, 80–90% ee on cyclic enones) were attained by Sammakia *et al.* [34] with a ligand based on the ferrocene framework (see Scheme 15.11). They noted that when the ferrocene was replaced by a phenyl group, very low enantiomeric excesses were obtained.

Catalytic systems with other than Grignards as primary organometallic were not investigated until very recently. It is already known that *diorganozinc* [35] or *trialkyl zincate* reagents [36] could undergo conjugate addition, with nickel catalysis seeming to be the most efficient [37]. Asymmetric conjugate addition of diethyl zinc to acyclic enones was reported with a catalytic amount of a chiral nickel complex [38] (Scheme 15.12). Ee's as high as 90% were obtained. With cyclic enones, no asymmetric induction was observed.

More recently copper catalysis has also been shown to be effective in the conjugate addition of organozinc reagents [39]. In 1993, we reported [31] the first enantioselective result with a catalytic amount of the chiral copper complex A* (see Scheme 15.10) and diethyl zinc (Scheme 15.13). The best solvent appeared to be a non-polar one such as toluene. Although the enantiomeric excess was moderate (32% ee) with cyclic enones, the asymmetric copper-catalysed conjugate addition of dialkyl zinc seemed to us to be the most promising way to do this important synthetic transformation. A most important discovery

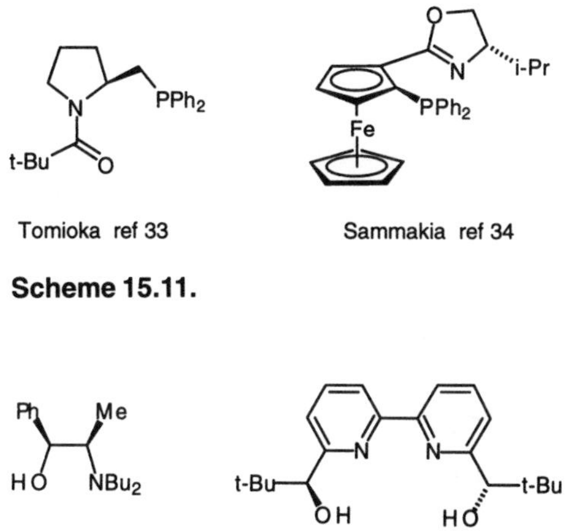

Tomioka ref 33   Sammakia ref 34

Scheme 15.11.

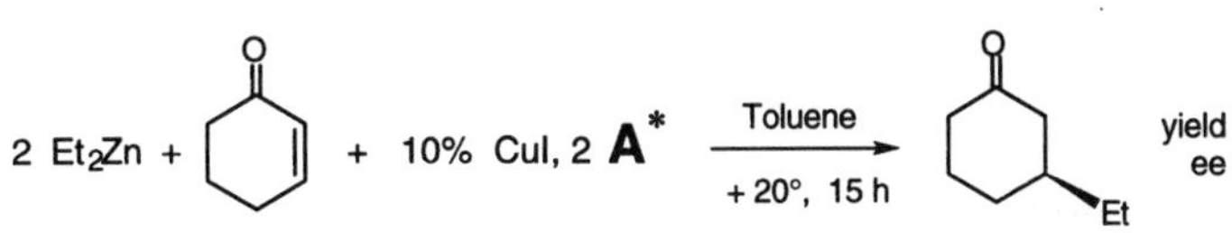

Soai ref 38a   Bolm ref 38b   Feringa ref 38c

Scheme 15.12.

2 Et$_2$Zn + [cyclohexenone] + 10% CuI, 2 **A*** $\xrightarrow[\text{+ 20}^\circ\text{, 15 h}]{\text{Toluene}}$ [3-ethylcyclohexanone] (S)   yield 70% ee 32%

Scheme 15.13.

was the fact that a ligand accelerating effect was observed when the copper source was changed from CuI to Cu(OTf)$_2$. The reaction of Scheme 15.13 was complete in 30 min at 0 °C instead of 15 h at room temperature! In the absence of any phosphorus ligand the reaction time (with Cu(OTf)$_2$) was 12 h at +20 °C [40]. This result matches Noyori's observation [41] that CuCN does not catalyse the conjugate addition of diethyl zinc, unless an equimolar (to Cu) amount of the sulfonamide of Scheme 15.14 was added.

We have investigated several phosphorus ligands, in toluene as well as in CH$_2$Cl$_2$ (the latter being somewhat better), with as little as 0.5% Cu(OTf)$_2$ and 0.5–1% chiral ligand [42]. Some of them are shown in Scheme 15.15. Although some of these ligands are very efficient in transition metal catalysis (such as BINAP with Rh or Ru), they gave low to moderate enantiomeric excesses in the conjugate addition to cyclic and acyclic enones. For example (R,R)-chiraphos gave 40% ee on cyclohexenone, whereas the tartrate-derived phosphite ligand was more efficient on acyclic benzalacetone (63% ee, the highest for this enone).

Based on our preliminary results, Feringa et al. greatly improved the efficiency of the ligand using binaphthol-based phosphoramidates [43]. A similar phosphite was also successfully employed by Pfaltz et al. [44] (Scheme 15.16). Both ligands gave excellent enantiomeric excess with cyclohexenone. With cyclopentenone and cycloheptenone, the phosphite ligand is more efficient (72–77% ee) [44] than the phosphoramidate ligand (15

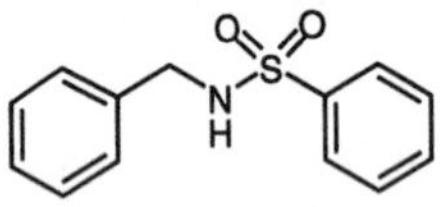

Scheme 15.14.

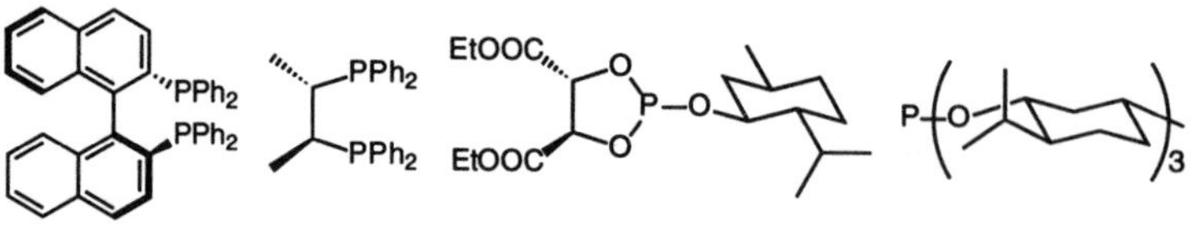

Scheme 15.15.

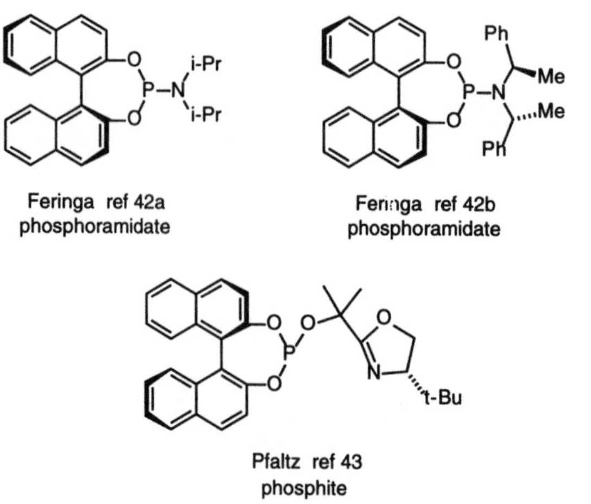

Feringa ref 42a
phosphoramidate

Feringa ref 42b
phosphoramidate

Pfaltz ref 43
phosphite

Scheme 15.16.

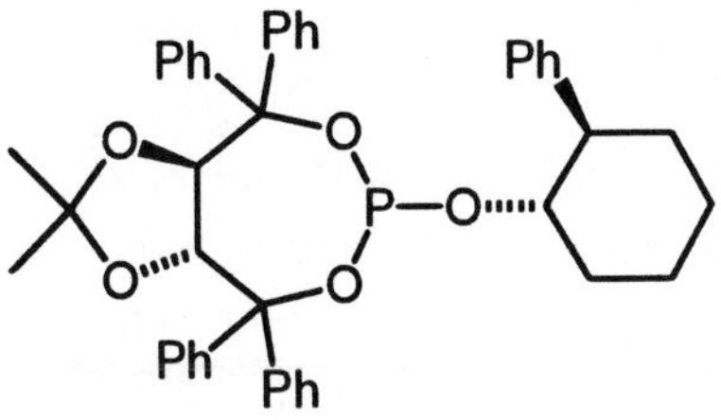

Scheme 15.17.

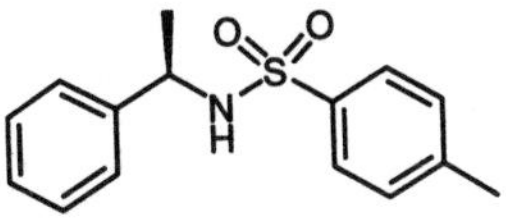

Scheme 15.18.

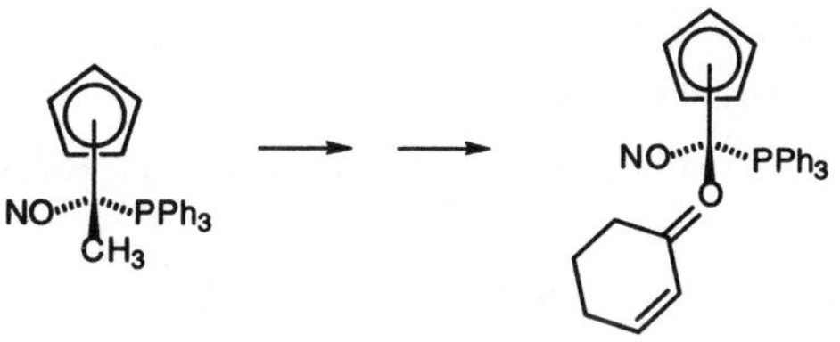

Scheme 15.19.

and 53% ee, respectively) [43]. With acyclic enones, only chalcone-type enones gave good enantiomeric excesses (50–90%), benzalacetone being a very poor case. In a very recent report [45], we have disclosed a new highly efficient ligand based on TADDOL (Scheme 15.17) which gave up to 96% ee on cyclohexenone.

An interesting observation was made in many of the above works. The absolute stereochemistry of the conjugate adduct may vary, or even be inversed, just by changing the copper salt catalyst! For example, Sewald and Wendish reported that diethyl zinc gave the (R)-ethyl cyclohexanone (30% ee) with CuCN, whereas with CuSPh the (S)-adduct (22% ee) was obtained [46]. Both experiments were performed with the same chiral sulfonamide (Scheme 15.17). In a similar way, we observed that CuI gave the (S)-adduct on cyclohexenone with ligand A* (Scheme 15.10), whereas Cu(OTf)$_2$ gave the (R)-adduct [32].

A last approach should be mentioned due to its originality. Because the Lewis acid-assisted conjugate addition is known [11], the idea arose to use a chiral Lewis acid (Scheme 15.19). A chiral rhenium complex was shown to be quite efficient. Although preformation of the Lewis acid–enone complex is needed, the enantiomeric excesses of the conjugate adducts on cyclohexenone and cyclopentenone are high (77–93%) [47]. Again, solvent effects (THF versus CH$_2$Cl$_2$) are important, with even an inversion of enantioselectivity in some cases.

All these results show that there is not yet a unique ligand of general applicability. There is no doubt that in the forthcoming years a breakthrough will occur with this enantioselective reaction, thus adding a new powerful tool to the palette of efficient catalytic synthetic chemistry.

3   The S$_N$2′ substitution reaction

The S$_N$2′ substitution reaction is potentially a very interesting one for asymmetric synthesis [48]. Soft nucleophiles such as malonate-type enolates are already known to be very efficient (>90% ee) under palladium catalysis with a variety of chiral ligands [49]. Performing this reaction with other nucleophilic species, particularly classical organometallics (R–MgX, R–Li, R–ZnX, etc.) would greatly enhance the scope of this reaction. Organocopper reagents (stoichiometric or catalytic) are among the best reagents for regioselective γ-allylic substitution [1]. However, the problem of its asymmetric version still remains a challenge.

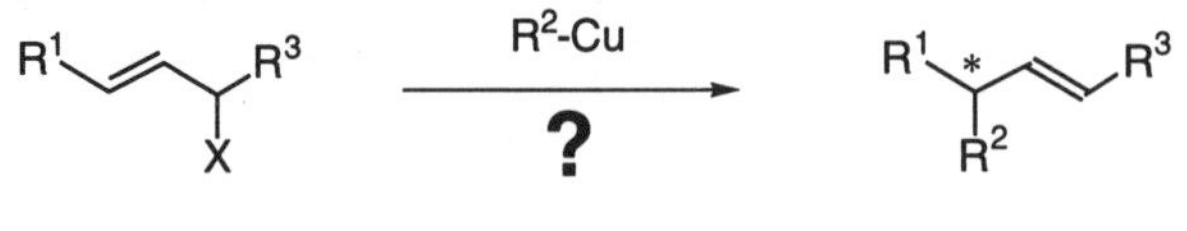

When R^3 is an alkyl group, the resulting stereochemistry is only a transfer of chirality. The stereoselectivity of the S$_N$2′ substitution in organocopper chemistry is purely an *anti* process (except for a single case [50]). However, when R^3 is H, the process could be made enantioselective. Only a few reports deal with this kind of reaction in the field of organocopper chemistry.

The example of chiral acetals disclosed in Scheme 15.2 is one of this class of reactions. The chirality is located on the leaving group. Enantioselectivities are high (85–95%) when the organocopper reagent (R–Cu) is aryl or alkenyl [6]. Three other interesting reports also deal with a chiral leaving group. Gais *et al.* [51] use chiral sulfoximines, Denmark and Marble [52] use chiral carbamates and Calo *et al.* use a chiral thiazoline [53] (Scheme 15.20). In the latter case both *E* and *Z* isomers react as well, affording the opposite enantiomers with excellent enantioselectivity.

These results attest to the viability of this approach. A catalytic process should rely on a different approach, where the chiral auxiliary lies on the metal side. Many unsuccessful

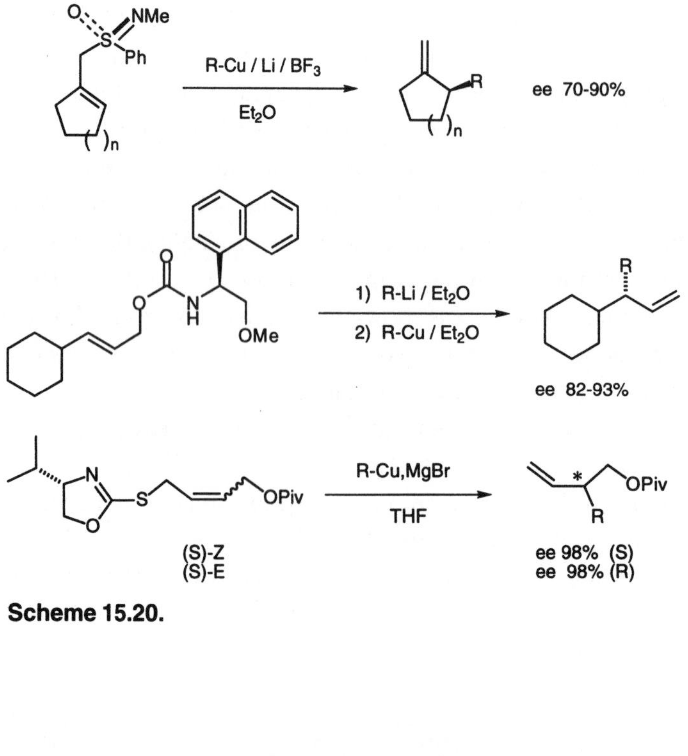

Scheme 15.20.

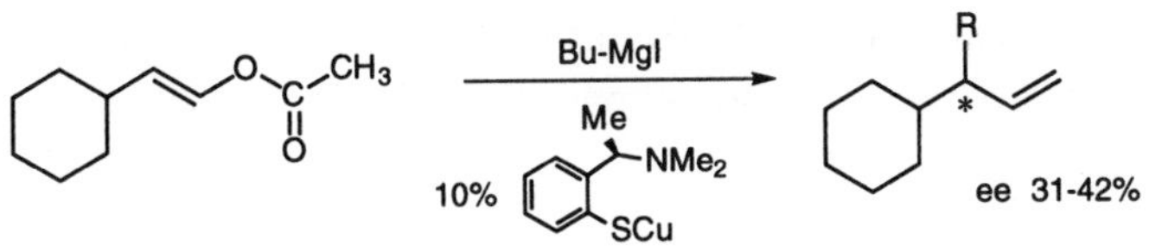

Scheme 15.21.

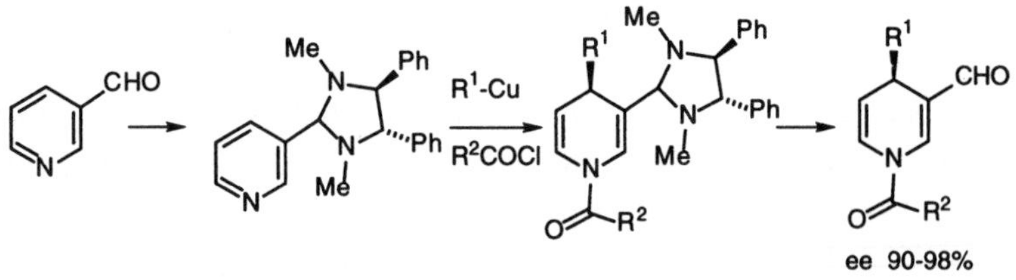

Scheme 15.22.

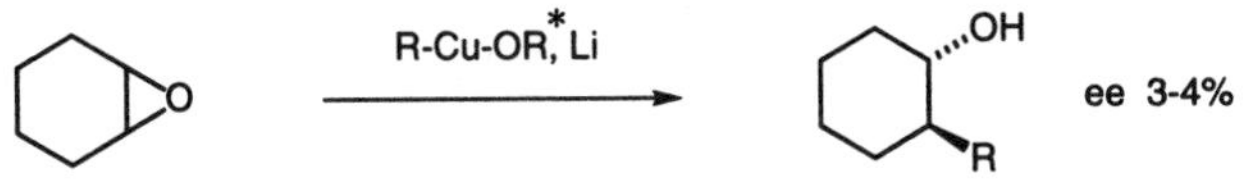

Scheme 15.23.

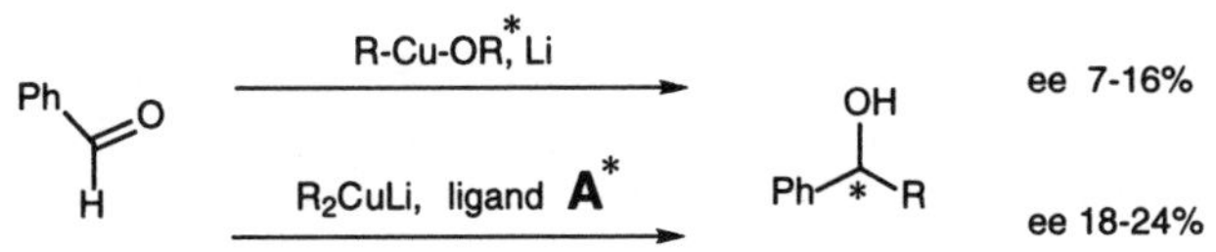

Scheme 15.24.

attempts have been performed in our laboratories and elsewhere. The only positive result was performed with van Koten's chiral copper thiolate (see Scheme 15.8) by van Klaveren *et al.* [54] (Scheme 15.21). Research in this area will hopefully disclose better results in the future.

4   Pyridinium salts and others

Among the other possible asymmetric transformations using organocopper reagents, we focused on the reaction of chiral pyridinium salts. It was known that pyridinium salts react readily with organocopper reagents to afford regioselectively the 4-substituted dihydropyridine [55]. When the reaction was carried out with 3-nicotinaldehyde chiral aminal derivative, an asymmetric reaction ensued with excellent enantioselectivity [56] (Scheme 15.22). Several synthetic applications of this reaction were found in the field of alkaloid natural products [56b, 57].

The ease of nucleophilic opening of epoxides by lithium diorganocuprates makes them the reagent of choice for this transformation [1]. Application to the desymmetrization of *meso* epoxides was attempted by Davies and Wollowitz [58] (Scheme 15.23). Unfortunately, the enantiomeric excess was very low.

Although lithium diorganocuprates do not react easily with the carbonyl functionality of ketones, they do so with aldehydes. Several attempts were made to add enantioselectively chirally modified cuprates to aldehydes [27b, 59] (Scheme 15.24). However, this reaction completely lost its interest because the asymmetric addition to aldehydes is now very successful with diorganozinc reagents.

5 Concluding remarks

Most of the work done in the field of asymmetric organocopper chemistry deals with conjugate addition. Although considerable progress was made in designing new efficient chiral ligands, there is a real need for new more general catalysts. The covalent chiral auxiliary approach still remains a viable route for α,β-ethylenic acid derivatives (esters and amides) as well as for other substrates such as pyridinium salts. As for the other possible asymmetric transformations, the field remains wide open for investigation in the future.

6 References

1 For recent reviews, see: Krause N. *Angew Chem Int Ed Engl* 1997; **36**: 187–204. Yamamoto Y. *Houben-Weyl Stereoselective Synthesis*, Vol. 4. Stuttgart: Thieme, 1995: 2041–57; Wipf P. *Synthesis* 1993: 537–57; Lipshutz BH, Sengupta S. *Org React* 1992; **41**: 135–631. Kozlowski JA. In: Trost BM, Fleming I, eds. *Comprehensive Organic Synthesis*, Vol. 4. Oxford: Pergamon Press, 1991: 169–98; Taylor RJK. *Synthesis* 1985: 364–92; Posner GH. *Org React* 1974; **22**: 253–400; Posner GH. *Org React* 1972; **19**: 1–113.

2 For a review covering the literature until 1992, see: Rossiter BE, Swingle NM. *Chem Rev* 1992; **92**: 771; Alexakis A. In: Taylor RJK, ed. *Organocopper Reagents, a Practical Approach*, Chapt. 8. Oxford: Oxford University Press, 1994: 159–83; for conjugate addition specifically, see also: Feringa BL, de Vries AHM. In: Doyle MD, ed. *Advances in Catalytic Processes*, Vol. 1. JAI, CT USA, 1995: 151–92; Noyori R. *Asymmetric Catalysis in Organic Synthesis*, New York: Wiley, 1994; Ojima I. *Catalytic Asymmetric Synthesis*, Weinheim: VCH, 1993.

3 For a review, see: van Koten G. *J Organomet Chem* 1990; **400**: 283–301.

4 Gerold A, Jastrzebski JTBH, Kronenburg CMP, Krause N, van Koten G. *Angew Chem Int Ed Engl* 1997; **36**: 755–7.

5 Perlmutter P. *Conjugate Addition Reaction in Organic Synthesis*, Tetrahedron Organic Chemistry Series No. 9 Oxford: Pergamon Press, 1992.

6 Alexakis A, Mangeney P, Ghribi A *et al. Pure Appl Chem* 1988; **60**: 49–56; Rakotoarisoa H, Guttierez Perez R, Mangeney P, Alexakis A. *Organometallics* 1996; **15**: 1957–602.

7 Posner GH. *Acc Chem Res* 1987; **20**: 72–8.

8 Scolastico C. *Pure Appl Chem* 1988; **60**: 1689–98.

9 Tamura R, Watabe KI, Ono N, Yamamoto Y. *J Org Chem* 1992; **57**: 4895–903.

10 (a) Corey EJ, Boaz NW. *Tetrahedron Lett* 1985; **26**: 6015–18, 6019–22; (b) Alexakis A, Berlan J, Besace Y. *Tetrahedron Lett* 1986; **27**: 1047–50; (c) Nakamura E, Matsuzawa S, Horiguchi Y, Kuwajima I. *Tetrahedron Lett* 1986; **27**: 4029–32.

11 Yamamoto Y. *Angew Chem Int Ed Engl* 1986; **25**: 947–59.

12 Alexakis A, Sedrani R, Mangeney P. *Tetrahedron Lett* 1990; **31**: 345–8.

13 Tomioka K, Suenaga T, Koga K. *Tetrahedron Lett* 1986; **27**: 369–72.

14 Oppolzer W, Löher H. *Helv Chim Acta* 1981; **64**: 2808–11.

15 Oppolzer W, Poli G, Kingma AJ, Starkemann C, Bernardinelli G. *Helv Chim Acta* 1987; **70**: 2201–14; Oppolzer W, Mills RJ, Pachinger W, Stevenson T. *Helv Chim Acta* 1986; **63**: 1542–5.

16 Helmchen G, Wegner G. *Tetrahedron Lett* 1985; **26**: 6051–4.

17 Bergdahl M, Nilsson M, Olsson T, Stern K. *Tetrahedron* 1991; **47**: 9691–702.

18 Posner GH, Whiten CE. *Tetrahedron Lett* 1973: 1815–18. Posner GH, Whiten CE, Sterling JJ. *J Am Chem Soc* 1973; **95**: 7788–800.

19 Corey EJ, Naef R, Hannon FJ. *J Am Chem Soc* 1986; **108**: 7114–16.
20 Rossiter BE, Eguchi M. *Tetrahedron Lett* 1990; **31**: 965–8; Rossiter BE, Eguchi M, Hernandez AE, Vickers D. *Tetrahedron Lett* 1991; **32**: 3973–6.
21 Tanaka K, Ushio H, Suzuki H. *J Chem Soc Chem Commun* 1990: 795–7; Tanaka K, Matsui J, Somemiya K, Suzuki H. *Synlett* 1994: 351–2.
22 Dieter RK, Tokles M, Lippard SJ. *J Am Chem Soc* 1987; **109**: 2040–46.
23 Villacorta GM, Rao CP, Lippard SJ. *J Am Chem Soc* 1988; **110**: 3175–82; Ahn KH, Klassen RB, Lippard SJ. *Organometallics* 1990; **9**: 3178–81.
24 Lambert F, Knotter DM, Janssen MD *et al. Tetrahedron: Asymm* 1991; **2**: 1097–100; Knotter DM, Grove DM, Smeets WJJ, Speck AL, van Koten G. *J Am Chem Soc* 1992; **114**: 3400–10.
25 Zhou Q-L, Pfaltz A. *Tetrahedron* 1994; **50**: 4467–79.
26 Spescha M, Rihs G. *Helv Chim Acta* 1993; **76**: 1219–30.
27 (a) Kretchmer RA. *J Org Chem* 1972; **37**: 2744–7; (b) Zweig JS, Luche J-L, Barreiro E, Crabbé P. *Tetrahedron Lett* 1975: 2355–8.
28 Langer W, Seebach D. *Helv Chim Acta* 1979; **62**: 1710–22.
29 Leyendecker F, Laucher D. *New J Chem* 1985; **9**: 13–19.
30 Alexakis A, Mutti S, Normant JF. *J Am Chem Soc* 1991; **113**: 6332–4.
31 Alexakis A, Frutos JC, Mangeney P. *Tetrahedron: Asymm* 1993; **4**: 2427–30.
32 Vastra J. Thèse de Doctorat, Université Pierre et Marie Curie, Paris, 1998.
33 (a) Kanai M, Koga K, Tomioka K. *Tetrahedron Lett* 1992; **33**: 7193–6; (b) Kanai M, Tomioka K. *Tetrahedron Lett* 1995; **36**: 4273–4; (c) Kanai M, Tomioka K. *Tetrahedron Lett* 1992; **33**: 4275–8.
34 Strangeland EL, Sammakia T. *Tetrahedron* 1997; **53**: 16503–10.
35 Reddy CK, Devasagayaraj A, Knochel P. *Tetrahedron Lett* 1996; **37**: 4495–8.
36 Suzuki M, Yanagisawa A, Noyori R. *J Am Chem Soc* 1988; **110**: 4718–26; Jansen JFGA, Feringa BL. *J Org Chem* 1990; **55**: 4168–75.
37 Greene AE, Lansard JP, Luche JL, Petrier C. *J Org Chem* 1984; **49**: 931–2.
38 (a) Soai K, Hayasaka T, Ugajin S, Yokoyama S. *Chem Lett* 1988: 1571–2; (b) Bolm C, Ewald M. *Tetrahedron Lett* 1990; **31**: 5011–12; (c) Jansen JFGA, Feringa BL. *Tetrahedron: Asymm* 1992; **3**: 581–2.
39 Nakamura E, Aoki S, Sekiya K, Oshino H, Kuwajima I. *J Am Chem Soc* 1987; **109**: 8056–66.
40 Alexakis A, Vastra J, Mangeney P. *Tetrahedron Lett* 1997; **38**: 7745–8.
41 Kitamura M, Miki T, Nakano K, Noyori R. *Tetrahedron Lett* 1996; **37**: 1253–7.
42 Alexakis A, Vastra J, Burton J, Mangeney P. *Tetrahedron: Asymm* 1997; **8**: 3193–6; Alexakis A, Burton J, Vastra J, Mangeney P. *Tetrahedron: Asymm* 1997; **8**: 3987–90.
43 (a) de Vries AHM, Meetsma A, Feringa BL. *Angew Chem Int Ed Engl* 1996; **35**: 2374–6; (b) Feringa BL, Pineschi M, Arnomd LA, Imbos R, de Vries AHM. *Angew Chem Int Ed Engl* 1997; **36**; 2620–23.
44 Knöbel AKH, Escher IH, Pfaltz A. *Synlett* 1997: 1429–31.
45 Alexis A, Vastra J, Burton J, Benhaim C, Mangeney P. *Tetrahedron Lett* 1998; **39**: 7869–82.
46 Wendish V, Sewald N. *Tetrahedron: Asymm* 1997; **8**: 1253–7.
47 Wang Y, Gladysz JA. *J Org Chem* 1995; **60**: 903–9.
48 Magid RM. *Tetrahedron* 1980; **36**: 1901–30.
49 Trost BM, van Vranken DL. *Chem Rev* 1996; **96**; 395–422.
50 Gallina C, Ciattini PG. *J Am Chem Soc* 1979; **101**: 1035–6; Goering HL, Kantner SS, Tseng CC. *J Org Chem* 1983; **48**: 715–21.
51 Gais H-J, Müller H, Bund J *et al. J Am Chem Soc* 1995; **117**: 2453–66.
52 Denmark SE, Marble LK. *J Org Chem* 1990; **55**: 1984–6.
53 Calo V, Fiandese V, Nacci A, Scilimati A. *Tetrahedron* 1994; **50**: 7283–92.
54 van Klaveren M, Personn ESM, del Villar A *et al. Tetrahedron Lett* 1995; **36**: 3059–62.
55 Piers S, Soucy S. *Can J Chem* 1974; **52**: 3563–4.
56 (a) Gosmini R, Mangeney P, Alexakis A, Commerçon M, Normant JF. *Synlett* 1991: 111–13; (b) Mangeney P, Gosmini R, Raussou S, Commerçon M, Alexakis A. *J Org Chem* 1994; **59**: 1877–88.

57 Raussou S, Gosmini R, Mangeney P, Alexakis A, Commerçon M. *Tetrahedron Lett* 1994; **35**: 5433–6; Raussou S, Urbain N, Mangeney P, Alexakis A. *Tetrahedron Lett* 1996; **37**: 1599–602; Mangeney P, Raussou S, Urbain N, Alexakis A. *Tetrahedron* 1998; **54**: 10349–62.

58 Davies SG, Wollowitz *Tetrahedron Lett* 1980; **21**: 4175–8.

59 Alexakis A, Koenig I. Diplôme d'Etudes Approfondies, Université P. & M. Curie, Paris, 1991.

16 Cyanocuprates: Controversial yet Useful Catalytic Reagents in Organic Synthesis for Today and Tomorrow

BRUCE H. LIPSHUTZ

Department of Chemistry, University of California, Santa Barbara, CA 93106, USA

1 Introduction

When a new organometallic reagent comes along, especially one that has the potential for immediate and widespread use for constructions of valuable carbon–carbon bonds, it is only natural for various groups worldwide to delve into the nature of such a species. Tools of physical organic chemistry are usually brought to bear on the problem, including theoretical studies, which provide at fairly high levels an external 'check' on the conclusions drawn from experiments. In an ideal world both are in harmony and under such circumstances all parties tend to assume that the problem is solved, more or less, at which point all then 'move on' to other challenges. So how is it that in 1999, 18 years after the arrival of 'higher order' (HO) cyanocuprates (1) [1], there has yet to appear an unequivocal conclusion regarding the structure of these otherwise extensively used synthetic reagents? Clearly, they are prepared today in the exact same fashion in which Gilman 'lower order' (LO) cuprates (2) have always been formulated [2], i.e. using two equivalents of an organolithium (RLi) plus one equivalent of copper cyanide (CuCN), rather than 2RLi and a copper(I) halide (CuX, X = I, Br, or Cl). But the observed chemistry of HO cuprates may be quite different from that of LO species: years of work by many groups have clearly shown that the cyanide ligand cannot always be equated with halide ion within a cuprate

cluster. One particularly informative example of innate differences between halocuprates 'R$_2$CuLi·LiX' (or 'R$_2$(X)CuLi$_2$') and cyanocuprates 'R$_2$Cu(CN)Li$_2$' can be seen in their simple oxidations upon treatment with oxidizing agents such as O$_2$ [3] or dinitrobenzene [4] (Scheme 16.1). Here, there is no external substrate with which a cuprate must interact, no issues with mixed cuprates about selectivity of ligand transfer: the reagent is the substrate, and vice versa. And yet, the ratios of products formed when mixed diarylcuprates (e.g. compound 3) are prepared and oxidized under otherwise identical ('kinetic') [5] conditions are nowhere close, nor are the yields comparable from such oxidations of compound 4 (Scheme 16.2) [6]. How can this be? Are not 'cuprates just cuprates'? Hardly! It has taken organometallic chemists decades to convince practitioners how different types of cuprates can lead to completely dissimilar products. Why this realization, and eventual acceptance, has come so slowly is not obvious, especially with the ongoing explosive growth in organopalladium chemistry [7]. Is there any debate as to the critical role that

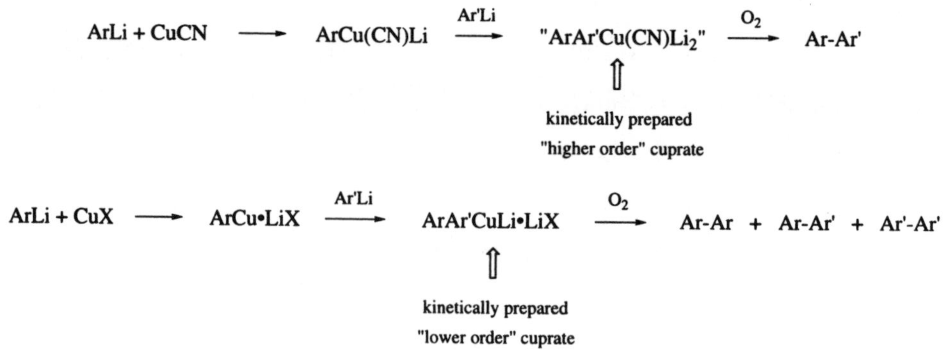

ArLi + CuCN $\longrightarrow$ ArCu(CN)Li $\xrightarrow{\text{Ar'Li}}$ "ArAr'Cu(CN)Li$_2$" $\xrightarrow{\text{O}_2}$ Ar-Ar'

⇑

kinetically prepared
"higher order" cuprate

ArLi + CuX $\longrightarrow$ ArCu•LiX $\xrightarrow{\text{Ar'Li}}$ ArAr'CuLi•LiX $\xrightarrow{\text{O}_2}$ Ar-Ar + Ar-Ar' + Ar'-Ar'

⇑

kinetically prepared
"lower order" cuprate

Scheme 16.1.

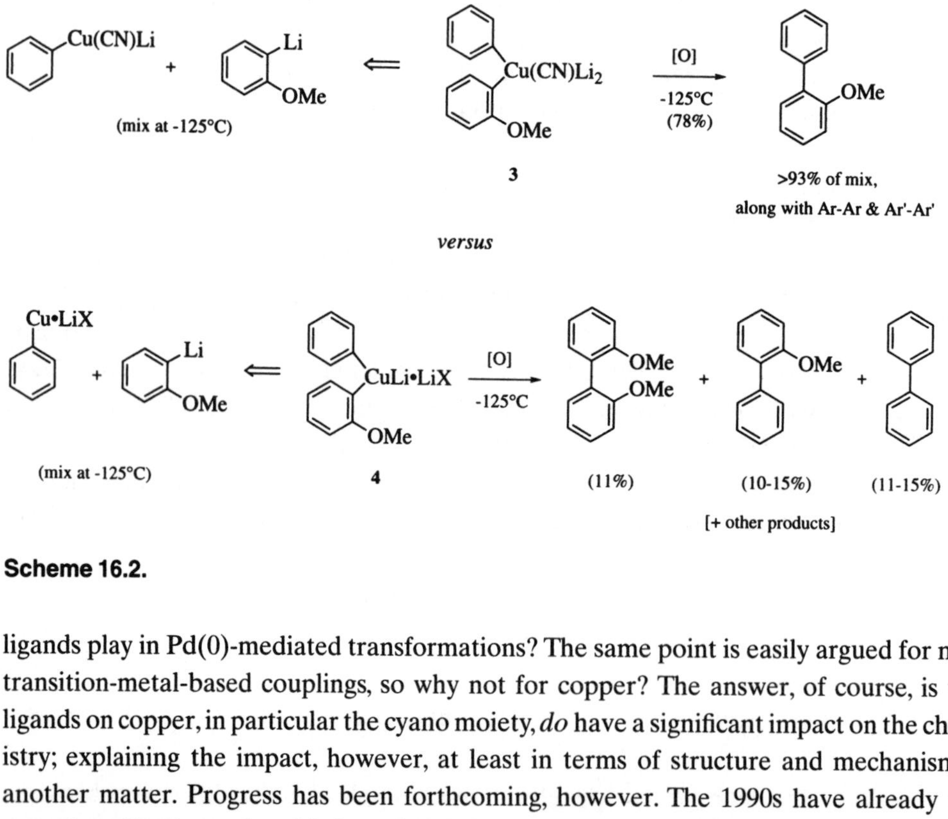

3

(mix at -125°C)

[O]
-125°C
(78%)

>93% of mix,
along with Ar-Ar & Ar'-Ar'

versus

4

(mix at -125°C)

[O]
-125°C

(11%)   (10-15%)   (11-15%)

[+ other products]

Scheme 16.2.

ligands play in Pd(0)-mediated transformations? The same point is easily argued for most transition-metal-based couplings, so why not for copper? The answer, of course, is that ligands on copper, in particular the cyano moiety, *do* have a significant impact on the chemistry; explaining the impact, however, at least in terms of structure and mechanism, is another matter. Progress has been forthcoming, however. The 1990s have already witnessed applications of sophisticated techniques such as EXAFS and XANES [8], NMR experiments applied to probing specified relationships between heteroatom-labeled reagents [9], theoretical treatments that can accommodate transition metal ions [10], cryoscopic measurements [11] and many attempts at X-ray analyses which have only recently been solved.

2   Cuprate-catalysed couplings: doing more with less copper

So with the HO cuprate debate likely to wind down now that hard X-ray data is available [12], what will be the focus of organocopper chemists for the future? From the author's perspective, two themes stand out, although they are not mutually exclusive. On the one hand, the notion that organocopper-induced couplings can afford chiral, non-racemic

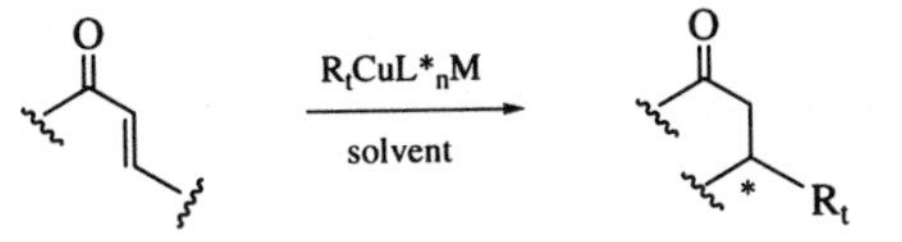

Scheme 16.3.

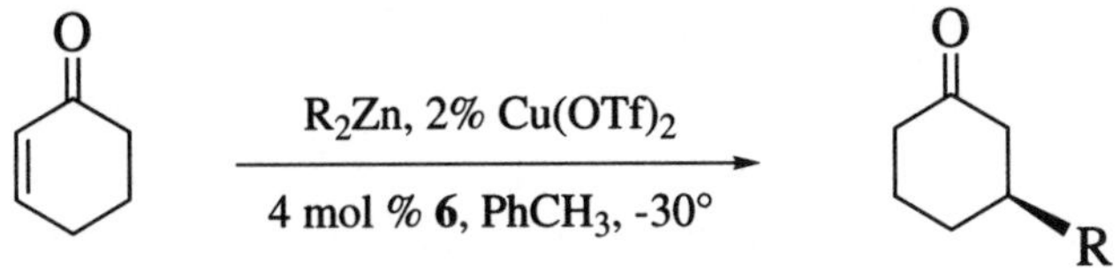

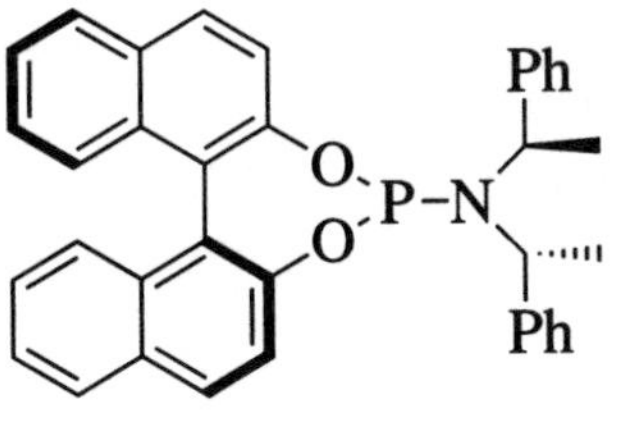

Michael adducts (**5**), where the asymmetry in the product is induced by the reagent, will continue to be a worthy goal [13]. Much work has already appeared and some guidelines have been espoused [13], although true substrate generality among stoichiometric chiral cuprate 1,4-additions is still lacking. Very recently, however, a considerable advance has been recorded based on the finding that a tailor-made, non-racemic phosphine (**6**) can combine with small percentages of a Cu(I) salt in the presence of organozinc reagents (R_2Zn) to give very high enantiomeric excesses (ee) of conjugate addition products from cyclohexenones (Scheme 16.3) [14]. These very impressive results underscore not only the prospects for an eventual, even more general, solution to this 'grail' of organocopper chemistry, but also the attractiveness of using copper in a strictly catalytic mode. Interestingly, when all of the highly valued carbon–carbon and carbon–heteroatom bond-forming reactions of copper now employed [15] can be carried out via copper catalysis, the field will have come full circle, dating back to the CuCl-catalysed Grignard couplings with enones first reported by Kharasch and Tawney in 1941 [16].

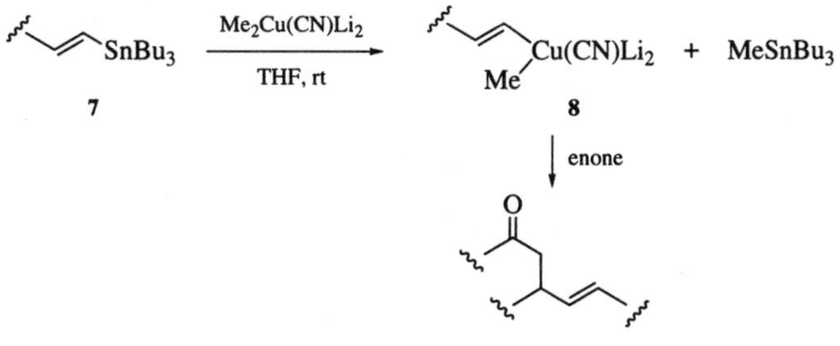

In recognition of the times, which seem to be squarely focused on catalysis, we have developed cuprate-catalysed processes based on the facility with which alkyl ligands on copper undergo exchange (i.e. transmetalation) with ligands on various other organometallics, in particular with zirconium and zinc reagents. The former was an outgrowth of our initial contribution to Searle's development of stoichiometric cuprate transmetalations of vinyl stannanes (**7**) [17] to mixed reagents (**8**), from which it was predicted that the corresponding zirconocenes (**9**) (Scheme 16.4) would be far more Lewis acidic and hence highly prone to ligand exchange with cyanocuprates. Several useful processes catalytic in Cu(I) have arisen from this concept [18], including a cuprate-catalysed 1,4-addition of vinyl zirconocenes [19]. Application of this one-pot hydrozirconation/transmetalation/Michael addition sequence to prostanoid syntheses could be achieved by quenching the intermediate enolate derived from a 4-silyloxy-substituted cyclopentenone (**10**) (Scheme 16.4) with an electrophile (**11**). A method for 1,4-delivery of alkyl ligands from zirconium to carbon via copper, using 10 mol.% of a copper(I) salt, is also available [20].

Organozinc reagents R_tZnX are especially valuable given their remarkable tolerance of electrophilic functionality within the reagents themselves [21]. Although well known to undergo conversion to lower order cyanocuprates $R_tCu(CN)ZnX$ upon exposure to CuCN [22] (Equation (16.1)), the $(ZnX)^+$ gegenion is not an effective Lewis acid, and

$$R_tZnX \quad + \quad CuCN \quad \xrightarrow{\text{THF}} \quad R_tCu(CN)^- (ZnX)^+$$

not a good Lewis acid
within cuprates

$$(16.1)$$

dramatically lowers reactivity (relative to Li$^+$) in Michael addition schemes. As a means of both increasing their reactivity and converting to an overall process that is catalytic in copper, a transmetalation route has been developed for arriving at lithiocuprate (**12**) using catalytic quantities of HO cuprate $Me_2Cu(CN)Li_2$ (Scheme 16.5). Mixed organozinc reagents (**13**) and a higher order cyanocuprate ($Me_2Cu(CN)Li_2$, 5 mol.%) in the presence of TMS-Cl combine to effect (via compound **12**) efficient conjugate delivery of the desired ligand R_t relatively quickly and under very mild conditions (Scheme 16.6) [23].

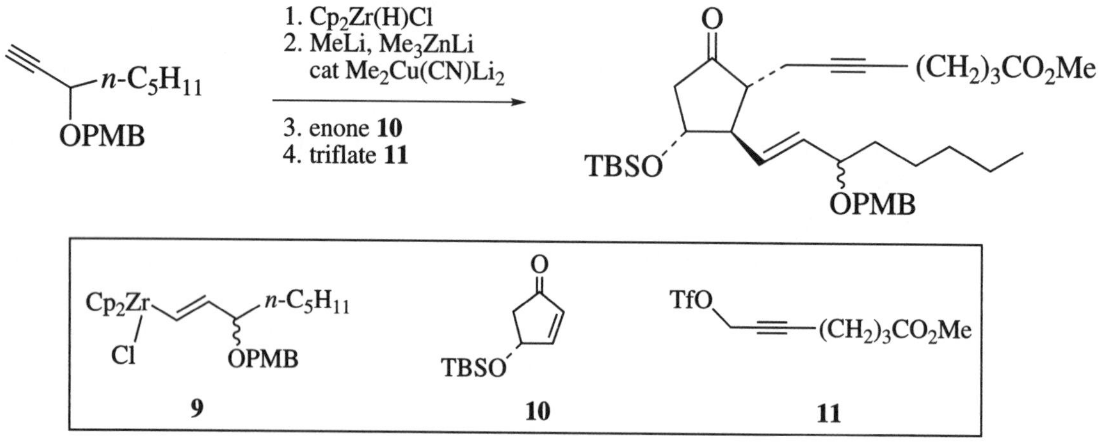

Scheme 16.4.

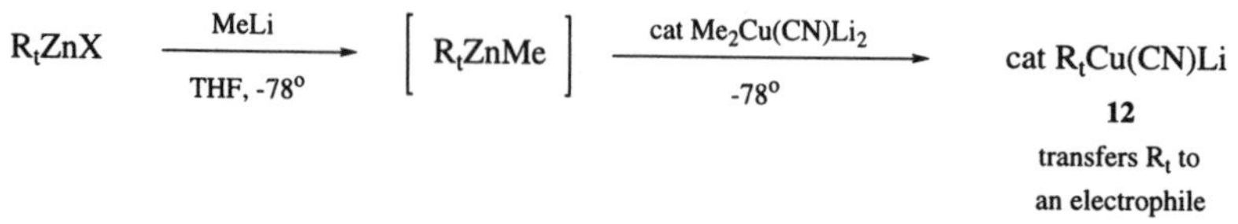

Scheme 16.5.

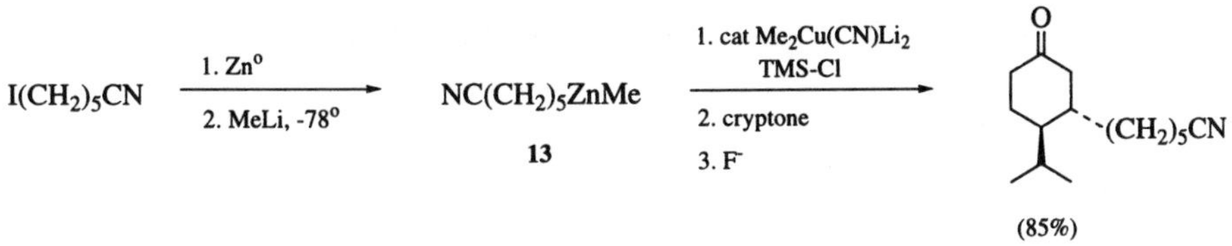

$$I(CH_2)_5CN \xrightarrow[\text{2. MeLi, -78°}]{\text{1. Zn°}} NC(CH_2)_5ZnMe \xrightarrow[\substack{\text{2. cryptone} \\ \text{3. F}}]{\substack{\text{1. cat Me}_2\text{Cu(CN)Li}_2 \\ \text{TMS-Cl}}} \quad (85\%)$$

13

Scheme 16.6.

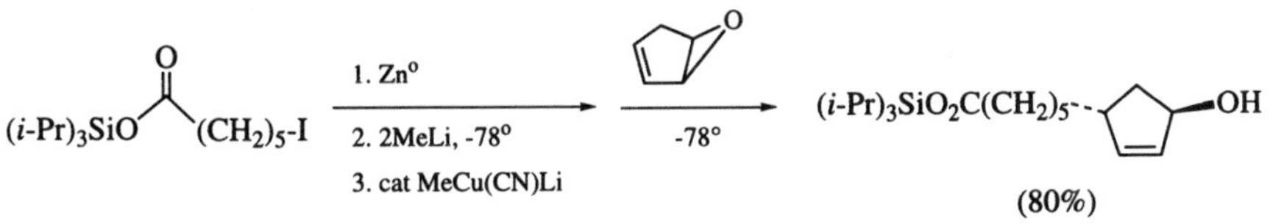

$$(i\text{-Pr})_3SiO\text{-C(=O)-}(CH_2)_5\text{-I} \xrightarrow[\substack{\text{2. 2MeLi, -78°} \\ \text{3. cat MeCu(CN)Li}}]{\text{1. Zn°}} \xrightarrow{-78°} (i\text{-Pr})_3SiO_2C(CH_2)_5\text{-}\cdots\text{-OH} \quad (80\%)$$

Scheme 16.7.

The powerful combination of a mixed zincate (i.e. $R_t ZnMe_2Li$)/catalytic cyanocuprate can be applied to selected alkylations. Unfortunately, simple epoxides are not yet acceptable substrates owing to their highly competitive opening by halide ions present in solution (recall that $R_t ZnMe_2Li$ comes from $R_t ZnX + 2MeLi$, thereby generating LiX as a byproduct). Allylic oxiranes, however, react cleanly at low temperatures, alkylating the intermediate cuprates in a Normant sense [24] to give allylic alcohol products via *anti*-S_N2' addition (Scheme 16.7) [25].

Cyanocuprate-catalysed alkylations of vinyl triflates are also under study. Although displacements by Gilman cuprates of unactivated educts of this type were reported years ago [26a], as were related couplings on aryl triflates mediated by HO cyanocuprates [26b], they normally require several equivalents of stoichiometric reagent, as well as higher-than-usual temperatures for full conversion. Thus, attempts to convert this important olefin synthesis to a process that is catalytic in copper has not yet been successful. However, activated vinyl triflates (**14**) react quickly at low temperatures using 5 mol.% of the 'cuprate in a bottle' [(2-Th)Cu(CN)Li] [27a], where the valuable transferable ligand R_t is brought in via mixed zinc species **15** (Equation 16.2) [27b].

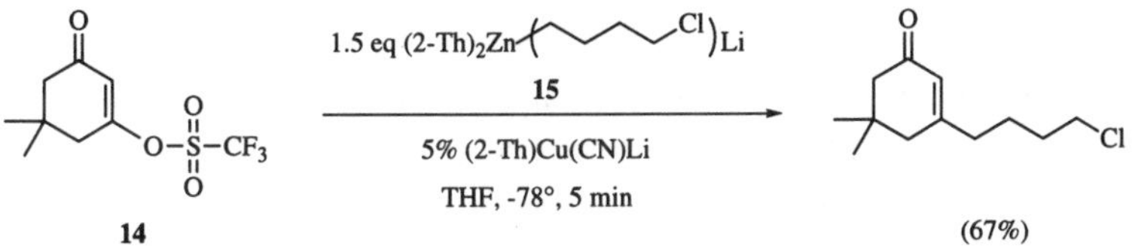

$$\xrightarrow[\substack{\text{5\% (2-Th)Cu(CN)Li} \\ \text{THF, -78°, 5 min}}]{\substack{\text{1.5 eq (2-Th)}_2\text{Zn}\left(\sim\sim\text{Cl}\right)\text{Li} \\ \textbf{15}}} \quad (67\%)$$

14

(16.2)

The ability of zinc to 'soften' the presence of a harder $R_t Li$ in the form of a zincate should readily apply to groups R_t that are not carbon–lithium bound. One example currently being developed involves a silyl group, most notably the phenyldimethylsilyl moiety, which makes up Fleming disilylcyanocuprate reagents $(PhMe_2Si)_2Cu(CN)Li_2$ (**16**) (Scheme 16.8) [28]. This cuprate, and its predecessor Gilman version $(PhMe_2Si)_2CuLi$ [30], have been well received by virtue of the availability of precursor $PhMe_2SiLi$ (from $PhMe_2SiCl + Li°$) [30], and the ease with which this aryl-substituted silicon residue can be exchanged for a hydroxyl group with maintenance of C_{sp3} stereochemistry [31]. Couplings are usually carried out with homocuprates (**16**) and therefore not only sacrifice one of the $PhMe_2Si$ groups but involve stoichiometric amounts of copper. The mixed zincate option (i.e.

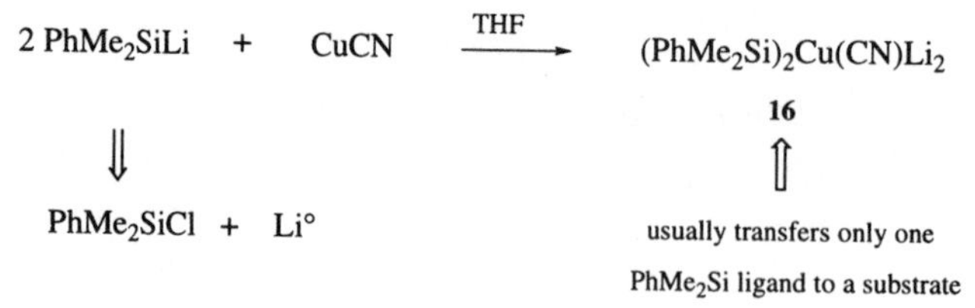

$$2 \; PhMe_2SiLi \quad + \quad CuCN \quad \xrightarrow{\;THF\;} \quad (PhMe_2Si)_2Cu(CN)Li_2$$

16

$$\Downarrow \qquad\qquad\qquad\qquad \Uparrow$$

$$PhMe_2SiCl \; + \; Li° \qquad\qquad \text{usually transfers only one}$$

$$\text{PhMe}_2\text{Si ligand to a substrate}$$

Scheme 16.8.

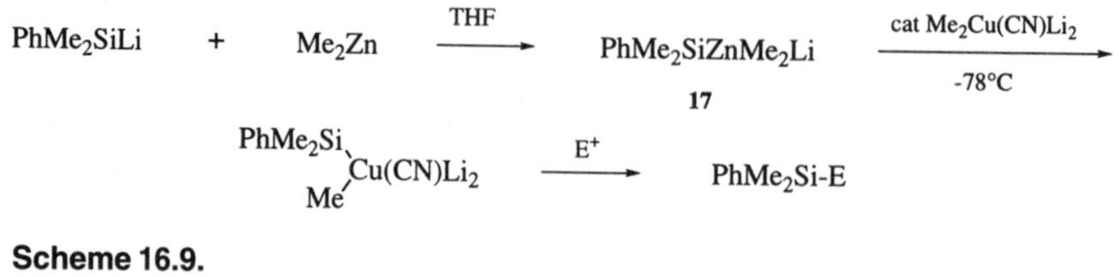

$$PhMe_2SiLi \quad + \quad Me_2Zn \quad \xrightarrow{\;THF\;} \quad PhMe_2SiZnMe_2Li \quad \xrightarrow[-78°C]{cat \; Me_2Cu(CN)Li_2}$$

17

$$\underset{Me}{\overset{PhMe_2Si}{\diagdown}}Cu(CN)Li_2 \quad \xrightarrow{\;E^+\;} \quad PhMe_2Si\text{-}E$$

Scheme 16.9.

PhMe$_2$SiZnMe$_2$Li, **17**) [32] provides an immediate benefit in that only one PhMe$_2$Si moiety need be invested (Scheme 16.9). The other player in this transmetalation scheme, Me$_2$Cu(CN)Li$_2$, remains the same as previously utilized for carbon ligand transfers (*vide supra*). Likewise, it is needed to the extent of only 5 mol.%. Conjugate additions typified by the reaction with isophorone (Equation (16.3)) appear to proceed quite readily at low temperatures, although the significant drop in quantity of Cu(I) does slow the rate of conversion to the 1,4-adduct [33]. None the less, with yields still high, there is another bonus to this approach. Because Cu(I) is catalytic, the resulting intermediate must retain zinc, which for 1,4-additions translates into a zinc enolate. Previous work [34] had shown that such

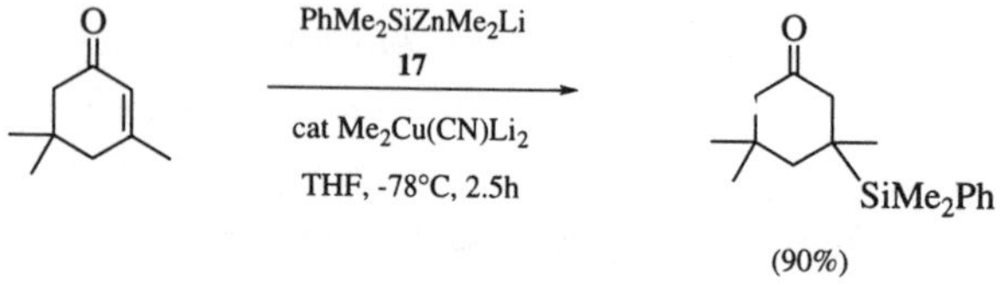

$$\text{(16.3)}$$

species readily participate in subsequent aldol events [35] (as well as alkylations with reactive electrophiles), the sequence thus becoming a net three-component coupling (3-CC) [34, 36]. One example of this multiple bond-forming method is illustrated in Scheme 16.10 [33].

A related 3-CC begins with a copper-catalysed, stereodefined silylzincation of terminal alkynes [37], rather than a Michael addition. Once complete, the presence of copper(I) can then lead to an *in situ*-derived vinylcuprate via a second transmetalation, in this case involving the intermediate *Z*-mixed zincate **18**. With the necessary ingredients already in the pot, a subsequent coupling can occur, e.g. upon introduction of an enone. Indeed, this tandem 1,2- followed by 1,4- sequence leads to potentially valuable vinylsilanes (**19**) bearing carbonyl functionality (Scheme 16.11) [33].

Another type of cuprate reaction ripe for development of a catalytic version is that of 1,4-hydride delivery. The reagent perhaps most commonly used in this regard is [CuH·PPh$_3$]$_6$ [38a], regaled years ago as 'Reagent of the Year' [39]. While its properties are very attractive (e.g. tolerant of moisture, commercially available, stable at room temperature, etc.), it would find even greater utility if its stoichiometric usage could be modified in favor of a *convenient* catalytic cycle for copper [38b].

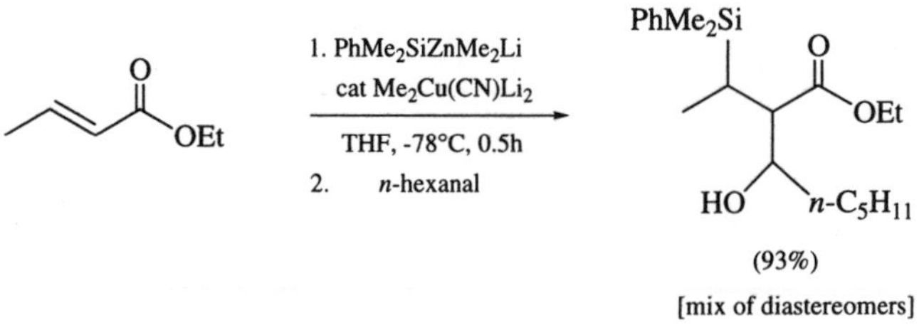

Scheme 16.10.

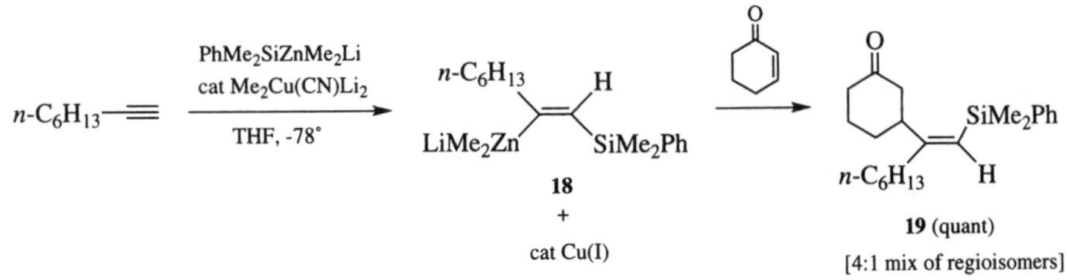

Scheme 16.11.

A decade ago, we described an alternative method for conjugate hydride delivery involving the *in situ* generation of copper hydride from equivalent amounts of a copper(I) halide salt (CuX, X = I) solubilized by LiCl, plus Bu$_3$SnH [40]. The reagent was presumed to exist as its halocuprate, X(H)CuLi (X = I or Cl; Equation (16.4)). At the time, this

$$\text{CuX·2LiCl} + \text{Bu}_3\text{SnH} \xrightarrow[\text{rt}]{\text{THF}} \text{"H(X)CuLi"} + \text{Bu}_3\text{SnX} \qquad (16.4)$$
$$(\text{X = I})$$

seemed like a reasonable stoichiometric approach involving transmetalation to X(H)CuLi and Bu$_3$SnX; the former species then goes on to react with Michael acceptors, such as compound **20** (Equation (16.5)). What has only recently occurred to us, as we look toward a potential catalytic variant of this process, is that it is already poised to be catalytic as is! What went unappreciated at the time is the potential for Bu$_3$SnX to capture the

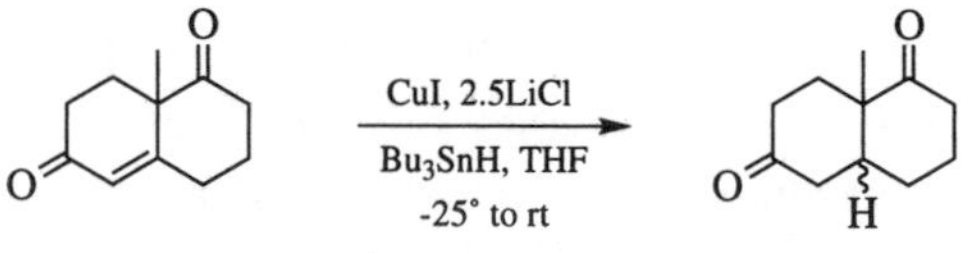

(16.5)

20 (84%)

initially formed enolate (**21**) [41], thereby liberating copper for re-entry into the catalytic cycle in the presence of stoichiometric Bu$_3$SnH (Scheme 16.12). In a test case conducted recently, use of 10 mol.% CuI·LiCl together with added Ph$_3$P gave rise to the 1,4-reduction product **22** in 80% gas chromatographic (GC) yield (Equation (16.6)) (B.H. Lipshutz and J. Keith, unpubl. obs.). The rest of the material was present as starting enone, and hence further work remains so that full consumption of educt is realized. Even better, though, would be a further modified procedure that also reduces the amount of tin present, or even avoids it altogether. For this purpose, we are currently screening various silicon hydrides.

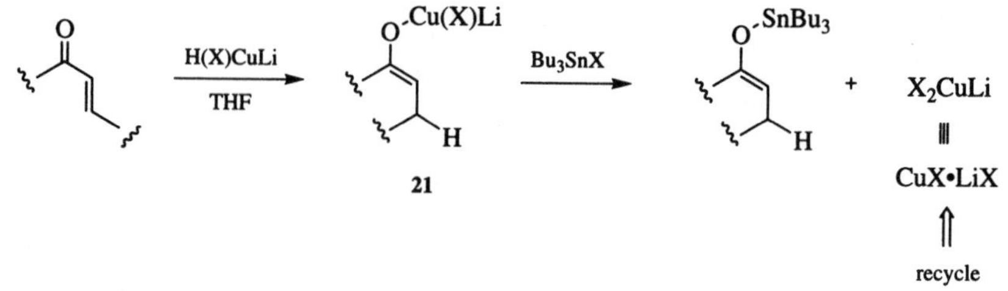

Scheme 16.12.

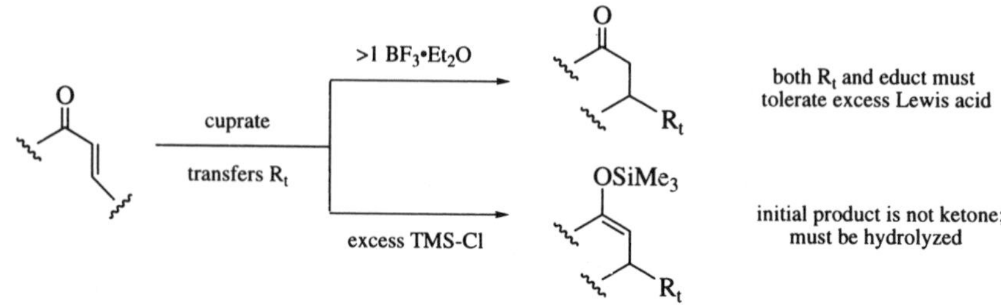

both R$_t$ and educt must tolerate excess Lewis acid

initial product is not ketone; must be hydrolyzed

Scheme 16.13.

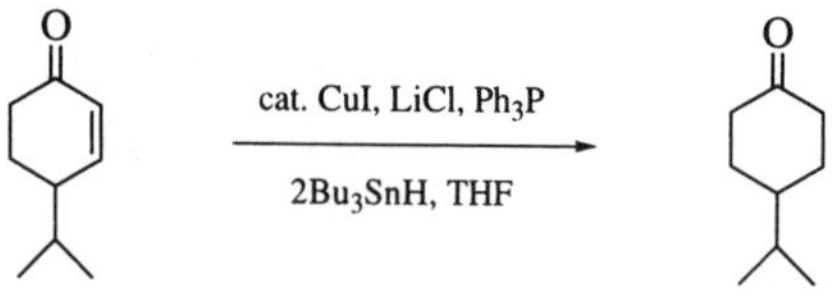

(16.6)

22
(80%, by GC)

Along with the accent on catalysis at the copper level, there is an associated issue that could well have an impact on the extent of usage of any newly developed catalytic process. Given that rates of couplings usually drop relative to those involving the corresponding stoichiometric cuprate, what additives are available for accelerating these catalytic couplings? The answers, perhaps surprising given the rich history of organocopper chemistry, are not completely satisfying because they highlight a far more general problem in this field for which no solution exists. There are essentially two classes of additives commonly relied upon for boosting cuprate reactivity (Scheme 16.13): BF$_3$·Et$_2$O, a good Lewis acid used in excess that is effective for various organocopper (RCu) [42] and cuprate Michael-type reactions, ultimately affording product ketones via intermediate boron enolates [43]; and trialkylsilyl halides, particularly Me$_3$SiCl, which, while not a Lewis acid toward most carbonyl-containing substrates [44], has a dramatic impact on most cuprate 1,4-additions [45]. However, the product (silyl enol ether) must be hydrolysed or cleaved by other means (e.g. fluoride ion) in order to arrive at the carbonyl product. Thus, there is no *general* additive known that is tolerant of acid-labile groups within the cuprate or substrate (e.g. a ketal), that affords the carbonyl product directly or that accelerates a cuprate reaction of any type, especially one that involves catalytic amounts of copper(I).

Very recently, while developing the copper-catalysed reactions of mixed silyl zincates (*vide supra*), we considered the potential for a rare-earth or lanthanide reagent (e.g. Sc(OTf)$_3$ or Yb(OTf)$_3$, respectively) to serve as a catalyst for silyl cuprate 1,4-additions. It was speculated that a metal of either type, known to strongly associate with an sp^2 carbonyl

Wanted: new additives for accelerating cuprate reactions

One possiblity: rare earths [*e.g.*, Sc(OTf)$_3$] and/or lanthanides [*e.g.*, Yb(OTf)$_3$]

Example:

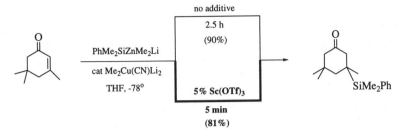

Scheme 16.14.

oxygen [46], might also undergo transmetalation with a cuprate or zincate after initial complexation and subsequent net 1,4-addition by silicon. The additive was not likely to react with a zincate or cuprate under mild (−78 °C) conditions, and its association with the nitrile ligand in the cuprate [47] was improbable when presented with a carbonyl moiety. The test case came with the hindered enone isophorone (*vide supra*), which had been observed to give the 1,4-adduct in 90% yield over the course of 2.5 h using a zincate in THF at 0.1 M, along with 3 mol.% Me$_2$Cu(CN)Li$_2$ (Scheme 16.14). The other control experiment of interest involved the stoichiometric Fleming cuprate (PhMe$_2$Si)$_2$Cu(CN)Li$_2$, which, at the same 0.1 M concentration as used for the zincate (i.e. 33 times more concentrated than when used catalytically), was found to consume educt in 15 min under otherwise identical conditions. The real surprise came when the reaction involving zincate and catalytic cuprate was run in the presence of 5 mol.% Sc(OTf)$_3$: it was over in less than 5 min! Thus, with an effective cuprate concentration of 0.003 M (0.1 M in zincate × 0.03 equivalents of cuprate), *the rare earth additive accounts for a reaction now three times faster than that using stoichiometric cuprate* [33].

Of course, this is only a single example, and many questions come up surrounding this result that are now demanding our attention. But it is experiments such as these, and other fascinating, exciting and curious observations [48,49] that are suggestive of new relationships between reagents, that await further scrutiny and eventual understanding. Ultimately, these Cu(I)-dependent reactions, along with those to be discoverd, will produce even more powerful synthetic technologies in the 21st century.

3 References

1 Lipshutz BH, Wilhelm RS, Floyd DM. *J Am Chem Soc* 1981; **103**: 7672.

2 Gilman H, Jones RG, Woods LA. *J Org Chem* 1952; **17**: 1630.

3 Whitesides GM, Fischer WF, San Filippo J, Bashe RW, House HO. *J Am Chem Soc* 1969; **91**: 4871.

4 Bertz SH, Gibson CP. *J Am Chem Soc* 1986; **109**: 8286.

5 Lipshutz BH, Siegmann K, Garcia E, Kayser F. *J Am Chem Soc* 1993; **115**: 9276.

6 Lipshutz BH, Kayser F, Siegmann K. *Tetrahedron Lett* 1993; **34**: 6689.

7 Tsuji J. *Palladium Reagents and Catalysts*. Chichester: Wiley, 1995.

8 Stemmler T, Penner-Hahn JE, Knochel P. *J Am Chem Soc* 1993; **115**: 348; Stemmler T, Barnhart TM, Penner-Hahn JE *et al. J Am Chem Soc* 1995; **117**: 12489; Barnhart TM, Huang H, Penner-Hahn JE. *J Org Chem* 1995; **60**: 4310.

9 Bertz SH. *J Am Chem Soc* 1991; **113**: 5470; 1990; **112**: 4031; Cabezas JA, Oehlschlager AC. *J Am Chem Soc* 1997; **119**: 3878.

10 Synder JP, Spangler DP, Behling JR, Rossiter BE. *J Org Chem* 1994; **59**: 2665; Synder JP, Bertz SH. *J Org Chem* 1995; **60**: 4312; Snyder JP, Tipsword GE, Spangler DP. *J Am Chem Soc* 1992; **114**: 1507.

11 Gerold A, Jastrzebski JTBH, Kronenburg CMP, Krause N., van Koten G, *Angew Chem Int Ed Engl* 1997; **36**: 755.

12 Kronenburg CMP, Jastrzebski JTBH, Spek AL *et al. J Am Chem Soc* 1998; **120**: 9688; Boche G, Bosold F, Marsch M *et al. Angew Chem Int Ed* 1998; **37**: 1684; Krause N, ibid, 1999; **38**: 79.

13 Rossiter BE, Swingle NM. *Chem Rev* 1992; **92**: 771.

14 Feringa BL, Pineschi M, Arnold LA, Imbos R, de Vries AHM. *Angew Chem Int Ed Engl* (In press.)

15 Lipshutz BH, Sengupta S. *Org React* 1992; **41**: 135; Lipshutz BH. In: Schlosser M, ed. *Organometallics in Synthesis: A Manual*. Chichester: Wiley, 1994: 283–382.

16 Kharasch MS, Tawney PO. *J Am Chem Soc* 1941; **63**: 2308.

17 Behling JR, Babiak KA, Ng JS *et al. J Am Chem Soc* 1988; **110**: 2461.

18 Yoshifuji M, Loots M, Schwartz J. *Tetrahedron Lett* 1977: 1303.

19 Lipshutz BH, Wood MR. *J Am Chem Soc* 1993; **115**: 12625.

20 Wipf P. *Synthesis* 1993: 537.

21 Knochel P, Singer RD. *Chem Rev* 1993; **93**: 2117.

22 Knochel P, Yeh MCP, Berk SC, Talbert J. *J Org Chem* 1988; **53**: 2390; Knochel P. *Synlett* 1995: 393.

23 Lipshutz BH, Wood MR, Tirado R. *J Am Chem Soc* 1995; **117**: 6126.

24 Cahiez G, Alexakis A, Normant JF. *Synthesis* 1978: 528; see also: Marino JP, Jaen JC. *J Am Chem Soc* 1982; **104**: 3165; Wender PA, Erhardt JM, Letendre LJ. *J Am Chem Soc* 1981; **103**: 2114; Ziegler FE, Cady MA. *J Org Chem* 1981; **46**: 122.

25 Lipshutz BH, Woo K, Gross T, Buzard DJ, Tirado R. *Synlett* 1997: 477.

26 (a) McMurry JE, Scott WJ. *Tetrahedron Lett* 1980; **21**: 1303; 1983; **24**: 979; (b) McMurry JE, Scott WJ. *Tetrahedron Lett* 1983; **24**: 2723.

27 (a) Lipshutz BH, Koerner M, Parker DA. *Tetrahedron Lett* 1987; **28**: 945; (b) Lipshutz BH, Vivian R. ibid, submitted.

28 Fleming I, Newton TW. *J Chem Soc Perkin Trans I* 1984: 1805.

29 Ager DJ, Fleming I. *J Chem Soc Chem Commun* 1978: 177; Ager DJ, Fleming I, Patel SK. *J Chem Soc Perkin Trans I* 1981: 2520.

30 Fleming I. In: Taylor RJK, ed. *Organocopper Reagents, A Practical Approach*. Oxford: Oxford University Press, 1994: 257–92.

31 Fleming I, Henning R, Plaut H. *J Chem Soc Chem Commun* 1984: 29; Fleming I, Sanderson PEJ. *Tetrahedron Lett* 1987; **28**: 4299; Tamao K, Kawachi A, Ito Y. *J Am Chem Soc* 1992; **114**: 3989.

32 Crump RANC, Fleming I, Urch CJ. *J Chem Soc Perkin Trans I* 1994: 701.

33 Lipshutz BH, Sclafani J, Takanami T. *J Am Chem Soc* 1998; **120**: 4021.

34 Lipshutz BH, Wood MR. *J Am Chem Soc* 1994; **116**: 11689.

35 Evans DA, McGee LR. *Tetrahedron Lett* 1980; **20**: 3975.

36 Lipshutz BH, Gross T, Buzard DJ, Tirado R. *J Chin Chem Soc* 1997; **44**: 1.

37 Wakamatsu K, Nonaka T, Okuda Y *et al. Tetrahedron* 1986; **42**: 4427.

38 (a) Mahoney WS, Brestensky DM, Stryker JM. *J Am Chem Soc* 1988; **110**: 291; Koenig TM, Daeuble JF, Brestensky DM, Stryker JM. *Tetrahedron Lett* 1990; **31**: 3237; (b) for a catalytic version using high pressures of molecular hydrogen, see: Mahoney WS, Stryker JM. *J Am Chem Soc* 1989; **111**: 8818.

39 For example, see: *J Am Chem Soc* 1991; **113**, issue no. 11, back cover.

40 Lipshutz BH, Ung CS, Sengupta S. *Synlett* 1989: 64.

41 Suzuki M, Yanagisawa A, Noyori R. *J Am Chem Soc* 1988; **110**: 4718.

42 Yamamoto Y. *Angew Chem Int Ed Engl* 1986; **25**: 947; Yamamoto Y, Yamamoto S, Yatagai H, Ishihara Y, Maruyama K. *J Org Chem* 1982; **47**: 119; Lipshutz BH, Ellsworth EL, Dimock SH. *J Am Chem Soc* 1990; **112**: 5869.

43 Lipshutz BH, Ellsworth EL, Siahaan TJ. *J Am Chem Soc* 1989; **111**: 1351.

44 Lipshutz BH, Aue DH, James B. *Tetrahedron Lett* 1996; **37**: 8471.

45 Corey EJ, Boaz NW. *Tetrahedron Lett* 1985; **26**: 6015, 6019; Alexakis A, Berlan J, Besace Y. *Tetrahedron Lett* 1986; **27**: 1047; Nakamura E, Matsuzawa S, Horiguchi Y, Kuwajima I. *Tetrahedron Lett* 1986; **27**: 4029.

46 Kobayashi S. *Synlett* 1994: 689; Ishihara K, Karumi Y, Yamamoto H. *Synlett* 1986: 839.

47 Lipshutz BH, Ellsworth EL, Siahaan T. *J Am Chem Soc* 1988; **110**: 4834.

48 Kitamura M, Miki T, Nakano K, Noyori R. *Tetrahedron Lett* 1996; **37**: 5141.

49 Alexakis A, Vastra J, Mangeney P. *Tetrahedron Lett* 1997; **38**: 7745.

17 Palladium-catalysed Oxidation of Alkenes

TAKAHIRO HOSOKAWA* and SHUN-ICHI MURAHASHI†

*Department of Environmental Systems Engineering, Faculty of Engineering, Kochi University of Technology, Tosayamada, Kochi 782-8502, Japan

†Department of Chemical Science and Engineering, Graduate School of Engineering Science, Osaka University, Machikaneyama 1–3, Toyonaka, Osaka 560-8531, Japan

1 Introduction

Catalysis by palladium complexes is classified into two categories by their oxidation states of either +2 or 0 [1, 2]. Compared with the outstanding progress made by palladium(0) catalysts in organic synthesis, the use of palladium(II) for the creation of efficient catalysts remains undeveloped. This is largely due to difficulties in retaining the oxidation state of +2 during catalytic reactions. However, efforts to enhance the catalytic efficiency of palladium have had significant consequences in the chemistry of palladium(II). This chapter thus deals with the conceptual advances in this field, focusing on the oxypalladation of alkenes, which is the representative reaction of palladium(II) catalysts.

Palladium(II) basically acts as a Lewis acid due to its cationic nature to activate functionalities such as alkenes. Activation of alkenes via π-coordination to palladium(II) allows nucleophilic attack on the functionality. Typically, water reacts with terminal alkenes in the presence of the $PdCl_2$ catalyst to give the corresponding methyl ketones (Scheme 17.1), which represents an invaluable functionalization step in organic synthesis [3]. In the early 1970s, an intramolecular version of the reaction was first introduced by us with alkenes bearing an OH group such as allylphenols [4, 5], which provided a useful entry to the synthesis of oxygen-containing heterocycles [6]. This approach was advanced by Hegedus et al. to a synthetic method for nitrogen-containing heterocycles from alkenes bearing intramolecular nitrogen nucleophiles [7]. Described herein are intramolecular oxypalladations of alkenes directing attention towards the nature of the catalytically active species consisting of palladium and copper.

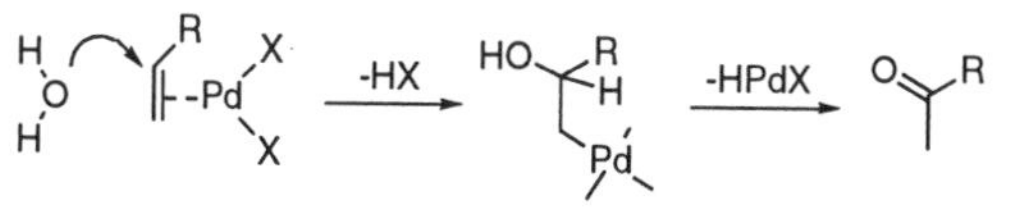

Scheme 17.1.

2 Intramolecular oxypalladation of alkenes by palladium–copper bimetallic systems

In the case of 1-alkenes containing OH groups, intramolecular nucleophilic attack of the OH on the alkene leads to the oxypalladation adduct **1** (Scheme 17.2), where the Pd–C bond is formed at the less sterically hindered C-1 position. Subsequent β-palladium hydride elimination gives oxygen-containing heterocycles bearing an alkene in the more stable position via equilibration. The resulting PdHCl species is reduced to Pd(0) and HX, when palladium(II) complexes are used as reagents. Representative examples of the stoichiometric reaction [4, 5, 8–10] are shown in Equations (17.1)–(17.5) and it is noteworthy that a tetronomycin precursor can be synthesized by this cyclization (Equation (17.5)).

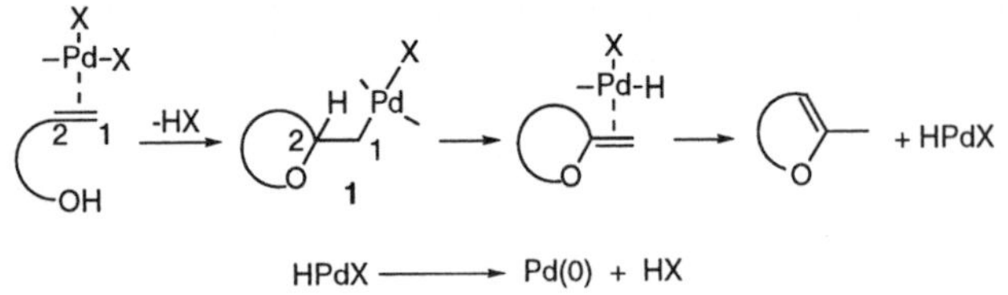

$$HPdX \longrightarrow Pd(0) + HX$$

Scheme 17.2.

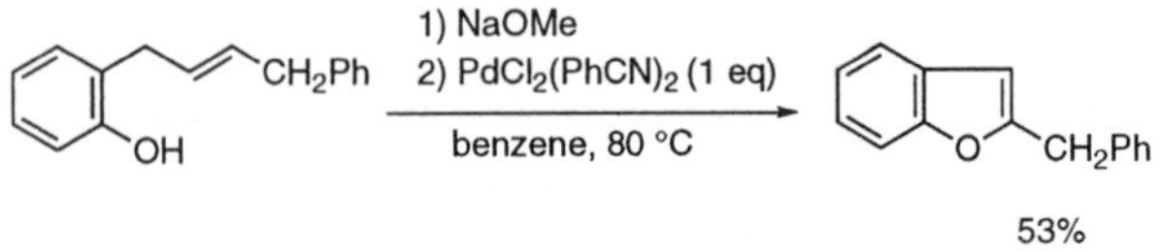

(17.1)[4]

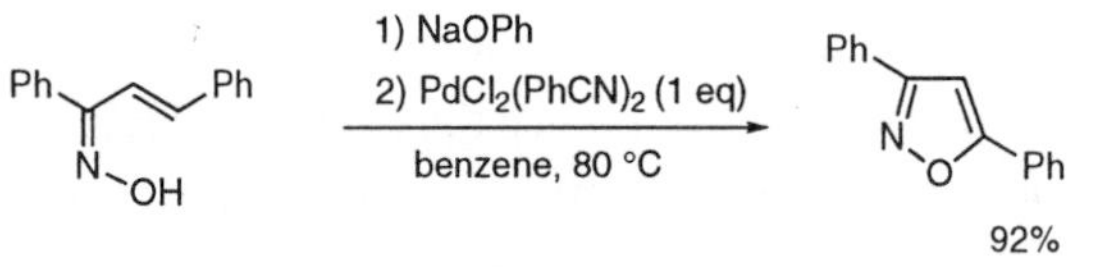

53%

(17.2)[5a]

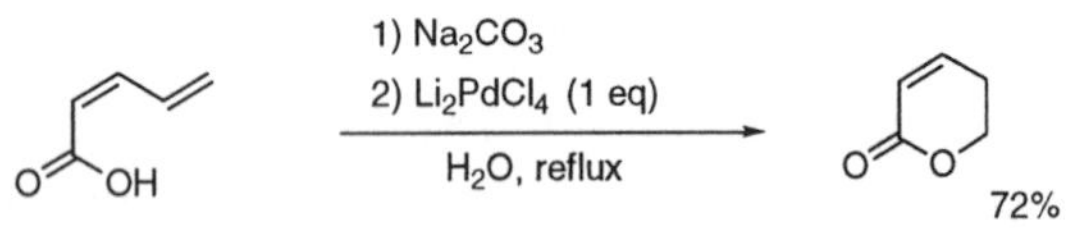

92%

(17.3)[8]

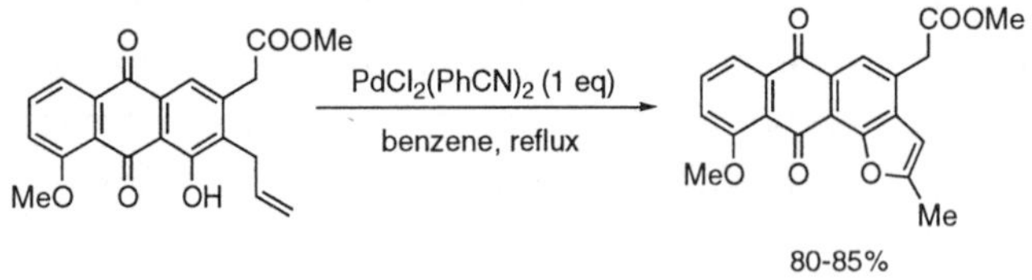

72%

(17.4)[9]

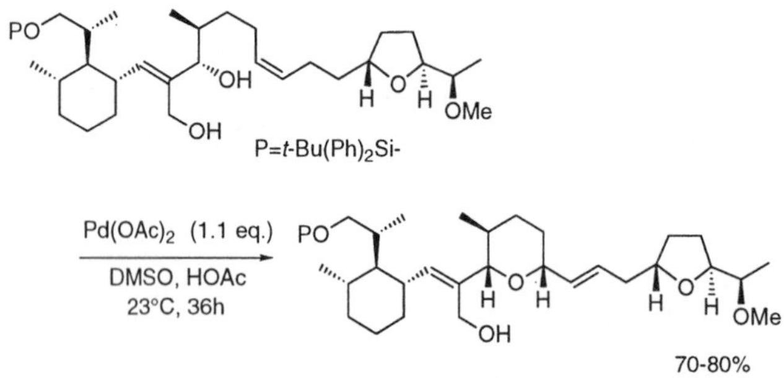

80-85%

(17.5)[10a]

70-80%

For catalytic reactions of this type, either a combination of copper salts and O_2 or *p*-benzoquinone is usually employed. Considering that the stoichiometric reaction gives rise to Pd(0) species (Scheme 17.2), the function of copper salts such as $CuCl_2$ is commonly understood to reoxidize Pd(0) into Pd(II) (Equations (17.6) and (17.7)). This understanding is, of course, based on the well-known Wacker catalysis for the production of acetaldehyde from ethylene [11]. However, if Pd(0) is not formed in the *catalytic reaction*, the redox catalysis of Equations (17.6) and (17.7) does not stand. Before approaching this subject, let us first review some examples of the catalytic cyclizations of hydroxyalkenes.

$$Pd(0) \ + 2\,CuX_2 \ \longrightarrow \ PdX_2 \ + 2\,CuX \tag{17.6}$$

$$2\,CuX \ + \ 1/2\,O_2 \ + \ 2\,HX \ \longrightarrow \ 2\,CuX_2 \ + \ H_2O \tag{17.7}$$

Illustrated in Scheme 17.3 is the oxypalladation intermediate having two possibilities of β-Pd–H elimination. This intermediate generally produces energetically favorable vinyl substitution rather than *exo*-methylene substitution. Thus, the γ,δ-unsaturated alcohol **2** can be converted catalytically to 2-vinyltetrahydrofuran **3** stereoselectively by using Pd(OAc)$_2$ catalyst together with Cu(OAc)$_2$ co-catalyst under O$_2$ (Equation (17.8)) [12]. Homoallylphenol **4** is similarly transformed to 2-vinylchroman **5**, which corresponds to a vitamin E moiety (Equation (17.9)) [13] (T. Hosokawa, T. Kono and S.-I. Murahashi, unpubl. data).

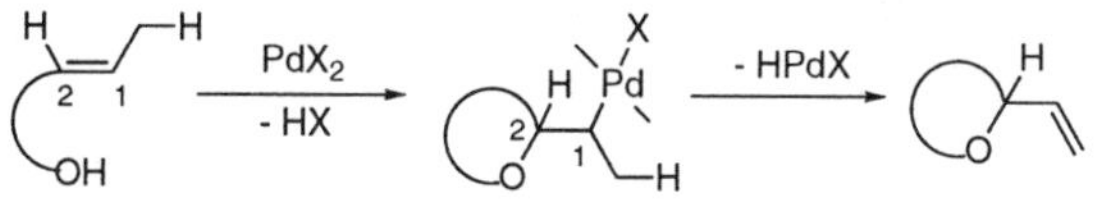

Scheme 17.3.

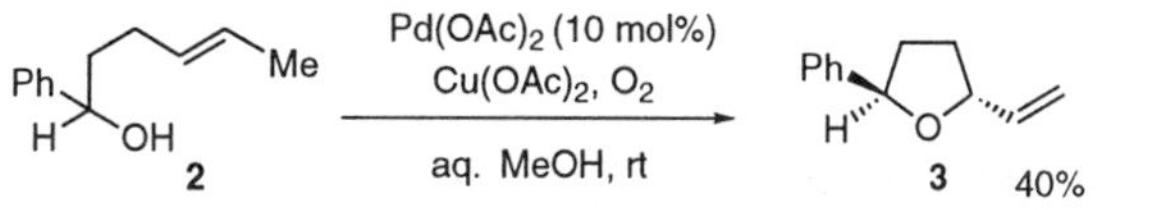

(17.8)

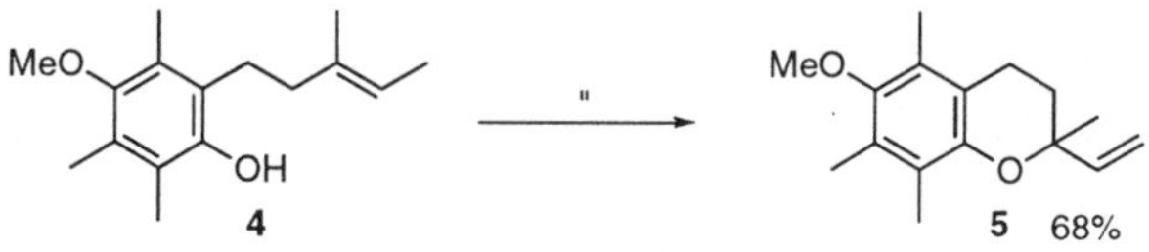

(17.9)

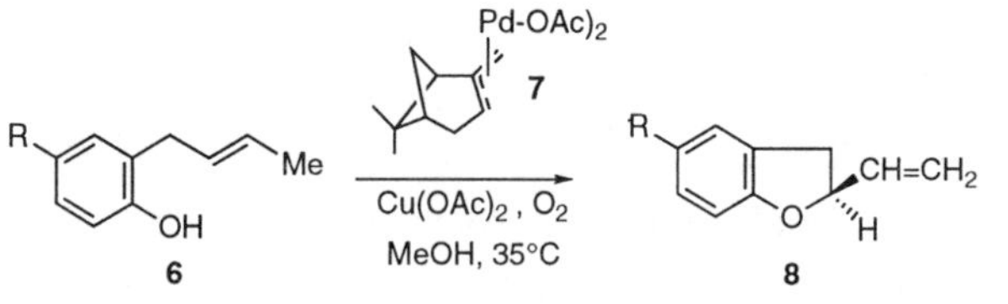

(17.10)

R= OMe: 44% (26 %ee)

Among these cyclizations, an intriguing example is the asymmetric cyclization of 2-allylphenols **6** using optically active [(η^3-pinene)Pd(OAc)]$_2$ (**7**) as a catalyst together with Cu(OAc)$_2$ and O$_2$ (Equation (17.10)), which was the first asymmetric Wacker-type oxidation developed by us in 1978 [14a]. Although the enantioselectivity of 2-vinyl-2,3-dihydrobenzofurans (**8**) formed by the cyclization of 2-(2-butenyl)phenols (**6**) is not high (e.g. R = OCH$_3$, 26% enantiomeric excess, ee), the [α]$_D$ value of product **8** (R = H) was constant, irrespective of the reaction times. This observation is important because it indicates that the pinanyl ligand is retained by Pd(II) throughout the reaction [14].

Shown in Scheme 17.4 is a catalytic cycle of the asymmetric reaction. Oxypalladation intermediate **9** is formed via intramolecular attack of the phenoxy group on the Pd(II)-coordinated olefin in compound **6**. The acetate ligand of palladium is removed as acetic acid. The cyclized product **8** arises from Pd–H elimination from the methyl group of intermediate **9**. At this stage, it is possible for the resulting Pd–H species **10** to decompose into Pd(0). However, because the pinanyl ligand is retained on palladium, no such decomposition takes place. Thus, the catalysis proceeds via oxygenation of the Pd–H bond in species

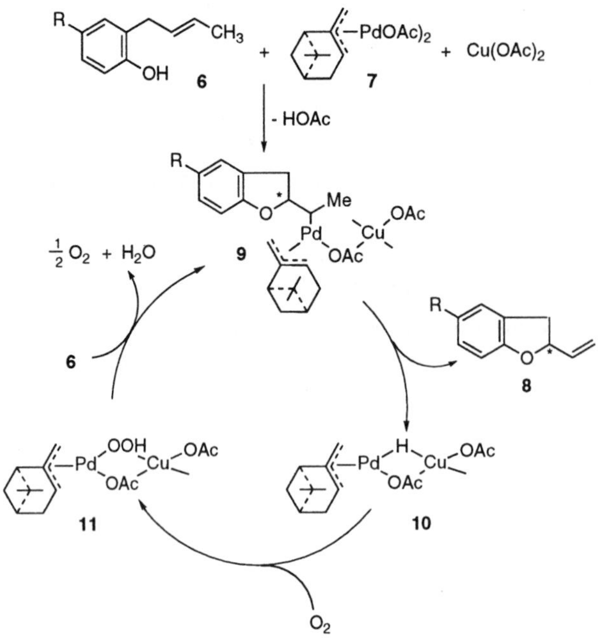

Scheme 17.4.

10 by O_2, leading to (η^3-pinene)palladium(II) hydroperoxide **11**. Species **10** and **11** must be Pd–Cu bimetallic complexes linked with μ-acetate and a peroxo ligand, because the reactivity and enantioselectivity in this reaction are influenced not only by the amount of $Cu(OAc)_2$ used relative to catalyst **7**, but also by the steric and electronic factors of the carboxylate ligands of Cu(II) (e.g. $ClCH_2COO$) [14b]. Coordination of the substrate to Pd(II) in species **11** by cleaving the acetate bridge, followed by oxypalladation with the loss of HOOH, leads to intermediate **9**, thus completing the catalytic cycle.

Our finding that the valency of Pd(II) does not change during the reaction is conceptually important because it provides the view that employing effective chiral ligands retained at Pd(II) will enhance the enantioselection of the Wacker-type oxidation. In fact, a remarkable enantioselectivity (97% ee) has been attained recently by Hayashi and Uozumi in the cyclization of 2-(2,2-dimethyl-2-butenyl)phenol (Equation (17.11)) [15] where (S,S)-2,2'-bis[4-isopropyloxazolyl]-1,1'-binaphthyl is utilized as the chiral ligand (L*) of $Pd(OCOCF_3)_2$ catalyst, and p-benzoquinone as the co-oxidant. The chiral ligand was intact during the reaction, and recovered quantitatively after the reaction. The chiral bis(oxazoline) ligand based on 1,1'-binaphthyl backbone is essential for high enantioselectivity; for higher reactivity, $OCOCF_3$ as the ligand and MeOH as the solvent are required, respectively.

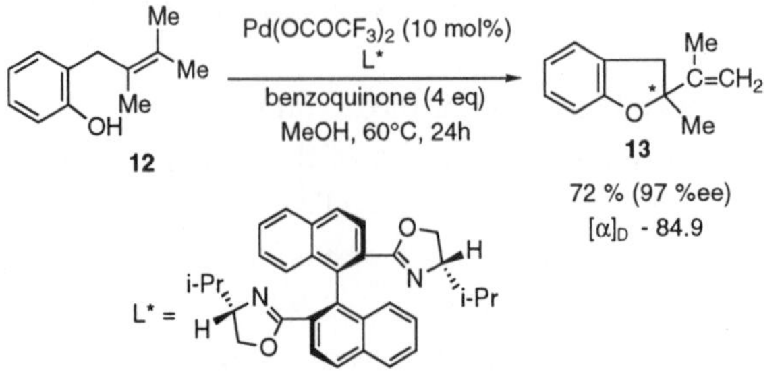

72 % (97 %ee)
$[\alpha]_D$ - 84.9

(17.11)

Let us again shift our attention to Scheme 17.4, where copper acetate as a co-catalyst participates as a stabilizing ligand of Pd–OOH species. The hydroperoxide ligand simply acts as a leaving group for the catalysis. The O-atom transfer from Pd–OOH species to alkenes seems to take place. However, in all cases mentioned so far, intramolecular nucleophilic attack to alkenes predominates. If no intra- or intermolecular nucleophiles exist, the O atom of the reactive OOH group could be transferred to alkenes. A requirement of this expectation is that the ability of capturing O_2 by the Pd–Cu bimetallic system must be highly enhanced. Furthermore, it is essential to use ligands that stabilize Pd–Cu complexes under O_2. Hexamethylphosphoramide (HMPA) is a non-oxidizable ligand and makes a variety of metal complexes readily soluble in organic solvents [16]. Thus, the choice of HMPA as a ligand has allowed us to develop a novel catalyst system for the oxidation of alkenes by O_2 alone.

3 Oxidation of alkenes with O_2 by palladium–copper bimetallic system

When the catalyst system of $PdCl_2/CuCl/HMPA$ is used in anhydrous 1,2-dichloromethane, the oxygenation of dec-1-ene by O_2 proceeds catalytically to give compounds **14** and **15** even in the absence of external nucleophiles such as ROH or H_2O (Equation (17.12)) [17]. In this reaction, 0.5 mol of O_2 is again consumed for the production of 1 mol of products (compounds **14** + **15**), indicating that two O atoms from molecular oxygen are incorporated into the substrates. The rate of reaction reaches a maximum when the $CuCl/PdCl_2$ ratio is 1 : 3, suggesting involvement of a Pd–Cu bimetallic species.

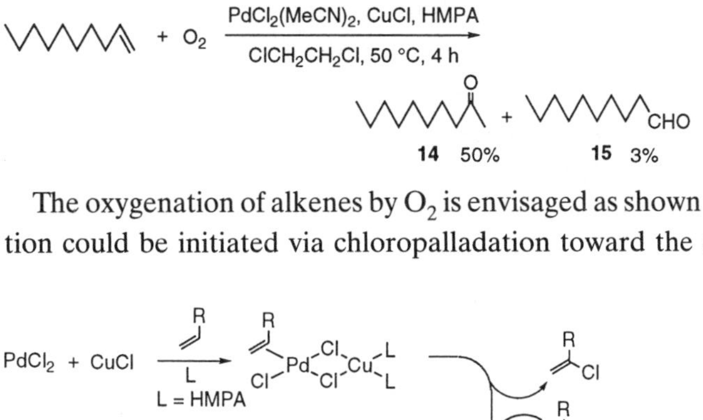

$$\text{(17.12)}$$

14 50%   **15** 3%

The oxygenation of alkenes by O_2 is envisaged as shown in Scheme 17.5, where the reaction could be initiated via chloropalladation toward the alkenes followed by Cl–Pd–H

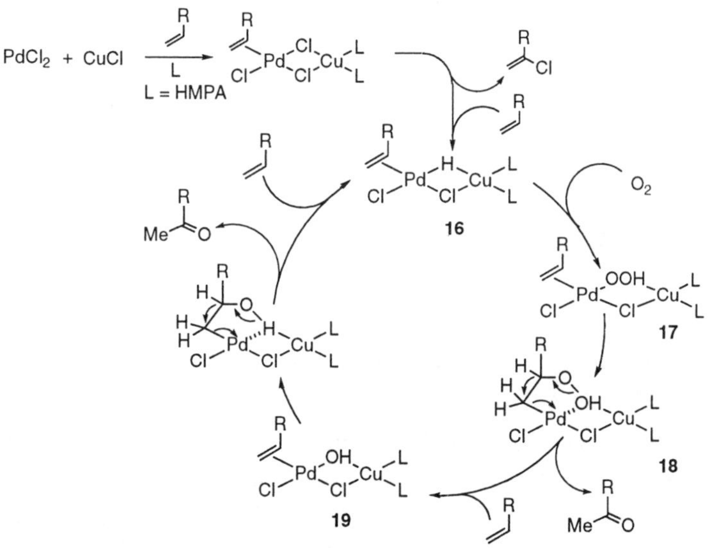

Scheme 17.5.

elimination. Molecular oxygen reacts with Pd–H species **16** stabilized by CuCl, giving rise to Pd–OOH **17** from which hydroperoxypalladation [18] takes place to give compound **18**. Cleavage of the Pd–C bond in the species **18**, formally shown by arrow lines, leads to methyl ketone and a Pd–OH species. Coordination of another alkene to the palladium gives Pd–OH species **19**, and subsequent hydroxypalladation affords another methyl ketone and Pd–H species **16**, thus completing the catalytic cycle. Two O atoms of O_2 are thus incorporated into the alkene.

A remarkable feature of this catalyst system can be seen in the regioselective transformation of the *N*-allylamide **20** into the aldehyde **21**, while the methyl ketone **22** becomes the major product in the presence of water (Equation (17.13)) [19]. Among various oxidations of alkenes, no such complete reversal of the regioselectivity in the presence or absence of water has been reported so far.

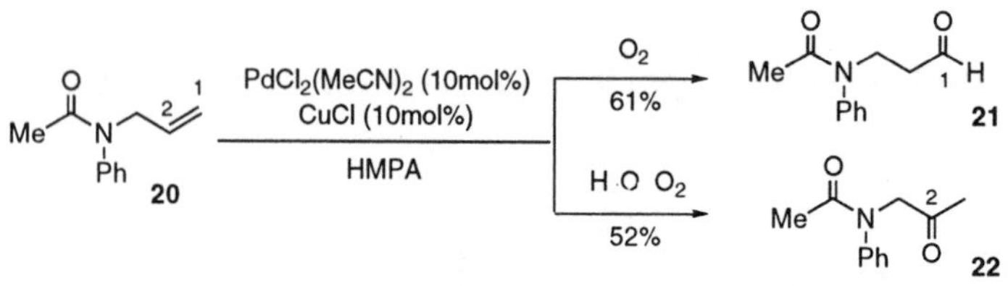

$$
\text{(reaction scheme: } 20 \xrightarrow[\text{CuCl (10mol\%)}]{\text{PdCl}_2(\text{MeCN})_2 \text{ (10mol\%)}, \ \text{HMPA}} \begin{cases} O_2, \ 61\% \rightarrow 21 \\ H_2O_2, \ 52\% \rightarrow 22 \end{cases}
$$

(17.13)

Formation of aldehydes and their derivatives via attack at the terminal olefinic carbon (C-1) is one of the important processes currently attracting interest in synthetic organic chemistry [20]. In the usual Wacker oxidation, the attacking position of the water nucleophile is altered from the C-2 carbon to the terminal C-1 carbon, depending on the functionality present in olefinic substrates [21]. With alkenes bearing electron-withdrawing groups, oxygen nucleophiles such as alcohols attack at the C-1 carbon to give acetals of aldehyde precursors [22].

The regioselective O atom transfer to the terminal olefinic carbon of *N*-allylamides shown in Equation (17.13) must be attained via the chelating Pd–OOH species **23** as shown in Scheme 17.6. When water is present in the reaction system, coordination of H_2O to Pd(II) takes place to interfere in the chelation of the amide carbonyl. Attack of water to the alkene in a Markovnikov manner results in the formation of methyl ketone **22**. The present result appears to justify the existence of Pd–OOH species in this oxidation.

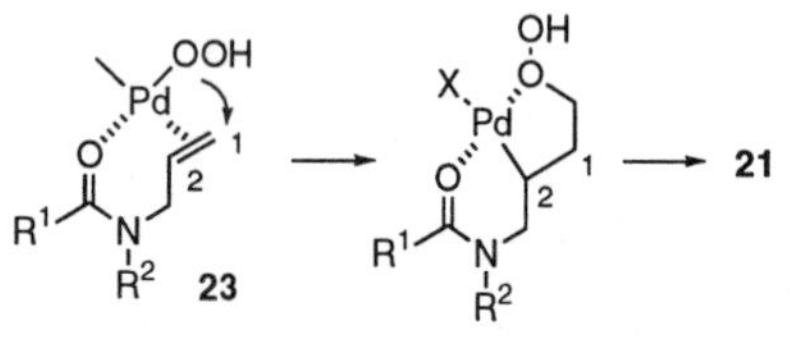

$$
\text{23} \longrightarrow \longrightarrow \text{21}
$$

Scheme 17.6.

In order to substantiate our view that the catalysis of palladium(II) with copper is exerted by a Pd–Cu hydroperoxide species $PdX(\mu\text{-OOH})(\mu\text{-X})CuX$ (= OAc, Cl) such as species **11** (Scheme 17.4) or species **17** (Scheme 17.5) arising from the reaction of $PdX(\mu\text{-H})(\mu\text{-X})CuX$ formed *in situ* with O_2, we have succeeded in isolating Pd–Cu bimetallic complexes. Thus, μ_4-oxo-bridged complex $Pd_6Cu_4Cl_{12}O_4(\text{HMPA})_4$ **(24)** and polymeric complex $[(PdCl_2)_2CuCl_2(\text{HMPA})_2]_n$ **(25)** were obtained from the reaction of $PdCl_2(\text{MeCN})_2$, CuCl and HMPA (1:1:2) in $ClCH_2CH_2Cl$ under O_2 (Equation (17.14)) [23].

$$PdCl_2(MeCN)_2 + CuCl + HMPA \xrightarrow[\text{ClCH}_2\text{CH}_2\text{Cl}]{O_2} (PdCl_2)_6(CuO)_4(HMPA)_4 + [(PdCl_2)_2CuCl_2(HMPA)_2]_n \qquad (17.14)$$

$$\textbf{24} \qquad\qquad\qquad \textbf{25}$$

Structures of these bimetallic complexes revealed by X-ray analyses are given in Fig. 17.1, where Cu and Pd atoms are linked to each other by a μ-Cl ligand in both cases. Noteworthy is that an O atom derived from O_2 is incorporated as an μ_4-O-bridged form in the cluster complex **24** [24]. Complex **24** readily undergoes O atom transfer to alkenes, and either bimetallic complex **24** or **25** serves as a catalyst for the oxygenation of alkenes with O_2. The catalytic efficiency, when compared with that using the catalyst system of $PdCl_2(MeCN)_2/$ $CuCl/HMPA/O_2$ itself, is not high. Consequently, it appears that the bimetallic complexes themselves are not real catalysts for this oxidation. However, this result demonstrates that a Pd–Cu species containing an O atom unequivocally participates as the catalyst in the reaction shown in Equations (17.12) and (17.13).

In the ketonization of terminal alkenes with water by $PdCl_2$–CuCl catalyst under O_2 (Equation (17.15)), N,N-dimethylformamide (DMF) is commonly used as the solvent [3, 25].

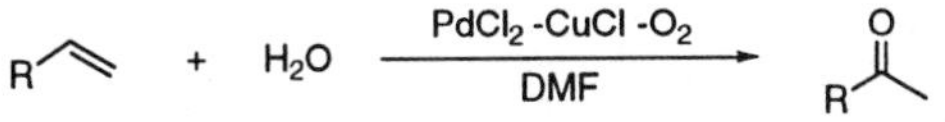

$$R\diagdown\diagup + H_2O \xrightarrow[\text{DMF}]{PdCl_2\text{ -CuCl -}O_2} R\overset{O}{\diagdown}\diagup \qquad (17.15)$$

The role of this solvent has been considered to lie in dissolving $PdCl_2$ and CuCl as well as hydrophobic higher terminal alkenes into solution. In line with the context of this chapter, it is reasonably expected that the catalyst constitutes a Pd–Cu–DMF complex. Indeed, we have also succeeded in the isolation of Pd–Cu complexes of $(PdCl_2)_x(CuO)_y(DMF)_z$ (**26**) and $[(PdCl_2)_2CuCl_2(DMF)_4]_n$ (**27**), as shown in Equation (17.16) [26].

$$PdCl_2(MeCN)_2 + CuCl + DMF \xrightarrow[\text{ClCH}_2\text{CH}_2\text{Cl}]{O_2} (PdCl_2)_x(CuO)_y(DMF)_z + [(PdCl_2)_2CuCl_2(DMF)_4]_n \qquad (17.16)$$

$$\textbf{26} \qquad\qquad \textbf{27}$$

Relevant to this section is that isolation of Pd–Cu bimetallic complexes that catalyse the oxidation of alkenes has been attained by the reaction of $Cu(OAc)_2$ and $PdCl_2$(2-hydroxy-

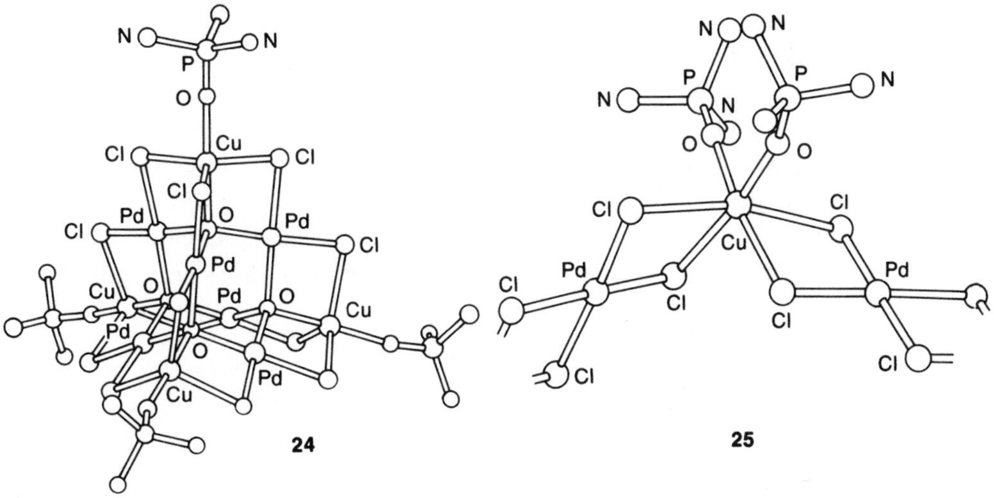

24

25

Figure 17.1. The molecular structures of; (**24**) $(PdCl_2)_6(CuO)_4(HMPA)_4$ and (**25**) $[(PdCl_2)_2CuCl_2(HMPA)_2]_n$.

pyridine)$_2$ [27]. In the PdCl$_2$-catalysed oxidation of cyclopentene into cyclopentanone with CuCl$_2$ and O$_2$ in EtOH, a Pd—OOH species has been reported to be formed via the reaction of Pd—H and O$_2$ [28], and several catalytic reactions involving Pd—OOH species also have been documented [29, 30].

4   Intramolecular oxypalladation of alkenes by palladium(II) in the absence of copper salts

In Scheme 17.4 or 17.5, the copper salt simply acts as the stabilizing ligand of Pd—H. No decomposition of the species into Pd(0) takes place. In view of the fact that the Pd—H is formally converted into Pd—OOH by O$_2$ alone, the reaction is expected to be catalytic with palladium even without using copper co-catalysts. In fact, the cyclization of 2-(2-cyclohexenyl)phenol **28** with Pd(OAc)$_2$ catalyst does proceed catalytically in the presence of O$_2$ alone (Equation (17.17)) [31], and 0.5 mol of O$_2$ is consumed for the production of 1 mol of cyclized products (**29 + 30 + 31**).

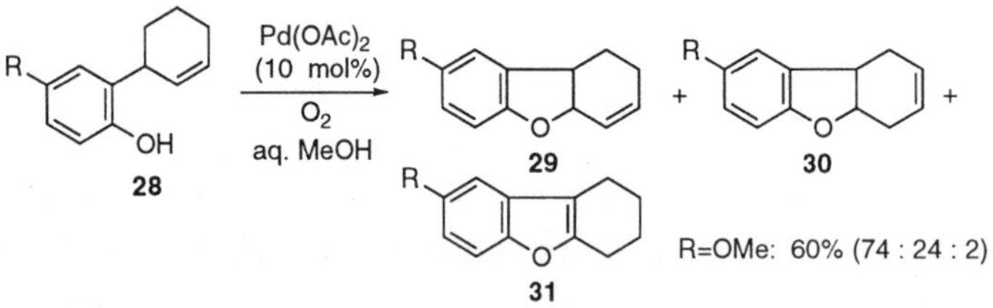

$$ (17.17) $$

Although the use of O$_2$ alone as an oxidant in the cyclization was demonstrated in 1978 [31], this technique has recently become a versatile tool for the synthesis of various heterocycles from functionalized alkenes [32–36], where DMSO is crucial as the solvent. The uniqueness of this solvent was originally found by Semmelhack in 1989 in the stoichiometric cyclization of compound **32** leading to tetrahydropyran **33**, where the mode of the β-Pd—H elimination process was controlled by the coordination ability of DMSO to Pd (Equation (17.18)) [37].

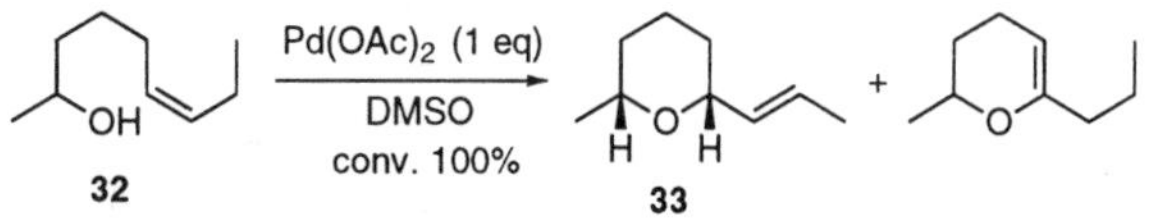

$$ (17.18) $$

Larock and Hightower thereafter have disclosed that when Pd(OAc)$_2$ catalyst is used along with NaOAc in DMSO, the cyclization of alkenoic acids to unsaturated lactones proceeds catalytically only by the aid of O$_2$ (Equations (17.19)–(17.21)) [32]. The use of excess amounts of Cu(OAc)$_2$ (2 equiv) leads to a poor yield of lactone **35** or **37**, while reducing the amount of Cu(OAc)$_2$ (10 mol%) or omitting it results in higher yields (86–90%).

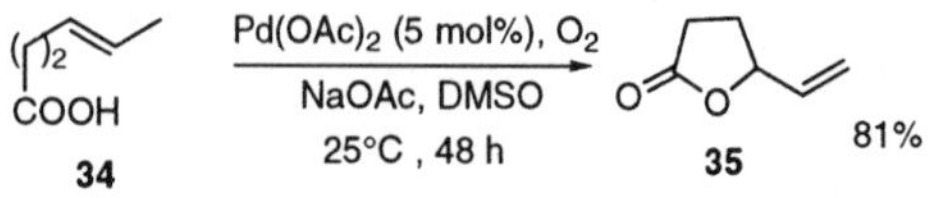

$$ (17.19) $$

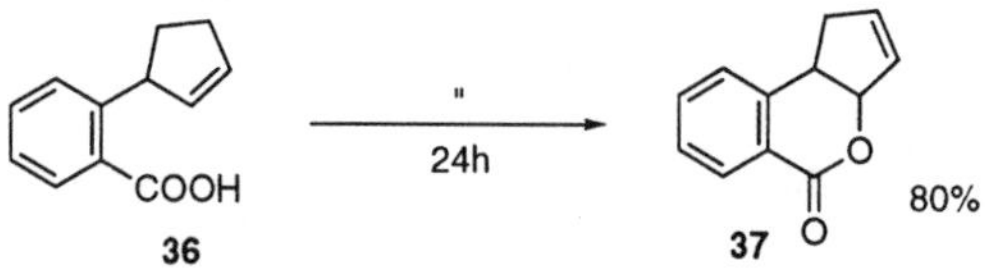

$$ (17.20) $$

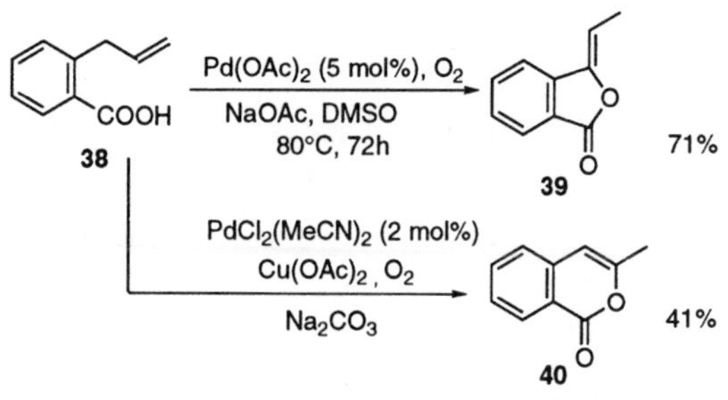

$$\text{(17.21)}$$

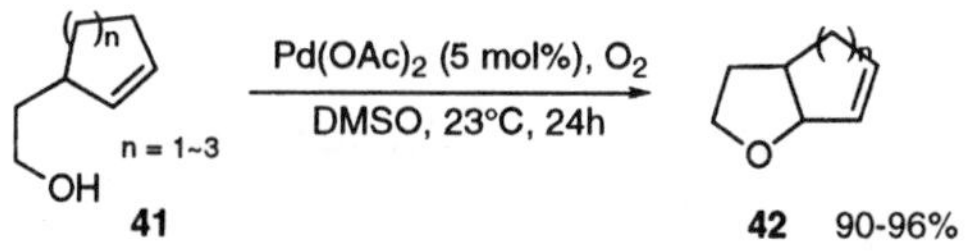

$$\text{(17.22)}$$

This observation indicates that the role of copper as the co-catalyst is not so straightforward as expressed in Equations (17.6) and (17.7). In the cyclization of *o*-allylbenzoic acid (**38**), the five-membered phthalide **39** is formed by the use of $Pd(OAc)_2$ as a catalyst [32], whereas $PdCl_2$ leads to the six-membered 3-methylisocoumarin **40** [38]. Alkenyl alcohols **41** are converted into tetrahydrofurans **42** even in the absence of NaOAc [34] (Equation (17.22)). In this case, the rate of reaction is not affected by the use of excess $Cu(OAc)_2$. Independent studies on the $Pd(OAc)_2/DMSO/O_2$ system by Hiemstra and Speckamp have indicated that methyl glyoxylate adducts of *N*-Boc-protected allylic amines (such as compound **43**) are catalytically cyclized into oxazolidines (such as compound **44**) by the aid of O_2 alone (Equation (17.23)) [39].

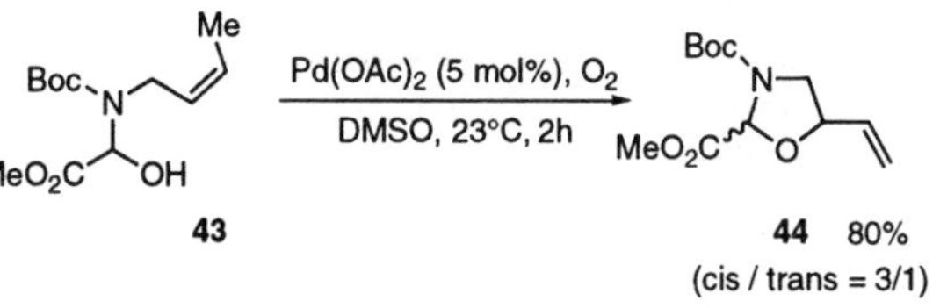

$$\text{(17.23)}$$

Elucidation of the nature of palladium species involved in the $Pd(OAc)_2/DMSO/O_2$ system by transmission electron microscopy (TEM) have shown that the resulting HPdOAc is reduced to a Pd(0) cluster that bears DMSO as a ligand, and the cluster is then reoxidized to $Pd(OAc)_2$ by O_2 and HOAc [40].

The catalyst system of $Pd(OAc)_2/DMSO/O_2$ effects the dehydrosilylation of silyl enol ethers, such as **45** → **46** (Equation (17.24)), in which the involvement of the Pd–OOH species is noted [41]. The catalytic dehydrosilylation of silyl enol ethers is also effected by the catalyst system of $PdCl_2(MeCN)_2/CuCl/HMPA/O_2$ (Equation (17.24)), and indirect evidence for the involvement of Pd–OOH species in this reaction has been put forward [17].

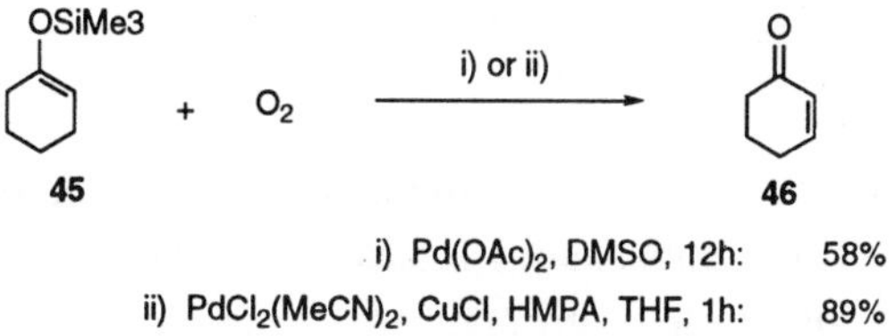

$$\text{(17.24)}$$

i) $Pd(OAc)_2$, DMSO, 12h: 58%

ii) $PdCl_2(MeCN)_2$, CuCl, HMPA, THF, 1h: 89%

5   Acetalization of alkenes

Intramolecular oxypalladation of alkenes in the presence of alcohols or water is briefly noted here from a synthetic viewpoint. In the case of an oxypalladation intermediate leading to *exo*-methylene substitution, attack of nucleophiles to the alkene gives acetals via a process formally represented in Scheme 17.7. The catalysis is thought to be operative by the oxygenation of the Pd—H bond by O_2. Under the conditions using CuCl and O_2, (2S,3S)-2-allylhydroxy-butyrate **47** is converted catalytically to the acetal **48** by using $PdCl_2(MeCN)_2$ catalyst and MeOH (Equation (17.25)) [42]. Hydroxyalkenes derived from (R)-(−)- or (S)-(+)-phenethylamine and/or (S)-(+)-lactic acid or (1R,2S)-(−)-ephedrine serve as the starting substrates for the synthesis of optically active tetrahydro-1,4-oxazines with high diastereoselectivity. A representative example is the transformation of compound **49** into compound **50** (Equation (17.26)) [43]. In this case, excess amounts of copper salts, even in the absence of O_2, make the reaction catalytic. When water is used in place of alcohols, hemiacetals are formed by $PdCl_2/CuCl/O_2$ (Equation (17.27)) [44].

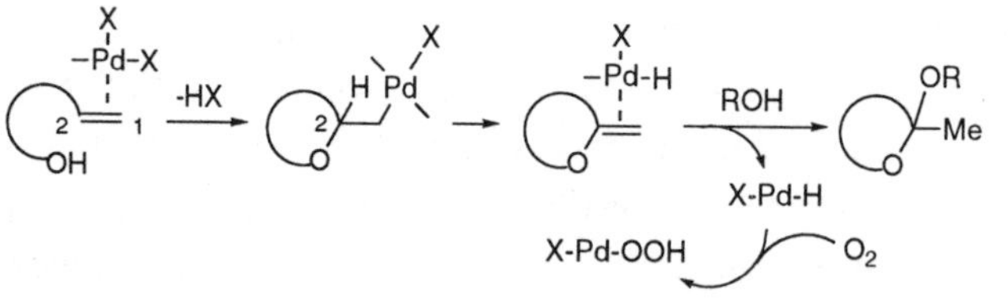

Scheme 17.7.

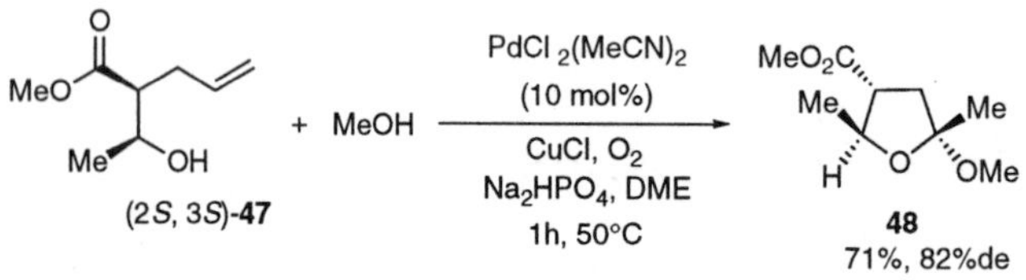

(17.25)

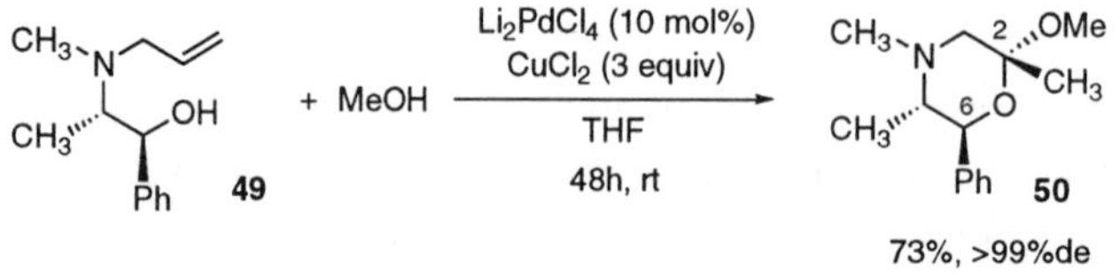

(17.26)

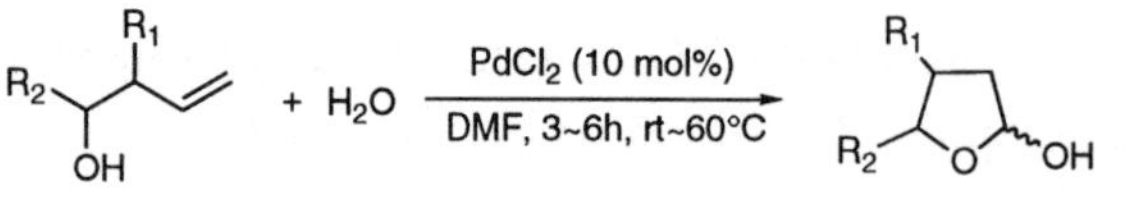

(17.27)

$R_1 = CH_3, R_2 = n\text{-}C_6H_{13}$ 64% $(CuCl\text{-}O_2)$
$R_1 = CH_3COO(CH_2)_7 \text{-}, R_2 = C_5H_{11}$ 74% (*p*-benzoqinone)

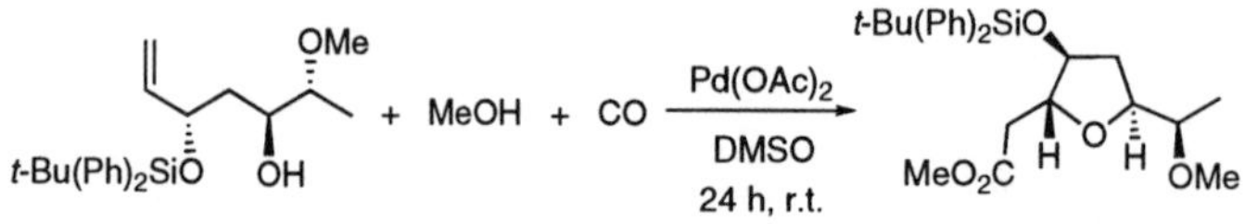

(17.28)

Insertion of CO into the Pd—C bond of oxypalladation intermediate **1** (shown in an earlier section, Scheme 17.2) and subsequent attack by alcohols leads to oxygen-containing heterocycles bearing ester groups. A unique example of this oxypalladation–carbonylation developed by Semmelhack, which was utilized for the synthesis of tetronomycine, is given in Equation (17.28) [10a]. Insertion of alkenes into the Pd—C bond of compound **1**, a process called intramolecular oxypalladation–olefination of hydroxyalkenes, also constitutes a synthetic method for oxygen-containing heterocycles bearing an olefinic group. In this reaction, molecular oxygen in combination with CuCl serves as the oxidant [45].

6　Conclusion

Catalytic oxidations involving oxypalladation as a key step are highly indispensable routes to synthesizing various organic compounds such as oxygen-atom-containing heterocycles. In a catalytic system of $PdX_2/CuX_2/O_2$ (X = OAc, Cl, etc.), a Pd—H species coupled with copper salts reacts with O_2, and the resulting Pd—OOH species acts as an active catalyst. Of importance in the catalysis is that the valency of palladium(II) does not change during the reaction. Such a conceptual advance has provided a clue for enhancing the synthetic utility of palladium(II), which includes future catalytic, asymmetric reactions of alkenes with various nucleophiles.

7　References

1　Tsuji J. *Palladium Reagents and Catalysts, Innovation in Organic Synthesis*. New York: Wiley, 1995: 19–124; Heck RF. *Palladium Reagents in Organic Syntheses*. New York: Academic, 1985; Trost BM, Verhoeven TR. In: Wilkinson G, ed. *Comprehensive Organometallic Chemistry*, Vol. 8. Oxford: Pergamon Press, 1982: 854–983; Davies SG. *Organotransition Metal Chemistry, Application to Organic Synthesis*. Oxford: Pergamon, 1982; Collman JP, Hegedus LS. *Principles and Applications of Organotransition Metal Chemistry*. California: University Science Books, 1980; Maitlis PM. *The Organic Chemistry of Palladium*, Vol. II. New York: Academic Press, 1971: 77–108. For palladium(0)-catalysed allylic substitution, see: O'Donnel MJ, Chen N, Zhou C *et al.* *J Org Chem* 1997; **62**: 3962–75, and references cited therein.

2　For palladium(II)-catalysed reactions, see: Heumann A, Jens K-J, Réglier M. In: Karlin KD, ed. *Progress in Inorganic Chemistry*, Vol. 42. New York: Wiley, 1994: 483–576; Henry PM. *Palladium Catalyzed Oxidation of Hydrocarbons*. Dordrecht: Reidel, 1980: 41–84.

3　Tsuji J. *Synthesis* 1984: 369–84; Tsuji J, Nagashima H, Nemoto H, *Org Synth* 1984; **62**: 9–13; Tsuji J, Nogami J, Mandai T. *J Synth Org Chem Jpn* 1989; **47**: 649–59.

4　Hosokawa T, Maeda K, Koga K, Moritani I. *Tetrahedron Lett* 1973; **14**: 739–40; Hosokawa T, Ohkata H, Moritani I. *Bull Chem Soc Jpn* 1975; **48**: 1533–9.

5　(a) Maeda K, Hosokawa T, Murahashi S-I, Moritani I. *Tetrahedron Lett* 1973; **14**: 5075–6; (b) Hosokawa T, Shimo N, Maeda K, Sonoda A, Murahashi S-I. *Tetrahedron Lett* 1976; **17**: 383–6.

6　Hosokawa T, Murahashi S-I. *Heterocycles* 1992; **33**: 1079–100; Hosokawa T, Murahashi S-I. *Acc Chem Res* 1990; **23**: 49.

7　Hegedus LS, Allen GF, Waterman EL. *J Am Chem Soc* 1976; **98**: 2674; Hegedus LS. In: Trost BM, Fleming I, Semmelhack MF, ed. *Comprehensive Organic Synthesis*, Vol. 4. Oxford: Pergamon Press, 1991: 559–63; Hegedus LS. *Angew Chem Int Ed Engl* 1988; **27**: 1113–26; Hegedus LS. *Tetrahedron* 1984; **40**: 2415–34; Harayama H, Abe A, Sakado T *et al.* *J Org Chem* 1997; **62**: 2113–22.

8　Izumi T, Kasahara A. *Bull Chem Soc Jpn* 1975; **48**: 1673–4; Kasahara A, Izumi T, Sato K, Maemura M, Hayasaka T. *Bull Chem Soc Jpn* 1977; **50**: 1899.

9　Pearlman BA, McNamara JM, Hasan I *et al.* *J Am Chem Soc* 1981; **103**: 4248–5.

10　(a) Semmelhack MF, Epa WR, Cheung AWH *et al.* *J Am Chem Soc* 1994; **116**: 7455; (b) Kraus GA, Li J, Gordon SM, Jensen JH. *J Org Chem* 1995; **60**: 1154–9; (c) Semmelhack MF, Kim CR, Dobler

W, Meier M. *Tetrahedron Lett* 1989; **37**: 4925–8; (d) Semmelhack MF, Zhang N. *J Org Chem* 1989; **54**: 4483–5; (d) McCormick M, Monahan III R, Soria J, Goldsmith D, Liotta D. *J Org Chem* 1989; **54**: 4485–7.

11 Smidt J, Hafner W, Jira R *et al. Angew Chem Int Ed Engl* 1962; **1**: 80–8; *Angew Chem* 1959; **71**: 176; March J. *Advanced Organic Chemistry*, 4th edn. New York: Wiley, 1992: 1196–8; Cotton FA, Wilkinson G. *Advanced Inorganic Chemistry*, 5th edn. New York: Wiley, 1988: 1276–7; Carey FA, Sundberg RT. *Advanced Organic Chemistry*, Part B, 3rd edn. New York: Plenum Press, 1990: 415–17.

12 Hosokawa T, Hirata M, Murahashi S-I, Sonoda A. *Tetrahedron Lett* 1976: 1821–4.

13 Hosokawa T, Kono T, Shinohara T, Murahashi S-I. *J Organomet Chem* 1989: **370**: C13–C16.

14 (a) Hosokawa T, Miyagi S, Murahashi S-I, Sonoda A. *J Chem Soc Chem Commun* 1978: 687–8; (b) Hosokawa T, Uno T, Inui S, Murahashi S-I. *J Am Chem Soc* 1981; **103**: 2318–23; (c) Hosokawa T, Okuda C, Murahashi S-I. *J Org Chem* 1985; **50**: 1282–7.

15 Uozumi Y, Kato K, Hayashi T. *J Am Chem Soc* 1997; **119**: 5063–4.

16 Hou Z, Kobayashi K, Yamazaki H. *Chem Lett* 1991: 265–8, and references cited therein.

17 Hosokawa T, Nakahira T, Takano M, Murahashi S-I. *J Mol Catal* 1992; **74**: 489–98.

18 Mimoun H. *Angew Chem Int Ed Engl* 1982; **21**: 734–50; Roussel H, Mimoun H. *J Org Chem* 1980; **45**: 5387–90.

19 Hosokawa T, Aoki S, Takano M *et al. J Chem Soc Chem Commun* 1991: 1559–60.

20 Wenzel TT. *J Chem Soc Chem Commun* 1993: 862–4; Feringa BL. *J Chem Soc Chem Commun* 1986: 909; Kiers NH, Feringa BL, van Leeuwen PWNM. *Tetrahedron Lett* 1992; **33**: 2403–6; also, see: Kiers NH, Feringa BL, Kooijman H, Spek AL, van Leeuwen PWNM. *J Chem Soc Chem Commun* 1992: 1169–70.

21 Kang S-K, Jung K-Y, Chung J-K, Namkoong E-Y, Kim T-H. *J Org Chem* 1995; **60**: 4678–9; Pellissier H, Michellys P-Y, Santelli M. *Tetrahedron Lett* 1994; **35**: 6481–4; Mori M, Yatanabe Y, Kagechika K, Shibasaki M. *Heterocycles* 1989; **29**: 2089–92; Bose A, Krishnan L, Wagle DR, Manhas MS. *Tetrahedron Lett* 1986; **27**: 5955–8.

22 Hosokawa T, Yamanaka T, Itotani M, Murahashi S-I. *J Org Chem* 1995; **60**: 6159–67; also, see: Lai J-Y, Shi XX, Dai L-X. *J Org Chem* 1992; **57**: 3485–7; and Igarashi S, Haruta Y, Ozawa M *et al. Chem Lett* 1989: 737–40.

23 Hosokawa T, Takano M, Murahashi S-I. *J Am Chem Soc* 1996; **118**: 3900–1; also, see: Hosokawa T, Takano M, Murahashi S-I *et al. J Chem Soc Chem Commun* 1994: 117–18.

24 Palladium complexes containing μ_4-oxo linkage are rare. Recently, a $Pd_4(\mu_4\text{-O})$ complex of formula $Pd_2(Pd_4(dmp)_4(\mu\text{-Cl})_2(\mu_4\text{-O})$ (dmp = dipivaloylmethanato) has been reported to be formed by the reaction of $Pd_2(dmp_2)(\mu\text{-OMe})_2$ with O_2; see: Zhang Y, Puddephatt RJ, Manojlovic-Muir L, Muir KW. *J Chem Commun* 1996: 2599–600.

25 Fahey DR, Zuech EA. *J Org Chem* 1974; **39**: 3276; Clement WHM, Selwitz CM. *J Org Chem* 1964; **29**: 241–3.

26 Hosokawa T, Nomura T, Murahashi S-I. *J Organomet Chem* 1998; **551**: 387–9.

27 Higashijima M, Masunaga T, Kojima Y, Watanabe E, Wada K. *Mitsubishi Kasei R&D Rev* 1994; **8**: 14–19; *Science and Technology*. Tokyo: Kodansha, 1995: 319–22.

28 Takehira K, Hayakawa T, Orita H, Shimizu M. *J Mol Catal* 1989; **53**: 15–21; also, see: Takehira K, Oh IH, Martinez VC *et al. J Mol Catal* 1987; **42**: 237–46; Bressan M, Morandini F, Morvillo A, Rigo P. *J Organomet Chem* 1985; **280**: 139–46; and Bégeault JM, Faraji M, Martin C. *N J Chim* 1987; **11**: 337–43.

29 Miyamoto M, Minami Y, Ukaji Y, Kinoshita H, Inomata K. *Chem Lett* 1994: 1149–52; Kataoka H, Watanabe K, Goto K. *Tetrahedron Lett* 1990; **31**: 4181–4; Kataoka H, Watanabe K, Miyazaki K *et al. Chem Lett* 1990: 1705.

30 Zargarian D, Alper H. *Organometallics* 1991; **10**: 2914–21.

31 Hosokawa T, Miyagi S, Murahashi S-I, Sonoda A. *J Org Chem* 1978; **43**: 2752–7.

32 Larock RC, Hightower TR. *J Org Chem* 1993; **58**: 5298–300.

33 Annby U, Stenkula M, Andersson C-M. *Tetrahedron Lett* 1993; **34**: 8545–8.

34 Rönn M, Bäckvall J-E, Andersson PG. *Tetrahedron Lett* 1995; **36**: 7749–52.

35 Larock RC, Hightower TR, Hasvold LA, Peterson KP. *J Org Chem* 1996; **61**: 3584–5.

36 Jabre-Truffert S, Waegell B. *Tetrahedron Lett* 1997; **38**: 835–6; Ronn M, Andersson PG, Bäckvall J-E. *Tetrahedron Lett* 1997; **38**: 3603–6.

37 Semmelhack MF, Kim CR, Dobler W, Meier M. *Tetrahedron Lett* 1989; **30**: 4925–8.

38 Korte DE, Hegedus LS, Wirth RK. *J Org Chem* 1977; **42**: 1329–36.

39 van Benthem RATM, Hiemstra H, Michels JJ, Speckamp WN. *J Chem Soc Chem Commun* 1994: 357–9; van Benthem RATM, Hiemstra H, Speckamp WN. *J Org Chem* 1992; **57**: 6083–5; van Benthem RATM, Hiemstra H, Longarela GR, Speckamp WN. *Tetrahedron Lett* 1994; **35**: 9281–4.

40 van Benthem RATM, Hiemstra H, van Leeuwen PWNM, Geus JW, Speckmap WN. *Angew Chem Int Ed Engl* 1995; **34**: 457–60; also, see: Schmid G, Harms M, Malm J-O *et al. J Am Chem Soc* 1993; **115**: 2046–8.

41 Larock RC, Hightower TR, Kraus GA, Hahn P, Zeng D. *Tetrahedron Lett* 1995; **36**: 2423–6.

42 Hosokawa T, Nakajima F, Iwasa S, Murahashi S-I. *Chem Lett* 1990: 1387–90.

43 Lai J-Y, Shi X-X, Gong Y-S, Dai L-X. *J Org Chem* 1993; **58**: 4775–7.

44 Nogami J, Ogawa H, Miyamoto S *et al. Tetrahedron Lett* 1988; **29**: 5181–4.

45 Semmelhack MF, Epa WR. *Tetrahedron Lett* 1993; **34**: 7205–8.

18 New Strategies for Hydrogen Peroxide Activation

ILYA I. MOISEEV

N. S. Kurnakov Institute of General and Inorganic Chemistry, Russian Academy of Sciences, Leninsky Prospekt 31, 117907, Moscow GSP-1, Russia

1 Introduction

Peroxide oxidations catalysed by transition metal complexes constitute an area of intensive studies. The peculiarities of the oxidations of alkanes, alkenes, arenes, phenols, organic arsines, stibines, amines and sulfides were reviewed in [1–19]. Despite these intensive studies, the mechanism of catalyst action in the polar reactions still remains a matter of contention.

The aim of this chapter is to call the reader's attention to the new pathways for oxidant action in peroxidic oxidation [20]. Reactions in which both the peroxo ligand and the substrate under oxidation can be coordinated to a common metal atom will be discussed in this chapter. In such a reaction the oxidation state of the central metal ion is not changed in the act of oxidation. The metal ion seemingly plays the role of a mediator, contributing to the electron transfer from a coordinated substrate to the coordinated oxidant.

2 Hydrogen peroxide decomposition: singlet dioxygen formation

Hydrogen peroxide (H_2O_2) decomposition to form triplet dioxygen and water is a thermodynamically favorable process ($\Delta G^o_{298} = -56\,\text{kcal}\,\text{mol}^{-1}$, $\Delta H^o_{298} = -47\,\text{kcal}\,\text{mol}^{-1}$) and is catalysed by a number of metals. Hydrogen peroxide decomposition to form singlet dioxygen, although being thermodynamically less favored ($\Delta H^o_{298} = -24\,\text{kcal}\,\text{mol}^{-1}$), could also proceed.

Hydrogen peroxide is quite stable for a long time if protected against light in the absence of any catalysts in neutral aqueous solution. The H_2O_2 molecule is both a weak one-electron reductant and a one-electron oxidant. Decomposition into H_2O and $^3O_2$ is observed in basic pyridine solution as a consequence of an initiation reaction of the type described by Equation (18.1) [21].

$$HO_2^- + HOOH \rightarrow O_2^- + H_2O + HO^{\cdot} \tag{18.1}$$

Hypochlorite ion is known to oxidize H_2O_2 in alkaline solution, giving rise to singlet oxygen formation [22, 23]. According to mechanistic studies [23–25], the reaction involves the intermediacy of a chlorohydroperoxy anion as the active species, as shown in Equations (18.2)–(18.4).

$$HOO^- + Cl-OH \rightarrow HOO-Cl + HO^- \tag{18.2}$$

$$HO^- + HOO-Cl \rightarrow {}^-OO-Cl + H_2O \tag{18.3}$$

$$^-\overset{\frown}{O}-O-\overset{\frown}{C}l \rightarrow {}^1O_2 + Cl^- \tag{18.4}$$

The intramolecular redox reaction (18.4), involving heterolysis of the $Cl–O_2^-$ bond, is responsible for singlet dioxygen generation.

Carboxylic peracids undergo two-electron dismutation in alkaline solution to yield singlet dioxygen, as exemplified by the decomposition of diisoperoxyphthalic acid (Equation (18.5)) [26]. The oxygen–oxygen bond in the percarboxylic group, $RCOO^{\delta-}$—$^{\delta+}OH$,

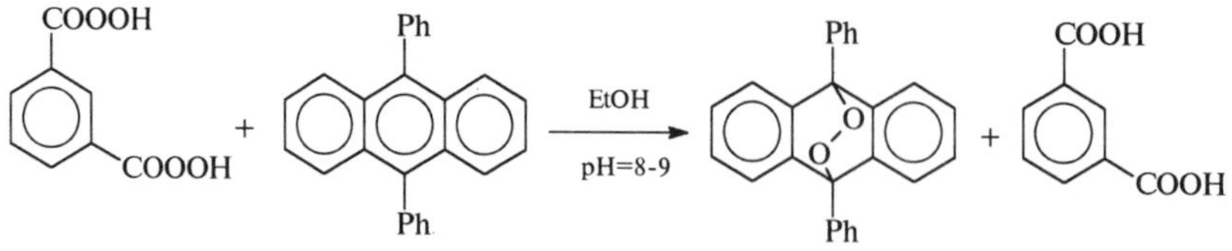

$$(18.5)$$

seems to be polar enough to make the disproportionation of two percarboxylic acids possible (Equation (18.6)).

$$(18.6)$$

The idea of trioxide formation via nucleophilic metathesis of polar peroxo compounds (see Equation (18.6)) has been discussed in the literature with peroxycarboximic acids as an example [27]. Subsequent decomposition of the intermediate monoacyl hydrotrioxide is expected to give rise to singlet dioxygen (Equation (18.7)). Thus, benzoic acid and singlet

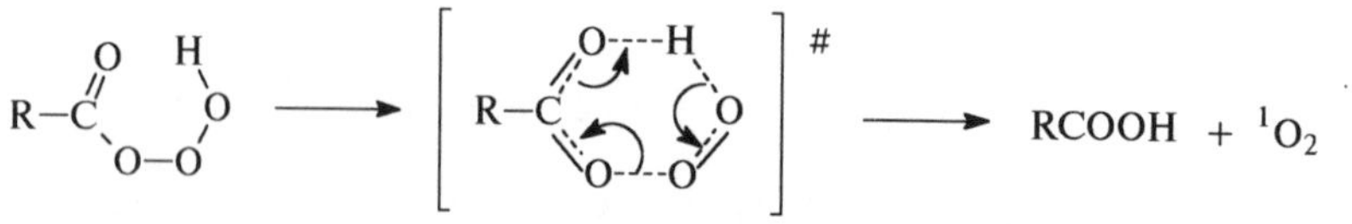

$$(18.7)$$

oxygen (96%) were reported to be the main decomposition products of benzoyl hydrotrioxide prepared from benzaldehyde and ozone [28].

Singlet dioxygen formation under metal complex catalysis was first observed when 9,10-diphenylanthracene and some other scavengers of singlet dioxygen were oxidized with t-BuOOH/Mo(VI) systems (e.g. see Equation (18.8)) [29].

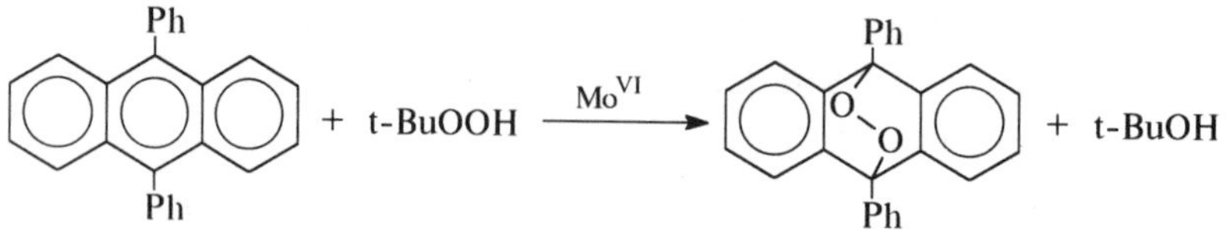

$$(18.8)$$

9,10-Endoperoxide formation (Equations (18.5) and (18.8)) is a typical reaction indicating intermediate $^1O_2$ involvement [29].

Another example of singlet dioxygen transfer from a peroxide compound to an organic substrate was observed with the $V(V)/H_2O_2/AcOH$ system [11–13,30,31]. Thus, 9,10-anthracene endoperoxide intermediate formation was detected with NMR and GC–MS techniques in the case of the anthracene reaction with H_2O_2 in acetic acid solution containing V(V) compounds at room temperature [30]. 9,10-Dimethylanthracene was converted into 9,10-dimethyl-9,10-epidioxy anthracene with c 70–90% yield under the same oxidation conditions (Equation (18.9)) [31].

$$\text{(18.9)}$$

The absence of active free radical species in the V(V)/AcOH/H$_2$O$_2$ system should be emphasized [30–32]. In this respect the system is different from those based on OCl$^-$/H$_2$O$_2$ reagents, which produce various free radicals capable of inducing autoxidative reactions [23].

A systematic study of H$_2$O$_2$ decomposition in alkaline aqueous solution showed that many mineral oxides, hydroxides or oxoanions, including vanadate and molybdate salts, can catalyse the reaction described by Equation (18.10) [33].

$$H_2O_2 + HO_2^- \longrightarrow HO^- + H_2O + {}^1O_2 \qquad \text{(18.10)}$$

The most thoroughly studied catalyst is molybdate ion MoO$_4^{2-}$ [34–39]. The kinetic study showed that oxotriperoxomolybdate MoO(O$_2$)$_3^{2-}$ is the main precursor of singlet dioxygen (see Equation (18.11)) [39].

$$\longrightarrow MoO_3(O_2)^{2-} + {}^1O_2 \qquad \text{(18.11)}$$

A study of the bleaching of phenolphthalein by H$_2$O$_2$ catalysed by Mo(VI) in high pH solution suggests Mo(O$_2$)$_4^{2-}$ and W(O$_2$)$_4^{2-}$ to be the active species. In the range of $(0.5–5) \times 10^{-3}$ M concentration of K$_2$MoO$_4$ and Na$_2$MoO$_4$, oxidation with singlet dioxygen was found to contribute to the reaction hand in hand with the direct oxidation by peroxo complexes. No influence of the free radical scavenger N-tert-butyl-α-phenylnitrone on the reaction kinetics was observed [37].

Another candidate for singlet dioxygen catalytic formation is Cr(VI) [39]. However, the participation of singlet dioxygen in Cr(VI) oxidations has not been firmly established, despite extensive studies [40].

The dismutation of two peroxo groups to form singlet dioxygen (see Scheme 18.2, step A; and Equation (18.11)) is an example of an inner-sphere redox process in which the oxidation state of the central atom seems to remain unchanged in all stages of the reaction. Simultaneous inner-sphere transfer of two electrons or two, fast, one-electron consecutive transfers between reacting peroxo groups are the critical requirements for singlet dioxygen formation. Vacant metal orbitals can be used to mediate such a redox process between coordinated reactants. Therefore, the data available can be considered as suggesting that vacant d-levels of a catalyst (V(V) and Mo(VI) ions with d^o non-bonding electron configuration) can be used for electron transfer in the redox reactions between coordinated ligands, e.g. two peroxo groups, a peroxo group and an acac$^-$ anion or a peroxo group and a trioxo dianion (see Sections 3–5).

There is evidence implying that such a mediator mechanism can be possible in the cases when d-orbitals of the central atom are partially filled. Thus, Mn(III) porphyrin-like complexes have been found to catalyse the oxidation of naphthalene into naphthoquinone via an intermediate endoperoxide pathway [41, 42]. Oxidation of naphthalene and its derivatives into corresponding 1,4-naphthoquinones and 1-naphthols by AcOOH catalysed with PMnCl (P = 3,5-octanitrophthalocyanine) in CH_3CN at 20 °C has been shown to proceed via thermally unstable intermediate(s) formation. On the basis of chemical reactions of the intermediates, 1,4-endoperoxide and 2,3-epoxy-1,2,3,4-tetrahydronaphthalene-1,4-endoperoxide were proposed to be involved in the oxidation.

Oxygenation of rubrene and 9,10-diphenylanthracene with H_2O_2 induced by Fe(II) $(CH_3CN)_4(ClO_4)_2$ in dry CH_3CN gives rise to the corresponding 9,10-endoperoxides with yields in the range 70–85 % [43]. No evidence for free radical oxidation reaction was found; iron(II) complexes do not undergo oxidation into iron(III) compounds under the conditions of the dioxygenations [43].

Free singlet dioxygen molecules are rather unstable species in solutions undergoing quenching by interaction with solvent molecules. Metal complexes with the $^1O_2$ molecule as a ligand can be expected to be more stable. In this context, oxidations in the $V(V)/H_2O_2/AcOH$ system seems to be of interest.

3   Decomposition of peroxides catalysed with V(V) in acetic acid solution

Vanadium(V) is known to form a number of peroxo derivatives in aqueous solutions [44] in which H_2O_2 is relatively stable.

Vanadium(V) oxo- and peroxo complexes in aqueous solution

$[HV(O)_4]^{2-}$	$[HV(O)_3(O_2)]^{2-}$	$[V(O)_2(O_2)_2]^{3-}$	$[VO(O_2)_3]^{3-}$	$[V(O_2)_4]^{3-}$
$[H_2V(O)_4]^-$	$[H_2V(O)_3(O_2)]$	$[HV(O)_2(O_2)]^{2-}$	$[HVO(O_2)_3]$	
$[H\{V(O)_4\}_2O]^{3-}$	$[VO(O_2)]^+$	$[H_2V(O)_2(O_2)_2]^-$		
$[V(O)_2]^+$				

However, in an acetic acid solution V(V) compounds were found to catalyse the decomposition of both H_2O_2 and AcOOH.

3.1   *Decomposition of peracetic acid catalysed with V(V) in acetic acid solution [30,32]*

Decomposition of peracetic acid catalysed with V(V) compounds proceeds according to Equation (18.12).

$$AcOOH \xrightarrow[\text{AcOH}]{\text{V(V)}} AcOH + 1/2 O_2 \qquad (18.12)$$

The kinetic curves for the AcOOH decomposition have no induction period. The conventional inhibitors of free radical chain reactions, such as tris-(*tert*-butyl)phenol and *p*-benzoquinone, do not affect the rates of the decomposition (see Fig. 18.1).

Correspondingly, neither the relatively stable phenoxyl radical nor the acyl and peroxyl radicals, which could be formed as chain carriers, were detected in the reaction solutions. These facts allow one to rule out a free radical chain mechanism for this reaction. No influence of strong acids (H_2SO_4, $HClO_4$) on the reaction rate was detected, suggesting that the protonation of AcOOH is not important (see Fig. 18.1).

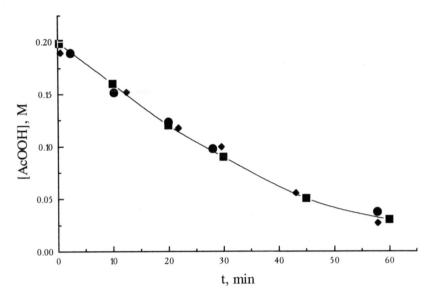

Figure 18.1. The AcOOH concentration vs time plot: (●) no additives; (■) $[HClO_4] = 10^{-2}\,M$; (◆) tris-(*tert*-butyl)phenol, $10^{-3}\,M$.

The observed kinetic equation suggests that the formation of the complex between catalyst and substrate precedes the limiting stage of the reaction (see Equation (18.13)).

$$W = k \frac{K[AcOOH][V]_\Sigma}{1 + K[AcOOH] + K_x[X]} \qquad (18.13)$$

Water and alcohols (X) can occupy the vacancies in the coordination sphere of vanadium ion and therefore retard the reaction.

In this context, the kinetic equation (Equation (18.13)) and the absence of evidence for participation of free radicals imply that the peroxide molecule decomposition proceeds via a polar mechanism. The shape of the kinetic curve (Fig. 18.1; Equation (18.13)) suggests a mechanism of equilibrium complex formation between V(V) and the peracid molecule (see Scheme 18.1) and a subsequent slow step in which some transformation of the coordinated AcOOH molecule should take place.

Such a transformation can involve the transfer of an acyl group from the bound AcOOH molecule to the neighboring hydroxylic group to form a coordinated peroxo ligand HO_2^- or O_2^{2-} and acetic acid or acetate ion. Thus, this rearrangement gives rise to a peroxo ligand that can be oxidized easier than the initial peracetic acid.

The complex with a coordinated O_2^{2-} ligand formed in step 'a' is assumed to undergo fast inner- or outer-sphere oxidation by another AcOOH molecule, leading to the observed products.

3.2 *Decomposition of H₂O₂ catalysed with V(V) in acetic acid solution*

The main gaseous product of the H_2O_2 decomposition catalysed with V(V) in acetic acid solution is dioxygen (Equation (18.14)).

$$H_2O_2 \xrightarrow[\text{AcOH}]{V(V)} H_2O + 1/2\,O_2 \qquad (18.14)$$

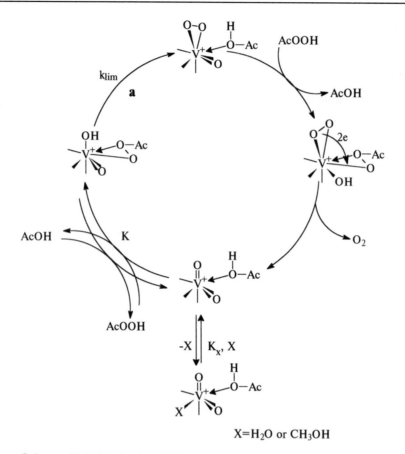

$X = H_2O$ or CH_3OH

Scheme 18.1. Mechanism of the AcOOH decomposition.

Throughout this century dioxygen has been believed to be the sole product of the decomposition of H_2O_2 catalysed with transition metal compounds. However, the formation of ozone according to Equation (18.15) is in fact thermodynamically allowed ($\Delta G_0^{298} = -188$ kJ mol^{-1}), along with dioxygen (either in ground or in exited state, e.g. $^1\Delta_g$). The decomposition of H_2O_2 catalysed with V(V) compounds in glacial acetic acid and especially in CF_3COOH was found to be accompanied by ozone formation [45].

$$3H_2O_2 \xrightarrow[\text{CF}_3\text{COOH or AcOH}]{\text{V(V)}} 3H_2O + O_3 \tag{18.15}$$

Ozone was identified by its UV spectra (a band with maximum absorption at 253 nm) and by its reactions with alkenes to form ozonides. The yield of ozone was no more than 1% (v/v) of the total gaseous products in the V(V)/H_2O_2/AcOH system.

The rate of the H_2O_2 decomposition reaction in acetic acid solution obeys the kinetic equation (Equation (18.16)).

$$-\frac{d[H_2O_2]}{d\tau} = \frac{k_{H_2O_2}[H_2O_2][NH_4VO_3]}{[H_2O]} \tag{18.16}$$

The ESR signal corresponding to the complex of V(V) with superoxide anion [V(V)(O$_2^-$)] was observed in the course of the H_2O_2 decomposition in the absence of peracetic acid. However, no induction period in kinetic curves was observed [11,30,32]. Conventional inhibitors of chain radical reactions did not affect the reaction rate (see Fig. 18.2).

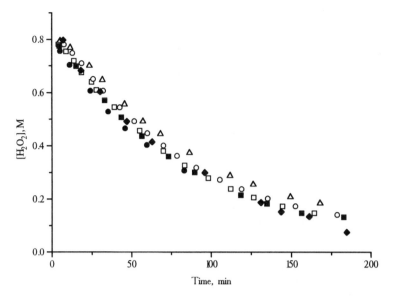

Figure 18.2. The H_2O_2 concentration vs time plot in the presence of inhibitors ($[V(V)] = 10^{-2}$ M, $[H_2O_2]_0 = 1$ M, 30 °C, AcOH): (■) no additives; (□) Pyrex glass powder; (●) [BHT] = 2×10^{-3} M; (○) p-nitroso-N,N-dimethylaniline additions, [Inhibitor] = 2×10^{-3} M; (△) [Anthracene] = 2×10^{-3} M; (◆) [Anthraquinone] = 2×10^{-3} M.

Table 18.1. Dependences of the rate of O_2 evolution and the concentration of $V(V)(O_2^-)$ on the reaction conditions

Influencing factors	O_2 evolution rate	$[V(V)(O_2^-)]$
Wall material	Non-sensitive	Sensitive
Surface/volume ratio	Non-sensitive	Sensitive
H_2O presence	Sensitive	Non-sensitive
Anthracene presence	Sensitive	Non-sensitive

Thus, all the kinetic data available, including both the kinetic equation and the insensitivity of the reaction rate to the presence of inhibitors, imply that the H_2O_2 decomposition does not involve the $V(V)(O_2^-)$ radical as a chain carrier. Moreover, this conclusion was supported by analysis of the reaction conditions influencing the reaction rate and the complex radical concentration (see Table 18.1).

As seen in Table 18.1, factors such as the wall material and surface/volume ratio exert different effects on the concentration of the paramagnetic complex and the reaction rate.

Based on the results obtained, free radical mechanisms of H_2O_2 decomposition in AcOH solution in the presence of vanadium compounds can be ruled out. The reaction under question is a polar process rather than a free radical chain reaction.

According to the [51]V-NMR data, at least two peroxo groups are coordinated with the V(V) atom in AcOH solution containing H_2O_2, and these diperoxo complexes dominate under the conditions of the kinetic experiments to which Equation (18.16) refers.

This fact and the first-order H_2O_2 concentration in Equation (18.16) can be explained in the framework of a scheme involving the reaction of the diperoxo complex with H_2O_2 as the slow step. In this slow step a complex with three peroxo groups is assumed to form as a transition state or intermediate species (see Scheme 18.2). In such a complex containing

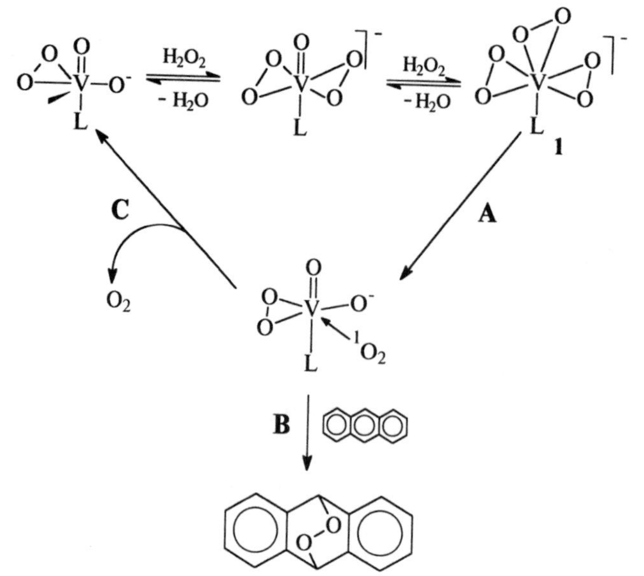

Scheme 18.2. Mechanism of the H_2O_2 decomposition and transfer of singlet dioxygen to the anthracene molecule.

three peroxo ligands (see complex **1** in Scheme 18.2), an inner-sphere electron transfer can supposedly take place to afford dioxygen and a V(V) monoperoxo complex.

This conclusion was substantiated by experiments with the anthracene additives to the dihydro-9,10-epidioxoanthracene that was found to be formed due to trapping of the singlet dioxygen molecule by the anthracene molecule (see Section 3.3). In the solution containing both AcOOH and H_2O_2, the kinetics of the reaction obeys a simple second-order equation (see Equation (18.17)).

$$W = -\frac{d[H_2O_2]}{d\tau} = -\frac{d[AcOOH]}{d\tau} = k[AcOOH][V]_\Sigma \qquad (18.17)$$

The absence of the Michaelis-type retardation of the reaction rate in these solutions is the result of fast formation of peroxo complexes from V(V) and H_2O_2. The same reason can be used to explain a remarkable increase in the rate of peracetic acid decomposition in the presence of H_2O_2 (see Fig. 18.3).

The decomposition of both peroxides was found to be accelerated by small additions of either H_2O_2 to AcOOH or AcOOH to H_2O_2. The rate of decomposition of H_2O_2 is equal to that of AcOOH if both are present. The reaction rate drops dramatically after the deficiency in peroxide was exhausted (see Fig. 18.4).

All these facts indicate that peroxide decomposition in solutions containing both AcOOH and H_2O_2 can be interpreted as the oxidation of H_2O_2 by peracetic acid, as shown in Equation (18.18).

$$H_2O_2 + AcOOH \xrightarrow[\text{AcOH}]{\text{V(V)}} AcOH + H_2O + O_2 \qquad (18.18)$$

In the H_2O_2/HOOAc/V(V)/AcOH system a new pathway becomes available for peroxide decomposition. In the H_2O_2/V(V)AcOH system peroxo ligands are formed rapidly with the participation of H_2O_2, and dioxygen is formed via the oxidation of one peroxo ligand by another, according to Equation (18.19).

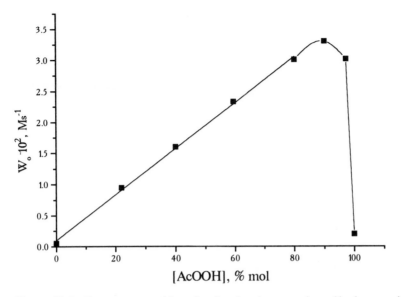

Figure 18.3. Rate vs composition plot showing the synergism of hydroperoxide decomposition ([V(V)] = 2×10^{-6} M, 20 °C).

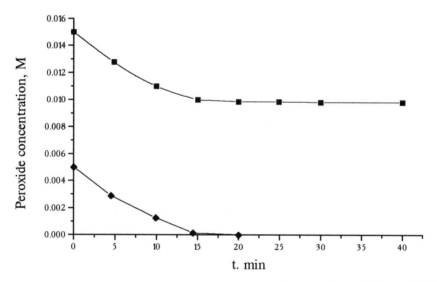

Figure 18.4. Concentrations vs time plot for the co-decomposition of H₂O₂ and AcOOH ([V(V)] = 2×10^{-6} M, 20 °C): (■) H₂O₂ decomposition (measured by permanganatometry); (◆) AcOOH decomposition (measured by iodometry).

$$O_2^{2-} + O_2^{2-} \longrightarrow O_2 + 2\, O^{2-} \tag{18.19}$$

In the AcOOH/V(V)/AcOH system the peroxo ligand is formed in a slow step that involves rearrangement of the coordinated peracetic acid or its anion. Further oxidation of the peroxo group by another AcOOH molecule is a comparatively fast reaction.

In the H₂O₂/HOOAc/V(V)/AcOH system, the formation of the peroxo group is facilitated due to the presence of H₂O₂ and the oxidation of this group proceeds as fast as in the AcOOH/V(V)/AcOH system. This peculiarity of the H₂O₂/HOOAc/V(V)/AcOH system is demonstrated by Scheme 18.3.

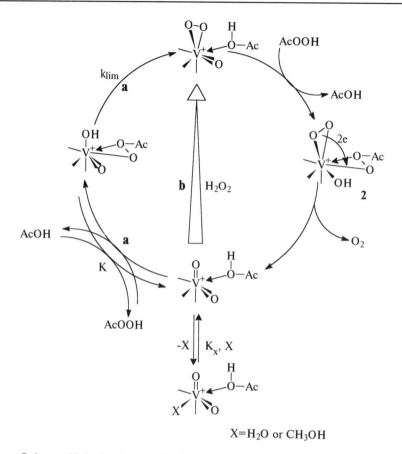

Scheme 18.3. Oxidation of H_2O_2 by AcOOH.

The peroxo ligand forms rapidly in step b and undergoes oxidation by its reaction with peracetic acid via complex **2**. In this complex a two-electron transfer between the coordinated O_2^{2-} ligand and the coordinated or free peracetic acid gives rise to free or coordinated dioxygen. It is seen that the reaction (e.g. Equation (18.19)) between the diamagnetic species should give rise to free singlet dioxygen or its complex if all other components of the reaction are diamagnetic compounds. To prove this mechanism of hydroperoxide decomposition, the oxidation of anthracene was studied.

3.3   *Oxidation reactions with the V(V)/H₂O₂/AcOH system*

Anthracene and 2-ethylanthracene have been found to be converted selectively into the corresponding anthraquinones (the yields are close to 100%) by reaction with the V(V)/H₂O₂/AcOH system [30,32]. No anthrone (**3**), bianthrone or anthracene cation radical (**4**) was detected under the conditions of anthracene oxidation with the V(V)/H₂O₂/AcOH system, which is consistent with the reaction not involving one-electron transfer from the substrate molecule to the active oxidant species (see Scheme 18.4).

Both the H₂O₂ decomposition and anthracene oxidation are unaffected by the presence of such conventional inhibitors of free radical reactions as *p*-benzoquinone and tris(*tert*-butyl)phenol, suggesting that the reactions proceed via polar ways [11,12,30,32].

According to the kinetic data, H₂O₂ decomposition catalysed by V(V) compounds in an

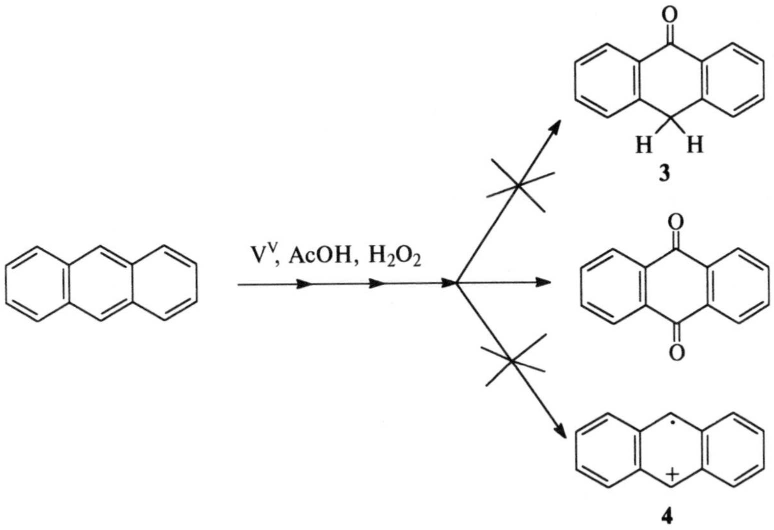

Scheme 18.4. Selective oxidation of anthracene to anthraquinone in the V(V)/AcOH/H₂O₂ system.

AcOH solution involves the formation of the triperoxo complex $V(O_2)_3^{2-}$ as a key interme-diate [30–32]. This complex is believed to form a complex with singlet oxygen molecule as a ligand (see Scheme 18.2, step A) in the slow step.

The coordinated singlet dioxygen is captured by an anthracene molecule to form the intermediate 9,10-dihydro-9,10-epidioxyanthracene (see Scheme 18.2, step B). The subse-quent oxidation of this short-lived intermediate by H₂O₂ affords anthraquinone (Equation (18.20)).

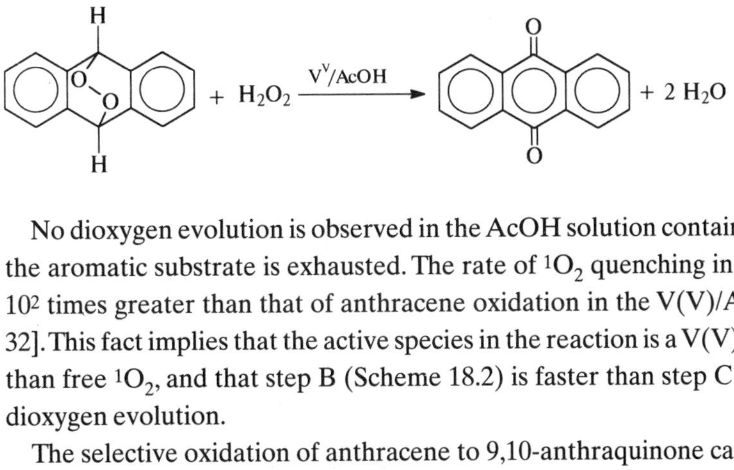

$$\text{(18.20)}$$

No dioxygen evolution is observed in the AcOH solution containing anthracene until all the aromatic substrate is exhausted. The rate of $^1O_2$ quenching in the AcOH solution is c 10^2 times greater than that of anthracene oxidation in the V(V)/AcOH/H₂O₂ system [30, 32]. This fact implies that the active species in the reaction is a V(V) peroxo complex rather than free $^1O_2$, and that step B (Scheme 18.2) is faster than step C and responsible for the dioxygen evolution.

The selective oxidation of anthracene to 9,10-anthraquinone can be rationalized within a scheme involving the intermediate formation of 9,10-dihydro-9,10-epidioxy anthracene, which intercepts the singlet dioxygen molecule from the V(V)/$^1O_2$ complex (see Scheme 18.2, step B); the consequent transformations of the primary endoperoxide finally afford anthraquinone (see Equation (18.20)) [32].

A different pathway for the oxidation of aromatic substrates to quinones has been put forward [46]. A series of Re(VII) compounds (Re₂O₇, CH₃ReO₃, EtReO₃, η⁵-C₅H₅ReO₃ and η⁵-(C₅H₄Me)ReO₃) have been shown to catalyse oxidation reactions of different sub-strates with H₂O₂ [46, 47]. The oxidation of 2,3-dimethylnaphthalene in acetic acid with

85% aqueous H_2O_2 yielded 2,3-dimethyl-1,4-naphthoquinone [46]. The highest activity (75% for 4 h at c 20 °C) was observed with CH_3ReO_3 (catalyst concentration 0.1 M, starting substrate concentration 5 M). The reaction has been rationalized as involving epoxidation of the naphthalene derivative by analogy with [48, 49], as shown by Scheme 18.5.

Slow decomposition of methylrhenium diperoxide $CH_3ReO(\eta^2O_2)_2H_2O$ in aqueous solution to yield dioxygen and methylrhenium trioxide has been observed [50] (see Equation (18.21)). In view of this fact, the transfer of an $^1O_2$ molecule from rhenium diperoxo complexes to naphthalene substrates seems to be possible.

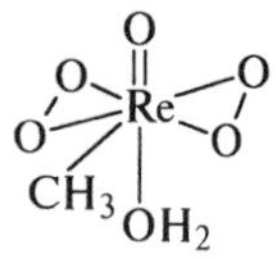

$$CH_3ReO_3 + O_2 \qquad (18.21)$$

Another redox reaction involving coordinated singlet dioxygen as the oxidant could be the oxidation of alkenes in the $V(V)/H_2O_2/AcOH$ system. Vanadium oxide and vanadic acid are capable of forming catalysts for the synthesis of glycols by reaction of olefins with H_2O_2 in both aqueous solution and organic non-aqueous solvents [1–3,6,51–58]. Sodium vanadate is used, under neutral and basic conditions, for selective epoxidation of olefins with H_2O_2. Bromination and hydroxybromination of olefins with H_2O_2 and KBr, catalysed by NH_4VO_3 in a two-phase H_2O/CH_2Cl_2 or $CHCl_3$ system, mimicking the hydrophilic and hydrophobic parts of vanadium-dependent bromoperoxidase, has been described [59, 60]. The available data suggest that the reaction of olefins with H_2O_2 catalysed by vanadium(V) compounds is sensitive to solution composition and pH values. By analogy [1–3, 6, 51–58], the $V(V)/H_2O_2/AcOH$ system was expected to yield glycols or their esters. However, experiments with hex-1-ene, 3-methylpent-1-ene, 4-methylpent-1-ene, cyclohexene, styrene and *trans*-1,2-diphenylethylene showed (see Table 18.2) that all the alkenes other than cyclohexene undergo >C=C< double bond cleavage to form the corresponding aldehydes (Equation (18.22)) [31].

$$R^1HC=CR^2H + 2H_2O_2 \rightarrow R^1CHO + R^2CHO \qquad (18.22)$$

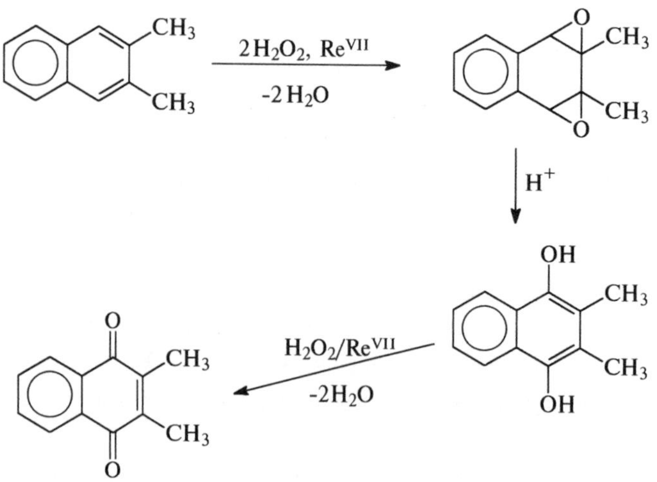

Scheme 18.5. Hypothetical mechanism of the 2,3-dimethylnaphthalene oxidation in an acetic acid solution by H_2O_2 catalysed with CH_3ReO_3 (MTO).

Table 18.2. Oxidation of olefins by H_2O_2 catalysed with V(V) compounds in an AcOH solution

No.	Substrate (M)	$[H_2O_2]_0$ (M)	$[VO(acac)_2]$ $\times 10^2$ (M)	Conversion of substrate (%)	Yield of RCHO (%)	$[H_2O_2]^a$ $\times 10^2$ (M) (Time of reaction) (h)
1	3-Methylpentene-1 (0.092)	0.5	1.02	<5	1.7	0.4 (1.7)
2	4-Methylpentene-1 (0.10)	0.5	0.88	<5	2.1	0.5 (1.8)
3	Hexene-1 (0.10)	0.5	1.00	<5	1.3	0.5 (1.8)
4	Hexene-1 (0.10)	1.0	0.95	6.4	2.9	0.8 (1.8)
5	Cyclohexene (0.10)	1.0	1.01	<5	<0.1	0.4 (2.4)
6	*Trans*-1,2-diphenylethylene (0.10)	0.97	1.09	72	51	0.6 (2.0)
7	*Trans*-1,2-diphenylethylene (0.089)	0.5	0.92	48	38	0.8 (1.8)
8	Styrene (0.10)	1.0	1.06	65	40	0.4 (1.8)
9	Styrene (0.10)	0.5	1.20	41	29	0.4 (1.8)

^{a} H_2O_2 concentration after the reaction is accomplished.

In no case were epoxides, glycols or their esters detected among the products by GC–MS techniques, contrary to expectations based on the above-mentioned literature data. Epoxides and vicinal diols are known to undergo oxidative cleavage by reaction with V(V)/TBHP complexes in organic solvents (Equation (18.23)).

$$\underset{\underset{\text{OH \ OH}}{|\quad\;|}}{\overset{\diagdown}{\diagup}C-C\overset{\diagup}{\diagdown}} + 2ROOH \xrightarrow{\;V^V\;} 2 \overset{\diagdown}{\diagup}C=O + 2ROH \qquad (18.23)$$

No benzaldehyde was found among the products of the oxidation of styrene oxide in the V(V)/H_2O_2/AcOH system [31]. Moreover, styrene oxide and ethylene glycol were found to be rather stable, undergoing very slow oxidation in the H_2O_2/AcOH/V(V) system. These facts suggest that neither epoxides nor glycols are the intermediates of the reactions in question.

Neither epoxycyclohexane nor the corresponding glycol and its ester were found among the products of the cyclohexene oxidation by H_2O_2 in acetic acid solution containing V(V); also, no adipic dialdehyde was observed [31]. Rather low conversion of cyclohexene was observed. The main direction of the reaction is allylic oxidation to yield cyclohexene-3-ol and cyclohex-2-enone, similar to the cycloxehene oxidation catalysed with V_2O_5 in acetone solution (Equation (18.24)) [61–63].

$$\text{(cyclohexene)} + H_2O_2 \xrightarrow{V_2O_5} \text{(2-cyclohexenol, OH)} + \text{(2-cyclohexenone, O)} \quad (18.24)$$

Acyclic aliphatic alkenes, e.g. hex-1-ene, 3-methylpent-1-ene and 4-methylpent-1-ene, can be split oxidatively to form the corresponding aldehydes according to Equation (18.22). However, the reactivity of these alkenes towards the $V(V)/H_2O_2/AcOH$ system was found to be rather low, contrary to the oxidation in organic or aqueous solutions resulting in the formation of epoxy derivatives or the corresponding glycols [1–3,6,51–60]. Unlike aliphatic alkenes, both styrene and *trans*-1,2-diphenylethylene were found to undergo >C=C< bond oxidative cleavage smoothly under very mild conditions. At 20 °C, the conversions are 50–70% after *c* 2 h (see Table 18.2). The enhanced reactivity of aryl-substituted alkenes as compared to the alkyl-substituted alkenes suggests that the active oxidant is an electrophilic agent.

Both the oxidation of anthracenes and the oxidative cleavage of the >C=C< bonds in alkyl- and aryl-substituted alkenes can be rationalized by assuming a mechanistic scheme involving the intermediate formation of V(V) complexes containing singlet dioxygen as a ligand and the interaction of these complexes with substrates to form the corresponding endoperoxides (see Equation (18.20)) or dioxetanes (see Equation (18.25)).

$$\underset{H}{\overset{R_1}{>}}C=C\underset{R_2}{\overset{H}{<}} + V^V[^1O_2] \longrightarrow \underset{H}{\overset{R_1}{>}}\underset{O-O}{\overset{|\;\;|}{C-C}}\underset{R_2}{\overset{H}{<}} + V^V \quad (18.25)$$

The decomposition of the intermediate dioxetane could give rise to the observed products (see Equation (18.26)) [64].

$$\underset{H}{\overset{R_1}{>}}\underset{O-O}{\overset{|\;\;|}{C-C}}\underset{R_2}{\overset{H}{<}} \longrightarrow R_1\underset{O}{\overset{\|}{C}}-H + R_2\underset{O}{\overset{\|}{C}}-H \quad (18.26)$$

The lack of any 'normal' products, e.g. epoxy compounds, glycols, etc., implies the absence of species responsible for their formation in the $H_2O_2/AcOH/V(V)$ system. Correspondingly, it is not surprising that the reactivity of alkenes in this system differs markedly from that in the 'normal' epoxidation systems, e.g. V(V) complexes in aqueous solution and organic solvents [1–3, 6, 51–60].

3.4 *Interaction between singlet dioxygen and superoxide radical anion coordinated with V(V) ion [65]*

Singlet dioxygen is known to oxidize free superoxide anion in solution according to reaction (18.27) [66, 67].

$$^1O_2 + O_2^- = O_2^- + {}^3O_2 + 22\,\text{kcal} \quad (18.27)$$

In this section we describe the first observations concerning the influence of $^1O_2$ on the shape of the ESR spectra of superoxide anion coordinated to V(V). The coordinated superoxide radical anion $V(V)(O_2^-)$ was detected in the $V(V)/H_2O_2/AcOH$ system in the

Table 18.3. The ESR spectrum parameters of $V(V)(O_2^-)$ complex in the $H_2O_2/V(V)/AcOH$ system $[V(V)] = 10^{-2}\,M, [H_2O_2]_0 = 1\,M, [H_2O]_0 = 2.3\,M, 20\,°C$

Purging gas	Time[a] (min)	$g \pm 0.0003$	$(a_v \pm 0.2) \times 10^{-4}$ (cm⁻¹)	$\alpha_v \pm 0.2$[b] (G)	$(a_{Mn} \pm 0.2) \times 10^{-4}$ (cm⁻¹)	Error[c]
Ar	12.0	2.0013	4.4	2.6	79.9	0.016
	18.8	2.0013	4.4	2.5	80.1	0.021
	40.0	2.0012	4.4	2.4	80.2	0.029
	72.0	2.0012	4.4	2.7	80.1	0.024
O₂	16.5	2.0012	4.5	4.0	80.2	0.018
	22.0	2.0013	4.5	4.0	80.2	0.016
	33.0	2.0013	4.5	3.8	80.2	0.016
	46.5	2.0013	4.5	3.8	80.2	0.020
	54.8	2.0012	4.5	3.8	80.2	0.017
No gas bubbling	3.5	2.0111	4.7	6.0	80.2	0.014
	8.5	2.0111	4.7	6.3	80.2	0.014
	35.0	2.0110	4.7	6.2	80.2	0.014
	48.0	2.0109	4.6	7.2	80.2	0.012
	67.0	2.0108	4.6	8.1	80.2	0.012

[a] The start of run is the time of H_2O_2 addition to the catalyst solution.
[b] The linewidth was estimated according to $\Delta H = \alpha + \beta m_1 + \gamma m_1^2$, where m_1 is the projection of nuclear spin on the direction of external magnet field and α is a parameter accounting for all the effects of broadening, which are equal for all HFS lines. Parameters β and γ are about two orders of magnitude smaller than α and are not given here.
[c] Root-mean-square deviation.

course of the H_2O_2 catalytic decomposition [30, 32]. An octet with $g = 2.01125 \pm 0.00005$ G and $a_V = 0.00044\,cm^{-1}$ was observed,* these values coinciding with those for the known vanadium complex with coordinated superoxide anion [68–70] (Fig. 18.5; Table 18.3). The integral intensity of the signal observed reduces as H_2O_2 is decomposed (Figs 18.2 and 18.6). To the best of our knowledge, this is the first time that rather high (10^{-3}–10^{-4} M) steady-state concentrations of complexes with coordinated superoxide ion $V(V)(O_2^-)$ have been observed.

The shape of the spectra was found to change drastically, depending on the gas under which H_2O_2 was decomposed and the spectra were recorded. Thus, the strong and well-resolved octet was observed only when the ESR cell was purged with Ar or CO₂ (see Fig. 18.5a and Table 18.3). Under the triplet dioxygen flow (Fig. 18.5a), the spectral parameters (g factor and hyperfine structure constant) were unchanged but the lines of the octet were broadened by c 1.5 G (see Fig. 18.5b and Table 18.3). The observed broadening can be attributed to the dipole–dipole interaction between the $V(V)(O_2^-)$ species and the $^3O_2$

* The ESR spectra were recorded using a Radiopan SE/X-2542 radiospectrometer (9450 MHz, 0.5 G modulation amplitude) at 255–293 K. 1,1-Diphenyl-2-picrylhydrazyl (DPPH), $g = 2.0036$, was used as the internal field marker. Methods of treatment and simulation of ESR data are described elsewhere [71]. The intensity of the $V(V)(O_2^-)$ signal was determined by using third and fourth components of the Mn/MgO ESR signal as a standard.

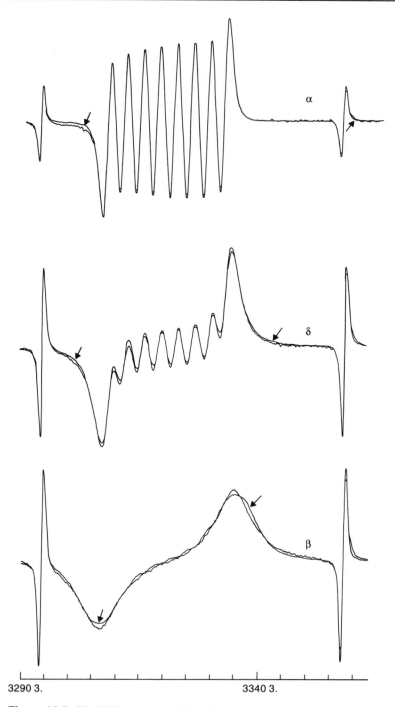

3290 3.   3340 3.

Figure 18.5. The ESR spectra of V(V) (O_2^-) in the H_2O_2/V(V)/AcOH system ([V(V)] = 10^{-2} M, $[H_2O_2]_0$ = 1 M, 20 °C): (a) under Ar purging; (b) under O_2 purging; (c) without any gas purging. The arrows indicate estimated curves. The two lines edging the spectra at the high- and low-frequency fields are the third and fourth components of the Mn/MgO ESR signal, correspondingly.

molecule. A dramatic change in the spectrum shape was observed in experiments without any gas purging. In this case, the lines of the spectrum are broadened so much that the hyperfine structure of the spectrum is practically unresolved (see Fig. 18.5c and Table 18.3). The spectrum lines become broadened in *c* 1 min after the Ar flow has been stopped. This

Relative integral intensity

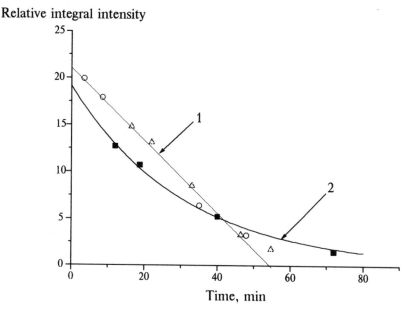

Figure 18.6. Relative integral intensity of the V(V)(O$_2^-$) ESR signal vs time plot ([V(V)] = 10^{-2} M, [H$_2$O$_2$]$_0$ = 1 M, 20 °C): (1) without any gas purging; (2) under Ar purging.

suggests the presence of a volatile, very effective relaxant agent accumulating rapidly during the H$_2$O$_2$ decomposition. Minor side products of the H$_2$O$_2$/AcOH decomposition, such as CO$_2$, CH$_3$OH and CH$_3$OOAc, cannot serve as relaxant agents because the addition of these compounds has been proved to have no effect on the shape of the spectra.

The observed broadening (c 6–8 G, see Table 18.3) is much more pronounced than that normally observed for the ^{3}O$_2$ dipole–dipole interaction, which is usually close to 1.5 G. For instance, we found that the spectrum of 2,2,6,6-tetramethyl-4-hydropiperidine-1-oxyl is broadened by c 1 G under ^{3}O$_2$ purging.

On the other hand, the V(V)/H$_2$O$_2$/AcOH system is known to generate complexes containing the ^{1}O$_2$ molecule as a ligand, i.e. V(V)(^{1}O$_2$) (see Section 3.2) [30–32]. In the absence of the ^{1}O$_2$ trapping agent the complex is decomposed, evolving ^{1}O$_2$, which is quenched via ^{1}O$_2$/solvent interaction. Therefore, singlet dioxygen can be expected to occur in the V(V)/H$_2$O$_2$/AcOH system in the steady-state concentration. Under a constant gas flow, the concentration of dissolved ^{1}O$_2$ should be lower than in the absence of purging, so the observed changes in the shape of the V(V)(O$_2^-$) spectra can be rationalized from the perspective of ^{1}O$_2$/V(V)(O$_2^-$) interaction.

The analogous unusual broadening of the ESR signal of 2,2,6,6-tetramethyl-4-hydroxypiperidine-1-oxyl has been observed but not discussed during the sensitized oxidation of the corresponding amine by singlet dioxygen [72]. In general, both free ^{1}O$_2$ and its V(V) complex should be taken into consideration as possible relaxant agents. Fast equilibrium between these species would be expected (see Equation (18.28)).

$$(^1O_2)V(V) \rightleftharpoons {}^1O_2 + V(V) \tag{18.28}$$

An inert gas flow can affect the steady-state concentration of ^{1}O$_2$ and, as a consequence, the concentration of the V(V)(^{1}O$_2$) complex. If both the ^{1}O$_2$ molecule and superoxide radical anion were bound to the same vanadium atom, a spin or electron exchange between the ligands could be expected, as has been observed [73] (e.g. according to Equation (18.29)).

$$({}^1O_2)V(V)(O_2^-) \rightleftharpoons (O_2^-)V(V)({}^1O_2) \tag{18.29}$$

The observed change in the shape of the ESR spectra could be attributed to such an exchange. However, the oxidation of free O_2^- by a free $^1O_2$ molecule is a diffusion-controlled reaction in Me_2SO ($7 \pm 6 \times 10^9 \, M^{-1} \, s^{-1}$) and CH_3CN ($1.6 \times 10^9 \, M^{-1} \, s^{-1}$) solutions [66]. The reaction between the coordinated ligands O_2^- and $^1O_2$, resulting in a complex containing the $^3O_2$ molecule and superoxide radical anion (Equation (18.30)), should be an even faster reaction.

$$({}^1O_2)V(V)(O_2^-) \rightleftharpoons (O_2^-)V(V)^3O_2 \tag{18.30}$$

An enormously high broadening of the O_2^- ESR signal in such a complex containing both O_2^- and $^3O_2$ would be expected. The triplet dioxygen molecule $^3O_2$ should be a very poor ligand for the V(V) ion being expelled from the complex (Equation (18.31)).

$$(O_2^-)V(V)({}^3O_2) \rightleftharpoons (O_2^-)V(V) + ({}^3O_2) \tag{18.31}$$

However, the coincidence of the experimental ESR spectrum and the curve simulating the ESR spectrum under approximation of homogeneous broadening suggests the occurrence of only one type of paramagnetic V(V) complex (Fig. 18.7). All attempts to record the signals of V(IV) complexes in the $V(V)/H_2O_2/AcOH$ system failed. The assumption of the coexistence of a complex with a well-resolved ESR signal, $V(V)(O_2^-)$, and that with a broadened spectrum, $(O_2^-)V(V)({}^3O_2)$, is in disagreement with the experimental data. Thus, superposition of a well-resolved ESR signal and a broadened one in $1:1$ ratio cannot be described by the model of homogeneous broadening of the spectrum of a single species (the deviations exceed 4%) (see Fig. 18.7(a)).

The interaction between a $V(V)(O_2^-)$ complex and a free singlet dioxygen molecule cannot be ruled out. A quenching of the $^1O_2$ molecule by $V(V)(O_2^-)$ complex according to Equation (18.32) could be an effective way for the electron spin exchange and spin relaxation because of the gain in free energy at $^1\Delta({}^1O_2)$ into a $^3\Sigma({}^3O_2)$ transition (c 23 kcal).

$$\begin{array}{cccc} \uparrow & \uparrow\downarrow & \downarrow & \uparrow\uparrow \\ V(V)(O_2^-) + {}^1O_2 &\rightarrow& V(V)(O_2^-) + {}^3O_2 \end{array} \tag{18.32}$$

As a result, the ESR lines are broadened. The same mechanism can be operative in the 2,2,6,6-tetramethyl-4-hydroxypiperidine-1-oxyl/$^1O_2$ system [72]. Note that complex formation between the components of the latter system is impossible, so reaction (18.32) seems to be a general interaction between singlet dioxygen and the free oxyl radical. However, in the case of the $V(V)(O_2^-)/^1O_2$ system, an additional avenue for the $V(V)(O_2^-)/^1O_2$ interaction appears to be possible.

It is easy to see that the presence of dissolved $^1O_2$ affects not only the shape of the ESR signal but also the kinetics of the $V(V)(O_2^-)$ disappearance (Fig. 18.6). Under Ar flow, the decrease in the integral ESR intensity is described by a first-order kinetic equation (Fig. 18.6, curve 2), while in the absence of Ar flow the kinetics are zero order (Fig. 18.6, curve 1). Thus, singlet dioxygen can participate in the disappearance of $V(V)(O_2^-)$ as well. A redox reaction between the $V(V)(O_2^-)$ complex and the $^1O_2$ molecule can be expected.

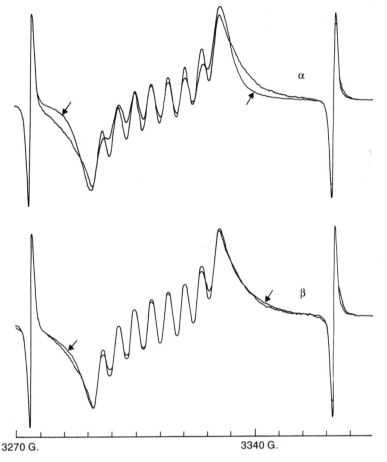

α

β

3270 G.   3340 G.

Figure 18.7. Simulated superposition of the experimental well-resolved ESR signal (see Fig. 18.5a) and the broadened ESR signal (see Fig. 18.5c) in the 1:1 ratio. The arrows indicate: (a) estimated curve under assumption of homogeneous broadening of the single species spectrum (root-mean-square deviation is 4.5%); (b) estimated curve under assumption of the coexistence of two complexes that have equal g factors and hyperfine structure constants but differ in their linewidths (root-mean-square deviation is 1.6%).

An electron transfer from the V(V)(O$_2^-$) complex to the ^{1}O$_2$ molecule in an AcOH solution should result in the formation of HO$_2^-$ (see Equation (18.33)).

$$V(O_2^-) + {}^1O_2 \xrightarrow[H^+]{k_{33}} V({}^1O_2) + HO_2^- \qquad (18.33)$$

A reaction between HO$_2^-$ and the V(V) complex will restore the V(V)(O$_2^-$) complex (Equation (18.34), $k_{34} = 10^5$ M^{-1} s^{-1} in H$_2$O solution [68]).

$$V(V) + HO_2^- \xrightarrow{k_{34}} V(O_2^-) + H^+ \qquad (18.34)$$

However, the HO$_2^-$ radicals should undergo disproportionation according to reaction (18.35) more rapidly ($k_{35} = 10^6$ M^{-1} s^{-1} in H$_2$O solution [74]).

$$2HO_2^- \xrightarrow{k_{35}} H_2O_2 + O_2 \qquad (18.35)$$

The net result of reactions (18.33), (18.34) and (18.35) is the dismutation of the complex with superoxide anion as a ligand. If this is the case, the system under question can be considered as the first example of the O$_2^-$ dismutase not containing the Zn(II)/Cu(II) couple.

4   Ozone formation

Combinations of O atoms with dioxygen molecules constitute the basis for many processes of ozone formation, such as the photolysis of O_2 or NO_2, electrolysis, electrical discharge, ionizing radiation and UHF [75]. Another high energy source for ozone formation can be the recombination of certain peroxy radicals [76]. Chemical ways to form ozone that do not involve high energy species are rather rare, normally yielding only small quantities of ozone. Thus, small detectable amounts of ozone were reported to occur upon the action of H_2O_2 on selenic acid, an acid on potassium peroxydisulfate, fluorine on aqueous KOH and dioxygen on thin Al films [77]. Ozone is evolved through the action of strong mineral acids on peroxides [78]. Recently, evidence for ozone formation was obtained by reacting an iron(III) porphyrin/NO_2^- adduct with dioxygen and an alkene [77].

A peculiarity of the V(V)-catalysed decomposition of H_2O_2 in a CF_3COOH solution is that this reaction yields a substantial amount of ozone (no less than 10–15% (v/v) of the total gaseous products), i.e. the amount of H_2O_2 that decomposes according to Equation (18.36) to give ozone (c 15%) is comparable to the amount of H_2O_2 that decomposes to give dioxygen (Equation (18.37)) (c 85%) [12, 13, 45].

$$3H_2O_2 \xrightarrow{\text{V(V)/H}_2\text{O}_2\text{/CF}_3\text{COOH}} 3H_2O + O_3 \tag{18.36}$$

$$2H_2O_2 \xrightarrow{\text{V(V)/H}_2\text{O}_2\text{/CF}_3\text{COOH}} 2H_2O + O_2 \tag{18.37}$$

Reaction (18.36) is the first example of catalytic H_2O_2 in O_3 transformation. As we have mentioned above (see Section 3.2), the formation of small amounts of ozone has also been observed in the decomposition of H_2O_2 in acetic acid solutions containing V(V) [45].

It may seem that the comparison of the yields of ozone in the decomposition of H_2O_2 in acetic and trifluoroacetic acid solutions in the presence of 10^{-5}–10^{-3} M V(V) supports the hypothesis that O_3 is formed with the participation of the HO^+ ion. For example, the increase in the yield of O_3 on going from CH_3COOH to CF_3COOH could have been explained by the fact that CF_3COOH is a strong acid (unlike CH_3COOH) and can protonate H_2O_2 or the peroxide group coordinated to vanadium to give a free or coordinated HO^+ ion or similar species. In this case, ozone would actually result from the reaction of these species with H_2O_2 or with the oxygen evolved. Contrary to this suggestion, the presence of 2 M H_2SO_4 or 0.7 M $HClO_4$ in acetic acid completely suppresses the evolution of O_3 (Table 18.4): this rules out the possibility of the HO^+ involvement in the formation of O_3. In the presence of 0.2–2 M CF_3COONa in trifluoroacetic acid solution, the yield of O_3 was not changed considerably. These data imply that the formation of ozone is not associated with protonation of H_2O_2 or vanadium.

Table 18.4. The formation of ozone in the H_2O_2/V(V)/RCOOH system (10 °C, [V] = 10^{-3} M)

No.	RCOOH	Compound added (M)	Yield of O_3 (%)
1	CH_3COOH	—	~1
2	CH_3COOH	$HClO_4$ (0.7)	~0
3	CH_3COOH	H_2SO_4 (2.0)	~0
4	CF_3COOH	—	~15
5	CF_3COOH	CF_3COONa (0.2–2.0)	~15

A thermodynamically favorable disproportionation of RO$_2^{\cdot}$ to yield ^{1}O$_2$ and R$_2$O$_2$ is known [28,79,80]. In the case of HO$_2^{\cdot}$, i.e. when R = H, the reaction can be depicted by Equation (18.38).

$$2HO_2^{\cdot} \longrightarrow H_2O_2 + {}^1O_2 \qquad\qquad \Delta G°_{298} \approx -11\,kcal\,mol^{-1} \qquad\qquad (18.38)$$

Disproportionation of HO$_2^{\cdot}$ radicals with the formation of ozone is even more thermodynamically favorable (see Equation (18.39)).

$$2HO_2^{\cdot} \longrightarrow H_2O + O_3 \qquad\qquad \Delta G°_{298} \approx -23\,kcal\,mol^{-1} \qquad\qquad (18.39)$$

Thus, ozone can arise in catalytic systems for the decomposition of hydroperoxides that are able to generate free radicals. For example, the evolution of large quantities of ozone was reported in a patent [81] in which a system consisting of an aqueous solution of an Fe(II) salt, an ammonium or alkali metal salt of heptamolybdenic acid and concentrated H$_2$O$_2$ is described. The absence of any experimental details (concentrations of the starting compounds, the yield of O$_3$, the reaction temperature, etc.) makes it impossible to compare this system with the V(V)/H$_2$O$_2$/CF$_3$COOH system.

A small amount of ozone (0.4–0.8%) has been produced [82] in the decomposition of peroxyacetic acid initiated by Co(II) compounds. The formation of ozone in this catalytic system may be due to the disproportionation depicted by a square termination reaction (Equation (18.39)). In principle, this mechanism could occur in the V(V) system under consideration, being seemingly supported by the existence of a V(V)(O$_2^{\cdot-}$) complex radical in the solutions [11–13]. However, the absence of V(IV) signals in the ESR spectra recorded during decomposition of H$_2$O$_2$ in CF$_3$COOH solution casts some doubt on the free radical mechanism of the formation of ozone in the H$_2$O$_2$/V(V)/CF$_3$COOH system. Although the occurrence of reaction (18.39) under the conditions of the catalysis by vanadium complexes cannot be ruled out completely, all of the facts and considerations outlined suggest that the polar pathways must be responsible of the formation of ozone during the decomposition of H$_2$O$_2$ in CF$_3$COOH solution containing V(V) complexes.

A possible intermediate for the ozone formation reaction could be a complex with trioxo dianion O$_3^{2-}$. Dihydrogen trioxide (H$_2$O$_3$) and its derivatives, such as (RO)$_3$PO$_3$ and HO$_3$SiR$_3$, are well known [28, 83–86]. The kinetic and ^{19}F-NMR study of H$_2$O$_2$ into ozone transformation showed the reaction depicted by Equation (18.36) to involve perfluoroperacetic acid (CF$_3$COOOH) and a V(V) peroxo complex [12,13]. For instance, the reaction does not take place in the absence of CF$_3$COOOH or if its coordination to the V(V) center is retarded, e.g. as a consequence of F$^-$ ion presence in the reaction solution. These facts imply that the trioxide dianion forms by reacting the perfluoroperacetic acid with a peroxo ligand (Equation (18.40)).

$$\underset{\substack{|\\H}}{\overset{\substack{O\\\|}}{CF_3C\!O\!-\!O}} + O_2^{2-} \longrightarrow CF_3COO^- + HO_3^- \qquad\qquad (18.40)$$

Similar to the reaction described by Equation (18.6), this reaction can be considered as a nucleophilic substitution at the OH group of the CF$_3$COOOH molecule with the O$_2^{2-}$ dianion as a nucleophile and the perfluoroacetate ion CF$_3$COO$^-$ as a leaving group. The stabilization of this leaving group with the V(V) metal ion can contribute to the energetics of the inner-sphere analogue of reaction (18.40), which represents oxidation of a peroxo

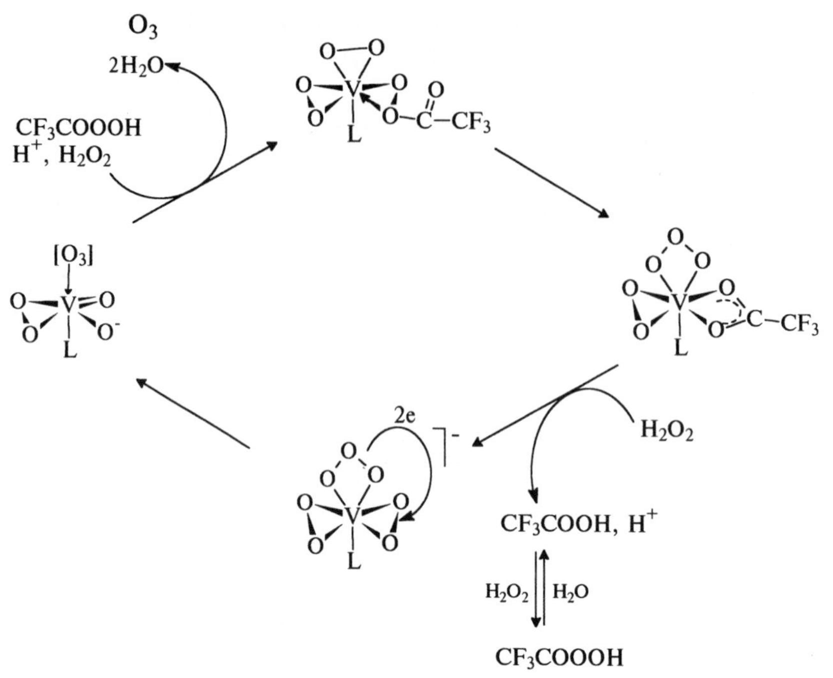

Scheme 18.6. Mechanism of the H_2O_2 decomposition to form ozone.

ligand into a trioxo ligand with perfluoroperacetate anion coordinated to V(V) and is depicted schematically by Equation (18.41).

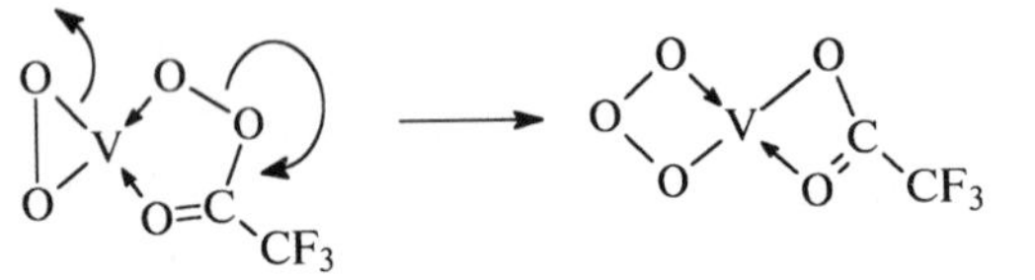

$$\text{(18.41)}$$

The coordinated trioxo dianion O_3^{2-} should lose two electrons to afford free ozone. It is most likely that the coordinated peroxo group O_2^{2-} would act as the acceptor of the electrons (Equation (18.42)).

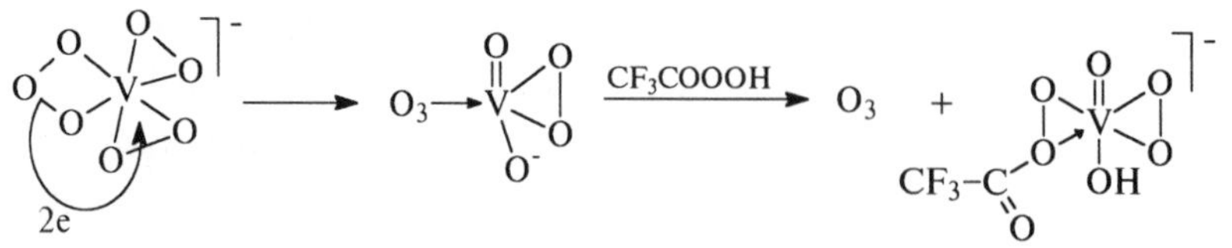

$$\text{(18.42)}$$

Hypothetical mechanistic considerations concerning the catalytic conversion of H_2O_2 into ozone are represented by Scheme 18.6.

Hydrogen peroxide is decomposed vigorously at the diffusion-controlled rate at 5–20 °C in the present of c 10^{-3} M V(V) compounds in CF_3COOH solution. Many substrates that are hard to oxidize undergo fast oxidation under mild conditions with this system.

4.1   Oxidation reactions in the V(V)/H_2O_2/CF_3COOH system

The H_2O_2/CF_3COOH/V(V) catalytic system showed great activity in a variety of oxidations. Alkanes, electronegatively substituted arenes and perfluoroalkenes were found to react with hydroperoxides in CF_3COOH solution in the presence of V(V) complexes under mild conditions (see Scheme 18.7).

Thus, cyclohexane was oxidized by the $H_2O_2/CF_3COOH/V(V)$ system to give cyclohexanol and its ester as main products (Equation (18.43)). The limited conversion of cyclohexane reaches 85–98% at the ratio $[H_2O_2]_0/[C_6H_{12}]_0 = 6$–7, although in a solution of HOAc maximal conversion of cyclohexane was 8–10% when this ratio was 5 [11–13].

$$(18.43)$$

The oxidation reactions of perfluoroalkenes (perfluorooctene-1 and -2, and perfluorocyclohexene) were found to give rise to the fluoroanhydrides of the corresponding perfluorocarboxylic acids smoothly with 45–100% yield [12, 13] (Equation (18.44)).

$$R^1_F\text{-}CF = CF\text{-}R^2_F + 2H_2O_2 \longrightarrow R^1_F\text{-}CF = O + O = CF\text{-}R^2_F + 2H_2O \qquad (18.44)$$

Oxalic acid or its fluoroanhydride (c 10–30%) were found among the reaction products of hexafluorobenzene and benzene oxidations [12, 13]. The rates of the individual oxidation reactions were commensurate with the rate of H_2O_2 decomposition and controlled by the diffusion of the reagents. There are five oxidizing species potentially capable of the observed reactions: $^1O_2$, $^3O_2$, O_3, coordinated superoxide anion V(V)(O_2^-) and peroxo complexes of type V(V)(O_2)$_n$. However $^1O_2$, $^3O_2$ and coordinated superoxide anion V(V)(O_2^-) have been shown to be inactive in the reactions under consideration. A detailed study, including comparison of the oxidation reactions under discussion and ozonization reactions of the same substrates, showed that the only oxidants responsible for the catalytic oxidations observed are vanadium complexes of type **5–8** [12, 13].

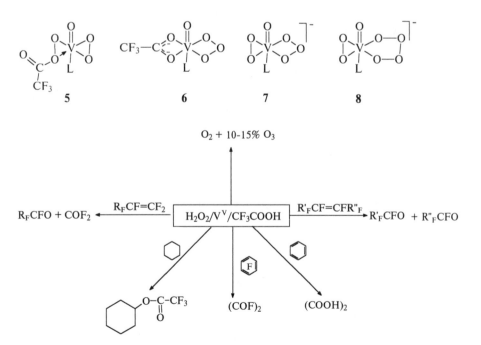

Scheme 18.7. Oxidation reactions in the $V(V)/H_2O_2/CF_3COOH$ system.

A detailed analysis led to the conclusion that complex **7** is responsible for ozone formation and perfluorooct-1-ene oxidation. Complex **5** oxidizes cyclohexane and complex **6** reacts with aromatic compounds. Complex **8** reacts with perfluorocyclohexene and internal perfluoroalkenes. The rates of interconversions ($\mathbf{5 \to 6 \to 7 \to 8}$) are smaller than those for the reactions of the intermediates with corresponding substrates. All the complexes under discussion are capable of dioxygen evolution and their concentrations obey steady-state requirements [12, 13].

5   Redox reactions between coordinated enolate anion of 1,3-dicarbonyl compounds and the peroxoligand [87, 88]

As with the dismutation of two peroxo groups of V(V) to form singlet dioxygen (see Scheme 18.2), acetylacetone oxidation catalysed with Mo(VI) in AcOH solution is another example of an inner-sphere redox process in which the oxidation state of the central atom seems to remain unchanged in all stages of the reaction. Compounds W(VI) and Mo(VI) are known to form peroxo complexes by reacting with H_2O_2 in aqueous media or CH_3CN solution (Equation (18.45)).

$$MO_4^{2-} + nH_2O_2 \rightleftharpoons O_mM(O_2)_n^{2-} + nH_2O \tag{18.45}$$

where $m + n = 4$ and M = W or Mo.

Analogously, acetylacetone and its derivatives are known to react with M(VI) complexes to form complex enolate anions [89] (Equation (18.46)).

$$OWF_4 + Hacac \xrightarrow{CH_3CN} W(O)F_3(acac) \tag{18.46}$$

The Hacac molecule was found to undergo oxidation in both a solution of tungsten(VI) and hydrogen peroxide (or in solutions of peroxo complexes of tungsten(VI)) and in analogous solutions of Mo(VI) complexes. Thus, when Hacac was added to an aqueous $SrMoO_6$ solution, rapid warming up of the mixture occurred and CO_2 was evolved, along with dioxygen formed by decomposition of the peroxomolybdate. On completion of the reaction, acetic acid, but not Hacac, was detected in the solution by GLC. The Hacac molecule also underwent oxidation by 20–30% aqueous H_2O_2 in the presence of Na_2MoO_4. Stoichiometric experiments showed that 2 mol of acetic acid and 1 mol of CO_2 are formed for each H_2O_2 molecule consumed, in accordance with Equation (18.47).

$$H_3C-\underset{\underset{O}{\|}}{C}-CH_2-\underset{\underset{O}{\|}}{C}-CH_3 + 4H_2O_2 \xrightarrow{Na_2MoO_4} 2CH_3COOH + CO_2 + 4H_2O \tag{18.47}$$

Analogously, ethyl acetoacetate was oxidized by H_2O_2 in aqueous Na_2MoO_4 solution according to Equation (18.48).

$$H_3C-\underset{\underset{O}{\|}}{C}-CH_2-\underset{\underset{O}{\|}}{C}-C_2H_5 + 4H_2O_2 \xrightarrow{Na_2MoO_4}$$
$$\longrightarrow 2CH_3COOH + C_2H_5OH + CO_2 + 4H_2O \tag{18.48}$$

Methyl acetylacetone also entered into reaction with H_2O_2 in the presence of MoO_4^{2-}. However, no CO_2 was found among the reaction products of this compound: 3 mol of acetic acid were formed per 1 mol of methyl acetylacetone oxidized, according to Equation (18.49).

$$H_3C-\underset{\underset{O}{\|}}{C}-CH(CH_3)-\underset{\underset{O}{\|}}{C}-C_2H_5 + 3\,H_2O_2 \xrightarrow{\ \text{Na}_2\ \text{oO}_4\ } 3CH_3COOH + 2H_2O \tag{18.49}$$

Thus, H$_2$O$_2$ reacts with all 1,3-dicarbonyl compounds bearing at least one H atom at the central C atom by splitting the C–C bond between this C atom and the acetyl groups. Remarkably, no oxidations were observed when 2,2-dimethyl acetylacetone (**9**) or dimethylmalonate (**10**) was added to CH$_3$CN or an aqueous solution of H$_2$O$_2$/Na$_2$MoO$_4$ or H$_2$O$_2$/WOF$_4$.

$$
\begin{array}{cc}
\underset{\underset{O}{\|}}{\overset{}{H_3C-C}}-\underset{\underset{CH_3}{|}}{\overset{\overset{CH_3}{|}}{C}}-\underset{\underset{O}{\|}}{\overset{}{C}}-CH_3 & \quad MeO-\underset{\underset{O}{\|}}{C}-CH_2-\underset{\underset{O}{\|}}{C}-OMe \\[2ex]
\textbf{9} & \textbf{10}
\end{array}
$$

Compounds **9** and **10** do not undergo enolization and are incapable of forming complexes with Mo(VI) and W(VI) ions.

To rationalize the observed data, it is essential to assume that both participants of the reaction, the peroxy group and 1,3-diketonate, are coordinated to a metal atom. Thus, when a solution of [WOF$_3$(acac)] containing excess of [WOF$_4$MeCN]$^-$ and [W$_2$O$_2$F$_9$]$^-$ to ensure the absence of free Hacac is mixed with a solution of [WOF$_3$(O$_2$)MeCN]$^-$ and [WOF$_2$(O$_2$)(MeCN)$_2$]$^-$ not containing free H$_2$O$_2$, warming of the mixture and changes in the ^{19}F-NMR spectra were observed, indicating the oxidation of Hacac. When Hacac is oxidized in aqueous peroxymolybdate solutions, the reaction is always preceded by complete dissolution of the Hacac, probably as a result of coordination to a metal ion because the solubility of Hacac in water is significantly below the concentration used in the experiments. Thus, any mechanisms involving participation of free 1,3-diketones and H$_2$O$_2$ can be excluded from consideration.

The reaction of tris(β-ketoenolato)cobalt(III) complexes with protic acids results in a ligand–metal electron transfer with the formation of a variety of β-ketoenolyl radicals that were detected by ESR spectroscopy [90]. Free-radical displacement of an H atom in a coordinated 1,3-diketonate has been observed in the reactions of acetylacetonates with N-bromosuccinimide [91]. The reactions (Equations (18.47)–(18.49)) could have been formally assumed to involve the attack on the coordinated 1,3-diketonate of HO$^{\cdot}$ or HO$_2^{\cdot}$ radicals generated under the reaction conditions. However, this mechanism appears quite unlikely, because the reactions proceed smoothly in aqueous ethanol, which reacts effectively with free oxyl radicals. Moreover, such strong free-radical acceptors as stable iminoxyl radicals, p-benzoquinone or 2,4,6-tri-$tert$-butylphenol did not affect the course of the reaction. No free O- or C-centered free radicals were detected with the ESR technique in the course of the reaction.

One might have expected intermediate formation of CH$_2$O and/or HCOOH if Hacac or coordinated acac$^-$ were oxidized according to the Baeyer–Villiger reaction (e.g. see Equation (18.50)).

$$AcCH_2Ac \xrightarrow[-H_2O]{H_2O_2} AcOCH_2Ac \xrightarrow[-H_2O]{H_2O_2} (AcO)_2CH_2 \xrightarrow[-2HOAc]{H_2O} CH_2O$$

$$\xrightarrow[-H_2O]{H_2O_2} HCOOH \xrightarrow[-2H_2O]{H_2O_2} CO_2 \tag{18.50}$$

However, joint oxidation of Hacac and CH_2O or HCOOH showed the acetylacetone to react much more rapidly. This fact and the absence of HCOOH in the oxidation products of Hacac (GLC data) refute the possibility that $AcCH_2Ac$ oxidation can proceed according to the Baeyer–Villiger reaction. The ^{1}H-NMR spectra of Hacac and Na_2MoO_3 showed the intermediate product of the reaction to be complex **11**.

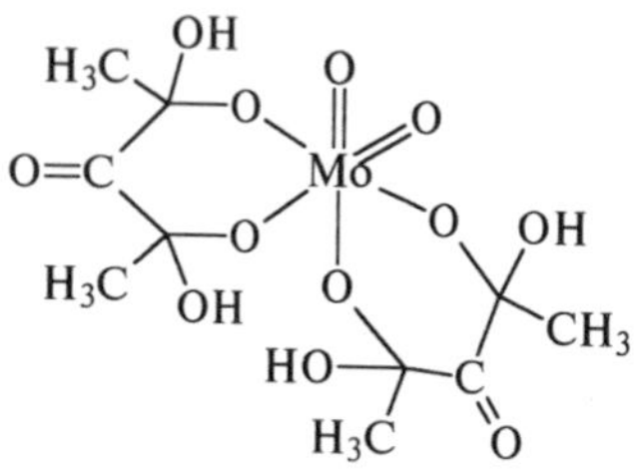

11

In the ^{1}H-NMR spectra, all the observed intermediate products have chemically equivalent CH_3 groups. If the Hacac were oxidized according to a Baeyer–Villiger mechanism or through a step of epoxidation of the C=C bond of the enol form of acac$^-$ ligand, then signals of equal intensity from non-equivalent CH_3 groups would be observed at H_2O_2/Hacac mole ratios of <2. The absence of the inequality of CH_3 groups suggests that at the first stages of reaction (Equation (18.47)) only the C^3 atom of the Hacac molecule is involved into the reaction.

All the data available suggest the oxidation of the acac$^-$ anion to involve two-electron removal or two successive steps each of one electron removal. The removal of an electron pair from the HOMO orbital of the acac$^-$ anion would convert this ligand into a carbocation (Equation (18.51)).

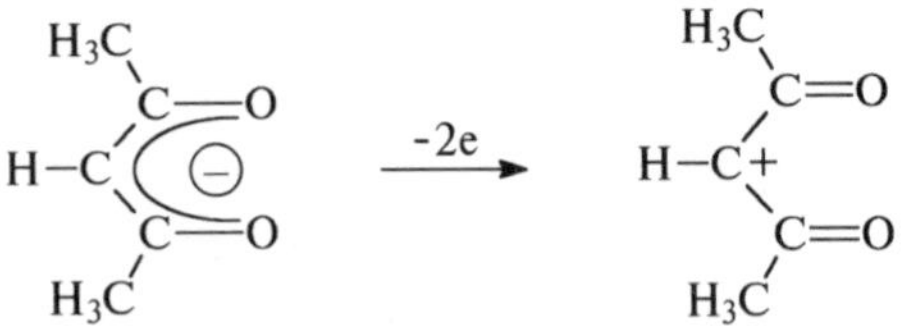

$$\tag{18.51}$$

It is apparent that this transformation should be impossible without the nucleophilic assistance of a solvent molecule attacking the C^3 atom when the redox reaction (Equation (18.51)) occurs. Thus, the reaction between $SrMoO_4$ and Hacac in dry CH_3CN does not take place until at least small amounts of water are added to the solution. After the addition of the water, rapid oxidation of Hacac was observed, suggesting that oxidation of the coordinated acac$^-$ ligand proceeds according to Equation (18.52).

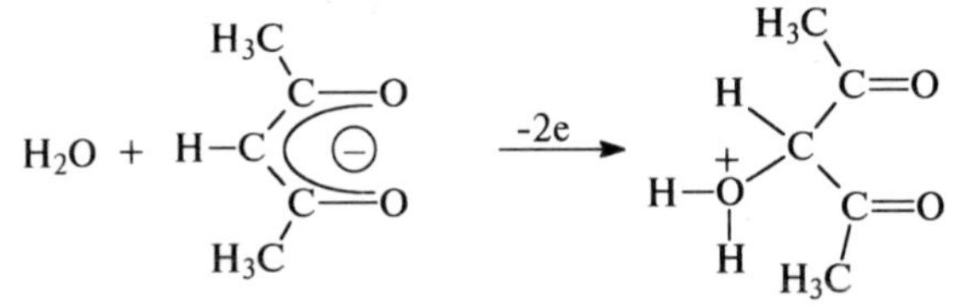

$$\text{(18.52)}$$

12

The elusive cation **12** after deprotonation can form 3-hydroxy-2,4-pentadione, which, after enolization and deprotonation, affords anion **13** (Equation (18.53)) capable of coordinating to Mo(VI).

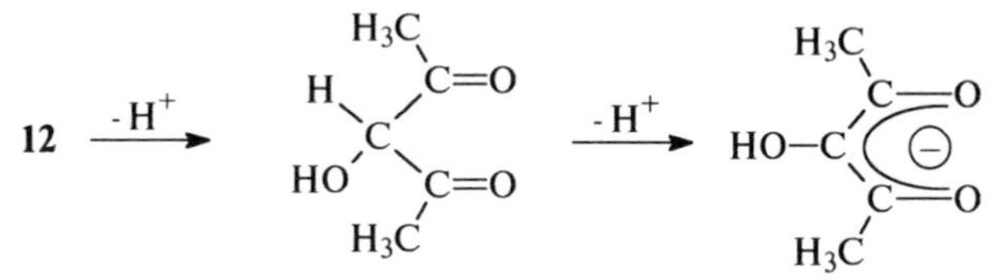

$$\text{(18.53)}$$

13

The two-electron oxidation of anion **13** should give cation **14** and, after its deprotonation, 2,3,5-pentatrione **15a** (Equation (18.54)).

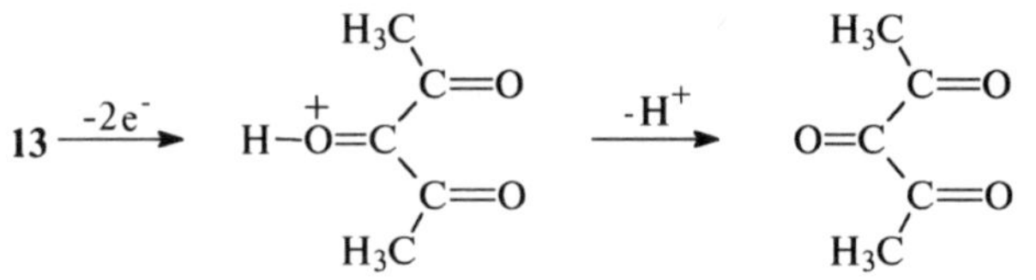

$$\text{(18.54)}$$

14   **15a**

A detailed study [88] showed the signal at 1.95 ppm in the spectra of the reaction mixture to belong to the dihydrate of the triketone **15a**, i.e. tetrahydroxy-3-pentanone **15b** (Equation (18.55)).

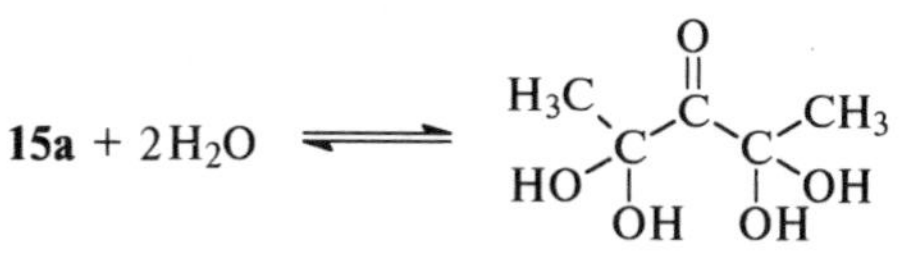

$$15a + 2H_2O \rightleftharpoons \qquad \text{(18.55)}$$

15b

The independently synthesized **15b** was found to form a complex with Mo(VI), which gave AcOH and CO_2 upon treatment with H_2O_2, presumably via the Baeyer–Villiger mechanism. Coordinated peroxo groups are to accept the electrons that are removed from the acac$^-$ anion and related species in reactions (18.51)–(18.53). Transfer of an electron from the 1,3-diketonate to O_2^{2-} would necessarily lead to the appearance of an electron in the antibonding σ* orbital of the O—O^{2-} fragment and consequently to decomposition of this fragment. In place of it, double-bonded O^{2-} and a coordinated O$^-$· or ·OH radical, arising from protonation of O$^-$, must appear in the metal complex. The lack of any evidence for free radicals taking part in the reaction suggests that it involves two- rather than one-electron transfer between the coordinated ligands (Equation (18.56)).

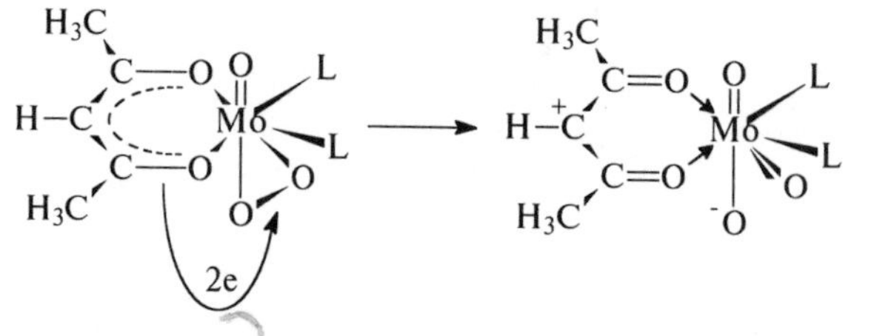

$$(18.56)$$

Such a redox reaction can occur as a one-step process or as two successive acts with the second one being faster than the first. The whole reaction can be depicted as shown in Scheme 18.8.

The tendency of Mo and W atoms to form 'oxometal' groups, M=O, must favor such a conversion. According to the mechanistic scheme under discussion, the metal atom is assumed to play the role of a mediator transferring the electrons from the coordinated ligand under oxidation to the coordinated oxidant, peroxo group by using its vacant d-orbitals.

Oxidation of the acetylacetonato ligand was also observed in the system $TiO(acac)_2/H_2O_2$ in $CDCl_3$ solution at 25 °C [92]. Hence, titanium(IV), a metal ion with d^0 con-

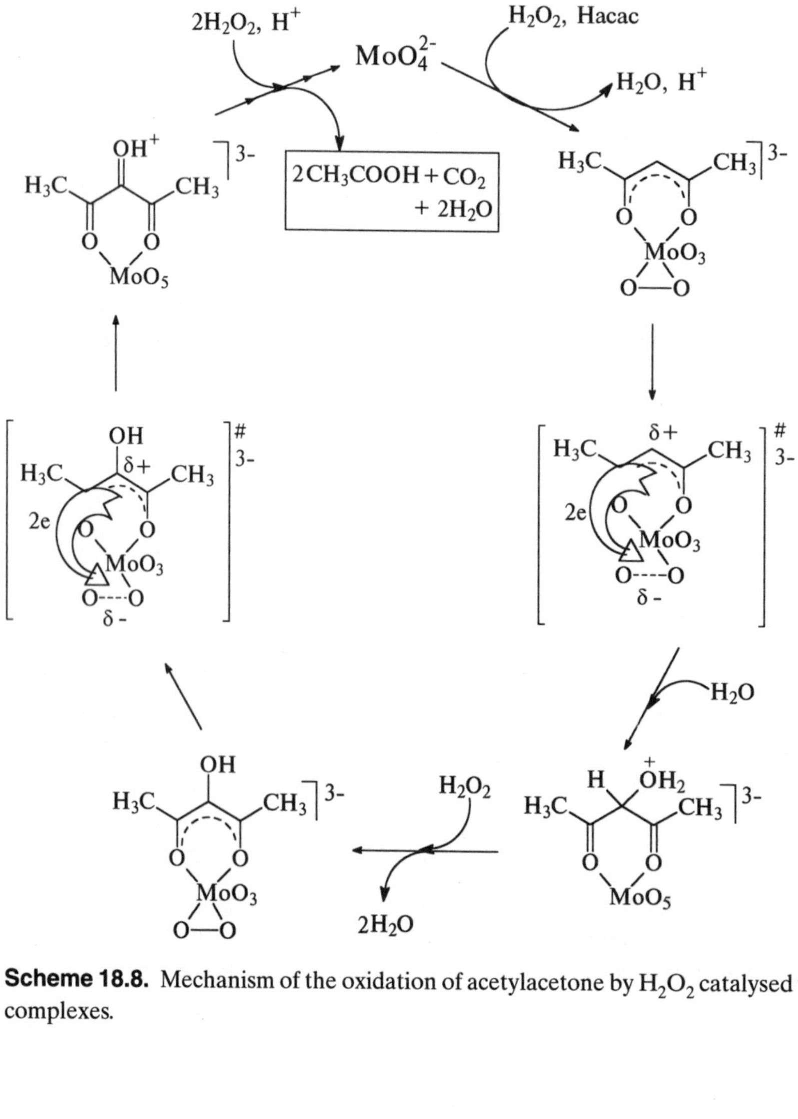

Scheme 18.8. Mechanism of the oxidation of acetylacetone by H_2O_2 catalysed with Mo(VI) complexes.

figuration, can serve as an electron transfer mediator like molybdenum(VI) and tungsten(VI) metal ions.

6 Concluding remarks

Hydrogen peroxide and organic peroxides have been widely used in organic synthesis since the discovery of Fenton chemistry. Being tremendously enlarged, the approach based on the oxidation with peroxides encompasses different reactions catalysed by metal complexes, such as epoxidation reactions after Halcon/ARCO processes, Milas dihydroxylation, Baeyer–Villiger oxidation and many other reactions of practical interest.

The diversity of reaction pathways in an H_2O_2/metal complex/solvent system may be attributed to the different modes of peroxide molecule coordination as well as to the variety of peroxide intermediates. Intermediate complexes containing such ligands as dioxo dianion O_2^{2-}, a singlet dioxygen molecule, trioxo dianion O_3^{2-} or an ozone molecule should be taken into consideration in parallel with free oxyl or peroxyl radicals.

A metal atom is able to stabilize a leaving group in reactions involving one oxygen atom transfer from a peroxidic oxidant to a substrate molecule. Mixing of HOMO's substrate and LUMO's oxidant orbitals with vacant d-orbitals on the central metal atom is believed to be a factor in reactions involving inner-sphere electron transfer. In such redox reactions the metal atom serves as an electron mediator and its oxidation state is not changed in the reaction.

7 Acknowledgments

Grants from INTAS (no. 94/1515), the Russian Foundation for Basic Research (no. 96-03-34101a) and the Russian Government (SS-5 no. 97-2-15/2) are gratefully acknowledged. Helpful comments and suggestions by Professor Roger A. Sheldon of Delft TU are greatly appreciated.

8 References

1 Sheldon RA, Kochi JK. *Metal-Catalysed Oxidations of Organic Compounds*, Chapt. 9. New York: Academic Press, 1981: 424.

2 Sheldon RA. In: Ugo R, ed. *Aspects of Homogeneous Catalysis*, Vol. 4. Dordrecht: Reidel, 1981: 1.

3 Tolstikov GA. *Reaktsii gidroperekisnogo okisleniya* [*Reactions of Hydroperoxide Oxidation*]. Moscow: Nauka, 1976 (in Russian).

4 Sheldon RA. In: Herrmann WA, ed. *Topics in Current Chemistry*, Vol. 164. Berlin: Springer-Verlag, 1993: 21.

5 Franz G, Sheldon RA. In: *Ullmann's Encyclopedia of Industrial Chemistry*, Vol. A18. Weinheim: VCH, 1991: 261.

6 Strukul G, ed. *Catalytic Oxidations with Hydrogen Peroxide as Oxidant*. Dordrecht: Kluwer Academic Publishers, 1992.

7 Weigert W, ed. *Wasserstoffperoxid und seine Derivate. Chemie und Anwendung*. Heidelberg: Hüthig, 1978: 127.

8 Drago RS. *Coord Chem Rev* 1992; **117**: 185.

9 Sheldon RA. In: Sheldon RA, van Santen RA, eds. *Catalytic Oxidation. Principles, Applications*. Singapore: World Scientific, 1995: 1, 151, 175, 234.

10 Dear K. In: Herrmann WA, ed. *Topics in Current Chemistry*, Vol. 164. Berlin: Springer-Verlag, 1993: 115.

11 Moiseev II, Shishkin DI, Gekhman AE. *New J Chem* 1989; **13**: 683.

12 Moiseeva NI, Gekhman AE, Moiseev II. *Gazz Chim Ital* 1992; **122**: 187.

13 Gekhman AE, Moiseeva NI, Moiseev II. *Russ Chem Bull* 1995; **44**: 584.

14 Murahashi SI, Naota T. *Zh Org Khim* 1996; **32**: 223 (in Russian).

15 Murahashi SI, Naota T, Miyaguchi N, Noda S. *J Am Chem Soc* 1996; **118**: 2509.

16 Murahashi SI, Naota T, Oda Y, Hirai N. *Synlett* 1995: 733.

17 Murahashi SI, Naota T. *Stud Org Chem* 1998; **33**: 283.

18 Shestakov AF, Shilov AE. *J Mol Catal* 1996; **A105**: 1.

19 Moiseev II. *J Mol Catal* 1997; **A127**: 1.

20 Moiseeva NI, Gekhman AE, Moiseev II. *J Mol Catal* 1997; **A117**: 39.

21 Roberts JL, Morrison MM, Sawyer DT. *J Am Chem Soc* 1978; **100**: 329.

22 Seliger HH. *Anal Biochem* 1960; **1**: 60.

23 Frimer AA. In: Patai S, ed. *The Chemistry of Functional Groups. Peroxides.* Chichester: Wiley–Interscience, 1983: 202.

24 Connick RE. *J Am Chem Soc* 1947; **69**: 1509.

25 Cahill AE, Taube H. *J Am Chem Soc* 1952; **74**: 2312.

26 Murray RW. In: Wasserman HH, Murray RW, eds. *Singlet Oxygen.* New York: Academic Press, 1979: 59.

27 Plesnicar B. In: Patai S, ed. *The Chemistry of Functional Groups. Peroxides.* Chichester: Wiley–Interscience, 1983: 522.

28 Plesnicar B. In: Patai S, ed. *The Chemistry of Functional Groups. Peroxides.* Chichester: Wiley–Interscience, 1983: 483.

29 Hamann HJ, Dahlmann J, Höft E. *Oxid Commun* 1980; **1**: 183.

30 Makarov AP, Gekhman AE, Nekipelov VM *et al. Bull Acad Sci USSR Div Chem Sci* 1985; **34**: 544 (English Translation).

31 Gekhman AE, Moiseeva NI, Moiseev II. *Dokl Phys Chem* 1996; **349**: 165 [Translation from *Dokl Acad Nauk* 1996; **349**: 53].

32 Makarov AP, Gekhman AE, Nekipelov VM, Talsi EP *et al. Bull Acad Sci USSR Div Chem Sci* 1985; **34**: 1764 (English Translation).

33 Aubry JM. *J Am Chem Soc* 1985; **107**: 5844.

34 Aubry JM, Cazin B. *Inorg Chem* 1988; **27**: 2013.

35 Boeme K, Brauer H-D. *Inorg Chem* 1992; **31**: 3468.

36 Aubry JM, Cazin B, Duprat I. *J Org Chem* 1989; **54**: 726.

37 Thompson KM, Griffith WP, Spiro M. *J Chem Soc Faraday Trans* 1994; **90**: 1105.

38 Nardello V, Marko J, Vermeersch G, Aubri JM. *Inorg Chem* 1995; **34**: 4950.

39 Muzart J. *Chem Rev* 1992; **92**: 113.

40 Ait Ajjou AN, Ait-Mohand S, Richard C, Sabo-Etienne S, Muzart J. *New J Chem* 1996; **20**: 571.

41 Kalia OL, Luk'yanets EA. In: *Abstracts of the 6th International Symposium on the Activation of Dioxygen, Homogeneous Catalytic Oxidation*, 4–19 April 1996, Noordwijkerhout, The Netherlands, p. 48.

42 Barkanova SV, Kalia OL, Luk'yanets EA. In: *Abstracts of the 6th International Symposium on the Activation of Dioxygen, Homogeneous Catalytic Oxidation*, 4–19 April 1996, Noordwijkerhout, The Netherlands, p. 150.

43 Sobkowiak A, Tung H-C, Sawyer DT. In: Lippard SJ, ed. *Prog Inorg Chem* 1992; **40**: 292.

44 Howarth OW, Rickards RE. *J Chem Soc Dalton Trans* 1979: 1388.

45 Gekhman AE, Moiseeva NI, Blumberg EA, Moiseev II. *Izv Acad Nauk SSSR Ser Khim* 1985: 2653 (in Russian).

46 Herrmann WA, Correia JDG, Kühn FE, Artus GRJ, Romro CC. *Chem Eur J* 1996; **2**: 168.

47 Herrmann WA. *J Organomet Chem* 1995; **500**: 149.

48 Crandall JK, Zucco M, Kirsch RS, Coppert DM. *Tetrahedron Lett* 1991; **32**: 5441.

49 Adam W, Shimizu M. *Synthesis* 1994; **6**: 560.

50 Abu-Omar MM, Hansen PJ, Espenson JH. *J Am Chem Soc* 1996; **118**: 1966.

51 Conte V, Di Furia F, Modena G. In: Ando W, ed. *Organic Peroxides.* Chichester: Wiley, 1992: 559.

52 Drago RS. *Coord Chem Rev* 1992; **117**: 185.

53 Mimoun H. In: Patai S, ed. *The Chemistry of Functional Groups. Peroxides*. Chichester: Wiley–Interscience, 1983: 463.

54 Mimoun H. *Pure Appl Chem* 1981; **53**: 2389.

55 Mimoun H. *Angew Chem Int Ed Engl* 1982; **21**: 734.

56 Mimoun H, Saussine L, Daire E *et al. J Am Chem Soc* 1983; **105**: 3101.

57 Di Furia F, Modena G. *Pure Appl Chem* 1982; **54**: 1853.

58 Di Furia F, Modena G. *Rev Chem Intermed* 1985; **6**: 51.

59 Andersson M, Conte V, Di Furia F, Moro S. *Tetrahedron Lett* 1995; **36**: 2675.

60 Conte V, Di Furia F, Moro S. *Tetrahedron Lett* 1994; **35**: 7429.

61 Milas N. *J Am Chem Soc* 1937; **59**: 2342.

62 Milas N, Sussman S. *J Am Chem Soc* 1936; **58**: 1302; 1937; **59**: 2345.

63 Eisenbraun EJ, Bader AR, Polacheck JW, Reif E. *J Org Chem* 1963; **28**: 2057.

64 Adam W. In: Weigert W, ed. *Wasserstoffperoxid und seine Derivate. Chemie und Anwendung*. Heidelberg: Hüthig, 1978: 127.

65 Gekhman AE, Moiseeva NI, Minin VV, Larin GM, Miseev II. *Mendeleev Commun* 1997; **No 6**: 221.

66 Guiraud HJ, Foote CS. *J Am Chem Soc* 1976; **98**: 1984.

67 Darmanyan AP, Foote CS, Jardon P. *J Phys Chem* 1995; **99**: 11854.

68 Samuni A, Czapski G. *Isr J Chem* 1969; **91**: 4673.

69 Setaka M, Kirino Y, Ozawa T, Kwan T. *J Catal* 1969; **15**: 209.

70 Berdnikov VM, Schastnev PV. *Kinet Kat* 1975; **16**: 83 (in Russian).

71 Larin GM, Zvereva GA, Minin VV, Rakitin YV. *Zh Neorg Khim* 1988; **33**: 2011 (in Russian).

72 Ivanov VB, Slyapintoch VYa, Khvostach OM, Shapiro AB, Rozantsev EG. *J Photochem* 1976; **4**: 313.

73 Wertz JE, Bolton JR. *Electron Spin Resonance. Elementary Theory, Practical Applications*. New York: McGraw-Hill, 1972.

74 Denisov ET. *Konstanty skorosti gomoliticheskikh zhidkofaznykh reaktsyi* [*Rate Constants of Homolytic Liquid-phase Reactions*]. Moscow: Nauka, 1971: 237 (in Russian).

75 Razumovskii SD, Zaikov GE. *Ozone, Its Reactions with Organic Compounds*. New York: Elsevier, 1984.

76 Horvath M, Bilitzky L, Huffner J. *Ozone*. New York: Elsevier, 1985.

77 Castro CE. *J Am Chem Soc* 1996; **118**: 3984, and references therein.

78 Melikoff P, Pissarjevsky L. *Z Anorg Chem* 1898; **18**: 59.

79 Russel GA. *J Am Chem Soc* 1957; **79**: 3871.

80 Howard JA. *Adv Free-Rad Chem* 1972; **4**: 49.

81 Faber M. *Patent BRD 3512439u*, 1986; *Chem Abstr* 1986; **105**: 229262w.

82 Levush SS, Garbuzyuk IA. *Dokl Akad Nauk Ukr SSR B* 1984: 37 (in Russian).

83 Cerkovnik J, Plesnicar B. *J Am Chem Soc* 1993; **115**: 12169.

84 Plesnicar B, Cerkovnik J, Koller J, Kovac F. *J Am Chem Soc* 1991; **113**: 4946.

85 Thompson QE. *J Am Chem Soc* 1961; **83**: 845.

86 Plesnicar B. In: Ando W, ed. *Organic Peroxides*. Chichester: Wiley, 1992: 479.

87 Moiseev II, Bochkareva VA, Gekhman AE, Kokunov YV, Buslaev YA. *Dokl Acad Nauk SSSR* 1977; **233**: 375 (in Russian).

88 Gekhman AE, Nekipelov VM, Sahkarov SG, Zamaraev KI, Moiseev II. *Izv Akad Nauk SSSR Ser Khim* 1986: 1242 (in Russian).

89 Buslaev YA, Kokunov YV, Bochkareva VA, Schustorovich EM. *Zh Neorg Khim* 1972; **17**: 3184 (in Russian).

90 Diversi P, Forte C, Franceschi M *et al. J Chem Soc Chem Commun* 1992: 1345.

91 Collman JP, Moss RA, Maltz H, Heindel CC. *J Am Chem Soc* 1961; **83**: 531.

92 Talsi EP, Babushkin DE. *J Mol Catal* 1996; **A106**: 179.

19 Organorhenium(VII) Oxides in Catalysis

WOLFGANG A. HERRMANN and FRITZ E. KÜHN

Anorganisch-chemisches Institut der Technischen Universität München, Lichtenbergstraße 4, D-85747 Garching b. München, Germany

1 Introduction

The entire class of organorhenium oxides is younger than 25 years. For a considerable part of this time these complexes have been widely regarded as curiosities, but the picture has changed drastically during recent years. Today, not only are an amazing number of these complexes known and easy to synthesize, but they also have very interesting applications in both catalysis and material sciences. Especially in the catalytic field, the use of organorhenium(VII) oxides is in rapid progress and therefore prototypal for organometal oxides in general.

2 Synthesis of organorhenium(VII) oxides

The first organorhenium(VII) oxides, namely $(CH_3)_3ReO_2$ and $((CH_3)_3CCH_2)_3ReO_2$, were prepared in 1975 by Mertis and Wilkinson [1]. In 1979 Beattie and Jones reported the preparation of CH_3ReO_3 (MTO) [2]. A few other complexes were published during the following years but the breakthrough for catalytical applications came in 1987 when we developed an efficient synthesis starting from dirhenium heptaoxide and tetramethyltin (Equation (19.1)). This enabled us to conduct the synthesis of MTO within a few hours and in amounts of up to 10 g [3].

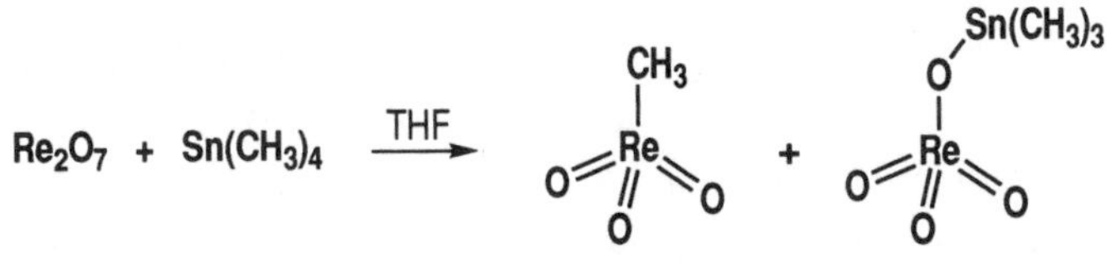

$$Re_2O_7 + Sn(CH_3)_4 \xrightarrow{THF} \quad \text{MTO, 1} \quad + \quad \tag{19.1}$$

The method was improved further in 1992, when it proved possible to reach nearly quantitative yields by the addition of carboxylic acid anhydrides to the reaction mixture and additionally by utilizing the trimethylstannyl perrhenate, formed as a by-product in the original synthesis (see Equation (19.1)) [4, 5]. Recently, the latter method could be extended for other perrhenates as starting materials so that the moisture-sensitive Re_2O_7 can be replaced by starting materials that are more convenient to handle [6]. This synthetic progress was accompanied by the discovery of a plethora of derivatives and catalytic applications of organorhenium(VII) oxides. The use of these complexes in catalysis, however, is dominated by MTO itself [7–12].

I (Re^{VII})   II (Re^{VII})   III (Re^{VII})

3   Oxidation catalysis

3.1   *Synthesis and behavior of methylrhenium(VII) peroxo complexes*

Oxidation catalysis is probably the most thoroughly examined field of application of organorhenium(VII) oxides. The catalytic activity of MTO and some of its derivatives in the oxidation of olefins was discovered soon after these complexes were accessible in higher amounts [13]. However, the breakthrough in understanding the role of organorhenium(VII) oxides of type I in oxidation catalysis was the isolation and characterization of the reaction product of MTO with H_2O_2. According to Equation (19.2) a bisperoxo complex of stoichiometry $(CH_3)Re(O_2)_2O$ (3) is formed.

$$\tag{19.2}$$

1   2   3

In the solid state it is isolated as an adduct with a donor ligand L (L = H_2O, **3a**; L = $O=P(N(CH_3)_2)_3$, **3b**) [14,15], which is lost in the gas phase. The structures of compounds **3** (electron diffraction) [8], **3a** and **3b** (X-ray diffraction) were determined; the structure of the ligand-free complex (**3**) is known from the gas phase. Complex **3** has a trigonal bipyramidal structure, with significant shorter Re—O and Re—C bond distances than in compounds **3a** and **3b**. This is very likely due to the increased Lewis acidity of the free complex **3**. If each peroxo group is considered as a coordinative unit, the molecular structure of complexes **3a** and **3b** can also be described as trigonal bipyramidal. The adduct **3a** melts at 56 °C and can be sublimed at room temperature in an oil pump vacuum, whereas adduct **3b** melts at 65 °C and decomposes at *c* 75 °C. Both are explosive.

Experiments using ^{17}O-NMR showed that adducts **3a** and **3b** exchange their ligand L rapidly in solution [8,9,14] and are in equilibrium with the uncoordinated form (**3**). The terminal oxygen of adduct **3a** is observed at $\delta(^{17}O) = 762$ ppm and the peroxo groups at $\delta(^{17}O) = 422$ and 362 ppm, respectively; in adduct **3b** the terminal oxygen is observed at $\delta(^{17}O) = 774$ ppm. Only the terminal oxygen atom is involved in an oxygen exchange with water [14,15].

Experiments with the isolated bis(peroxo)complex (**3a**) have shown that it is an active species in oxidation catalysis, e.g. in the oxidation of olefins [15,16]. *In situ* experiments show that the reaction of MTO with one equivalent of H_2O_2 leads to a monoperoxo complex (**2**) according to Equation (19.2). Complex **2** is also catalytically active in certain oxidation processes, as discussed below. The decomposition of complexes **2** and **3** in solution was also examined [17]. In the presence of H_2O_2 they decompose to methanol and perrhenate at a rate that is dependent on $[H_2O_2]$ and $[H_3O]^+$. The complex peroxide and pH dependences are explained by two possible pathways: attack of either hydroxide on complex **2**, or HO_2^- on MTO. The bisperoxo complex (**3**) decomposes much more slowly to yield O_2 and MTO [17]. Mechanistic studies in the gas phase showed that high-valent rhenium compounds (e.g. derivative **3**) and high-valent chromium compounds are oxygen-transfer reagents, whilst iron derivatives usually catalyse the decomposition of H_2O_2 [18]. The activation parameters for the coordination of H_2O_2 to MTO have also been determined. They indicate a mechanism involving nucleophilic attack. The protons lost in converting H_2O_2 to a coordinated O_2^{2-} ligand are transferred to one oxide oxygen, which remains on the ligand as well as the aqua ligand L. The rate of this reaction is not pH dependent [19]. The photochemical and photophysical behaviors of complex **3** have also been closely examined [20].

The amazing versatility of MTO as an oxidation catalyst precursor is demonstrated by the fact that complexes **2** and **3** perform catalytic oxidation of alkenes [13–17, 21–32] and aromatic compounds [33–40]; Baeyer–Villiger oxidation and Dakins reaction [41–43]; oxidation of sulfur compounds [44–48], phosphines, arsines and stilbines [49], amines and other organonitrogen compounds [50–55] and halide ions [56,57]; oxygen insertion into C–H bonds [58]; and oxidation of alkynes [59] and certain metal carbonyl complexes [60,61].

3.2 Oxidation of alkenes

One of the most intensively examined catalytic processes using organorhenium(VII) oxides as catalysts is olefin epoxidation [13–17, 21–32]. Usually <85 wt.% of H_2O_2 is used and MTO is typically used in concentrations of 0.2–0.4 mol.%. Turnover numbers up to 2000 (mol mol^{-1} catalyst) and turnover frequencies of c 1200 (mol mol^{-1} catalyst h^{-1}) can be achieved.

The catalytic MTO/H_2O_2 system is active even at low temperature, e.g. at –30 °C. The reactions between complex **3** and alkenes are about one order of magnitude faster in semi-aqueous solvents (e.g. 85% H_2O_2) than in methanol. The rate constants for the reaction of complex **3** with aliphatic alkenes correlate closely with the number of alkyl groups on the olefinic carbons. The reactions become significantly slower when electron-attracting groups, such as –OH, –CO, –Cl and –CN, are present. Interestingly, the MTO/H_2O_2 system also has been employed successfully for the oxidation of [60]fullerene to 1,2-epoxy[60]fullerene [26].

Based on Equation (19.2), two catalytic pathways may be described corresponding to the concentration of H_2O_2 used. If 85% H_2O_2 is used, the described equilibrium is on the right-hand side of Equation (19.2) and only compound **3** seems to be responsible for the epoxidation activity (Scheme 19.1).

When a solution with 30 wt.% or less H_2O_2 is used, the monoperoxo complex **2** is also responsible for the epoxidation process and a second catalytic cycle is involved, as shown in Scheme 19.2.

For both cycles a concerted mechanism is suggested in which the electron-rich double bond of the alkene attacks a peroxidic oxygen of compound **3**, according to Schemes 19.1 and 19.2. It has been deduced from experimental data that the system might show a spiro arrangement [13,16,21,26]. The selectivity towards epoxides can be enhanced by the

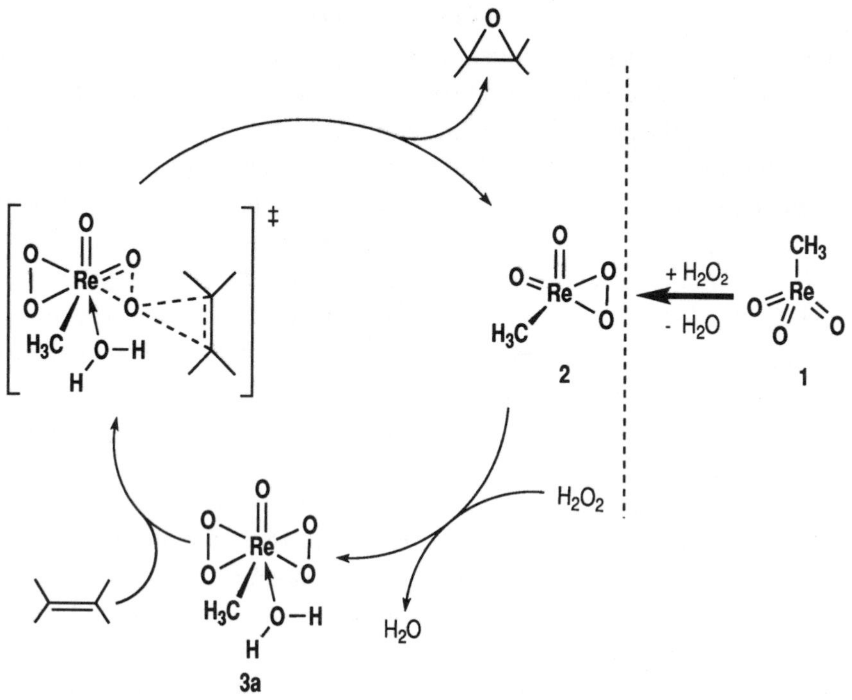

Scheme 19.1.

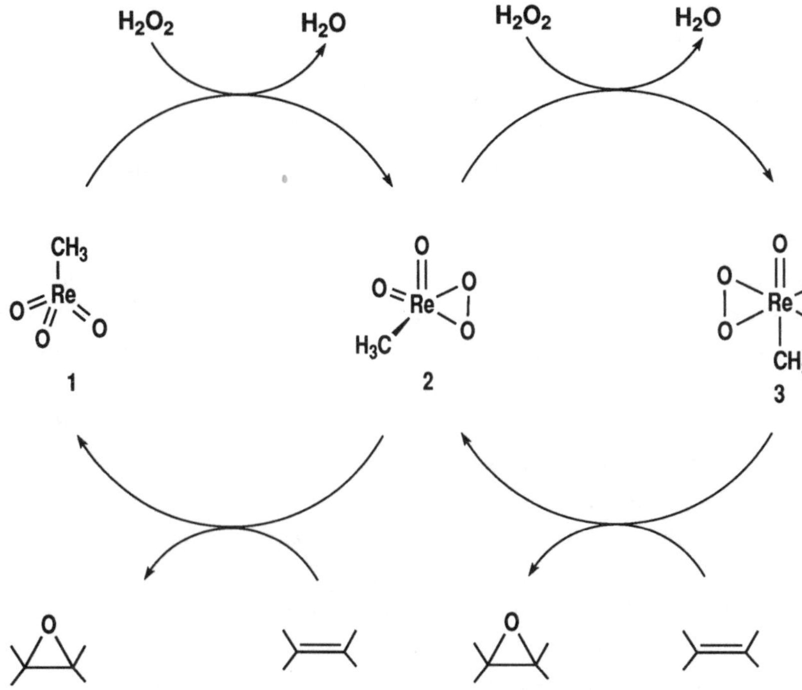

Scheme 19.2.

addition of Lewis bases such as quinuclidine, pyridine or 2,2′-bipyridine to the system [16, 28–31]. Lewis acids catalyse ring-opening reactions and diol formation. These reactions are suppressed by the addition of Lewis bases. It has been shown also that the selectivity towards epoxides is at least partially dependent on the pK_b values of the Lewis bases used.

The lower the pK_b values, the higher the selectivity [28]. There is recent evidence that an excess of certain N-base ligands, especially under two-phase conditions, leads to accelerated reactions [28,30–32]. Several functionalized epoxides can be prepared in high yields by this method [31]. Another possibility for enhancing the selectivity towards epoxides is the use of the urea–H_2O_2 adduct, which enables oxidation to be carried out in water-free solutions, thus avoiding the formation of any diols and other side reactions. In the case of the oxidation of chiral allyl alcohols, high diastereoselectivities have been achieved [23,26]. Lewis base adducts of some derivatives of MTO, e.g. (cyclopropyl)trioxorhenium(VII), have shown even higher activity in the olefin epoxidation than MTO itself [16]. However, the more difficult synthesis and, especially, the reduced stability of these complexes have prevented widespread use up to now.

3.3 Oxidation of aromatic compounds

The mechanistic aspects of the oxidation of aromatics have not been examined in great detail so far. It seems clear, however, that complex 3 plays an important role as a catalyst [33–40]. Noteworthy is the high regioselectivity, most notably in the industrially interesting synthesis of vitamin K_3 (see Equation (19.3)).

$$\tag{19.3}$$

Because water is an inhibitor, concentrated (85 wt.%) H_2O_2 is preferred. Alternatively, commercially available 35% H_2O_2 in acetic anhydride can be employed; a considerable regioselectivity is obtained with this system. The conversion is higher for electron-rich arenes (nearly 100%) and selectivities of >85% have been reached [33]. Hydroxy-substituted arenes can be oxidized by aqueous H_2O_2 (85 wt.%) in acetic acid, to afford the corresponding p-quinonones in isolated yields of up to 80% [36]. Recently it has been shown that using a mixture of acetic acid and acetic anhydride further improves the product yield [40]. Anisole was also found to undergo selective oxidation with the MTO/H_2O_2 system to yield o- and p-methoxyphenols. There is no need to use a solvent in this case. It is assumed that the active species in this system might be a highly electrophilic species [34].

3.4 Baeyer–Villiger oxidation and Dakins reaction

It has been shown that compound 3 also acts as an active species in the Baeyer–Villiger oxidation of ketones (Equation (19.4)) and in the Dakins reaction [41–43].

$$\tag{19.4}$$

It is slightly surprising that the MTO/H_2O_2 system exhibits this activity because these oxidations involve nucleophilic attack at the carbonyl group, which contrasts with all the preceding examples where the substrates attacked the electrophilic Re–peroxo complexes, e.g. in the olefin epoxidation. Nevertheless, compound **3** reacts stoichiometrically with cyclobutanone in the absence of H_2O_2 as shown in Equation (19.4). This reversed behavior may be due to substrate binding to rhenium.

The reaction was found to be strongly solvent dependent by means of the mechanistic probe for oxygen-transfer reactions, which is thiantrene-5-oxide [41,42]. Donor solvents such as acetonitrile seem to enhance the nucleophilicity of the peroxo groups. It has been suggested that the double bond of the enol form (the major tautomer) attacks a peroxo oxygen of peroxorhenium complex **2** or **3**. This reaction affords a 2-hydroxy-1,3-dicarbonyl intermediate that can be detected by ^{1}H-NMR. This hydroxy intermediate is susceptible to cleavage via Baeyer–Villiger oxidation to yield carboxylic acids as final products. Low H_2O_2 concentrations are sufficient and no H_2O_2 decomposition is observed at temperatures up to 70 °C. This is an advantage of the catalytic MTO/H_2O_2 system over the known transition metal catalysts containing V, Mo, Mn or Os. However, the nucleophilic character of the peroxidic atoms in complex **3** is not as pronounced as in Pt or Ir peroxo complexes that react with CO_2 or SO_2 to give isolable cycloaddition products [41,42]. In the case of complex **3**, turnover frequencies of 18 000 (mol mol^{-1} catalyst h^{-1}) are obtained for cyclobutanone, whereas in other cases turnover numbers up to 100 are usual [43]. Cycloketones can be converted into lactones even below room temperature (15 °C) by diluted H_2O_2 (10 wt.%).

3.5   Oxidation of sulfur compounds

Organic sulfides can be oxidized to the corresponding sulfoxides by hydrogen peroxide in the presence of MTO (Equation (19.5)) [44]. Both complexes **2** and **3** seem to be active in this reaction but kinetic results indicate that complex **2** might be more active than complex **3**. The kinetic results suggest to a mechanism that involves the nucleophilic attack of the sulfur atom on a coordinated peroxide oxygen, because electron-donating substituents have accelerating effects. The first reports on the scope and selectivity of the reaction [45] have been re-examined recently. Using ethanol as solvent, the MTO/H_2O_2 system can be used to oxidize dialkyl, diaryl and alkyl aryl sulfides to sulfoxides ($R_2S/H_2O_2 = 1:1.1$) or sulfones ($R_2S/H_2O_2 = 1:2.2$) with excellent yields and selectivities even in the presence of oxidatively sensitive functions on the sulfide side-chain [44–48]. Functional groups in the side-chain of the sulfide, e.g. carbon double bonds, are not affected under the reaction conditions: the sulfur atom is selectively oxidized [44].

$$R-S-R \xrightarrow{MTO/H_2O_2} R-\overset{O}{\underset{}{S}}-R \xrightarrow{MTO/H_2O_2} O=\overset{O}{\underset{R}{S}}\cdots R \qquad (19.5)$$

This mild reactivity has also been demonstrated in the oxidation of thioether Fischer carbene complexes [46]. Oxidation of thiophene and its derivatives has been achieved with this system. However, the rate constants are two to four orders of magnitude below those reported for the 'aliphatic' sulfides where the S atom is not part of an aromatic heterocyclic ring [47]. Oxidation studies of coordinated thiolates were carried out on the model Co(III) complex $[(en)_2Co(SCH_2CH_2NH_2)]^{2+}$. The thiolato complex is first oxidized to a sulfenato complex $[(en)_2Co(S(O)CH_2CH_2NH_2)]^{2+}$, which then is oxidized more slowly to the sulfinato complex $[(en)_2Co(S(O)_2CH_2CH_2NH_2)]^{2+}$. The second step is c 1500 times slower than the first. From these kinetic examinations, complex **2** seems to be the catalytically active species that is attacked by the nucleophilic S atom [48].

3.6 Oxidation of phosphines, arsines and stilbines

Tertiary phosphines, triaryl arsines and triarylstilbines are converted to their oxides R_3EO (E = P, As, Sb) by MTO/H_2O_2. Kinetic studies lead to the assumption that complexes **2** and **3** have similar catalytic activities in all cases. The kinetic data support a mechanism involving nucleophilic attack of the substrate at the rhenium peroxides. The proposed catalytic cycle is given in Scheme 19.3 [49].

3.7 Oxidation of amines

The MTO/H_2O_2 system also catalyses the oxidation of anilines (Equation (19.6)). The major product of the oxidation of aniline is nitrosobenzene. For 4-substituted N,N-dimethylanilines, the N-oxide is the only oxidation product. Electron-withdrawing substituents inhibit the reaction.

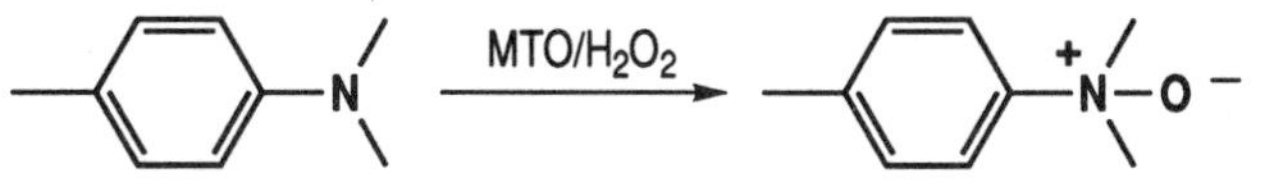

(19.6)

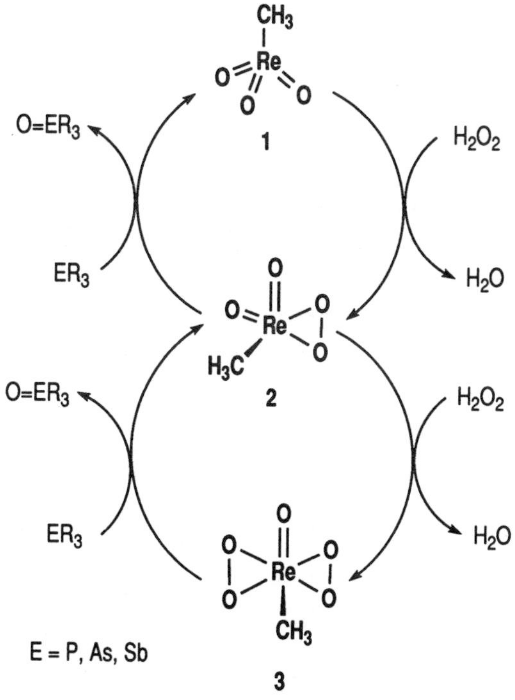

Scheme 19.3.

Kinetic results suggest that both compounds **2** and **3** are involved in the oxidation process [50]. It is suggested that the rate-determining step is the nucleophilic attack of the nitrogen lone-pair electrons of the anilines on a peroxidic oxygen of the catalyst. Electron-donating groups attached to the nitrogen atom of aniline increase the rate constant. In the case of $ArNH_2$ derivatives, the oxidation proceeds c 50 times faster than without catalyst [50]. In general, the reactions are facile and high yielding at or below room temperature [51]. Furthermore, not only anilines but a broad variety of other aromatic and aliphatic amines are oxidized to the corresponding *N*-oxides [52–55]. The oxidation of secondary amines affords nitrones in very good yields. Both H_2O_2 and the urea–H_2O_2 complex can be used together with MTO. Benzylamines are selectively oxidized to oximes.

It is noteworthy in this context that the *N*-oxides also form adducts with MTO [29]. While aliphatic *N*-oxides form temperature-sensitive adducts with MTO and are inactive in the catalytic epoxidation of olefins, aromatic *N*-oxide adducts of MTO are both stable and catalytically active. In some cases their selectivity in olefin epoxidation is even higher than with N-base adducts of MTO (see above).

3.8   Oxidation of halide ions

Another application of the MTO/H_2O_2 system is the catalytic oxidation of chloride and bromide ions in acidic aqueous solutions. The chloride oxidation steps are three to four orders of magnitude slower than the corresponding bromine oxidation steps. Both compounds **2** and **3** have been shown to be active catalysts in this process. In both cases the catalysed reactions were c 10^5 times faster than the uncatalysed reactions under similar conditions. In a first step HOX is formed and then HOX reacts with X⁻ to form X_2. When H_2O_2 is used in excess, the reaction yields O_2 [56, 57].

3.9   Oxidation of C—H bonds

The MTO/H_2O_2 system also catalyses the insertion of oxygen into a variety of activated and unactivated C—H bonds with yields varying from good to excellent. Alcohols or ketones are formed, as shown in Equations (19.7) and (19.8). Suitable substrates proved that the reaction is stereospecific with retention of configuration. Alcohols, e.g. ethanol and *t*-butanol, are used as solvents, the reaction temperatures range between 40 and 60 °C and nearly quantitative yields have been obtained. However, the reaction times are generally longer than those used for most epoxidations, ranging between 10 and 72 h [58].

(19.7)

(19.8)

3.10 *Oxidation of alkynes*

Internal alkynes yield carboxylic acids and α-diketones when oxidized with the MTO/H_2O_2 system. Rearrangement products were observed only for aliphatic alkynes. Terminal alkynes give carboxylic acids, derivatives thereof and α-keto acids as the major products. The yields of these products vary with the solvent used [59].

3.11 *Oxidation of metal carbonyls*

The organorhenium(VII) oxide MTO and some of its derivatives catalyse the oxidation of metal carbonyls to metal oxides with H_2O_2 [60, 61]. This reaction runs at room temperature and yields of up to 90% are obtained. However, only organometal carbonyls with oxidation-resistant organic groups can be oxidized, e.g. (pentamethylcyclopentadienyl) tricarbonyl rhenium(I) (Equation (19.9)) [61]. In all other cases the organic ligand is also oxidized, leading to decomposition of the product complex.

$$
\begin{array}{ccc}
\text{(Cp}^*\text{Re}^{I}(CO)_3) & \xrightarrow[\text{[MTO]}]{H_2O_2/Et_2O} & \text{(Cp}^*\text{Re}^{VII}O_3) \quad (+ \; CO/CO_2)
\end{array}
\tag{19.9}
$$

$R^{1,2}$ = alkyl

4 Aldehyde olefination and related reactions

Aldehydes or strained cycloketones, treated with aliphatic diazoalkanes in the presence of an equimolar amount of a tertiary phosphine and MTO as catalyst, afford an olefinic coupling product in good yields at room temperature according to Equation (19.10) [62].

$$
R^1R^2C=N_2 + O=CHR^3 + P(C_6H_5)_3 \xrightarrow{\text{cat.}} R^1R^2C=CHR^3 + N_2 + O=P(C_6H_5)_3
\tag{19.10}
$$

The *trans* selectivity is between 60 and 95%, depending on the substrate, and the yields are *c* 85%. The *trans* selectivity is higher when lower catalyst concentrations are used, but with catalyst concentrations below 5 mol.% significant amounts of azines (RCH=N–N=CHR') are formed as undesired byproducts. However, the advantage of this method over Tebbe–Grubbs coupling is that it does not require the use of a stoichiometric amount of an organometallic coupling reagent. It has been shown by *in situ* NMR spectroscopy, and by isolation and crystallization of the reaction product of MTO with tertiary phosphines, that Re(V) species act as catalysts in this process. Therefore, ligand- or solvent-stabilized methyldioxorhenium(V) seems to be the key compound in the olefination reaction cycle represented in Scheme 19.4 [62–66].

Methyldioxorhenium generated *in situ* is also able to abstract oxygen atoms successively from the notoriously sluggish oxidation agent, perchlorate [64]. These observations explain the catalytic behavior: the Re–O bond is weaker than the Ti–O bond in the case of

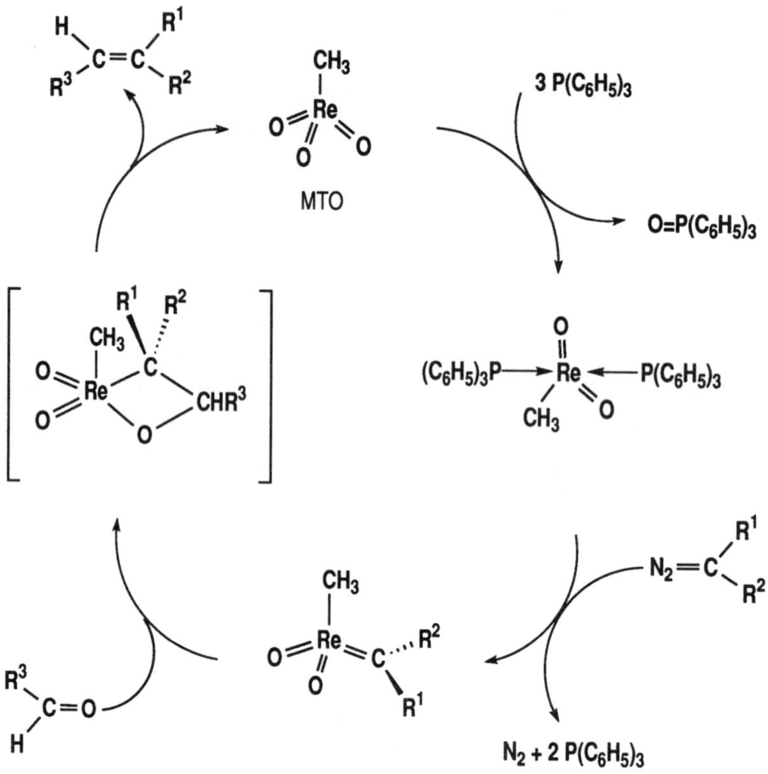

Scheme 19.4.

Tebbe–Grubbs reagents. Only the former is amenable to reductive elimination by phosphines as phosphine(V) oxides. The deoxygenation of epoxides, sulfoxides, N-oxides, triphenylarsine and triphenylstilbine oxide at room temperature is also catalysed by MTO with PPh$_3$ as oxygen acceptor. A possible reaction mechanism involves phosphine attack at a compound formed between MTO and the epoxide or other oxygen-donor compounds [63]. It is noteworthy in this context that MTO was also found to be a good catalyst for the oxidation of tertiary phosphines by molecular oxygen at room temperature. Again, a rhenium(V) intermediate containing the ligand-stabilized methyldioxorhenium seems to be involved. Methyldioxorhenium is also produced from MTO and hydrophosphorus acid in acidic aqueous medium. Oxygen donors such as sulfoxides and pyridine N-oxides oxidize methyldioxorhenium to MTO, as also do metal oxo complexes, e.g. VO^{2+}(aq) and MnO$_4^-$ [67]. The pentamethylcyclopentadienyl rhenium(V) complex, which is analogous to methyldioxorhenium, has also been described [68].

Other organorhenium(VII) oxides form less active catalysts for aldehyde olefination. On the other hand, taking into account the findings that the catalyst is an Re(V) species derived from MTO, it was found that trichlorooxo rhenium(V) and its derivatives are very active aldehyde olefination catalyst precursors, even more active than MTO [69, 70]. Olefin yields of nearly 100% are reached, and the *trans* selectivity is considerably higher than 90%. In contrast to the oxidation catalysis with organorhenium(VII) oxides, the presence of a rhenium–carbon bond seems not to be crucial for this process.

5 Olefin metathesis

The system Re_2O_7/Al_2O_3 is an effective heterogeneous catalyst for carrying out olefin metathesis under mild conditions (Equation (19.11)) and its activity can be increased further by the addition of tetraalkyl tin compounds [71–74], an observation that triggered the synthesis of MTO in our group.

$$(19.11)$$

Because tin-containing co-catalysts are essential for the metathesis of functionalized olefins [71–74] it was soon discovered that MTO and other rhenium(VII) oxides of type I supported on acidic metal oxides form metathesis catalysts that are active without additives even for functionalized olefins [75]. Standard supports are Al_2O_3–SiO_2 or Nb_2O_5 and the activity is related to the surface acidity [76, 77]. A high metathesis activity is observed when MTO is chemisorbed on the surface. No evidence for a surface carbene species was obtained, but there appears to be a correlation between the catalytic activity and the presence of an alkyl fragment on the surface [75–77]. The surface-fixed catalyst is significantly more sensitive to water than the free MTO, being rapidly decomposed to methane, ethylene and perrhenate in the presence of moisture. Free MTO can neither be removed from the surface with solvents nor sublimed out of the carrier at temperatures up to 150 °C. At higher temperatures decomposition accompanied by predominantly methane and ethylene evolution is observed [75]. The carbene species that is implied in the metathesis catalytic cycle seems not to originate from the methyl group of MTO [77]. The MTO supported on niobia has also been used for several other catalytic applications recently [78]. These include reactions of ethyl diazoacetate (equivalent to carbene transfer processes) and selective oxidations that utilize H_2O_2 (see above).

It was also possible to encapsulate MTO in zeolite, maintaining its metathesis activity. Both IR and EXAFS data indicate that the structure of MTO remains unchanged and is anchored by hydrogen bridges to the zeolite oxygens [79, 80]. There is evidence in the case of zeolite Y that the loading level of MTO corresponds to four molecules per supercage. The still intact molecules of MTO can either hydrogen bond to Brønsted acid sites or interact with extra-framework cationic sites in the supercages of $M_{56}Y$, where M = H, Na, Rb. The favored anchoring interaction is found to be $CH_3ReO_3 \cdots$ MOZ, in which the oxygen end of the bond in MTO binds preferentially to supercage cation or proton sites. The guest molecules associate through Re=O $\cdots$ Re interactions. The formation of these aggregates of MTO is induced by the high Lewis acidity and the intense electric fields associated with the anchoring cations or proton sites, respectively. Adsorption of water causes the de-aggregation of the guest molecules. Thermal treatment around 120 °C is found to yield methane and water, together with the formation of an intrazeolite cluster species containing Re—Re bonds [81].

The MTO supported on Al_2O_3–SiO_2 catalyses in particular the self-metathesis of allyl aldehydes, ethers, silanes and unsaturated carboxylates and nitriles, but also the ethenolysis of olefins with internal double bonds [82]. The catalyst system is also suitable for the metathesis of simple open-chain and cyclic olefins. Otherwise frequent side reactions such as double bond isomerization and olefin dimerization are insignificant. Ring-opening polymerization is catalysed by the homogeneous catalyst MTO/R_nAlCl_{3-n} ($R = CH_3$, C_2H_5, $n = 1, 2$). As in the case of the heterogeneous olefin metathesis, the reaction can be performed at room temperature [75]. Several functionalized diolefins cyclize to hydroazulenes via olefin metathesis in the presence of MTO. These conversions are regioselective and occur with high diastereoselectivity [83]. However, certain ruthenium-based catalysts show superior performance, especially in the building of strained ring systems and with α, β-unsaturated carbonyl groups in the neighborhood of the reaction centers [84].

6 Other catalytic applications

6.1 *Diels–Alder reaction*

There are some other catalytic applications of organorhenium(VII) oxides, mainly of MTO, that have not yet been examined in great detail. For instance, MTO enhances the Diels–Alder reactivity of unsaturated C=C compounds, the standard case of which is given in Equation (19.12) [85].

$$(19.12)$$

The organorhenium(VII) oxide MTO proves to be an efficient and effective catalyst in this reaction when the dienophile is an α,β-unsaturated ketone or aldehyde. It is especially active in water: the isolated yields are usually >90%. Kinetic studies show that the reaction rate is proportional to the catalyst concentration. The desirability of MTO as a Diels–Alder catalyst stems from a combination of favorable properties: the tolerance for many substrates, the inertness to air and oxygen, the use of aqueous medium and the absence of product inhibition. The initial step appears to be the coordination of the carbonyl oxygen to the rhenium center. Steric crowding around rhenium inhibits reactions of the larger dienophiles [86].

The use of MTO allows the combination of the Diels–Alder reaction with a subsequent epoxidation reaction. Besides MTO, several other Re(VII) oxides catalyse the Diels–Alder reaction owing to their Lewis acidity. In the case of Re_2O_7, turnover frequencies of 4000 (mol mol^{-1} catalyst h^{-1}) are reached at room temperature using 0.05 mol.% catalyst. In the specific example of Equation (19.12) the *endo* selectivity is 98% [85].

6.2 *Alkoxylations*

In the presence of MTO the catalytic alkoxylation of cyclohexene oxide with secondary and tertiary alcohols can be performed. This catalyst is known to cause disproportionation of epoxides, yielding olefins and diols. Fourier transform infrared spectroscopy indicated the formation of an active intermediate composed of MTO and an oxirane, which evolves as depicted in Scheme 19.5. The carbocationic species **4** must be regarded as highly reactive with respect to nucleophilic compounds [87].

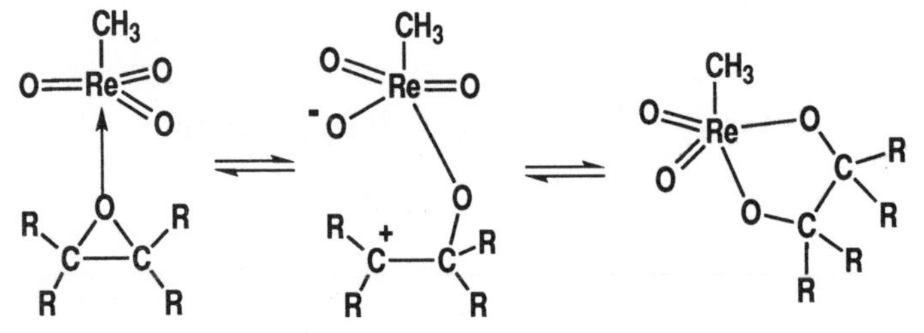

Scheme 19.5.

6.3 Conversion of amines to aziridines and related reactions

A third very recent application is the conversion of aromatic imines to aziridines by catalytic amounts of MTO with ethyl diazoacetate, as shown in Equation (19.13).

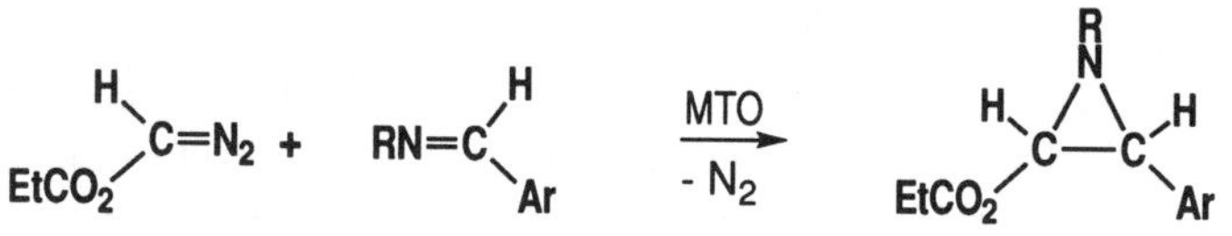

$$\text{(19.13)}$$

R = Ph, Bu, Hex

The same type of cycloadditions occur between ethyl diazoacetate and olefins or carbonyl compounds, in the presence of catalytic amounts of MTO, to produce cyclopropanes and epoxides, respectively. The mechanism of these transformations is unknown but a metal–carbene intermediate seems plausible, thus making MTO the first high-valent oxo complex for carbene transfer. The presence of an oxyethylidene-type active species **5**, analogous to species **2**, is proposed [88].

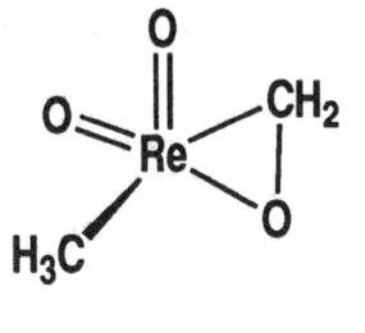

5

Under mild conditions and in the absence of other substrates, ethyl diazoacetate is converted to a 9:1 mixture of diethyl maleate and diethyl fumarate. In the presence of alcohols, α-alkoxyethyl acetates were obtained in good yield. An electron-donating group in the *para* position of phenols favors the formation of α-ethyl acetates. In the presence of an oxygen source such as an epoxide, ethyl diazoacetate and azibenzil are converted to an oxalic acid monoethyl ester and to benzil. Epoxide is converted to an olefin [89].

6.4 Dehydration, amination and disproportionation of alcohols

The Re(VII) oxide MTO has also been claimed to be the first transition metal complex to catalyse the direct solvent-independent formation of ethers from alcohols. Aromatic alcohols give better yields than aliphatic alcohols, and reactions between different alcohols have been used to prepare asymmetric ethers. Also catalysed by MTO is the dehydration

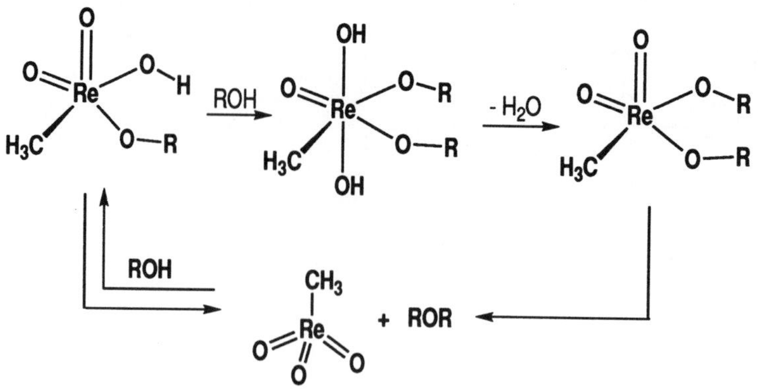

Scheme 19.6.

of alcohols to form olefins at room temperature. When primary or secondary amines, respectively, are used as the limiting reagents, direct amination of alcohols gives the expected secondary or tertiary amines in yields of *c* 95%. Disproportionation of alcohols to carbonyl compounds and alkanes is also observed for aromatic alcohols in the presence of MTO as catalyst. Hypothetical rhenium compounds that might act as intermediates in these processes are given in Scheme 19.6 [90].

7 Conclusions

Organorhenium(VII) oxides have found numerous applications in catalysis during recent years. Methyltrioxorhenium(VII), a water-soluble, air- and temperature-stable metal alkyl, has become the most versatile oxidation catalyst in organic chemistry. Its outstanding reactivity in oxidation catalysis is due to the highly Lewis acidic and sterically unsaturated rhenium(VII) center and the thermally and chemically very stable rhenium–methyl bond. Methyltrioxorhenium(VII) and its derivatives have also been applied successfully as catalysts in olefin metathesis, aldehyde olefination and several other organic reactions. However, in spite of a rich body of chemistry and applications now available, this area is far from mature. We particularly believe that organometal oxides and derivatives in rhenium's neighborhood of the Periodic Table will still improve the catalytic strength of this class of compounds.

8 References

1 Mertis K, Wilkinson G. *J Chem Soc Dalton Trans* 1976: 1488.
2 Beattie IR, Jones, PJ. *Inorg Chem* 1979; **18**: 2318.
3 Herrmann WA, Kuchler JG, Felixberger JK, Herdtweck E, Wagner W. *Angew Chem* 1988: **100**: 420; *Angew Chem Int Ed Engl* 1988; **27**: 394.
4 Herrmann WA, Kühn FE, Fischer RW, Thiel WR, Romão CC. *Inorg Chem* 1992; **31**: 4431.
5 Herrmann WA, Thiel WR, Kühn FE *et al. J Inorg Chem* 1993; **32**: 5188.
6 Herrmann WA, Kratzer R, Fischer RW. *Angew Chem* 1997; **109**: 2767; *Agnew Chem Int Ed Engl* 1997; **36**: 2652.
7 Herrmann WA. *Angew Chem* 1988; **100**: 1269; *Angew Chem Int Ed Engl* 1988; **27**: 394.
8 Herrmann WA. *J Organomet Chem* 1995; **500**: 149.
9 Herrmann WA, Kühn FE. *Acc Chem Res* 1997; **30**: 169.

10 Romão CC, Kühn FE, Herrmann WA. *Chem Rev* 1997; **97**: 3197.

11 Schmidt B. *J Prakt Chem* 1997; **339**: 439.

12 Gable KP. *Adv Organomet Chem* 1997; **41**: 127.

13 Herrmann WA, Fischer RW, Marz DW. *Angew Chem* 1991; **109**: 1706; *Angew Chem Int Ed Engl* 1991; **30**: 1638.

14 Herrmann WA, Correia JDG, Artus GRJ, Fischer RW, Romão CC. *J Organomet Chem* 1996; **520**: 139.

15 Herrmann WA, Fischer RW, Scherer W, Rauch MU. *Angew Chem* 1993; **106**: 1209; *Angew Chem Int Ed Engl* 1993; **32**: 1157.

16 Herrmann WA, Fischer RW, Rauch MU, Scherer W. *J Mol Catal* 1994; **86**: 243.

17 Abu-Omar MM, Hansen PJ, Espenson JH. *J Am Chem Soc* 1996; **118**: 4966.

18 Schröder D, Fiedler A, Herrmann WA, Schwarz H. *Angew Chem* 1995; **107**: 2714; *Angew Chem Int Ed Engl* 1995; **34**: 2636.

19 Pestovsky O, van Eldik R, Huston P, Espenson JH. *J Chem Soc Dalton Trans* 1995: 133.

20 Hatzopoulos I, Brauer HD, Geisberger M, Herrmann WA. *J Organomet Chem* 1996; **520**: 201.

21 Al-Ajlouni A, Espenson JH. *J Am Chem Soc* 1995; **117**: 9243.

22 Herrmann WA. In: Herrmann WA, ed. *Organic Peroxygen Chemistry*, Vol. 164. Berlin: Springer Verlag, 1993: 130.

23 Boehlow TR, Spilling CD. *Tetrahedron Lett* 1996; **37**: 2717.

24 Murray RW, Singh M, Williams BL, Moncrieff HM. *Tetrahedron Lett* 1995; **36**: 2437.

25 Al-Ajlouni AM, Espenson JH. *J Org Chem* 1996; **61**: 3969.

26 Adam W, Mitchell CM. *Angew Chem* 1996; **108**: 578; *Angew Chem Int Ed Engl* 1996; **35**: 533.

27 Murray RW, Iyanar K. *Tetrahedron Lett* 1997; **38**: 335.

28 Herrmann WA, Kühn FE, Mattner MR *et al. J Organomet Chem* 1997; **538**: 203.

29 Herrmann WA, Correia JDG, Rauch MU, Artus GRJ, Kühn FE. *J Mol Catal* 1997; **118**: 33.

30 Sharpless KB, Rudolph J, Redding KL, Chiang JP. *J Am Chem Soc* 1997; **119**: 6189.

31 Copéret C, Adolfsson H, Sharpless KB. *Chem Commun* 1997: 1565.

32 Herrmann WA, Ding H, Kratzer R *et al. J Organomet Chem* 1998; **555**: 293.

33 Adam W, Herrmann WA, Lin J *et al. Angew Chem* 1994; **106**: 2545; *Angew Chem Int Ed Engl* 1994; **33**: 2475.

34 Karasevich El, Nikitin AV, Rubailo VL. *Kinet Katal* 1994; **35**: 810, 878.

35 Yamazaki S. *Chem Lett* 1995: 127.

36 Adam W, Herrmann WA, Lin J, Saha-Möller CR. *J Org Chem* 1994; **59**: 8281.

37 Adam W, Herrmann WA, Saha-Möller CR, Shimizu M. *J Mol Catal* 1995; **97**: 15.

38 Schuchardt U, Mandelli D, Shul'pin GB. *Tetrahedron Lett* 1996; **37**: 6487.

39 Kühn FE, Haider JJ, Herdtweck E *et al. Inorg Chim Acta* 1998; **279**: 44.

40 Herrmann WA, Haider JJ, Fischer RW. *J Mol Catal* 1999; **138**: 115.

41 Herrmann WA, Fischer RW, Correia JDG. *J Mol Catal* 1994; **94**: 213.

42 Abu-Omar MM, Espenson JH. *Organometallics* 1996; **15**: 3543.

43 Fischer RW, PhD Thesis, Technical University Munich, 1994.

44 Yamazaki S. *Bull Chem Soc Jpn* 1996; **69**: 2955.

45 Adam W, Mitchell CM, Saha-Möller CR. *Tetrahedron* 1994; **50**: 13121.

46 Beddoes RL, Painter JE, Quayle P, Patel P. *Tetrahedron Lett* 1996; **37**: 9385.

47 Brown KN, Espenson JH. *Inorg Chem* 1996; **35**: 7211.

48 Huston P, Espenson JH, Bakac A. *Inorg Chem* 1993; **32**: 4517.

49 Abu-Omar MM, Espenson JH. *J Am Chem Soc* 1995; **117**: 272.

50 Zhu Z, Espenson JH. *J Org Chem* 1995; **60**: 1326.

51 Vassell KA, Espenson JH. *Inorg Chem* 1994; **33**: 5491.

52 Goti A, Nannelli L. *Tetrahedron Lett* 1996; **37**: 6025.

53 Murray RW, Iyanar K, Chen I, Wearing T. *J Org Chem* 1996; **61**: 8099.

54 Yamazaki S. *Bull Chem Soc Jpn* 1997; **70**: 877.

55 Murray RW, Iyanar K, Chen I, Wearing T. *Tetrahedron Lett* 1996; **37**: 809.

56 Espenson JH, Pestovsky O, Huston P, Staudt S. *J Am Chem Soc* 1994; **116**: 2869.

57 Hansen PJ, Espenson JH. *Inorg Chem* 1995; **34**: 5389.
58 Murray RW, Iyanar K, Chen L, Wearing JT. *Tetrahedron Lett* 1995; **36**: 6415.
59 Zhu Z, Espenson JH. *J Org Chem* 1995; **60**: 7728.
60 Thiel WR, Fischer RW, Herrmann WA. *J Organomet Chem* 1993; **459**: C9.
61 Herrmann WA, Correia JDG, Kühn FE, Artus GRJ, Romão CC. *Chem J Eur* 1996; **2**: 168.
62 Herrmann WA, Wang M. *Angew Chem* 1991; **103**: 1709; *Angew Chem Int Ed Engl* 1991; **30**: 1641.
63 Zhu Z, Espenson JH. *J Mol Catal* 1995; **103**: 87.
64 Abu-Omar MM, Espenson JH. *Inorg Chem* 1995; **34**: 6239.
65 Herrmann WA, Roesky PW, Scherer W, Kleine M. *Organometallics* 1994; **13**: 4536.
66 Herrmann WA, Felixberger JK, Kuchler JG, Herdtweck E. *Z Naturforsch* 1990; **45b**: 874.
67 Abu-Omar MM, Appelman EH, Espenson JH. *Inorg Chem* 1996; **33**: 7751.
68 Gable KP, Juliette JJJ, Gartman MA. *Organometallics* 1995; **14**: 3138.
69 Herrmann WA, Rauch MU, Roesky PW. *J Organomet Chem* 1996; **511**: 299.
70 Roesky PW. PhD Thesis, Technische Universität München, 1994.
71 Moulijn JA, Boelhouwer C. *J Chem Soc Chem Commun* 1979: 330.
72 Verkuijlen E, Kapsteijn F, Mol JC, Boelhouwer C. *J Chem Soc Chem Commun* 1977: 198.
73 Mol JC, Woerlee EFG. *J Chem Soc Chem Commun* 1979: 330.
74 Mol JC. *J Mol Catal* 1982; **15**: 35.
75 Herrmann WA, Wagner W, Flessner UN, Volkhardt U, Komber H. *Angew Chem* 1991; **103**: 1704; *Angew Chem Int Ed Engl* 1991; **30**: 1636.
76 Buffon R, Auroux A, Lefebvre F *et al. J Mol Catal* 1992; **76**: 287.
77 Buffon R, Choplin A, Leconte M *et al. J Mol Catal* 1992; **72**: L7.
78 Zhu Z, Espenson JH. *J Mol Catal* 1997; **121**: 139.
79 Borvornwattanought A, Bein T. *J Phys Chem* 1992; **96**: 9446.
80 Huber C, Wu GC, Möller K, Bein T. *Abstr Pap Am Chem Soc* 1995; **209**: 172.
81 Malek A, Ozin G. *Adv Mater* 1995; **7**: 160.
82 Flessner UN. PhD Thesis, Technische Universität München, 1992.
83 Junga H, Blechert S. *Tetrahedron Lett* 1993; **34**: 3731.
84 Schneider M, Junga H, Blechert S. *Tetrahedron* 1995; **51**: 13003.
85 Correia JDG. PhD Thesis, Technische Universität München, 1996.
86 Zhu Z, Espenson JH. *J Am Chem Soc* 1997; **119**: 3501.
87 Kholopov AB, Nikitin AV, Rubailo VL. *Kinet Katal* 1995; **36**: 101, 111.
88 Zhu Z, Espenson JH. *J Org Chem* 1995; **60**: 7090.
89 Zhu Z, Espenson JH. *J Org Chem* 1996; **61**: 324.
90 Zhu Z, Espenson JH. *J Am Chem Soc* 1996; **118**: 9901.

20 Ruthenium Catalysts as Innovative Architects for Molecule Building from Alkynes

P.H. DIXNEUF and C. BRUNEAU

Laboratoire de Chimie de Coordination et Catalyse, UMR 6409, CNRS-Université de Rennes, Campus de Beaulieu, F-35042 Rennes, France

1 Introduction

During the last decade ruthenium catalysts have created a variety of new activation processes allowing the design of new molecule architectures. For instance, ruthenium catalysts can equally promote selective oxidation reactions [1] or activate inert C—H bonds of arenes and alkenes [2, 3]. On the other hand, ruthenium–carbene species now allow access to new polymers via mild ring-opening metathesis reactions of cyclic olefins with tolerance of functional groups [4,5]. The same catalysts applied to non-conjugated dienes have opened the route to cyclic olefins via ring-closing metathesis [6, 7].

It is probably in the field of catalytic transformations of alkynes that ruthenium complexes have shown their power and their versatility to selectively produce carbon–heteroatom or carbon–carbon bonds, via activation processes that were not observed with other metal catalysts [8]. Ruthenium(II) complexes containing electron-rich ligands able to favor oxidative couplings have been shown to selectively couple the C≡C bond of alkynes with a variety of unsaturated bonds. Thus, they have created new routes to dienes [9], cyclobutenes [10] and butenolides [11]. The coupling of alkynes with allylic alcohols offers access to unsaturated ketones or aldehydes [12,13]. Ruthenium systems have been shown recently to constitute excellent promoters for the intramolecular coupling of C≡C and C= C bonds of 1,6- or 1,7-enynes, taking place with skeleton rearrangement [14,15], and for the insertion of CO in Pauson–Khand-type reactions [16].

Simple *regioselective* intermolecular additions to the C≡C bond of terminal alkynes for the formation of a C—O or C—C bond involve initial activation processes that are difficult to control. However, these simple combinations offer access to elemental functional building blocks. The breakthrough for the anti-Markovnikov addition to alkynes was brought about by the general ability of metal complexes to activate terminal alkynes into metal–vinylidene species (M(HC≡C—R) → M=C=CHR) bearing an electrophilic C1 atom bonded to the metal center [17]. Ruthenium complexes were the most efficient to transform this stoichiometric activation into a catalytic reaction.

Since the discovery of the one-step transformation of terminal alkynes into alkenyl carbamates via an unexpected anti-Markovnikov addition [18], ruthenium–vinylidene species have been recognized as key catalytic species in the transformation of terminal alkynes with regioselective addition to the terminal carbon atom. Thus, butatrienes [19], enynes [20], aromatic polycycles [21], unsaturated ketones [22], enol ethers [23] and enol esters [24] can be prepared from catalytic anti-Markovnikov addition to terminal alkynes via vinylidene intermediates with C—C or C—O bond formations.

It is noteworthy that all these catalytic reactions involving an addition to a ruthenium–vinylidene intermediate result from the combination of two substrates with atom economy [24, 25].

Here we present general catalytic additions to terminal alkynes that are promoted and regioselectively controlled by ruthenium complexes, modulated by simple modification of the ancillary ligands. It will be shown that:

1 Markovnikov additions are catalysed by Ru(arene)X$_2$(L) and [Ru(O$_2$CR)(CO)$_2$(L)]$_2$ complexes and create O–C2 linkages.

2 Anti-Markovnikov additions with initial formation of an O–C1 bond are exclusively promoted by tailor-made Ru(methallyl)$_2$(diphosphine) catalysts.

3 The carbon–carbon linkage of C≡C and C=C bonds can be performed by RuCl(L$_2$)(C$_5$Me$_5$) derivatives.

All these catalyst precursors are easy to prepare and handle for the selective transformation of simple alkynes and propargylic alcohols under very mild and clean conditions.

2 Regioselective Markovnikov addition of O–H bonds to terminal alkynes

2.1 Synthesis of enol esters and recovery of the catalyst

The activation of terminal alkynes with *electrophilic* ruthenium(II) complexes allows the regioselective Markovnikov addition of carboxylic acids and N-protected amino acids to the triple bond, as exemplified by Equations (20.1) and (20.2). The best catalyst appears to be the air-stable readily available complex RuCl$_2$(PPh$_3$)(p-cymene) (**I**) at 80–100 °C in toluene, which leads to the nearly quantitative formation of enol esters [26, 27].

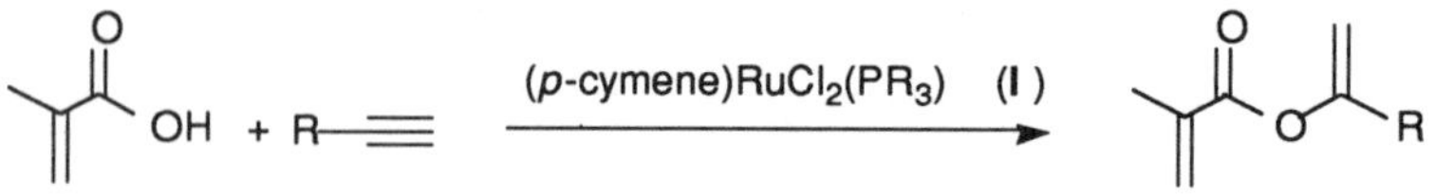

$$\text{(20.1)}$$

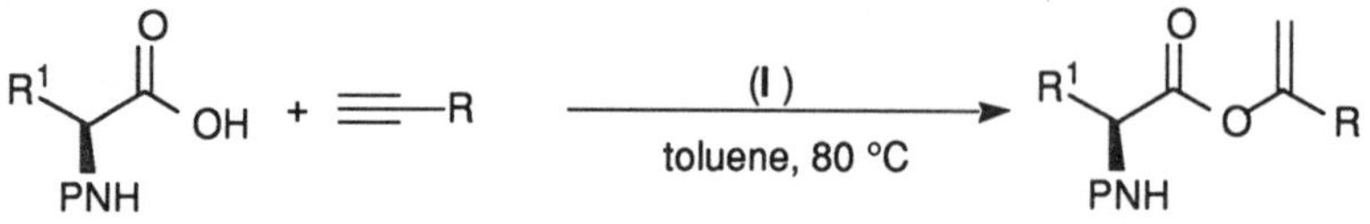

$$\text{(20.2)}$$

Thus, this catalytic reaction offers environmental advantages over the mercury(II)-assisted synthesis of enol esters. The functional enol esters formed in one step can be used as monomers for radical-initiated polymerization. They are also efficient acylating reagents of alcohols, amines, amino acids or amides, releasing only a neutral ketone as the byproduct [28].

However, the addition of oxalic acid or mandelic acid to alkynes is not efficiently promoted by catalyst **1** but our attempts to produce an enol formate led us to discover, by accident, an excellent catalyst precursor [Ru(O$_2$CH)(CO)$_2$(PPh$_3$)]$_2$ (**II**) for the Markovnikov addition of functional carboxylic acids to alkynes under mild conditions (60 °C) [29]. From mandelic acid, a subsequent catalytic addition to the enol C=C bond leads to the stereoselective formation of dioxolanones (Equation (20.3)) [30]. In addition, catalyst **2** appears to allow Markovnikov addition to a variety of functional alkynes and diynes.

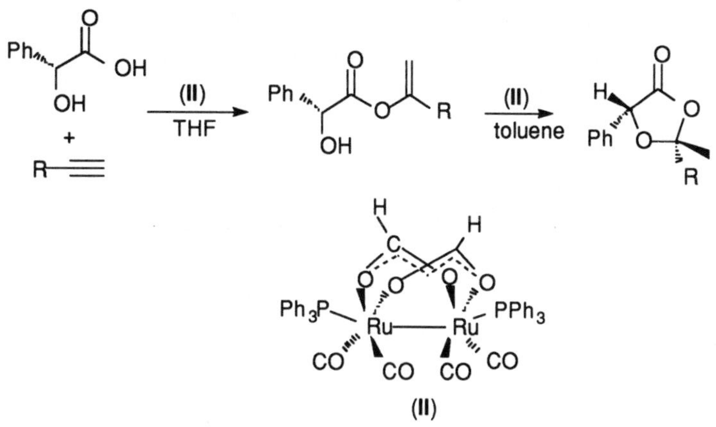

(20.3)

It is noteworthy that catalyst **II** was used as model for the recovery and re-use of the ruthenium catalyst. Catalyst **III** containing the polymeric phosphine $Ph_2P(CH_2CH_2)_nCH_2CH_3$ ($n = 50$) was insoluble at 25 °C and performed the quantitative addition of diacids to alkynes (Equation (20.4)) in a toluene homogeneous solution at 100 °C. After cooling, it was completely recovered unchanged by filtration and re-used six times without loss of activity for the production of a bifunctional precursor of polyamide [31].

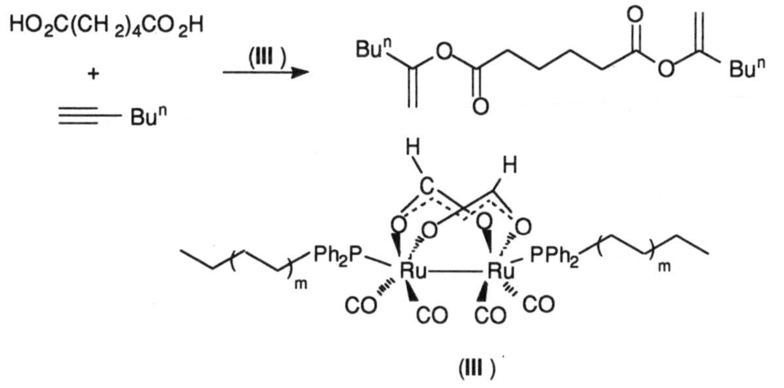

(20.4)

The mechanism of Markovnikov addition of the carboxylic acid O–H bond to alkynes has not been fully established, but we have shown by cross-experiments that the co-ordinated carboxylate $R^1CO_2^-$ of the catalyst precursors $Ru(O_2CR^1)(Cl)(arene)$ or $[Ru(O_2CR^1)(CO)_2(PPh_3)]_2$ did not add to the alkyne. Only the external carboxylate from the free R^2CO_2H adds to the triple bond. We concluded that the C–O bond formation did not occur via insertion of the C≡C bond into the $(R^1CO_2)O$–Ru bond but via external addition of the free carboxylate to the electrophilically activated coordinated alkyne.

2.2 Synthesis of functional dienes

The Markovnikov addition of carboxylic acids to conjugated enynes catalysed by the complex $RuCl_2(PPh_3)(p\text{-cymene})$ (**I**) actually constitutes a simple way to generate functional 1,3-dienes in high yields (Equation (20.5)).

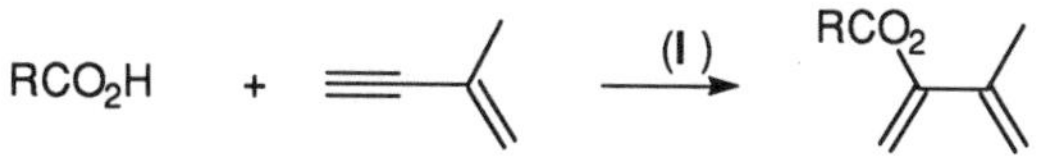

(20.5)

Similarly, the activation of (E)-3-methylpent-2-en-4-yn-l-ol with the catalyst $RuCl_2(PPh_3)(p\text{-cymene})$ (**I**) in the presence of carboxylic acids leads to the formation of hydroxylated dienyl esters (Equation (20.6)) [32].

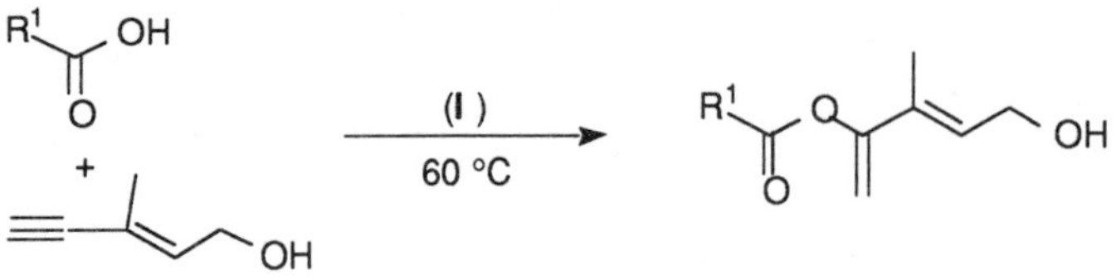

$$(20.6)$$

These dienes actually offer access to a variety of functional dienes (or their related α,β-unsaturated ketones) by modification of the hydroxy group. Thus, it can easily be transformed into a leaving group upon esterification or carboxylation to produce allylic esters or carbonates. Their activation by palladium(0) complexes provides a three-step access to functionalized dienes from the starting enynol (Equation (20.7)).

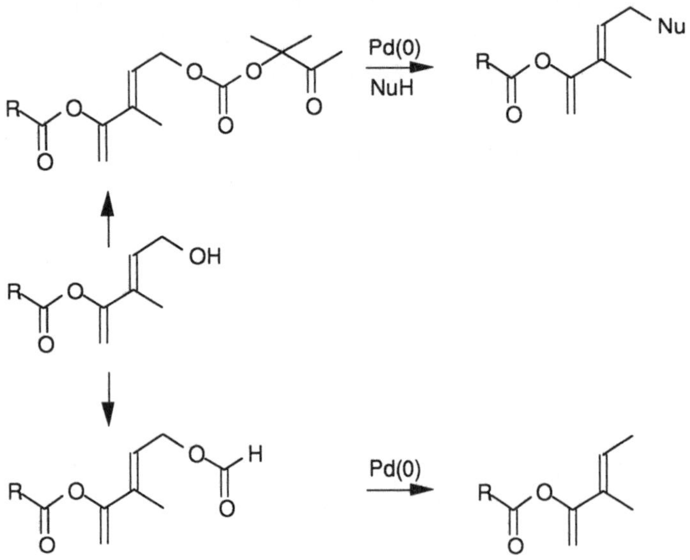

$$(20.7)$$

2.3 Synthesis of furans: intramolecular addition of O–H to a terminal carbon–carbon triple bond

By contrast, activation of the Z isomer, the (Z)-3-methylpent-2-en-4-yn-l-ol, leads to the formation of furans, showing that *intramolecular* addition of the O–H group in a suitable position with respect to the HC≡C group is favored as compared to *intermolecular* addition of the carboxylic acid to the C≡CH group (Equation (20.8)) [33].

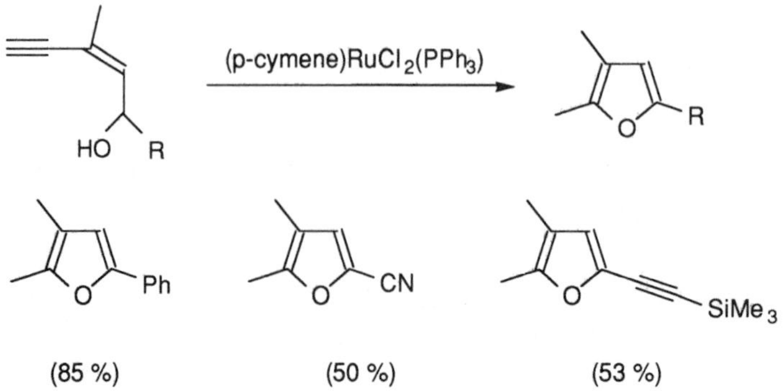

$$(20.8)$$

(85 %)   (50 %)   (53 %)

This catalytic formation of furans can be carried out from a variety of substituted enynols with different R groups under mild and neutral conditions, and thus presents advantages over the preparation of furans promoted by strong bases.

Very efficient new ruthenium catalysts have been found recently for the catalytic formation of furans: 1,4,5,6-tetrahydropyrimidine complexes (**IV**) (arene = C_6Me_6 or p-cymene) operate at 80 °C and binuclear catalysts (**V**) ($n = 0, 1$) operate at room temperature, with furans obtained quantitatively (Equation (20.9) [34] and Equation (20.10) [35]).

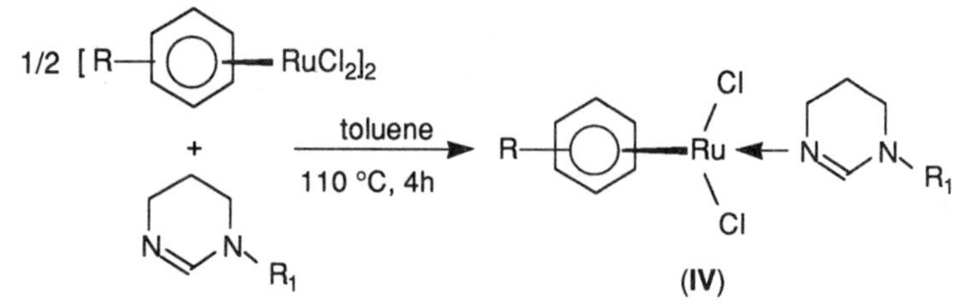

$$(20.9)$$

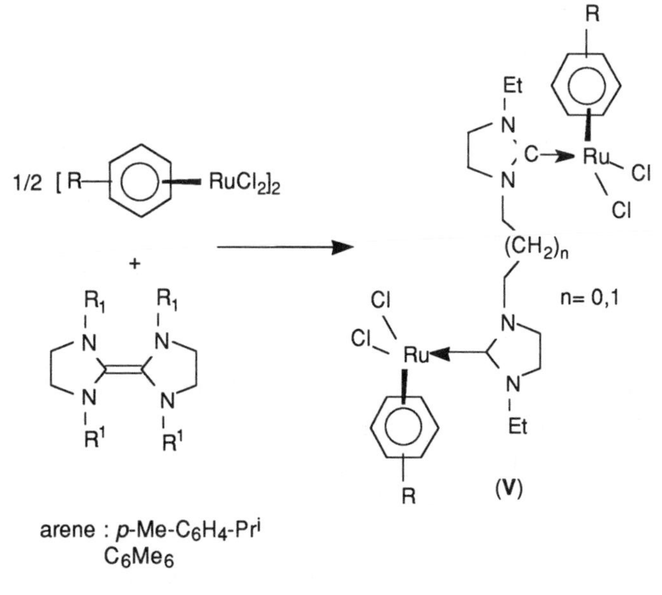

$$(20.10)$$

arene : p-Me-C_6H_4-Pr^i
C_6Me_6

The mechanism of the reaction corresponds to the Markovnikov addition of the O–H group to the C≡CH bond, but presumably in this case the regioselectivity is also controlled by the length of the chain between the two groups (Scheme 20.1) [36]. This ruthenium-catalysed reaction seems to be limited to terminal C≡C–H, but this restriction has been overtaken recently with the use of palladium catalysts [37].

3   Anti-Markovnikov addition of carboxylates to terminal alkynes

Whereas the above reactions correspond to the expected Markovnikov addition to the C≡C bond activated by an electrophilic reagent (the ruthenium(II) catalyst), the challenge was to perform regioselective anti-Markovnikov addition to the C≡CH bond of terminal alkynes. This was achieved for a variety of alkyl-, aryl- and functional terminal acetylenes and led to the formation of (Z)-alk-l-en-l-yl esters, dienes or α,β-unsaturated aldehydes, according to the nature of the alkyne. Anti-Markovnikov addition requires specific ruthenium catalysts containing a diphosphine ligand, the

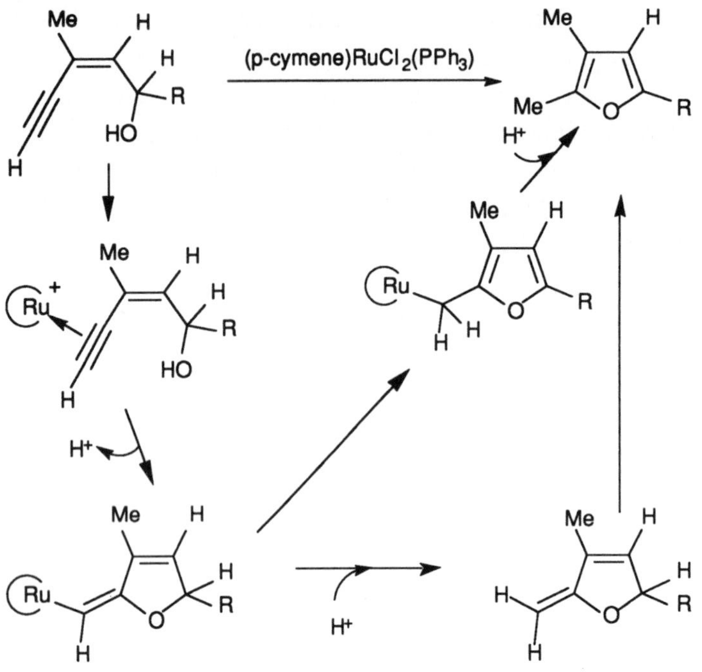

Scheme 20.1.

nature of which has to be modulated according to the nature of the functional alkyne.

3.1 Synthesis of alkenyl carbamates

We discovered anti-Markovnikov addition to terminal alkynes when we studied the catalytic synthesis of alkenyl carbamates in one step from a terminal alkyne, a secondary amine and CO_2 in the presence of ruthenium(II) catalysts, especially the $RuCl_2(PR_3)$(arene) derivatives of type **I** (Equation (20.11)) [38].

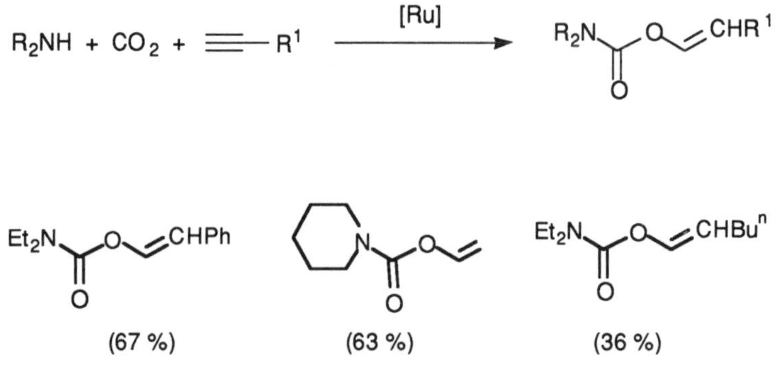

$$R_2NH + CO_2 + \equiv\!\!-R^1 \xrightarrow{\text{[Ru]}} R_2N\!\!-\!\!C(=O)\!\!-\!\!O\!\!-\!\!CHR^1$$

(67 %) (63 %) (36 %)

(20.11)

To explain the regioselective addition of the in situ formed ammonium carbamate $R_2NCO_2^- \; {}^+H_2NR_2$ only to the terminal carbon of the triple bond, we postulated a vinylidene–ruthenium intermediate Ru=C=CHR [38]. Indeed, such vinylidene complexes are readily obtained from a variety of metal complexes and terminal alkynes and the carbon atom linked to the metal is the electrophilic site of the M=C=CHR moiety [17]. We showed that $RuCl_2(PR_3)$(arene) derivatives readily produced ruthenium–vinylidene

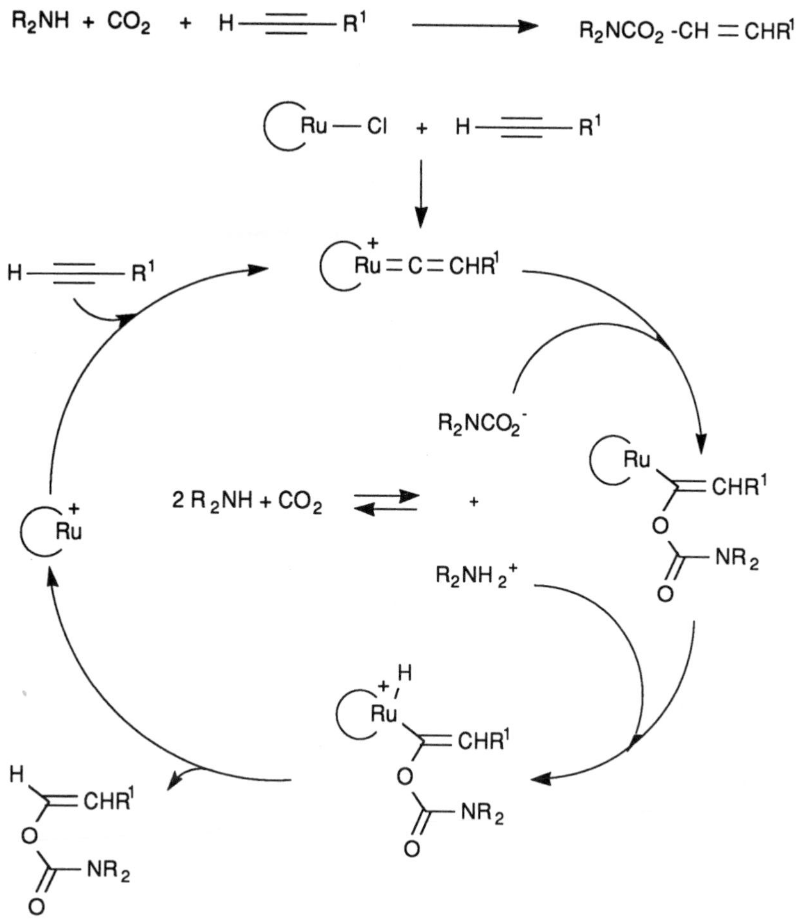

$$R_2NH + CO_2 + H \equiv R^1 \longrightarrow R_2NCO_2\text{-}CH = CHR^1$$

Scheme 20.2.

species with terminal alkynes when a vacant site could be generated under mild conditions [39]. The catalytic cycle shown in Scheme 20.2 accounts for the direct formation of alkenyl carbamates.

3.2 *Synthesis of (Z)-alk-1-en-1-yl esters*

As shown before, the $RuCl_2(PR_3)(arene)$ complexes are good catalysts for the addition of the carboxylate to the C(2) atom of the terminal triple bond. In order to reverse the addition of carboxylates to the triple bond of terminal alkynes, we studied the catalytic activity of a series of ruthenium catalysts possessing labile allylic ligands and an electron-releasing bidentate diphosphine ligand that can be modulated. Complexes of the type $Ru(\eta^3\text{-}CH_2C(Me)=CH_2)_2(Ph_2P(CH_2)_nPPh_2)$ were prepared with $n = 1$ (**VI**), 2 (**VII**), 3 (**VIII**) and 4 (**IX**) on the basis of the synthesis of bis-allylruthenium derivatives from $(cod)Ru(\eta^3\text{-}CH_2C(Me)=CH_2)_2$ [40]. The above four catalysts (**6–9**) have been evaluated for the activation of hex-1-yne and phenylacetylene towards a variety of carboxylic acids and only the catalyst $(dppb)Ru(methallyl)_2$ (**9**) led to a remarkable regioselective anti-Markovnikov addition. Moreover, a high stereoselectivity (>95%) was observed as the (*Z*)-alk-1-en-1-yl esters corresponding to the *trans*-addition to the C≡C bond were obtained in excellent yields (Equation (20.12)) [41].

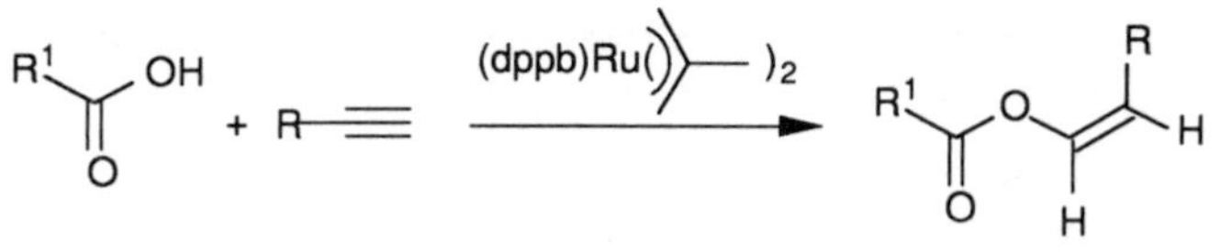

$$\text{(20.12)}$$

R = Bu^n ; R^1 = Ph (95 %), Ph$_2$CH (97 %), Bu^{t}O$_2$CNHCH(CH$_2$Ph) (97 %)

R = Ph ; R^1 = Bu^n (92 %), CF$_3$ (90 %), MeOCH$_2$ (96 %), 2-HO-C$_6$H$_4$ (94 %)

For the addition of acids to isopropenylacetylene (Equation (20.13)) and 4-methoxybut-3-en-1-yne (Equation (20.14)), the same complex (**IX**) was shown to be the only efficient catalyst [40], offering a new route to a variety of functional butadiene derivatives.

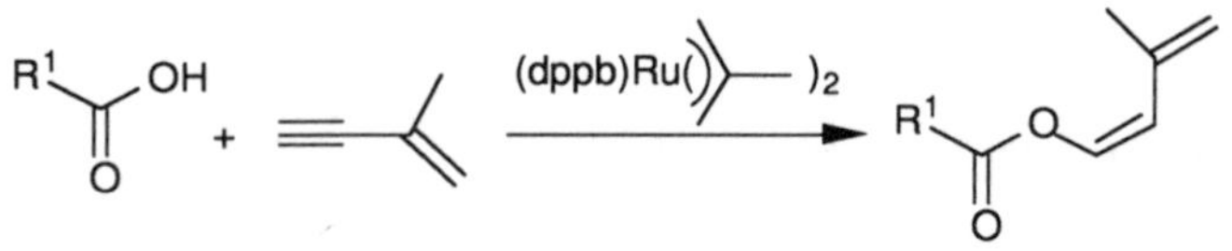

$$\text{(20.13)}$$

R^1= Ph (92%), n-C$_4$H$_9$ (77%)

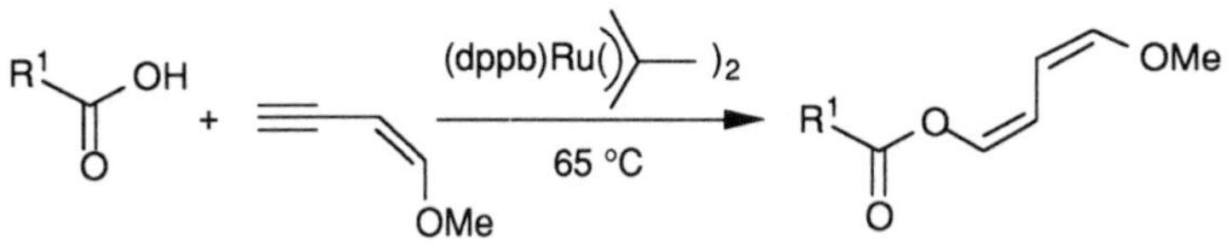

$$\text{(20.14)}$$

R^1= Ph (81%), PhCH$_2$(76%),
MeOCH$_2$ (69%)

By contrast, catalyst **IX** was not efficient for alkynes with a bulky substituent. For instance, the C(1) addition to trimethylsilylacetylene was achieved with catalyst (dppe)Ru(methallyl)$_2$ (**VII**), which led to good yields of the anti-Markovnikov addition products (Equation (20.15)) [42] without cleavage of the carbon–silicon bond.

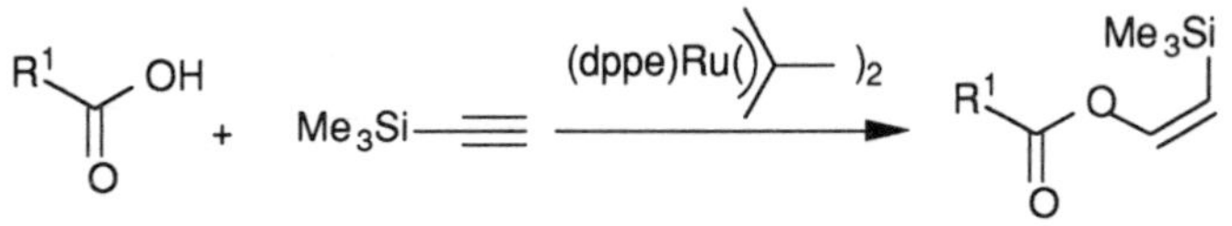

$$\text{(20.15)}$$

R^1 = Ph (88 %), PhCH$_2$ (96 %), (L)-Bu^{t}O$_2$CNHCH(Me) (75 %)

Complex **VII** also appeared to be by far the most efficient catalyst for the addition to a variety of disubstituted 3-methoxyprop-1-ynes, and the corresponding (Z)-alkenyl esters were obtained with excellent regio- and stereoselectivities (Equation (20.16)) [43].

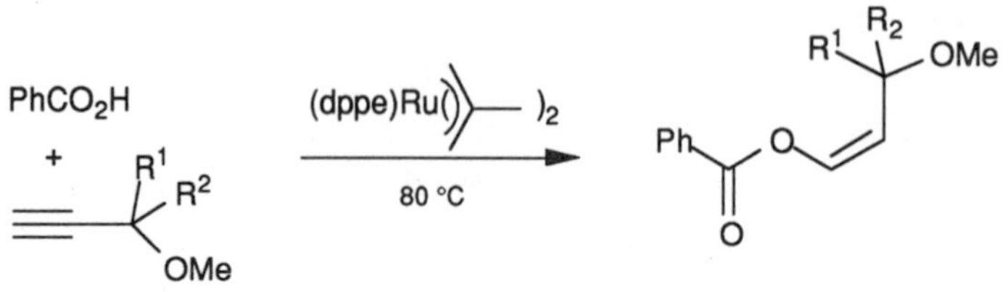

$$\text{(20.16)}$$

R^1 = Me ; R^2 = Me (86 %), R^2 = Et (73 %), R^2 = Ph (95 %)

It is noteworthy that alk-1-en-1-yl esters represent protected aldehydes directly generated from terminal alkynes. They are actually key precursors for access to enamines [44] or cyanohydrin esters [45], as shown in Scheme 20.3.

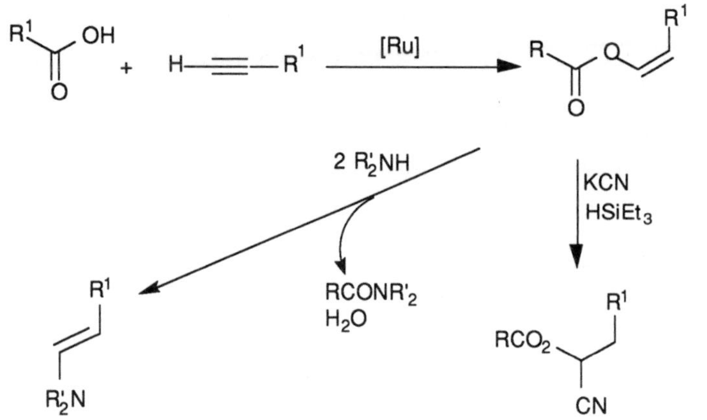

Scheme 20.3.

3.3   *Isomerization of prop-2-yn-1-ols into α,β-unsaturated aldehydes via a transient anti-Markovnikov addition of benzoic acid*

The reaction of benzoic acid with prop-2-yn-1-ols in the presence of the catalyst **7** at 50 °C selectively gave anti-Markovnikov addition of the benzoate to the C≡C bond to produce 3-hydroxyprop-1-en-1-yl benzoates. The heating of these benzoates in the presence or the absence of catalyst **7** led to the elimination of benzoic acid and the formation of the corresponding α,β-unsaturated aldehydes (Equation (20.17)). The overall transformation corresponding to the formal isomerization of the alkynol could be achieved in one step by treatment of the prop-2-yn-2-ol with one equivalent of benzoic acid in the presence of 5 mol.% of (dppe)Ru(methallyl)$_2$ (**VII**) followed by distillation under reduced pressure [46]. Scheme 20.4 accounts for the successive steps of this reaction [46].

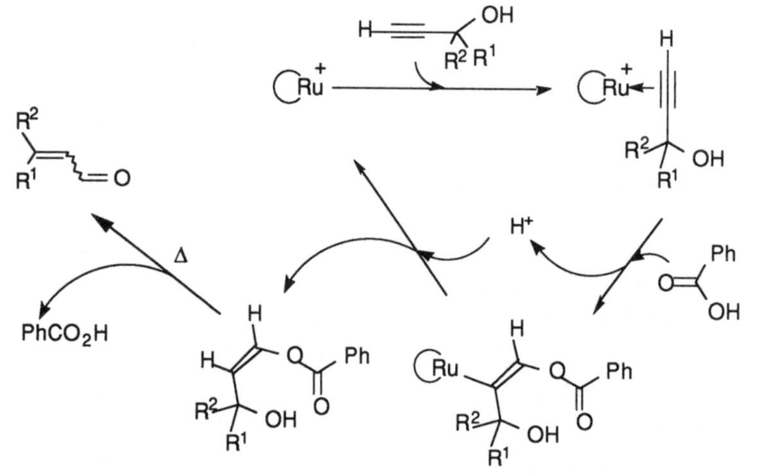

Scheme 20.4.

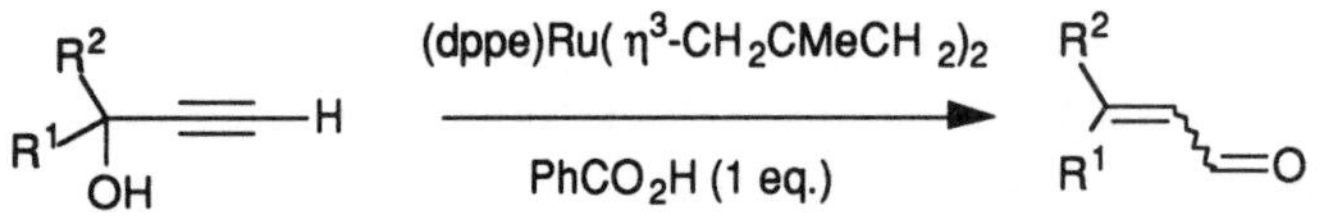

$$\underset{OH}{\overset{R^2}{\underset{R^1}{\bigg\rangle}}}\!\!=\!\!\!=\!\!-H \quad \xrightarrow[\text{PhCO}_2\text{H (1 eq.)}]{\text{(dppe)Ru}(\eta^3\text{-CH}_2\text{CMeCH}_2)_2} \quad \underset{R^1}{\overset{R^2}{\bigg\rangle}}\!\!=\!\!\!\diagdown\!\!=\!\!O \qquad (20.17)$$

This isomerization of prop-2-yn-1-ols into α,β-unsaturated aldehydes is an alternative to the similar isomerization promoted by metal oxides in the presence of acids [47].

It must be noted that this isomerization reaction contrasts with the reaction of the same prop-2-yn-1-ols with carboxylic acids promoted by the catalyst $[\text{Ru}(\text{O}_2\text{CH})(\text{CO})_2(\text{PPh}_3)]_2$ (**II**), which allows direct access to stable β-oxopropyl esters corresponding to Markovnikov addition to the C≡C bond followed by transesterification (Equation (20.18)) [32,48].

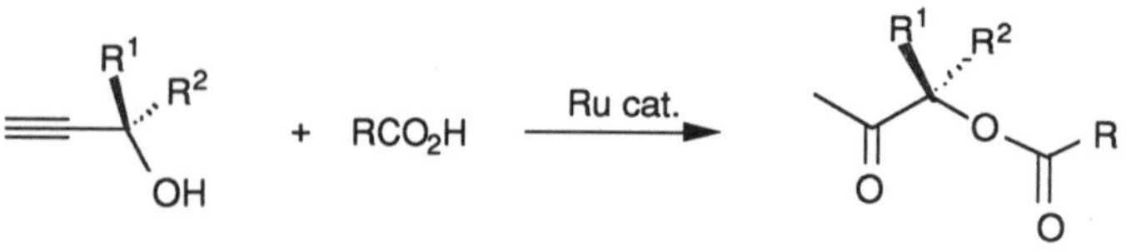

$$\equiv\!\!\!=\!\!\!\underset{OH}{\overset{R^1}{\underset{}{\bigg\rangle}}}\!\!\!R^2 \;+\; \text{RCO}_2\text{H} \quad \xrightarrow{\text{Ru cat.}} \quad \underset{O}{\overset{R^1\;\;R^2}{\bigg\rangle}}\!\!\!\diagdown\!\!O\!\!\diagdown\!\!\underset{O}{\overset{}{R}} \qquad (20.18)$$

The main difference arises from the initial step, which is the addition of carboxylate either at C(1) or C(2) of the triple bond and is controlled by the nature of the ruthenium catalyst.

Our study of the anti-Markovnikov addition of carboxylic acids to a variety of alkynes promoted by catalyst precursors of the type $(\text{Ph}_2\text{P}(\text{CH}_2)_n\text{PPh}_2)\text{Ru}(\eta^3\text{-CH}_2\text{C}(\text{Me})\text{CH}_2)_2$ reveals two classes of alkynes:

1 The less bulky alkynes, such as HC≡CBu^n, HC≡CPh, HC≡CC(Me)=CH$_2$ and HC≡C–CH=CH–OMe, for which the reaction is promoted by catalyst **IX** ($n = 4$).

2 The more bulky alkynes, such as HC≡C–SiMe$_3$, HC≡CCR$_2$(OMe) and HC≡C–CR$_2$(OH), for which the reaction is promoted by catalyst **VII** ($n = 2$).

The ruthenium site in the $(\text{Ph}_2\text{PCH}_2\text{CH}_2\text{CH}_2\text{CH}_2\text{PPh}_2)\text{Ru}$ moiety in catalyst **IX** is more sterically hindered than in the $(\text{Ph}_2\text{PCH}_2\text{CH}_2\text{PPh}_2)\text{Ru}$ moiety in catalyst **VII**. Indeed, the P–Ru–P angle is larger in catalyst **IX** than in catalyst **VII**, so the (PPh$_2$) phenyl groups in catalyst **IX** are expected to sterically hinder the ruthenium site, and if a vinylidene is formed with the bulky alkynes the external approach of the carboxylate to carbon C1 is likely to be disfavored compared to access to a similar moiety formed from complex **VII**. The hypothesis is made that the steric hindrance around the ruthenium site controls the regioselectivity of the addition at carbon C1.

4 Regioselective formation of carbon–carbon bonds by coupling C≡C and C=C bonds

4.1 *Catalytic synthesis of γ,δ-unsaturated aldehydes*

Ruthenium(II) complexes are able to favor the regioselective coupling of C≡C and C=C bonds. The coupling of terminal alkynes with α,β-enones and water, or allyl alcohol derivatives, can selectively lead to diketones or unsaturated ketones, respectively, in the presence of the catalyst precursor $\text{RuCl}(\text{PPh}_3)_2(\text{C}_5\text{H}_5)\text{NH}_4\text{PF}_6$ [49,50]. By contrast, the related $\text{RuCl}(\text{cod})(\text{C}_5\text{Me}_5)$ complex promotes the coupling of alkynes with olefins to produce cyclobutene derivatives [10].

We have shown that the catalyst precursor $\text{RuCl}(\text{cod})(\text{C}_5\text{Me}_5)$ (**X**), containing a labile cyclooctadiene ligand and a bulky electron-releasing C_5Me_5 group, is able to catalyse the coupling of alkynes with allyl alcohol and produce γ,δ-unsaturated aldehydes (Equation (20.19)) [13]. The branched isomer (**a**), corresponding to C—C coupling of the substituted C≡C carbon atom, is the major isomer. The reaction can be performed in water, or better without solvent, to give the highest regioselectivity **a/b** = 4/1 [13]. This regioselectivity in the branched isomer is remarkable because in previous mixed coupling of C≡C and C=C bonds, in the presence of the less bulky and electron-rich complex RuCl(cod)(C$_5$H$_5$), the major isomer was the linear one [12].

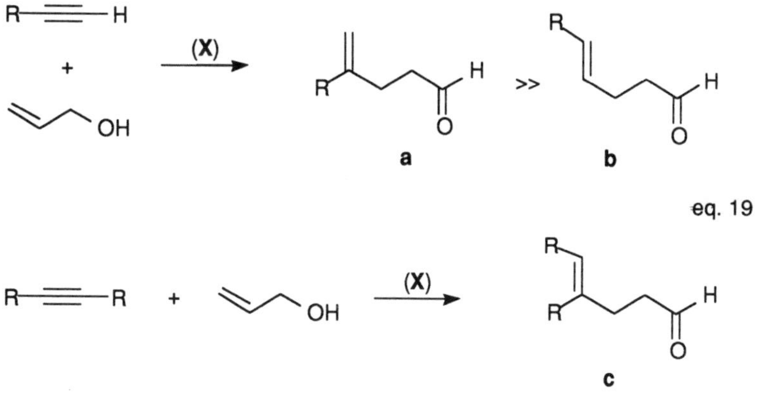

eq. 19

(20.19)

Symmetrically disubstituted alkynes lead to only one aldehyde (**c**) in 90–95% yield (Equation (20.19)) [13], showing that this reaction is not specific to terminal alkynes. Therefore, a vinylidene–ruthenium intermediate cannot be involved in the mechanism and an oxidative coupling of both C≡C and C=C bonds should be considered.

4.2   Catalytic synthesis of 5-methylenetetrahydropyrans

The same RuCl(cod)(C$_5$Me$_5$) catalyst precursor (**X**) promotes the oxidative coupling of prop-2-yn-1-ols and allyl alcohol and leads to tetrahydropyran derivatives. The hemiacetal and the mixed acetal can be prepared selectively, depending on the reaction temperature (Equation (20.20)) [51]. The disubstituted (R^1, R^2 ≠ H) propargylic alcohols favor the formation of the branched isomers, observed as the sole reaction products.

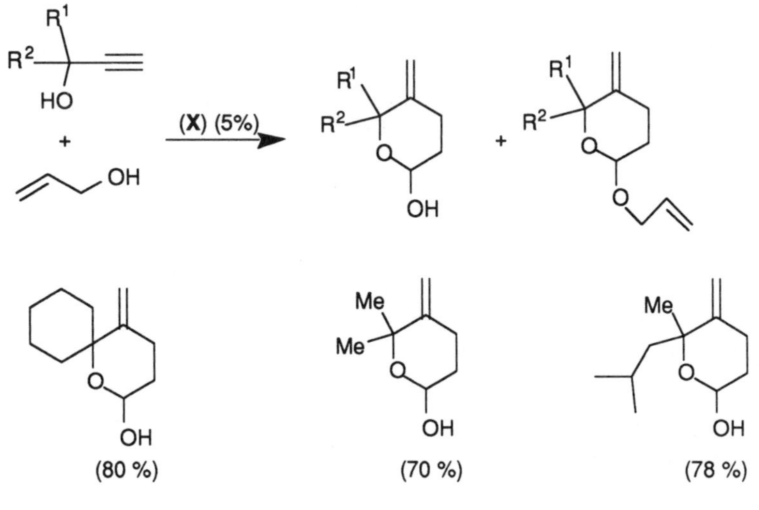

(20.20)

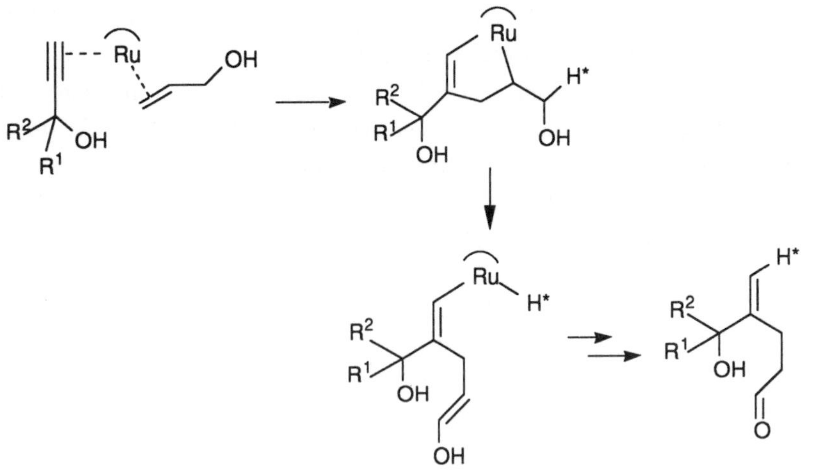

Scheme 20.5.

Allenylidene intermediates containing the $Ru=C=C=CR_2$ moiety are easily produced by activation of prop-2-yn-1-ols with a variety of ruthenium complexes via vinylidene formation and loss of water [52, 53]. The mechanism involving such an intermediate has been ruled out by the reaction with a substituted propargyl alcohol derivative, which cannot generate an allenylidene species. It leads to the formation of the hemiacetal and the hydroxy unsaturated aldehyde, corresponding to the two regioisomers of the C–C coupling with the C≡C bond (Equation (20.21)).

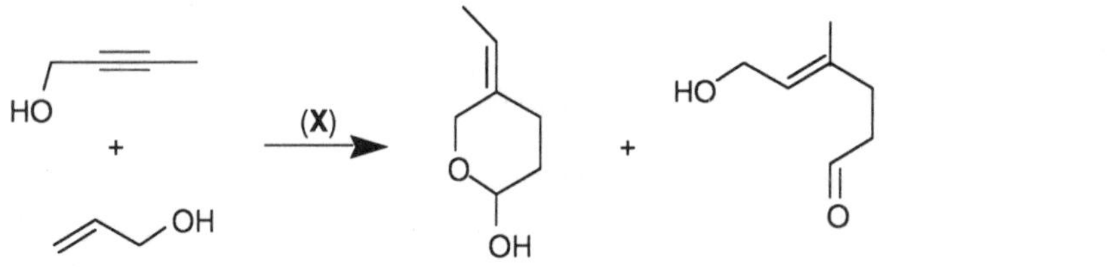

$$(20.21)$$

To explain both the formation of γ,δ-unsaturated aldehydes and that of 5-methylene-tetrahydropyrans, the mechanism displayed in Scheme 20.5 could be proposed.

This mechanism is based on the oxidative coupling of both C≡C and C=C bonds, β-elimination of an exocyclic hydrogen atom and the formation of an γ,δ-unsaturated aldehyde, which can lead to cyclization with the hydroxy group of propynylic alcohols.

The above two reactions show examples of regioselective C–C bond formation from two unsaturated substrates to produce high value chemicals with atom economy.

5   Conclusion

A variety of ruthenium complexes that are easy to make and handle have been used to promote the activation of terminal alkynes and the regioselective formation of C–O and C–C bonds with *electrophilic* and *electron-rich* ruthenium(II) complexes. Thus, these ruthenium precursors provide smart catalysts because by slight modifications of only one ancillary ligand around the ruthenium site the activity of the catalyst can be created and the regioselectivity of the coupling highly modified.

These catalytic reactions show that ruthenium complexes are able to build a variety of molecule architectures from alkynes via methods that were not observed with other transition metal catalysts.

The electrophilic $(arene)RuCl_2(L)$ (**I**) or $[Ru(O_2CH)(CO)_2(PPh_3)]_2$ (**II**) complexes control the regioselective formation of C—O bonds by addition of the oxygen atom to the C2 carbon of terminal alkynes.

On the other hand, ruthenium catalysts **VII** and **IX** containing a chelating diphosphine are able to combine molecules, specifically by C—O bond formation involving the C1 carbon atom of terminal alkynes. This regioselectivity is reached only when slight modifications of the chelating ligand are brought to fit the nature of the alkynes.

The electron-rich ruthenium(II) catalyst $RuCl(cod)(C_5Me_5)$ (**X**) is the most suitable for the design of C—C bonds, by selective coupling of C≡C and C=C bonds.

One can predict that the ruthenium catalysts that are now able to perform a variety of new combinations of an alkyne with a simple substrate will be the new architects of this decade.

6 References

1 Murahashi SI, Saito T, Hanaoka H *et al. J Org Chem* 1993; **58**: 2929; Murahashi SI, Naota T. *Synthesis* 1993: 433; Barf GA, van den Hoek D, Sheldon RA. *Tetrahedron* 1993; **52**: 12971.

2 Murai S, Kakiuchi F, Sekine S *et al. Nature (London)* 1993; **366**: 529; Kakiuchi F, Tanaka Y, Sato T, Chatani N, Murai S. *Chem Lett* 1995: 679; Harris PWR, Woodgate PD. *J Organomet Chem* 1996; **506**: 339; Sonoda M, Kakiuchi F, Kamatasu A, Chatani N, Murai S. *Chem Lett* 1996: 109. Kakiuchi F, Yamauchi M, Chatani N, Murai S. *Chem Lett* 1996: 111.

3 Trost BM, Imi K, Davies I. *J Am Chem Soc* 1995; **117**: 5371; Kakiuchi F, Yamamoto Y, Chatani N, Murai S. *Chem Lett* 1995: 681.

4 Demonceau A, Noels AF, Saive E, Hubert AJ. *J Mol Catal* 1992; **76**: 123; Stumpf AW, Saive E, Demonceau A, Noels AF. *J Chem Soc Chem Commun* 1995: 1127; Wache S, Hermann NA, Artus G, Nuyken O, Wolf D. *J Organomet Chem* 1995; **491**: 181.

5 France MB, Paciello RA, Grubbs RH. *Macromolecules* 1993; **26**: 4739; Nguyen ST, Grubbs RH, Ziller JW. *J Am Chem Soc* 1993; **115**: 9858; Wu Z, Nguyen ST, Grubbs RH, Ziller JW. *J Am Chem Soc* 1995; **117**: 5503; Schwab P, France MB, Ziller JW, Grubbs RH. *Angew Chem Int Ed Engl* 1995; **34**: 2039; Mohr B, Lynn DM, Grubbs RH. *Organometallics* 1996; **15**: 4317.

6 Fu GC, Nguyen ST, Grubbs RH. *J Am Chem Soc* 1993; **115**: 9856; Kim SH, Bowden N, Grubbs RH. *J Am Chem Soc* 1994; **116**: 10801; Grubbs RH, Miller SJ, Fu GC. *Acc Chem Res* 1995; **28**: 446.

7 Schmalz HG. *Angew Chem Int Ed Engl* 1995; **34**: 1833; König B, Horn C. *Synlett* 1996: 1013; Fürstner A, Langemann K. *J Org Chem* 1996; **61**: 3942; Fürstner A, Müller T. *Synlett* 1997: 1010; Huwe CM, Kiehl OC, Blechert S. *Synlett* 1996: 65.

8 Trost BM. *Chem Ber* 1996; **129**: 1313.

9 Mitsudo TA, Shang SW, Nagao M, Watanabe Y. *J Chem Soc Chem Commun* 1991: 598; Trost BM, Indolese A. *J Am Chem Soc* 1993; **115**: 4361; Trost BM, Imi K, Indolese A. *J Am Chem Soc* 1993; **115**: 8831; Trost BM, Indolese AF, Müller TJJ, Treptow B. *J Am Chem Soc* 1995; **117**: 615.

10 Mitsudo TA, Naruse H, Kondo T, Ozaki Y, Watanabe Y. *Angew Chem Int Ed Engl* 1994; **33**: 580.

11 Trost BM, Müller TJJ. *J Am Chem Soc* 1994; **116**: 4985; Trost BM, Müller TJJ, Martinez J. *J Am Chem Soc* 1995; **117**: 1888.

12 Trost BM, Martinez JA, Kulawiec JA, Indolese AF. *J Am Chem Soc* 1993; **115**: 10402.

13 Dérien S, Dixneuf PH. *J Chem Soc Chem Commun* 1994: 2551; Dérien S, Jan D, Dixneuf PH. *Tetrahedron* 1995; **52**: 5511.

14 Chatani N, Morimoto T, Muto T, Murai S. *J Am Chem Soc* 1994; **116**: 6049.

15 Kinoshita A, Mori M. *Synlett* 1994: 1020.

16 Kondo T, Suzuki N, Okada T, Mitsudo TA. *J Am Chem Soc* 1997; **119**: 6187; Morimoto T, Chatani N, Fukumoto Y, Murai S. *J Org Chem* 1997; **62**: 3762.

17 Bruce MI. *Chem Rev* 1991; **91**: 197.

18 Mahé R, Dixneuf PH, Lécolier S. *Tetrahedron Lett* 1986; **27**: 6333.

19 Wakatsuki Y, Yamazaki H, Kumegawa N, Satoh T, Satoh JY. *J Am Chem Soc* 1991; **113**: 9604.

20 Bianchini C, Peruzzini M, Frediani P. *J Am Chem Soc* 1991; **113**: 5453; Slugovc C, Mereiter K, Zobetz E, Schmid R, Kirchner K. *Organometallics* 1996; **15**: 5275; Yi CS, Liu N. *Organometallics* 1996; **15**: 3986.

21 Merlic CA, Pauly ME. *J Am Chem Soc* 1996; **118**: 11319.

22 Trost BM, Kulawiec RJ. *J Am Chem Soc* 1992; **114**: 5579; Trost BM, Kulawiec RJ, Hammes A. *Tetrahedron Lett* 1993; **34**: 587; Trost BM, Flygare JA. *J Org Chem* 1994; **59**: 1078; *J Am Chem Soc* 1992; **114**: 5476.

23 Gemel C, Trimmel G, Slugovc C *et al. Organometallics* 1996; **15**: 3998.

24 Bruneau C, Dixneuf PH. *J Chem Soc Chem Commun* 1997: 507.

25 Trost BM. *Angew Chem Int Ed Engl* 1995; **34**: 259.

26 Ruppin C, Dixneuf PH. *Tetrahedron Lett* 1986; **27**: 6323.

27 Bruneau C, Neveux M, Ruppin C, Kabouche Z, Dixneuf PH. *Synlett* 1991: 755.

28 Neveux M, Bruneau C, Dixneuf PH. *J Chem Soc Perkin Trans 1* 1991; 1197: Kabouche Z, Bruneau C, Dixneuf PH. *Tetrahedron Lett* 1991; **32**: 5359; Seiller B, Heins D, Bruneau C, Dixneuf PH. *Tetrahedron* 1995; **51**: 10901; Seiller B, Bruneau C, Dixneuf PH. *Synlett* 1995: 707.

29 Neveux M, Bruneau C, Lécolier S, Dixneuf PH. *Tetrahedron* 1993; **49**: 2629.

30 Neveux M, Seiller B, Hagedorn F, Bruneau C, Dixneuf PH. *J Organomet Chem* 1993; **451**: 133.

31 Lavastre O, Bebin P, Marchaland O, Dixneuf PH. *J Mol Catal A* 1996; **108**: 29.

32 Bruneau C, Kabouche Z, Neveux M, Seiller B, Dixneuf PH. *Inorg Chim Acta* 1994; **222**: 155.

33 Seiller B, Bruneau C, Dixneuf PH. *J Chem Soc Chem Commun* 1994: 493.

34 Cetinkaya B, Alici B, Ozdemir I, Bruneau C, Dixneuf PH. *J Organomet Chem* 1999 (In press.)

35 Cetinkaya B, Ozdemir I, Bruneau C, Dixneuf PH. *J Mol Catal* 1997; **118**: L1.

36 Seiller B, Bruneau C, Dixneuf PH. *Tetrahedron* 1995; **51**: 13089.

37 Bartolo G, Salerno G. *J Chem Soc Chem Commun* 1997: 1083.

38 Mahé R, Sasaki Y, Bruneau C, Dixneuf PH. *J Org Chem* 1989; **54**: 1518.

39 Le Bozec H, Ouzzine K, Dixneuf PH. *Organometallics* 1991; **10**: 2768.

40 Doucet H, Höfer J, Derrien N, Bruneau C, Dixneuf PH. *Bull Soc Chim Fr* 1996; **133**: 939.

41 Doucet H, Höfer J, Bruneau C, Dixneuf PH. *J Chem Soc Chem Commun* 1993: 850.

42 Doucet H, Martin-Vaca B, Bruneau C, Dixneuf PH. *J Org Chem* 1995; **60**: 850.

43 Doucet H, Derrien N, Kabouche Z, Bruneau C, Dixneuf PH. *J Organomet Chem* 1998; **551**: 151.

44 Doucet H, Bruneau C, Dixneuf PH. *Synlett* 1997: 807.

45 Doucet H. PhD Thesis, Rennes, 1994.

46 Picquet M, Bruneau C, Dixneuf PH. *J Chem Soc Chem Commun* 1997: 1201.

47 Chabardès P. *Tetrahedron Lett* 1988; **48**: 6253; Nasaka K, Kusama H, Hayashi Y. *Chem Lett* 1991: 1413; Lorber CY, Osborn JA. *Tetrahedron Lett* 1996; **37**: 853.

48 Darcel C, Bruneau C, Dixneuf PH, Roberts SM. *Tetrahedron* 1997; **53**: 9241.

49 Trost BM, Portnoy M, Kurihara H. *J Am Chem Soc* 1997; **119**: 836.

50 Trost BM, Dyker G, Kulawiec RJ. *J Am Chem Soc* 1990; **112**: 7809.

51 Dérien S, Gomez Vicente B, Dixneuf PH. *J Chem Soc Chem Commun* 1997: 1405.

52 Selegue JP. *Organometallics* 1982; **1**: 217; Touchard D, Guesmi S, Bouchaib M *et al. Organometallics* 1996; **15**: 2579.

53 Le Bozec H, Ouzzine K, Dixneuf PH. *Organometallics* 1991; **10**: 2768; Le Bozec H, Pilette D, Dixneuf PH. *New J Chem* 1990; **14**: 793; Le Bozec H, Ouzzine K, Dixneuf PH. *J Chem Soc Chem Commun* 1989: 219; Pilette D, Moreau S, Le Bozec H *et al. J Chem Soc Chem Commun* 1994: 409; Haquette P, Pirio N, Touchard D, Toupet L, Dixneuf PH. *J Chem Soc Chem Commun* 1993: 163; Touchard D, Haquette P, Daridor A, Toupet L, Dixneuf PH. *J Am Chem Soc* 1994; **116**: 11157; Touchard D, Pirio N, Dixneuf PH. *Organometallics* 1995; **14**: 4920.

21 Palladium-catalysed Asymmetric Hydrosilylation of Olefins

TAMIO HAYASHI

Department of Chemistry, Graduate School of Science, Kyoto University, Sakyo, Kyoto 606-8502, Japan

1 Introduction

Asymmetric reactions catalysed by transition metal complexes containing enantiomerically pure chiral ligands have attracted significant interest owing to their synthetic utility [1]. Catalytic asymmetric synthesis requires ideally only one molecule of the chiral catalyst in order to produce a large quantity of an optically active substance. Some of the catalytic asymmetric reactions have developed so well that they are useful practically for the synthesis of key intermediates for biologically active compounds. Representatives are rhodium- or ruthenium-catalysed asymmetric hydrogenation of olefins and ketones and asymmetric epoxidation and dihydroxylation of olefins [1]. However, other types of reactions remain not so useful in terms of enantioselectivity and catalytic activity, although attempts have been made to apply a number of catalytic reactions to asymmetric synthesis.

We reported the first example of catalytic asymmetric hydrosilylation in 1971, when the reaction of 2-phenylpropene with methyldichlorosilane in the presence of a chiral phosphine–platinum catalyst gave 5% enantiomeric excess (ee) [2]. At that time asymmetric hydrosilylation was not as attractive as asymmetric hydrogenation because organosilicon compounds of such low % ee have no practical use. In 1977 Tamao reported the oxidative cleavage of a carbon–silicon bond into a carbon–oxygen bond, which proceeds with retention of configuration at the stereogenic carbon center [3]; with this efficient oxidation, asymmetric hydrosilylation became a method of forming optically active alcohols from olefins, i.e. overall catalytic asymmetric functionalization of olefins. The catalytic asymmetric synthesis of optically active alcohols through asymmetric hydrosilylation has been reported, but asymmetric hydrosilylation has been typically one of low enantioselectivity [1,4].

It is well documented that the hydrosilylation of olefins is catalysed by platinum, rhodium and nickel complexes [5]. On the other hand, rather surprisingly, little attention has been paid to the use of palladium catalysts for hydrosilylation [6] except for the reaction of 1,3-dienes and styrene derivatives [5]. We have concentrated our efforts into the use of palladium catalysts for asymmetric hydrosilylation. It was found that palladium complexes coordinated with bisphosphine ligands such as 1,4-bis(diphenylphosphino)butane (dppb) did not catalyse the hydrosilylation of olefins at all, even upon elevation to a high temperature, while hydrosilylation took place under mild conditions with monodentate phosphine ligands. It follows that efficient chiral monodentate phosphine ligands are required for catalytic asymmetric synthesis to be viable. Although a number of chiral phosphine ligands have been prepared and used for transition-metal-catalysed reactions, most of them are bisphosphines, which are in general anticipated to be effective in constructing a chiral environment by chelate coordination to a metal [1]. On the other hand, there have been reported only a limited number of monodentate chiral phosphine ligands, probably because they have been described as being of little practical use [7]. We have chosen the chiral binaphthyl skeleton as the basic structure of the monodentate phosphine ligand because in the case of using axially chiral binaphthyl compounds to construct an effective chiral template for asymmetric reactions there are numerous examples documented in the literature [1]: they are 2-(diphenylphosphino)-2′-methoxy-1,1′-binaphthyl (MeO-MOP, **1a**) and its derivatives. Here we describe the use of MOP ligands for the palladium-catalysed asymmetric hydrosilylation of olefins, which proceeds with high enantioselectivity as well as with high catalytic activity and provides a new efficient route to optically active alcohols.

2   Preparation of MOP ligands

Morgans and co-workers have reported [8] the selective monophosphinylation of 2,2′-bis(trifluoromethanesulfonyloxy)-1,1′-binaphthyl (**2**) with diphenylphosphine oxide in the presence of a palladium catalyst to give a high yield of 2-diphenylphosphinyl-2′-trifluoromethanesulfonyloxy-1,1′-binaphthyl (**3**), which attracted our attention as a versatile starting compound for the preparation of chiral monophosphine ligands. The triflate group on compound **3** was considered to be a convenient functionality for the introduction of various types of functional groups onto the binaphthyl ring. The conversion of compound **3** into 2-(diphenylphosphino)-2′-methoxy-1,1′-binaphthyl (MeO-MOP, **1a**) was achieved [9,10] in high yield by the three-step sequence shown in Scheme 21.1. Thus, the triflate (*S*)-**3** was hydrolysed with aqueous sodium hydroxide to give a 99% yield of alcohol, and its phenolic hydroxy group was alkylated by treatment with methyl iodide in the presence of potassium carbonate in acetone to give a 99% yield of the methyl ether (*S*)-**4a**. Reduction of the phosphine oxide with trichlorosilane and triethylamine in refluxing xylene led to (*S*)-MeO-MOP (**1a**) in 97% yield. The overall yield from 2,2′-dihydroxy-1,1′-binaphthyl was *c* 90%. Similar phosphines containing a benzyl ether and an isopropyl ether, (*S*)-**1b** and (*S*)-**1c**, were also prepared by alkylation of the phenol oxygen with benzyl bromide and isopropyl iodide, respectively, followed by reduction of the phosphine oxide.

The trifluoromethanesulfonyloxy group on the 2′ position can be replaced by an alkyl group by nickel-catalysed cross-coupling with a Grignard reagent. Introduction of an ethyl group on (*S*)-**3** with ethylmagnesium bromide followed by the reduction with trichlorosilane gave (*R*)-**1d** in 64% overall yield (Scheme 21.2). A cyano group also can be introduced at the 2′ position of compound **3** in quantitative yield by nickel-catalysed cyanation

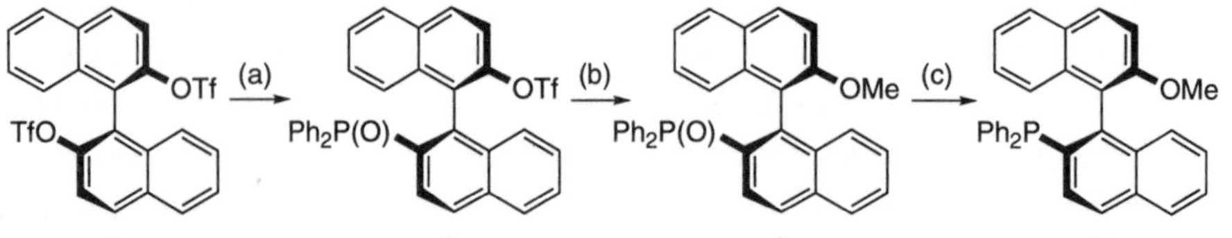

(a) Ph$_2$POH (2 eq), Pd(OAc)$_2$ (5 mol %), dppb (5 mol %), *i*-Pr$_2$NEt (4 eq), DMSO, 100 °C, 12 h (**3**, 95%). (b) (i) 3N NaOH, 1,4-dioxane, methanol. (ii) MeI (4 eq), K$_2$CO$_3$ (4 eq), acetone, reflux, 3 h (**4a**, 99%). (c) Et$_3$N (20 eq), HSiCl$_3$ (5 eq), xylene, 120 °C, 5 h, (**1a**, 97%).

Scheme 21.1. Preparation of MeO-MOP (**1a**).

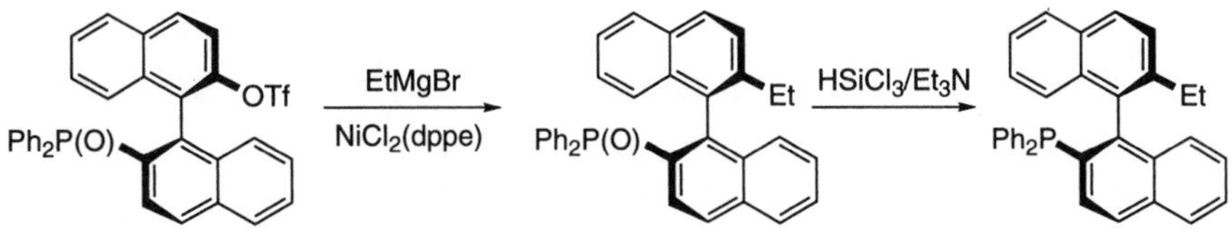

Scheme 21.2. Preparation of Et-MOP (**1d**).

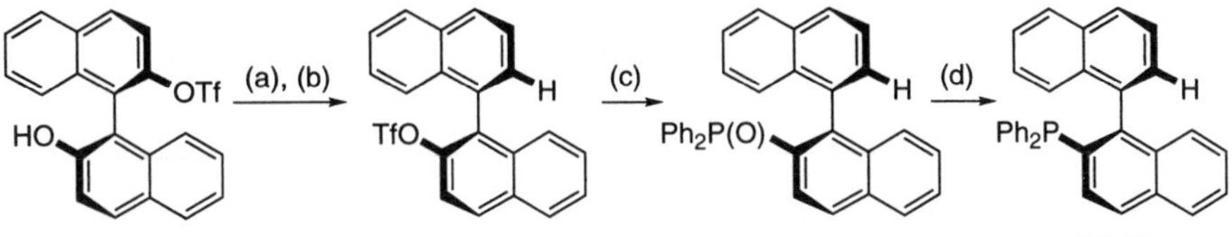

(a) H$_2$ (1 atm), 10% Pd-C, *i*-Pr$_2$NEt (2 eq) (100%). (b) Tf$_2$O (1.2 eq), pyridine (2.5 eq) (92%).
(c) Ph$_2$POH (2 eq), Pd(OAc)$_2$ (5 mol %), dppb (5 mol %), *i*-Pr$_2$NEt (4 eq), DMSO, 100 °C, 8 h
(88%). (d) Et$_3$N (20 eq), HSiCl$_3$ (5 eq), toluene, 100 °C, 16 h, (**1j**, 90%).

Scheme 21.3. Preparation of H-MOP (**1j**).

with potassium cyanide to give compound **1e** after reduction of the phosphine oxide [11]. Reduction of the cyano group with diborane followed by methylation with formaldehyde/ formic acid gave aminophosphine **1f**. The MOP ligands **1g** and **1h**, which contain an ester and carboxylic acid, respectively, were prepared through palladium-catalysed monocarbonylation of bis(triflate) **2**, giving 2-methoxycarbonyl-2′-trifluoromethanesulfonyloxy-1,1′-binaphthyl [11].

The MOP derivative (*R*)-**1j** bearing no substituent at the 2′ position, which is needed to evaluate the steric and/or electronic effects of various functional groups in other MOP derivatives, was prepared starting from (*S*)-2-hydroxy-2′-trifluoromethanesulfonyloxy-1, 1′-binaphthyl (**5**) by a sequence of reactions including palladium-catalysed phosphinylation and reduction of the resulting phosphine oxide (Scheme 21.3) [11].

The enantiomerically pure monophosphine containing the biphenanthryl skeleton, MOP-phen (**6**), was also prepared by a sequence of reactions from 3,3′-dihydroxy-4,4′-biphenanthryl that are essentially the same as those for the binaphthyl analog **1a** [12].

We have found a new catalytic method for the preparation of enantiomerically pure axially chiral biaryls: an enantioposition-selective substitution reaction of one of the two enantiotopic triflate groups on achiral biaryl ditriflates (Scheme 21.4) [13]. One of the monosubstitution products was readily converted into the enantiomerically pure monophosphine ligand. Thus, the reaction of 1-[2,6-bis[[(trifluoromethyl)sulfonyl]oxy] phenyl]naphthalene (**7a**) with phenylmagnesium bromide in the presence of 5 mol.% of PdCl$_2$[(*S*)-phephos], where phephos stands for 2-(dimethylamino)-1-(diphenylphosphino)-3-phenylpropane [14], and 1 equiv of lithium bromide at −30 °C for 48 h gave an 87% yield of axially chiral monophenylated biaryl (*S*)-**8a** in 93% ee and a 13% yield of diphenylated biaryl **9a**. The biaryl **8a** is readily made enantiomerically pure with high recovery by simple recrystallization. The enantiomerically pure monotriflate (*S*)-**8a** was subjected to palladium-catalysed diphenylphosphinylation [8] followed by reduction of

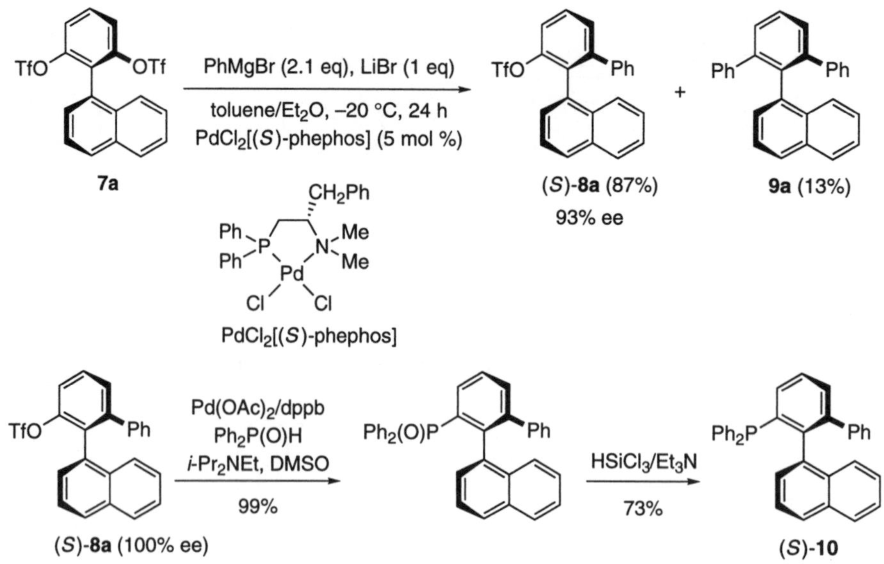

Scheme 21.4. Preparation of MOP ligand **10** via catalytic asymmetric cross-coupling.

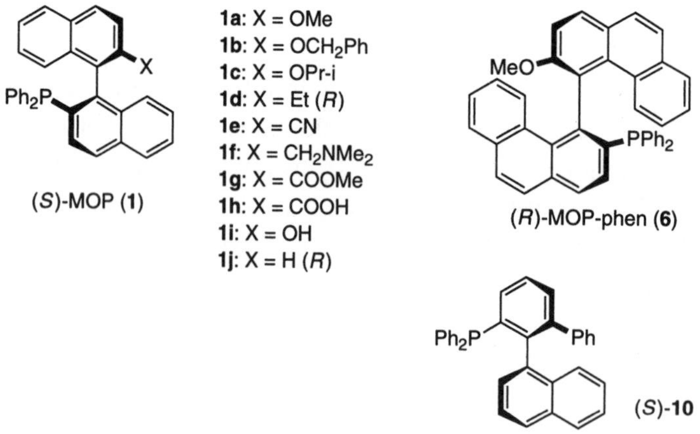

Figure 21.1. MOP ligands prepared and used for catalytic asymmetric reactions.

diphenylphosphine oxide with trichlorosilane and triethylamine, giving the new axially chiral triarylmonophosphine (*S*)-**10**. The MOP ligands prepared and used for catalytic asymmetric reactions are shown in Fig. 21.1.

The crystal structure of *trans*-[PdCl₂{(*R*)-MeO-MOP}₂] is shown in Fig. 21.2 [15]. The complex has a square-planar geometry with two phosphorus atoms and two chlorine atoms, where the MOP ligand coordinates to palladium with the phosphorus atom as a monodentate ligand. The phosphorus atoms or chlorine atoms are *trans* to each other. It should be noted that the naphthyl ring having a methoxy group plays an important role in the construction of the chiral environment of the palladium. Thus, the naphthyl ring A (A′) points toward the vicinity of palladium, while the methoxy group is located in the side opposite palladium. The conformation of the naphthyl group where the C2′ substituent is well removed from the palladium center is interesting. The phenyls B (B′) and C (C′) are situated below and above the plane around the palladium atom. These structural features

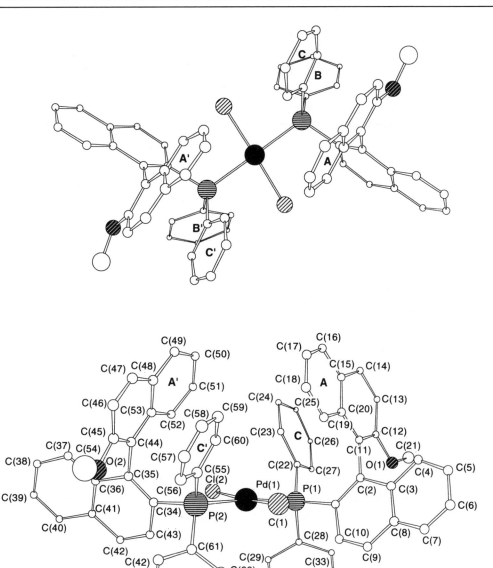

Figure 21.2. Molecular structure for *trans*-PdCl$_2$\{(*R*)-MeO-MOP\}$_2$·Et$_2$O. Ether molecule is omitted for simplicity.

are very different from those commonly observed in complexes coordinated with chiral bidentate bis(phosphino) ligands, such as BINAP [16].

3   Asymmetric hydrosilylation of simple terminal olefins

Hydrosilylation of terminal olefins catalysed by platinum, rhodium or nickel complexes is known to proceed with anti-Markovnikov selectivity to 1-silylalkanes [5]. In order to develop a catalyst that possesses high catalytic activity, is highly regioselective in giving 2-silylalkanes and is highly enantioselective in the hydrosilylation, we examined several types of phosphine–palladium catalysts for the reaction of 1-hexene (**11a**) with trichlorosilane. Palladium complexes coordinated with a chelating bis(phosphine), 1,4-bis-(diphenylphosphino)butane (dppb), 2,3-bis(diphenylphosphino)butane (chiraphos) or 2,2′-bis(diphenylphosphino)-1,1′-binaphthyl (BINAP) did not catalyse the hydrosilylation at all, even upon elevation of the reaction temperature to 80 °C. However, the reaction took place at 40 °C with monodentate phosphine ligands, although the chemical yields in forming hexylsilanes were low. For example, the reaction in the presence of 0.1 mol.% of a palladium–triphenylphosphine catalyst (P/Pd = 2 : 1) at 40 °C for 24 h gave 12% yield of the hydrosilylation products consisting of 2-hexylsilane (**12a**) and its 1-isomer (**12′a**) in a ratio of 9 : 91, the hydrosilylation being accompanied by isomerization of 1-hexene into internal olefins. The regioselectivity forming 2-silylhexane (**12a**) was increased to some extent by the use of sterically more bulky monophosphine ligands, namely pentafluorophenyl(diphenyl)phosphine and tris(2-methylphenyl)phosphine, giving compound **12a** with 15% and 22% regioselectivity, respectively, although the low chemical yield (<20%) was still the plague of this reaction. It is reasonable to expect that a monodentate phosphine ligand generates a palladium catalyst that is more active for hydrosilylation than a chelating bis(phosphine) ligand. The former can form a square-planar palladium(II) intermediate $PdH(SiCl_3)L(CH_2{=}CHR)$ (L = monophosphine) that offers a coordination site for the activation of the olefin, while the latter cannot. Studies of the effects of monodentate phosphine ligands on the catalytic activity and the regioselectivity forming 1-alkylsilanes or 2-alkylsilanes in palladium-catalysed hydrosilylation revealed that (*S*)-2-(diphenylphosphino)-2′-methoxy-1,1′-binaphthyl (MeO-MOP, **1a**) is a unique ligand for the hydrosilylation, its palladium complex exhibiting both high catalytic activity and unusually high regioselectivity in forming 2-alkylsilanes, and, moreover, high enantioselectivity [9, 15]. The predominant formation of 2-alkylsilanes from aliphatic 1-olefins has never before been observed with any transition metal catalysts. Mechanistic studies using deuterated olefins suggested that the catalytic cycle includes both Pd(1-alkyl)L(silyl) and Pd(2-alkyl)L(silyl) intermediates, which are in equilibrium with one another, and that the MOP ligand can accelerate reductive elimination of 2-silylalkane from the 2-alkylpalladium intermediate (Y. Uozumi, K. Kitayama and T. Hayashi, unpubl. data).

The results obtained for the asymmetric hydrosilylation of 1-alkenes (**11**) with trichlorosilane [9,15] are summarized in Scheme 21.5. The hydrosilylation products,

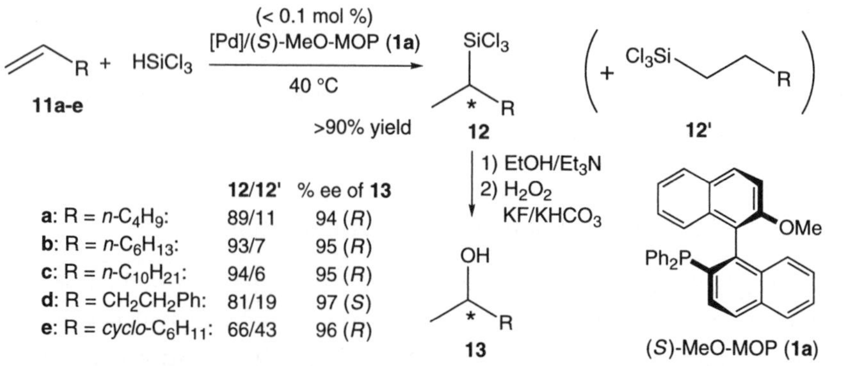
(< 0.1 mol %)
[Pd]/(S)-MeO-MOP (1a)
R + HSiCl3
40 °C
11a-e
>90% yield
SiCl3
R
12
Cl3Si
R
12'
1) EtOH/Et3N
2) H2O2
KF/KHCO3
OH
R
13
OMe
Ph2P
(S)-MeO-MOP (1a)
12/12' % ee of 13
a: R = n-C4H9: 89/11 94 (R)
b: R = n-C6H13: 93/7 95 (R)
c: R = n-C10H21: 94/6 95 (R)
d: R = CH2CH2Ph: 81/19 97 (S)
e: R = cyclo-C6H11: 66/43 96 (R)

Scheme 21.5. Palladium-catalysed asymmetric hydrosilylation of 1-alkenes with MeO-MOP (**1a**).

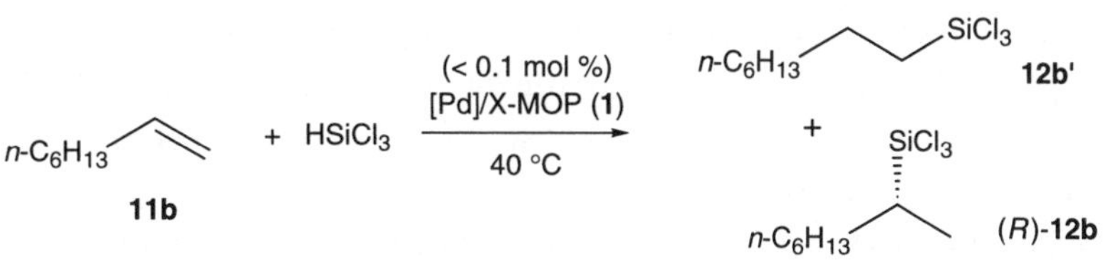

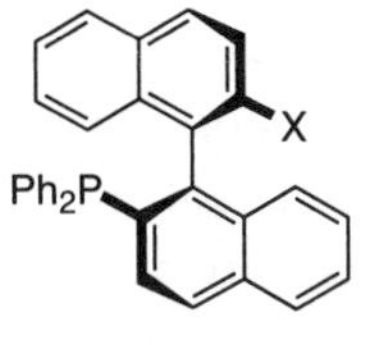

X-MOP

X-MOP	yield %	12b : 12b'	% ee
(*S*)-MeO-MOP (**1a**)	83	93 : 7	95 (*R*)
(*S*)-PhCH$_2$O-MOP (**1b**)	85	80 : 20	95 (*R*)
(*S*)-*i*-PrO-MOP (**1c**)	88	90 : 10	91 (*R*)
(*R*)-Et-MOP (**1d**)	80	90 : 10	93 (*R*)

Scheme 21.6. Palladium-catalysed asymmetric hydrosilylation of 1-octene with MOP ligands (**1a–d**).

namely 2-alkyl(trichloro)silanes (**12**), were readily converted into optically active 2-alkanols (**13**) by treatment with EtOH/Et$_3$N followed by oxidation of the resulting (triethoxy)silanes with hydrogen peroxide in the presence of a fluoride anion [3]. The terminal olefins, namely 1-hexene (**11a**), 1-octene (**11b**), 1-dodecene (**11c**), 4-phenyl-1-butene (**11d**) and vinylcyclohexane (**11e**), were transformed efficiently into the corresponding optically active alcohols (**13**) with enantioselectivities ranging between 94% and 97% ee by the catalytic hydrosilylation–oxidation procedure, the selectivity being highest for the enantioface selection of simple terminal olefins. The regioselectivity in forming 2-(silyl)alkanes is surprisingly high for the terminal olefins **11a–d** substituted with a primary alkyl group. Lower regioselectivity was observed with vinylcyclohexane (**11e**), which is substituted with a sterically bulky group on the double bond. Asymmetric hydrosilylation of 4-pentenyl benzoate and 1,5-heptadiene gave corresponding 2-alkanols of 90% ee and 87% ee, respectively, with the ester carbonyl and the internal double bond remaining intact [15]. It should be noted that the palladium–MOP complex is highly active as a catalyst, hydrosilylation taking place with a mere 0.01 mol.% of the catalyst material.

The high selectivity was also observed with MOP ligands **1b**, **1c** and **1d**, which have other substituents than methoxy at the 2′ position [15] (Scheme 21.6). Thus, the hydrosilylation of 1-octene (**11b**) with MOP ligands substituted with benzyloxy or isopropoxy gave >91% enantioselectivity and >80% branch selectivity, suggesting that the steric bulkiness of the 2′-substituents has little influence on the present asymmetric hydrosilylation. The presence of an alkoxy group at the 2′ position is not essential for the high selectivity. Replacement of the alkoxy group by an alkyl group did not affect the selectivity. The lack of influence of the 2′-substituents on the stereoselectivity is ascribed to the conformation of the ligand on palladium, where the 2′-substituent is located far away from the metal (see Fig. 21.2).

4 Asymmetric hydrosilylation of cyclic olefins

Asymmetric synthesis through a selective monofunctionalization of enantiotopic positions is arguably one of the most attractive strategies for one-step construction of multiple stereogenic carbon centers [17]. In spite of the impressive development of enantioface selective asymmetric reactions catalysed by transition metal complexes, the enantio-position selective approach is yet to be developed [1]. We have applied the MOP/

palladium-catalysed hydrosilylation to the catalytic asymmetric functionalization of a meso bicyclo[2.2.1] system [18] because the optically active bicyclo[2.2.1]heptane derivatives represented by norbornanol are indispensable as versatile chiral building blocks for the synthesis of a variety of important compounds.

The hydrosilylation of norbornene (**14**) with trichlorosilane took place at 0 °C in the presence of 0.01 mol.% of the MOP–palladium catalyst to give a quantitative yield of *exo*-2-(trichlorosilyl)norbornane (**15**) as a single product (Scheme 21.7). Direct oxidation of compound **15** with hydrogen peroxide in the presence of a large excess of potassium fluoride and potassium bicarbonate gave *exo*-2-norbornanol (**16**) in yields of >90%, which was shown to be the (1*S*,2*S*,4*R*)-isomer by its optical rotation (93% ee). The hydrosilylation carried out at –20 °C raised the enantiomeric excess to 96% ee. The trichlorosilane (**15**) also can be converted into (1*S*,2*R*,4*R*)-*endo*-2-bromonorbornane (**17**) by treatment with an excess of potassium fluoride followed by bromination of the resulting pentafluorosilicate with *N*-bromosuccinimide [19]. Bicyclo[2.2.2]octene, a diester of norbornenedicarboxylic acid, and 2,5-dihydrofuran derivatives [20] also were successfully subjected to asymmetric hydrosilylation–oxidation under similar reaction conditions to give the corresponding optically active alcohols with enantioselectivity in excess of 92%.

It is remarkable that the monofunctionalization of norbornadiene (**18**) giving *exo*-5-trichlorosilyl-2-norbornene (**19a**) is effected by the palladium–MOP catalyst with high chemo- and enantioselectivity [18] (Scheme 21.8). It is in striking contrast to the reaction catalysed by chloroplatinic acid or palladium–triphenylphosphine, which gives nortricyc-

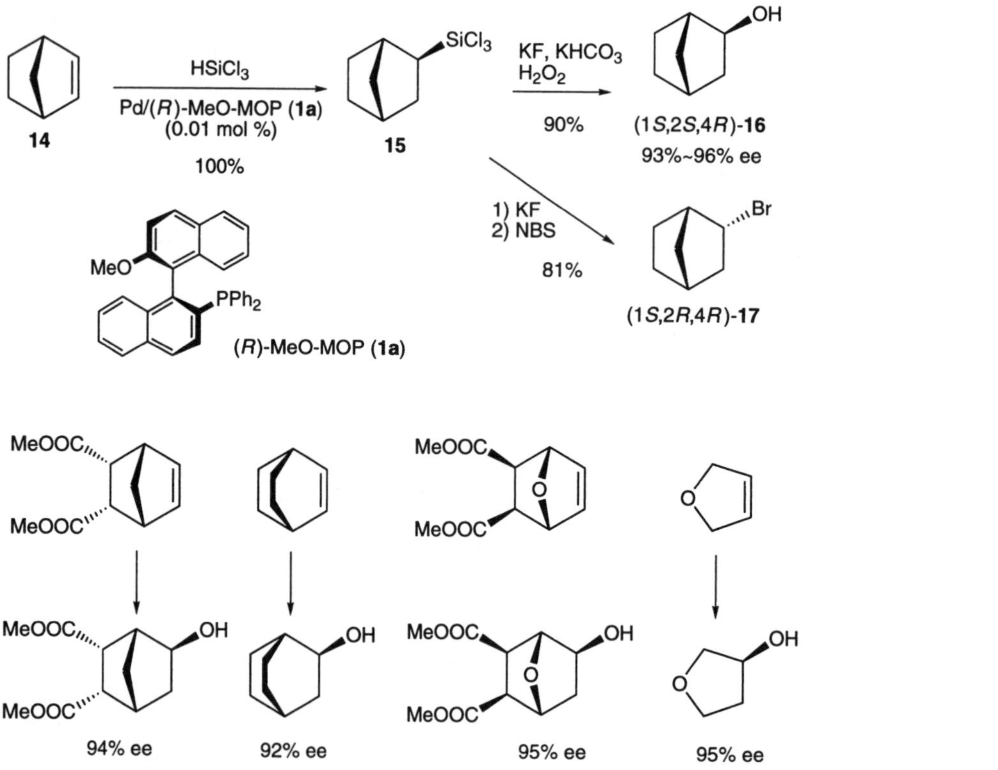

Scheme 21.7. Palladium-catalysed asymmetric hydrosilylation of norbornene and its derivatives with MeO-MOP (**1a**).

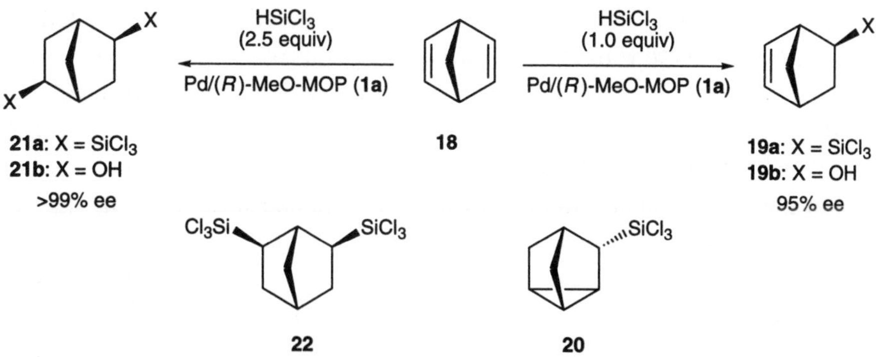

Scheme 21.8. Palladium-catalysed asymmetric hydrosilylation of norbornadiene with MeO-MOP (**1a**).

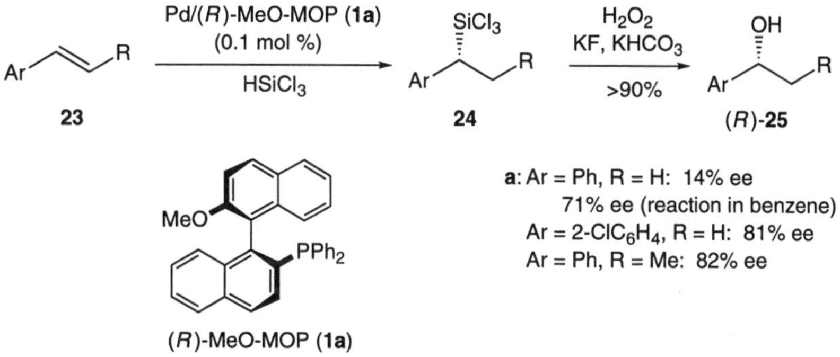

Scheme 21.9. Palladium-catalysed asymmetric hydrosilylation of styrenes with MeO-MOP (**1a**).

lene (**20**) as a major product. Thus, the reaction of compound **18** with 1.0 equiv of trichlorosilane and the palladium–MOP catalyst (0.1 mol.%) followed by the hydrogen peroxide oxidation gave (1*R*,4*R*,5*S*)-*exo*-5-hydroxy-2-norbornene (**19b**) with 95% ee. Reacting compound **18** with 2.5 equiv of trichlorosilane induced enantioselective hydrosilylation in both double bonds, thus giving a 78% yield of chiral disilylnorbornane (**21a**) and the meso isomer (**22**) in a ratio of 18:1. The oxidation of compound **21a** gave the diol (1*R*,2*S*,4*R*,5*S*)-**21b** with >99% ee, the high purity attained being due to the expected double stereoselection.

5   Asymmetric hydrosilylation of styrenes

Although simple terminal olefins such as 1-octene and cyclic olefins such as norbornene have been converted efficiently into the corresponding secondary alkyl alcohols with over 90% enantioselectivity by use of the palladium catalyst coordinated with MeO-MOP (**1a**), such high selectivity has not been observed in the hydrosilylation of styrene derivatives (Scheme 21.9) [21]. Thus, the palladium-catalysed hydrosilylation of styrene (**23a**) with trichlorosilane in the presence of MeO-MOP (**1a**) ligand under standard conditions (without solvent), followed by oxidation, gave 1-phenylethanol (**25a**) of only 14% ee. Use of benzene as solvent for the hydrosilylation improved the enantioselectivity to 71%. Although this value is the highest for the hydrosilylation of styrene at this moment, it is still low compared with the selectivity observed in the reaction of simple terminal olefins. For substituted styrenes such as *o*-chlorostyrene or β-methylstyrene, an enantioselectivity of *c* 80% was observed with the MeO-MOP ligand [22].

We have examined MOP ligands where the methoxy group at the 2′ position in MeO-MOP is replaced by several groups for their enantioselectivity in the palladium-catalysed asymmetric hydrosilylation of styrene (**23a**) (Scheme 21.10). The enantiomeric purities and absolute configuration of alcohol **25a** obtained with Et-MOP (**1d**), CN-MOP (**1e**), CO₂Me-MOP (**1g**) and HO-MOP (**1i**) are 18% ee (*R*), 26% ee (*R*), 30% ee (*S*) and 34% ee (*S*), respectively. These results suggest that the electronic nature of the substituent is not a decisive factor in the enantioselection, because all of the MOPs substituted with methoxy, hydroxy, methoxycarbonyl, cyano and ethyl groups show low enantioselectivity irrespective of their electron-withdrawing or electron-donating character. It turned out that H-MOP (**1j**), which has the same 1,1′-binaphthyl skeleton as MeO-MOP but lacks the methoxy group, is particularly effective for the palladium-catalysed hydrosilylation of styrene [23]. Hydrosilylation of styrene (**23a**) with trichlorosilane in the presence of 0.1 mol.% of H-MOP–palladium catalyst, generated *in situ* by mixing [PdCl(π-C₃H₅)]₂ and (*S*)-H-MOP (**1j**) at 0 °C for 12 h, gave a quantitative yield of 1-phenyl-1-trichlorosilylethane (**24a**) as a single regioisomer, which was converted into (*R*)-1-phenylethanol (**25a**) in 97% yield by oxidative cleavage of the carbon–silicon bond. The enantiomeric excess determined by HPLC analysis with a chiral stationary-phase column was 93% ee. The hydrosilylation carried out at −10 °C raised the enantiomeric excess slightly to 94% ee. The monophosphine (*S*)-**10** that was prepared through catalytic asymmetric cross-coupling was as effective as (*S*)-H-MOP (**1j**) for the hydrosilylation of styrene, giving (*R*)-**25a** of 91% ee [13]. Neither of the ligands (*S*)-**1j** or (*S*)-**10** has any substituent at the 2′ position. It follows that the small size of the hydrogen at the 2′ position in H-MOP (**1j**) is important for high enantioselectivity. The dihedral angle between the two naphthyl rings in the binaphthyl skeleton, which is controlled by the steric bulkiness of the 2′-substituent, is presumably related to the enantioselectivity.

The H-MOP–palladium complex also catalysed the asymmetric hydrosilylation of styrene derivatives substituted on the phenyl ring (derivatives **23b–e**) and β-alkyl-

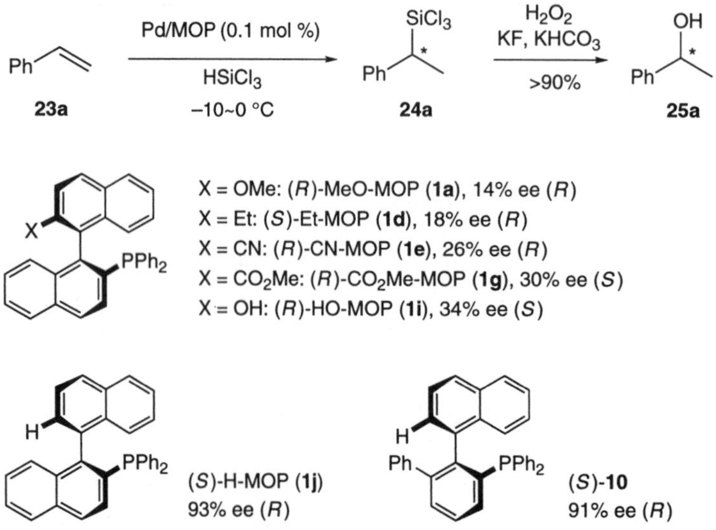

Scheme 21.10. Palladium-catalysed asymmetric hydrosilylation of styrene with several MOP ligands.

substituted styrenes **23f** and **23g** to give the corresponding benzylic alcohols (*R*)-**25b–g** [23] (Scheme 21.11). The enantioselectivity is high for the styrenes containing electron-withdrawing groups on the phenyl, *p*-chlorostyrene and *p*-trifluoromethylstyrene, giving the corresponding alcohols in >94% ee, but unfortunately it is not so high for those containing electron-donating groups on the phenyl ring. The enantiomeric purity of alcohol **25c** obtained by the asymmetric hydrosilylation of *p*-methoxystyrene is 61% ee. The reaction of β-substituted styrenes proceeded with perfect regioselectivity to give the corresponding benzylic alcohols of *c* 90% ee. Interestingly, H-MOP–palladium catalyst was less enantioselective and/or less active than MeO-MOP–palladium for the hydrosilylation of non-styrene-type olefins such as 1-octene and norbornene.

The H-MOP ligand was further modified for higher enantioselectivity in the asymmetric hydrosilylation of styrenes. Several functional groups, including methoxy and trifluoromethyl groups, were introduced on the phenyl rings of the diphenylphosphino group. The representative results are shown in Scheme 21.12 (Y. Uozumi, K. Kitayama, S. Hirate and T. Hayashi, unpubl. data). It was found that the palladium catalyst co-ordinated with (*S*)-2-bis[(3,5-trifluoromethyl)phenyl]phosphino-1,1′-binaphthyl (**26**) is more catalytically active and more enantioselective than others. The hydrosilylation of styrene with trichlorosilane in the presence of 0.1 mol.% of the palladium–**26** catalyst at 0 °C was completed in 1 h to give the hydrosilylation product in 97% ee. The reaction carried out at −10 °C increased the enantioselectivity up to 98%. The high catalytic activity can

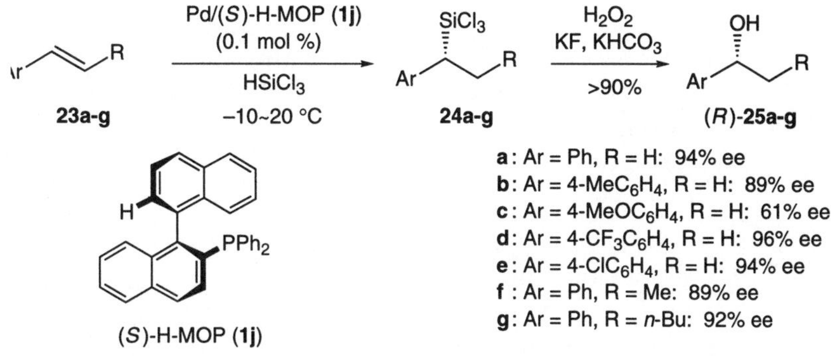

Scheme 21.11. Palladium-catalysed asymmetric hydrosilylation of styrenes with H-MOP (**1j**).

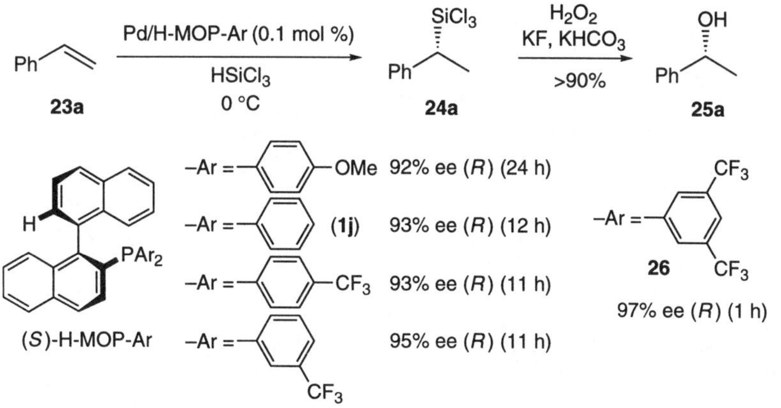

Scheme 21.12. Palladium-catalysed asymmetric hydrosilylation of styrene with modified H-MOP ligands.

be ascribed to the electron-withdrawing character of two trifluoromethyl groups on the phenyl, which will accelerate the reductive elimination step in the catalytic cycle of hydrosilylation.

6 Asymmetric hydrosilylation of 1,3-dienes

Palladium-catalysed hydrosilylation of 1,3-dienes is one of the important synthetic methods for allylic silanes, and considerable attention has been paid to their asymmetric synthesis by this catalytic method. Unfortunately, the binaphthyl monophosphine MeO-MOP (**1a**) or H-MOP (**1j**) is not as effective as a chiral ligand for the asymmetric hydrosilylation of 1,3-dienes as it is for other types of prochiral olefins shown above, where >90% enantioselectivity is usually observed. We found that MOP-phen (**6**), which is the 4,4′-biphenanthryl analog of MeO-MOP, shows higher enantioselectivity than others in the hydrosilylation of cyclic 1,3-dienes to give optically active allylic silanes (Scheme 21.13) [24,25].

The reaction of cyclopentadiene (**27a**) with trichlorosilane in the presence of MOP–phen-palladium catalyst proceeded at 20 °C in a 1,4-fashion to give a quantitative yield of (*R*)-3-(trichlorosilyl)cyclopentene (**28a**). The enantiomeric purity was determined to be 80% ee by HPLC analysis of (cyclopent-2-enyl)(phenyl)methanol (**29a**), which was obtained in 92% yield by treatment of the allylsilane **28a** with benzaldehyde in DMF according to Kobayashi's procedure [26]. Much lower enantioselectivity was observed in

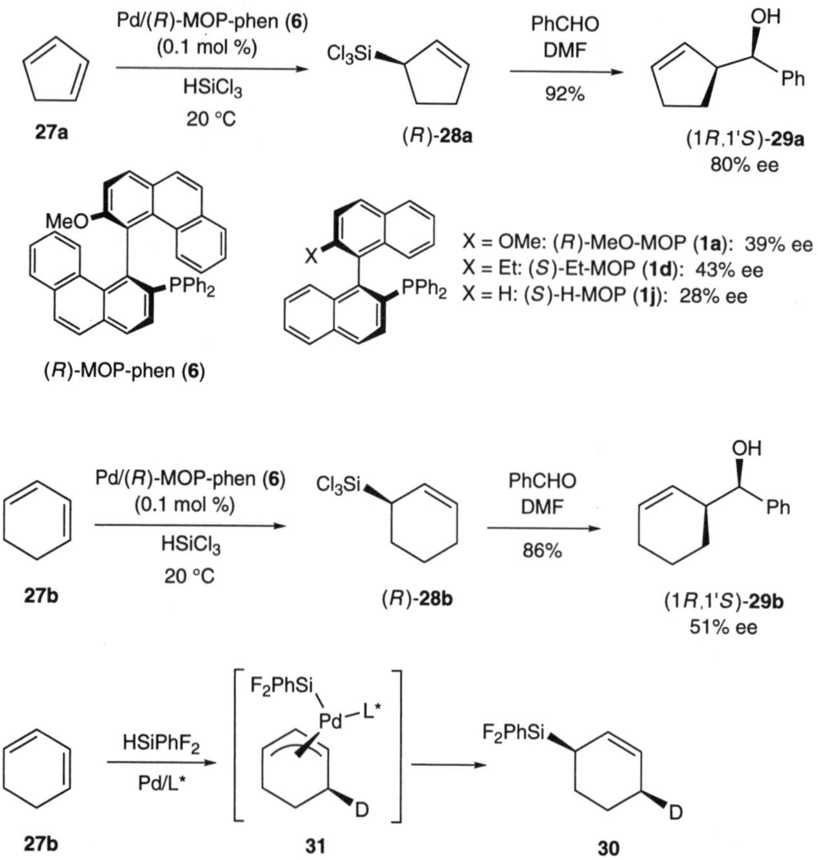

Scheme 21.13. Palladium-catalysed asymmetric hydrosilylation of 1,3-dienes with MOP-phen (**6**).

the hydrosilylation of compound **27a** with MOP ligands whose basic structure is the binaphthyl skeleton. Thus, MeO-MOP (**1a**), Et-MOP (**1d**) and H-MOP (**1j**) gave the allylsilane **28a** in 39% ee, 43% ee and 28% ee, respectively. In the asymmetric hydrosilylation of 1,3-cyclohexadiene (**27b**), MOP-phen (**6**) also exhibited higher enantioselectivity than MeO-MOP (**1a**). The $S_{E'}$ allylation of benzaldehyde was demonstrated to proceed through a six-membered cyclic transition state by the stereochemical outcome in the reaction of the allylsilane (R)-**28b** to form the homoallyl alcohol (1R,1'S)-**29b**. The use of phenyldifluorosilane in place of trichlorosilane did not improve the enantioselectivity, but the reaction with deuterium-labelled silane ($DSiF_2Ph$) gave us significant insight into the mechanism of palladium-catalysed hydrosilylation of 1,3-dienes. The reaction of 1,3-cyclohexadiene (**27b**) with $DSiF_2Ph$ gave cis-3-(phenyldifluorosilyl)-6-deuteriocyclohexene (**30**) as a single isomer without any diastereo- or regioisomers, demonstrating that 1,4-cis-addition is an exclusive pathway. The π-allylpalladium intermediate (**31**), which is formed by the addition of palladium hydride on a PdH(Si)L* species to the diene and has the silyl group located at the trans position to the π-allyl carbon next to the deuterated carbon, rapidly undergoes reductive elimination to form compound **30** before trans–cis isomerization of intermediate **31** can occur. It follows that the stereochemical outcome is determined in the enantioselective addition of palladium hydride to the diene.

7 Concluding remarks

The axially chiral MeO-MOP ligand was also found to be very useful for several other asymmetric reactions where chelating bisphosphine–metal complexes cannot be used because of their low catalytic activity or low selectivity towards a desired reaction pathway. Examples are:

1 Palladium-catalysed asymmetric 1,4-hydroboration of 1,3-enynes with catecholborane to form axially chiral allenylboranes [27, 28].

2 Palladium-catalysed asymmetric reduction of allylic esters with formic acid, which proceeds with high regioselectivity to give less-substituted olefins with high enantioselectivity [29–31].

3 Palladium-catalysed alkylation of allylic esters with soft carbon nucleophiles, which produces allylic alkylation products with high enantioselectivity [32].

   The MOP ligands can be modified on the side-chain of the 2' position as well as the biaryl skeleton according to the demand of the reaction type, and they are expected to find great utility in other types of catalytic asymmetric reactions where the use of monodentate phosphine ligands is essential or favorable for steric and/or electronic reasons.

8 Acknowledgments

The author wishes to thank Dr Uozumi and his co-workers, whose names are listed in the references, for their invaluable contributions.

9 References

1 For recent reviews: Brunner H. *Synthesis* 1988: 645–54; Brunner H. *Top Stereochem* 1988; **18**: 129–247; Noyori R, Kitamura M. In: Scheffold R, ed. *Modern Synthetic Methods*, Vol. 5. 115–98 New York: Springer-Verlag, 1989; Ojima I, Clos N, Bastos C. *Tetrahedron* 1989; **45**: 6091–939;

Ojima I. *Catalytic Asymmetric Synthesis*. New York: VCH, 1993; Noyori R. *Asymmetric Catalysis in Organic Synthesis*. New York: Wiley, 1994.

2   Yamamoto K, Hayashi T, Kumada M. *J Am Chem Soc* 1971; **93**: 5301.

3   Tamao K. In: Larson GL, ed. *Advances in Silicon Chemistry*, Vol. 3. London: JAI Press, 1996: 1–62; Fleming I. *Chemtracts—Organic Chemistry* 1996; **9**: 1–64; Tamao K. In: Sakurai H, ed. *Organosilicon and Bioorganosilicon Chemistry*. Chichester: Ellis Horwood, 1985: 231–42; Tamao K, Ishida N, Tanaka T, Kumada M. *Organometallics* 1983; **2**: 1694–6; Tamao K, Ishida N. *J Organomet Chem* 1984; **269**: C37–9; Tamao K, Nakajo E, Ito Y. *J Org Chem* 1987; **52**: 4412–14.

4   Hayashi T, Tamao K, Katsuro Y, Nakae I, Kumada M. *Tetrahedron Lett* 1980; **21**: 1871; Hayashi T, Kabeta K. *Tetrahedron Lett* 1985; **26**: 3023–6, and references cited therein.

5   For a review: Ojima I. In: Patai S, Rappoport Z, eds. *The Chemistry of Organic Silicon Compounds*. Chichester: Wiley, 1989: 1479–526.

6   Tsuji J, Hara M, Ohno K. *Tetrahedron* 1974; **30**: 2143–6.

7   Examples of optically active monophosphine ligands: (*S*)-(*o*-Methoxyphenyl)-cyclohexyl-methylphosphine ((*S*)-CAMP): Knowles WS, Sabacky MJ, Vineyard BD. *J Chem Soc Chem Commun* 1972: 10–11. Neomentyldiphenylphosphine: Morrison JD, Burnett RE, Aguiar AM, Morrow CJ, Phillips C. *J Am Chem Soc* 1971; **93**: 1301–3.

8   Kurz L, Lee G, Morgans Jr D, Waldyke MJ, Ward T. *Tetrahedron Lett* 1990; **31**: 6321–4.

9   Uozumi Y, Hayashi T. *J Am Chem Soc* 1991; **113**: 9887–8.

10 Uozumi Y, Tanahashi A, Lee S-Y, Hayashi T. *J Org Chem* 1993; **58**: 1945–8.

11 Uozumi Y, Suzuki N, Ogiwara A, Hayashi T. *Tetrahedron* 1994; **50**: 4293–302.

12 Hayashi T, Iwamura H, Uozumi Y, Matsumoto Y, Ozawa F. *Synthesis* 1994: 526–32.

13 Hayashi T, Niizuma S, Kamikawa T, Suzuki N, Uoumi Y. *J Am Chem Soc* 1995; **117**: 9101–2.

14 Hayashi T, Konishi M, Fukushima M *et al. J Org Chem* 1983; **48**: 2195–202.

15 Uozumi Y, Kitayama K, Hayashi T, Yanagi K, Fukuyo E. *Bull Chem Soc Jpn* 1995; **68**: 713–22.

16 For example, the structure of PdCl$_2${(*R*)-BINAP} has been reported: Ozawa F, Kubo A, Matsumoto Y, Hayashi T *et al. Organometallics* 1993; **12**: 4188–96.

17 For reviews: Morrison JD, ed. *Asymmetric Synthesis*, Vols 1–5. London: Academic Press, 1983–1985; Nógrádi M. *Stereoselective Synthesis*. New York: Weinheim, 1987.

18 Uozumi Y, Lee S-Y, Hayashi T. *Tetrahedron Lett* 1992; **33**: 7185–8.

19 Tamao K, Yoshida J, Murata M, Kumada M. *J Am Chem Soc* 1980; **102**: 3267–9.

20 Uozumi Y, Hayashi T. *Tetrahedron Lett* 1993; **34**: 2335–8.

21 For the asymmetric hydrosilylation of styrene with other chiral ligands: Yamamoto K, Kiso Y, Ito R, Tamao K, Kumada M. *J Organomet Chem* 1981; **210**: 9–17; Okada T, Morimoto T, Achiwa K. *Chem Lett* 1990: 999–1002; Marinetti A. *Tetrahedron Lett* 1994; **35**: 5861–4; Marinetti A, Ricard L. *Organometallics* 1994; **13**: 3956–62.

22 Uozumi Y, Kitayama K, Hayashi T. *Tetrahedron Asymm* 1993; **4**: 2419–22.

23 Kitayama K, Uozumi Y, Hayashi T. *J Chem Soc Chem Commun* 1995: 1533–4.

24 Kitayama K, Tsuji H, Uozumi Y, Hayashi T. *Tetrahedron Lett* 1996; **37**: 4169–72.

25 For the asymmetric hydrosilylation of 1,3-dienes with other chiral ligands: Hayashi T, Hengrasmee S, Matsumoto Y. *Chem Lett* 1990: 1377–80, and references cited therein; Okada T, Morimoto T, Achiwa K. *Chem Lett* 1990: 999–1002; Marinetti A, Ricard L. *Organometallics* 1994; **13**: 3956–62; Hatanaka Y, Goda K, Yamashita F, Hiyama T. *Tetrahedron Lett* 1994; **35**: 7981–2.

26 Kobayashi S, Nishio K. *J Org Chem* 1994; **59**: 6620–8.

27 Matsumoto Y, Naito M, Hayashi T. *Organometallics* 1992; **11**: 2732–4.

28 Matsumoto Y, Naito M, Uozumi Y, Hayashi T. *J Chem Soc Chem Commun* 1993: 1468–9.

29 Tsuji J, Yamakawa T. *Tetrahedron Lett* 1979: 613; Tsuji J, Shimizu I, Minami I. *Chem Lett* 1984: 1017–20; Tsuji J, Minami I, Shimizu I. *Synthesis* 1986: 623–7; Mandai T, Matsumoto T, Kawada M, Tsuji J. *J Org Chem* 1992; **57**: 1326; Oshima M, Shimizu I, Yamamoto A, Ozawa F. *Organometallics* 1991; **10**: 1221–3.

30 Hayashi T, Iwamura H, Naito M *et al. J Am Chem Soc* 1994; **116**: 775–6.

31 Hayashi T, Iwamura H, Uozumi Y. *Tetrahedron Lett* 1994; **35**: 4813–16.

32 Hayashi T, Kawatsura M, Uozumi Y. *Chem Commun* 1997: 561–2.

22 New Organic Synthesis Based upon Palladium-catalysed Activation of Silicon–Silicon σ-Bonds

MICHINORI SUGINOME and YOSHIHIKO ITO

Department of Synthetic Chemistry and Biological Chemistry, Graduate School of Engineering, Kyoto University, Kyoto 606-8501, Japan

1 Introduction

Synthesis of new organosilicon compounds has attracted much attention from the viewpoint of utility in organic synthesis as well as synthesis of new silicon-containing materials [1]. Hence, the development of new and efficient methodologies for the preparation of organosilicon compounds is desirable. Stereo- and regioselective silicon–carbon bond-forming reactions have been exploited by means of transition metal catalysts: for example, hydrosilane is one organosilicon compound that is accessible as a starting silicon source and has been used widely for the synthesis of organosilicon compounds, particularly through transition-metal-catalysed hydrosilation of carbon–carbon multiple bonds [2]. Recently, dehydrogenative condensation of hydrosilanes catalysed by transition metals also has been noted [3].

Silicon–carbon bond formation has been accomplished by the addition of the silicon–silicon σ-bond of disilanes across carbon–carbon multiple bonds [4]. The high-lying HOMO and low-lying LUMO corresponding to σ and σ* orbitals of the Si—Si bonds enable interaction of the σ-bonds with palladium complexes. Unlike hydrosilanes, however, organodisilanes have rarely found synthetic applications, probably due to lack of effective catalysts for activation of the Si—Si bond of disilanes. Recently, palladium complexes bearing isonitrile ligands in place of conventional phosphine ligands have been found by the authors' group to provide remarkable catalytic activation of Si—Si bonds. The palladium–isonitrile complexes have been widely applicable for the new synthesis of organosilicon compounds that are otherwise inaccessible.

This chapter describes new syntheses of organosilicon compounds based on activation of the Si—Si σ-bond by palladium–isonitrile complexes.

2 σ-Bond metathesis

Palladium-catalysed Si—Si σ-bond metathesis, i.e. cleavage and recombination of the silicon–silicon bonds, was reported in the 1970s [5]. Regardless of the requirement of high temperature in the non-catalysed reaction, the palladium-catalysed metathesis reaction proceeded under relatively mild conditions [6]. Although the reaction of acyclic disilanes resulted only in disproportionation of the Si—Si bonds, cyclic disilanes underwent σ-bond metathesis leading to new silicon–silicon bond formation. The cyclodimerization reaction of cyclic disilanes suggested the potent usefulness of Si—Si bond metathesis as a strategy for the efficient synthesis of organosilicon macrocycles that are otherwise difficult to synthesize [5, 7]. Although the detailed mechanism of the metathesis reaction has not been elucidated, interaction of the silicon–silicon σ-bond with the bis(organosilyl)palladium complex initially formed as a catalyst may be crucially involved.

On the other hand, it is also known that cyclic disilanes undergo ring-opening polymerization in the presence of certain palladium catalysts, for which a different mechanism from that of ring-enlargement oligomerization may be proposed [8, 9].

New syntheses of organosilicon compounds are described herein on the basis of Si–Si σ-bond metathesis catalysed by palladium–isonitrile complexes.

2.1 *Ring-enlargement oligomerization*

Ring-enlargement dimerization of a five-membered cyclic disilane (**1**) was reported using a palladium–phosphine catalyst (Equation (22.1)) [5]. Although the bis(organosilyl)palladium(II) complex, which may be generated from oxidative addition on a palladium(0) complex, was presumed to be the key intermediate, attempts at the stoichiometric reaction of cyclic disilanes with tetrakis(triphenylphosphine)palladium(0) complex did not result in the formation of the corresponding bis(organosilyl)palladium(II) complexes. In sharp contrast to this, bis(*t*-alkyl isocyanide)palladium(0) complex **2** readily underwent oxidative addition of the Si–Si bond of the cyclic disilane (Equation (22.2)) [10, 11]. The rapid reaction with 1,2-disilacyclopentane (**1**) gave an air-sensitive six-membered cyclic bis(organosilyl)bis(isonitrile)palladium(II) complex (**3**) at room temperature quantitatively, whose structure was established by single-crystal X-ray analysis. The requisite use of the *tert*-alkyl isonitrile ligand for the formation of the bis(organosilyl)palladium complex (**3**) may be relevant to the finding that an addition of two molar amounts of phosphine ligands to complex **3** resulted in reductive elimination with the formation of an Si–Si bond.

$$ (22.1) $$

$$ (22.2) $$

2a (R = *t*-Bu)
  b (R = 1-adamantyl)

3a (R = *t*-Bu)
  b (R = 1-adamantyl)

The highly reactive complex **2** showed remarkable catalytic activity in the reaction of cyclic disilane **1**. The catalytic activity of complex **2** for the ring-enlargement oligomerization of cyclic disilane **1** is particularly remarkable [11, 12]. Thus, the reaction of cyclic disilane **1** in the presence of complex **2** (0.01 equiv) at 50 °C in an NMR sample tube fitted with a rubber septum proceeded slowly to give a mixture of oligomeric cyclic products, including the cyclic dimer (Equation (22.3)). After 6 days, separation of the mixture by gel permeation chromatography furnished cyclic oligomers 4_n up to the 30-membered hexamer 4_6 (Table 22.1). Interestingly, the cyclic dimer, which was the major product in the

Table 22.1. Ring-enlargement oligomerization of compound **1** in the presence of palladium–isonitrile complexes

Entry	Catalyst[a] (0.01 equiv)	Temp. (°C)	4_2	4_3	4_4	4_5	4_6	4_7	4_8	Total
1	$(t\text{-BuNC})_2\text{Pd}$	50	6	29	20	8	3			66%
2	$(t\text{-BuNC})_2\text{Pd}$ + t-BuNC (0.01)	80	14	22	12	4	2			53%
3	$(t\text{-BuNC})_2\text{Pd}$ + t-BuNC (0.02)	80	13	18	7	2				41%
4	$(1\text{-AdNC})_2\text{Pd}$	80	6	15	16	9	6	5	4	62%
5	$(t\text{-HepNC})_2\text{Pd}$	80	22	5						27%
6	$(i\text{-PrNC})_2\text{Pd}$	50	25	20	5					50%
7	$(\text{XyNC})_2\text{Pd}$	50	42	4						46%

[a] 1-Ad: 1-adamantyl; t-Hep: 1,1,2,2-tetramethylpropyl; Xy: 2,6-dimethylphenyl.

Pd(PPh₃)₄-catalysed reaction, was obtained in only 6% yield, and cyclic trimer **4₃**, tetramer **4₄** and pentamer **4₅** were predominantly produced in the Pd(CNBu-t)₂-catalysed reaction (entry 1). The oligomerization was significantly retarded by addition of a small amount (1 or 2 equiv based on palladium) of t-butyl isocyanide, although the reaction at 80 °C gave a mixture of the cyclic oligomers in nearly the same distribution (entries 2 and 3). The substituents of the isonitriles critically affect the catalytic activity. Thus, the use of aryl isocyanides and *sec*-alkyl isonitrile failed to promote the ring-enlargement reaction, resulting in the formation of the dimer as a major product (entries 6 and 7). Among the *tert*-alkyl isonitriles examined, 1-adamantyl isonitrile gave a result comparable to that for t-butyl isocyanide (entry 4), although the bulkier 1,1,2,2-tetramethylpropyl isonitrile gave the cyclic dimer as a major product in poor yield (entry 5).

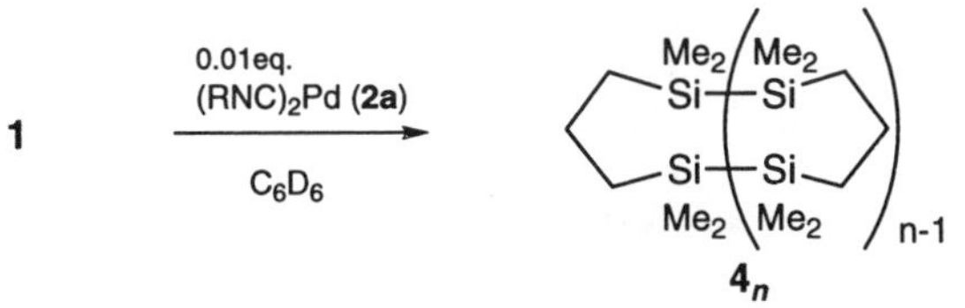

(22.3)

The yields of the oligomers were greatly improved by carrying out the reaction in a sealed tube under higher concentration (Equation (22.4)). Thus, in the presence of Pd(CNBu-t)₂, a mixture of oligomers **4ₙ** was obtained in 93% total yield, consisting of the dimer **4₂** (3%), trimer **4₃** (32%), tetramer **4₄** (34%), pentamer **4₅** (14%), hexamer **4₆** (6%), heptamer **4₇** (3%) and octamer **4₈** (1%) after 2 days.

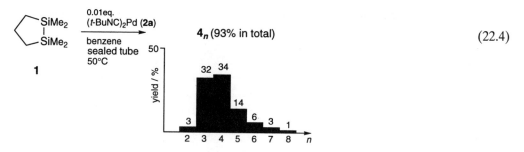

$$4_n \text{ (93\% in total)} \tag{22.4}$$

The remarkable feature of the reaction is the exclusive formation of the cyclic oligomers without the accompanying formation of any acyclic oligomers or polymers. To investigate the reaction mechanism, the reaction of each isolated oligomer was carried out in the presence of Pd(CNBu-t)$_2$ at 50 °C. Interestingly, the cyclic dimer 4_2 afforded a mixture of the cyclic oligomers 4_n in the distribution, similar to that of the reaction with cyclic disilane **1**. Furthermore, a stoichiometric reaction of the dimer 4_2 with Pd(CNBu-t)$_2$ gave a six-membered cyclic bis(silyl)palladium(II) complex (**3**), presumably because the four silicon atoms in the dimer are suitably arranged for simultaneous interaction with the palladium, leading to intramolecular metathesis [11,13]. These results suggest that compounds **1** and 4_2 serve equally as starting compounds in the ring-enlargement oligomerization through the generation of the key bis(silyl)palladium(II) intermediate (**3**). In contrast, cyclic trimer 4_3 and tetramer 4_4 underwent de-oligomerization and oligomerization only to the $(n-1)$-mer and $(n+1)$-mer to a lesser extent (Equation (22.5)). In the stoichiometric reaction with Pd(CNBu-t)$_2$, these higher oligomers did not give the six-membered palladium complex (**3**). These findings imply that the cyclooligomerization did not proceed by σ-bond metathesis between the two higher oligomers produced, but between the cyclic oligomer and the five-membered cyclic disilane through the six-membered cyclic bis(silyl)palladium(II) complex (**3**).

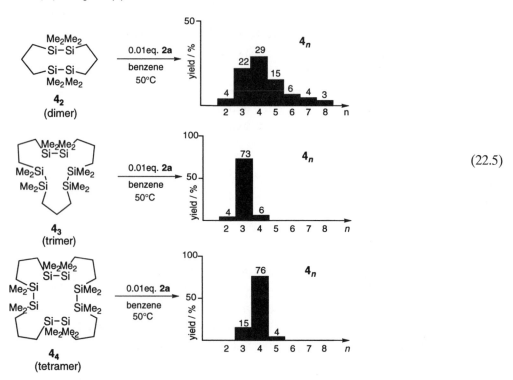

$$\tag{22.5}$$

A schematic illustration is presented for the mechanism of the ring-enlargement oligomerization of compound **1** (Equation (22.6)). Intermediate **3**, produced by oxidative addition of compound **1** onto Pd(CNBu-t)$_2$, may undergo oxidative addition of the Si–Si bond of compound **1** or the oligomers to give a tetrakis(silyl)palladium(IV) intermediate (**5**). Subsequent double reductive elimination produces the higher oligomers. Although the catalytic cycle may be reversible, monomer **1** is completely consumed to produce higher oligomers, because oligomers higher than the cyclic trimer hardly undergo the de-oligomerization by intramolecular metathesis.

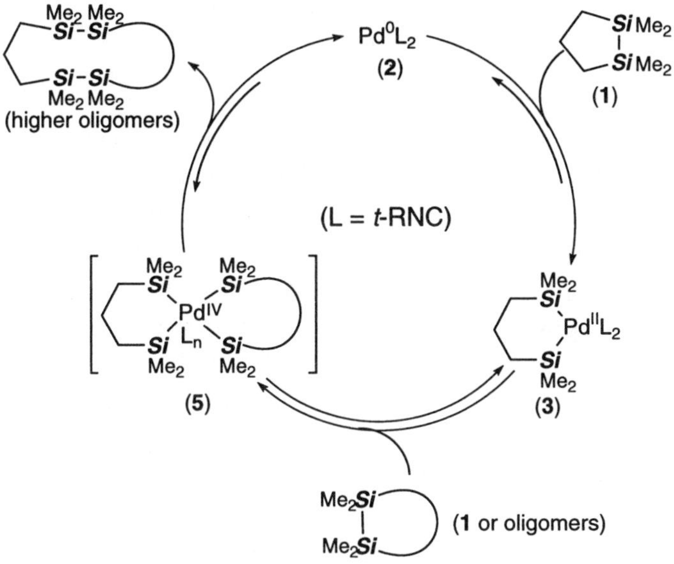

$$(22.6)$$

As to the formation of the palladium(IV) intermediate (**5**), the related bis(iso-nitrile)tetrakis(organosilyl)platinum(IV) complex (**6**) was isolated in the reaction of a platinum(0)-t-butyl isonitrile complex with bis(disilanyl)dithiane (**7**) [14], which under-went intramolecular Si–Si σ-bond metathesis in the reaction with Pd(CNBu-t)$_2$ to give a four-membered cyclic bis(organosilyl)palladium(II) complex (**8**) (Equation (22.7)) [13].

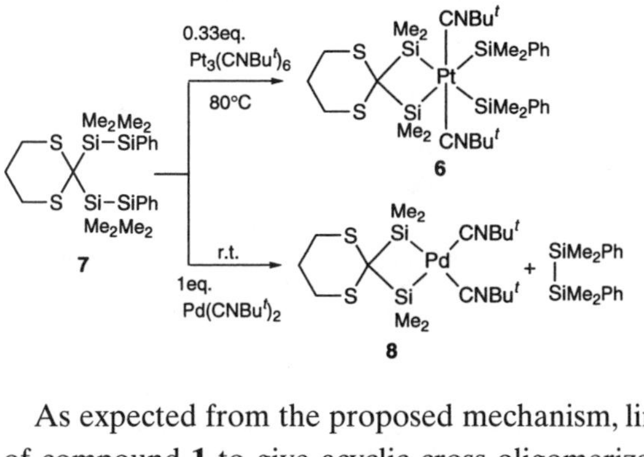

$$(22.7)$$

As expected from the proposed mechanism, linear disilane **9** also underwent insertion of compound **1** to give acyclic cross-oligomerization products **10**$_n$ along with the cyclic oligomers **4**$_n$ derived from the ring-enlargement oligomerization of compound **1** (Equation (22.8)) [11]. The insertion of compound **1** into the Si–Si bonds of the linear disilane occurred preferentially over the ring-enlargement oligomerization of compound **1** (67:33 ratio of the yields based on compound **1**). Furthermore, digermane **11** also underwent the insertion of compound **1** to give a mixture of oligomers 12_n ($n = 1$–6) bearing the dimethylphenylgermyl groups at both ends (Equation (22.9)).

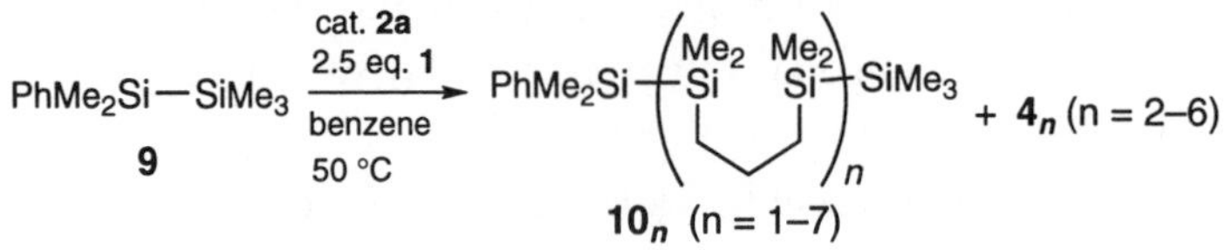

$$(22.8)$$

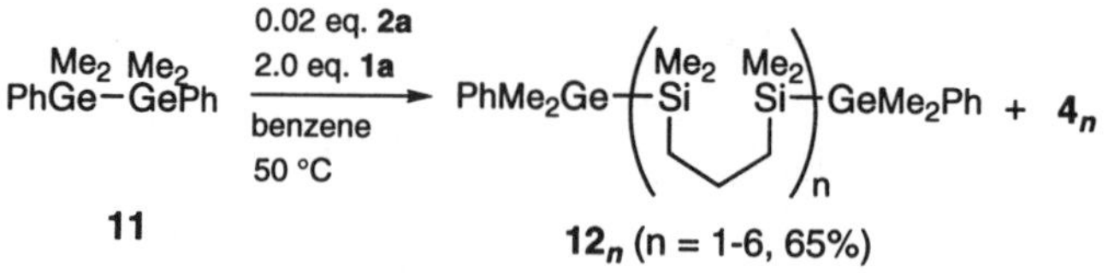

$$(22.9)$$

The cyclic oligomers 4_n thus far prepared, which contain the Si–Si linkages in the ring, were elaborated further by virtue of the high but controllable reactivities of the Si–Si bonds toward transition metal complex catalysts, as demonstrated by insertion reactions of unsaturated molecules into the Si–Si bonds [11]. Thus, in the presence of Pd(OAc)$_2$ catalyst, 2,6-diisopropylphenyl isocyanide was inserted into all the Si–Si bonds of the trimer 4_3 and tetramer 4_4 to give 18-membered triimine **13a** and 24-membered tetraimine **13b**, respectively (Equations (22.10) and (22.11)) [15–17]. The cyclic oligomers also underwent insertion of oxygen atoms by oxidation of the Si–Si linkages with trimethylamine oxide in refluxing benzene [18]. The reactions of the trimer and tetramer cleanly proceeded without ring-opening, affording tri(disiloxane) **14a** and tetra(disiloxane) **14b**, respectively, in quantitative yields.

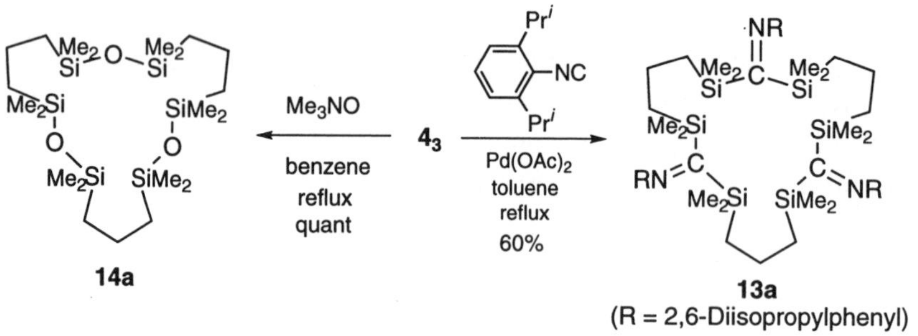

$$(22.10)$$

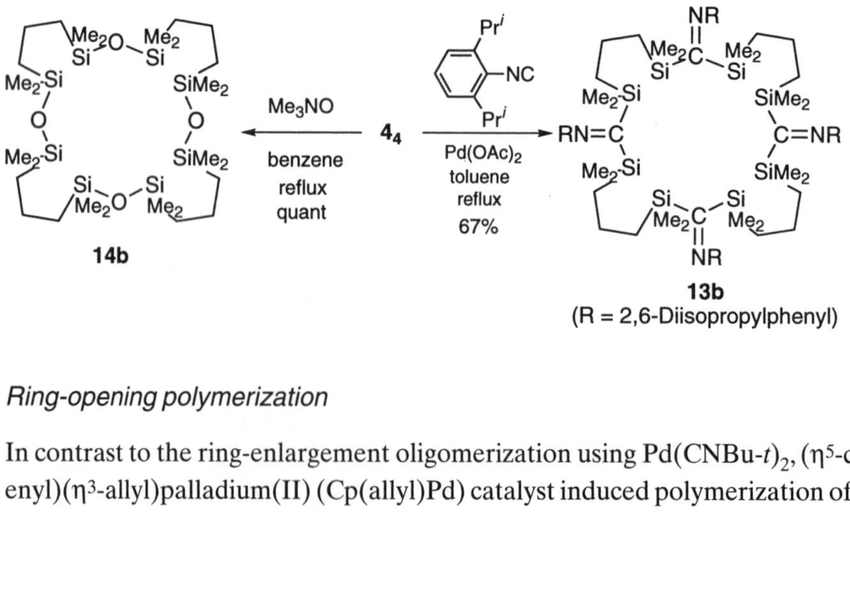

$$(22.11)$$

2.2 Ring-opening polymerization

In contrast to the ring-enlargement oligomerization using Pd(CNBu-t)$_2$, (η^5-cyclopentadienyl)(η^3-allyl)palladium(II) (Cp(allyl)Pd) catalyst induced polymerization of compound **1** to give polymers **15** with very high molecular weights (51%, Mn > 500000), accompanied only by the cyclic dimer **4₂** (37%) without the formation of any higher cyclic oligomers (Equation (22.12)) [11,12]. Although the structure of the polymer, i.e. cyclic or acyclic, could not be determined by spectroscopic measurements, it may be presumed that a different mechanism from the above-mentioned ring-enlargement oligomerization was involved. In fact, a stoichiometric reaction of compound **1** with Cp(allyl)Pd produced a ring-opening product (**16**) in which the Cp and the allyl group were attached to each silicon atom of compound **1**, suggesting that a ring-opening mechanism could be responsible for the formation of the polymer (Equation (22.13)) [19].

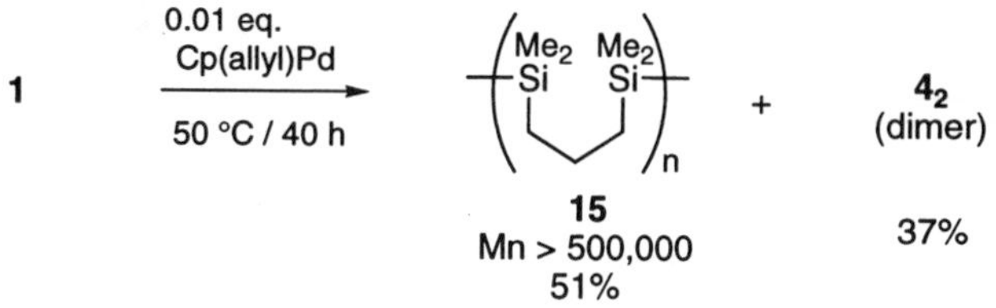

$$(22.12)$$

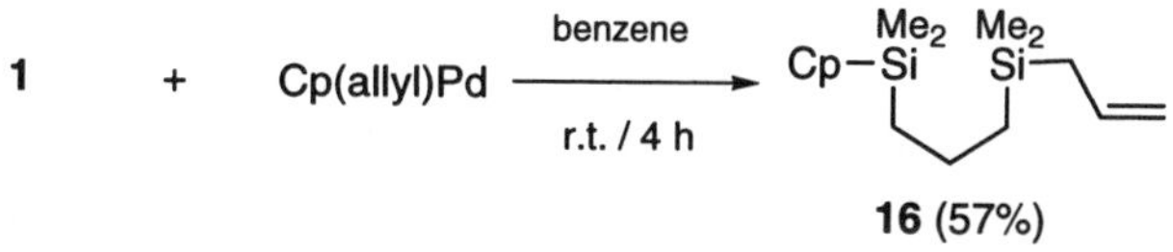

$$(22.13)$$

A plausible polymerization mechanism, in which palladium(IV) complexes were crucially involved, is illustrated in Equation (22.14). Initially, oxidative addition of the Si–Si bond of compound **1** onto Cp(allyl)Pd is immediately followed by reductive elimination of the Si–allyl bond to give Cp(organosilyl)palladium(II) complex **A**. Complex **A**, which undergoes further reductive elimination to give compound **16** in the absence of excess of compound **1**, may react with compound **1** to afford (Cp)tris(organosilyl)palladium(IV) intermediate **B**. The formation of a new Si–Si bond through reductive elimination from the palladium(IV) intermediate **B** to produce complex **A′** constitutes a propagation step for the ring-opening polymerization. The polymerization is terminated by reductive elimination of the Si–Cp bond from Cp(organosilyl)palladium(II) complex **A″**.

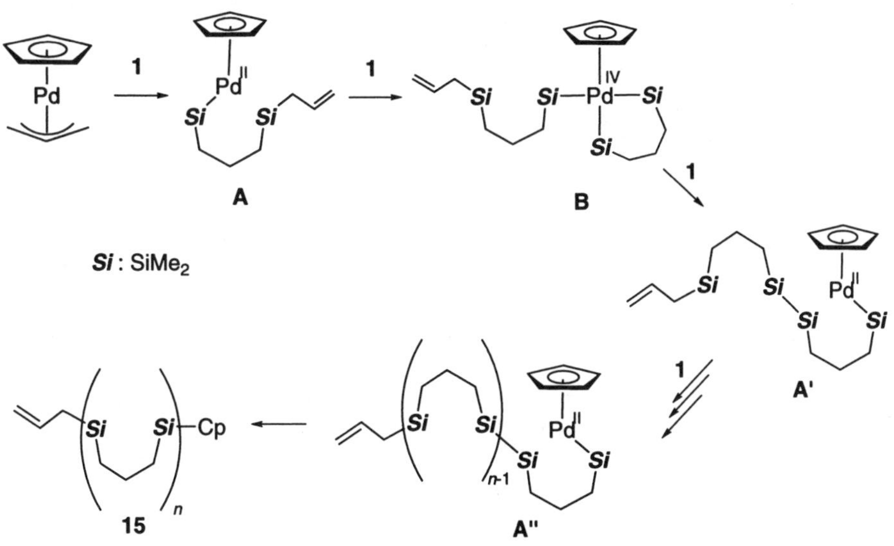

$$(22.14)$$

The palladium(IV) intermediacy may be confirmed by the formation of a closely related palladium(IV) complex (**17**) in a reaction of Cp(allyl)Pd with spirotrisilane (**18**), in which each five-membered ring contains a Si–Si bond (Equation (22.15)) [20]. In fact, the (Cp)tris(organosilyl)palladium(IV) complex (**17**) catalysed the ring-opening polymerization, being suggestive of the involvement of the palladium(IV) intermediate.

$$(22.15)$$

Unfortunately, the ring-enlargement oligomerization and ring-opening polymerization are only applicable to 1,1,2,2-tetramethyl-1,2-disilacyclopentane (**1**).

3 Bis-silylation

The remarkable activity of the palladium–isonitrile complexes mentioned so far can be attributed primarily to an effective formation of the bis(silyl)palladium(II) intermediate through oxidative addition of the Si–Si bond [21]. Based on activation of the silicon–silicon bond of disilanes by the palladium(0)–isonitrile complex, new methodologies for stereo- and regioselective silicon–carbon bond formation have been developed.

Palladium-catalysed bis-silylation, i.e. addition of the Si–Si bond across carbon–carbon multiple bonds, involves the bis(silyl)palladium(II) complex as a key intermediate. Extensive studies on the bis-silylation of alkynes with palladium–phosphine catalysts manifested that electronegative hetero-substituents such as fluorine, chlorine and alkoxy groups on the silicon atoms of disilane were crucially important for high yields of the bis-silylation [22–24]. Peralkylated disilanes, e.g. hexamethyldisilane, are provided with much lower reactivity in the phosphine-palladium-catalysed bis-silylation of alkynes [24, 25].

It is noted that the isonitrile–palladium(0) catalyst exhibited remarkable catalytic activity even in the bis-silylation of alkynes with unreactive disilanes [26]. Among the isonitriles examined, *t*-alkyl isocyanides were most effective, as described in the ring-enlargement oligomerization. Furthermore, the palladium complex catalyst for the bis-silylation could be generated *in situ* by mixing palladium(II) precursors, e.g. Pd(OAc)$_2$ and Pd(acac)$_2$, with *t*-alkyl isonitrile (4–15 equiv to Pd), instead of the use of the unstable bis(isonitrile)palladium(0) complexes isolated. For example, reactions of hexamethyldisilane (**19**) with phenylacetylene and 1-octyne afforded the corresponding (*Z*)-bis(trimethylsilyl)alkenes (**20**) in high yields in the presence of the catalyst prepared from Pd(OAc)$_2$ (2 mol.%) and 1,1,3,3-tetramethylbutyl isocyanide (30 mol.%) under reflux in toluene (Equation (22.16)).

$$(22.16)$$

3.1 *Intramolecular bis-silylation of alkynes*

With the highly efficient palladium catalyst in hand, we examined intramolecular bis-silylation of carbon–carbon multiple bonds, which may serve as useful methodologies in organic synthesis. The intramolecular bis-silylation with high regio-, chemo- and stereo-selectivities can provide efficient synthesis of stereodefined organosilicon compounds, whose elaboration is useful for stereoselective organic synthesis.

First of all, the reaction of alkynes tethered to disilanyl groups by a three-atom chain, including the silyl ether linkage, was examined [27]. The disilanyl ethers (**21**) were readily prepared in good yields by the reaction of homopropargylic alcohols with chlorodisilanes in the presence of amines in THF or DMF. The intramolecular bis-silylation of the disilanyl ethers proceeded effectively in the presence of the palladium–isonitrile catalyst to give 5-*exo* ring-closure products (**22**) in high yields (Equation (22.17)). Noteworthy was that intermolecular bis-silylation, which may lead to the formation of oligomeric products, was never observed and that internal alkynes, which hardly underwent the intermolecular bis-silylation, gave the cyclized product in high yields under conditions identical to those for internal alkynes. Furthermore, carbon–carbon triple bonds conjugated with an ester and a vinyl group similarly underwent the bis-silylation exclusively at the triple bonds to give the corresponding five-membered cyclic products.

$$(22.17)$$

The intramolecular bis-silylation was retarded by bulky substituents at the alkynyl carbons; phenyl- and trimethyl-substituted alkynes require higher temperatures to afford the corresponding bis-silylation products in high yields. Although bis-silylacetylene (**23**) failed to cyclize even at 200 °C in the presence of the palladium–isonitrile catalyst, the reaction was accelerated by high pressure (1×10^9 Pa) to give tetrakis(organosilyl)ethene (**24**) in good yield (Equation (22.18)) [28]. It is interesting to note that, in contrast to the twisted structure of tetrakis(trimethylsilyl)ethene, alkene **24** exhibited a non-twisted structure around the C=C bond due to less steric repulsion between the vicinal silyl groups, as revealed by X-ray crystal structure.

$$(22.18)$$

Intramolecular bis-silylation of a disilanyl ether of 4-pentyn-1-ol, which has a longer tether, provided a corresponding 6-*exo* ring-closure product (Equation (22.19)). However, the reaction was less effective than the bis-silylation giving 5-*exo* products, resulting in the formation of a mixture of (*Z*)- and (*E*)-alkenes **26** (88:12) in 71% yield.

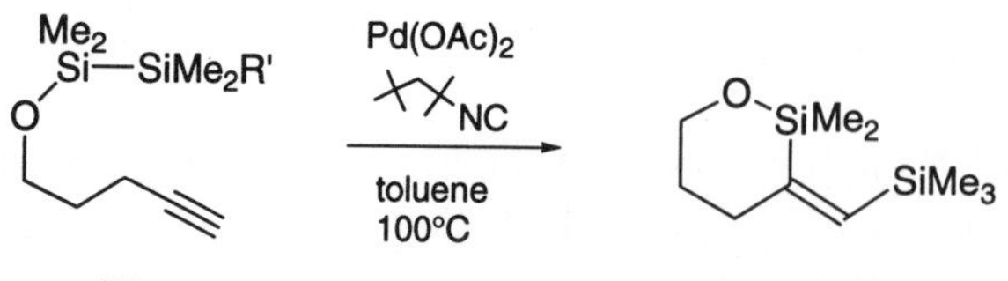

$$(22.19)$$

25   **26** (Z/E = 88/12)

On the other hand, disilanyl propargyl ethers underwent the bis-silylation smoothly (Equation (22.20)) [29]. An eight-membered ring product (**30**) was obtained in the reaction of primary propargyl ether (**27**), presumably through the 4-*exo* cyclization followed by dimerization of the resultant strained 1,2-oxasiletane. The exclusive formation of four-membered products (**31**) could be confirmed in the reaction of secondary propargyl ethers (**28**), although they were unstable toward moisture and underwent cleavage of the Si–O bonds. Interestingly, a stable four-membered cyclic oxasiletane (**32**) was isolated in the reaction of tertiary propargyl ether (**29**) in high yields.

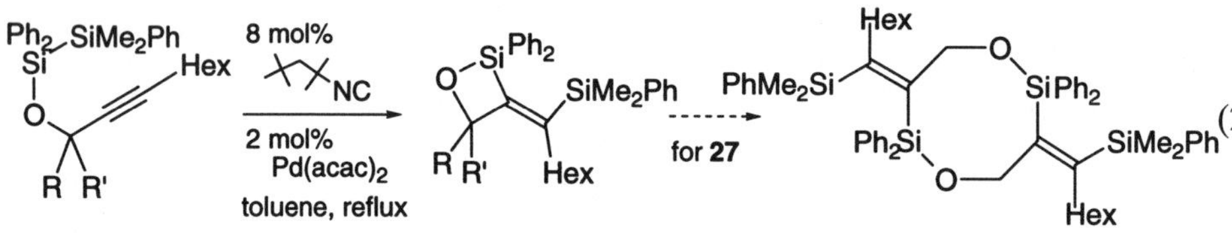

$$(22.20)$$

27 (R = R' = H)   **31** (R = Me, R' = H)   **30**
28 (R = Me, R' = H)   **32** (R = R' = Me)
29 (R = R' = Me)

Although the eight-membered dimer (**30**) was stable toward organometallic reagents such as organolithium and organomagnesium, the four-membered cyclic (**31**) underwent nucleophilic attack onto the silicon atom in the ring. On reacting with methylmagnesium bromide, a ring-opening reaction at the Si–O bond of compound **31** proceeded at low temperature to afford the corresponding allylic alcohol (**33**) (Equation (22.21)). In contrast, the reaction of compound **31** with *n*-BuLi gave allenylsilane (**34**) in high yield. The formation of the allenylsilane could be rationalized by the initial attack of BuLi to the ring silicon, followed by Peterson-type elimination of the BuPh$_2$SiO$^-$ species from the resultant lithium alkoxide (Equation (22.22)) [30]. It should be noted that the β-elimination proceeded highly effectively, while closely related allene formation via the nucleophilic Peterson-type elimination was reported to be difficult to furnish a satisfactory result [31].

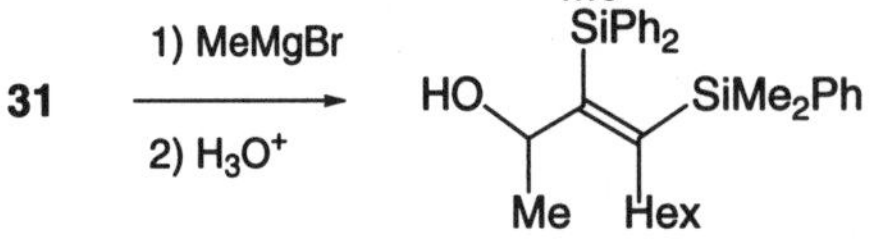

$$(22.21)$$

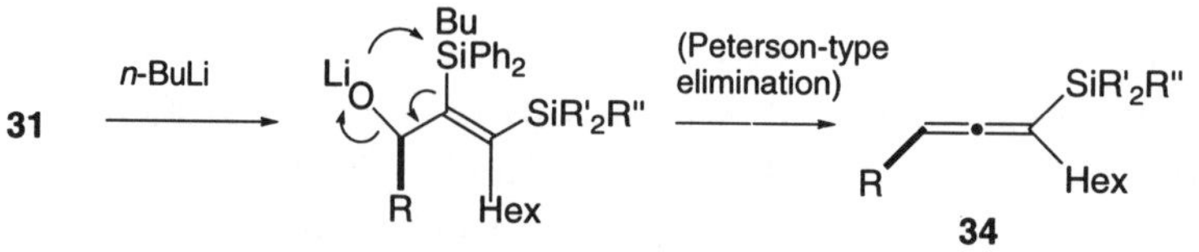

33

$$(22.22)$$

34

It is quite attractive that synthetically useful allenylsilanes were synthesized in one flask with manipulative simplicity [32]. The intramolecular bis-silylation reaction starting with propargylic alcohols could be applied generally to the synthesis of allenylsilanes having a variety of substituents (Equation (22.23)) [29].

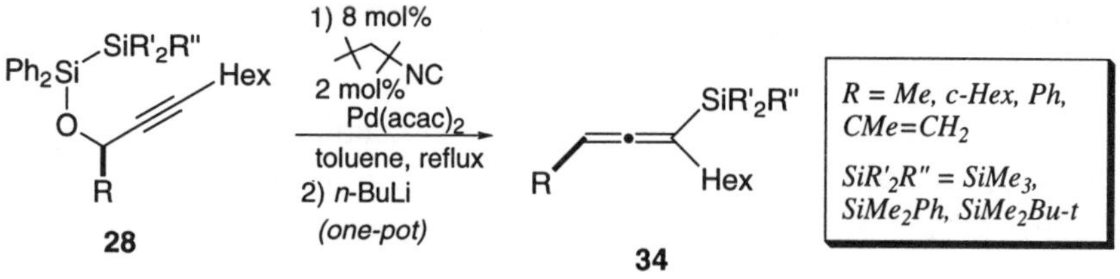

(22.23)

Enantio-enriched allenylsilanes having a chirality solely on the allene axis, which have never been prepared [33], may provide useful synthetic application from the viewpoint of stereoselective synthesis. A disilanyl ether ((R)-**28**) prepared from enantio-enriched propargyl alcohol of 96.7% ee produced allenylsilane (**34**) on treatment with the palladium–isonitrile catalyst followed by treatment with BuLi (Equation (22.24)). The allenylsilane showed a specific rotation of −13.2, although the enantiomeric excess could not be determined as it stood. The high enantiomeric excess of the allenylsilane was revealed by the enantiomeric excess of the homopropargyl alcohol (**35**), which was obtained diastereoselectively by the reaction of compound **34** with an aldehyde in the presence of TiCl$_4$ [34].

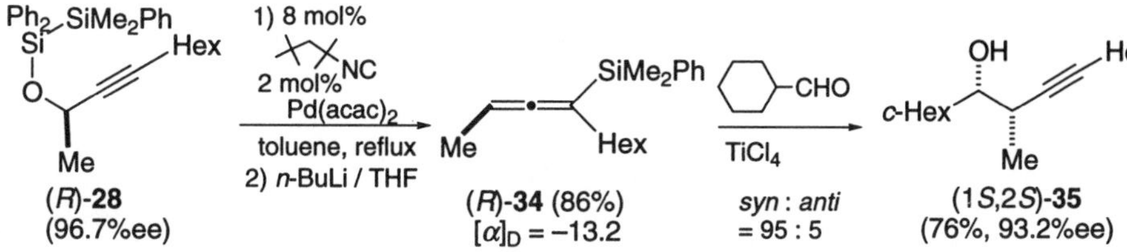

(22.24)

The highly stereospecific formation of the enantio-enriched allenylsilane was rationalized by the *cis*-addition of the Si–Si bond across the carbon–carbon triple bond and the subsequent *syn*-elimination of BuPh$_2$SiO$^-$. The mechanism suggests that allenylsilane (**34**) with an *R*-configuration should be formed from the starting disilanyl ether (**28**) of *R*-configuration (Equation (22.25)). Furthermore, the absolute configuration of the homopropargylic alcohol (**35**), i.e. (1*S*,2*S*)-configuration, indicates that reaction of compound **34** with the aldehyde occurs with high stereoselectivity at the π-face *anti* to the silyl group of the allenylsilane.

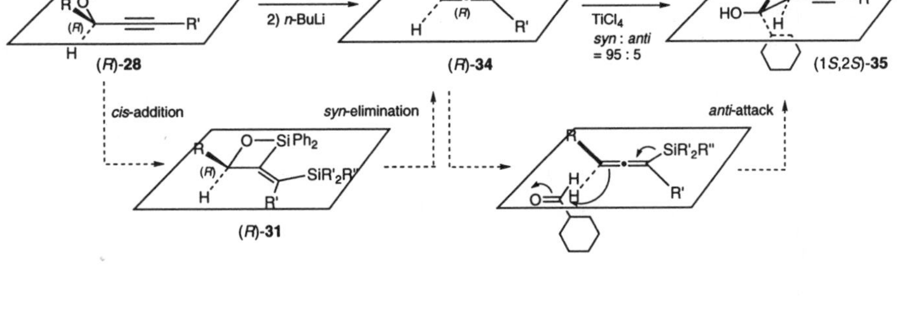

(22.25)

3.2 *Intramolecular bis-silylation of alkenes*

In spite of the extensive study on the alkyne bis-silylation, bis-silylation of isolated carbon–carbon double bonds has scarcely been achieved even by using the activated disilanes [35, 36]. The isonitrile–palladium catalyst was also ineffective for the intermolecular reactions of disilanes with alkenes. However, intramolecular bis-silylation of disilanyl ethers (**36**) prepared from homoallylic alcohols proceeded in the presence of the palladium–isonitrile catalyst under even milder conditions than those for bis-silylation of the corresponding alkynes, giving 5-*exo* ring-closure products (**37**) in high yields (Equation (22.26)) [37,38]. Stereoselectivities of the intramolecular bis-silylation of homoallylic ethers having substituents at the tethers were especially noteworthy; those having a substituent α to the C=C bond (**37a**) provided, 3,4-*trans*-oxasilolane, and those having a substituent β to the C=C bond (**37b**) provided 3,5-*cis*-oxasilolane in high diastereoselectivities. The stereochemical outcome suggests that the intramolecular bis-silylation proceeds through a well-ordered chair-like bis(silyl)palladium(II) intermediate (**38**), which was formed via oxidative addition of the Si–Si bond onto the palladium(0)–isonitrile complex. Presumably, the bis-silylation takes place preferentially from the chair-like conformer **38** in which the α- or β-substituent occupies the equatorial position to afford the 3,4-*trans*- or 3,5-*cis*-oxasilolane, respectively, in high diastereoselectivities. In good agreement with this mechanism, bulkier substituents, e.g. phenyl groups, at the silicon proximal to the ether oxygen improve the stereoselectivities presumably by the increased steric demands in the chair-like conformation.

The phenyl substitution at the proximal silicon also made possible the intramolecular bis-silylation of vicinally disubstituted carbon–carbon double bonds (Equation (22.27)) [39]. Intramolecular bis-silylation of (*E*)- and (*Z*)-alkenes (**39**) proceeded stereospecifically to give the corresponding products **40a** and **40b**, in which the Si–Si bonds added to the C=C bonds in a *cis* fashion. The corresponding pentamethyldisilanyl ethers were completely reluctant to undergo bis-silylation even at high temperature. Interestingly, diastereoselectivities of the bis-silylation of the internal alkenes having α-substituents (compound **41a**) were largely improved to provide 3,4-*trans*-oxasilolanes exclusively, while those for the β-substituted substrates (compound **41b**) were moderate (Equation (22.28)).

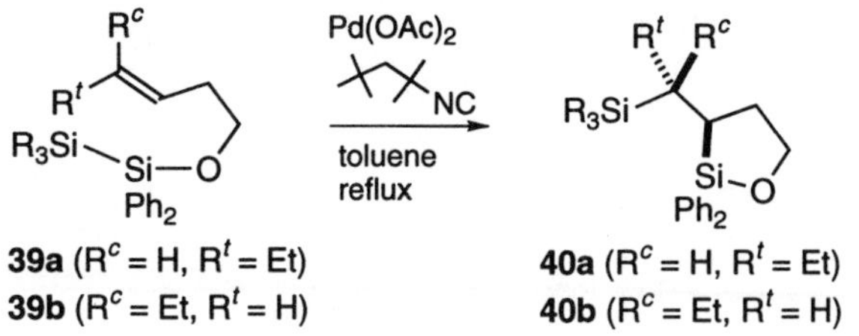

39a (R^c = H, R^t = Et)
39b (R^c = Et, R^t = H)

40a (R^c = H, R^t = Et)
40b (R^c = Et, R^t = H)

(22.27)

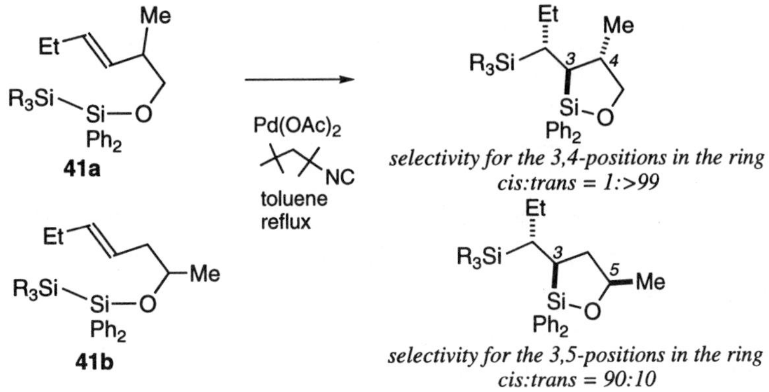

41a

selectivity for the 3,4-positions in the ring
cis:trans = 1:>99

(22.28)

41b

selectivity for the 3,5-positions in the ring
cis:trans = 90:10

The bis-silylation was successfully applied to the stereoselective synthesis of 1,2,4-polyols by oxidative transformation of the resultant oxasilolanes. Thus, the silicon–carbon bonds of the oxasilolanes were oxidized under the Tamao conditions by use of basic hydrogen peroxide with a fluorine source, after cleavage of the silicon–phenyl linkages of the terminal dimethylphenylsilyl group (Equation (22.29)) [40, 41]. The oxidative transformation proceeded with retention of the stereochemistry at the carbon atoms attached to the silicon atoms to provide the stereo-defined polyols.

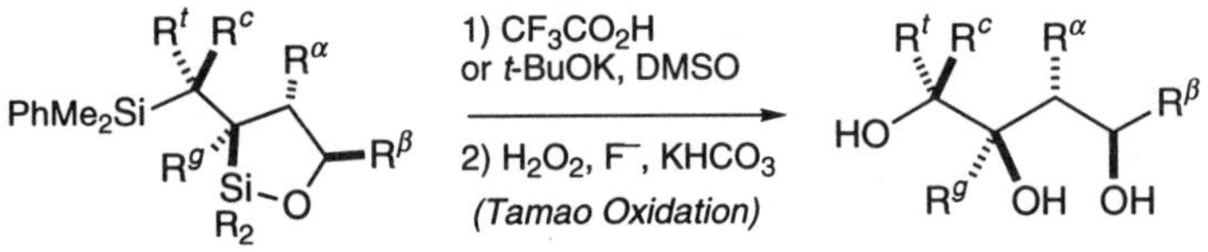

(22.29)

Intramolecular bis-silylation of 1,4-dienes (42) with one stereogenic center, which gives the corresponding trisubstituted oxasilolane with two additional stereogenic centers on the ring, was developed on the basis of the remarkable stereoselectivities as well as the ready accessibility to the polyols (Equation (22.30)) [42]. The attempt at the bis-silylation using compound 42a in the presence of the isonitrile–palladium catalyst prepared from 1,1,3,3-tetramethylbutyl isocyanide and Pd(OAc)$_2$ resulted in the formation of two of the four possible diastereomers, i.e. (R*,R*,R*)- and (R*,S*,R*)-43a, in a 6:4 ratio. This result indicates that the cyclization occurred with high diastereoselectivity for the 3,5-diastereoselection (diastereofacial selection) but with low 3,4-diastereoselection (diastereotopic group selection). To improve the diastereotopic group selection, the reactions of compounds 42b–e with disilanyl groups having various substituents proximal to the ether oxygen were examined. As shown in Equation (22.30), the selectivity for (R*,R*,R*)-43 was increased as the substituents of the disilanyl groups became bulkier. Thus, in the case of i-Pr-substituted 42c the selectivity reached a 92:8 ratio, although the reaction was too sluggish to obtain a reasonable yield of compound 43c. It was found that use of *i*-Bu-substituted disilanyl ether **42e** was preferable with respect to chemical yield and diastereoselectivity.

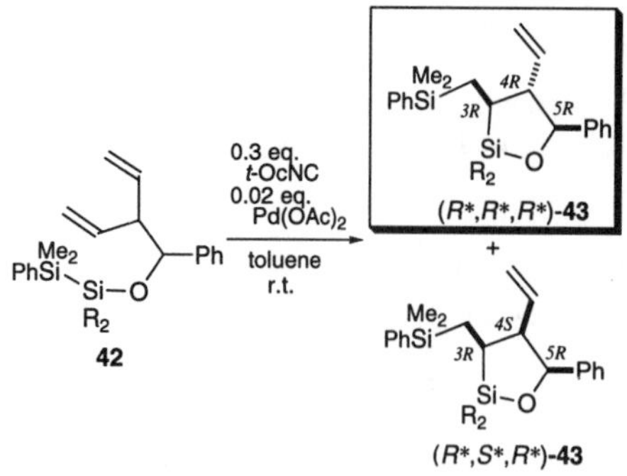

42	R	Product 43 (yield / %)	Ratio (%)[a]
a	Me	98	59 : 41
b	Et	82	75 : 25
c	*i*-Pr	27	92 : 8
d	Ph	87	83 : 17
e	*i*-Bu	90	88 : 12

[a]Ratios of (R^*,R^*,R^*) : (R^*,S^*,R^*).

$$(22.30)$$

As mentioned above, the high selectivity may be attributed to the well-ordered, chair-like bis(silyl)palladium(II) intermediate, which was formed through oxidative addition of the silicon–silicon bond onto the isonitrile–palladium(0) complex (Equation (22.31)). Presumably, the strong tendency of R groups to occupy the equatorial position renders more favorable the conformers **eq-eq** and **eq-ax**, which are responsible for the formation of (R^*,R^*,R^*)-**43** and (R^*,S^*,R^*)-**43**, respectively. Furthermore, the bulky substituents at the proximal silicon make the insertion of C=C via **eq-ax** unfavorable, resulting in selective reaction through **eq-eq** to give (R^*,R^*,R^*)-**43** predominantly.

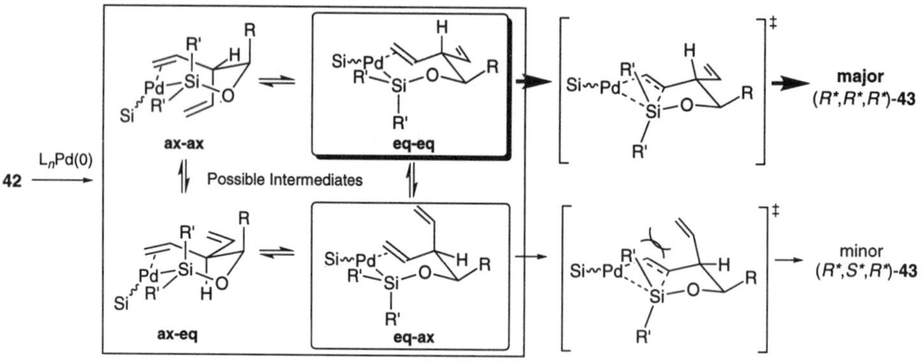

$$(22.31)$$

The stereoselective bis-silylation could be applied to the total synthesis of antifungal metabolite (−)-avenaciolide (**44**), which has the three contiguous stereogenic centers, (R,R,R), equivalent to those in the bis-silylation product **43** [43,44]. As shown in the retro-synthetic analysis (Equation (22.32)), (R)-3-vinyl-1-dodecen-4-ol (**45**) was chosen as the starting material and, at the final step, the *exo*-methylene group was introduced by the reported method [45]. The optically active (>98% ee) compound **45** was successfully prepared by enantioselective γ-pentadienylation of nonanal using a pentadienylborane reagent with Corey's chiral auxiliary, which was used for enantioselective allylation [46]. Use of (pentadienyl)diisopinocampheylborane showed lower selectivity (90% ee) for the reaction of nonanal [47]. Intramolecular bis-silylation of the optically active compound (R)-**46** gave a 90 : 10 mixture of (R,R,R)-**47** and (R,S,R)-**47**, from which the desired (R,R,R) isomer was isolated by column chromatography in nearly pure form (Equation (22.33)). The compound obtained, (R,R,R)-**47**, was transformed into alcohol (**48**) through a rhodium-catalysed hydroboration–oxidation sequence in high yield [48]. The Si—C bonds were not affected by the alkaline hydrogen peroxide oxidation. After trityl protection of the hydroxy group, followed by cleavage of the Si–Ph bonds, the remaining Si–C bonds were subjected to hydrogen peroxide oxidation in the presence of tetrabutylammonium fluoride to afford triol **49** in good yield. Final elaboration of triol **49** for the total synthesis of (–)-avenaciolide (**44**) involved stepwise oxidation with appropriate protection and deprotection followed by *exo*-methylenation of the resultant bis-lactone (**52**) according to Equation (22.33). The transformation of racemic bis-lactol **51** to compound **52** has been reported [49]. The structure of (–)-avenaciolide (**44**) was identified by comparison with spectroscopic data as well as with the specific reported rotation [50].

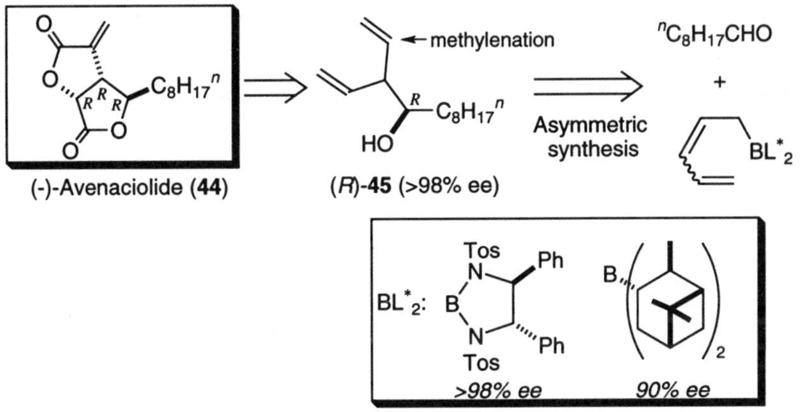

(22.32)

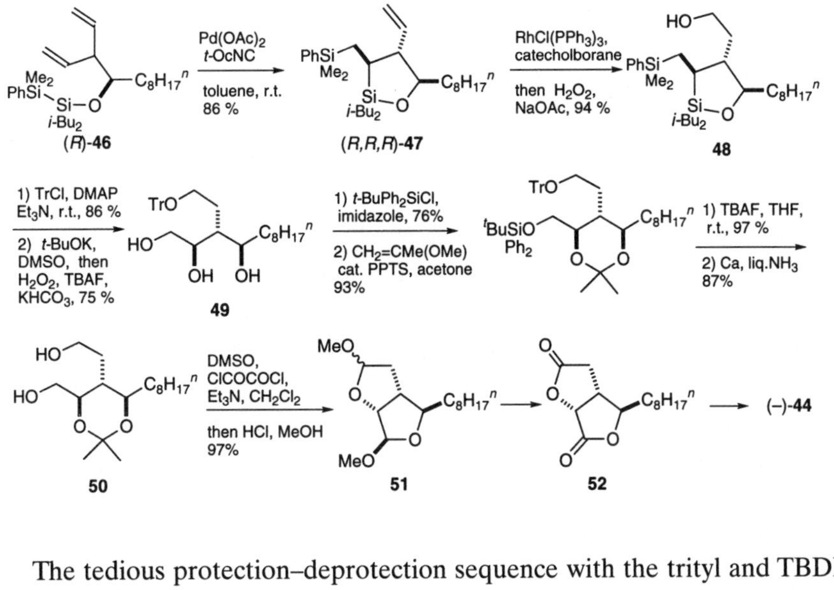

(22.33)

The tedious protection–deprotection sequence with the trityl and TBDPS groups was skipped by using trisilanyl ether ((*R*)-**53**), which underwent selective 5-*exo* ring-closure at the internal Si–Si bond in the presence of the palladium catalyst (Equation (22.34)) [51]. (–)-Avenaciolide was successfully synthesized through selective oxidation at the silyl group, hydroboration–oxidation and subsequent oxidation at the disilanyl groups, which was a synthetic equivalent to the hydroxy group as reported previously [52]. The successful total synthesis of (–)-avenaciolide demonstrated efficient construction of multiple stereogenic centers by the diastereoselective bis-silylation reactions.

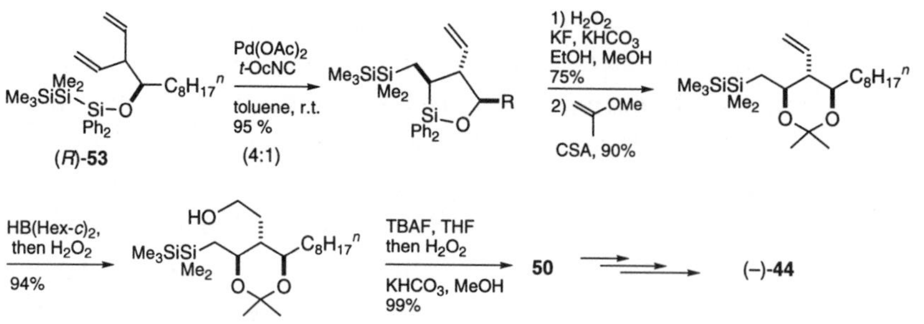

(22.34)

Enantioselective intramolecular bis-silylation was also possible by using optically active *t*-alkyl isonitrile ligands on a palladium catalyst [53]. We designed and synthesized rigid, bicyclic isonitriles (**54a–d**) with various substituents β to the isocyano group, taking into consideration the linear structure of isonitriles, which might make it difficult to create an effective asymmetric environment in the vicinity of the metal center (Equation (22.35)).

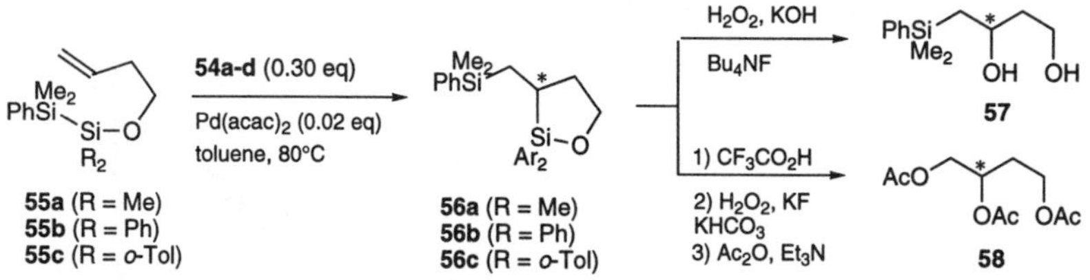

Optically pure isocyanides used

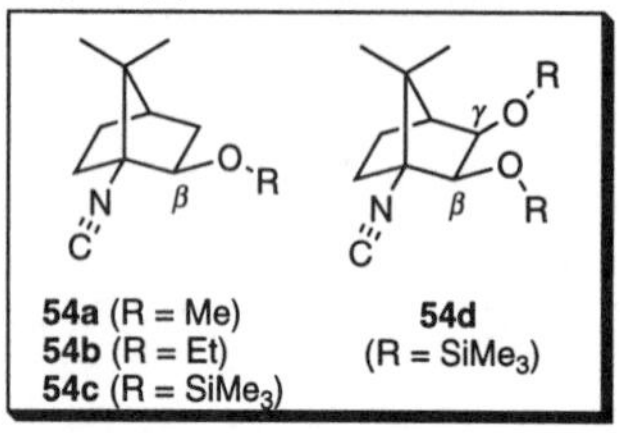

55	Isonitrile (54)	%Yield of 56 or 57	%ee (config)
a (Me)	c	72	4 (*R*)
b (Ph)	a	79	14 (*S*)
b	b	85	20 (*S*)
b	c	83	33 (*S*)
c (o-Tol)	c	74	55 (*S*)
c	d	87	64 (*S*)

(22.35)

Reactions of disilanyl ethers (**55a–c**) in the presence of compounds **54a–c** and Pd(acac)$_2$ were examined. Unlike the bis-silylation reactions of compound **55** with ordinary *t*-alkyl isonitriles, which proceeded at room temperature in the presence of palladium(II) acetate, the palladium-catalysed bis-silylation with the optically active isonitriles was sluggish at room temperature but proceeded at 80 °C at a reasonable reaction rate. For determination of the enantiomeric excesses and absolute configuration, the bis-silylation product **56a** was transformed into diol **57** and triol **58** by oxidative cleavage of the Si–C bonds. Results are shown in the table in Equation (22.35), indicating that the combination of the bulky substituents on both the isonitrile and the disilanyl group is preferable to attain high enantioselectivity. Use of modified isonitrile **54d**, in which the β-substituent effect may be reinforced by the 'buttressing effect' of the γ-substituent, improved the enantioselectivity in the bis-silylation of compound **55c**. The highest enantioselectivity (78% ee) attained so far has been for the reaction of compound **55d** in the presence of compound **54d** (Equation (22.36)). From the optically active compound **56d** thus obtained, the optically active 2-methyl-1,2,4-butanetriol was synthesized. It should be worth mentioning that this is the first example using optically active isonitriles as chiral, spectator ligands in catalytic asymmetric synthesis, and they provide a new possibility to serve as effective chiral ligands.

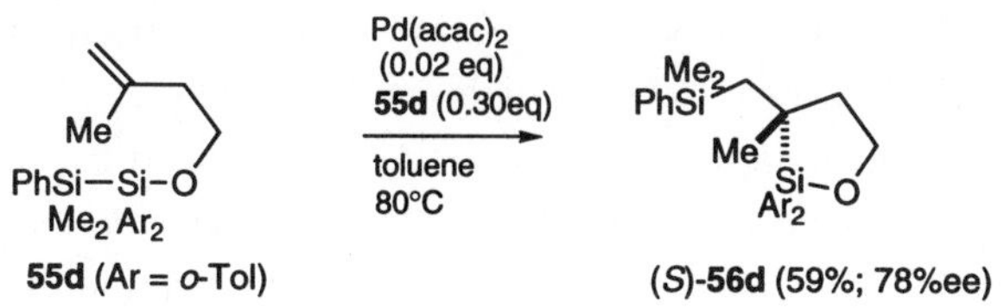

(22.36)

Here, attention is turned to the intramolecular bis-silylation reactions proceeding with 4-*exo* cyclization. The intramolecular bis-silylation of 4-disilanyl-1-butenes (**59**) underwent the 4-*exo* cyclizations in high yields with moderate selectivities; α-methyl (**59a**) and β-methyl (**59b**) derivatives selectively gave the corresponding disubstituted *trans*- and *cis*-silacyclobutanes (**60a** and **60b**), respectively (Equation (22.37)) [38]. Attempts at the intramolecular bis-silylation of allylic alcohols, however, gave more complicated results [54]. Thus, the disilanyl ether (*E*)-**61a** prepared from the corresponding allylic alcohol and 1-chloro-2,2-dimethyl-1,1,2-triphenyldisilane was heated for 2h in the presence of Pd(acac)$_2$ (2 mol.%) and 1,1,3,3-tetramethylbutyl isocyanide (8 mol.%) under reflux in toluene (Equation (22.38)). Unexpectedly, the reaction gave (*E*)-allylsilane **62a** (49%) and six-membered cyclic siloxane **63a** (46%), which were separated and isolated by column chromatography. Presumably, instability of the four-membered oxasiletane **64a**, which may be formed initially by 4-*exo* cyclization, led to a disproportionation reaction to give these products. Noteworthy is that the relative stereochemistry of the three consecutive stereo-centers in compound **63a**, as well as the *trans* geometry of the carbon–carbon double bond in compound **62a**, were completely controlled. The stereochemical outcome indicates obviously that the intramolecular bis-silylation of allyl alcohols takes place with highly stereoselective 4-*exo* cyclization to give *trans*-oxasiletane (**64a**).

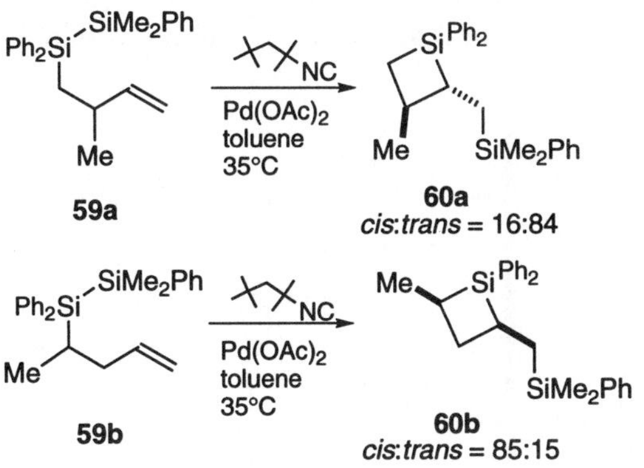

(22.37)

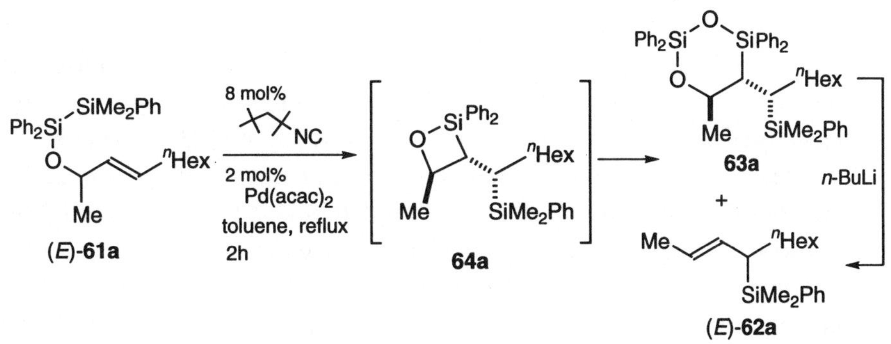

(22.38)

Treatment of the six-membered cyclic siloxane (**63a**) with *n*-butyllithium in THF at 0 °C led to the formation of (*E*)-allylsilane (**62a**) in high yield (Equation (22.38)). The transformation of compound **63a** to compound **62a** may be rationalized by cleavage of the silicon–oxygen bonds of compound **63a** followed by Peterson-type *syn*-elimination. The selective formation of compound **62a** from compound **63a** indicated that the bis-silylation followed by treatment with BuLi is useful for the stereoselective synthesis of allylsilanes, which have been used widely as useful synthetic tools in organic synthesis [55]. In fact, the palladium-catalysed bis-silylation followed by treatment with *n*-butyllithium was carried out in one flask to afford the allylsilane (*E*)-**62a** in 93% yield. It should be remarked that the (*E*)-allylsilane **62a** was obtained also from (*Z*)-**61a** in 83% yield according to the one-pot procedure with phenyllithium (Equation (22.39)).

$$(22.39)$$

The exclusive formation of the (*E*)-allylsilanes from either (*E*)- or (*Z*)-allylic alcohols indicates highly stereoselective formation of the *trans*-four-membered oxasiletanes by the intramolecular bis-silylation reactions. The high diastereoselection may originate in the conformation of the bis(silyl)palladium(II) intermediates, which undergo intramolecular insertion of the carbon–carbon double bonds (Equation (22.40)). It is presumed that the insertion reaction proceeds through the '*exo*' complex, which involves less steric repulsion than the '*endo*' complex. The high diastereoselection in the intramolecular bis-silylation could lead to the stereoselective formation of the unstable *trans*-oxasiletanes (**64**), whose disproportionation gave the (*E*)-allylsilanes (**62**) and six-membered cyclic **63**.

$$(22.40)$$

Use of enantio-enriched allylic alcohols, which are readily available by asymmetric synthesis, e.g. Sharpless kinetic resolution, was prompted by the highly selective formation of (*E*)-allylsilanes [56]. The palladium-catalysed reaction of (*R*,*E*)-**61a** (99.7% ee) followed by treatment with *n*-butyllithium gave (*E*)-**62a** in 90% yield (Equation (22.41)). The enantiomeric excess of the allylsilane was determined to be 99.1% ee by derivatization of (*E*)-**62a** with 9-BBN followed by oxidation with basic hydrogen peroxide, which gave *anti*-β-silyl alcohol (**65**) with excellent regio- and stereoselectivity [57,58]. This result showed unambiguously that the synthesis of allylsilanes proceeded with highly stereoselective 1,3-

chirality transfer from optically active allylic alcohols according to the mechanism involving the exclusive formation of the *trans*-oxasiletane.

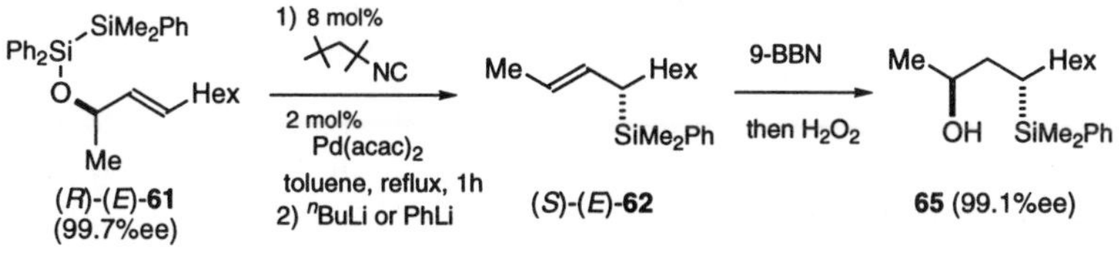

$$(22.41)$$

Various enantio-enriched allylsilanes ((*E*)-**62**) were synthesized from highly enantio-enriched disilanyl ethers ((*E*)-**61a**) of allylic alcohols by bis-silylation followed by treatment with organolithium reagents (Equation (22.42)). Highly enantio-enriched allylsilanes having various silyl groups as well as alkyl substituents at the allylic carbons were obtained stereoselectively. In all cases, nearly complete 1,3-chirality transfer took place without formation of the corresponding (*Z*)-allylsilanes. It should be noted that (*R*,*Z*)-**61a** provided (*R*,*E*)-**62a** stereoselectively (Equation (22.43)).

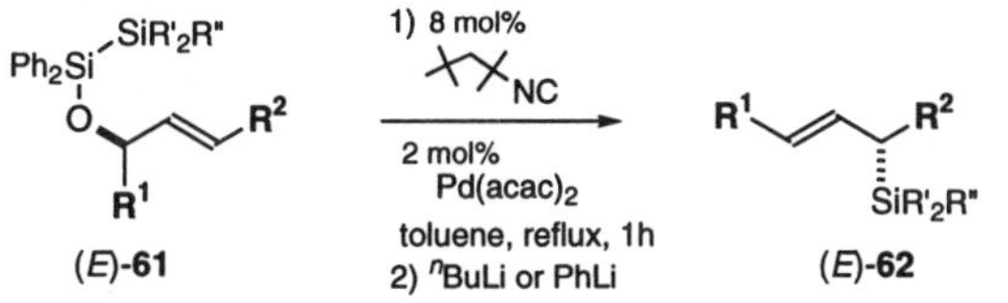

$$(22.42)$$

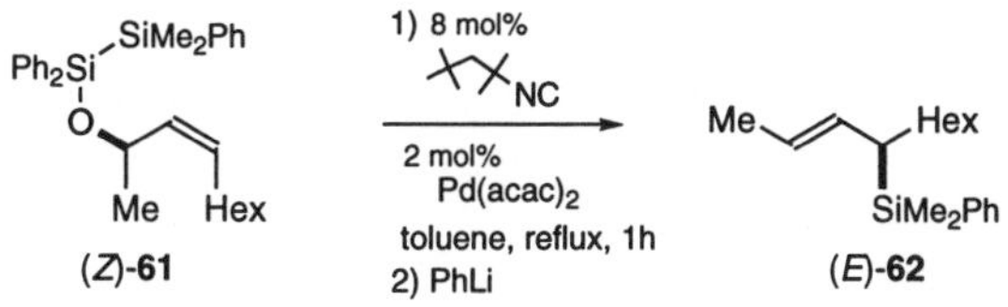

$$(22.43)$$

Finally, the mechanism for the disproportionation of four-membered oxasiletane **64** is briefly discussed. When the reaction of (*R*,*E*)-**61a** was carried out in refluxing hexane, an eight-membered product (**66**) was obtained as a single stereoisomer in high yield (Equation (22.44)) [59]. Similarly, (*R*,*Z*)-**61a** provided compound **66′** stereoselectively in the reaction in hexane under reflux. The structures of compounds **66** and **66′**, which may be formed through dimerization of the corresponding four-membered oxasiletanes (**64**), were established by single-crystal X-ray analyses, revealing that the bis-silylation proceeded in high diastereoselectivity as well as stereospecific *cis*-addition. On heating in toluene under reflux in the absence of the palladium catalyst, the eight-membered rings **66** and **66′** underwent a ring contraction reaction to six-membered ring **63** with the formation of (*E*)-allylsilanes (**62a**) (Equation (22.45)).

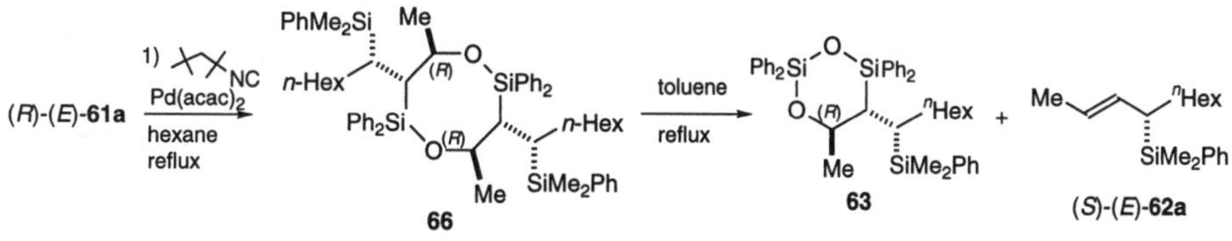

$$(22.44)$$

$$(22.45)$$

The new synthesis of allylsilanes has the advantages of a general preparative method for highly enantio-enriched allylsilanes with selective (*E*)-geometry as well as manipulative simplicity [60].

4   Concluding remarks

The new palladium catalyst for the activation of silicon–silicon bonds of disilanes has opened up new synthetic chemistry for organosilicon compounds that hitherto have been difficult to synthesize. The Si–Si σ-bond metathesis reactions offer new entry into organosilicon macrocycles through the palladium-catalysed formation of silicon–silicon bonds. On the other hand, the bis-silylation reactions find wide application in organic synthesis through the stereoselective formation of silicon–carbon bonds. It should be emphasized that the use of the palladium(0)–isonitrile catalyst is crucial to promote these reactions effectively.

5   References

1   Michl J, Ed. *Chem Rev* 1995; **95**: 1135–673.
2   Marciniec B. *Comprehensive Handbook on Hydrosilylation*. Oxford: Pergamon Press, 1992.
3   West R. In: Abel EW, Stone FGA, Wilkinson G, eds. *Comprehensive Organometallic Chemistry II*, Vol. 2. Oxford: Pergamon Press, 1995: 77–110.
4   Sharma HK, Pannell KH. *Chem Rev* 1995; **95**: 1351–74.
5   Tamao K, Hayashi T, Kumada M. *J Organomet Chem* 1976; **114**: C19–21.
6   Sakurai H, Kamiyama Y, Nakadaira Y. *J Organomet Chem* 1977; **131**: 147–52.
7   Kusukawa K, Kabe Y, Ando W. *Chem Lett* 1993: 985–8.
8   Suzuki M, Obayashi T, Amii H, Saegusa T. *Polym Prep Jpn* 1991; **40**: 355.
9   Suzuki M, Obayashi T, Saegusa T. *J Chem Soc Chem Commun* 1993: 717–18.
10   Suginome M, Oike H, Park S-S, Ito Y. *Bull Chem Soc Jpn* 1996; **69**: 289–99.
11   Suginome M, Oike H, Shuff PH, Ito Y. *Organometallics* 1996; **15**: 2170–78.
12   Suginome M, Oike H, Ito Y. *J Am Chem Soc* 1995; **117**: 1665–6.
13   Suginome M, Oike H, Ito Y. *Organometallics* 1994; **13**: 4148–50.
14   Suginome M, Oike H, Shuff PH, Ito Y. *J Organomet Chem* 1996; **521**: 405–8.
15   Ito Y, Nishimura S, Ishikawa M. *Tetrahedron Lett* 1987; **28**: 1293–5.
16   Ito Y, Matsuura T, Murakami M. *J Am Chem Soc* 1988; **110**: 3692–3.
17   Ito Y, Suginome M, Matsuura T, Murakami M. *J Am Chem Soc* 1991; **113**: 8899–908.
18   Ishikawa M, Hatano T, Hasegawa Y *et al. Organometallics* 1992; **11**: 1604–18.
19   Uchimaru Y, Tanaka Y, Tanaka M. *Chem Lett* 1995: 164.
20   Suginome M, Kato Y, Takeda N, Oike H, Ito Y. *Organometallics* 1998; **17**: 495–7.
21   Pan Y, Mague JT, Fink MJ. *Organometallics* 1992; **11**: 3495–7.
22   Okinoshima H, Yamamoto K, Kumada M. *J Organomet Chem* 1975; **86**: C27–30.
23   Sakurai H, Kamiyama Y, Nakadaira Y. *J Am Chem Soc* 1975; **97**: 931–2.
24   Watanabe H, Kobayashi M, Higuchi K, Nagai Y. *J Organomet Chem* 1980; **186**: 51–62.
25   Yamashita H, Catellani M, Tanaka M. *Chem Lett* 1991: 241–4.
26   Ito Y, Suginome M, Murakami M. *J Org Chem* 1991; **56**: 1948–51.

27 Murakami M, Oike H, Sugawara M, Suginome M, Ito Y. *Tetrahedron* 1993; **49**: 3933–46.

28 Murakami M, Suginome M, Fujimoto K, Ito Y. *Angew Chem Int Ed Engl* 1993; **32**: 1473–5.

29 Suginome M, Matsumoto A, Ito Y. *J Org Chem* 1996; **61**: 4884–5.

30 Peterson DJ. *J Org Chem* 1968; **33**: 780–84.

31 Chan TH, Mychajlowskij W, Ong BS, Harpp DN. *J Org Chem* 1978; **43**: 1526–32.

32 Danheiser RL, Carini DJ, Basak A. *J Am Chem Soc* 1981; **103**: 1604–6.

33 Marshall JA, Tang Y. *J Org Chem* 1994; **59**: 1457–64.

34 Danheiser RL, Carini DJ, Kwasigroch CA. *J Org Chem* 1986; **51**: 3870–78.

35 Hayashi T, Kobayashi T, Kawamoto AM, Yamashita H, Tanaka M. *Organometallics* 1990; **9**: 280–81.

36 Ozawa F, Sugawara M, Hayashi T. *Organometallics* 1994; **13**: 3237–43.

37 Murakami M, Andersson PG, Suginome M, Ito Y. *J Am Chem Soc* 1991; **113**: 3987–8.

38 Murakami M, Suginome M, Fujimoto K *et al. J Am Chem Soc* 1993; **115**: 6487–98.

39 Suginome M, Matsumoto A, Nagata K, Ito Y. *J Organomet Chem* 1995; **499**: C1–3.

40 Tamao K. Ishida N, Tanaka T, Kumada M. *Organometallics* 1983; **2**: 1694–6.

41 Tamao K, *Adv Silicon Chem* 1996; **3**: 1–62.

42 Suginome M, Yamamoto Y, Fujii K, Ito Y. *J Am Chem Soc* 1995; **117**: 9608–9.

43 Ohrui H, Emoto S. *Tetrahedron Lett* 1975; **16**: 3657–60.

44 Anderson RC, Fraser-Reid B. *J Am Chem Soc* 1975; **97**: 3870–71.

45 Parker WL, Johnson F. *J Org Chem* 1973; **38**: 2489–96.

46 Corey EJ, Yu C-M, Kim SS. *J Am Chem Soc* 1989; **111**: 5495–6.

47 Brown HC, Jadhav PK. *J Am Chem Soc* 1989; **111**: 5495–6.

48 Männig D, Nöth H. *Angew Chem Int Ed Engl* 1985; **24**: 878–9.

49 Schreiber SL, Hoveyda AH. *J Am Chem Soc* 1984; **106**: 7200–2.

50 Burke SD, Pacofsky GJ, Piscopio AD. *J Org Chem* 1992; **57**: 2228–35.

51 Suginome M, Matsunaga S, Iwanami T, Matsumoto A, Ito Y. *Tetrahedron Lett* 1996; **37**: 8887–90.

52 Suginome M, Matsunaga S, Ito Y. *Synlett* 1995: 941–2.

53 Suginome M, Nakamura H, Ito Y. *Tetrahedron Lett* 1997; **38**: 555–8.

54 Suginome M, Matsumoto A, Ito Y. *J Am Chem Soc* 1996; **118**: 3061–2.

55 Fleming I, Dunogués J, Smithers R. In: Kende AS, ed. *Organic Reactions*, Vol. 37. New York: Wiley, 1989: 57–575.

56 Gao Y, Hanson RM, Klunder JM *et al. J Am Chem Soc* 1987; **109**: 5765–80.

57 Fleming I, Lawrence NJ. *J Chem Soc Perkin Trans 1* 1992: 3309–26.

58 Suginome M, Iwanami T, Matsumoto A, Ito Y. *Tetrahedron: Asymm* 1997; **8**: 859–62.

59 Suginome M, Iwanami T, Ohmori H, Matsumoto A, Ito Y. (Manuscript in preparation.)

60 Hayashi T, Konishi M, Kumada M. *J Am Chem Soc* 1982; **104**: 4963–5.

23 Catalytic Reactions in Organoboron Compounds

AKIRA SUZUKI

Department of Chemical Technology, Kurashiki University of Science and the Arts, Kurashiki 712-8505, Japan

1 Introduction

Useful C—C bond-forming reactions employing organoboron compounds are divided into two categories, one involving an organic radical from an organoboron compound in the presence of a catalytic amount of oxygen, and another including an intramolecular shift of an organic group on boron to an electron-deficient carbon atom, that generally need no catalysts [1]. In the hydroboration reaction of alkenes and alkynes, which provides the general and most convenient method for preparing organoboron compounds, usually catalysts are not required [2]. Recently, in some of the reactions using organoboron compounds or even in the hydroboration reaction, the necessity of catalysts has been demonstrated. This review mentions such catalytic reactions in organoboron compounds.

2 Catalytic hydroboration of alkenes and alkynes with catecholborane

Borane derivatives that contain groups other than alkyl are potentially of great value in synthesis because the reactivity of borane itself can be greatly modified. Among such derivatives, catecholborane (1,3,2-benzodioxaborole) (**1**) is useful as a hydroboration reagent [3], although the reaction is generally rather inert. The most obvious application of catecholborane derivatives involves hydrolytic removal of the catechol moiety to give the simple alkyl- or alkenyldihydroxyboranes (boronic acids) [4]. For the synthesis of stereo-defined or functionalized alkenylboronic acids and their esters, the hydroboration of a terminal alkyne with catecholborane is important (Equation (23.1)). The hydroboration with catecholborane can be carried out under mild conditions by using palladium, rhodium or nickel catalysts [5].

$$RC\equiv CH + H\text{-}B \quad \longrightarrow \quad \text{(alkenylboronate)} \quad \longrightarrow \quad R\text{-CH=CH-}B(OH)_2 \tag{23.1}$$

1

It is also observed that such a catalytic hydroboration can direct the course of the addition of borane toward different regioselectivity, compared with that in an uncatalysed reaction. For instance, it is known that monohydroboration of common 1,3-dienes without catalysts mainly occurs at the terminal double bond to give homoallylic boranes as major products. On the other hand, the catalytic 1,4-hydroboration of 1,3-butadiene, isoprene, myrcene, 2,3-dimethyl-1,3-butadiene and 1,3-cyclohexadiene with catecholborane in the presence of $Pd(PPh_3)_4$ or $Rh_4(CO)_{12}$ proceeds extremely regio- and stereoselectively to afford (*Z*)-allylic boronates as sole products. Thus, 1,3-butadiene gives 2-[(*Z*)-crotyl]-1,3,2-benzodioxaborole in a yield of 87% with almost 100% pure (*Z*)-configuration. Additionally, the erythro selectivity in the addition to benzaldehyde supports the (*Z*)-configuration (Equation (23.2)) [6].

$$
\text{(23.2)}
$$

87% (Z-, almost 100%)   81%

Arase and co-workers reported that the conventional hydroboration of thioacetylenes with several dialkylboranes gives high yields of vinylborane intermediates, with high preference for the addition of a boron atom at the carbon adjacent to the alkylthio group [7]. Although our preliminary results also indicated that such an addition of a boron atom occurs preferentially at the carbon α to sulfur, the regioselectivity of addition is highly dependent upon substituents in the thioacetylenes and dialkylboranes employed, and complete control of regioselectivity is generally difficult. However, we have found that the catalytic hydroboration of thioacetylenes with catecholborane in the presence of Ni, Pd or Rh catalysts results in a complete reversal of the regiochemical preference [5b]. The representative results of hydroboration of 1-(ethylthio)-1-propyne with catecholborane in the presence of 3 mol.% of catalysts (Equation (23.3)) are summarized in Table 23.1, the reactions of which are conducted in benzene at room temperature for 5 h.

$$
\text{(23.3)}
$$

The extremely high efficiency of nickel complexes having bidentate ligands over that of monodentate complexes is unlikely to correspond to the substitution of a strongly chelated diphosphine ligand with alkynes, which is generally speculated as being the coordination step in the mechanism of hydrosilylation [8] and hydroboration [5d] of alkenes catalysed by monophosphine–transition metal complexes. Most probably, the coordination and insertion of alkyne may proceed through a saturated five-coordinated Ni(II) complex without elimination of a phosphine ligand. Such a five-coordinated intermediate can be stabilized best by electron-rich nickel rather than palladium and rhodium [9].

However, we have found that $NiCl_2$(dppe) and $NiCl_2$(dppp) are the most effective catalysts for the hydroboration of thioacetylenes with catecholborane. These nickel catalysts do not allow the hydroboration and cross-coupling reactions to take place in the

Table 23.1. Effect of catalysts on hydroboration of 1-(ethylthio)-1-propyne with catecholborane (see Eqn 23.3)

Entry	Catalyst	Yield (%)	Ratio 2:3
1	$Pd(PPh_3)_4$	54	98:2
2	$PdCl_2(PPh_3)_4$	29	66:34
3	$PdCl_2$(dppf)	69	96:4
4	$RhCl(PPh_3)_4$	40	58:42
5	$RhCp(PPh_3)_4$	12	91:9
6	$NiCl_2(PPh_3)_4$	19	52:48
7	$NiCl_2$(dppe)	100	>99:1
8	$NiCl_2$(dppp)	97	98:2
9	none	20	50:50

same flask because nickel complexes are not effective for the cross-coupling reaction of compound **2** with organic halides. Thereafter, it has been found that palladium hydroboration is slower than the nickel-catalysed reaction, but phosphine-based palladium complexes such as $Pd(PPh_3)_4$ and $PdCl_2(dppf)$ are sufficiently effective to permit the preparation of quantitative yields of compound **2** after an overnight reaction period. Thus, a variety of vinylic sulfides are synthesized stereoselectively by the catalytic hydroboration of thioacetylenes with catecholborane in the presence of palladium catalysts followed by cross-coupling of the resulting boron derivatives with organic halides by using the same catalyst (Equation (23.4)). Vinylic sulfides thus prepared are useful intermediates in organic synthesis. For example, the reaction of compound **5** with (2-methylpropenyl)magnesium bromide in the presence of $NiCl_2(dppp)$ gives compound **6** readily in 78% yield (Equation (23.4)) [5c].

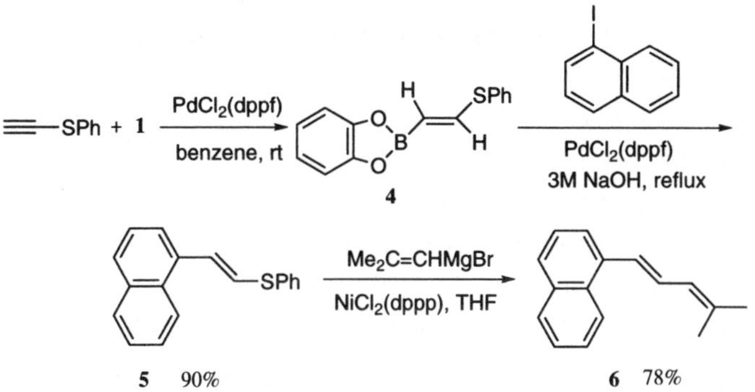

$$(23.4)$$

Rhodium-mediated hydroboration of allylic alcohol derivatives, such as compound **7**, followed by oxidation gives predominantly *syn*-2-methyl-1,3-diol, whereas conventional hydroboration of the same substrate without catalyst affords mostly *anti*-products (Equation (23.5)) [10]. It is proposed that mixing of s*-orbitals involved in bonding at the asymmetric center with p*-orbitals of the alkene lowers the LUMO involved in complexation of rhodium, and this could control diastereofacial selectivities in the catalysed hydroboration.

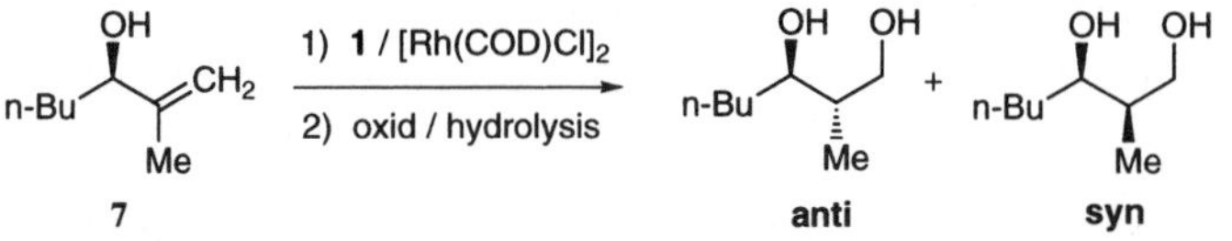

$$(23.5)$$

Knorr and Merola have reported [11] that oxidative addition of the B—H bond in catecholborane to $Ir(cyclooctene)(PMe_3)_3Cl$ produces mer-$Ir(H)(BO_2C_6H_4)(Cl)(PMe_3)_3$ (**8**), which is characterized by single-crystal X-ray diffraction (Equation (23.6)). Compound **8** reacts with alkynes to form vinyliridium complexes and catalyses the hydroboration of alkynes with catecholborane [11].

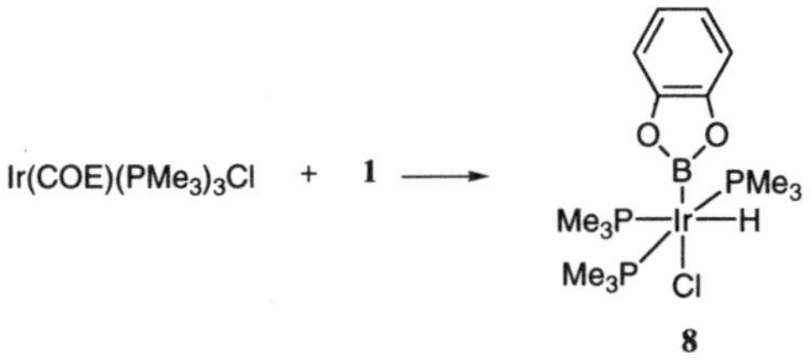

$$(23.6)$$

Most recently, a comprehensive review on hydroborations catalysed by transition metal complexes has been described by Beletskaya and Pelter [12].

3   Addition of heteroatom–boron bond to alkynes

Recently, we have reported a general and convenient method for the stereo- and regioselective syntheses of conjugated alkadienes, alkenynes, arylated alkenes and other olefinic compounds by the cross-coupling reaction of 1-alkenylboron compounds with a wide variety of unsaturated organic halides and related compounds [4]. In connection with the development of such coupling reactions, the importance of stereoselective synthesis of alkenyl halides has increased. As one of the new types of these syntheses, one can expect a halometallation reaction. Actually, we have found that B-halo-9-borabicyclo[3.3.1]nonane derivatives (B-X-9-BBN, X = Br or I) react readily with 1-alkynes stereo-, regio- and chemoselectively to give corresponding haloboration adducts, which are very useful intermediates for the synthesis of various vinylic halide derivatives [13]. Such a haloboration reaction proceeds smoothly without any catalysts.

The addition of boron–heteroatom bonds to alkynes is an attractive route for the synthesis of functionalized alkenylboron compounds. However, the reaction usually does not take place due to the high boron–heteroatom bond energy and low Lewis acidity of boron. The copper(I)- or cobalt(II)-catalysed addition of Si–B (silylboration) [14] and Sn–B compounds (stannylboration) [14,15] to alkynes has been observed. Additionally, it has been discovered recently that the palladium(0)-catalysed addition of 9-(organothio)-9-BBN derivatives to 1-alkynes (thioboration) provides (β-(organothio)vinyl)boranes in high yields [16].

3.1   *Addition of Si–B (silylboration)*

Although several compounds containing a B–Si bond were synthesized [17, 18], no examples of synthetic utility were given. Utimoto and co-workers have reported that the silylboron compound $PhMe_2SiBEt_3Li$ adds to triple bonds effectively to give silylboration adducts under good control of regio- and stereoselectivity [14]. For the reaction of 1-alkynes, it is observed that CuI, $CuBr \cdot SMe_2$ and $CoCl_2(PPh_3)_2$ catalyse the reaction, while $Pd(PPh_3)_4$, $RuCl_2(PPh_3)_3$ and $NiCl_2(PPh_3)_2$ are not effective. In particular, the Co catalyst gives the best result in terms of reaction yield and selectivity. The addition proceeds in *cis* fashion exclusively to produce compound **9** (Equation (23.7)). Weak acids such as water, methanol and phenol react readily with the intermediary vinylboron derivatives (**9**) to yield corresponding vinylsilanes (**10**) (Equation (23.7)). The reaction was extended by the same group to stannylboration of acetylenes. However, the result was not excellent like the case of silylboration. Thereafter, Sharma and Oehlschlager [15] have found a modified method that is mentioned in Section 3.2.

$$RC \equiv CH \ + \ PhMe_2SiBEt_3Li \ \xrightarrow{\ CoCl_2(PPh_3)_2\ }$$

(23.7)

3.2 Addition of Sn—B (stannylboration)

The stanylboron 'ate' complex (**11**) has been shown to react with 1-alkynes in the presence of one equivalent of CuBr·SMe$_2$ to afford the corresponding addition products (**12**) regio- and stereoselectively, whereas phenylacetylene gives the reverse result, namely products **13** mainly (Equation (23.8)) [15]. From these addition derivatives, corresponding vinylic tin compounds are prepared.

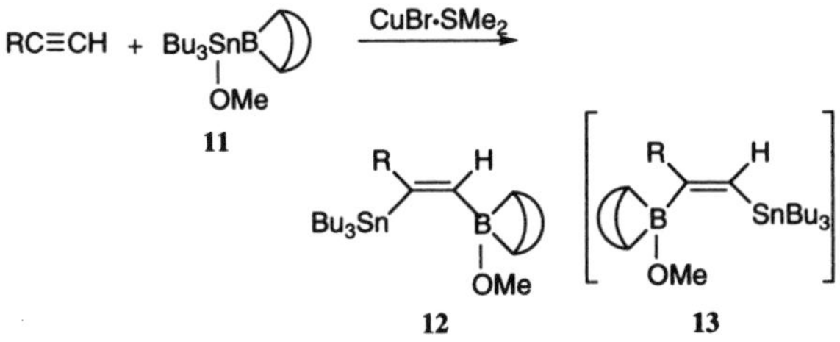

$$(23.8)$$

3.3 Addition of S—B (thioboration)

A novel palladium(0)-catalysed addition reaction of 9-(organothio)-9-BBN derivatives (**14**) to terminal alkynes has been discovered to give (β-(organothio)-vinyl)boranes (**15**) regio- and stereoselectively in high yields (Equation (23.9)) [16]. Among the catalysts examined, it had been shown that Pd(PPh$_3$)$_4$ is the best, and the reaction proceeds efficiently in THF solvent. Compound RhCl(PPh$_3$)$_3$ also exhibits some catalytic activity, but PdCl$_2$(PPh$_3$)$_2$, CuI and CoCl(PPh$_3$)$_3$ are ineffective. (Organothio)boranes (**14**) are readily prepared by the reaction of boranes or haloboranes with thiols (Equation (23.10)). By selecting catalysts, various different types of organothioboron compounds are obtained in almost quantitative yields, as shown in Table 23.2 (T. Ishiyama, N. Miyaura and A. Suzuki, unpubl. data).

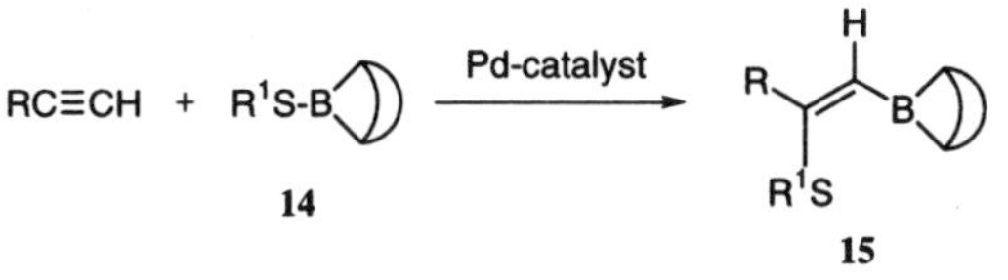

$$(23.9)$$

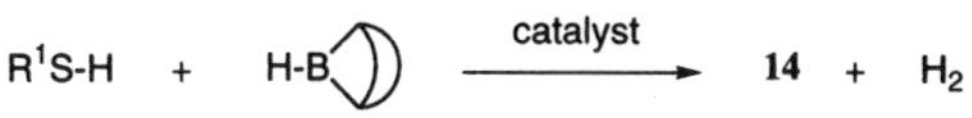

$$(23.10)$$

The addition of methanol immediately induces a very fast protodeboronation of compound **15** to give 2-organothio-1-alkenes (**16**) readily in excellent yields (Equation (23.11)).

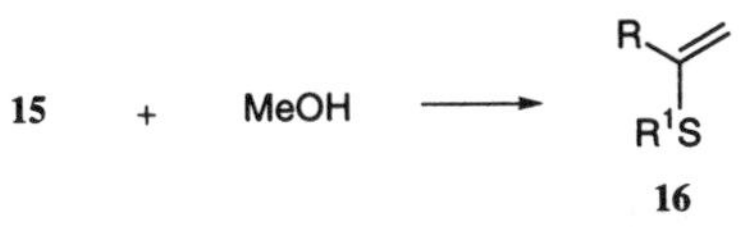

$$(23.11)$$

Like other related reactions catalysed by transition metals, the thioboration reaction may involve three steps: oxidative addition of a B—S bond to a palladium(0) complex, insertion of an alkyne and reductive elimination to sulfides.

As mentioned in Section 5, we have reported the palladium-catalysed cross-coupling reaction of 1-alkenyl-, aryl and alkylboron compounds with various organic halides. The

Table 23.2. Synthesis of (9-organothio)-9-BBN derivatives **11** (see Eqn 23.10)

14	Yield (%)[a]		
	Catalyst: None	Pd(PPh$_3$)$_4$	Pt(PPh$_3$)$_4$
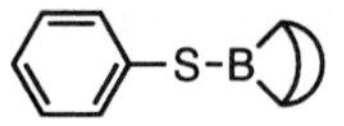	95	–	–
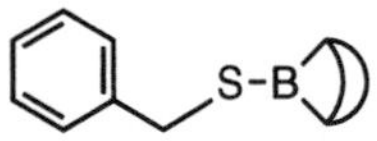	86	–	–
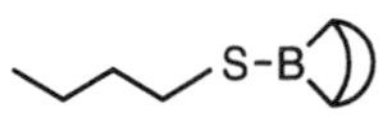	89	–	–
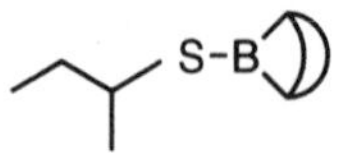	42	96	98
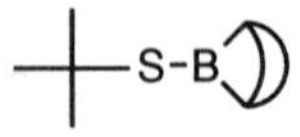	10	32	96

All reactions are carried out at 25 °C in THF (20 mL), by using the catalyst (0.1 mmol), thiol (10 mmol), and 9-BBN (10 mmol).
[a] Isolated yields based on thiols.

reaction of (*E*)- and (*Z*)-1-alkenylboron derivatives takes place with ease in the presence of base and palladium catalyst, thus allowing the ready preparation of isomerically pure alkenes or alkadienes. The usefulness of the reaction is demonstrated in the one-pot synthesis of vinylic sulfides by sequential thioboration and cross-coupling reactions, both of which are catalysed by a common palladium(0) catalyst; one such example is shown in Equation (23.12).

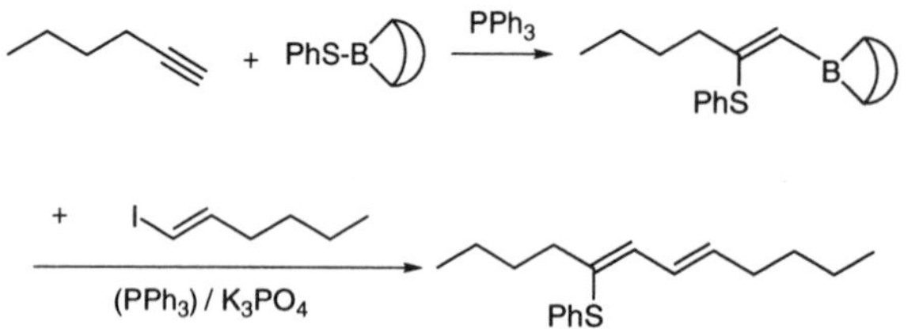

77% (purity, 99%)

$$(23.12)$$

1-Alkenyl sulfides are valuable precursors for the preparation of carbonyl compounds by mercury(II)-promoted hydrolysis [19], the synthesis of 1-alkenyl sulfoxides [20] as the dienophiles of cycloaddition or the acceptors for Michael addition, and the synthesis of stereodefined alkenes by the cross-coupling reaction with Grignard reagents [21]. However, reported methods using the reaction of carbonyl compounds with (1-(methylthio)alkyl)-phosphonate [22] and ((alkylthio)methyl)trimethylsilane [23] unfortunately lead to a mixture of (*E*)- and (*Z*)-isomers. On the other hand, the thioboration–cross-coupling sequence described above gives expected sulfides stereoselectively in high yields. The cross-coupling reaction of 2-(phenylthio)-1-bromo-1-alkenes with alkyl- or 1-alkenylboron compounds is also known to provide such sulfides selectively [24].

It has been reported recently that the catalytic hydroboration of 1-(alkylthio)-1-alkynes with catecholborane selectively produces (β-(alkylthio)vinyl)boronates (**17**) in almost quantitative yields (Equation (23.13)) [5b]. The ready availability of such boron reagents by catalytic thioboration or hydroboration may now offer a more flexible and reliable route to such stereodefined alkenyl sulfides in combination with numerous reactions reported in organoboron chemistry.

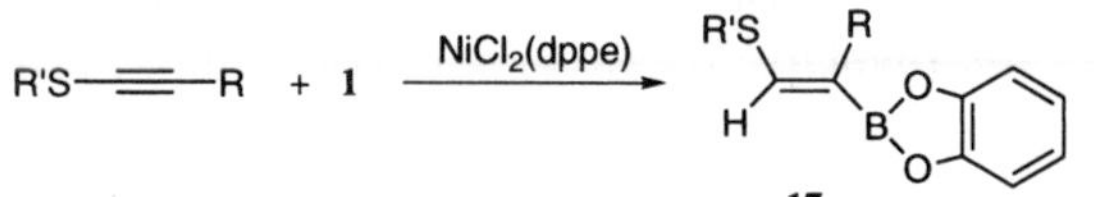

$$(23.13)$$

4 Diboration to alkynes

The addition of diboron compounds (X_2B-BX_2) to unsaturated hydrocarbons, first discovered by Schlesinger in 1954 [25], is an attractive and straightforward method to introduce two boryl units into organic molecules. Although diboron tetrahalides (X = Cl, F) have been used extensively for this transformation, their synthetic use has been severely limited because of instability and unavailability of the reagents. Tetrakis(alkoxo)- and tetrakis(amido)-diborons (X = RO, R_2N) [26], which are readily available in large quantity by the Wurtz coupling of the corresponding haloboranes with sodium or potassium metal, are rather stable; however, the compounds are reported to be quite inert to alkenes and alkynes due to the low Lewis acidity of the boron centers conjugated with oxygen or nitrogen atoms and the high B–B bond energy.

Considerable attention has been focused on the transition-metal-catalysed additions of metal reagents, including magnesium, aluminum, silicon, zinc, germanium and tin compounds, to unsaturated hydrocarbons [16]. Recently, we have discovered a novel platinum(0)-catalysed addition reaction of the pinacol ester of diboron (18) to both terminal and internal alkynes to afford *cis*-bis(boryl)alkenes (19) (Equation (23.14)) [27].

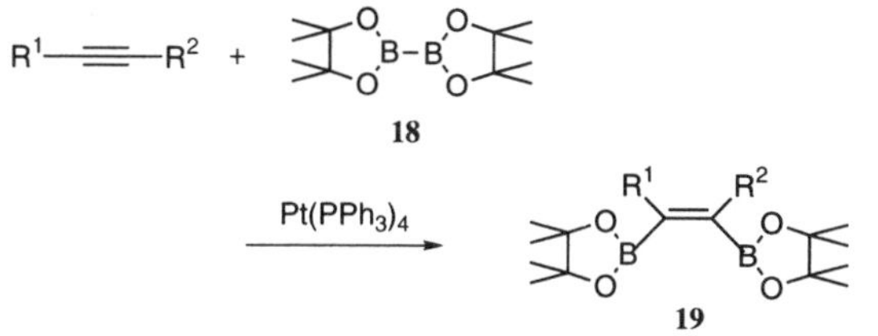

$$(23.14)$$

For instance, the reaction of 1-octyne is shown to be catalysed efficiently by platinum(0) complexes. The expected addition product is produced in high yield (>90%) by heating the mixture at 80°C in DMF for 24h in the presence of a catalytic amount of $Pt(PPh_3)_4$ (3 mol.%). The addition of tetrakis(methoxo)diboron to 1-octyne under similar conditions gives the corresponding addition product in equally high yield.

Gallery Chemical Co., Evans City, PA, has established a novel synthetic industrial process to form compound 18 from trichloroborane (Equation (23.15)) [28].

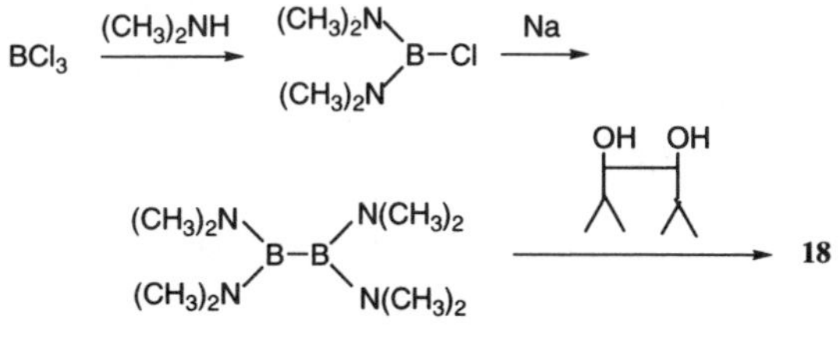

$$(23.15)$$

The reaction is catalysed by platinum(0) complexes but not by divalent complexes. The reaction proceeds through *syn* addition of the B—B bond to alkynes, giving the isomerically pure compound **19** in Fig. 23.1. These results strongly suggest a mechanism involving B—B bond activation by oxidative addition of compound **18** to the platinum(0) complex to form a bis(boryl)platinum(II) intermediate (**20**), followed by alkyne insertion to the B—Pt bond to give compound **21** and finally reductive elimination of compound **21** to provide compound **19**. A similar oxidative addition of B—H [29], Si—Si [30,31], and Si—Sn [32] compounds to low-valent transition metal complexes has been studied extensively and postulated as a key step in the catalytic hydroboration and addition of disilanes or silyl-stannanes to alkenes and alkynes. Monitoring of the reaction mixture of compound **18** and Pt(PPh$_3$)$_4$ in toluene by multinuclear NMR spectroscopy reveals the formation of a new Pt(II) species. Finally, a single crystal of compound **20** suitable for X-ray analysis was obtained in 82% yield by treatment of Pt(PPh$_3$)$_4$ with 20 equiv of compound **18** in hexane at 80 °C for 2 h, followed by recrystallization from hexane/toluene (Equation (23.16)). The results of an X-ray diffraction analysis of the compound confirm the structure to be compound **20**.

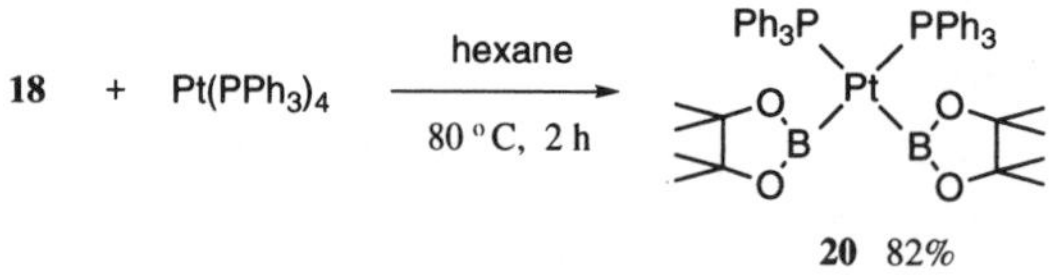

$$(23.16)$$

The isolated compound **20** exhibits high reactivity for the insertion of alkynes. On the other hand, compound **20** provides no addition products to alkenes such as 1-octene, thus limiting this reaction only to alkynes, which is in contrast to the bis(boryl)rhodium(III) complex giving diboration products with styrene [33].

On the basis of these findings, we propose the catalytic cycle shown in Fig. 23.1.

The diboration adducts of alkynes are useful intermediates in organic synthesis. For instance, compound **22** reacts with iodobenzene at 90 °C in dioxane in the presence of 3 mol.% Pd(PPh$_3$)$_4$ as catalyst and aqueous 3 M KOH to give (*Z*)-1,2-diphenyl-1-decene as the sole product (Equation (23.17)) [27a].

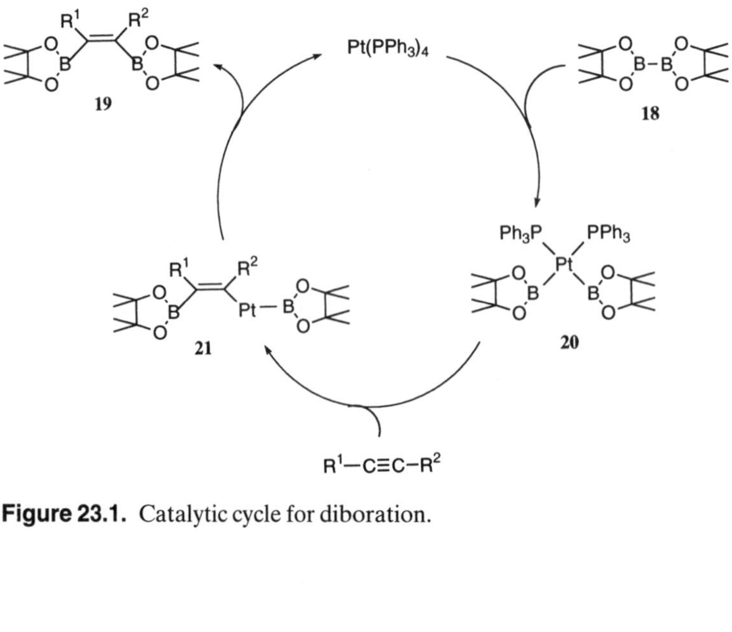

Figure 23.1. Catalytic cycle for diboration.

$$(23.17)$$

5 Cross-coupling reactions of organoboron compounds with organic halides

The cross-coupling reaction of organometallic compounds with organic halides and related derivatives provides a convenient synthetic methodology for C—C bond formation [4]. Although organoboron compounds are readily obtained by the hydroboration of alkenes and alkynes or from the other organometallic compounds, the B—C bond is observed to be almost completely covalent, and organic groups on boron are weakly nucleophilic, thus limiting the use of such organoboron reagents for ionic reactions. The coordination of negatively charged bases to the boron atom has been recognized to be an efficient way of increasing its nucleophilicity to transfer the organic group on boron to the adjacent positive center (1,2-migration reaction). However, intermolecular transfer reactions such as the Grignard-like reaction are relatively rare. The cross-coupling reaction of organoboron compounds with organic halides in the presence of a catalytic amount of palladium complexes and bases has been found to proceed smoothly, and such a reaction has proved to be quite a general technique for a wide range of selective carbon–carbon bond formation [4]. Many organometallic reagents undergo similar cross-coupling reactions, but much attention has been focused recently on the use of organoboronic acids in laboratories and industries, because they are convenient reagents that are generally thermally stable and inert to water and oxygen, thus allowing their handling without special precautions. In this section, the palladium-catalysed cross-coupling reaction of organoboron compounds with organic halides and their synthetic applications are briefly mentioned.

5.1 *Coupling of 1-alkenylboron derivative*

5.1.1 WITH 1-ALKENYL HALIDES: SYNTHESIS OF CONJUGATED ALKADIENES

The stereo- and regioselective syntheses of conjugated alkadienes are of great importance in organic chemistry in themselves, as well as in their utilization in other reactions, such as the Diels–Alder reaction. A number of new methods for the preparation of conjugated dienes and polyenes have been developed by utilizing various organometallic compounds. Among these procedures, the most promising ones are perhaps those based on the direct cross-coupling reaction of stereodefined alkenylmetals with stereodefined haloalkenes in the presence of transition metal catalysts. Although the representative 1-alkenylmetal reagents undergo a similar type of coupling with haloalkenes, there are several limitations when one wishes, for example, to obtain unsymmetrical dienes without homo-coupling and highly functionalized dienes. Also, stoichiometric conditions relative to metal reagents and halides are difficult to use in such coupling reactions. Thus, the use of 1-alkenylboronic acids or their esters has been noted recently.

The first observation of the preparation of conjugated dienes was discovered in 1978 (Equation (23.18)) [34a,b]. High yields of the diene (22) are obtained when relatively strong bases such as sodium ethoxide and hydroxide are used together with a phosphine-based Pd complex such as $Pd(PPh_3)_4$ or $PdCl_2(PPh_3)_2$. The use of a palladium catalyst without phosphine ligands or of a weak base (KOAc or Et_3N) leads to a tendency for contamination by an undesired head-to-tail coupling product (**23**) [35].

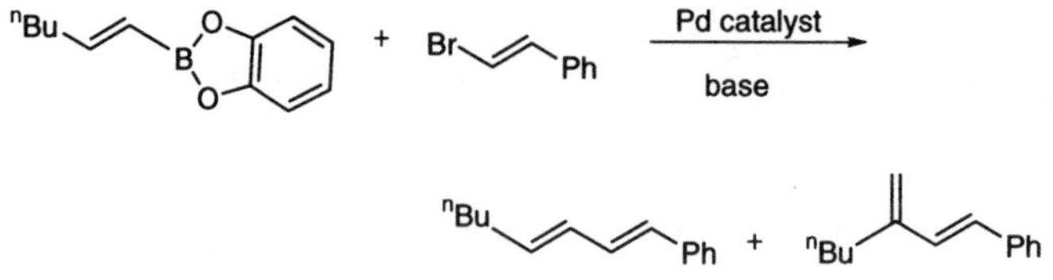

$$(23.18)$$

Although disiamyl- or dicyclohexylborane is a selective and efficient hydroboration reagent of alkynes, the 1-alkenyldialkylboranes thus prepared, especially (Z)-1-alkenylboron derivatives, give relatively poor yields of coupling products (c 50%) [36]. Fortunately, in such cases, the difficulty can be overcome by using 1-alkenylboronic acids or their esters. For instance, the yield and stereoselectivity on the cross-coupling of (Z)-1-hexenylboronic acid diisopropyl ester with iodobenzene are shown in Table 23.3 [34d].

Oxidation of the two boron—sp^3 carbon bonds with trimethylamine N-oxide prior to coupling solves the difficulty arising from B—C bond protonolysis and contamination of the coupling products with an alkyl group [37].

The conjugated alkadiene synthetic method by the cross-coupling of vinylic boranes has been applied to the synthesis of many natural products that have such diene structures, including bombykol and related isomers [38] and the sex pheromones of insects such as the European grape wine moth [39], the red bollworm moth [40] and the Egyptian cotton leafworm [41]. Many syntheses of alkadienes and trienes, such as unsaturated fatty acid amides [42], alkenylsilanes [43], *gem*-difluoroalkenes [44], cyclic alkenes [45], *trans*-(C10)-allofarnesene [46], trisporol B [47] and vinylsulfides [24b], are reported using Pd-catalysed cross-coupling.

The coupling rate enhancement has been realized by Kishi by using aqueous TlOH in place of sodium or potassium alkoxide or hydroxide. The cross-coupling between an (E)-1-alkenylboronic acid and a (Z)-iodoalkene stereoselectively furnished the C75—C76 bond

Table 23.3. Reaction of (Z)-BuCH=CHBX$_2$ with PhI[a]

-BX$_2$	Yield (%)[b]	Isomeric purity (%)
-B(Sia)$_2$	58	>94
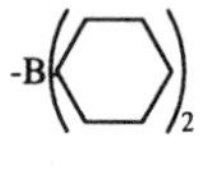 -B⟨C$_6$H$_{10}$⟩$_2$	49	>83
-B(OPr^i)$_2$	98	>97

[a] A mixture of Pd(PPh$_3$)$_4$ (3 mol.%), 2M NaOEt in EtOH (2 equiv), PhI (1 equiv) and (Z)-BuCH=CHBX$_2$ (1.1 equiv) in benzene was refluxed for 3h.
[b] Yields of (Z)-BuCH=CHPh.

formation of palytoxin at room temperature [48]. Roush, Nicolaou and Evans have also demonstrated the efficiency of thallium hydroxide on the synthesis of an aglycon of antibiotic kijanimicin [49], chlorothricolide [50], (5Z,8Z,10E,12R,14Z)-12-hydroxy-5,8,10,14-icosatetraenoic acid [(12R)-HETE] [51] and a macrolide antibiotic rutamycin B [52] (Scheme 23.1).

Hydroboration of enynes provides 1,3-alkadienylboron derivatives. The coupling of such compounds with haloalkenes allows a short synthesis of conjugated trienes; for example, the synthesis of leukotriene B4 is shown in Equation (23.19) [53, 54].

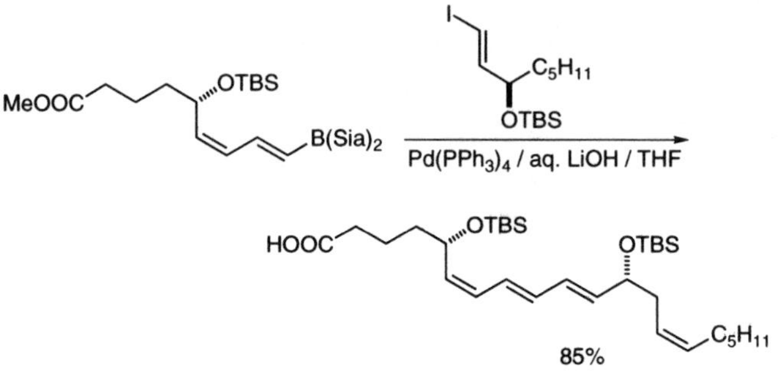

$$(23.19)$$

A combination of 1-alkenylboronates and 1-halo-1,3-alkadienes is expected to lead to the same trienes, but this combination is generally not recommended because of the synthetic problems of unstable dienyl halides and the side reaction eliminating hydrogen halides with base to produce the corresponding enyne. However, the thallium base allows this combination for synthesis of the conjugated pentaene (24) in Equation (23.20) [55].

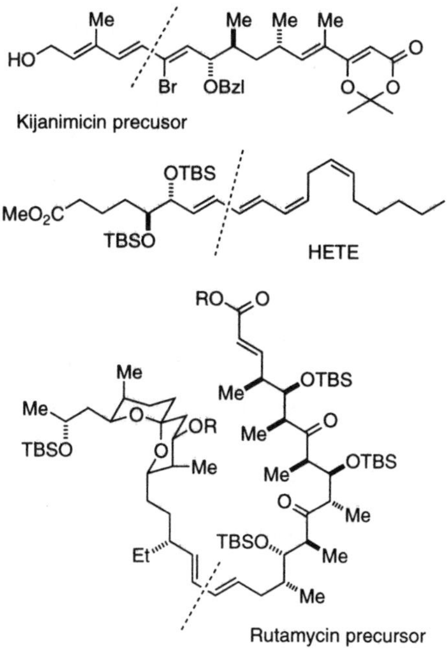

Scheme 23.1.

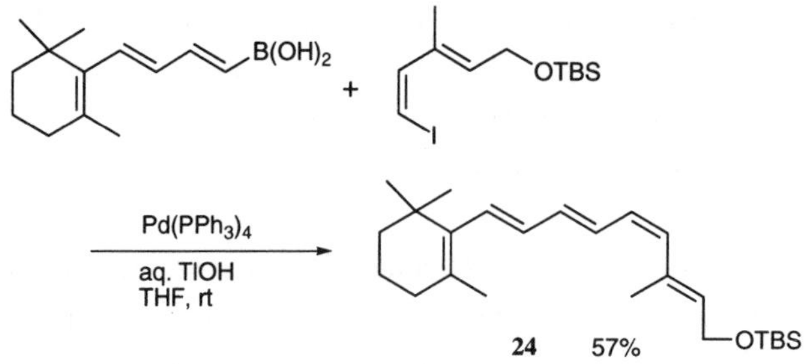

$$(23.20)$$

24   57%

β-Halo-α,β-unsaturated ketones and esters are highly susceptible to S_N2 displacement at the carbon attached to halogen, thus strong bases are undesirable for such substrates [56–58]. However, relatively weak bases such as sodium acetate and even triethylamine are effective when the reaction is conducted in alcoholic solvents [56]. Sodium acetate suspended in methanol and aqueous or solid carbonate in methanol give best results for haloenones (Equation (23.21)) [56] and haloesters [57], respectively.

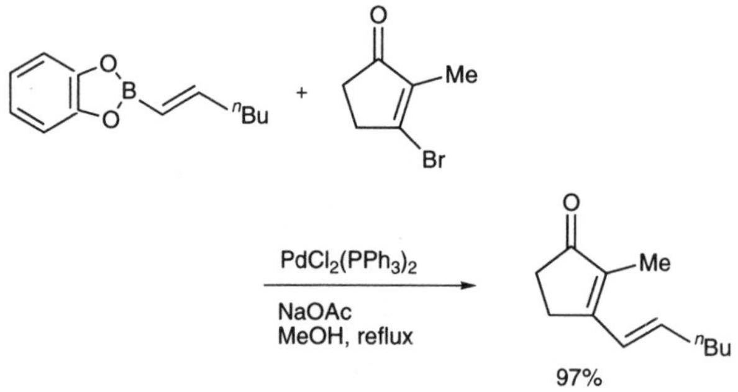

$$(23.21)$$

97%

The Pd(0)-catalysed cross-coupling of boronic acids or esters conducted with a water-soluble catalyst in the presence of organic base under mild conditions allows the production of functionalized dienes in good yields [59]. General synthesis of retinoids and arotinoids [60] and the synthesis of *trans*-1,2-dicyclopropyl alkenes [61] have been reported recently.

The zirconocene-mediated preparation of 1,3-, 1,4- and 2,3-dibora-1,3-butadienes, followed by their Suzuki coupling, provides an interesting synthesis of conjugated alkadines [62].

5.1.2   WITH 1-ALKYNYL HALIDES

Conjugated enynes are of importance in themselves and in their utilization for the synthesis of stereodefined conjugated alkadines. The cross-coupling reaction of 1-alkenyl(disiamyl)-boranes with 1-bromo-1-alkynes provides corresponding conjugated enynes in high yields (Equation (23.22)) [34].

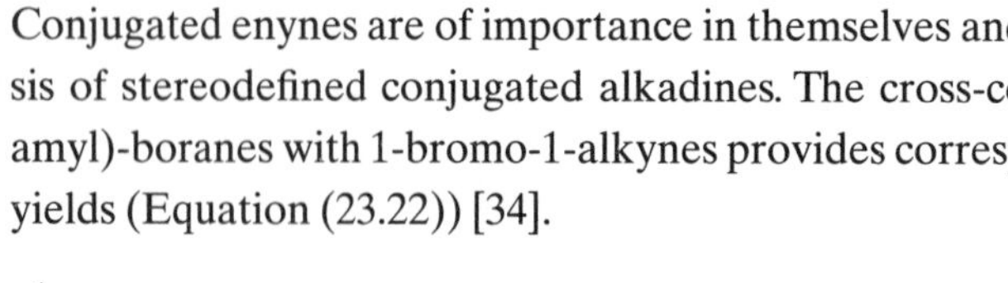

$$(23.22)$$

5.1.3   WITH ARYL HALIDES

Cross-coupling between 1-alkenylboronates and haloarenes is useful for alkenylation of haloarenes (Equation (23.23)) [63].

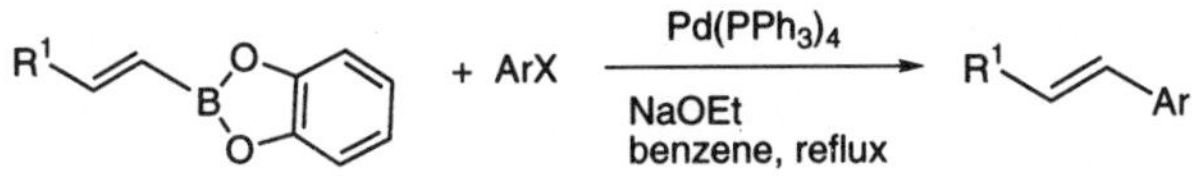

$$(23.23)$$

The relative reactivity of halobenzenes appears to be Ph > p-ClC$_6$H$_4$Br > PhBr > o-MeC$_6$H$_4$Br > o-MeOC$_6$H$_4$Br [63]. The order of reactivity is in good agreement with the substituent effect in the oxidative addition of aryl halides to palladium(0) complex [64], and presumably the substituents accelerate the transmetallation rate in the same order. The procedure, involving a hydroboration–coupling sequence, gives new access to the HGM-CoA reductase inhibitor NK-104 [65], as indicated in Equation (23.24).

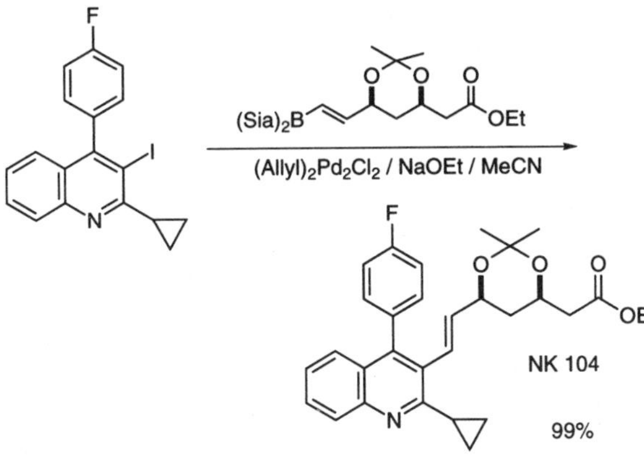

$$(23.24)$$

The efficient synthesis of a part of the antiparasitic agent milbemycin β3 has been described recently by Marko (Equation (23.25)) [66]. In the report, the unique role played by thallium carbonate in the Pd(0)-catalysed reaction is discussed. While Kishi showed that such a coupling is brought to fruition by using TlOH [48a], Suzuki utilized the corresponding Tl$_2$CO$_3$ to promote some alkyl–aryl/alkyl–vinyl coupling reactions [74,134]. Because the presence of an ester group in the aromatic fragment precluded the use of TlOH, Marko used TlOEt. Disappointingly, a mediocre yield of product (12%) was obtained. However, in the presence of Tl$_2$CO$_3$, a smooth reaction took place to give a good overall yield (68%) (Equation (23.25)).

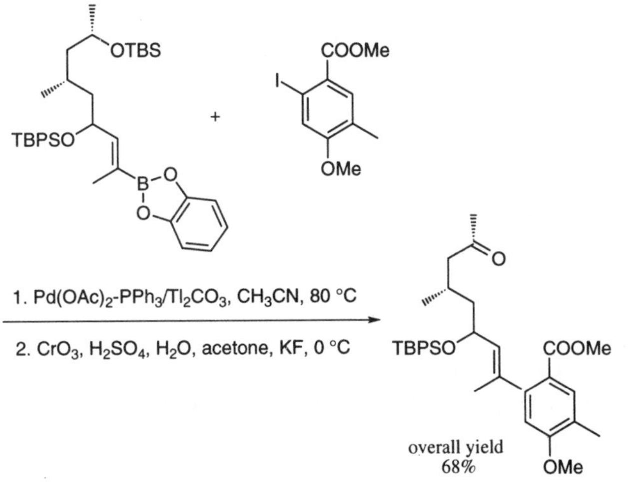

$$(23.25)$$

Polymer-bound aryl halides react with a variety of arylboronic acids under the Suzuki coupling conditions, followed by cleavage of the resin part by different electrophiles to give *ipso*-substitution products in good yields [67].

A novel, stereospecific synthesis of 1,2-*trans*-disubstituted cyclopropanes has been demonstrated, based on the palladium(0)-mediated cross-coupling of cyclopropylboronic esters with aryl halides [68].

5.2   *Coupling of arylboron compounds with aryl halides: synthesis of biaryls*

The first synthetic approach to prepare biaryls by employing a coupling reaction of arylboron derivatives with aryl halides in the presence of palladium catalysts and bases was reported in 1981 (Equation (23.26)) [69]. After this discovery, various modifications have been made for the reaction conditions. A combination of $Pd(PPh_3)_4$ or $PdCl_2(PPh_3)_4$ and aqueous Na_2CO_3 in dimethoxyethane (DME) works satisfactorily in most cases [70, 71].

$$\text{Ph}-B(OH)_2 + X-\text{Ar}(Z) \xrightarrow[\substack{\text{aq. } Na_2CO_3 \\ \text{benzene, reflux}}]{Pd(PPh_3)_4} \text{Ph}-\text{Ar}(Z) \qquad (23.26)$$

Combinations with other bases such as Et_3N [72], $NaHCO_3$ [70], Cs_2CO_3 [73], Tl_2CO_3 [74] and K_3PO_4 [75], with or without Bu_4NCl [76] and 18-crown-6 ether [73], also have been used. The reaction is successful for aryl triflates and iodo- and bromoarenes. Chlorobenzene derivatives are generally quite inert to oxidative addition, but some π-deficient heteroaryl chlorides give coupling products [77]. The reaction proceeds more rapidly in homogeneous conditions (aqueous base in DME), but reasonable yields are also obtained under heterogeneous conditions. For example, K_2CO_3 suspended in toluene works well for base-sensitive reactants [78]. As mentioned above, the coupling reactions take place nicely in aqueous conditions. Beletskaya and co-workers have reported that $PdCl_2$- and $Pd(OAc)_2$-catalysed cross-coupling reactions of arylboric acids with aryl halides in aqueous solutions give corresponding addition products in good yields [79], and also demonstrated the same type of coupling of symmetric diaryliodonium salts with sodium tetraphenylborate in an aqueous Na_2CO_3 solution to produce coupling products in 96–98 % yields [80]. For base-sensitive reactants, the extremely mild conditions using CsF or Bu_4NF allow the synthesis of variously functionalized biaryls (Equation (23.27)) [81].

$$\text{Ph}-B(OH)_2 \quad + \quad \text{Br}-\text{Ar}-CH_2COCH_3$$

$$\xrightarrow[\text{CsF / DME at 100 °C}]{Pd(PPh_3)_4} \text{Ph}-\text{Ar}-CH_2COCH_3 \qquad (23.27)$$

$$85\%$$

Phosphine-based palladium catalysts are generally used because they are stable on prolonged heating; however, an extremely high coupling reaction rate can be achieved sometimes by using palladium catalysts without phosphine ligands such as $Pd(OAc)_2$, $((\eta^3\text{-}C_3H_5)PdCl)_2$ and $Pd_2(dba)_3C_6H_6$ [82, 83]. Phosphine-free palladiums are approximately an order of magnitude more active than $ArPd^{II}I\cdot(PPh_3)_2$, both of which are, in turn, markedly more active than $Pd(PPh_3)_4$ (Equation (23.28)).

(23.28)

Catalyst:

Pd(PPh$_3$)$_4$ (8 h, 23%); PhPdI(PPh$_3$)$_2$ (0.33 h, 53%); Pd(OAc)$_2$ (0.75 h, 98%)

Although the steric hindrance of aryl halides is not a major factor for the formation of substituted biaryls, low yields resulted when using *ortho*-substituted arylboronic acids and esters. For instance, the reaction with mesitylboronic acid proceeds only slowly because of steric hindrance during the transmetallation to palladium(II) halide. The addition of strong bases such as aqueous NaOH or Ba(OH)$_2$, both in benzene and DME, exerts a remarkable effect on the acceleration of the coupling rate (Equation (23.29)) [84–86]. Weak bases give better results for less hindered arylboronic acids, and the order of reactivity for mesitylboronic acids corresponds to the basic strength: Ba(OH)$_2$ > NaOH > K$_3$PO$_4$ > Na$_2$CO$_3$ > NaHCO$_3$ [84].

(23.29)

ArX: 2-MeOC$_6$H$_4$I (80%), 2-ClC$_6$H$_4$I (94%), 2-bromonaphthalene (86%)

Even if there is no large steric hindrance, the reaction under aqueous conditions gives undesirable results due to competitive hydrolytic deboronation [87]. The rate for the cleavage of XC$_6$H$_4$B(OH)$_2$ with water at pH 6.7 is shown as follows (relative to phenylboronic acid): 2,6-dimethoxy (125), 2-F (77), 2-Cl (59), 2-MeO (11), 4-MeO (4.2), 2-Me (2.5), 3-F (2.3), 3-Me (2), 4-F (1.7) [88]. For example, the coupling of 2-formylphenylboronic acid with 2-iodotoluene at 80 °C using an aqueous Na$_2$CO$_3$ in DME gives only 54% of the biaryl together with benzaldehyde (39%). The yield can be improved to 89% by using the corresponding ester of boronic acid and anhydrous K$_3$PO$_4$ suspended in DMF. Other examples are shown in Equation (23.30) [84].

(23.30)

ArX: iodomesitylene (73%), 2-MOMOC$_6$H$_4$I (85%), 2-MeO$_2$CC$_6$H$_4$Br (63%)

An aryl–aryl exchange between the palladium center and phosphine ligands in palladium(II) complexes is enhanced by electron-donating substituents [89]. Consequently, the synthesis of biaryls substituted with electron-donating groups often results in contamination of coupling products with the aryl group on phosphine ligand. Tris(2-methoxyphenyl)phosphine is effective in reducing the formation of such a byproduct while maintaining a high yield of the desired product [90].

The cross-coupling reaction of arylboronic acids is largely unaffected by the presence of water, tolerates a broad range of functionality and yields non-toxic byproducts. The reaction offers the additional great advantage of being insensitive to the presence of *ortho*-functional groups or heteroaromatic rings. Gronowitz has shown that unsymmetrically substituted dithienyls [70, 91] and thienylpyridines [92] can be synthesized regioselectively by the cross-coupling reaction of thienylboronic acids. Arylation of 5-bromonicotinate is demonstrated by Thompson [93]. Diethyl(3-pyridyl)borane employed by Terashima is a unique air-stable reagent for heteroarylation (Equation (23.31)) [76].

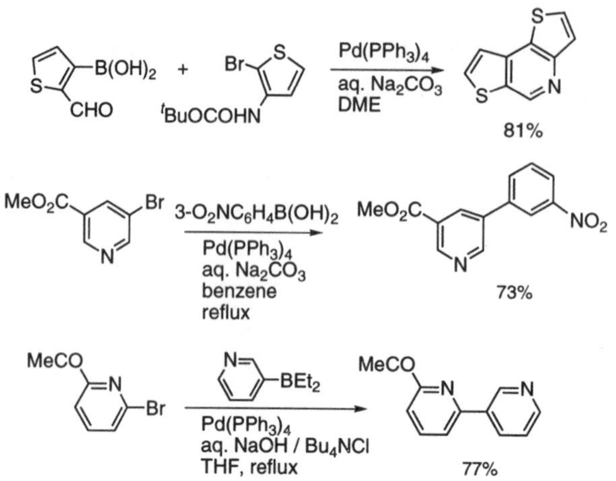

(23.31)

Most recently, other syntheses of bi(hetero)aryls have been reported [94].

The ready availability of *ortho*-functionalized arylboronic acids by a directive *ortho*-metallation–boronation sequence provides a synthetic link to the cross-coupling. Snieckus has amply demonstrated that the sequence has considerable scope for the synthesis of unsymmetrical biaryls, heterobiaryls and terphenyls (Equation (23.32)) [95].

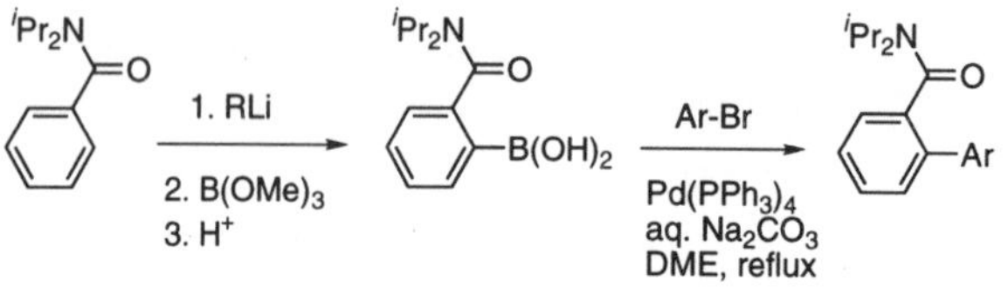

(23.32)

The utility of the sequence has been applied recently in industry, e.g. the industrial scale synthesis of a non-peptide angiotensin II receptor antagonist is reported (Equation (23.33)) [96].

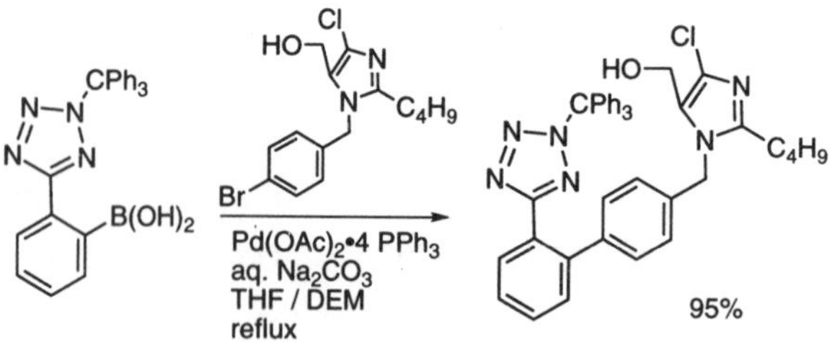

(23.33)

In consequence, the reaction has been used extensively in the synthesis of natural and unnatural products and pharmaceuticals, such as saddle-shaped host compounds [97], ferrocene derivatives [98], bis-cyclometallating N-C-N hexadentated ligands [99], helically chiral ligands [100], michellamine [86], biphenomycin A [101], vancomycin [102], receptor molecules for oxo acids [103], leukotriene B4 receptor antagonist [104], hemispherand [105], fascaplysin and streptonigrin alkaloids [106], ungerimine and hippadine alkaloids [107], other biaryls [108, 109] and others [110].

Aromatic rigid-rod polymers play an important role in a number of diverse technologies, including high-performance engineering materials, conducting polymers and non-linear optical materials. The cross-coupling reaction of aryldiboronic acids and dihaloarenes for the synthesis of poly(p-phenylenes) was first reported by Schlüter [111]. The method has been applied extensively to monodispersed aromatic dendrimers [112], water-soluble poly(p-phenylenes) [113], planar poly(p-phenylenes) fixed with ketoimine bonds [114], poly(phenylenes) fused with polycyclic aromatics [115] and non-linear optical materials [116].

The cross-coupling reactions using polymer-bound derivatives have become of interest [117]. In particular, it is attractive that microwave-assisted palladium-catalysed coupling of aryl- or heteroarylboronic acids with iodo- or bromo-substituted benzoic acids, anchored to TentaGel S RAM, provides high isolated yields of coupled products after a reaction time of 3.8 min (45 W) [117a].

Palladacycles have been reported as efficient catalysts for aryl–aryl coupling reactions [118].

Instead of palladium catalysts, the Ni(0)-catalysed Suzuki-type cross-coupling reaction of various aryl sulfonates, including mesylate, with arylboronic acids in the presence of K_3PO_4 has been observed [119]. The mechanism is proposed to be similar to that for the Pd-catalysed coupling suggested by Suzuki [34b].

5.3 *Coupling of alkylboron derivatives*

Although alkylmagnesium, -zinc, -tin and -aluminum reagents have been used successfully for the cross-coupling reaction with organic halides [4], the reaction of alkylborane derivatives is particularly useful when one wishes to start from alkenes via hydroboration. Also, bases, as well as palladium catalysts, are essential for the success of the coupling reaction [120–123]. A combination of $PdCl_2(dppf)$ and aqueous NaOH in THF works nicely for most cases. Strong bases accelerate the coupling reaction, but more weak bases and aprotic conditions are desirable for functionalized alkylboranes or organic halides. The reaction can be carried out by powdered K_2CO_3 or K_3PO_4 suspended in DMF at 50 °C in the presence of $PdCl_2(dppf)$ catalyst [120, 121]. The $Pd(PPh_3)_4$ catalyst works well when aqueous NaOH in benzene or K_3PO_4 in dioxane is used [120]. The characteristic features of both catalysts are that $PdCl_2(dppf)$ is used well in polar solvents (e.g. THF and DMF) but $Pd(PPh_3)_4$ gives good results in non-polar solvents such as benzene and dioxane.

One of the primary alkyl groups in trialkylboranes participates in the coupling, and the reaction of organoboranes with secondary alkyl is very slow [120]. Thus, representative hydroboration reagents such as 9-BBN, disiamylborane, dicyclohexylborane and borane can be used as hydroboration reagents for terminal alkenes. However, 9-BBN is most accessible owing to its ease of use, high selectivity on hydroboration and high reactivity on the cross-coupling reaction.

The hydroboration–coupling approach for the construction of carbon skeletons affords several advantages. The high stereoselectivity of hydroboration provides a stereodefined alkyl center on boron. For instance, as shown in Equation (23.34) the hydroboration occurs chemoselectively at the less hindered C20–C21 double bond. In addition, the alkyl group thus constructed can be cross-coupled readily with alkenyl or aryl halides under mild conditions [120].

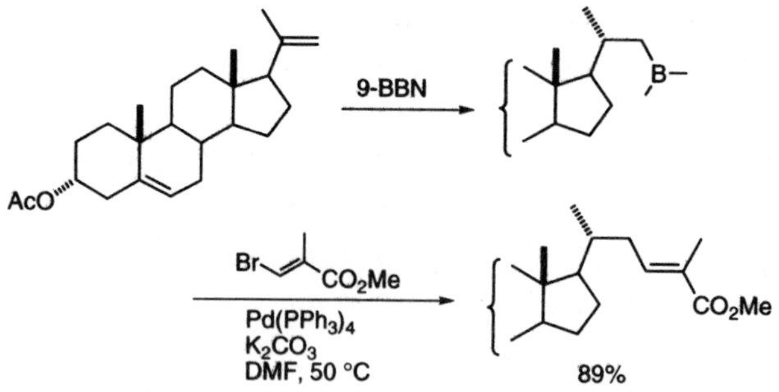

(23.34)

The procedure has been used in a variety of syntheses of natural products: e.g. dihydroxyserrulatic acid [124], the aggregation pheromone of *Cathartus quadricollis* (quadrilure) [125] and aza-C-disaccharides [126]. Epothilone A has been reported most recently by Danishefsky and his group [127b], as a promising anticancer agent (Equation (23.35)).

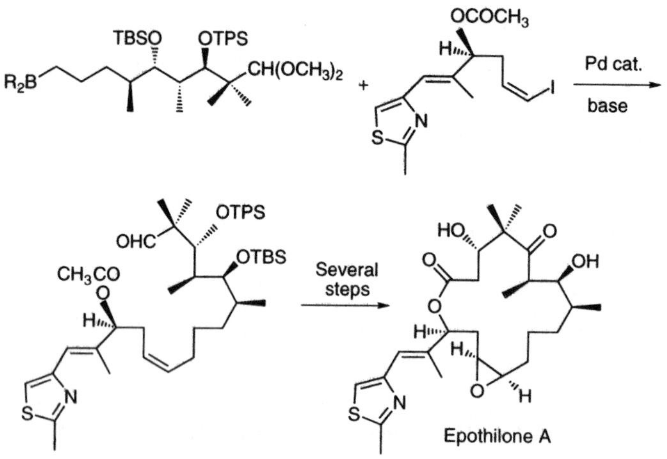

(23.35)

A three-step, three-component synthesis of PGEl is achieved by utilization of the cross-coupling reaction of the 9-alkyl-9-BBN derivative with an α-iodoenone (Equation (23.36)). It is recognized that cesium carbonate in the presence of water rapidly accelerates the reaction carried out at room temperature [128].

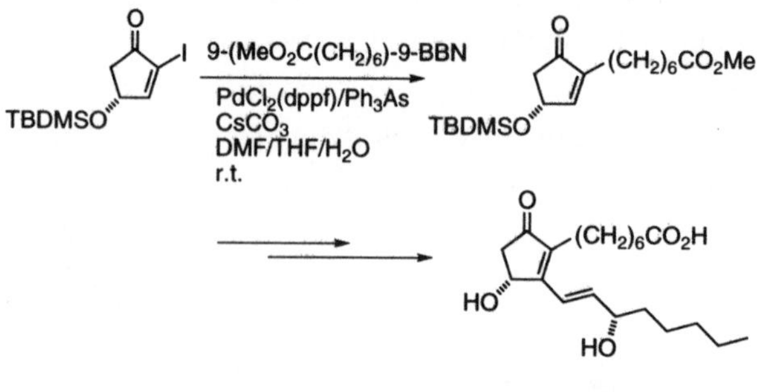

(23.36)

9-Methyl and 9-(trimethylsilylmethyl)-9-BBN derivatives are easily synthesized by the reaction of the corresponding lithium reagents with 9-methoxy-9-BBN. Unfortunately, such compounds are spontaneously flammable in air, making them particularly hazardous to handle for the treatment. However, selective oxidation with anhydrous trimethylamine *N*-oxide converts them to air-stable borinate esters, which are efficient reagents for the methylation [129,130] of haloalkenes (Equation (23.37)) or for syntheses of allylic and propargylic silanes [131].

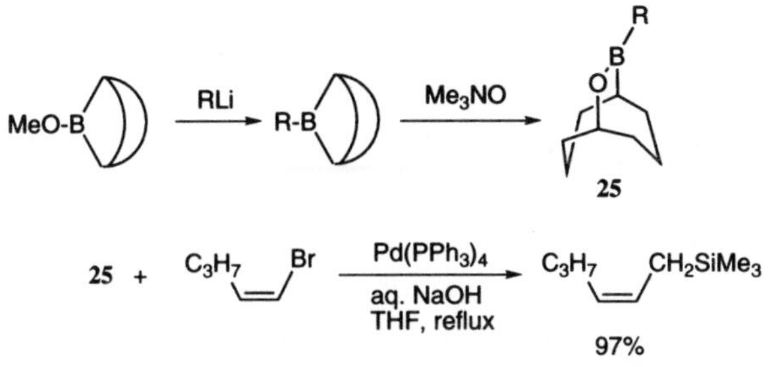

$$(23.37)$$

Beletskaya has reported that the reaction of tetraethylammonium tetrakis(trimethyl-silylmethyl)borate with aryl halides in THF containing $PdCl_2(dppf)$ gives 72–98% of the corresponding trimethylsilyl arenes. Much lower yields are obtained in reactions with lithium tetrakis(trimethylsilylmethyl)borate or tris(trimethylsilylmethyl)borane [132].

The intramolecular cross-coupling proceeds smoothly when the cyclization results in the formation of either five- or six-membered rings [120,133]. The hydroboration of the terminal double bond with 9-BBN is faster than that of the halogenated double bond, e.g. (the relative rate): 2-methyl-1-pentene (196); 1-hexene (100); (Z)-1-bromo-1-butene (0.011). Thus, the hydroboration–coupling approach provides a new route for stereodefined exocyclic alkenes; one such example is indicated in Equation (23.38).

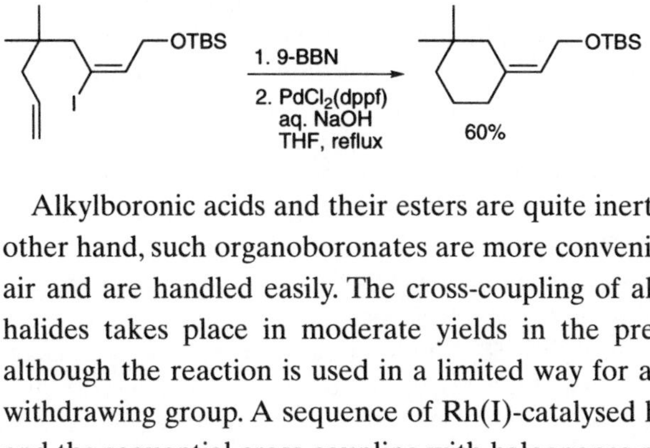

$$(23.38)$$

Alkylboronic acids and their esters are quite inert under the above conditions. On the other hand, such organoboronates are more convenient to use because they are stable in air and are handled easily. The cross-coupling of alkylboronates with 1-alkenyl or aryl halides takes place in moderate yields in the presence of Tl_2CO_3 and $PdCl_2(dppf)$, although the reaction is used in a limited way for activated halides having an electron-withdrawing group. A sequence of Rh(I)-catalysed hydroboration [5a] of 1-hexen-5-one and the sequential cross-coupling with haloenones produces the corresponding coupling products in 62–69% yields [134] (Equation (23.39)) [134].

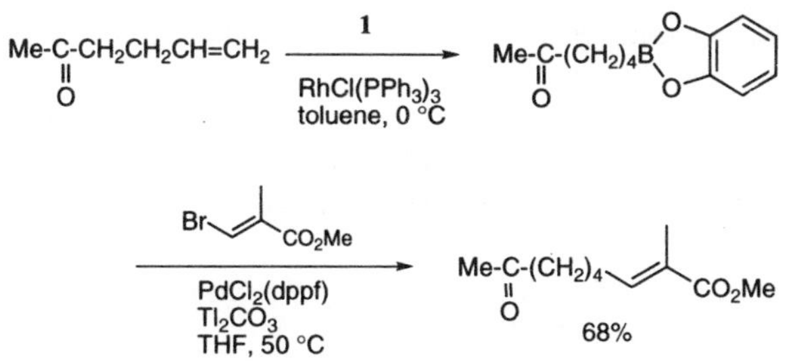

$$(23.39)$$

6 Conclusion

There are three major ways in which organoboron compounds find use:

1 The organic group attached to boron is transferred to a heteroatom to provide functionalized organic compounds. The heteroatom that replaces boron can be hydrogen, oxygen, nitrogen, sulfur, selenium, halogen or a heavy metal.

2 Carbon–carbon bond formation by transfer of an organic group from boron to a carbon atom; this is probably the most important type of application.

3 A number of synthetically useful reactions involve boron-bound groups other than simple organic units.

Most of these applications of organoboron compounds without catalysts have been studied in detail. In contrast, the catalytic reactions have not been investigated much but interest is growing. The continued study of this essentially novel area may provide new chemical developments of major interest and utility.

7   References

1 Suzuki A. *Acc Chem Res* 1982; **15**: 178–84; Pelter A, Smith K, Brown HC. *Borane Reagents*. London: Academic Press, 1988: 194.

2 Brown HC. *Hydroboration*. New York: Benjamin, 1962; Brown HC, Kramer GW, Levy AB, Midland MM. *Organic Synthesis via Boranes*. New York: Wiley, 1975.

3 For a review of catecholborane and the further reactions of its derivatives, see: Lane CF, Kabalka GW. *Tetrahedron* 1976; **32**: 981–90.

4 Miyaura N, Suzuki A. *Chem Rev* 1995; **95**: 2457–83.

5 (a) Männig D, Nöth H. *Angew Chem Int Ed Engl* 1985; **24**: 878–9; (b) Gridnev ID, Miyaura N, Suzuki A. *Organometallics* 1993; **12**: 589–92; (c) Gridnev ID, Miyauira N, Suzuki A. *J Org Chem* 1993; **58**: 5351–4; for a review, see: (d) Burgess K, Ohlmeyer MJ. *Chem Rev* 1991; **91**: 1179–91.

6 Satoh M, Nomoto Y, Miyaura N, Suzuki A. *Tetrahedron Lett* 1989; **30**: 3789–92.

7 Hoshi M, Masuda Y, Arase A. *Bull Chem Soc Jpn* 1990; **63**: 447–52.

8 Bergens SH, Noheda P, Whelan J, Bosnich B. *J Am Chem Soc* 1992; **114**: 2128–35.

9 Yamamoto A. *Organotransition Metal Chemistry*. New York: Wiley, 1986: 201.

10 Burgess K, Ohlmeyer MJ. *J Org Chem* 1988; **53**: 5178–9; Burgess K, Cassidy J, Ohlmeyer MJ. *J Org Chem* 1991; **56**: 1020–27.

11 Knorr JR, Merola JS. *Organometallics* 1990; **9**: 3008.

12 Beletskaya I, Pelter A. *Tetrahedron* 1997; **53**: 4957–5026.

13 Suzuki A. *Pure Appl Chem* 1986; **58**: 629–38; Suzuki A. In: Hermanek S, ed. *Boron Chemistry*. New Jersey: World Scientific, 1987: 273–91; Suzuki A. In: Oae S, ed. *Reviews on Heteroatom Chemistry*, vol. 1. Tokyo: MYU, 1988: 291–303; Suzuki A. In: Siebert W, ed. *Advances in Boron Chemistry*. London: The Royal Society of Chemistry, 1997: 167–70.

14 Nozaki K, Nonaka K, Tückmantel W, Oshima K, Utimoto K. *Tetrahedron Lett* 1986; **27**: 2007–10.

15 Sharma S, Oehlschlager AC. *Tetrahedron Lett* 1988; **29**: 261–4.

16 Ishiyama T, Nishijima K, Miyaura N, Suzuki A. *J Am Chem Soc* 1993; **115**: 7219–25.

17 Biffar W, Nöth H. *Angew Chem* 1980; **92**: 65–6.

18 Biffar W, Nöth H. *Chem Ber* 1982; **115**: 934–45.

19 Trost BM, Hiroi K, Kurozumi S. *J Am Chem Soc* 1975; **97**: 438–40.

20 Durst T. In: Barton DHR, ed. *Comprehensive Organic Chemistry*. Oxford: Pergamon, 1979; **3**: 121–56.

21 Fiandanese V, Marchese G, Naso F, Ronzini L. *J Chem Soc Perkin Trans. 1* 1985: 1115–19.

22 Corey EJ, Shulman JI. *J Org Chem* 1970; **35**: 777–80.

23 Carey FA, Court AS. *J Org Chem* 1972; **37**: 939–43.

24 (a) Ishiyama T, Miyaura N, Suzuki A. *Chem Lett* 1987: 25–28; (b) Hoshino Y, Ishiyama T, Miyaura N, Suzuki A. *Tetrahedron Lett* 1988; **29**: 3983–6; (c) Ishiyama T, Miyaura N, Suzuki A. *Org Synth* 1992; **71**: 89–95.

25 Urry G, Kerrigan J, Parsons TD, Schlesinger HI. *J Am Chem Soc* 1954; **76**: 5299–301.

26 Nöth H. *Z Naturforsch* 1984; **39b**: 1463.
27 (a) Ishiyama T, Miyaura N, Suzuki A. *J Am Chem Soc* 1993; **115**: 11018–19; (b) Ishiyama T, Matsuda N, Murata M *et al. Organometallics* 1996; **15**: 713–20.
28 Stinson SC. *Chem Eng News* 1996; **Nov. 25**: 41–7.
29 Westcott SA, Marder TB, Baker RT, Calabrese JC. *Can J Chem* 1993; **71**: 930–6.
30 Ozawa F, Sugawara M, Hayashi T. *Organometallics* 1994; **13**: 3237–43.
31 Yamashita H, Kobayashi T, Hayashi T, Tanaka M. *Chem Lett* 1990: 1447–50.
32 Murakami M, Yoshida T, Kawanami S, Ito Y. *J Am Chem Soc* 1995; **117**: 6408–9.
33 Baker RT, Calabrese JC, Westcott SA, Nguyen P, Marder TB. *J Am Chem Soc* 1993; **115**: 4367–8.
34 (a) Miyaura N, Yamada K, Suzuki A. *Tetrahedron Lett* 1979; **20**: 3437–40; (b) Miyaura N, Yamada K, Suginome H, Suzuki A. *J Am Chem Soc* 1985; **107**: 972–80; (c) Miyaura N, Suzuki A. *Org Synth* 1990; **68**: 130–6; (d) Miyaura N, Satoh M, Suzuki A. *Tetrahedron Lett* 1986; **27**: 3745–8; (e) Satoh M, Miyaura N, Suzuki A. *Chem Lett* 1986: 1329–32.
35 Miyaura N, Suzuki A. *J Organomet Chem* 1981; **213**: C53–6.
36 Miyaura N, Sugionome H, Suzuki A. *Tetrahedron Lett* 1981; **22**: 127–30.
37 Soderquist JA, Colberg JC. *Tetrahedron Lett* 1994; **35**: 27–8.
38 Miyaura N, Suginome H, Suzuki A. *Tetrahedron* 1983; **39**: 3271–7.
39 Cassani G, Massardo P, Piccardi P. *Tetrahedron Lett.* 1983; **24**: 2513–16.
40 Rossi R, Carpita A, Quirici MG. *Tetrahedron* 1981; **37**: 2617–23.
41 Björkling F, Norin T, Unelius CR, Miller RB. *J Org Chem* 1987; **52**: 292–4.
42 Carpita A, Neri D, Rossi R. *Gazz Chim Ital* 1987; **117**: 503–5.
43 Roush WR, Warmus JS, Works AB. *Tetrahedron Lett* 1993; **34**: 4427–30.
44 Ichikawa J, Minami T, Sonoda T, Kobayashi H. *Tetrahedron Lett* 1992; **33**: 3779–82; Ichikawa J, Ikeura C, Minami T. *Synlett* 1992: 739–40.
45 Negishi E, Noda Y, Lamaty F, Vawter E. *Tetrahedron Lett* 1990; **31**: 4393–6.
46 Miyaura N, Suginome H, Suzuki A. *Bull Chem Soc Jpn* 1982; **55**: 2221–3.
47 Miyaura N, Satoh Y, Hara S, Suzuki A. *Bull Chem Soc Jpn* 1986; **59**: 2029–31.
48 (a) Uenishi J, Beau J, Armstrong RW, Kishi Y. *J Am Chem Soc* 1987; **109**: 4756–8; (b) Armstrong RW, Beau J, Cheon SH *et al. J Am Chem Soc* 1989; **111**: 7525–30.
49 Roush WR, Brown BB. *J Org Chem* 1993; **58**: 2162–72.
50 Roush WR, Sciotti RJ. *Tetrahedron Lett* 1992; **33**: 4691–4; Roush WR, Sciotti RJ. *J Am Chem Soc* 1994; **116**: 6457–8.
51 Nicolaou KC, Ramphal JY, Petasis NA, Serhan CN. *Angew Chem Int Ed Engl* 1991; **30**: 1100–16.
52 Evans DA, Ng HP, Rieger DL. *J Am Chem Soc* 1993; **115**: 11446–59.
53 Kobayashi Y, Okamoto S, Shimazaki T, Ochiai Y, Sato F. *Tetrahedron Lett* 1987; **28**: 3959–62; Avignon-Tropis M, Treihou M, Lebreton J, Pougny JR. *Tetrahedron Lett* 1989; **30**: 6335–6.
54 Kobayashi Y, Shimazaki T, Sato F. *Tetrahedron Lett* 1987; **28**: 5849–52; Kobayashi Y, Shimizu T, Taguchi T, Sato F. *J Org Chem* 1990, **55**: 5324–35.
55 de Lera AR, Torrado A, Iglesias B, Lopez S. *Tetrahedron Lett* 1992; **33**: 6205–8.
56 Satoh N, Ishiyama T, Miyaura N, Suzuki A. *Bull Chem Soc Jpn* 1987; **60**: 3471–3.
57 Yanagi T, Oh-e T, Miyaura N, Suzuki A. *Bull Chem Soc Jpn* 1989; **62**: 3892–5.
58 Kaga H, Ahmed Z, Gotoh K, Orito K. *Synlett* 1994: 607–8.
59 Genet JP, Linquist A, Blart E *et al. Tetrahedron Lett* 1995; **36**: 1443–6.
60 Torrado A, Lopez S, Alvarez R, de Lera AR. *Synthesis* 1995: 285–93.
61 Charette AB, Giroux A. *J Org Chem* 1996; **61**: 8718–19.
62 Desurmont G, Klein R, Uhlenbrock S *et al. Organometallics* 1996; **15**: 3323–8.
63 Miyaura N, Suzuki A. *J Chem Soc Chem Commun* 1979: 866.
64 Stille JK, Lau KSY. *Acc Chem Res* 1977; **10**: 434–42.
65 Miyachi N, Yanagawa Y, Iwasaki H, Ohara Y, Hiyama T. *Tetrahedron Lett* 1993; **34**: 8267–70.
66 Marko IE, Murphy F, Dolan S. *Tetrahedron Lett* 1996; **37**: 2507–10.
67 Han Y, Walker SD, Young RN. *Tetrahedron Lett* 1996; **37**: 2703–6.
68 Hildebrand JP, Marsden P. *Synlett* 1996: 893–4.
69 Miyaura N, Yanagi Y, Suzuki A. *Synth Commun* 1981; **11**: 513–19.

70 Gronowitz S, Bobosik V, Lawitz K. *Chem Scr* 1984; **23**: 120–2.

71 Alo BI, Kandil A, Patil PA *et al. J Org Chem* 1991; **56**: 3763–8.

72 Müller W, Lowe DA, Neijt H *et al. Helv Chim Acta* 1992; **75**: 855–64.

73 Katz HE. *J Org Chem* 1987; **52**: 3932–4.

74 Hoshino Y, Miyaura N, Suzuki A. *Bull Chem Soc Jpn* 1988; **16**: 3008–10.

75 Coleman RS, Grant EB. *Tetrahedron Lett* 1993; **34**: 2225–8.

76 Ishikura M, Kamada M, Terashima M. *Synthesis* 1984: 936–8.

77 Janietz D, Bauer M. *Synthesis* 1993: 33–4; Achab A, Guyot M, Potier P. *Tetrahedron Lett* 1993; **34**: 2127–30.

78 Shieh WC, Carlson JA. *J Org Chem* 1992; **57**: 379–81.

79 Bumagin NA, Bykov VV, Beletskaya IP. *Dokl Akad Nauk SSSR* 1990; **315**: 1133–6.

80 Bumagin NA, Luzikova EV, Sukhomlinova LI, Tolstaya TP, Beletskaya IP. *Izv Akad Nauk Ser Khim* 1995: 394.

81 Wright SW, Hageman DL, McClure LD. *J Org Chem* 1994; **59**: 6095–7.

82 Wallow TI, Navak BM. *J Org Chem* 1994; **59**: 5034–7.

83 Marck G, Villiger A, Buchecker R. *Tetrahedron Lett* 1994; **35**: 3277–80.

84 Watanabe T, Miyaura N, Suzuki A. *Synlett* 1992: 207–10.

85 Guillier F, Nivoliers F, Godard A, Marsais F, Queguiner G. *Tetrahedron Lett* 1994; **35**: 6489–92.

86 Kelly TR, Garcia A, Lang F *et al. Tetrahedron Lett* 1994; **35**: 7621–4.

87 Muller D, Fleury JP. *Tetrahedron Lett* 1991; **32**: 2229–32; Fukuyama Y, Kiriyama Y, Kodama M. *Tetrahedron Lett* 1993; **34**: 7637–8.

88 Kuivila HG, Reuwer JF, Mangravite JA. *J Am Chem Soc* 1964; **86**: 2666–70.

89 Kong KC, Cheng CH. *J Am Chem Soc* 1991; **113**: 6313–15; Segelstein BE, Butler TW, Chenard BL. *J Org Chem* 1995; **60**: 12–13.

90 O'Keefe DF, Donnock MC, Marcuccio SM. *Tetrahedron Lett* 1992; **33**: 6679–80.

91 Gronowitz S, Lawitz K. *Chem Scr* 1983; **22**: 265; Yang Y, Hörnfeldt AB, Gronowitz S. *Chem Scr* 1988, 28: 275.

92 Gronowitz S, Lawitz K. *Chem Scr* 1984; **24**: 5–6.

93 Thompson WJ, Gaudino J. *J Org Chem* 1984; **49**: 5237–43; Thompson WJ, Jones JH, Lyle PA, Thies JE. *J Org Chem* 1988; **53**: 2052–5.

94 Chang CK, Bag N. *J Org Chem* 1995; **60**: 7030–32; Jones K, Keenan M, Hilbbert F. *Synlett* 1996: 509–10; D'Alessio R, Rossi A. *Synlett* 1996: 513–14; Boland GM, Donnelly DMX, Finet JP, Rea MD. *J Chem Soc Perkin Trans 1* 1996: 2591–7.

95 For reviews, see: Snieckus V. *Chem Rev* 1990; **90**: 879–933; Snieckus V. *Pure Appl Chem* 1994; **66**: 2155–8.

96 Larsen DR, King AO, Chen CY *et al. J Org Chem* 1994; **59**: 6391–4.

97 Schwartz EB, Knobler CB, Cram DJ. *J Am Chem Soc* 1992; **114**: 10775–84.

98 Knapp R, Rerhahn M. *J Organomet Chem* 1993; **452**: 235–40.

99 Beley M, Chodorowski S, Collin JP, Sauvage JP. *Tetrahedron Lett* 1993; **34**: 2933–6.

100 Judice JK, Keipert SJ, Cram DJ. *J Chem Soc Chem Commun* 1993: 1323.

101 Schmidt U, Leitenberger V, Meyer R, Greisser H. *J Chem Soc Chem Commun* 1992: 951–3.

102 Brown AG, Grimmin MJ, Edwards PD. *J Chem Soc Perkin Trans 1* 1992: 123–30.

103 Manabe K, Okamura K, Date T, Koga K. *J Org Chem* 1993; **58**: 6692–700.

104 Sawyer JS, Baldwin RF, Sofia MI *et al. J Med Chem* 1993; **36**: 3982–4.

105 Ostaszewski R, Verboom W, Reihoudt DN. *Synlett* 1992: 354–6.

106 Rocca P, Marsais F, Godard A, Queguiner G. *Tetrahedron Lett* 1993; **34**: 7917–18; Rocca P, Cochennee C, Marsais F *et al. J Org Chem* 1993; **58**: 7832–8.

107 Siddiqui MA, Snieckus V. *Tetrahedron Lett* 1990; **31**: 1523–6.

108 Maddaford SP, Keay BA. *J Org Chem* 1994; **59**: 6501–3; Song ZZ, Wong HNC. *J Org Chem* 1994; **59**: 33–41; Carrera GM, Sheppard GS. *Synlett* 1994: 93–4; Feldman KS, Campbell RF. *J Org Chem* 1995; **60**: 1924–5.

109 Benbow JW, Martinez BL. *Tetrahedron Lett* 1996; **49**: 8829–32; Bahl A, Grahn W, Standler S *et al. Angew Chem Int Ed Engl* 1995; **34**: 1485–8.

110 Zhou X, Tse MK, Wan TSM, Chan KS. *J Org Chem* 1996; **61**: 3590–3; Galda P, Rehahn M. *Synthesis* 1996: 614–20; Harre K, Enkelmann V, Schulze M, Bunz UHF. *Chem Ber* 1996; **129**: 1323–5; Setayesh S, Bunz UHF. *Organometallics* 1996; **15**: 5470–2.

111 Rehahn M, Schlüter AD, Wegner G, Feast W. *J Polym* 1989; **30**: 1054–9.

112 Miller TM, Neenan TX, Zayas R, Bair HE. *J Am Chem Soc* 1992; **114**: 1018–25.

113 Wallow TI, Novak BM. *J Am Chem Soc* 1991; **113**: 7411–12.

114 Tour JM, Lamba JJS. *J Am Chem Soc* 1993; **115**: 4935–6.

115 Goldfinger MB, Swager TM. *J Am Chem Soc* 1994; **116**: 7895–6.

116 Wulff G, Schmidt H, Witt H, Zentel R. *Angew Chem Int Ed Engl* 1994; **33**: 188–91.

117 (a) Larhed M, Linderberg G, Hallberg A. *Tetrahedron Lett* 1996; **37**: 8219–22; (b) Guiles JW, Johnson SG, Murray WV. *J Org Chem* 1996; **61**: 5169–71; (c) Piettre SR, Baltzer S. *Tetrahedron Lett* 1997; **38**: 1197–200.

118 Beller M, Fischer H, Herrmann WA, Öfele K, Brossmer C. *Angew Chem Int Ed Engl* 1995; **34**: 1848–9.

119 Percee V, Bae JY, Hill DH. *J Org Chem* 1995; **60**: 1060–5.

120 Miyaura N, Ishiyama I, Sasaki H, Ishikawa M, Satoh M, Suzuki A. *J Am Chem Soc* 1989; **111**: 314–21.

121 Ishiyama T, Miyaura N, Suzuki A. *Synlett* 1991: 687–8.

122 Nomoto Y, Miyaura N, Suzuki A. *Synlett* 1992: 727–9.

123 Abe S, Miyaura N, Suzuki A. *Bull Chem Soc Jpn* 1992; **65**: 2863–5.

124 Uemura M, Nishimura H, Minami T, Hayashi T. *J Am Chem Soc* 1991; **113**: 5402–10.

125 Mori K, Puapoomchareon P. *Liebigs Ann Chem* 1990: 159–62.

126 Johnson CR, Miller MW, Golebiowski A, Sundram H, Ksebati MB. *Tetrahedron Lett* 1994; **35**: 8991–4.

127 (a) Borman S. *Chem Eng News* 1996; **Dec. 23**: 24–26; (b) Danishefsky SJ, Balog A, Meng D *et al*. *Angew Chem Int Ed Engl* 1996; **35**: 2801–3.

128 Johnson CR, Braun MP. *J Am Chem Soc* 1993; **115**: 11014–15.

129 Soderquist JA, Santiago B. *Tetrahedron Lett* 1990; **31**: 5541–2.

130 Moore WR, Schatzman GL, Jarvi ET, Gross RS, McCarthy JR. *J Am Chem Soc* 1992; **114**: 360–1.

131 Soderquist JA, Santiago B, Rivera I. *Tetrahedron Lett* 1990; **31**: 4981–4.

132 Bumagin NA, Bykov VV, Artamkina GA, Beletskaya IP. *Dokl Akad Nauk* 1995; **340**: 493–4.

133 Soderquist JA, Leon G, Colberg JC, Martinez I. *Tetrahedron Lett* 1995; **36**: 3119–22.

134 Sato M, Miyaura N, Suzuki A. *Chem Lett* 1989: 1405–8.

24 Mechanism and Synthesis in Catalytic Hydroboration

J.M. BROWN, H. DOUCET, E. FERNANDEZ, H.E. HEERES, M.W. HOOPER,
D.I. HULMES, F.I. KNIGHT, T.P. LAYZELL and G.C. LLOYD-JONES

Dyson Perrins Laboratory, South Parks Road, Oxford OX1 3QY, UK

1 Overview

Hydroboration has proved to be one of the most useful reactions in organic synthesis, because it provides a route from alkenes to many different types of C–X and C–R bonds and hence a formal addition of H–R or H–X to alkenes or alkynes. For this reason alone, the development of catalytic hydroboration, based on the initial observation by Ito and co-workers of the B–H activation of a cyclic boronate ester by Wilkinson's catalyst (Fig. 24.1) [1] and Hawthorne's observation of the Rh-catalysed B–H addition from a carborane to butyl acrylate [2], is of potential significance. The chapter, which essentially identified the possibility of catalytic hydroborations relevant to organic synthesis, was due to Mannig and Nöth in 1985, ten years after the initial stoichiometric chemistry described by Ito. In this chapter the catalytic addition of catecholborane to alkenes was demonstrated clearly, and rhodium complexes were shown to be superior to those of other transition metals [3]. These results provided the precedent on which most of the field has been based subsequently. Soon afterwards, Burgess reported the asymmetric addition of catecholborane to alkenes catalysed by various diphosphine–rhodium complexes [4]. The field has been reviewed several times, comprehensively by Burgess and Ohlmeyer in 1991 and then recently by Beletskaya and Pelter, with the emphasis on asymmetric hydroboration [5,6].

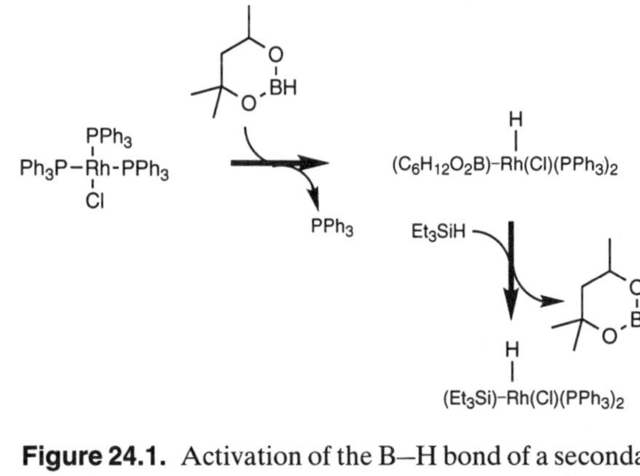

Figure 24.1. Activation of the B–H bond of a secondary boronic acid by Wilkinson's catalyst.

Despite the initial promise, the level of advance in the last ten years has been slower than would have been expected, and a lot of momentum has shifted to the newer alternatives of catalysed diboration or silaboration, which can in principle offer difunctional products from alkenes [7,8]. Several factors can be identified, namely:

1 Many attempts to provide superior hydroborating agents to catecholborane have been attempted but none have been particularly successful.

2 With existing catalysts there are limitations of alkene structure and, in particular, only lightly substituted reactants give successful results.

465

3 Boranes are unstable under the reaction conditions with BH_3 among the disproportionation products, so that slow reactions can be in competition with direct hydroboration.

4 As in hydrosilylation, ligand-free catalysts are more reactive so that decomposition of the catalyst during the course of reaction can result in loss of selectivity or even a change of reaction course.

5 The boronate ester products from catecholborane addition to alkenes have limited synthetic utility; oxidation to the corresponding alcohol, which occurs with near-complete retention of configuration, is the only direct reaction of interest.

None of these are insuperable obstacles and in fact they encourage further research; a problem can equally well be viewed as a challenge! The remainder of this review presents work largely carried out at Oxford on catalytic hydroboration, with the aim of extending the usefulness of the reaction through better understanding of the basic processes.

2   Reagents for asymmetric hydroboration (and boration)

In asymmetric catalysis, there are few examples where stereochemical control is derived from the reagent rather than the catalyst. This is not surprising, because the specific advantage of catalysis is the mass amplification of chirality initially residing only in the catalyst. Only when a reagent is readily available in enantiomerically pure form does reagent-controlled asymmetric catalysis become a reasonable objective.

Our first experiments were conducted with the readily available ephedrineboranes [9]. In fact, the asymmetric hydroboration of simple styrenes was modestly successful, yielding product in 76% enantiomeric excess (ee) and good regioselectivity in the best experiments with pseudoephedrine-derived reagent. Styrenes are a favorable substrate for the catalysed hydroboration reaction because a high level of reactivity is combined with good regioselectivity towards the branched isomer, which introduces a stereogenic center. An interesting constraint was that the results were very dependent on the catalyst ligand, with the more rigid 1,1′-bis-(diphenylphosphino)ferrocene being the more enantioselective (Table 24.1), albeit less regioselective than the corresponding 1,4-diphenylphosphinobutane-derived rhodium complex [10].

Further efforts to improve this level of asymmetric induction were unsuccessful, and the obvious step of modifying the nitrogen side-chain of ephedra was undertaken. Of the *N*-alkyl derivatives prepared from nor-ψ-ephedrine, the *i*-propyl derivative **1** of Fig. 24.2 was the one subject to most study.

When the standard hydroboration reaction conditions were applied (20 °C in THF), the reaction with borane **1** was substantially slower and much less clean. Worse still, it was quite irreproducible, and initially perplexing. When the reaction was deliberately run in air it was much faster but with no trace of the direct hydroboration product formed. This observation suggested that oxidation of the diphosphine might be occurring, and, as expected, when catalysis with a phosphine-free rhodium complex was attempted, the result was a clean conversion of the alkene reactant but exactly 50% partitioned to alkane and 50% to vinylborane. This same conversion could be realized with other vinylarenes, but not more generally. It is very rapid and efficient, complete reaction being observed with less than 0.1% catalyst, preferably an alkenerhodium chloride. This provides a direct high-yielding synthesis of vinylboronic acids from vinylarenes (Fig. 24.2), although it sacrifices one equivalent of the alkene for each equivalent of product formed [11].

At first this abrupt change in chemoselectivity was quite startling, although competitive

Table 24.1. Reagent-controlled asymmetric hydroboration with secondary boranes derived from ephedrines [9]

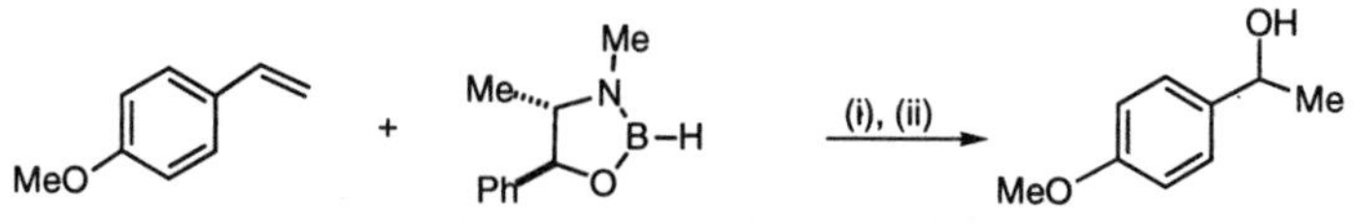

(i) 1 mol% LRh(C$_7$H$_8$)OSO$_2$CF$_3$, thf, 20 °C; (ii) H$_2$O$_2$, OH⁻

Ligand	Product: Config. (ee) selectivity
[ClRh(PPh$_3$)$_3$]	S (55%) 53
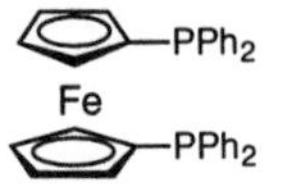 Ferrocene diphosphine (–PPh$_2$, Fe, –PPh$_2$)	S (76%) 82
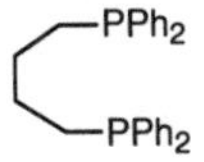 Diphosphine (–PPh$_2$, –PPh$_2$)	R (6%) 96

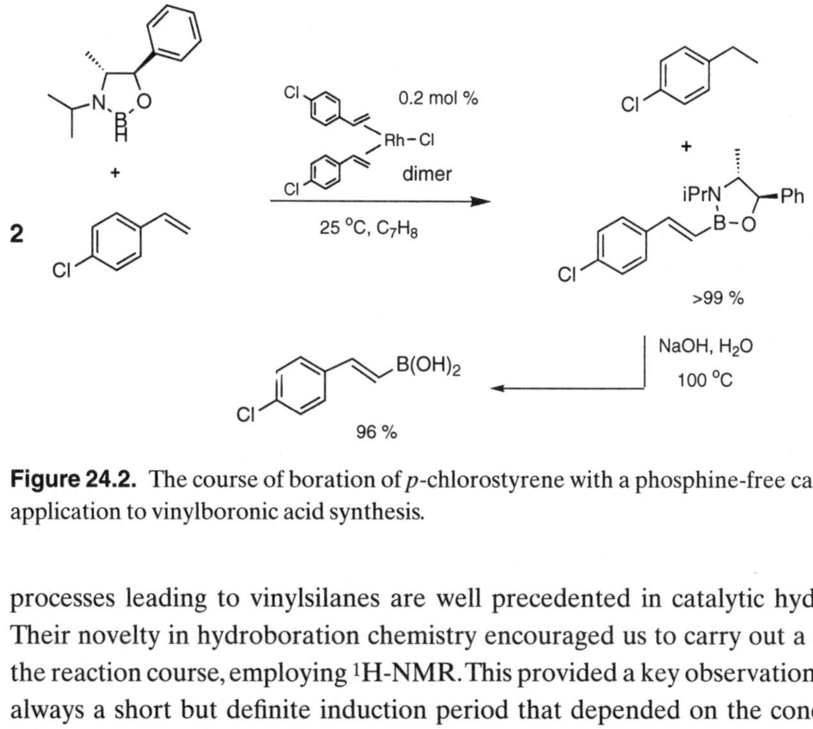

Figure 24.2. The course of boration of *p*-chlorostyrene with a phosphine-free catalyst and its application to vinylboronic acid synthesis.

processes leading to vinylsilanes are well precedented in catalytic hydrosilylation [12]. Their novelty in hydroboration chemistry encouraged us to carry out a detailed study of the reaction course, employing ¹H-NMR. This provided a key observation in that there was always a short but definite induction period that depended on the concentration of the reagents, alkene and catalyst. The catalyst is not present initially in its active form and, considering the likely events, an alkene-ligated rhodium hydride seemed the obvious possibility. It was gratifying to observe that this led naturally to a consistent mechanistic hypothesis [11] in which the addition of catecholborane to this intermediate promoted the reductive elimination of alkane. The resulting rhodium boryl then inserted in coordinated alkene, and the alkyl underwent a β-elimination to produce the vinylborane and regenerate the rhodium hydride. This catalytic cycle is summarized in the main pathway of Fig. 24.3. Further chemical evidence for the hydride mechanism came in experiments using isotopically labeled *p*-methoxystyrene. From these it was demonstrated that both internal and

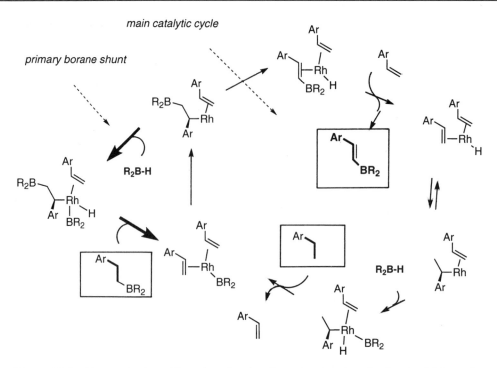

Figure 24.3. The mechanism of boration, showing the activation via a rhodium hydride catalyst generated *in situ*. The main path leads only to alkane and vinylborane and fully describes the ephedrineborane observations. With catecholborane, a further sequence is necessary to explain the formation of primary borane, as shown.

external exchange of deuterium within the alkene occurred. This is consistent with the exclusive addition of a Rh–H (or Rh–D) entity so that the rhodium is exclusively bound to the benzylic position of the chain and hydride is transferred exclusively to the terminal position. The reverse of this addition leads to isotopic scrambling because of free rotation of the (partially deuterated) methyl group, and also to intermolecular exchange because the rhodium can exchange hydrogen isotopes in the course of this process (Fig. 24.4).

There was a further surprise when the same catalyst was utilized for the hydroboration of *p*-methoxystyrene using catecholborane. The reaction was extraordinarily fast, several hundred turnovers being complete soon after mixing—the colour of the sample consistently changed from orange to brown on completion, probably due to formation of colloidal rhodium after the alkene was spent—but the product mixture was complex. In addition to the alkane and vinylborane, significant amounts of the terminal borane were formed, but none of the branched isomer was favored in the case of the diphosphine catalysts. This observation could be explained by invoking a branch in the catalytic cycle, such that the borylated alkylrhodium intermediate of Fig. 24.4 reacts further with catecholborane and undergoes a reductive elimination to provide the linear borane. This shunt from the main pathway could dominate the catalytic cycle completely, in principle, if the addition reaction that initiates it was sufficiently fast. The practical outcome is that up to 50% of the linear borane is formed under these conditions. Significantly, the proportion increases as the initial concentration of catecholborane is increased, and the data obtained from a series of experiments can be correlated with the model by employing reasonable rate constants. There have been sporadic observations of linear phenylethylborane formation in

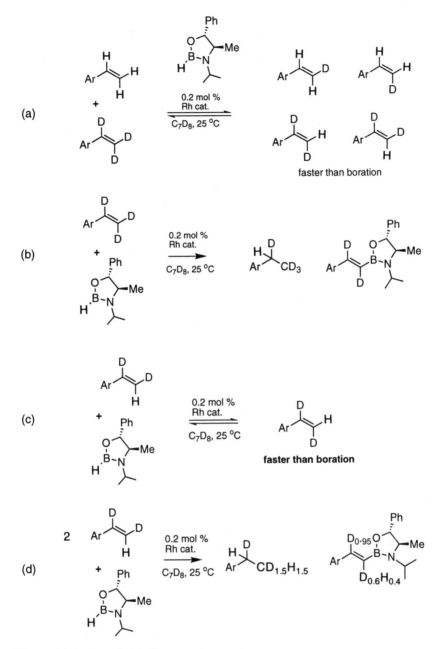

Figure 24.4. Isotopic labeling experiments that support the mechanistic pathway described in Fig. 24.3 indicating both inter- and intramolecular exchange processes.

the catalysed hydroboration of styrene, and it can be assumed that the mechanism proposed here operates in at least some of those cases. The reaction could provide a benchmark for new boranes in catalysis; the faster the rate of oxidative addition of the reagent, the greater the proportion of primary borane in the hydroboration of vinylarenes.

A further test of the rhodium hydride model shown in Fig. 24.3 is in its application to hydrosilylation. With triethylsilane, both the main cycle and the shunt were operative, with the latter at least as important as for catecholborane when a phosphine-free rhodium catalyst was employed. Hence, there appears to be a common model for M–H additions when this type of catalyst is involved. In contrast to hydroboration, desaturative additions had

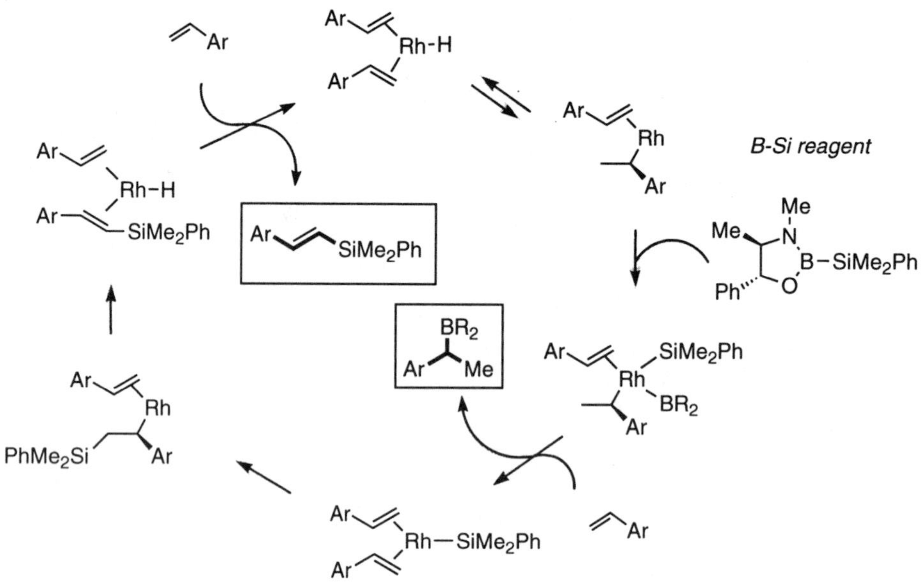

Figure 24.5. The observed disproportionation on attempted silaboration and its relationship to the boration pathway.

been well known in hydrosilylation [12] but the common mechanistic framework requiring a metal-hydride-based catalytic cycle had not been appreciated before our work.

We had hoped to extend this work to silylboration, and synthesized the ephedrine-based reagent shown in Fig. 24.5 in order to test this possibility. Reaction proved to be much slower than with catecholborane or ephedrineborane, but was again achieved using phosphine-free rhodium catalysts. A single reaction pathway was involved, which led to the exclusive formation of a mixture of the vinylsilane and secondary borane with >50 turnovers. The latter compound was a 50 : 50 mixture of diastereomers, so that the secondary alcohol liberated by H_2O_2 oxidation is racemic. The mechanism that best explains this observation is similar to that required to produce vinylboranes in Fig. 24.3, but the complete absence of the opposite products from this mechanism — the vinylborane and benzylsilane — is instructive. We assume that the intermediate formed by addition of the borasilane to the Rh benzyl has only one reaction pathway, which is to eliminate the benzylborane and not benzylsilane. This tells us something about the relative ease of the two elimination processes (assuming that the Rh intermediate has comparable access to the two pathways and lacks stereoelectronic constraints). An explanation may lie in the mechanism of elimination for the two cases; C—B elimination proceeds through a migration of the benzyl group from Rh to B, giving a transiently stable rhodium borane adduct. Because the silane first produced from C—Si is electron sufficient at silicon, this route is not as facile owing to migration and dissociation from rhodium, coupled with a higher energy process.

There would of course be a greater incentive to achieve a catalysed addition of B—Si to an alkene, and recent work where a Pt catalyst is employed indicates that this is a realizable goal [13].

3   Asymmetric hydroboration with phosphinamine ligands

At this stage the project was generating interesting mechanistic information but the original goal of a useful contribution to asymmetric hydroboration remained elusive. The most effective case of asymmetric hydroboration was reported from Hayashi's laboratory, and involved catecholborane addition/peroxide oxidation in the case of *para*-substituted styrenes, employing Rh–BINAP catalysts. The resulting secondary alcohols were produced in highest enantiomer excess (90–96%) when the reaction was conducted at −78 °C, and this was attenuated appreciably under ambient conditions [14].

Around that time we had been engaged in the synthesis of a novel class of ligands with axial chirality based on the binaphthyl framework [15]. The parent QUINAP (2-(diphenylphosphino)-1-(1′-isoquinolyl)naphthalene) is less bulky than BINAP in the region of the isoquinoline, which replaces one of the diphenylphosphinonaphthalene moieties (cf. Fig. 24.6, where the projections are based on the X-ray structures of complexes of the two ligands with extraneous backbone atoms excised). Because BINAP complexes are pre-eminent for asymmetric hydrogenation, and secondary borane reagents are necessarily more sterically demanding than dihydrogen, we wondered whether QUINAP ligands would be effective in rhodium-complex-catalysed hydroboration. This proved to be the case, although the limitation with regard to vinylarenes being the only effective reactants still remains.

The synthetic route to phosphinoisoquinolines permits several structural modifications [16]. In the course of early hydroboration work with this class of catalyst, it was noted that electron-rich vinylarenes were hydroborated with higher enantiomeric excesses than their

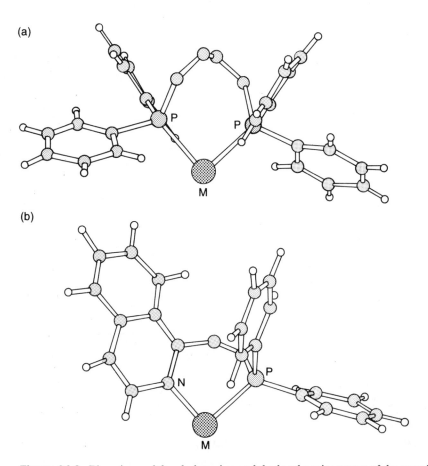

Figure 24.6. Plan views of the chelate ring and the local environment of the coordination sphere for: (a) a BINAP metal complex; (b) a QUINAP metal complex taken from X-ray structures. The lower steric crowding in (b) should be apparent.

electron-poor counterparts. This encouraged the speculation that there was a ligand electronic effect involved. In order to test this, several new diarylphosphine analogs were synthesized; the most interesting proved to be di(2-furyl)phosphine. In a comparison between this electron-rich ligand and its parent in rhodium-catalysed hydroboration, a consistent pattern emerges — the furylphosphine gives higher enantiomeric excesses with electron-withdrawing groups on the aromatic ring and the phenylphosphine works better with electron-releasing groups on the ring. A selected set of results is displayed in Table 24.2 [17].

The observation that substituted vinylarenes were hydroborated in high enantiomeric excess is of greater synthetic significance. Among the results obtained, a preparatively useful hydroboration of dihydronaphthalene leads to 2-tetralol in 96% ee. Related examples are shown in Table 24.3. The reaction works equally well in toluene or THF, although the sensitivity of the latter solvent to strong Lewis acids makes it less suitable when the catalytic turnover is slow and there is a danger of borane production through catecholborane

Table 24.2. Asymmetric hydroborations with phosphinoisoquinoline rhodium complexes [17]. All reactions occur with >95% chemo- and regioselectivity, yields 70–80% on a 0.5 mmolar scale

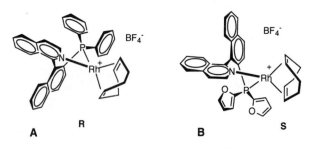

(i) 1 mol% LRh(C_8H_{12})BF$_4$ in situ, C_7H_8, 20°C; (ii) H$_2$O$_2$, OH$^-$

Reactant	Product with A, ee	Product with B, ee
Electron-rich alkenes		
R = *p*-Me	*R*, 89%	*S*, 88%
R = *m*-Me	*R*, 86%	*S*, 81%
R = *o*-Me	*R*, 92%	*S*, 90%
R = 2,4 di-Me	*R*, 94%	*S*, 93%
R = *p*-OEt	*R*, 94%	*S*, 78%
Electron-poor alkenes		
R = *p*-Cl	*R*, 78%	*S*, 82%
R = *m*-Cl	*R*, 63%	*S*, 89%
R = *o*-Cl	*R*, 55%	*S*, 69%
R = *m*-F	*R*, 67%	*S*, 77%
R = *m*-CF$_3$	*R*, 37%	*S*, 83%

Table 24.3. Asymmetric hydroborations with phosphinoisoquinoline rhodium complexes [17]. All reactions occur with ≥96% chemo- and regioselectivity. [a] yield on 1 g scale

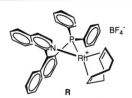

(i) 1 mol% $LRh(C_8H_{12})BF_4$ in situ, C_7H_8, 20°C; (ii) H_2O_2, OH^-

Reactant	e.e. of R-alcohol	yield
Ph Me (Z)	93%	80%
Ph Me (E)	95%	80%
Ph Ph	91%	86%
MeO—⬡—Me	97%	84%
indene	76%	75%
dihydronaphthalene	96%	78% [95%][a]
acenaphthylene	86%	67%
chromene	90%	82%

degradation. The enantiomeric excess seems to be comparable whether the alkene is prochiral ((E)- or (Z)-1-phenylpropene) or achiral (e.g. acenaphthene). This implies that the enantioselection is controlled by the specificity of borane transfer to the coordinated alkene, rather than recognition of the correct face of the alkene in the coordination step. For the moment, these represent the best results available in asymmetric hydroboration; Togni's ferrocenylphosphine/pyrazole catalysts give high enantioselectivity in the asymmetric hydroboration of styrene but at the expense of regioselectivity, the primary isomer being a significant component of the product [18].

Tables 24.2 and 24.3 indicate the strengths and weaknesses of the procedure: the limitations of substrate structure, where the most hindered side-chain giving good results is that of dihydronaphthalene and the most hindered ring substituent consistent with good enantioselectivity is *o*-methyl. When more hindered alkenes are employed, the reaction is appreciably slower and side-products including the alkane and primary borane (leading to the primary alcohol on oxidative workup) appear to the extent of >5%, which is not the usual observation.

4   Selectivity in asymmetric hydroboration

In the case of asymmetric hydrogenation, a combination of X-ray crystallography, NMR and kinetics has given us a detailed picture of the reaction mechanism [19]. Part of the reason why this has been possible is the stability of advanced intermediates in the catalytic cycle, particularly the key intermediate alkylrhodium hydride. Attempts to locate intermediates on the hydroboration pathway were less successful, and it appears that the true catalytic species are quite transient, at least when catecholborane is employed. The nature of the catalytic cycle has been addressed in MO calculations by both Schleyer's and Morokuma's group [20], although their theoretical cycle is promoted by an RhCl species. It is likely, however, that under experimental conditions this initial species undergoes reduction [Rh–Cl to Rh–H or Rh+ to Rh–H]. Some evidence for this stems from our unpublished observations that the extent of competing hydrogenation increases linearly with the mol.% of ClRh(PPh$_3$)$_3$ employed, as if the first cycle was one of hydrogenation and not hydroboration. This is also observed in NMR experiments (catecholborane plus styrene plus Wilkinson's catalyst) at low temperatures, where the first turnover leads to hydrogenation rather than hydroboration.

In the absence of direct information on reactive intermediates, a modeling approach to the source of enantioselectivity in asymmetric hydroboration was employed. In this, X-ray structural data on the QUINAP–Pd complex were combined with Chem 3D structures of the borane and alkene. It was assumed that the hydride transfer preceded B–C bond formation, and therefore Rh–H and coordinated C=C were coplanar. The hydride was further placed in an axial position, and the second axial position left unspecified—in practice this may be a further hydride, a catecholboronate or other ligating species. Given these constraints and the known relationship between the ligand configuration and that of the secondary alcohol derived from oxidation of the product borane, the optimum structure appears to be that shown in Fig. 24.7(a), although an alternative possibility represented in Fig. 24.7(b) is also consistent with the available evidence but more constrained sterically. The observation of electronic effects on enantiomeric excess may arise from the potential π-stack formed between the arene of the substrate and a proximate P–aryl group in the complex. Overall, this represents no more than a working model, of course, and like all such models that are based on the X-ray structures of stable complexes it suffers from a significant drawback. The metal–ligand fragment is assumed to be a rigid entity, providing an immutable framework within which the reactant and reagent are assembled. But in practice the bond lengths and more particularly bond angles within the coordination sphere are much more prone to distortion than the light atom connections, which are relatively hard. Analysis of the X-ray structures of a family of palladium complexes of QUINAP-type ligands leads to the conclusion:

(a)

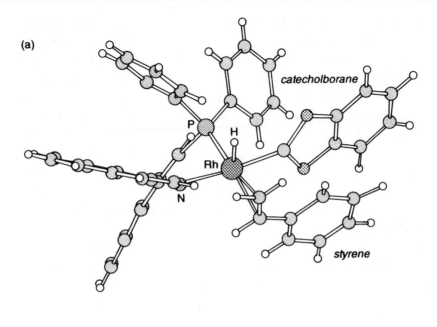

(b)

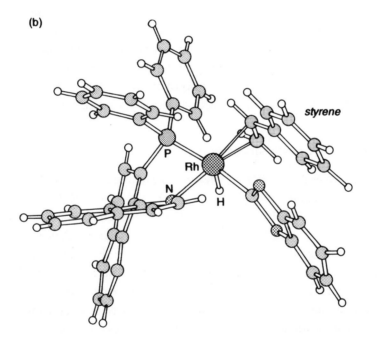

Figure 24.7. A possible model for the reactive intermediate in asymmetric hydroboration that predicts the correct configuration of the product and minimizes the steric strain: (a) with alkene coordinated *trans* to P; (b) with alkene coordinated *trans* to N.

Strain due to molecular crowding in an organometallic complex can be most readily accommodated by dispersal away from hard H—H and C—H repulsions into distortion of the coordination sphere; in this, angle bending strain is preferred over bond stretching strain.

This cautionary note suggests that attempts to rationalize asymmetric catalysis by treating it solely as an exercise in ligand design are doomed to failure, because all the interactions involved in the critical complex and its transition state must be taken into account fully at the outset [21].

Returning to the asymmetric hydroboration model, it can be seen readily that a styrene may bind in eight different ways once the ligand is positioned (*trans* to P or N, with each offering four different sites for the location of the phenyl group). It is probable that the alkene binding is not in itself so specific, but the requirement for a specific geometry in the hydride-transfer step, and also a requirement for specificity in catecholborane binding, provide further constraints that lead to the observed high enantiomeric excesses. For alkenes that lack a second handle enabling chelation to the metal, asymmetric organometallic catalysis of addition reactions is relatively rare, being limited to hydroformylation [22] and hydrosilylation [23] in a fairly narrow range of examples. The number of degrees of freedom that are normally encountered in such cases, and that need to be constrained before the enantioselectivity can be controlled, provide a clear reason for this observation.

A further question that needs to be addressed in catalytic hydroboration is the unique role of catecholborane. This has never been explained adequately and AM1 calculations on this and related secondary boranes do not provide an unequivocal answer. Part of the reason resides in the fact that catecholborane is monomeric and, being coplanar, it is spatially undemanding so that addition to the metal is unimpeded. This cannot be the whole story, and the unusual electronic nature of the borane was pointed out in the first paper of Mannig and Nöth. In essence the five-ring heterocycle is aromatic and possesses significant B—O double bond character, but in addition B—H should have enhanced acidity after the zwitterionic model of thiamine. Taken together, these factors provide for high reactivity, although the reagent is far from ideal; disproportion to BH_3 and the catechol-bridged boronate ester $B_2(C_6H_4O_2)_3$ competes with slow catalytic processes [24].

5   Manipulation of the carbon–boron bond

A severe limitation in the synthetic application of catalytic hydroboration is the relatively narrow range of reactions of the boronate esters with electrophiles. Only perborate or alkaline H_2O_2 oxidation works satisfactorily. None the less, this is a very useful transformation, providing the corresponding alcohols (usually secondary) with near-complete stereochemical control in the displacement step. At that stage the overall transformation is a Markovnikov hydration of the alkene, and provides a useful alternative to asymmetric syntheses of arylcarbinols from ketones via hydrosilylation or oxazaborolidine reduction [25, 26]. In contrast, a wide range of synthetically useful products are available from the more electrophilic trialkylboranes [27].

Our first attempts to extend the synthetic utility of catecholboronates involved attempted amination with a reagent related to hydrogen peroxide in its mechanism of action. Among electrophilic aminating agents, the most suitable thus appeared to be the nucleophilic substituted hydroxyamines $RN(M)OSiMe_3$, successfully applied in organocuprate chemistry [28]. Our initial experiments were very promising in that a mixture of the secondary amine RHNCHMeAr and the alcohol HOCHMeAr was obtained, with the enantiomeric purity of the amine and alcohol identical within experimental error. But despite considerable variation in experimental conditions, the ratio of the two could never be improved above 60:40. Because the silicon is mobile between the N and O in the anionic reagent, and simple *O*-alkylhydroxylamines do not work, an impasse was eventually reached. It was established through ^{13}C-labeling experiments that the first addition step of reagent to borane is irreversible at low temperatures and appears to produce both aminated and alkoxylated boranes in comparable quantities. These decompose along separate pathways at c 0 °C, giving borylated amine and alcohol, respectively [29].

A superior solution to the problem was discovered adventitiously. When an attempt was made to transmetallate the catecholboronate ester with diethylzinc, the expected benzylzinc reagent was not formed. Instead, a precipitate of zinc catecholate was formed, leaving a solution containing only the benzyldiethylborane. The same reaction took place using two moles of methylmagnesium chloride, albeit with more difficulty in controlling the outcome. In effect, we had rediscovered a forgotten observation of Cabiddu [30], who demonstrated the following reaction of catecholboronate esters for primary catecholboronates:

$$RBO_2C_6H_4 + 2\ R'MgBr \rightarrow R'_2RB + MgO_2C_6H_4$$

This observation stands in contrast to the rather difficult and stepwise conversions of aliphatic boronates, where considerable synthetic dexterity is required and stepwise displacement of the B—O bonds is preferred [27].

With a route to trialkylboranes on hand, the only impediment to further synthetic advance is the chemoselectivity of alkyl group transfers from boron to the added reagent. Fortunately, the self-limiting feature of asymmetric hydroboration proves to be a virtue; the benzyl group that possesses the stereogenic center is the one that migrates preferentially in the cases tested so far. The first and most obvious experiment was to attempt asymmetric amination, having failed to provide a satisfactory synthesis directly from the catecholboronate. It was found that fresh hydroxylamine-O-sulfonic acid is perfectly satisfactory in this regard, and the amine is the only functional organic product formed in the reaction, as detailed in Table 24.4. On this basis, a one-pot synthesis of benzylic amines was developed from alkene precursors, through successive asymmetric hydroboration in THF, *in situ* reaction with methylmagnesium chloride and finally stirring with a small excess of solid H_2NOSO_3H. Yields are modest at 50–60%, but the enantiomer excesses are often 1–2% higher than in the case of H_2O_2 oxidation. Part of the reason for the lower yield may be the difficulty in obtaining perfect control in the step where MeMgCl is added, because ^{11}B-NMR evidence indicates that any excess leads to the formation of a tetrahedral borate R_4B^- [31]. It may well be that the dialkylzinc procedure provides a superior entry to the desired trialkylboranes.

In a similar manner, preliminary experiments indicate that the trialkylborane synthesis can be used to promote C—C bond-forming reactions [32]. The initial results are indicated in Fig. 24.8; these data again demonstrate the stereochemical robustness of the borane migrations in that disparate procedures operate with an acceptable degree of retention of configuration.

6   Future directions

The results to date indicate that catalytic hydroboration with rhodium complexes of our phosphinoisoquinoline ligand QUINAP proceed with good enantioselectivity, but only for those cases where the reactant is a vinylarene. Ring substituent patterns (other than ortho, ortho' disubstitution) are tolerated and, unlike the BINAP catalysts, β-substituents on the

Table 24.4. A one-pot procedure for hydroboration/amination with (*S*)-QUINAP rhodium complexes [31]

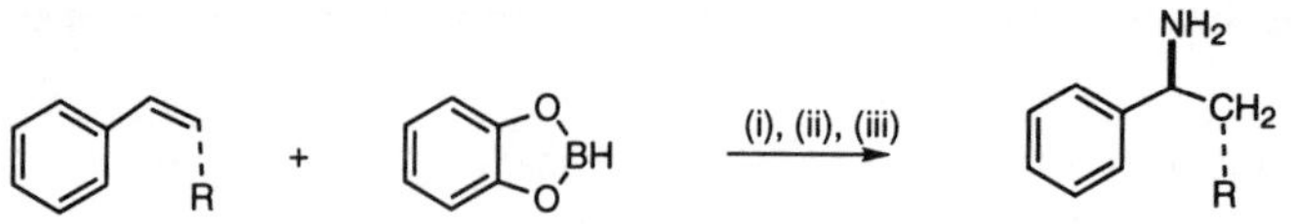

(i) 1 mol% LRh(C$_8$H$_{12}$)OSO$_2$CF$_3$, 20˚C, thf; (ii) 2 MeMgCl, thf; (iii) 3 equiv H$_2$NOSO$_3$H

Reactant	e.e. of *S*-amine	yield(chemoselectivity)
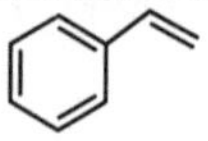	87%	56%(98%)
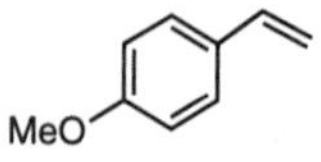	98%	54%(98%)
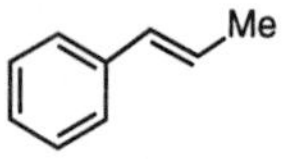	90%	50%(98%)
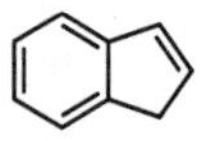	77%	61%(98%)
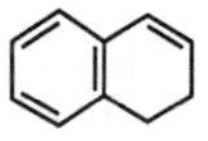	97%	51%(96%)
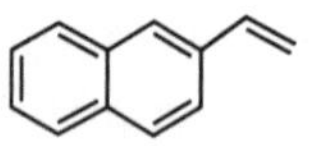	90%	62%(92%)
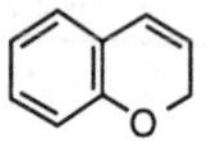	89%	64%(98%)

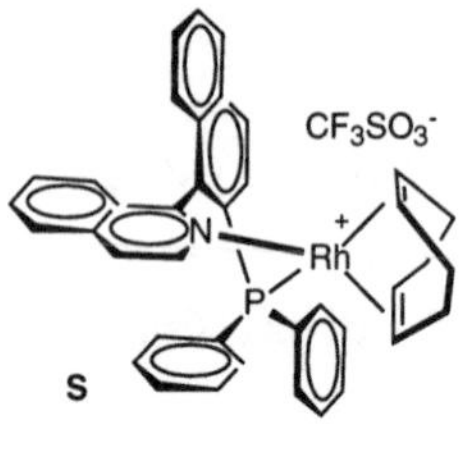

alkene incorporating a fused ring do not diminish the observed enantiomeric excess. The reaction occurs at an acceptable turnover rate at ambient temperature—indeed the selectivity is less at lower or higher temperature—and in preparative reactions a 1:500 catalyst/substrate ratio is effective. This still represents a rather limited range, given the power of uncatalysed hydroboration, and efforts to extend it will require new ligands for catalysis with modified electronic properties. Figure 24.9 shows an example that is at the limit of the range of present utility. The limitations of catecholboronate esters in synthesis appear to have been overcome by the clean Et$_2$Zn transformation observed that motivated the amination work. In extending this further it is worth noting that the literature reagents employed for borane transformations sometimes lack the stringency of modern synthetic

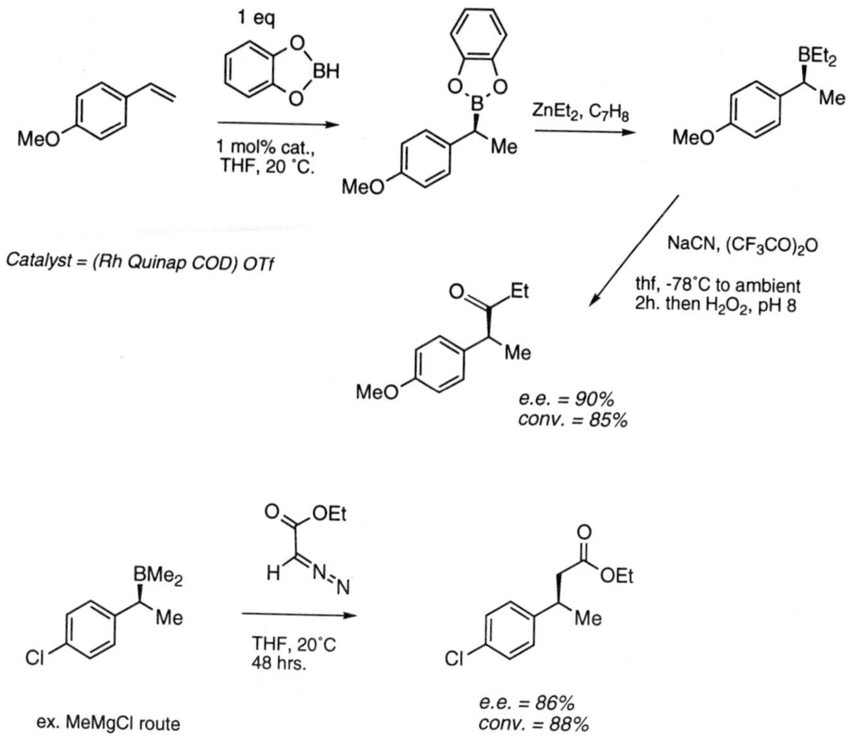

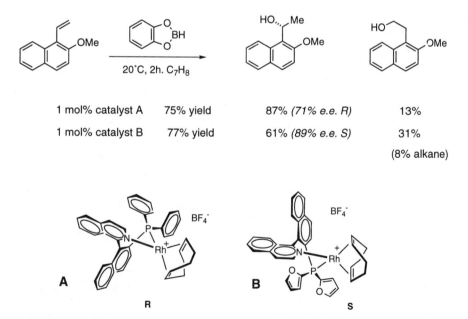

Figure 24.8. Applications of catalytic asymmetric hydroboration to C—C bond synthesis using known methodology for the C—B conversion steps.

Figure 24.9. Steric limitations in the asymmetric hydroboration reaction: some loss of enantiomeric excess and significant loss of chemoselectivity in a more hindered vinylarene.

practice, and the development of new methodology will be desirable. Finally, the impediment to progress brought about by the lack of good mechanistic information on asymmetric hydroboration will be challenged.

7 Acknowledgments

We thank EPSRC, LINK Asymmetric Synthesis, the Ramsay Memorial Trust, the Spanish Government and the UK chemical industry for support of aspects of this work. Johnson-Matthey kindly provided loans of rhodium salts.

8 References

1 Kono H, Ito K, Nagai Y. *Chem Lett* 1976: 1095.
2 Long JA, Marder TB, Behnken PE, Hawthorne MF. *J Am Chem Soc* 1984; **106**: 2979.
3 Mannig D, Noth H. *Angew Chem Int Ed Engl* 1985; **24**: 878–9.
4 Burgess K, Ohlmeyer MJ. *J Org Chem* 1988; **53**: 5178–9.
5 Burgess K, Ohlmeyer MJ. *Chem Rev* 1991; **91**: 1179–91.
6 Beletskaya I, Pelter A. *Tetrahedron* 1997; **53**: 4957–5026.
7 For diboration, see: Baker RT, Nguyen P, Marder TB, Westcott SA. *Angew Chem Int Ed Engl* 1995; **34**: 1336–8; Ishiyama T, Matsuda N, Miyaura N, Suzuki A. *J Am Chem Soc* 1993; **115**: 11018–119; Ishiyama T, Yamamoto M, Miyaura N. *Chem Commun* 1997: 689–90; Lawson YG, Lesley M, Marder TB, Norman NC, Rice CR. *Chem Commun* 1997: 2051–2; Ishiyama T, Matsuda N, Murata M *et al. Organometallics* 1996; **15**: 713–20; Lesley G, Nguyen P, Taylor NJ *et al. Organometallics* 1996; **15**: 5137–54; Iverson CN, Smith MR. *Organometallics* 1997; **16**: 2757–9; Marder TB, Norman NC, Rice CR. *Tetrahedron Lett* 1998; **39**: 155–8.
8 For silaboration, see: Suginome M, Nakamura H, Ito Y. *Chem Commun* 1996: 2777–8; Onozawa S, Hatanaka Y, Tanaka M. *Chem Commun* 1997: 1229–30.
9 Joshi NN, Srebnik M, Brown HC. *Tetrahedron Lett* 1989; **30**: 5551–4.
10 Brown JM, Lloyd-Jones GC. *Tetrahedron: Asymm* 1990; **1**: 869–72.
11 Brown JM, Lloyd-Jones GC. *J Am Chem Soc* 1994; **116**: 866–78.
12 For a recent example, see: LaPointe AM, Rix FC, Brookhart M. *J Am Chem Soc* 1997; **119**: 906–17, and references therein.
13 Suginome M, Nakamura H, Ito Y. *Angew Chem Int Ed Engl* 1997; **36**: 2516–18.
14 Hayashi T, Matsumoto Y, Ito Y. *Tetrahedron: Asymm* 1991; **2**: 601–12.
15 Alcock NW, Brown JM, Hulmes DI. *Tetrahedron: Asymm* 1993; **4**: 743–56.
16 Valk JM, Claridge T, Brown JM, Hibbs D, Hursthouse MB. *Tetrahedron: Asymm* 1995; **6**: 2597–610; Claridge T, Long JM, Brown JM, Hibbs D, Hursthouse MB. *Tetrahedron* 1997; **53**: 4035–50; Brown, JM, Doucet, H. *Tetrahedron: Asymm* 1997; **8**: 3775–984.
17 Brown JM, Doucet H, Fernandez E, Layzell TP. 1998 *Chem: Euro J*, 199 (in press).
18 Schnyder A, Hintermann L, Togni A. *Angew Chem Int Ed Engl* 1995; **34**: 931–3; Schnyder A, Togni A, Wiesli U. *Organometallics* 1997; **16**: 255–60.
19 Noyori, R. *Asymmetric Catalysis in Organic Chemistry.* New York: Wiley, 1994.
20 Dorigo AE, Schleyer P von R. *Angew Chem Int Ed Engl* 1995; **34**: 115–18; Musaev DG, Morokuma K. *J Phys Chem* 1996; **100**: 6509.
21 Brown JM, Hulmes DI, Long JM *et al. ECTOC Electronic Conference on Trends in Organometallic Chemistry*, HS Rzepa and C Leach (eds) Royal Society of Chemistry 1997: 28.
22 Nozaki K, Nanno T, Takaya H. *J Organomet Chem* 1997; **527**: 103–8. Horiuchi T, Shirakawa E, Nozaki K, Takaya H. *Organometallics* 1997; **16**: 2981–6, and earlier references cited therein.
23 Hayashi T. *Acta Chem Scand* 1996; **50**: 259–66; Uozumi Y, Kitayama K, Hayashi T, Yanagi K, Fukuyo E. *Bull Chem Soc Jpn* 1995; **68**: 713–22; Kitayama K, Uozumi Y, Hayashi T. *J Chem Soc Chem Commun* 1995: 1533–4.

24 Westcott SA, Blom HP, Marder TB, Baker RT, Calabrese JC. *Inorg Chem* 1993; **32**: 2175–82.
25 Langer T, Janssen J, Helmchen G. *Tetrahedron: Asymm* 1996; **7**: 1599–602, and earlier references cited therein.
26 Douglas AW, Tschaen DM, Reamer RA, Shi YJ. *Tetrahedron: Asymm* 1996; **7**: 1303–8; Jones GB, Heaton SB, Chapman BJ, Guzel M. *Tetrahedron: Asymm* 1997; **8**: 3625–36, and earlier references cited therein.
27 Brown HC, Singaram B. *Acc Chem Res* 1988; **21**: 287–93.
28 Casarini A, Dembech P, Lazzari D *et al. J Org Chem* 1993; **58**: 5620–23.
29 Knight FI, Brown JM, Lazzari D, Ricci A, Blacker AJ. *Tetrahedron* 1997; **53**: 11411–24.
30 Cabiddu S, Maccioni A, Secci M. *Gazz Chim Ital* 1972; **102**: 555.
31 Fernandez E, Hooper MW, Knight FI, Brown JM. *Chem Commun* 1997: 173–4.
32 Hooper MW, research in progress.

Index

Page references to figures appear in *italic* type and those for tables appear in **bold** type.